Numerical Recipes

The Art of Scientific Computing

William H. Press
Harvard-Smithsonian Center for Astrophysics

Brian P. Flannery
EXXON Research and Engineering Company

Saul A. Teukolsky
Department of Physics, Cornell University

William T. Vetterling
Polaroid Corporation

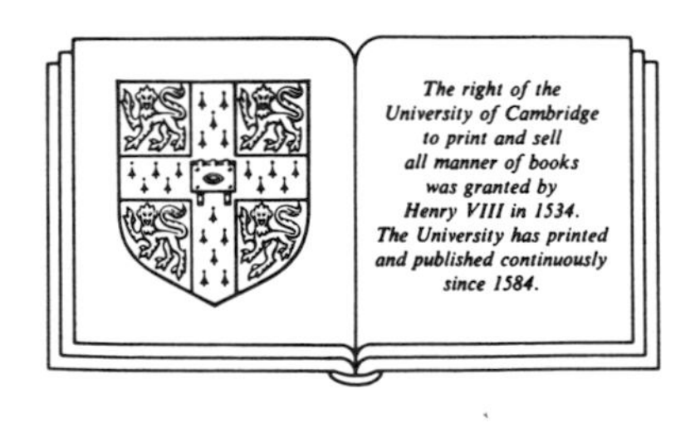
The right of the
University of Cambridge
to print and sell
all manner of books
was granted by
Henry VIII in 1534.
The University has printed
and published continuously
since 1584.

CAMBRIDGE UNIVERSITY PRESS

Cambridge

New York New Rochelle Melbourne Sydney

Published by the Press Syndicate of the University of Cambridge
The Pitt Building, Trumpington Street, Cambridge CB2 1RP
32 East 57th Street, New York, NY 10022 USA
10 Stamford Road, Oakleigh, Melbourne 3166, Australia

First published 1986
Reprinted 1986 (twice), 1987 (thrice), 1988 (twice)

Printed in the United States of America
Typeset in TEX

The computer programs in this book are available in several machine-readable formats, and in the **FORTRAN** and **Pascal** programming languages.

To purchase diskettes in IBM-compatible personal computer formats, use the order form at the back of the book or write to Cambridge University Press, 510 North Avenue, New Rochelle, NY 10801. Also available from Cambridge University Press are the *Numerical Recipes Example Books* and coordinated diskettes in **FORTRAN** or **Pascal**. These provide demonstration programs that illustrate the use of each subroutine and procedure in this book. They too may be ordered in the above manner.

Technical questions, corrections, and requests for information on other available formats and additional software products should be directed to Numerical Recipes Software, P.O. Box 243, Cambridge, MA 02238. Inquiries regarding OEM and site licenses should also be directed to this address.

Library of Congress Cataloging in Publication Data
Main entry under title:
Numerical recipes.
Bibliography: p.
Includes index.
1. Numerical analysis–Computer programs.
2. Science–Mathematics–Computer programs. 3. FORTRAN
(Computer program language) 4. PASCAL (Computer
program language) I. Press, William H.
QA297.N866 1986 001.64′2′0151 85-11397

British Library Cataloguing in Publication Data
Numerical recipes : the art of scientific computing.
1. Numerical analysis–Data processing
I. Press, William H.
519.4 QA297

ISBN 0 521 30811 9 Book
ISBN 0 521 30957 3 FORTRAN Example diskette
ISBN 0 521 30954 9 Pascal Example diskette
ISBN 0 521 30956 5 Example book in Pascal
ISBN 0 521 30955 7 Pascal diskette
ISBN 0 521 31330 9 Example book in FORTRAN
ISBN 0 521 30958 1 FORTRAN diskette

Contents

Preface

We call this book *Numerical Recipes* for several reasons. In one sense, this book is indeed a "cookbook" on numerical computation. However there is an important distinction between a cookbook and a restaurant menu. The latter presents choices among complete dishes in each of which the individual flavors are blended and disguised. The former – and this book – reveals the individual ingredients and explains how they are prepared and combined.

Another purpose of the title is to connote an eclectic mixture of presentational techniques. This book is unique, we think, in offering, for each topic considered, a certain amount of general discussion, a certain amount of analytical mathematics, a certain amount of discussion of algorithmics, and (most important) actual implementations of these ideas in the form of working computer routines. Our task has been to find the right balance among these ingredients for each topic. You will find that for some topics we have tilted quite far to the analytic side; this where we have felt there to be gaps in the "standard" mathematical training. For other topics, where the mathematical prerequisites are universally held, we have tilted towards more in-depth discussion of the nature of the computational algorithms, or towards practical questions of implementation.

We admit, therefore, to some unevenness in the "level" of this book. About half of it is suitable for an advanced undergraduate course on numerical computation for science or engineering majors. The other half ranges from the level of a graduate course to that of a professional reference. Most cookbooks have, after all, recipes at varying levels of complexity. An attractive feature of this approach, we think, is that the reader can use the book at increasing levels of sophistication as his/her experience grows. Even inexperienced readers should be able to use our most advanced routines as black boxes. Having done so, we hope that these readers will subsequently go back and learn what secrets are inside.

If there is a single dominant theme in this book, it is that practical methods of numerical computation can be simultaneously efficient, clever, and – important – clear. The alternative viewpoint, that efficient computational methods must necessarily be so arcane and complex as to be useful only in "black box" form, we firmly reject.

Our purpose in this book is thus to open up a large number of computational black boxes to your scrutiny. We want to teach you to take apart these black boxes and to put them back together again, modifying them to suit your specific needs. We assume that you are mathematically literate, i.e. that you have the normal mathematical preparation associated with an undergraduate degree in a physical science, or engineering, or economics, or a quantitative social science. We assume that you know how to program a computer. We do not assume that you have any prior formal knowledge of numerical analysis or numerical methods.

The scope of *Numerical Recipes* is supposed to be "everything up to, but not including, partial differential equations." We honor this in the breach:

First, we *do* have one introductory chapter on methods for partial differential equations (Chapter 17). Second, we obviously cannot include *everything* else. All the so-called "standard" topics of a numerical analysis course have been included in this book: linear equations (Chapter 2), interpolation and extrapolation (Chaper 3), integration (Chaper 4), nonlinear root-finding (Chapter 9), eigensystems (Chapter 11), and ordinary differential equations (Chapter 15). Most of these topics have been taken beyond their standard treatments into some advanced material which we have felt to be particularly important or useful. Some highlights are

- special linear systems (Vandermonde, Toeplitz, sparse)
- rational function interpolation and extrapolation
- integrals with singularities
- root-finding by the Van Wijngaarden-Dekker-Brent method
- Bulirsch-Stoer integration of differential equations with adaptive stepsize control.

Some other subjects that we cover in detail are not usually found in the standard numerical analysis texts. These include the evaluation of functions and of particular special functions of higher mathematics (Chapters 5 and 6); random numbers and Monte Carlo methods (Chapter 7); sorting (Chapter 8); optimization, including multidimensional methods (Chapter 10); Fourier transform methods, including FFT methods and maximum entropy methods (Chapter 12); two chapters on the statistical description and modeling of data (Chapters 13 and 14); and two-point boundary value problems, both shooting and relaxation methods (Chapter 16). In these chapters, some of the less usual, but highly useful, topics include

- working routines for statistical probability functions
- Bessel functions and modified Bessel functions
- random deviates from gamma, Poisson, and binomial distributions
- the Data Encryption Standard
- sorting for determining equivalence classes of data
- a complete implementation of linear programming
- an introduction to annealing methods of optimization
- routines for real Fourier transforms, for sine and cosine transforms
- linear prediction and linear predictive coding
- digital filtering
- entropy measures of statistical dependency
- robust statistical fitting

The programs in this book are included in both ANSI-standard `FORTRAN-77` and in `Pascal`. Available separately is a different version of this book, in `C`. We have more to say about programming languages, and the computational environment assumed by our routines, in §1.1 (Introduction). But first, an important formality:

Legal Matters

We make no warranties, express or implied, that the programs contained in this volume are free of error, or are consistent with

any particular standard of merchantability, or that they will meet your requirements for any particular application. They should not be relied on for solving a problem whose incorrect solution could result in injury to a person or loss of property. If you do use the programs in such a manner, it is at your own risk. The authors and publisher disclaim all liability for direct or consequential damages resulting from your use of the programs.

Several registered trademarks appear within the text of this book: DEC and VAX are trademarks of Digital Equipment Corporation. IBM is a trademark of International Business Machines Corporation. IMSL is a trademark of IMSL Inc. for its proprietary computer software. NAG refers to proprietary computer software of Numerical Algorithms Group (USA) Inc. Microsoft and MS are trademarks of Microsoft Corporation. UCSD Pascal and UCSD p-system are trademarks of the Regents of the University of California. TURBO Pascal refers to copyrighted computer software of Borland International.

Like artistic or literary compositions, computer programs can be protected by copyright. Generally it is an infringement for you to copy into your computer a program from a copyrighted source. It is also not a friendly thing to do, since it deprives the program's author of compensation for his or her creative effort. Although this book and its programs are copyrighted, we specifically authorize you, a reader of the book, to make one machine-readable copy of each program for your own use. You may wish to consider purchasing this book's collected programs in one of the machine-readable formats that are available (see front of book). Distribution of the machine-readable programs (either as copied by you or as purchased) to any other person is not authorized.

Copyright does not protect ideas, but only the expression of those ideas in a particular form. In the case of a computer program, the ideas consist of the programs's methodology and algorithm, including the sequence of processes adopted by the programmer. The expression of those ideas is the program source code and its derived object code.

If you analyze the ideas contained in a program, and then express those ideas in your own distinct implementation, then that new program implementation belongs to you. That is what we have done in assembling the programs n this book (except for those entirely of our own devising). When programs in is book are said to be "based" on programs published in copyright sources, mean that the ideas are the same. The expression of these ideas as source is our own. We believe that no material in this book infringes on an g copyright.

fact that ideas are legally "free as air" in no way supersedes the uirement that ideas be credited to their known originators. When n this book are based on known sources, whether copyrighted or domain, published or "handed-down," we have attempted to ribution. Unfortunately, the lineage of many programs in com- is often unclear. We would be grateful to readers for new o

corrected information regarding attributions, which we will attempt to incorporate in subsequent printings.

Acknowledgments

Many colleagues have been generous in giving us the benefit of their numerical and computational experience, in providing us with programs, in commenting on the manuscript, or in general encouragement. We particularly wish to thank George Rybicki, Douglas Eardley, Philip Marcus, Stuart Shapiro, Paul Horowitz, Bruce Musicus, Irwin Shapiro, Stephen Wolfram, Henry Abarbanel, Larry Smarr, Richard Muller, John Bahcall, and A.G.W. Cameron.

We also wish to acknowledge two individuals whom we have never met: Forman Acton, whose 1970 textbook *Numerical Methods that Work* (New York: Harper and Row) has surely left its stylistic mark on us; and Donald Knuth, both for his series of books on *The Art of Computer Programming* (Reading, Mass.: Addison-Wesley), and for TeX, a computer typesetting language which immensely aided production of this book.

Research by the authors on computational methods was supported in part by the U.S. National Science Foundation.

Corrections in This Printing

Our publisher has kindly allowed us to make corrections from time to time, as this book is reprinted. In the third printing (1986), corrections were made on pages 19, 83, 87, 100*, 121, 161*, 162*, 175*, 180*, 232, 241, 247*, 264*, 265*, 305*, 310*, 398*, 400*, 401, 406*, 412*, 432, 466, 475*, 496*, 513*, 526*, 527*, 554*, 555, 597*, 606*, 612, 616, 703*, 709*, 711*, 717*, 718*, 730*, 733*, 734*, 755*, 756*, 757*, 760*, 766*, 770*, 775*, 778*, 787*.

In the fifth printing (1987), corrections were made on pages 33, 34, 35, 36*, 63*, 64*, 86, 134*, 138, 140, 175*, 176, 176*, 178*, 180*, 293, 299*, 304, 322, 399, 400*, 404*, 406*, 508*, 511, 559*, 565*, 567*, 571, 585*, 587*, 597*, 598*, 629, 682*, 684*, 689*, 702*, 709*, 710*, 711*, 741*, 767*, 779*, 782*, 783*, 784*, 785*.

In this printing, additional corrections or improvements have been mad on pages 28, 57, 73*, 118, 119, 129*, 140, 167, 215*, 220, 245, 258*, 265*, 30 321, 348, 400*, 416*, 468, 469*, 475*, 495, 522, 524, 528, 586, 587, 643, 6 687*, 732*, 733*, 734*, 736*, 737*, 738*, 749*, 756*, 765*, 766*, 777*

Asterisk (*) indicates that the correction is in program source cod of these corrections are quite minor; many affect only particular c We thank our readers for helping to find them.

Computer Programs by Chapter and Section

1.0	FLMOON	phases of the moon, calculated by date
1.1	JULDAY	Julian day number, calculated by date
1.1	BADLUK	Friday the 13th when the moon is full
1.1	CALDAT	calendar date, calculated from Julian day number
2.1	GAUSSJ	matrix inversion and linear equation solution, Gauss-Jordan
2.3	LUDCMP	linear equation solution, *LU* decomposition
2.3	LUBKSB	linear equation solution, backsubstitution
2.6	TRIDAG	linear equation solution, tridiagonal equations
2.7	MPROVE	linear equation solution, iterative improvement
2.8	VANDER	linear equation solution, Vandermonde matrices
2.8	TOEPLZ	linear equation solution, Toeplitz matrices
2.9	SVDCMP	singular value decomposition of a matrix
2.9	SVBKSB	singular value backsubstitution
2.10	SPARSE	linear equation solution, sparse matrix, conjugate-gradient method
3.1	POLINT	interpolation, polynomial
3.2	RATINT	interpolation, rational function
3.3	SPLINE	interpolation, construct a cubic spline
3.3	SPLINT	interpolation, evaluate a cubic spline
3.4	LOCATE	search an ordered table, bisection
3.4	HUNT	search an ordered table, correlated calls
3.5	POLCOE	polynomial coefficients from a table of values
3.5	POLCOF	polynomial coefficients from a table of values
3.6	POLIN2	interpolation, two-dimensional polynomial
3.6	BCUCOF	interpolation, two-dimensional, construct bicubic
3.6	BCUINT	interpolation, two-dimensional, evaluate bicubic
3.6	SPLIE2	interpolation, two-dimensional, construct two-dimensional spline
3.6	SPLIN2	interpolation, two-dimensional, evaluate two-dimensional spline
4.2	TRAPZD	integrate a function by trapezoidal rule
4.2	QTRAP	integrate a function to desired accuracy, trapezoidal rule
4.2	QSIMP	integrate a function to desired accuracy, Simpson's rule
4.3	QROMB	integrate a function to desired accuracy, Romberg adaptive method
4.4	MIDPNT	integrate a function by extended midpoint rule
	QROMO	integrate a function to desired accuracy, open Romberg

4.4	MIDINF	integrate a function on a semi-infinite interval
4.4	MIDSQL	integrate a function with a square-root singularity
4.4	MIDSQU	integrate a function with an inverse square-root singularity
4.4	MIDEXP	integrate a function which decreases exponentially
4.5	QGAUS	integrate a function by Gaussian quadratures
4.5	GAULEG	Gauss-Legendre weights and abscissas, compute
4.6	QUAD3D	integrate a function over a three-dimensional space
5.1	EULSUM	sum a series, Euler–van Wijngaarden algorithm
5.3	DDPOLY	polynomial, fast evaluation of specified derivatives
5.3	POLDIV	polynomials, divide one by another
5.6	CHEBFT	fit a Chebyshev polynomial to a function
5.6	CHEBEV	Chebyshev polynomial evaluation
5.7	CHINT	integrate a function already Chebyshev fitted
5.7	CHDER	derivative of a function already Chebyshev fitted
5.8	CHEBPC	polynomial coefficients from a Chebyshev fit
5.8	PCSHFT	polynomial coefficients of a shifted polynomial
6.1	GAMMLN	logarithm of gamma function
6.1	FACTRL	factorial function
6.1	BICO	binomial coefficients function
6.1	FACTLN	factorial function, logarithm
6.1	BETA	beta function
6.2	GAMMP	gamma function, incomplete
6.2	GAMMQ	gamma function, incomplete, complementary
6.2	GSER	gamma function, incomplete, series evaluation
6.2	GCF	gamma function, incomplete, continued fraction evaluation
6.2	ERF	error function
6.2	ERFC	error function, complementary
6.2	ERFCC	error function, complementary, concise routine
6.3	BETAI	beta function, incomplete
6.3	BETACF	beta function, incomplete, continued fraction evaluation
6.4	BESSJ0	Bessel function J_0
6.4	BESSY0	Bessel function Y_0
6.4	BESSJ1	Bessel function J_1
6.4	BESSY1	Bessel function Y_1
6.4	BESSJ	Bessel function J of integer order
6.4	BESSY	Bessel function Y of integer order
6.5	BESSI0	Modified Bessel function I_0
6.5	BESSK0	Modified Bessel function K_0
6.5	BESSI1	Modified Bessel function I_1
6.5	BESSK1	Modified Bessel function K_1
6.5	BESSI	Modified Bessel function I of integer order
6.5	BESSK	Modified Bessel function K of integer order

12.8	MEMCOF	power spectrum estimation, evaluate maximum entropy coefficients
12.8	EVLMEM	power spectrum estimation using maximum entropy coefficients
12.10	FIXRTS	roots of a polynomial, reflects inside unit circle
12.10	PREDIC	linear prediction using MEM coefficients
12.11	FOURN	Fourier transform (FFT) in multidimensions
13.1	MOMENT	moments of a data set, calculate
13.2	MDIAN1	median of a data set, calculate by sorting
13.2	MDIAN2	median of a data set, calculate iteratively
13.4	TTEST	Student's t-test for difference of means
13.4	AVEVAR	mean and variance of a data set, calculate
13.4	TUTEST	Student's t-test for means, with unequal variances
13.4	TPTEST	Student's t-test for means, with paired data
13.4	FTEST	F-test for difference of variances
13.5	CHSONE	chi-square test for difference between data and model
13.5	CHSTWO	chi-square test for difference between two data sets
13.5	KSONE	Kolmogorov-Smirnov test of data against model
13.5	KSTWO	Kolmogorov-Smirnov test between two data sets
13.5	PROBKS	Kolmogorov-Smirnov probability function
13.6	CNTAB1	contingency table analysis using chi-square
13.6	CNTAB2	contingency table analysis using entropy measure
13.7	PEARSN	correlation between two data sets, Pearson's
13.8	SPEAR	correlation between two data sets, Spearman's rank
13.8	CRANK	rank, replaces array elements by their
13.8	KENDL1	correlation between two data sets, Kendall's tau
13.8	KENDL2	contingency table analysis using Kendall's tau
13.9	SMOOFT	smooth data using FFT
14.2	FIT	fit data to a straight line, least squares
14.3	LFIT	linear least squares fit, general, normal equations
14.3	COVSRT	covariance matrix, sort, used by LFIT
14.3	SVDFIT	linear least squares fit, general, singular value decomposition
14.3	SVDVAR	variances from singular value decomposition
14.3	FPOLY	fit a polynomial, using LFIT or SVDFIT
14.3	FLEG	fit a Legendre polynomial, using LFIT or SVDFIT
14.4	MRQMIN	nonlinear least squares fit, Marquardt's method
14.4	MRQCOF	nonlinear least squares fit, used by MRQMIN
14.4	FGAUSS	fit a sum of Gaussians, using MRQMIN
14.6	MEDFIT	fit data to a straight line robustly, least absolute deviation
14.6	ROFUNC	fit data robustly, used by MEDFIT
15.1	RK4	integrate one step of ODEs, 4th order Runge-Kutta
15.1	RKDUMB	integrate ODEs by 4th order Runge-Kutta

15.2	RKQC	integrate one step of ODEs with accuracy monitoring
15.2	ODEINT	integrate ODEs with accuracy monitoring
15.3	MMID	integrate ODEs by modified midpoint method
15.4	BSSTEP	integrate ODEs, Bulirsch-Stoer step
15.4	RZEXTR	rational function extrapolation, used by BSSTEP
15.4	PZEXTR	polynomial extrapolation, used by BSSTEP
16.1	SHOOT	two-point boundary value problem, solve by shooting
16.2	SHOOTF	two-point boundary value problem, shooting to a fitting point
16.3	SOLVDE	two-point boundary value problem, solve by relaxation
16.3	BKSUB	backsubstitution, used by SOLVDE
16.3	PINVS	diagonalize a sub-block, used by SOLVDE
16.3	RED	reduce columns of a matrix, used by SOLVDE
16.4	SFROID	spheroidal functions, obtain using SOLVDE
16.4	DIFEQ	spheroidal matrix coefficients, used by SFROID
17.5	SOR	elliptic PDE solved by simultaneous overrelaxation method
17.6	ADI	elliptic PDE solved by alternating direction implicit method

Chapter 1. Preliminaries

1.0 Introduction

This book is supposed to teach you methods of numerical computing which are practical, efficient, and (insofar as possible) elegant. We presume throughout this book that you, the reader, have particular tasks that you want to get done. We view our job as educating you on how to proceed. Occasionally we may try to reroute you briefly onto a particularly beautiful side road; but by and large, we will travel with you along main highways that lead to practical destinations.

Throughout this book, you will find us fearlessly editorializing, telling you what you should and shouldn't do. This prescriptive tone results from a conscious decision on our part, and we hope that you will not find it irritating. We do not claim that our advice is infallible! Rather, we are reacting against a tendency, in the textbook literature of computation, to discuss every possible method that has ever been invented, without ever offering a practical judgment on relative merit. We do, therefore, offer you our practical judgments whenever we can. As you gain experience, you will form your own opinion of how reliable our advice is.

We presume that you are able to read computer programs in either `FORTRAN` or `Pascal`. In this edition of *Numerical Recipes*, the former language is the "default" that occurs in the main body of the text. However, we have translated all the routines in the book into working `Pascal`, and these translations are printed, by chapters, in the back of the book. `Pascal` programmers may need to do some extra flipping of pages back and forth but should not be otherwise penalized. (We intend that a future version of this book will be entirely in `Pascal`, redressing the balance. An version in `C` is already available.)

Typographically, when we include programs in the text, they look like this:

```
SUBROUTINE FLMOON(N,NPH,JD,FRAC)
   Our programs begin with an introductory comment summarizing their purpose and ex-
   plaining their calling sequence. This routine calculates the phases of the moon. Given
   an integer N and a code NPH for the phase desired (NPH= 0 for new moon, 1 for first
   quarter, 2 for full, 3 for last quarter), the routine returns the Julian Day Number JD, and
```

```
   the fractional part of a day FRAC to be added to it, of the Nth such phase since January,
   1900. Greenwich Mean Time is assumed.
PARAMETER (RAD=3.14159265/180.)
C=N+NPH/4.                      This is how we comment an individual line.
T=C/1236.85
T2=T**2
AS=359.2242+29.105356*C         You aren't really intended to understand this algorithm,
AM=306.0253+385.816918*C+0.010730*T2        but it does work!
JD=2415020+28*N+7*NPH
XTRA=0.75933+1.53058868*C+(1.178E-4-1.55E-7*T)*T2
IF(NPH.EQ.0.OR.NPH.EQ.2)THEN
    XTRA=XTRA+(0.1734-3.93E-4*T)*SIN(RAD*AS)-0.4068*SIN(RAD*AM)
ELSE IF(NPH.EQ.1.OR.NPH.EQ.3)THEN
    XTRA=XTRA+(0.1721-4.E-4*T)*SIN(RAD*AS)-0.6280*SIN(RAD*AM)
ELSE
    PAUSE 'NPH is unknown.'  This is how we will indicate error conditions.
ENDIF
IF(XTRA.GE.0.)THEN
    I=INT(XTRA)
ELSE
    I=INT(XTRA-1.)
ENDIF
JD=JD+I
FRAC=XTRA-I
RETURN
END
```

A few notes: For simplicity, we adopt the convention of handling all errors and exceptional cases by the PAUSE statement. In general, we do not intend that you continue program execution after a pause occurs, but FORTRAN allows you to do so — if you want to see what kind of wrong answer or catastrophic error results. In some applications, you will want to modify our programs to do more sophisticated error handling, for example to return with an error flag set.

Pascal readers who are new to FORTRAN will be baffled at this point by the fact that variables in FORTRAN are all *implicitly declared.* In brief, the rule is that variables which begin with the letters I,J,K,L,M,N are implicitly declared to be type INTEGER, while all other variables are implicitly declared to be type REAL. Operations between two integer variables are done with integer arithmetic; operations between two reals, or a real and an integer, are done with real arithmetic, and the result is of type REAL.

If you are a FORTRAN programmer who has never seen an IF-THEN-ELSE-ENDIF construction before, then you are undoubtedly a user of the 1966, rather than 1977, standard. You should have no difficulty in picking up a reading knowledge of FORTRAN-77 as you go through this book. For information on obtaining 1966 versions of the programs in this book, write to the address given on the reverse of the title page.

We will need to say more about programming languages and style in §1.1.

Computational Environment and Program Validation

We have attempted to make the programs in this book as generally useful as possible, not just in terms of the subjects covered, but also in terms of their degree of portability among very different computers. Specifically, we intend

that all the programs should work on both mainframe, multi-user computers and on personal computers.

As surrogates for the large number of computers in each category, we have used the DEC VAX-11/780 running VMS, and the IBM PC or XT running PC-DOS (version 2.0 or later). The FORTRAN programs in this book run without modification in VAX-11 Fortran, and with minor modifications in Microsoft FORTRAN (version 3.20 or later) on the IBM PC. A diskette containing these slightly modified programs is available for purchase, see back of book for information. The reason that modification is required is that Microsoft FORTRAN does not (at time of writing) implement the full FORTRAN-77 standard, but only a subset of it. The programs should run without modification on any compiler that implements the standard.

In Pascal, the programs in this book run, essentially without modification, in TURBO Pascal (Borland International) on an IBM PC or compatible, in VAX-11 Pascal on a DEC VAX, in UCSD Pascal on most machines, and in IBM Pascal/VS on mainframes. For further details turn to the section *Numerical Recipes in Pascal* at the back of this book. The programs should run without modification (except for implementation-dependent bindings of files to external file names) on any compiler that implements the ISO or IEEE standard. Diskettes containing the Pascal program set in machine-readable form are available.

In validating the programs, we have taken the program source code directly from the machine-readable form of the book's manuscript, so as to decrease the chance of propagating typographical errors. "Driver" or demonstration programs which we used as part of our validations are available separately as the *Numerical Recipes Example Book*, as well as in machine-readable form. If you plan to use more than a few of the programs in this book, or if you plan to use programs in this book on more than one different computer, then you may find it useful to obtain a copy of these demonstration programs.

Of course we would be foolish to claim that there are no bugs in our programs, and we do not make such a claim. We have been very careful. But if you find a bug, please document it and tell us!

The remaining two sections of this chapter review some basic concepts of programming (control structures, etc.) and of numerical analysis (roundoff error, etc.). Thereafter, we plunge into the substantive material of the book.

REFERENCES AND FURTHER READING:

Meeus, Jean 1982, *Astronomical Formulae for Calculators*, Second Edition, revised and enlarged (Richmond, Virginia: Willmann-Bell).

You will find references listed like this at the end of most sections of this book.

Because computer algorithms often circulate informally for quite some time before appearing in a published form, the task of uncovering "primary literature" is quite difficult. We have not attempted this, and we do not pretend to any degree of bibliographical completeness in this book. For topics where a substantial secondary literature exists (discussion in textbooks, reviews, etc.) we have

consciously limited our references to a few of the more useful secondary sources, especially those with good references to the primary literature. Where the existing secondary literature is insufficient, we give references to a few primary sources that are intended to serve as starting points for further reading, not as complete bibliographies for the field.

The references at the end of each section are arranged in order, roughly, according to their usefulness or relevance to that section. References relevant to more than one section are also repeated in the separate list at the back of the book.

1.1 Program Organization and Control Structures

The term *structured programming* has become a catch phrase that means many different things to many different people. Some people incorrectly think that structured programming means "break everything up into subroutines and include a lot of comments." Others opine that structured programming is possible in some computer languages (e.g. `Pascal`) and not in others (e.g. `FORTRAN`). This, too, is wrong, although some languages certainly do make structured programming relatively easier or more difficult.

We sometimes like to point out the close analogies between computer programs, on the one hand, and written poetry or written musical scores, on the other. All three present themselves as visual media, symbols on a two-dimensional page or computer screen. Yet, in all three cases, the visual, two-dimensional, *frozen-in-time* representation communicates (or is supposed to communicate) something rather different, namely a process that *unfolds in time.* A poem is meant to be read; music, played; a program, executed as a sequential series of computer instructions.

In all three cases, the target of the communication, in its visual form, is a human being. The goal is to transfer to him/her, as efficiently as can be accomplished, the greatest degree of understanding, in advance, of how the process *will* unfold in time. In poetry, this human target is the reader. In music, it is the performer. In programming, it is the program user.

Now, you may object that the target of communication of a program is not a human but a computer, that the program user is only an irrelevant intermediary, a lackey who feeds the machine. This is perhaps the case in the situation where the corporate executive pops the floppy disk into his personal computer and feeds it a black-box program in binary executable form. The computer doesn't care whether that program is "structured" or not! We envision, however, that you, the readers of this book, are in quite a different situation. You need, or want, to know not just *what* a program does, but also *how* it does it, so that you can tinker with it and modify it to your particular application. You need others to be able to see what you have done, so that they can criticize or admire. In such cases, the targets of a program's communication are surely human, not machine.

We are not yet done drawing the analogies. Music, poetry, and programming, all three as symbolic constructs of the human brain, are found to be naturally structured into hierarchies that have many different nested levels. Sounds (phonemes) form small meaningful units (morphemes) which in turn form words; words group into phrases, which group into sentences; sentences make paragraphs, and these are organized into higher levels of meaning. Notes form musical phrases, which form themes, counterpoints, harmonies, etc.; which form movements, which form concertos, symphonies, and so on.

The structure in programs is equally hierarchical, if not so universally recognized. At a low level is the ASCII character set. Then, constants, identifiers, operands, operators. Then program statements, like `A(J+1)=B+C/3.0`.

At the next level, the terminology becomes vague, there being no standard vocabulary in use. Statements frequently come in "groups" or "blocks" which make sense only taken as a whole. For example,

```
SWAP=A(J)
A(J)=B(J)
B(J)=SWAP
```

makes immediate sense to any programmer as the exchange of two variables, while

```
SUM=0.0
ANS=0.0
N=1
```

is very likely to be an initialization of variables prior to some iterative process. This level of hierarchy in a program is usually evident to the eye. The more compulsive of programmers put in comments at this level, e.g., "Initialize" or "Exchange variables."

The next level is that of *control structures.* These are things like DO-loops, IF's, and so on. They are so important that we will come back to them just below. Then we have the level of subroutines or procedures, and the whole "global" organization of the computational task to be done. In the musical analogy, we are now at the level of movements and complete works. Organization at this level is so closely linked to *what* the composer or programmer hopes to accomplish, rather than *how* it is accomplished, that we can offer few, if any, general principles. At this level, *you* are supposed already to be the expert.

Control Structures

An executing program unfolds in time, but not strictly in the linear order in which the statements are written. Program statements which affect the order in which statements are executed, or which affect whether statements are executed, are called *control statements.* Control statements never make useful sense by themselves. They only make sense in the context of the groups or blocks of statements that they in turn control. If you think of those blocks as paragraphs containing sentences, then the control statements are perhaps

best thought of as the indentation of the paragraph and the punctuation between the sentences, not the words within the sentences.

We can now say what the goal of structured programming is. It is *to make program control manifestly apparent in the visual presentation of the program.* You see that this goal has nothing at all to do with how the computer sees the program. As already remarked, computers don't care whether you use structured programming or not. Human readers, however, *do* care. You yourself will also care, once you discover how much easier it is to perfect and debug a well-structured program than one whose control structure is obscure.

You accomplish the goals of structured programming in two complementary ways. First, you acquaint yourself with the small number of essential control structures which occur over and over again in programming, and which are therefore given convenient representations in most programming languages. You should learn to think about your programming tasks, insofar as possible, exclusively in terms of these standard control structures. In writing programs, you should get into the habit of representing these standard control structures in consistent, conventional ways.

"Doesn't this inhibit *creativity*?" our students sometimes ask. Yes, just as Mozart's creativity was inhibited by the sonata form, or Shakespeare's by the metrical requirements of the sonnet. The point is that creativity, when it is meant to communicate, does *well* under the inhibitions of appropriate restrictions on format.

Second, you *avoid*, insofar as possible, control statements whose controlled blocks or objects are difficult to discern at a glance. This means, in practice, that *you must try to avoid statement labels and* `GOTO`*'s.* It is not the `GOTO`'s which are dangerous (although they do interrupt one's reading of a program); the statement labels are the hazard. In fact, whenever you encounter a statement label while reading a program, you will soon become conditioned to get a sinking feeling in the pit of your stomach. Why? Because the following questions will, by habit, immediately spring to mind: Where did control come *from* in a branch to this label? It could be anywhere in the routine! What circumstances resulted in a branch to this label? They could be anything! Certainty becomes uncertainty, understanding dissolves into a morass of possibilities.

Some languages, notably 1966 `FORTRAN` and to a lesser extent `FORTRAN-77`, *require* statement labels in the construction of certain standard control structures. We will see this in more detail below. This is a demerit for these languages, but we cannot ask you to bear the guilt for their deficiencies. In such cases, you must use labels as required. But you should never branch to them independently of the standard control structure. If you must branch, let it be to an additional label, one which is not masquerading as part of a standard control structure.

We call labels that are part of a standard construction and never otherwise branched to *tame labels.* They do not interfere with structured programming in any way, except possibly typographically as distractions to the eye.

Some examples are now in order to make these considerations more concrete (see Figure 1.1.1).

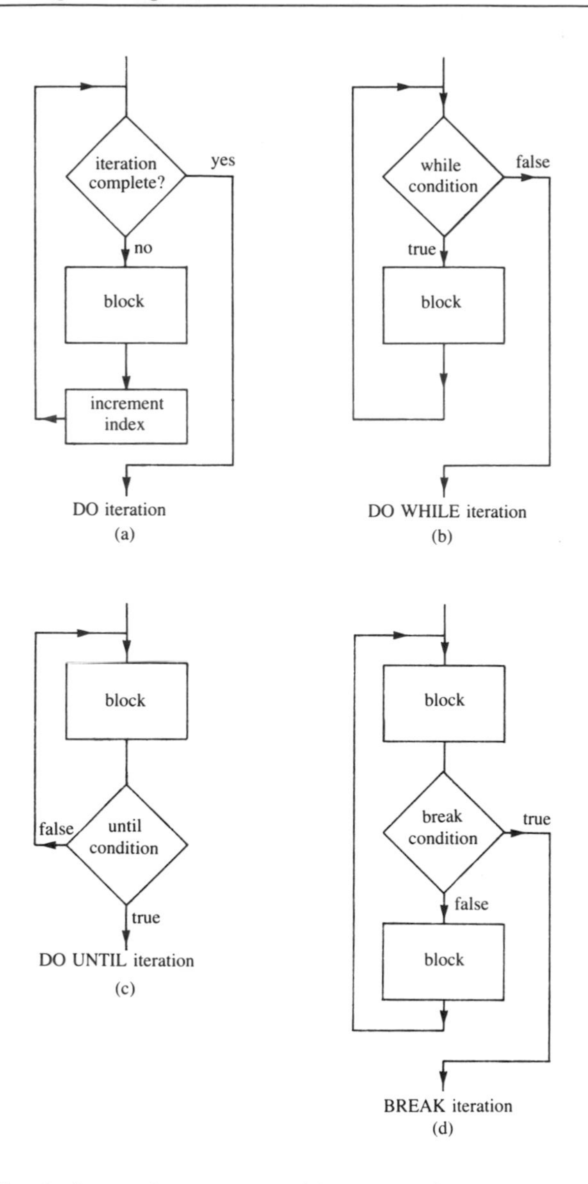

Figure 1.1.1. Standard control structures used in structured programming: (a) DO iteration; (b) DO WHILE iteration; (c) DO UNTIL iteration; (d) BREAK iteration; (e) IF structure; (f) obsolete form of DO iteration found in **FORTRAN-66**, where the block is executed once even if the iteration condition is initially not satisfied.

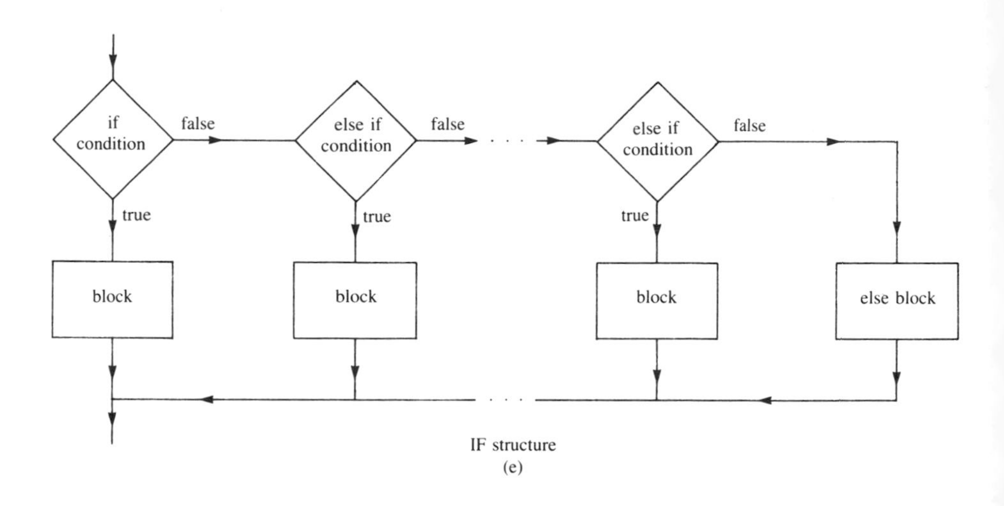

IF structure
(e)

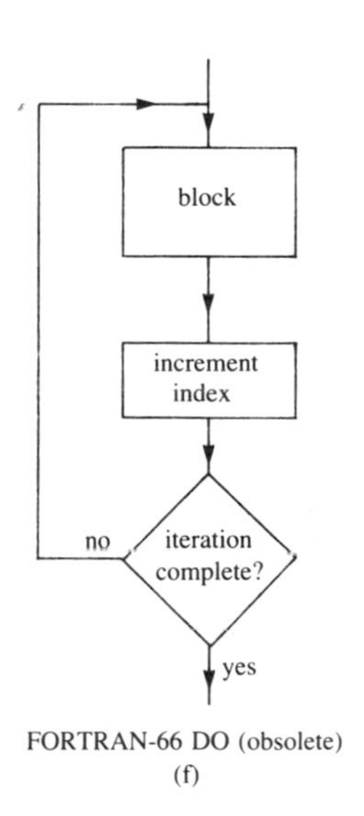

FORTRAN-66 DO (obsolete)
(f)

Figure 1.1.1. Standard control structures used in structured programming (see caption on previous page).

Catalog of Standard Structures

Iteration. In FORTRAN, simple iteration is performed with a DO-loop, for example

```
      DO 10 J=2,1000
          B(J)=A(J-1)
          A(J-1)=J
10    CONTINUE
```

Notice how we always indent the block of code that is acted upon by the control structure, leaving unindented the structure itself. In Pascal the structure would be

```
FOR j := 2 TO 1000 DO BEGIN
    b[j] := a[j-1];
    a[j-1] := j
END;
```

The statement label 10 in the above FORTRAN example is a tame label. Some better nonstandard FORTRAN's, notably Digital Equipment Corporations's VAX-11 FORTRAN, provide a language extension that obviates the tame label,

```
DO J=2,1000
    B(J)=A(J-1)
    A(J-1)=J
ENDDO
```

In fact, we consider it a terrible mistake that the American National Standard for FORTRAN-77 (ANSI X3.9–1978) failed to provide an ENDDO or equivalent construction. This mistake by the people who write standards, whoever they are, has presented us with a painful quandary in preparing this book: Do we stick to the standard, and clutter our programs with tame labels? Or do we adopt a nonstandard FORTRAN like VAX-11?

We think that standards, even imperfect standards, are terribly important and highly necessary in a time of rapid expansion in the use of computers. Therefore, not without reluctance, we decided to live within the ANSI FORTRAN-77 standard in all respects except one (see below). We also decided, however, to use typography to mitigate the standard's deficiencies, at least in the printed version of this book. Whenever a tame label occurs, we enclose it in a ruled box. That means that, when you are reading the programs, *you should never look inside the boxes.* The example above, in our adopted typographical format, is

```
DO[10] J=2,1000
    B(J)=A(J-1)
    A(J-1)=J
  [10]CONTINUE
```

Quickly you will come to recognize statements like [10]CONTINUE as, functionally, ENDDO's. Notice that the ruled boxes at the beginning and end are

matched in level of indentation. (The computer doesn't care about this, but your eye does.) A nested DO loop looks like this:

```
DO 12 J=1,20
    S(J)=0.
    DO 11 K=5,10
        S(J)=S(J)+A(J,K)
    11 CONTINUE
12 CONTINUE
```

IF structure. In this structure the FORTRAN-77 standard is exemplary, similar to Pascal, to Algol and other languages. Here is a working program which consists dominantly of IF control statements:

```
FUNCTION JULDAY(MM,ID,IYYY)
    In this routine JULDAY returns the Julian Day Number which begins at noon of the calendar
    date specified by month MM, day ID, and year IYYY, all integer variables. Positive year
    signifies A.D.; negative, B.C. Remember that the year after 1 B.C. was 1 A.D.
PARAMETER (IGREG=15+31*(10+12*1582))  Gregorian Calendar was adopted on Oct. 15, 1582.
IF (IYYY.EQ.0) PAUSE 'There is no Year Zero.'
IF (IYYY.LT.0) IYYY=IYYY+1
IF (MM.GT.2) THEN                  Here is an example of a block IF-structure.
    JY=IYYY
    JM=MM+1
ELSE
    JY=IYYY-1
    JM=MM+13
ENDIF
JULDAY=INT(365.25*JY)+INT(30.6001*JM)+ID+1720995   Uses built-in function for integer part.
IF (ID+31*(MM+12*IYYY).GE.IGREG) THEN   Test whether to change to Gregorian Calendar.
    JA=INT(0.01*JY)
    JULDAY=JULDAY+2-JA+INT(0.25*JA)
ENDIF
RETURN
END
```

(Astronomers number each 24-hour period, starting and ending at *noon*, with a unique integer, the Julian Day Number. Julian Day Zero was a very long time ago; a convenient reference point is that Julian Day 2440000 began at noon of May 23, 1968. If you know the Julian Day Number that begins at noon of a given calendar date, then the day of the week of that date is obtained by adding 1 and taking the result modulo base 7; a zero answer corresponds to Sunday, 1 to Monday, ..., 6 to Saturday.)

Do-While iteration. Most good languages, except FORTRAN, provide for structures like the following Pascal example:

```
WHILE (n<1000) DO BEGIN
    n:=2*n;
    j:=j+1
END;
```

In fact, VAX-11 FORTRAN has

```
DO WHILE (N.LT.1000)
    N=2*N
    J=J+1
ENDDO
```

Within the FORTRAN-77 standard, however, the structure requires a tame label:

```
[17]IF (N.LT.1000) THEN
    N=2*N
    J=J+1
GOTO[17]
ENDIF
```

There are other ways of constructing a Do-While in FORTRAN, but we try to use the above format consistently. You will quickly get used to a statement like [17]IF as signaling this structure. Notice that the two final statements are not indented, since they are part of the control structure, not of the inside block.

Do-Until iteration. In Pascal this is rendered as

```
REPEAT
    n:=n DIV 2;                 Pascal's integer divide is DIV
    k:=k+1;
UNTIL (n=1);
```

In FORTRAN we will consistently write

```
[19]CONTINUE
    N=N/2
    K=K+1
IF (N.NE.1) GOTO[19]
```

Break iteration. In this case, you have a loop which is repeated indefinitely until some condition *tested somewhere in the middle of the loop* (and possibly tested in more than one place) becomes true. At that point you wish to exit the loop and proceed with what comes after it. Neither standard Pascal nor standard FORTRAN make this structure accessible without labels. We will try to avoid using the structure when we can. Sometimes, however, it is plainly necessary. We do not have the patience to argue with the designers of computer languages over this point. In FORTRAN we write

```
[13]CONTINUE
    [statements before the test]
    IF (···) GOTO[14]
    [statements after the test]
GOTO[13]
[14]CONTINUE
```

Here is a program that uses several different iteration structures. One of us was once asked, for a scavenger hunt, to find the date of a Friday the

Thirteenth on which the moon was full. This is a program which accomplishes that task, giving incidentally all other Friday the Thirteenth's as a by-product.

```
      PROGRAM BADLUK
      PARAMETER (TIMZON=-5./24.)            Time zone -5 is Eastern Standard Time.
      DATA IYBEG,IYEND /1900,2000/          The range of dates to be searched.
      WRITE (*,'(1X,A,I5,A,I5)') 'Full moons on Friday the 13th from',
     *      IYBEG,' to',IYEND
      DO 12 IYYY=IYBEG,IYEND                Loop over each year,
          DO 11 IM=1,12                     and each month.
              JDAY=JULDAY(IM,13,IYYY)       Is the thirteenth a Friday?
              IDWK=MOD(JDAY+1,7)
              IF(IDWK.EQ.5) THEN
                  N=12.37*(IYYY-1900+(IM-0.5)/12.)
        This value N is a first approximation to how many full moons have occurred since 1900.
        We will feed it into the phase routine and adjust it up or down until we determine that
        our desired 13th was or was not a full moon. The variable ICON signals the direction
        of adjustment.
                  ICON=0
1                 CALL FLMOON(N,2,JD,FRAC)          Get date of full moon N.
                  IFRAC=NINT(24.*(FRAC+TIMZON))     Convert to hours in correct time zone.
                  IF(IFRAC.LT.0)THEN                Convert from Julian Days beginning at noon
                      JD=JD-1                       to civil days beginning at midnight.
                      IFRAC=IFRAC+24
                  ENDIF
                  IF(IFRAC.GT.12)THEN
                      JD=JD+1
                      IFRAC=IFRAC-12
                  ELSE
                      IFRAC=IFRAC+12
                  ENDIF
                  IF(JD.EQ.JDAY)THEN                Did we hit our target day?
                      WRITE (*,'(/1X,I2,A,I2,A,I4)') IM,'/',13,'/',IYYY
                      WRITE (*,'(1X,A,I2,A)') 'Full moon ',IFRAC,
     *                    ' hrs after midnight (EST).'
        Don't worry if you are unfamiliar with FORTRAN's esoteric input/output statements; very
        few programs in this book do any input/output.
                      GOTO 2            Part of the break-structure, case of a match.
                  ELSE                  Didn't hit it.
                      IC=ISIGN(1,JDAY-JD)
                      IF(IC.EQ.-ICON) GOTO 2        Another break, case of no match.
                      ICON=IC
                      N=N+IC
                  ENDIF
                  GOTO 1
2                 CONTINUE
              ENDIF
11        CONTINUE
12    CONTINUE
      END
```

If you are merely curious, there were (or will be) occurrences of a full moon on Friday the Thirteenth (Eastern Standard Time) on: 3/13/1903, 10/13/1905, 6/13/1919, 1/13/1922, 11/13/1970, 2/13/1987, 10/13/2000, 9/13/2019, and 8/13/2049.

You may have noticed that, by its looping over the months and years, the program BADLUK avoids using any algorithm for converting a Julian Day Number back into a calendar date. A routine for doing just this is not very interesting structurally, but it is occasionally useful:

```
SUBROUTINE CALDAT(JULIAN,MM,ID,IYYY)
   Inverse of the function JULDAY given above. Here JULIAN is input as a Julian Day Number,
   and the routine outputs MM,ID, and IYYY as the month, day, and year on which the
   specified Julian Day started at noon.
PARAMETER (IGREG=2299161)          Cross-over to Gregorian Calendar
IF(JULIAN.GE.IGREG)THEN            produces this correction,
   JALPHA=INT(((JULIAN-1867216)-0.25)/36524.25)
   JA=JULIAN+1+JALPHA-INT(0.25*JALPHA)
ELSE                               or else no correction.
   JA=JULIAN
ENDIF
JB=JA+1524
JC=INT(6680.+((JB-2439870)-122.1)/365.25)
JD=365*JC+INT(0.25*JC)
JE=INT((JB-JD)/30.6001)
ID=JB-JD-INT(30.6001*JE)
MM=JE-1
IF(MM.GT.12)MM=MM-12
IYYY=JC-4715
IF(MM.GT.2)IYYY=IYYY-1
IF(IYYY.LE.0)IYYY=IYYY-1
RETURN
END
```

Other "standard" structures. Our advice is to avoid them. Every programming language has some number of "goodies" that the designer just couldn't resist throwing in. They seemed like a good idea at the time. Unfortunately they don't stand the *test* of time! Your program becomes difficult to translate into other languages, and difficult to read (because rarely used structures are unfamiliar to the reader). You can almost always accomplish the supposed conveniences of these structures in other ways. Try to do so with the above standard structures, which really *are* standard. If you can't, then use straightforward, that is unstructured, tests and GOTO's. This will introduce real (not tame) statement labels, whose very existence will warn the reader to give special thought to the program's control flow.

In FORTRAN we consider the ill-advised control structures to be

- assigned GOTO and ASSIGN statements
- computed GOTO statement
- arithmetic IF statement

In Pascal an example is

- CASE ... OF

Some Habits and Departures from ANSI Standard

All programmers have their individual habits or "tics". Mentioning a few of ours here will make it easier for you to read the programs in this book.

- Often, when a subroutine or procedure is to be passed some integer N, it needs to have a preset value for the largest possible value that will be passed. We habitually call this NMAX, and set it in a PARAMETER statement (in Pascal, a CONST declaration). When we say in a comment, "largest expected value of N," we do not mean to imply that the program will fail algorithmically for larger values, but only that NMAX must be altered.

- We habitually use M and N to refer to the logical dimensions of a matrix, MP and NP to refer to the physical dimensions. (These important concepts are detailed in §2.0 and Figure 2.0.1.)
- The variable DUM, when it occurs, is usually a temporary "dummy" without any particular mnemonic significance.
- A number represented by TINY, usually a parameter, is supposed to be much smaller than any number of interest to you, but not so small that it underflows. Its use is usually prosaic, to prevent divide checks in some circumstances.

The printed FORTRAN programs in this book, if typed into a computer exactly as written, violate the FORTRAN-77 standard in a few trivial ways. These violations are a matter of typographical convenience, to make the text more readable, and they are corrected in the machine-readable materials which are coordinated with this book. The violations are detailed as follows:

The FORTRAN-77 standard provides that variables internal to a subroutine are *not* guaranteed to be saved between successive calls to that subroutine unless the variable is declared global in a SAVE statement. (This allows stack-oriented implementations.) If this part of the standard were "enforced," a large fraction of all programs written by all living programmers would immediately cease to function. Only a few compilers *dare* to implement this part of the standard, therefore. According to the standard, however, the program line

```
SAVE
```

should be inserted as the second line of several of the subroutines in this book. (If you are overly cautious, you can insert SAVE in all subroutines.) In Pascal, incidentally, we do follow that language's standard: any variable that is to be saved is explicitly, globally defined, external to the procedure (see introduction to Pascal section). In case of doubt, the FORTRAN programmer can usefully look at the comments in the corresponding Pascal routine, to ascertain whether any variables need to be SAVEd.

Standard FORTRAN reads no more than 72 characters on a line and ignores input from column 73 onward. Longer statements are broken up onto "continuation lines." In the printed programs in this book, some lines contain more than 72 characters. The break to a continuation line is not always shown explicitly, but it should be inserted if you type the program into a computer.

In standard FORTRAN, columns 1 through 6 on each line are used variously for (i) statement labels, (ii) signaling a comment line, and (iii) signaling a continuation line. We simplify the format slightly: To the left of the "program left margin," an integer is a statement label (not a "tame label" as described above), an asterisk (*) indicates a continuation line, and a "C" indicates a comment line. Comment lines shown in this way are generally "commented-out program lines" that are separately explained in each instance.

REFERENCES AND FURTHER READING:

Kernighan, B.W. 1978, *The Elements of Programming Style* (New York: McGraw-Hill).

Yourdon, E. 1975, *Techniques of Program Structure and Design* (Englewood Cliffs, N.J.: Prentice-Hall).

Meissner, L.P. and Organick, E.I. 1980, *Fortran 77 Featuring Structured Programming* (Reading, Mass.: Addison-Wesley).

Hoare, C.A.R. 1981, "The Emperor's Old Clothes" in *Communications of the ACM*, vol. 24, no. 2 (Feb.), p. 75.

VAX-11 FORTRAN Language Reference Manual (AA-D034C-TE) 1982 (Maynard, Mass.: Digital Equipment Corporation).

IEEE Standard Pascal Computer Programming Language (ANSI/IEEE 7780 X3.97-1983) 1983 (New York: Wiley or IEEE).

Meeus, Jean 1982, *Astronomical Formulae for Calculators*, Second Edition, revised and enlarged (Richmond, Virginia: Willmann-Bell).

1.2 Error, Accuracy, and Stability

Although we assume no prior training of the reader in formal numerical analysis, we will need to presume a common understanding of a few key concepts. We will define these briefly in this section.

Computers store numbers not with infinite precision but rather in some approximation that can be packed into a fixed number of *bits* (binary digits) or *bytes* (groups of 8 bits). Almost all computers allow the programmer a choice among several different such *representations* or *data types.* Data types can differ in the number of bits utilized (the *wordlength*), but also in the more fundamental respect of whether the stored number is represented in *fixed point* (also called *integer*) or *floating point* (also called *real*) format.

A number in integer representation is exact. Arithmetic between numbers in integer representation is also exact, with the provisos that (i) the answer is not outside the range of (usually, signed) integers that can be represented, and (ii) that division is interpreted as producing an integer result, throwing away any integer remainder.

In floating point representation, a number is represented internally by a sign bit s (interpreted as plus or minus), an exact integer exponent e, and an exact positive integer mantissa M. Taken together these represent the number

$$s \times M \times B^{e-E} \tag{1.2.1}$$

where B is the base of the representation (usually $B = 2$, but sometimes $B = 16$), and E is the *bias* of the exponent, a fixed integer constant for any given machine and representation. An example is shown in Figure 1.2.1.

Several floating-point bit patterns can represent the same number. If $B = 2$, for example, a mantissa with leading (high-order) zero bits can be left-shifted, i.e. multiplied by a power of 2, if the exponent is decreased by a compensating amount. Bit patterns which are "as left-shifted as they can be" are termed *normalized.* Most computers always produce normalized

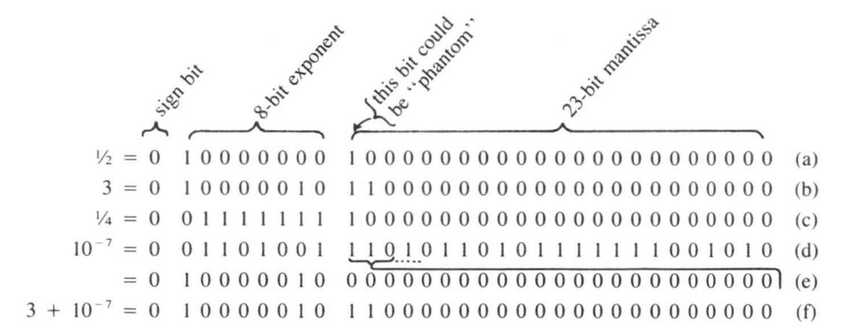

Figure 1.2.1. Floating point representations of numbers in a typical 32-bit (4-byte) format. (a) the number 1/2 (note the bias in the exponent); (b) the number 3; (c) the number 1/4; (d) the number 10^{-7}, represented to machine accuracy; (e) the same number 10^{-7}, but shifted so as to have the same exponent as the number 3; with this shifting, all significance is lost and 10^{-7} becomes zero; shifting to a common exponent must occur before two numbers can be added; (f) sum of the numbers $3+10^{-7}$, which equals 3 to machine accuracy. Even though 10^{-7} can be represented accurately by itself, it cannot accurately be added to a much larger number.

results, since these don't waste any bits of the mantissa and thus allow a greater accuracy of the representation. Since the high-order bit of a properly normalized mantissa (when $B = 2$) is *always* one, most computers don't store this bit at all, giving one extra bit of significance.

Arithmetic among numbers in floating-point representation is not exact, even if the operands happen to be exactly represented (i.e. have exact values in the form of equation 1.2.1). For example, two floating numbers are added by first right-shifting (dividing by two) the mantissa of the smaller (in magnitude) one, simultaneously increasing its exponent, until the two operands have the same exponent. Low-order (least significant) bits of the smaller operand are lost by this shifting. If the two operands differ too greatly in magnitude, then the smaller operand is effectively replaced by zero, since it is right-shifted to oblivion.

The smallest (in magnitude) floating point number which, when added to the floating point number 1.0, produces a floating point result different from 1.0 is termed the *machine accuracy* ϵ_m. A typical computer with $B = 2$ and a 32-bit wordlength has ϵ_m around 3×10^{-8}. Roughly speaking, the machine accuracy ϵ_m is the fractional accuracy to which floating point numbers are represented, corresponding to a change of one in the least significant bit of the mantissa. Pretty much any arithmetic operation among floating numbers should be thought of as introducing an additional fractional error of at least ϵ_m. This type of error is called *roundoff error.*

It is important to understand that ϵ_m is not the smallest floating-point number that can be represented on a machine. *That* number depends on how many bits there are in the exponent, while ϵ_m depends on how many bits there are in the mantissa.

Roundoff errors accumulate with increasing amounts of calculation. If, in the course of obtaining a calculated value, you perform N such arithmetic operations, you *might* be so lucky as to have a total roundoff error on the order of $\sqrt{N}\epsilon_m$, if the roundoff errors come in randomly up or down. (The square root comes from a random-walk.) However, this estimate can be very

badly off the mark for two reasons:

(i) It very frequently happens that the regularities of your calculation, or the peculiarities of your computer, cause the roundoff errors to accumulate preferentially in one direction. In this case the total will be of order $N\epsilon_m$.

(ii) Some especially unfavorable occurrences can vastly increase the roundoff error of single operations. Generally these can be traced to the subtraction of two very nearly equal numbers, giving a result whose only significant bits are those (few) low-order ones in which the operands differed. You might think that such a "coincidental" subtraction is unlikely to occur. Not always so. Some mathematical expressions magnify its probability of occurrence tremendously. For example, in the familiar formula for the solution of a quadratic equation,

$$x = \frac{-b + \sqrt{b^2 - 4ac}}{2a} \qquad (1.2.2)$$

the addition becomes delicate and roundoff-prone whenever $ac \ll b^2$. (In §5.5 we will learn how to avoid the problem in this particular case.)

Roundoff error is a characteristic of computer hardware. There is another, different, kind of error that is a characteristic of the program or algorithm used, independent of the hardware on which the program is executed. Many numerical algorithms compute "discrete" approximations to some desired "continuous" quantity. For example, an integral is evaluated numerically by computing a function at a discrete set of points, rather than at "every" point. Or, a function may be evaluated by summing a finite number of leading terms in its infinite series, rather than all infinity terms. In cases like this, there is an adjustable parameter, e.g. the number of points or of terms, such that the "true" answer is obtained only when that parameter goes to infinity. Any practical calculation is done with a finite, but sufficiently large, choice of that parameter.

The discrepancy between the true answer and the answer obtained in a practical calculation is called the *truncation error*. Truncation error would persist even on a hypothetical, "perfect" computer that had an infinitely accurate representation and no roundoff error. As a general rule there is not much that a programmer can do about roundoff error, other than to choose algorithms that do not magnify it unnecessarily (see discussion of "stability" below). Truncation error, on the other hand, is entirely under the programmer's control. In fact, it is only a slight exaggeration to say that clever minimization of truncation error is practically the entire content of the field of numerical analysis!

Most of the time, truncation error and roundoff error do not strongly interact with one another. A calculation can be imagined as having, first, the truncation error that it would have if run on an infinite-precision computer, "plus" the roundoff error associated with the number of operations performed.

Sometimes, however, an otherwise attractive method can be *unstable*. This means that any roundoff error that becomes "mixed into" the calcula-

tion at an early stage is successively magnified until it comes to swamp the true answer. An unstable method would be useful on a hypothetical, perfect computer; but in this imperfect world it is necessary for us to require that algorithms be stable – or if unstable that we use them with great caution.

Here is a simple, if somewhat artificial, example of an unstable algorithm: Suppose that it is desired to calculate all integer powers of the so-called "Golden Mean," the number given by

$$\phi \equiv \frac{\sqrt{5}-1}{2} \approx 0.61803398 \tag{1.2.3}$$

It turns out (you can easily verify) that the powers ϕ^n satisfy a simple recursion relation,

$$\phi^{n+1} = \phi^{n-1} - \phi^n \tag{1.2.4}$$

Thus, knowing the first two values $\phi^0 = 1$ and $\phi^1 = 0.61803398$, we can successively apply (1.2.4) performing only a single subtraction, rather than a slower multiplication by ϕ, at each stage.

Unfortunately, the recurrence (1.2.4) also has *another* solution, namely the value $-\frac{1}{2}(\sqrt{5}+1)$. Since the recurrence is linear, and since this undesired solution has magnitude greater than unity, any small admixture of it introduced by roundoff errors will grow exponentially. On a typical machine with 32-bit wordlength, (1.2.4) starts to give completely wrong answers by about $n = 16$, at which point ϕ^n is only down to 10^{-4}. The recurrence (1.2.4) is *unstable*, and cannot be used for the purpose stated.

We will encounter the question of stability in many more sophisticated guises, later in this book.

REFERENCES AND FURTHER READING:

Stoer, J., and Bulirsch, R. 1980, *Introduction to Numerical Analysis* (New York: Springer-Verlag), Chapter 1.

Johnson, Lee W., and Riess, R. Dean. 1982, *Numerical Analysis*, 2nd ed. (Reading, Mass.: Addison-Wesley), §1.3.

Chapter 2. Solution of Linear Algebraic Equations

2.0 Introduction

A set of linear algebraic equations looks like this:

$$
\begin{aligned}
a_{11}x_1 + a_{12}x_2 + a_{13}x_3 + \cdots + a_{1N}x_N &= b_1 \\
a_{21}x_1 + a_{22}x_2 + a_{23}x_3 + \cdots + a_{2N}x_N &= b_2 \\
a_{31}x_1 + a_{32}x_2 + a_{33}x_3 + \cdots + a_{3N}x_N &= b_3 \\
\cdots \qquad & \quad \cdots \\
a_{M1}x_1 + a_{M2}x_2 + a_{M3}x_3 + \cdots + a_{MN}x_N &= b_M
\end{aligned}
\tag{2.0.1}
$$

Here the N unknowns x_j, $j = 1, 2, \ldots, N$ are related by M equations. The coefficients a_{ij} with $i = 1, 2, \ldots, M$ and $j = 1, 2, \ldots, N$ are known numbers, as are the *right-hand side* quantities b_i, $i = 1, 2, \ldots, M$.

Nonsingular versus Singular Sets of Equations

If $N = M$ then there are as many equations as unknowns, and there is a good chance of solving for a unique solution set of x_j's. Analytically, there can fail to be a unique solution if one or more of the M equations is a linear combination of the others, a condition called *row degeneracy*, or if all equations contain certain variables only in exactly the same linear combination, called *column degeneracy*. (For square matrices, a row degeneracy implies a column degeneracy, and vice versa.) A set of equations that is degenerate is called *singular*. We will consider singular matrices in some detail in §2.9.

Numerically, at least two additional things can go wrong:

- While not exact linear combinations of each other, some of the equations may be so close to linearly dependent that roundoff errors in the machine render them linearly dependent at some stage in the solution process. In this case your numerical procedure will fail, and it can tell you that it has failed.

- Accumulated roundoff errors in the solution process can swamp the true solution. This problem particularly emerges if N is too large. The numerical procedure does not fail algorithmically. However, it returns a set of x's that are wrong, as can be discovered by direct substitution back into the original equations. The closer a set of equations is to being singular, the more likely this is to happen, since increasingly close cancellations will occur during the solution. In fact, the preceding item can be viewed as the special case where the loss of significance is unfortunately total.

Much of the sophistication of complicated "linear equation-solving packages" is devoted to the detection and/or correction of these two pathologies. As you work with large linear sets of equations, you will develop a feeling for when such sophistication is needed. It is difficult to give any firm guidelines, since there is no such thing as a "typical" linear problem. But here is a rough idea: Linear sets with N as large as 20 or 50 can be routinely solved in single precision (32 bit floating representations) without resorting to sophisticated methods, *if* the equations are not close to singular. With double precision (60 or 64 bits), this number can readily be extended to N as large as a few hundred, by which point the limiting factor is almost always machine time, not accuracy.

Even larger linear sets, N in the thousands, can be solved when the coefficients are sparse (that is, mostly zero), by methods which take advantage of the sparseness. We discuss this further in §2.10.

At the other end of the spectrum, one seems just as often to encounter linear problems which, by their underlying nature, are close to singular. In this case, you *might* need to resort to sophisticated methods even for the case of $N = 10$ (though rarely for $N = 5$). Singular value decomposition (§2.9) is a technique which can sometimes turn singular problems into nonsingular ones, in which case additional sophistication becomes unnecessary.

Matrices

Equation (2.0.1) can be written in matrix form as

$$\mathbf{A} \cdot \mathbf{x} = \mathbf{b} \tag{2.0.2}$$

Here the raised dot denotes matrix multiplication, $\mathbf{A}$ is the matrix of coefficients, and $\mathbf{b}$ is the right-hand side written as a column vector,

$$\mathbf{A} = \begin{bmatrix} a_{11} & a_{12} & \dots & a_{1N} \\ a_{21} & a_{22} & \dots & a_{2N} \\ & \dots & & \\ a_{M1} & a_{M2} & \dots & a_{MN} \end{bmatrix} \qquad \mathbf{b} = \begin{bmatrix} b_1 \\ b_2 \\ \dots \\ b_M \end{bmatrix} \tag{2.0.3}$$

By convention, the first index on an element a_{ij} denotes its row, the second index its column. A computer will store the matrix $\mathbf{A}$ as a two-dimensional array. However, computer memory is numbered sequentially by its address, and so is intrinsically one-dimensional. Therefore the two-dimensional array A will, at the hardware level, either be *stored by columns* in the order

$$a_{11}, a_{21}, \ldots, a_{M1}, a_{12}, a_{22}, \ldots, a_{M2}, a_{31}, \ldots, a_{1N}, a_{2N}, \ldots a_{MN}$$

or else *stored by rows* in the order

$$a_{11}, a_{12}, \ldots, a_{1M}, a_{21}, a_{22}, \ldots, a_{2M}, a_{13}, \ldots, a_{N1}, a_{N2}, \ldots a_{NM}$$

FORTRAN always stores by columns, and user programs are generally allowed to exploit this fact to their advantage. Pascal generally stores by rows, but user programs are discouraged from using this fact, the one exception being that whole rows A[i,j], j=1,..., M can be referenced as A[i]. Note one confusing point in the terminology, that a matrix which is stored by columns (as in FORTRAN) has its *row* (i.e. first) index changing most rapidly as one goes linearly through memory, the opposite of a car's odometer!

For most purposes you don't *need* to know what the order of storage is, since you reference an element by its two-dimensional address (a_{34} = A(3,4)). It is, however, *essential* that you understand the difference between an array's *physical dimensions* and its *logical dimensions.* When you pass an array to a subroutine or other procedure, you must, in general, tell the subroutine *both* of these dimensions. The distinction between them is this: It may happen that you have a 4×4 matrix stored in an array dimensioned as 10×10. This occurs most frequently in practice when you have dimensioned to the largest expected value of N, but are at the moment considering a value of N smaller than that largest possible one. In the example posed, the 16 elements of the matrix do not occupy 16 consecutive memory locations. Rather they are spread out among the 100 dimensioned locations of the array as if the whole 10×10 matrix were filled. Figure 2.0.1 shows an additional example.

If you have a subroutine to invert a matrix, its call might typically look like this

```
CALL MATINV(A,AI,N,NP)
```

Here the subroutine has to be told both the logical size of the matrix that you want to invert (here N=4), and the physical size of the array in which it is stored (here NP=10).

This seems like a trivial point, and we are sorry to belabor it. But it turns out that *most* reported failures of standard linear equation and matrix manipulation packages are due to user errors in passing inappropriate logical or physical dimensions!

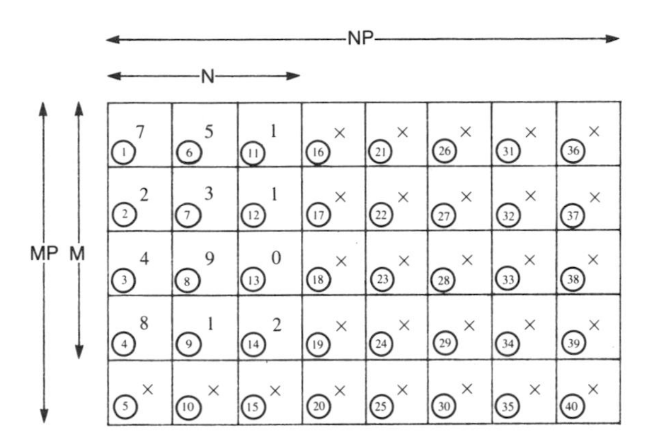

Figure 2.0.1. A matrix of logical dimension `M` by `N` is stored in an array of physical dimension `MP` by `NP`. Locations marked by "x" contain extraneous information which may be left over from some previous use of the physical array. Circled numbers show the actual ordering of the array in computer memory, not usually relevant to the programmer. Note, however, that the logical array does not occupy consecutive memory locations. To locate an `(I,J)` element correctly, a subroutine must be told `MP` and `NP`, not just `I` and `J`.

Tasks of Computational Linear Algebra

We will consider the following tasks as falling in the general purview of this chapter:

- Solution of the matrix equation $\mathbf{A} \cdot \mathbf{x} = \mathbf{b}$ for an unknown vector $\mathbf{x}$, where $\mathbf{A}$ is a square matrix of coefficients, raised dot denotes matrix multiplication, and $\mathbf{b}$ is a known right-hand side vector (§2.1–§2.3).
- Solution of more than one matrix equation $\mathbf{A} \cdot \mathbf{x}_j = \mathbf{b}_j$, for a set of vectors $\mathbf{x}_j$, $j = 1, 2, \ldots$, each corresponding to a different, known right-hand side vector $\mathbf{b}_j$. In this task the key simplification is that the matrix $\mathbf{A}$ is held constant, while the right-hand sides, the $\mathbf{b}$'s, are changed (§2.1–§2.3).
- Calculation of the matrix $\mathbf{A}^{-1}$ which is the matrix inverse of a square matrix $\mathbf{A}$, i.e. $\mathbf{A} \cdot \mathbf{A}^{-1} = \mathbf{A}^{-1} \cdot \mathbf{A} = \mathbf{1}$, where $\mathbf{1}$ is the identity matrix (all zeros except for ones on the diagonal). This task is equivalent, for an $N \times N$ matrix $\mathbf{A}$, to the previous task with N different $\mathbf{b}_j$'s ($j = 1, 2, \ldots N$), namely the unit vectors ($\mathbf{b}_j =$ all zero elements except for 1 in the jth component). The corresponding $\mathbf{x}$'s are then the columns of the matrix inverse of $\mathbf{A}$ (§2.1 and §2.4).
- Calculation of the determinant of a square matrix $\mathbf{A}$ (§2.5).

If $M < N$, or if $M = N$ but the equations are degenerate, then there are effectively fewer equations than unknowns. In this case there can either be no solution, or else more than one solution vector $\mathbf{x}$. In the latter event, the solution space consists of a particular solution $\mathbf{x}_p$ added to any linear

combination of (typically) $N-M$ vectors (which are said to be in the nullspace of the matrix $\mathbf{A}$). The task of finding the solution space of $\mathbf{A}$ is called

- Singular value decomposition of a matrix $\mathbf{A}$.

This subject is treated in §2.9.

In the opposite case there are more equations than unknowns, $M > N$. When this occurs there is, in general, no solution vector $\mathbf{x}$ to equation (2.0.1), and the set of equations is said to be *overdetermined.* It happens frequently, however, that the best "compromise" solution is sought, the one which comes closest to satisfying all equations simultaneously. If closeness is defined in the least squares sense, i.e., that the sum of the squares of the differences between the left and right-hand sides of equation (2.0.1) be minimized, then the overdetermined linear problem reduces to a (usually) solvable linear problem, called the

- Linear least-squares problem.

The reduced set of equations to be solved can be written as the $N \times N$ set of equations

$$(\mathbf{A}^T \cdot \mathbf{A}) \cdot \mathbf{x} = (\mathbf{A}^T \cdot \mathbf{b}) \tag{2.0.4}$$

where $\mathbf{A}^T$ denotes the transpose of the matrix $\mathbf{A}$. Equations (2.0.4) are called the *normal equations* of the linear least squares problem. There is a close connection between singular value decomposition and the linear least squares problem, and the latter is also discussed in §2.9. You should be warned that direct solution of the normal equations (2.0.4) is not generally the best way to find least-squares solutions.

Some other topics in this chapter include

- Iterative improvement of a solution (§2.7)
- Various special forms: tridiagonal (§2.6), Toeplitz (§2.8), Vandermonde (§2.8), sparse (§2.10)
- Strassen's "fast matrix inversion" (§2.11).

Standard Subroutine Packages

We cannot hope, in this chapter or in this book, to tell you everything there is to know about the tasks that have been defined above. In many cases you will have no alternative but to use sophisticated black-box program packages. Several good ones are available. LINPACK was developed at Argonne National Laboratories and deserves particular mention because it is published, documented, and available for free use. Packages available commercially include those in the IMSL and NAG libraries.

You should keep in mind that the sophisticated packages are designed with very large linear systems in mind. They therefore go to great effort to minimize not only the number of operations, but also the required storage. Routines for the various tasks are usually provided in several versions, corresponding to several possible simplifications in the form of the input coefficient matrix: symmetric, triangular, banded, positive definite, etc. If you have a

large matrix in one of these forms, you should certainly take advantage of the increased efficiency provided by these different routines, and not just use the form provided for general matrices.

There is also a great watershed dividing routines which are *direct* (i.e. execute in a predictable number of operations) from routines which are *iterative* (i.e. attempt to converge to the desired answer in however many steps are necessary). Iterative methods become preferable when the battle against loss of significance is in danger of being lost, either due to large N or because the problem is close to singular. We will say only a little about iterative methods in this book, in §2.10 and in Chapter 17. These methods are important, but beyond our scope. We will, however, discuss a technique which is on the borderline between direct and iterative methods, namely the iterative improvement of a solution that has been obtained by direct methods (§2.7).

REFERENCES AND FURTHER READING:

Golub, Gene H., and Van Loan, Charles F. 1983, *Matrix Computations* (Baltimore: Johns Hopkins University Press).

Dongarra, J.J., et al. 1979, *LINPACK User's Guide* (Philadelphia: Society for Industrial and Applied Mathematics).

Forsythe, George E., and Moler, Cleve B. 1967, *Computer Solution of Linear Algebraic Systems* (Englewood Cliffs, N.J.: Prentice-Hall).

Wilkinson, J.H., and Reinsch, C. 1971, *Linear Algebra*, vol. II of *Handbook for Automatic Computation* (New York: Springer-Verlag).

Westlake, Joan R. 1968, *A Handbook of Numerical Matrix Inversion and Solution of Linear Equations* (New York: Wiley).

NAG Fortran Library Manual Mark 8, 1980 (NAG Central Office, 7 Banbury Road, Oxford OX26NN, U.K.), chapter F01.

IMSL Library Reference Manual, 1980, ed. 8 (IMSL Inc., 7500 Bellaire Boulevard, Houston TX 77036), chapter L.

Johnson, Lee W., and Riess, R. Dean. 1982, *Numerical Analysis*, 2nd ed. (Reading, Mass.: Addison-Wesley), chapter 2.

Ralston, Anthony, and Rabinowitz, Philip. 1978, *A First Course in Numerical Analysis*, 2nd ed. (New York: McGraw-Hill), chapter 9.

Stoer, J., and Bulirsch, R. 1980, *Introduction to Numerical Analysis* (New York: Springer-Verlag), chapter 4.

2.1 Gauss-Jordan Elimination

For inverting a matrix, *Gauss-Jordan elimination* is about as efficient as any other method. For solving sets of linear equations, Gauss-Jordan elimination produces *both* the solution of the equations for one or more right-hand side vectors $\mathbf{b}$, and also the matrix inverse $\mathbf{A}^{-1}$. However, its principal weaknesses are (i) that it requires all the right-hand sides to be stored and manipulated at the same time, and (ii) that when the inverse matrix is *not* desired, Gauss-Jordan is three times slower than the best alternative technique for solving a

single linear set (§2.3). The method's principal strength is that it is as stable as any other direct method, perhaps even a bit more stable when full pivoting is used (see below).

If you come along later with an additional right-hand side vector, you can multiply it by the inverse matrix, of course. This does give an answer, but one that is quite susceptible to roundoff error, not nearly as good as if the new vector had been included with the set of right-hand side vectors in the first instance.

For these reasons, Gauss-Jordan elimination should usually not be your method of first choice, either for solving linear equations, or for matrix inversion. The decomposition methods in §2.3 are better. Why do we give you Gauss-Jordan at all? Because it is straightforward, understandable, solid as a rock, and an exceptionally good "psychological" backup for those times that something is going wrong and you think it *might* be your linear-equation solver.

Some people believe that the backup is more than psychological, that Gauss-Jordan elimination is an "independent" numerical method. This turns out to be mostly myth. Except for the relatively minor differences in pivoting, described below, the actual sequence of operations performed in Gauss-Jordan elimination is very closely related to that performed by the routines in the next two sections.

For clarity, and to avoid writing endless ellipses ($\cdots$) we will write out equations only for the case of four equations and four unknowns, and with three different right-hand side vectors that are known in advance. You can write bigger matrices and extend the equations to the case of $N \times N$ matrices, with M sets of right-hand side vectors, in completely analogous fashion. The routine implemented below is, of course, general.

Elimination on Column-Augmented Matrices

Consider the linear matrix equation

$$\begin{bmatrix} a_{11} & a_{12} & a_{13} & a_{14} \\ a_{21} & a_{22} & a_{23} & a_{24} \\ a_{31} & a_{32} & a_{33} & a_{34} \\ a_{41} & a_{42} & a_{43} & a_{44} \end{bmatrix} \cdot \left[\begin{pmatrix} x_{11} \\ x_{21} \\ x_{31} \\ x_{41} \end{pmatrix} \sqcup \begin{pmatrix} x_{12} \\ x_{22} \\ x_{32} \\ x_{42} \end{pmatrix} \sqcup \begin{pmatrix} x_{13} \\ x_{23} \\ x_{33} \\ x_{43} \end{pmatrix} \sqcup \begin{pmatrix} y_{11} & y_{12} & y_{13} & y_{14} \\ y_{21} & y_{22} & y_{23} & y_{24} \\ y_{31} & y_{32} & y_{33} & y_{34} \\ y_{41} & y_{42} & y_{43} & y_{44} \end{pmatrix} \right]$$

$$= \left[\begin{pmatrix} b_{11} \\ b_{21} \\ b_{31} \\ b_{41} \end{pmatrix} \sqcup \begin{pmatrix} b_{12} \\ b_{22} \\ b_{32} \\ b_{42} \end{pmatrix} \sqcup \begin{pmatrix} b_{13} \\ b_{23} \\ b_{33} \\ b_{43} \end{pmatrix} \sqcup \begin{pmatrix} 1 & 0 & 0 & 0 \\ 0 & 1 & 0 & 0 \\ 0 & 0 & 1 & 0 \\ 0 & 0 & 0 & 1 \end{pmatrix} \right] \qquad (2.1.1)$$

Here the raised dot ($\cdot$) signifies matrix multiplication, while the operator $\sqcup$ just signifies column augmentation, that is, removing the abutting parentheses and making a wider matrix out of the operands of the $\sqcup$ operator.

It should not take you long to write out equation (2.1.1) and to see that it simply states that x_{ij} is the ith component ($i = 1, 2, 3, 4$) of the vector

solution of the jth right-hand side ($j = 1, 2, 3$), the one whose coefficients are $b_{ij}, i = 1, 2, 3, 4$; and that the matrix of unknown coefficients y_{ij} is the inverse matrix of a_{ij}. In other words, the matrix solution of

$$[\mathbf{A}] \cdot [\mathbf{x}_1 \sqcup \mathbf{x}_2 \sqcup \mathbf{x}_3 \sqcup \mathbf{Y}] = [\mathbf{b}_1 \sqcup \mathbf{b}_2 \sqcup \mathbf{b}_3 \sqcup \mathbf{1}] \tag{2.1.2}$$

where $\mathbf{A}$ and $\mathbf{Y}$ are square matrices, the $\mathbf{b}_i$'s and $\mathbf{x}_i$'s are column vectors, and $\mathbf{1}$ is the identity matrix, simultaneously solves the linear sets

$$\mathbf{A} \cdot \mathbf{x}_1 = \mathbf{b}_1 \qquad \mathbf{A} \cdot \mathbf{x}_2 = \mathbf{b}_2 \qquad \mathbf{A} \cdot \mathbf{x}_3 = \mathbf{b}_3 \tag{2.1.3}$$

and

$$\mathbf{A} \cdot \mathbf{Y} = \mathbf{1} \tag{2.1.4}$$

Now it is also elementary to verify the following facts about (2.1.1):

- Interchanging any two *rows* of $\mathbf{A}$ and the corresponding *rows* of the $\mathbf{b}$'s and of $\mathbf{1}$, does not change (or scramble in any way) the solution $\mathbf{x}$'s and $\mathbf{Y}$. Rather, it just corresponds to writing the same set of linear equations in a different order.
- Likewise, the solution set is unchanged and in no way scrambled if we replace any row in $\mathbf{A}$ by a linear combination of itself and any other row, as long as we do the same linear combination of the rows of the $\mathbf{b}$'s and $\mathbf{1}$ (which then is no longer the identity matrix, of course).
- Interchanging any two *columns* of $\mathbf{A}$ gives the same solution set only if we simultaneously interchange corresponding *rows* of the $\mathbf{x}$'s and of $\mathbf{Y}$. In other words, this interchange scrambles the order of the rows in the solution. If we do this, we will need to unscramble the solution by restoring the rows to their original order.

Gauss-Jordan elimination uses one or more of the above operations to reduce the matrix $\mathbf{A}$ to the identity matrix. When this is accomplished, the right-hand side becomes the solutions set, as one sees instantly from (2.1.2).

Pivoting

In "Gauss-Jordan elimination with no pivoting," only the second operation in the above list is used. The first row is divided by the element a_{11} (this being a trivial linear combination of the first row with any other row — zero coefficient for the other row). Then the right amount of the first row is subtracted from each other row to make all the remaining a_{i1}'s zero. The first column of $\mathbf{A}$ now agrees with the identity matrix. We move to the second column and divide the second row by a_{22}, then subtract the right amount of the second row from rows 1,3, and 4, so as to make their entries in the second

column zero. The second column is now reduced to the identity form. And so on for the third and fourth columns. As we do these operations to **A**, we of course also do the corresponding operations to the **b**'s and to **1** (which by now no longer resembles the identity matrix in any way!).

Obviously we will run into trouble if we ever encounter a zero element on the (then current) diagonal when we are going to divide by the diagonal element. (The element that we divide by, incidentally, is called the *pivot element* or *pivot.*) Not so obvious, but true, is the fact that Gauss-Jordan elimination with no pivoting (no use of the first or third procedures in the above list) is numerically unstable in the presence of any roundoff error, even when a zero pivot is not encountered. You must *never* do Gauss-Jordan elimination (or Gaussian elimination, see below) without pivoting!

So what *is* this magic pivoting? Nothing more than interchanging rows (*partial pivoting*) or rows and columns (*full pivoting*), so as to put a particularly desirable element in the diagonal position from which the pivot is about to be selected. Since we don't want to mess up the part of the identity matrix that we have already built up, we can choose among elements that are both (i) on rows below (or on) the one that is about to be normalized, and also (ii) on columns to the right (or on) the column we are about to eliminate. Partial pivoting is easier than full pivoting, because we don't have to keep track of the permutation of the solution vector. Partial pivoting makes available as pivots only the elements already in the correct column. It turns out that partial pivoting is "almost" as good as full pivoting, in a sense that can be made mathematically precise, but which need not concern us here (for discussion and references, see Wilkinson 1965). To show you both variants, we do full pivoting in the routine in this section, partial pivoting in §2.3.

We have to state how to recognize a particularly desirable pivot when we see one. The answer to this is not completely known theoretically. It is known, both theoretically and in practice, that simply picking the largest (in magnitude) available element as the pivot is a very good choice. A curiosity of this procedure, however, is that the choice of pivot will depend on the original scaling of the equations. If we take the third linear equation in our original set and multiply it by a factor of a million, it is almost guaranteed that it will contribute the first pivot; yet the underlying solution of the equations is not changed by this multiplication! One therefore sometimes sees routines which choose as pivot that element which *would* have been largest if the original equations had all been scaled to have their largest coefficient normalized to unity. This is called *implicit pivoting.* There is some extra bookkeeping to keep track of the scale factors by which the rows would have been multiplied. (The routines in §2.3 include implicit pivoting, but the routine in this section does not.)

Finally, let us consider the storage requirements of the method. With a little reflection you will see that at every stage of the algorithm, *either* an element of **A** is predictably a one or zero (if it is already in a part of the matrix which has been reduced to identity form) *or else* the exactly corresponding element of the matrix which started as **1** is predictably a one or zero (if its mate in **A** has not been reduced to the identity form). Therefore the matrix **1**

does not have to exist as separate storage: the matrix inverse of **A** is gradually built up in **A** as the original **A** is destroyed. Likewise, the solution vectors **x** can gradually replace the right-hand side vectors **b** and share the same storage, since after each column in **A** is reduced, the corresponding row entry in the **b**'s is never again used.

Here is the routine for Gauss-Jordan elimination with full pivoting:

```
SUBROUTINE GAUSSJ(A,N,NP,B,M,MP)
    Linear equation solution by Gauss-Jordan elimination, equation (2.1.1) above. A is an
    input matrix of N by N elements, stored in an array of physical dimensions NP by NP. B
    is an input matrix of N by M containing the M right-hand side vectors, stored in an array
    of physical dimensions NP by MP. On output, A is replaced by its matrix inverse, and B is
    replaced by the corresponding set of solution vectors.
PARAMETER (NMAX=50)
DIMENSION A(NP,NP),B(NP,MP),IPIV(NMAX),INDXR(NMAX),INDXC(NMAX)
    The integer arrays IPIV, INDXR, and INDXC are used for bookkeeping on the pivoting. NMAX
    should be as large as the largest anticipated value of N.
DO 11 J=1,N
    IPIV(J)=0
  11 CONTINUE
DO 22 I=1,N                          This is the main loop over the columns to be reduced.
    BIG=0.
    DO 13 J=1,N                      This is the outer loop of the search for a pivot element.
        IF(IPIV(J).NE.1)THEN
            DO 12 K=1,N
                IF (IPIV(K).EQ.0) THEN
                    IF (ABS(A(J,K)).GE.BIG)THEN
                        BIG=ABS(A(J,K))
                        IROW=J
                        ICOL=K
                    ENDIF
                ELSE IF (IPIV(K).GT.1) THEN
                    PAUSE 'Singular matrix'
                ENDIF
              12 CONTINUE
        ENDIF
      13 CONTINUE
    IPIV(ICOL)=IPIV(ICOL)+1
    We now have the pivot element, so we interchange rows, if needed, to put the pivot element
    on the diagonal. The columns are not physically interchanged, only relabeled: INDXC(I),
    the column of the Ith pivot element, is the Ith column that is reduced, while INDXR(I) is
    the row in which that pivot element was originally located. If INDXR(I)≠INDXC(I) there
    is an implied column interchange. With this form of bookkeeping, the solution B's will end
    up in the correct order, and the inverse matrix will be scrambled by columns.
    IF (IROW.NE.ICOL) THEN
        DO 14 L=1,N
            DUM=A(IROW,L)
            A(IROW,L)=A(ICOL,L)
            A(ICOL,L)=DUM
          14 CONTINUE
        DO 15 L=1,M
            DUM=B(IROW,L)
            B(IROW,L)=B(ICOL,L)
            B(ICOL,L)=DUM
          15 CONTINUE
    ENDIF
    INDXR(I)=IROW                    We are now ready to divide the pivot row by the pivot element, located
    INDXC(I)=ICOL                        at IROW and ICOL.
    IF (A(ICOL,ICOL).EQ.0.) PAUSE 'Singular matrix.'
    PIVINV=1./A(ICOL,ICOL)
```

```
          A(ICOL,ICOL)=1.
          DO 16 L=1,N
            A(ICOL,L)=A(ICOL,L)*PIVINV
16        CONTINUE
          DO 17 L=1,M
            B(ICOL,L)=B(ICOL,L)*PIVINV
17        CONTINUE
          DO 21 LL=1,N                        Next, we reduce the rows...
            IF(LL.NE.ICOL)THEN                ...except for the pivot one, of course.
              DUM=A(LL,ICOL)
              A(LL,ICOL)=0.
              DO 18 L=1,N
                A(LL,L)=A(LL,L)-A(ICOL,L)*DUM
18            CONTINUE
              DO 19 L=1,M
                B(LL,L)=B(LL,L)-B(ICOL,L)*DUM
19            CONTINUE
            ENDIF
21        CONTINUE
22      CONTINUE                              This is the end of the main loop over columns of the reduction.
        DO 24 L=N,1,-1                        It only remains to unscramble the solution in view of the column
          IF(INDXR(L).NE.INDXC(L))THEN        interchanges. We do this by interchanging pairs of columns
            DO 23 K=1,N                       in the reverse order that the permutation was built up.
              DUM=A(K,INDXR(L))
              A(K,INDXR(L))=A(K,INDXC(L))
              A(K,INDXC(L))=DUM
23          CONTINUE
          ENDIF
24      CONTINUE
        RETURN                                And we are done.
        END
```

REFERENCES AND FURTHER READING:

Carnahan, Brice, Luther, H.A., and Wilkes, James O. 1969, *Applied Numerical Methods* (New York: Wiley), Example 5.2, p.282.

Bevington, Philip R. 1969, *Data Reduction and Error Analysis for the Physical Sciences* (New York: McGraw-Hill), Program B-2, p.298.

Westlake, Joan R. 1968, *A Handbook of Numerical Matrix Inversion and Solution of Linear Equations* (New York: Wiley).

Ralston, Anthony, and Rabinowitz, Philip. 1978, *A First Course in Numerical Analysis*, 2nd ed. (New York: McGraw-Hill), §9.3-1.

2.2 Gaussian Elimination with Backsubstitution

The usefulness of Gaussian elimination with backsubstitution is primarily pedagogical. It stands in between full elimination schemes such as Gauss-Jordan, and triangular decomposition schemes such as will be discussed in the next section. Gaussian elimination reduces a matrix not all the way to the identity matrix, but only halfway, to a matrix whose components on the diagonal and above (say) remain nontrivial. Let us now see what advantages accrue.

Suppose that in doing Gauss-Jordan elimination, as described in §2.1, we at each stage subtract away rows only *below* the then-current pivot element. When a_{22} is the pivot element, for example, we divide the second row by its value (as before), but now use the pivot row to zero only a_{32} and a_{42}, not a_{12} (see equation 2.1.1). Suppose, also, that we do only partial pivoting, never interchanging columns, so that the order of the unknowns never needs to be modified.

Then, when we have done this for all the pivots, we will be left with a reduced equation that looks like this (in the case of a single right-hand side vector):

$$\begin{bmatrix} a'_{11} & a'_{12} & a'_{13} & a'_{14} \\ 0 & a'_{22} & a'_{23} & a'_{24} \\ 0 & 0 & a'_{33} & a'_{34} \\ 0 & 0 & 0 & a'_{44} \end{bmatrix} \cdot \begin{bmatrix} x_1 \\ x_2 \\ x_3 \\ x_4 \end{bmatrix} = \begin{bmatrix} b'_1 \\ b'_2 \\ b'_3 \\ b'_4 \end{bmatrix} \tag{2.2.1}$$

Here the primes signify that the a's and b's do not have their original numerical values, but have been modified by all the row operations in the elimination to this point. The procedure up to this point is termed *Gaussian elimination.*

Backsubstitution

But how do we solve for the x's? The last x (x_4 in this example) is already isolated, namely

$$x_4 = b'_4/a'_{44} \tag{2.2.2}$$

With the last x known we can move to the penultimate x,

$$x_3 = \frac{1}{a'_{33}}[b'_3 - x_4 a'_{34}] \tag{2.2.3}$$

and then proceed with the x before that one. The typical step is

$$x_i = \frac{1}{a'_{ii}}\left[b'_i - \sum_{j=i+1}^{N} a'_{ij}x_j\right] \tag{2.2.4}$$

The procedure defined by equation (2.2.4) is called *backsubstitution.* The combination of Gaussian elimination and backsubstitution yields a solution to the set of equations.

The advantage of Gaussian elimination and backsubstitution over Gauss-Jordan elimination is simply that the former is faster in raw operations count: The innermost loops of Gauss-Jordan elimination, each containing one subtraction and one multiplication, are executed N^3 and N^2M times (where

there are N equations and M unknowns). The corresponding loops in Gaussian elimination are executed only $\frac{1}{3}N^3$ times (only half the matrix is reduced, and the increasing numbers of predictable zeros reduce the count to one-third), and $\frac{1}{2}N^2M$ times, respectively. Each backsubstitution of a right-hand side is $\frac{1}{2}N^2$ executions of a similar loop (one multiplication plus one subtraction). For $M \ll N$ (only a few right-hand sides) Gaussian elimination thus has about a factor three advantage over Gauss-Jordan. (We could reduce this advange to a factor 1.5 by *not* computing the inverse matrix as part of the Gauss-Jordan scheme.)

For computing the inverse matrix (which we can view as the case of $M = N$ right-hand sides, namely the N unit vectors which are the columns of the identity matrix), Gaussian elimination and backsubstitution at first glance require $\frac{1}{3}N^3$ (matrix reduction) $+\frac{1}{2}N^3$ (right-hand side manipulations) $+\frac{1}{2}N^3$ (N backsubstitutions) $= \frac{4}{3}N^3$ loop executions, which is more than the N^3 for Gauss-Jordan. However, the unit vectors are quite special in containing all zeros except for one element. If this is taken into account, the right-side manipulations can be reduced to only $\frac{1}{6}N^3$ loop executions, and, for matrix inversion, the two methods have identical efficiencies.

Both Gaussian elimination and Gauss-Jordan elimination share the disadvantage that all right-hand sides must be known in advance. The *LU* decomposition method in the next section does not share that deficiency, and also has an equally small operations count, both for solution with any number of right-hand sides, and for matrix inversion. For this reason we will not implement the method of Gaussian elimination as a routine.

REFERENCES AND FURTHER READING:

Ralston, Anthony, and Rabinowitz, Philip. 1978, *A First Course in Numerical Analysis*, 2nd ed. (New York: McGraw-Hill), §9.3-1.

Isaacson, Eugene, and Keller, Herbert B. 1966, *Analysis of Numerical Methods* (New York: Wiley), §2.1.

Johnson, Lee W., and Riess, R. Dean. 1982, *Numerical Analysis*, 2nd ed. (Reading, Mass.: Addison-Wesley), §2.2.1.

Westlake, Joan R. 1968, *A Handbook of Numerical Matrix Inversion and Solution of Linear Equations* (New York: Wiley).

2.3 LU Decomposition

Suppose we are able to write the matrix $\mathbf{A}$ as a product of two matrices,

$$\mathbf{L} \cdot \mathbf{U} = \mathbf{A} \tag{2.3.1}$$

where $\mathbf{L}$ is *lower triangular* (has elements only on the diagonal and below) and $\mathbf{U}$ is *upper triangular* (has elements only on the diagonal and above). For the

case of a 4×4 matrix $\mathbf{A}$, for example, equation (2.3.1) would look like this:

$$\begin{bmatrix} \alpha_{11} & 0 & 0 & 0 \\ \alpha_{21} & \alpha_{22} & 0 & 0 \\ \alpha_{31} & \alpha_{32} & \alpha_{33} & 0 \\ \alpha_{41} & \alpha_{42} & \alpha_{43} & \alpha_{44} \end{bmatrix} \cdot \begin{bmatrix} \beta_{11} & \beta_{12} & \beta_{13} & \beta_{14} \\ 0 & \beta_{22} & \beta_{23} & \beta_{24} \\ 0 & 0 & \beta_{33} & \beta_{34} \\ 0 & 0 & 0 & \beta_{44} \end{bmatrix} = \begin{bmatrix} a_{11} & a_{12} & a_{13} & a_{14} \\ a_{21} & a_{22} & a_{23} & a_{24} \\ a_{31} & a_{32} & a_{33} & a_{34} \\ a_{41} & a_{42} & a_{43} & a_{44} \end{bmatrix} \tag{2.3.2}$$

We can use a decomposition such as (2.3.1) to solve the linear set

$$\mathbf{A} \cdot \mathbf{x} = (\mathbf{L} \cdot \mathbf{U}) \cdot \mathbf{x} = \mathbf{L} \cdot (\mathbf{U} \cdot \mathbf{x}) = \mathbf{b} \tag{2.3.3}$$

by first solving for the vector $\mathbf{y}$ such that

$$\mathbf{L} \cdot \mathbf{y} = \mathbf{b} \tag{2.3.4}$$

and then solving

$$\mathbf{U} \cdot \mathbf{x} = \mathbf{y} \tag{2.3.5}$$

What is the advantage of breaking up one linear set into two successive ones? The advantage is that the solution of a triangular set of equations is quite trivial, as we have already seen in §2.2 (equation 2.2.4). Thus, equation (2.3.4) can be solved by *forward substitution* as follows,

$$\begin{aligned} y_1 &= \frac{b_1}{\alpha_{11}} \\ y_i &= \frac{1}{\alpha_{ii}} \left[b_i - \sum_{j=1}^{i-1} \alpha_{ij} y_j \right] \qquad i = 2, 3, \ldots, N \end{aligned} \tag{2.3.6}$$

while (2.3.5) can then be solved by *backsubstitution* exactly as in equations (2.2.2) – (2.2.4),

$$\begin{aligned} x_N &= \frac{y_N}{\beta_{NN}} \\ x_i &= \frac{1}{\beta_{ii}} \left[y_i - \sum_{j=i+1}^{N} \beta_{ij} x_j \right] \qquad i = N-1, N-2, \ldots, 1 \end{aligned} \tag{2.3.7}$$

Equations (2.3.6) and (2.3.7) total (for each right-hand side $\mathbf{b}$) N^2 executions of an inner loop containing one multiply and one add. If we have N

right-hand sides which are the unit column vectors (which is the case when we are inverting a matrix), then taking into account the leading zeros reduces the total execution count of (2.3.6) from $\frac{1}{2}N^3$ to $\frac{1}{6}N^3$, while (2.3.7) is unchanged at $\frac{1}{2}N^3$.

Notice that, once we have the LU decomposition of $\mathbf{A}$ we can solve with as many right-hand sides as we then care to, one at a time. This is a distinct advantage over the methods of §2.1 and §2.2.

Performing the LU Decomposition

How then can we solve for $\mathbf{L}$ and $\mathbf{U}$, given $\mathbf{A}$? First, we write out the i, j^{th} component of equation (2.3.1) or (2.3.2). That component always is a sum beginning with

$$\alpha_{i1}\beta_{1j} + \cdots = a_{ij}$$

The number of terms in the sum depends, however, on whether i or j is the smaller number. We have, in fact, the three cases,

$$i < j: \qquad \alpha_{i1}\beta_{1j} + \alpha_{i2}\beta_{2j} + \cdots + \alpha_{ii}\beta_{ij} = a_{ij} \tag{2.3.8}$$

$$i = j: \qquad \alpha_{i1}\beta_{1j} + \alpha_{i2}\beta_{2j} + \cdots + \alpha_{ii}\beta_{jj} = a_{ij} \tag{2.3.9}$$

$$i > j: \qquad \alpha_{i1}\beta_{1j} + \alpha_{i2}\beta_{2j} + \cdots + \alpha_{ij}\beta_{jj} = a_{ij} \tag{2.3.10}$$

Equations (2.3.8)–(2.3.10) total N^2 equations for the $N^2 + N$ unknown α's and β's (the diagonal being represented twice). Since the number of unknowns is greater than the number of equations, we are invited to specify N of the unknowns arbitrarily and then try to solve for the others. In fact, as we shall see, it is always possible to take

$$\alpha_{ii} \equiv 1 \qquad i = 1, \ldots, N \tag{2.3.11}$$

A surprising procedure, now, is *Crout's algorithm*, which quite trivially solves the set of $N^2 + N$ equations (2.3.8)–(2.3.11) for all the α's and β's by just arranging the equations in a certain order! That order is as follows:

- Set $\alpha_{ii} = 1$, $i = 1, \ldots, N$ (equation 2.3.11).

- For each $j = 1, 2, 3, \ldots, N$ do these two procedures: First, for $i = 1, 2, \ldots, j$, use (2.3.8), (2.3.9), and (2.3.11) to solve for β_{ij}, namely

$$\beta_{ij} = a_{ij} - \sum_{k=1}^{i-1} \alpha_{ik}\beta_{kj}. \tag{2.3.12}$$

(When $i = 1$ in 2.3.12 the summation term is taken to mean zero.) Second, for $i = j+1, j+2, \ldots, N$ use (2.3.10) to solve for α_{ij}, namely

$$\alpha_{ij} = \frac{1}{\beta_{jj}}\left(a_{ij} - \sum_{k=1}^{j-1} \alpha_{ik}\beta_{kj}\right) \tag{2.3.13}$$

Be sure to do both procedures before going on to the next j.

If you work through a few iterations of the above procedure, you will see that the α's and β's that occur on the right-hand side of equations (2.3.12) and (2.3.13) are already determined by the time they are needed. You will also see that every a_{ij} is used only once and never again. This means that the corresponding α_{ij} or β_{ij} can be stored in the location that the a used to occupy: the decomposition is "in place." [The diagonal unity elements α_{ii} (equation 2.3.11) are not stored at all.] In brief, Crout's method fills in the combined matrix of α's and β's,

$$\begin{bmatrix} \beta_{11} & \beta_{12} & \beta_{13} & \beta_{14} \\ \alpha_{21} & \beta_{22} & \beta_{23} & \beta_{24} \\ \alpha_{31} & \alpha_{32} & \beta_{33} & \beta_{34} \\ \alpha_{41} & \alpha_{42} & \alpha_{43} & \beta_{44} \end{bmatrix} \tag{2.3.14}$$

by columns from left to right, and within each column from top to bottom (see Figure 2.3.1).

What about pivoting? Pivoting (i.e. selection of a salubrious pivot element for the division in equation 2.3.13) is absolutely essential for the stability of Crout's method. Only partial pivoting (interchange of rows) can be implemented efficiently. However this is enough to make the method stable. This means, incidentally, that we don't actually decompose the matrix **A** into LU form, but rather we decompose a rowwise permutation of **A**. (If we keep track of what that permutation is, this decomposition is just as useful as the original one would have been.)

Pivoting is slightly subtle in Crout's algorithm. The key point to notice is that equation (2.3.12) in the case of $i = j$ (its final application) is *exactly the same* as equation (2.3.13) except for the division in the latter equation; in

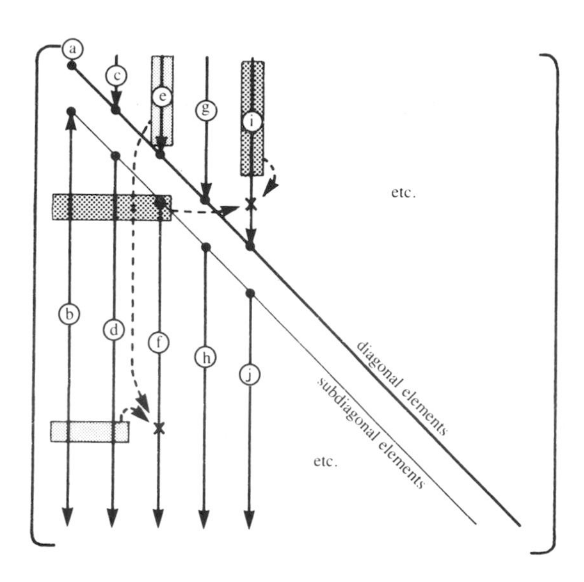

Figure 2.3.1. Crout's algorithm for *LU* decomposition of a matrix. Elements of the original matrix are modified in the order indicated by lower case letters: a, b, c, etc. Shaded boxes show the previously modified elements that are used in modifying two typical elements, each indicated by an "×".

both cases the upper limit of the sum is $k = j - 1$ $(= i - 1)$. This means that we don't have to commit ourselves as to whether the diagonal element β_{jj} is the one which happens to fall on the diagonal in the first instance, or whether one of the (undivided) α_{ij}'s below it in the column, $i = j + 1, \ldots, N$, is to be "promoted" to become the diagonal β. This can be decided after all the candidates in the column are in hand. As you should be able to guess by now, we will choose the largest one as the diagonal β (pivot element), then do all the divisions by that element *en masse.* This is *Crout's method with partial pivoting.* Our implementation has one additional subtlety: it initially finds the largest element in each row, and subsequently (when it is looking for the maximal pivot element) scales the comparison *as if* we had initially scaled all the equations to make their maximum coefficient equal to unity; this is the *implicit pivoting* mentioned in §2.1.

```
SUBROUTINE LUDCMP(A,N,NP,INDX,D)
```

Given an N × N matrix A, with physical dimension NP, this routine replaces it by the *LU* decomposition of a rowwise permutation of itself. A and N are input. A is output, arranged as in equation (2.3.14) above; INDX is an output vector which records the row permutation effected by the partial pivoting; D is output as ±1 depending on whether the number of row interchanges was even or odd, respectively. This routine is used in combination with LUBKSB to solve linear equations or invert a matrix.

```
PARAMETER (NMAX=100,TINY=1.0E-20)      Largest expected N, and a small number.
DIMENSION A(NP,NP),INDX(N),VV(NMAX)    VV stores the implicit scaling of each row.
D=1.                                   No row interchanges yet.
DO 12 I=1,N                            Loop over rows to get the implicit scaling information.
    AAMAX=0.
    DO 11 J=1,N
```

```
          IF (ABS(A(I,J)).GT.AAMAX) AAMAX=ABS(A(I,J))
11      CONTINUE
        IF (AAMAX.EQ.0.) PAUSE 'Singular matrix.'    No nonzero largest element.
        VV(I)=1./AAMAX                               Save the scaling.
12    CONTINUE
      DO 19 J=1,N                                    This is the loop over columns of Crout's method.
        DO 14 I=1,J-1                                This is equation 2.3.12 except for i = j.
          SUM=A(I,J)
          DO 13 K=1,I-1
            SUM=SUM-A(I,K)*A(K,J)
13        CONTINUE
          A(I,J)=SUM
14      CONTINUE
        AAMAX=0.                                     Initialize for the search for largest pivot element.
        DO 16 I=J,N                                  This is i = j of equation 2.3.12 and i = j+1...N of equation
          SUM=A(I,J)                                     2.3.13.
          DO 15 K=1,J-1
            SUM=SUM-A(I,K)*A(K,J)
15        CONTINUE
          A(I,J)=SUM
          DUM=VV(I)*ABS(SUM)                         Figure of merit for the pivot.
          IF (DUM.GE.AAMAX) THEN                         Is it better than the best so far?
            IMAX=I
            AAMAX=DUM
          ENDIF
16      CONTINUE
        IF (J.NE.IMAX)THEN                           Do we need to interchange rows?
          DO 17 K=1,N                                Yes, do so...
            DUM=A(IMAX,K)
            A(IMAX,K)=A(J,K)
            A(J,K)=DUM
17        CONTINUE
          D=-D                                       ...and change the parity of D.
          VV(IMAX)=VV(J)                             Also interchange the scale factor.
        ENDIF
        INDX(J)=IMAX
        IF(A(J,J).EQ.0.)A(J,J)=TINY
        IF(J.NE.N)THEN                               Now, finally, divide by the pivot element.
          DUM=1./A(J,J)                              If the pivot element is zero the matrix is singular (at least to the
          DO 18 I=J+1,N                                  precision of the algorithm). For some applications on singular
            A(I,J)=A(I,J)*DUM                            matrices, it is desirable to substitute TINY for zero.
18        CONTINUE
        ENDIF
19    CONTINUE                                       Go back for the next column in the reduction.
      RETURN
      END
```

Here is the routine for forward substitution and backsubstitution, implementing equations (2.3.6) and (2.3.7).

```
      SUBROUTINE LUBKSB(A,N,NP,INDX,B)
```

Solves the set of N linear equations $A \cdot X = B$. Here **A** is input, not as the matrix A but rather as its LU decomposition, determined by the routine **LUDCMP**. **INDX** is input as the permutation vector returned by **LUDCMP**. **B** is input as the right-hand side vector

B, and returns with the solution vector X. A, N, NP and INDX are not modified by this routine and can be left in place for successive calls with different right-hand sides B. This routine takes into account the possibility that B will begin with many zero elements, so it is efficient for use in matrix inversion.

```
DIMENSION A(NP,NP),INDX(N),B(N)
II=0                        When II is set to a positive value, it will become the index of the first
DO 12 I=1,N                    nonvanishing element of B. We now do the forward substitu-
    LL=INDX(I)                 tion, equation 2.3.6. The only new wrinkle is to unscramble
    SUM=B(LL)                  the permutation as we go.
    B(LL)=B(I)
    IF (II.NE.0)THEN
        DO 11 J=II,I-1
            SUM=SUM-A(I,J)*B(J)
11      CONTINUE
    ELSE IF (SUM.NE.0.) THEN
        II=I                A nonzero element was encountered, so from now on we will have to
    ENDIF                      do the sums in the loop above.
    B(I)=SUM
12 CONTINUE
DO 14 I=N,1,-1              Now we do the backsubstitution, equation 2.3.7.
    SUM=B(I)
    IF(I.LT.N)THEN
        DO 13 J=I+1,N
            SUM=SUM-A(I,J)*B(J)
13      CONTINUE
    ENDIF
    B(I)=SUM/A(I,I)         Store a component of the solution vector X.
14 CONTINUE
RETURN                      All done!
END
```

The LU decomposition in LUDCMP requires about $\frac{1}{3}N^3$ executions of the inner loops (each with one multiply and one add). This is thus the operation count for solving one (or a few) right-hand sides, and is a factor of 3 better than the Gauss-Jordan routine GAUSSJ which was given in §2.1, and a factor of 1.5 better than a Gauss-Jordan routine which does not compute the inverse matrix. For inverting a matrix, the total count (including the forward and backsubstitution as discussed following equation 2.3.7 above) is $(\frac{1}{3}+\frac{1}{6}+\frac{1}{2})N^3 = N^3$, the same as GAUSSJ.

To summarize, this is the preferred way to solve the linear set of equations $\mathbf{A}\cdot\mathbf{x}=\mathbf{b}$:

```
CALL LUDCMP(A,N,NP,INDX,D)
CALL LUBKSB(A,N,NP,INDX,B)
```

The answer $\mathbf{x}$ will be returned in B. Your original matrix $\mathbf{A}$ will have been destroyed.

If you subsequently want to solve a set of equations with the same $\mathbf{A}$ but a different right-hand side $\mathbf{b}$, you repeat *only*

```
CALL LUBKSB(A,N,NP,INDX,B)
```

not, of course, with the original matrix $\mathbf{A}$, but with A and INDX as were already returned from LUDCMP.

REFERENCES AND FURTHER READING:

Golub, Gene H., and Van Loan, Charles F. 1983, *Matrix Computations* (Baltimore: Johns Hopkins University Press), Chapter 4.

Dongarra, J.J., et al. 1979, *LINPACK User's Guide* (Philadelphia: Society for Industrial and Applied Mathematics).

Forsythe, George E., Malcolm, Michael A., and Moler, Cleve B. 1977, *Computer Methods for Mathematical Computations* (Englewood Cliffs, N.J.: Prentice-Hall), §3.3.

Forsythe, George E., and Moler, Cleve B. 1967, *Computer Solution of Linear Algebraic Systems* (Englewood Cliffs, N.J.: Prentice-Hall), Chapters 9 and 16.

IMSL Library Reference Manual, 1980, ed. 8 (IMSL Inc., 7500 Bellaire Boulevard, Houston TX 77036), Chapter L.

NAG Fortran Library Manual Mark 8, 1980 (NAG Central Office, 7 Banbury Road, Oxford OX26NN, U.K.), Chapters F01, F04.

Westlake, Joan R. 1968, *A Handbook of Numerical Matrix Inversion and Solution of Linear Equations* (New York: Wiley).

2.4 Inverse of a Matrix

Using the *LU* decomposition routines of the previous section, it is completely straightforward to find the inverse of a matrix column by column. (There is no better way to do it.)

```
DIMENSION A(NP,NP),Y(NP,NP),INDX(NP)
...
DO 12 I=1,N                        Set up identity matrix.
    DO 11 J=1,N
        Y(I,J)=0.
11  CONTINUE
    Y(I,I)=1.
12 CONTINUE
CALL LUDCMP(A,N,NP,INDX,D)         Decompose the matrix just once.
DO 13 J=1,N                        Find inverse by columns.
    CALL LUBKSB(A,N,NP,INDX,Y(1,J))
13 CONTINUE
```

It is necessary to recognize that FORTRAN stores two dimensional matrices by column, so that Y(1,J) is the address of the Jth column of Y. In Pascal you must create a unit column vector for this call to LUBKSB, then copy the result into the corresponding column of the matrix Y on return.

The matrix **Y** will now contain the inverse of the original matrix **A**, which will have been destroyed.

REFERENCES AND FURTHER READING:

Forsythe, George E., and Moler, Cleve B. 1967, *Computer Solution of Linear Algebraic Systems* (Englewood Cliffs, N.J.: Prentice-Hall), Chapter 18.

Dongarra, J.J., et al. 1979, *LINPACK User's Guide* (Philadelphia: Society for Industrial and Applied Mathematics).

Stoer, J., and Bulirsch, R. 1980, *Introduction to Numerical Analysis* (New York: Springer-Verlag), §4.2.

2.5 Determinant of a Matrix

The determinant of an LU decomposed matrix is just the product of the diagonal elements,

$$\det = \prod_{j=1}^{N} \beta_{jj} \tag{2.5.1}$$

We don't, recall, compute the decomposition of the original matrix, but rather a decomposition of a rowwise permutation of it. Luckily, we have kept track of whether the number of row interchanges was even or odd, so we just preface the product by the corresponding sign. (You now finally know what was the purpose of returning D in the routine LUDCMP §2.3.)

Calculation of a determinant thus requires one call to LUDCMP, with *no* subsequent backsubstitutions by LUBKSB.

```
DIMENSION A(NP,NP),INDX(NP)
...
CALL LUDCMP(A,N,NP,INDX,D)    This returns D as ±1.
DO 11 J=1,N
   D=D*A(J,J)
11 CONTINUE
```

D now contains the determinant of the original matrix A, which will have been destroyed.

For a matrix of any substantial size, it is quite likely that the determinant will overflow or underflow your computer's floating point dynamic range. In this case you can modify the loop of the above fragment to (e.g.) divide by powers of ten to keep track of the scale separately, or (e.g.) accumulate the sum of logarithms of the absolute values of the factors and the sign separately.

REFERENCES AND FURTHER READING:

Forsythe, George E., Malcolm, Michael A., and Moler, Cleve B. 1977, *Computer Methods for Mathematical Computations* (Englewood Cliffs, N.J.: Prentice-Hall), p.50.

Ralston, Anthony, and Rabinowitz, Philip. 1978, *A First Course in Numerical Analysis*, 2nd ed. (New York: McGraw-Hill), §9.11.

2.6 Tridiagonal Systems of Equations

The special case of a system of linear equations that is *tridiagonal*, that is, has nonzero elements only on the diagonal plus or minus one column, is one that occurs frequently. For tridiagonal sets, the procedures of LU decomposition, forward- and backsubstitution each take only $O(N)$ operations, and the whole solution can be encoded very concisely. The resulting routine TRIDAG is one which we will use in later chapters.

Naturally, one does not reserve storage for the full $N \times N$ matrix, but only for the nonzero components, stored as three vectors. The set of equations to be solved is

$$\begin{bmatrix} b_1 & c_1 & 0 & \cdots & & & \\ a_2 & b_2 & c_2 & \cdots & & & \\ & & & \cdots & & & \\ & & & \cdots & a_{N-1} & b_{N-1} & c_{N-1} \\ & & & \cdots & 0 & a_N & b_N \end{bmatrix} \cdot \begin{bmatrix} u_1 \\ u_2 \\ \cdots \\ u_{N-1} \\ u_N \end{bmatrix} = \begin{bmatrix} r_1 \\ r_2 \\ \cdots \\ r_{N-1} \\ r_N \end{bmatrix} \tag{2.6.1}$$

Notice that a_1 and c_N are undefined and are not referenced by the routine that follows.

```
SUBROUTINE TRIDAG(A,B,C,R,U,N)
    Solves for a vector U of length N the tridiagonal linear set given by equation (2.6.1). A,
    B, C and R are input vectors and are not modified.
PARAMETER (NMAX=100)           One vector of workspace, GAM is needed.
DIMENSION GAM(NMAX),A(N),B(N),C(N),R(N),U(N)
IF(B(1).EQ.0.)PAUSE            If this happens then you should rewrite your equations as a set of
BET=B(1)                            order N - 1, with u2 trivially eliminated.
U(1)=R(1)/BET
DO 11 J=2,N                    Decomposition and forward substitution.
    GAM(J)=C(J-1)/BET
    BET=B(J)-A(J)*GAM(J)
    IF(BET.EQ.0.)PAUSE         Algorithm fails; see below.
    U(J)=(R(J)-A(J)*U(J-1))/BET
11  CONTINUE
DO 12 J=N-1,1,-1               Backsubstitution.
    U(J)=U(J)-GAM(J+1)*U(J+1)
12  CONTINUE
RETURN
END
```

There is no pivoting in TRIDAG. It is for this reason that TRIDAG can fail (PAUSE) even when the underlying matrix is nonsingular: a zero pivot can be encountered even for a nonsingular matrix. In practice, this is not something to lose sleep about. The kinds of problems that lead to tridiagonal linear sets usually have additional properties which guarantee that the algorithm in TRIDAG will succeed. For example, if

$$|b_j| > |a_j| + |c_j| \qquad j = 1, \ldots, N \tag{2.6.2}$$

(called *diagonal dominance*) then it can be shown that the algorithm cannot encounter a zero pivot.

It is possible to construct special examples in which the lack of pivoting in the algorithm causes numerical instability. In practice, however, such instability is almost never encountered — unlike the general matrix problem where pivoting is essential.

The tridiagonal algorithm is the rare case of an algorithm that, in practice, is more robust than it appears to be in theory. Of course, should you ever encounter a problem for which TRIDAG fails, you can fall back on elimination with pivoting.

REFERENCES AND FURTHER READING:

Keller, Herbert B. 1968, *Numerical Methods for Two-Point Boundary-Value Problems* (Waltham, Mass.: Blaisdell), p.74.

Dahlquist, Germund, and Bjorck, Ake. 1974, *Numerical Methods* (Englewood Cliffs, N.J.: Prentice-Hall), Example 5.4.3, p.166.

Ralston, Anthony, and Rabinowitz, Philip. 1978, *A First Course in Numerical Analysis*, 2nd ed. (New York: McGraw-Hill), §9.11.

2.7 Iterative Improvement of a Solution to Linear Equations

Obviously it is not easy to obtain greater precision for the solution of a linear set than the precision of your computer's floating-point word. Unfortunately, for large sets of linear equations, it is not always easy to obtain precision equal to, or even comparable to, the computer's limit. In direct methods of solution, roundoff errors accumulate, and they are magnified to the extent that your matrix is close to singular. You can easily lose two or three significant figures for matrices which (you thought) were *far* from singular.

If this happens to you, there is a neat trick to restore the full machine precision, called *iterated improvement* of the solution. The theory is very straightforward (see Figure 2.7.1): Suppose that a vector $\mathbf{x}$ is the exact solution of the linear set

$$\mathbf{A} \cdot \mathbf{x} = \mathbf{b} \tag{2.7.1}$$

You don't, however, know $\mathbf{x}$. You only know some slightly wrong solution $\mathbf{x} + \delta\mathbf{x}$, where $\delta\mathbf{x}$ is the unknown error. When multiplied by the matrix $\mathbf{A}$, your slightly wrong solution gives a product slightly discrepant from the desired right-hand side $\mathbf{b}$, namely

$$\mathbf{A} \cdot (\mathbf{x} + \delta\mathbf{x}) = \mathbf{b} + \delta\mathbf{b} \tag{2.7.2}$$

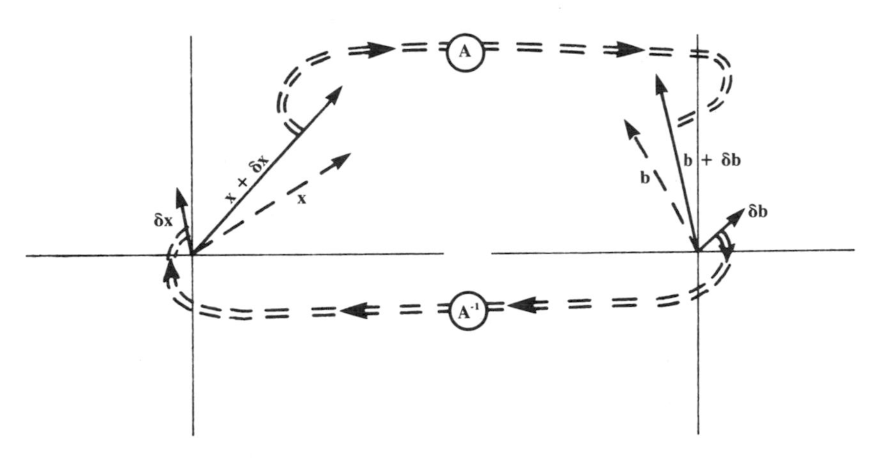

Figure 2.7.1. Iterative improvement of the solution to $\mathbf{A} \cdot \mathbf{x} = \mathbf{b}$. The first guess $\mathbf{x} + \delta\mathbf{x}$ is multiplied by $\mathbf{A}$ to produce $\mathbf{b} + \delta\mathbf{b}$. The known vector $\mathbf{b}$ is subtracted, giving $\delta\mathbf{b}$. The linear set with this right hand side is inverted, giving $\delta\mathbf{x}$. This is subtracted from the first guess giving an improved solution $\mathbf{x}$.

Subtracting (2.7.1) from (2.7.2) gives

$$\mathbf{A} \cdot \delta\mathbf{x} = \delta\mathbf{b} \tag{2.7.3}$$

But (2.7.2) can also be solved, trivially, for $\delta\mathbf{b}$. Substituting this into (2.7.3) gives

$$\mathbf{A} \cdot \delta\mathbf{x} = \mathbf{A} \cdot (\mathbf{x} + \delta\mathbf{x}) - \mathbf{b} \tag{2.7.4}$$

In this equation, the whole right-hand side is known, since $\mathbf{x}+\delta\mathbf{x}$ is the wrong solution that you want to improve. It is a good idea to calculate the right-hand side in double-precision (if available), since there will be a lot of cancellation in the subtraction of $\mathbf{b}$. Then, we need only solve (2.7.4) for the error $\delta\mathbf{x}$, then subtract this from the wrong solution to get an improved solution.

An important extra benefit occurs if we obtained the original solution by *LU* decomposition. In this case we already have the *LU* decomposed form of $\mathbf{A}$, and all we need do to solve (2.7.4) is compute the right-hand side and backsubstitute!

The code to do all this is concise and straightforward:

```
SUBROUTINE MPROVE(A,ALUD,N,NP,INDX,B,X)
   Improves a solution vector X of the linear set of equations A · X = B. The matrix A,
   and the vectors B and X are input, as is the dimension N. Also input is ALUD, the LU
   decomposition of A as returned by LUDCMP, and the vector INDX also returned by that
   routine. On output, only X is modified, to an improved set of values.
PARAMETER (NMAX=100)
DIMENSION A(NP,NP),ALUD(NP,NP),INDX(N),B(N),X(N),R(NMAX)
REAL*8 SDP
DO 12 I=1,N                    Calculate the right-hand side, accumulating the residual in double
   SDP=-B(I)                      precision.
   DO 11 J=1,N
```

```
          SDP=SDP+DBLE(A(I,J))*DBLE(X(J))
        11 CONTINUE
        R(I)=SDP
12    CONTINUE
      CALL LUBKSB(ALUD,N,NP,INDX,R)     Solve for the error term,
      DO 13 I=1,N                       and subtract it from the old solution.
        X(I)=X(I)-R(I)
13    CONTINUE
      RETURN
      END
```

You should note that the routine LUDCMP in §2.3 destroys the input matrix as it LU decomposes it. Since iterative improvement requires *both* the original matrix and its LU decomposition, you will need to copy $\mathbf{A}$ before calling LUDCMP. Likewise LUBKSB destroys $\mathbf{b}$ in obtaining $\mathbf{x}$, so make a copy of $\mathbf{b}$ also. If you don't mind this extra storage, iterated improvement is *highly* recommended: It is a process of order only N^2 operations (multiply vector by matrix, and backsubstitute – see discussion following equation 2.3.7); it never hurts; and it can really give you your money's worth if it saves an otherwise ruined solution on which you have already spent of order N^3 operations.

You can call MPROVE several times in succession if you want. Unless you are starting quite far from the true solution, one call is generally enough; but a second call to verify convergence can be reassuring.

REFERENCES AND FURTHER READING:

Johnson, Lee W., and Riess, R. Dean. 1982, *Numerical Analysis*, 2nd ed. (Reading, Mass.: Addison-Wesley), §2.3.4, p.55.

Golub, Gene H., and Van Loan, Charles F. 1983, *Matrix Computations* (Baltimore: Johns Hopkins University Press), p.74.

Dahlquist, Germund, and Bjorck, Ake. 1974, *Numerical Methods* (Englewood Cliffs, N.J.: Prentice-Hall), §5.5.6, p.183.

Forsythe, George E., and Moler, Cleve B. 1967, *Computer Solution of Linear Algebraic Systems* (Englewood Cliffs, N.J.: Prentice-Hall), Chapter 13.

Ralston, Anthony, and Rabinowitz, Philip. 1978, *A First Course in Numerical Analysis*, 2nd ed. (New York: McGraw-Hill), §9.5, p. 437.

2.8 Vandermonde Matrices and Toeplitz Matrices

In §2.6 the case of a tridiagonal matrix was treated specially, because that particular type of linear system admits a solution in only of order N operations, rather than of order N^3 for the general linear problem. When such particular types exist, it is important to know about them. Your computational savings, should you ever happen to be working on a problem which involves the right kind of particular type, can be considerable.

This section treats two special types of matrices which can be solved in of order N^2 operations, not as good as tridiagonal, but a lot better than the general case. (Other than the operations count, these two types having nothing in common.) Matrices of the first type, termed *Vandermonde matrices*, occur in some problems having to do with the fitting of polynomials, the reconstruction of distributions from their moments, and also other contexts. In this book, for example, a Vandermonde problem crops up in §3.5. Matrices of the second type, termed *Toeplitz matrices*, tend to occur in problems involving deconvolution and signal processing. In this book, a Toeplitz problem is encountered in §12.8.

These are not the *only* special types of matrices worth knowing about. The *Hilbert matrices*, whose components are of the form $a_{ij} = 1/(i + j - 1)$, $i, j = 1, \ldots, N$ can be inverted by an exact integer algorithm, and are very *difficult* to invert in any other way, since they are notoriously ill-conditioned (see Forsythe and Moler for details). The Sherman-Morrison and Woodbury formulas, discussed below in §2.10, can sometimes be used to convert new special forms into old ones. Westlake gives some other special forms. We have not found these additional forms to arise as frequently as the two that we now discuss.

Vandermonde Matrices

A Vandermonde matrix of size $N \times N$ is completely determined by N arbitrary numbers $x_1, x_2, \ldots, x_N$, in terms of which its N^2 components are the integer powers x_i^{j-1}, $i, j = 1, \ldots, N$. Evidently there are two possible such forms, depending on whether we view the i's as rows, j's as columns, or vice versa. In the former case, we get a linear system of equations that looks like this,

$$\begin{bmatrix} 1 & x_1 & x_1^2 & \cdots & x_1^{N-1} \\ 1 & x_2 & x_2^2 & \cdots & x_2^{N-1} \\ & & & \cdots & \\ 1 & x_N & x_N^2 & \cdots & x_N^{N-1} \end{bmatrix} \cdot \begin{bmatrix} c_1 \\ c_2 \\ \cdots \\ c_N \end{bmatrix} = \begin{bmatrix} y_1 \\ y_2 \\ \cdots \\ y_N \end{bmatrix} \tag{2.8.1}$$

Performing the matrix multiplication, you will see that this equation solves for the unknown coefficients c_i which fit a polynomial to the N pairs of abscissas and ordinates (x_j, y_j). Precisely this problem will arise in §3.5, and the routine given there will solve (2.8.1) by the method that we are about to describe.

The alternative identification of rows and columns leads to the set of equations

$$\begin{bmatrix} 1 & 1 & \cdots & 1 \\ x_1 & x_2 & \cdots & x_N \\ x_1^2 & x_2^2 & \cdots & x_N^2 \\ & & \cdots & \\ x_1^{N-1} & x_2^{N-1} & \cdots & x_N^{N-1} \end{bmatrix} \cdot \begin{bmatrix} w_1 \\ w_2 \\ \cdots \\ w_N \end{bmatrix} = \begin{bmatrix} q_1 \\ q_2 \\ \cdots \\ q_N \end{bmatrix} \tag{2.8.2}$$

Write this out and you will see that it relates to the *problem of moments*: Given the values of N points x_i, find the unknown weights w_i, assigned so as to match the given values q_j of the first N moments. (For more on this problem, consult von Mises.) The routine given in this section solves (2.8.2).

The method of solution of both (2.8.1) and (2.8.2) is closely related to Lagrange's polynomial interpolation formula, which we will not formally meet until §3.1 below. Notwithstanding, the following derivation should be comprehensible:

Let $P_j(x)$ be the polynomial of degree $N-1$ defined by

$$P_j(x) = \prod_{\substack{n=1 \\ (n \neq j)}}^{N} \frac{x - x_n}{x_j - x_n} = \sum_{k=1}^{N} A_{jk} x^{k-1} \tag{2.8.3}$$

Here the meaning of the last equality is to define the components of the matrix A_{ij} as the coefficients which arise when the product is multiplied out and like terms collected.

The polynomial $P_j(x)$ is a function of x generally. But you will notice that it is specifically designed so that it takes on a value of zero at all x_i with $i \neq j$, and has a value of unity at $x = x_j$. In other words,

$$P_j(x_i) = \delta_{ij} = \sum_{k=1}^{N} A_{jk} x_i^{k-1} \tag{2.8.4}$$

But (2.8.4) says that A_{jk} is exactly the inverse of the matrix of components x_i^{k-1}, which appears in (2.8.2), with the subscript as the column index. Therefore the solution of (2.8.2) is just that matrix inverse times the right hand side,

$$w_j = \sum_{k=1}^{N} A_{jk} q_k \tag{2.8.5}$$

As for the transpose problem (2.8.1), we can use the fact that the inverse of the transpose is the transpose of the inverse, so

$$c_j = \sum_{k=1}^{N} A_{kj} y_k \tag{2.8.6}$$

The routine in §3.5 implements this.

It remains to find a good way of multiplying out the monomial terms in (2.8.3), in order to get the components of A_{jk}. This is essentially a bookkeeping problem, and we will let you read the routine itself to see how it can be solved. One trick is to define a master $P(x)$ by

$$P(x) \equiv \prod_{n=1}^{N} (x - x_n) \tag{2.8.7}$$

work out its coefficients, and then obtain the numerators and denominators of the specific P_j's via synthetic division by the one supernumerary term. (See §5.3 for more on synthetic division.) Since each such division is only a process of order N, the total procedure is of order N^2.

You should be warned that Vandermonde systems are notoriously ill-conditioned, by their very nature. (As an aside anticipating §5.6, the reason is the same as that which makes Chebyshev fitting so impressively accurate: there exist high-order polynomials that are very good uniform fits to zero. Hence roundoff error can introduce rather substantial coefficients of the leading terms of these polynomials.) It is a good idea always to compute Vandermonde problems in double precision.

The routine for (2.8.2) which follows is due to G. Rybicki.

```
SUBROUTINE VANDER(X,W,Q,N)
    Solves the Vandermonde linear system sum_{i=1}^{N} x_i^{k-1} w_i = q_k (k = 1,...,N). Input consists
    of the vectors X and Q, each of length N; the vector W is output.
PARAMETER (NMAX=100,ZERO=0.0,ONE=1.0)
    NMAX is the maximum expected value of N. Make constants double precision if you convert
    program to double precision — which is a good idea.
DIMENSION X(N),W(N),Q(N),C(NMAX)
IF(N.EQ.1)THEN
    W(1)=Q(1)
ELSE
    DO 11 I=1,N                         Initialize array.
        C(I)=ZERO
11  CONTINUE
    C(N)=-X(1)                          Coefficients of the master polynomial are found by recursion.
    DO 13 I=2,N
        XX=-X(I)
        DO 12 J=N+1-I,N-1
            C(J)=C(J)+XX*C(J+1)
12      CONTINUE
        C(N)=C(N)+XX
13  CONTINUE
    DO 15 I=1,N                         Each subfactor in turn
        XX=X(I)
        T=ONE
        B=ONE
        S=Q(N)
        K=N
        DO 14 J=2,N                     is synthetically divided,
            K1=K-1
            B=C(K)+XX*B
            S=S+Q(K1)*B                 matrix-multiplied by the right-hand side,
            T=XX*T+B
            K=K1
```

```
        14 CONTINUE
          W(I)=S/T                and supplied with a denominator.
        15 CONTINUE
      ENDIF
      RETURN
      END
```

Toeplitz Matrices

An $N \times N$ Toeplitz matrix is specified by giving $2N-1$ numbers R_k, $k = -N+1, \ldots, -1, 0, 1, \ldots, N-1$. Those numbers are then emplaced as matrix elements constant along the (upper-left to lower-right) diagonals of the matrix:

$$\begin{bmatrix} R_0 & R_{-1} & R_{-2} & \cdots & R_{-N+2} & R_{-N+1} \\ R_1 & R_0 & R_{-1} & \cdots & R_{-N+3} & R_{-N+2} \\ R_2 & R_1 & R_0 & \cdots & R_{-N+4} & R_{-N+3} \\ \cdots & & & \cdots & & \\ R_{N-2} & R_{N-3} & R_{N-4} & \cdots & R_0 & R_{-1} \\ R_{N-1} & R_{N-2} & R_{N-3} & \cdots & R_1 & R_0 \end{bmatrix} \tag{2.8.8}$$

The linear Toeplitz problem can thus be written as

$$\sum_{j=1}^{N} R_{i-j} x_j = y_i \qquad (i = 1, \ldots, N) \tag{2.8.9}$$

where the x_j's, $j = 1, \ldots, N$, are the unknowns to be solved for.

The Toeplitz matrix is symmetric if $R_k = R_{-k}$ for all k. Levinson developed an algorithm for fast solution of the symmetric Toeplitz problem, by a *bordering method*, that is, a recursive procedure which solves the M-dimensional Toeplitz problem

$$\sum_{j=1}^{M} R_{i-j} x_j^{(M)} = y_i \qquad (i = 1, \ldots, M) \tag{2.8.10}$$

in turn for $M = 1, 2, \ldots$ until $M = N$, the desired result, is finally reached. The vector $x_j^{(M)}$ is the result at the Mth stage, and becomes the desired answer only when N is reached.

Levinson's method is well documented in standard texts (e.g. Robinson and Treitel). The useful fact that the method generalizes to the *nonsymmetric* case seems to be less well known. At some risk of excessive detail, we therefore give a derivation here, due to G. Rybicki.

In following a recursion from step M to step $M+1$ we find that our developing solution $x^{(M)}$ changes in this way:

$$\sum_{j=1}^{M} R_{i-j} x_j^{(M)} = y_i \qquad i = 1, \ldots, M \tag{2.8.11}$$

becomes

$$\sum_{j=1}^{M} R_{i-j} x_j^{(M+1)} + R_{i-(M+1)} x_{M+1}^{(M+1)} = y_i \qquad i = 1, \ldots, M+1 \tag{2.8.12}$$

By eliminating y_i we find

$$\sum_{j=1}^{M} R_{i-j} \left(\frac{x_j^{(M)} - x_j^{(M+1)}}{x_{M+1}^{(M+1)}} \right) = R_{i-(M+1)} \qquad i = 1, \ldots, M \tag{2.8.13}$$

or by letting $i \to M+1-i$ and $j \to M+1-j$,

$$\sum_{j=1}^{M} R_{j-i} G_j^{(M)} = R_{-i} \tag{2.8.14}$$

where

$$G_j^{(M)} \equiv \frac{x_{M+1-j}^{(M)} - x_{M+1-j}^{(M+1)}}{x_{M+1}^{(M+1)}} \tag{2.8.15}$$

To put this another way,

$$x_{M+1-j}^{(M+1)} = x_{M+1-j}^{(M)} - x_{M+1}^{(M+1)} G_j^{(M)} \qquad j = 1, \ldots, M \tag{2.8.16}$$

Thus, if we can use recursion to find the order M quantities $x^{(M)}$ and $G^{(M)}$ *and* the single order $M+1$ quantity $x_{M+1}^{(M+1)}$, then all of the other $x_j^{(M+1)}$ will follow. Fortunately, the quantity $x_{M+1}^{(M+1)}$ follows from equation (2.8.12) with $i = M+1$,

$$\sum_{j=1}^{M} R_{M+1-j} x_j^{(M+1)} + R_0 x_{M+1}^{(M+1)} = y_{M+1} \tag{2.8.17}$$

For the unknown order $M+1$ quantities $x_j^{(M+1)}$ we can substitute the previous order quantities in G since

$$G_{M+1-j}^{(M)} = \frac{x_j^{(M)} - x_j^{(M+1)}}{x_{M+1}^{(M+1)}} \tag{2.8.18}$$

The result of this operation is

$$x_{M+1}^{(M+1)} = \frac{\sum_{j=1}^{M} R_{M+1-j} x_j^{(M)} - y_{M+1}}{\sum_{j=1}^{M} R_{M+1-j} G_{M+1-j}^{(M)} - R_0} \tag{2.8.19}$$

The only remaining problem is to develop a recursion relation for G. Before we do that, however, we should point out that there are actually two distinct sets of solutions to the original linear problem for a nonsymmetric matrix, namely right-hand solutions (which we have been discussing) and left-hand solutions z_i. The formalism for the left-hand solutions differs only in that we deal with the equations

$$\sum_{j=1}^{M} R_{j-i} z_j^{(M)} = y_i \qquad i = 1, \ldots, M \tag{2.8.20}$$

Then, the same sequence of operations on this set leads to

$$\sum_{j=1}^{M} R_{i-j} H_j^{(M)} = R_i \tag{2.8.21}$$

where

$$H_j^{(M)} \equiv \frac{z_{M+1-j}^{(M)} - z_{M+1-j}^{(M+1)}}{z_{M+1}^{(M+1)}} \tag{2.8.22}$$

(compare with 2.8.14 – 2.8.15). The reason for mentioning the left-hand solutions now is that, by equation (2.8.21), the H_j satisfy exactly the same equation as the x_j except for the substitution $y_i \to R_i$ on the right-hand side. Therefore we can quickly deduce from equation (2.8.19) that

$$H_{M+1}^{(M+1)} = \frac{\sum_{j=1}^{M} R_{M+1-j} H_j^{(M)} - R_{M+1}}{\sum_{j=1}^{M} R_{M+1-j} G_{M+1-j}^{(M)} - R_0} \tag{2.8.23}$$

By the same token, G satisfies the same equation as z, except for the substitution $y_i \rightarrow R_{-i}$. This gives

$$G_{M+1}^{(M+1)} = \frac{\sum_{j=1}^{M} R_{j-M-1} G_j^{(M)} - R_{-M-1}}{\sum_{j=1}^{M} R_{j-M-1} H_{M+1-j}^{(M)} - R_0} \tag{2.8.24}$$

The same "morphism" also turns equation (2.8.16), and its partner for z, into the final equations

$$\begin{aligned} G_j^{(M+1)} &= G_j^{(M)} - G_{M+1}^{(M+1)} H_{M+1-j}^{(M)} \\ H_j^{(M+1)} &= H_j^{(M)} - H_{M+1}^{(M+1)} G_{M+1-j}^{(M)} \end{aligned} \tag{2.8.25}$$

Now, starting with the initial values

$$x_1^{(1)} = y_1/R_0 \qquad G_1^{(1)} = R_{-1}/R_0 \qquad H_1^{(1)} = R_1/R_0 \tag{2.8.26}$$

we can recurse away. At each stage M we use equations (2.8.23) and (2.8.24) to find $H_{M+1}^{(M+1)}, G_{M+1}^{(M+1)}$, and then equation (2.8.25) to find the other components of $H^{(M+1)}, G^{(M+1)}$. From there the vectors $x^{(M+1)}$ and/or $z^{(M+1)}$ are easily calculated.

The program below does this. It incorporates the second equation in (2.8.25) in the form

$$H_{M+1-j}^{(M+1)} = H_{M+1-j}^{(M)} - H_{M+1}^{(M+1)} G_j^{(M)} \tag{2.8.27}$$

so that the computation can be done "in place."

Notice that the above algorithm fails if $R_0 = 0$. In fact, because the bordering method does not allow pivoting, the algorithm will fail if any of the diagonal principal minors of the original Toeplitz matrix vanish. (Compare with discussion of the tridiagonal algorithm in §2.6.) If the algorithm fails, your matrix is not necessarily singular — you might just have to solve your problem by a slower and more general algorithm such as *LU* decomposition with pivoting.

The routine that implements equations (2.8.17)–(2.8.20) is also due to Rybicki. Note that the routine's `R(N+J)` is equal to R_j above, so that subscripts on the `R` array vary from 1 to $2N - 1$.

```
      SUBROUTINE TOEPLZ(R,X,Y,N)
```

Solves the Toeplitz system $\sum_{j=1}^{N} R_{(N+i-j)} x_j = y_i$ $(i = 1, \ldots, N)$. The Toeplitz matrix need not be symmetric. **Y** and **R** are input arrays of length N and 2*N-1 respectively. **X** is the output array, of length N.

```
      PARAMETER (NMAX=100)
      DIMENSION R(2*N-1),X(N),Y(N),G(NMAX),H(NMAX)
      IF(R(N).EQ.0.)  GO TO 99
      X(1)=Y(1)/R(N)                    Initialize for the recursion.
      IF(N.EQ.1)RETURN
      G(1)=R(N-1)/R(N)
      H(1)=R(N+1)/R(N)
      DO 15 M=1,N                       Main loop over the recursion.
        M1=M+1
        SXN=-Y(M1)                      Compute numerator and denominator for x,
        SD=-R(N)
        DO 11 J=1,M
          SXN=SXN+R(N+M1-J)*X(J)
          SD=SD+R(N+M1-J)*G(M-J+1)
11      CONTINUE
        IF(SD.EQ.0.)GO TO 99
        X(M1)=SXN/SD                    whence x.
        DO 12 J=1,M
          X(J)=X(J)-X(M1)*G(M-J+1)
12      CONTINUE
        IF(M1.EQ.N)RETURN
        SGN=-R(N-M1)                    Compute numerator and denominator for G and H,
        SHN=-R(N+M1)
        SGD=-R(N)
        DO 13 J=1,M
          SGN=SGN+R(N+J-M1)*G(J)
          SHN=SHN+R(N+M1-J)*H(J)
          SGD=SGD+R(N+J-M1)*H(M-J+1)
13      CONTINUE
        IF(SD.EQ.0..OR.SGD.EQ.0.)GO TO 99
        G(M1)=SGN/SGD                   whence G and H.
        H(M1)=SHN/SD
        K=M
        M2=(M+1)/2
        PP=G(M1)
        QQ=H(M1)
        DO 14 J=1,M2
          PT1=G(J)
          PT2=G(K)
          QT1=H(J)
          QT2=H(K)
          G(J)=PT1-PP*QT2
          G(K)=PT2-PP*QT1
          H(J)=QT1-QQ*PT2
          H(K)=QT2-QQ*PT1
          K=K-1
14      CONTINUE
15    CONTINUE                          Back for another recurrence.
      PAUSE 'never get here'
99    PAUSE 'Levinson method fails:  singular principal minor'
      END
```

REFERENCES AND FURTHER READING:

Golub, Gene H., and Van Loan, Charles F. 1983, *Matrix Computations* (Baltimore: Johns Hopkins University Press), Chapter 5 [also treats some other special forms].

Westlake, Joan R. 1968, *A Handbook of Numerical Matrix Inversion and Solution of Linear Equations* (New York: Wiley).

Forsythe, George E., and Moler, Cleve B. 1967, *Computer Solution of Linear Algebraic Systems* (Englewood Cliffs, N.J.: Prentice-Hall), §19.

von Mises, Richard. 1964, *Mathematical Theory of Probability and Statistics* (New York: Academic Press), p. 394 ff.

Levinson, N., Appendix B. of N. Wiener, 1949, *Extrapolation, Interpolation and Smoothing of Stationary Time Series* (New York: Wiley).

Robinson, E.A., and Treitel, S. 1980, *Geophysical Signal Analysis* (Englewood Cliffs, N.J.: Prentice-Hall), p. 163 ff.

2.9 Singular Value Decomposition

There exists a very powerful set of techniques for dealing with sets of equations or matrices that are either singular or else numerically very close to singular. In many cases where Gaussian elimination and LU decomposition fail to give satisfactory results, this set of techniques, known as *singular value decomposition* or *SVD*, will diagnose for you precisely what the problem is. In some cases, SVD will not only diagnose the problem, it will also solve it, in the sense of giving you a useful numerical answer, although, as we shall see, not necessarily "the" answer that you thought you should get.

SVD is also the method of choice for solving most *linear least squares* problems. We will outline the relevant theory in this section, but defer detailed discussion of the use of SVD in this application to Chapter 14, whose subject is the parametric modeling of data.

SVD methods are based on the following theorem of linear algebra, whose proof is beyond our scope: Any $M \times N$ matrix $\mathbf{A}$ whose number of rows M is greater than or equal to its number of columns N, can be written as the product of an $M \times N$ column-orthogonal matrix $\mathbf{U}$, an $N \times N$ diagonal matrix $\mathbf{W}$ with positive or zero elements, and the transpose of an $N \times N$ orthogonal matrix $\mathbf{V}$. The various shapes of these matrices will be made clearer by the following tableau:

$$\left(\begin{array}{c} \\ \\ \mathbf{A} \\ \\ \\ \end{array} \right) = \left(\begin{array}{c} \\ \\ \mathbf{U} \\ \\ \\ \end{array} \right) \cdot \left(\begin{array}{cccc} w_1 & & & \\ & w_2 & & \\ & & \cdots & \\ & & \cdots & \\ & & & w_N \end{array} \right) \cdot \left(\begin{array}{c} \mathbf{V}^T \end{array} \right) \tag{2.9.1}$$

The matrices **U** and **V** are each orthogonal in the sense that their columns are orthonormal,

$$\sum_{i=1}^{M} U_{ik} U_{in} = \delta_{kn} \qquad \begin{array}{l} 1 \le k \le N \\ 1 \le n \le N \end{array} \tag{2.9.2}$$

$$\sum_{j=1}^{N} V_{jk} V_{jn} = \delta_{kn} \qquad \begin{array}{l} 1 \le k \le N \\ 1 \le n \le N \end{array} \tag{2.9.3}$$

or as a tableau,

$$\left(\quad \mathbf{U}^T \quad \right) \cdot \left(\quad \mathbf{U} \quad \right) = \left(\quad \mathbf{V}^T \quad \right) \cdot \left(\quad \mathbf{V} \quad \right) = \left(\quad \mathbf{1} \quad \right) \tag{2.9.4}$$

Since **V** is square, it is also row-orthonormal, $\mathbf{V} \cdot \mathbf{V}^T = 1$.

The decomposition (2.9.1) can always be done, no matter how singular the matrix is, and it is "almost" unique. That is to say, it is unique up to (i) making the same permutation of the columns of **U**, elements of **W**, and columns of **V** (or rows of $\mathbf{V}^T$), or (ii) forming linear combinations of any columns of **U** and **V** whose corresponding elements of **W** happen to be exactly equal.

At the end of this section, we give a routine, `SVDCMP`, that performs SVD on an arbitrary matrix **A**, replacing it by **U** (they are the same shape) and returning **W** and **V** separately. The routine `SVDCMP` is based on a routine by Forsythe et al., which is in turn based on the original routine of Golub and Reinsch, found, in various forms, in Wilkinson and Reinsch, in LINPACK, and elsewhere. These references include extensive discussion of the algorithm used. As much as we dislike the use of black-box routines, we are going to ask you to accept this one, since it would take us too far afield to cover its necessary background material here. Suffice it to say that the algorithm is very stable, and that it is very unusual for it ever to misbehave. Most of the concepts that enter the algorithm (Householder reduction to bidiagonal form, diagonalization by QR procedure with shifts) will be discussed further in Chapter 11. Along with those already mentioned, another useful reference is Stoer and Bulirsch.

If you are as suspicious of black boxes as we are, you will want to verify yourself that SVDCMP does what we say it does. That is very easy to do: Generate an arbitrary matrix **A**, call the routine, and then verify by matrix multiplication that (2.9.1) and (2.9.4) are satisfied. Since these two equations are the only defining requirements for SVD, this procedure is (for the chosen **A**) a complete end-to-end check.

Now let us find out what SVD is good for.

SVD of a Square Matrix

If the matrix **A** is square, $N \times N$ say, then **U**, **V**, and **W** are all square matrices of the same size. Their inverses are also trivial to compute: **U** and **V** are orthogonal, so their inverses are equal to their transposes; **W** is diagonal, so its inverse is the diagonal matrix whose elements are the reciprocals of the elements w_j. From (2.9.1) it now follows immediately that the inverse of **A** is

$$\mathbf{A}^{-1} = \mathbf{V} \cdot [\text{diag}\ (1/w_j)] \cdot \mathbf{U}^T \tag{2.9.5}$$

The only thing that can go wrong with this construction is for one of the w_j's to be zero, or (numerically) for it to be so small that its value is dominated by roundoff error and therefore unknowable. If more than one of the w_j's have this problem, then the matrix is even more singular. So, first of all, SVD gives you a clear diagnosis of the situation.

Formally, the *condition number* of a matrix is defined as the ratio of the largest of the w_j's to the smallest of the w_j's. A matrix is singular if its condition number is infinite, and it is *ill-conditioned* if its condition number is too large, that is, if its reciprocal approaches the machine's floating point precision (for example, less than 10^{-6} for single precision or 10^{-12} for double).

For singular matrices, the concepts of *nullspace* and *range* are important. Consider the familiar set of simultaneous equations

$$\mathbf{A} \cdot \mathbf{x} = \mathbf{b} \tag{2.9.6}$$

where **A** is a square matrix, **b** and **x** are vectors. Equation (2.9.6) defines **A** as a linear mapping from the vector space **x** to the vector space **b**. If **A** is singular, then there is some subspace of **x**, called the nullspace, that is mapped to zero, $\mathbf{A} \cdot \mathbf{x} = 0$. The dimension of the nullspace (the number of linearly independent vectors **x** which can be found in it) is called the *nullity* of **A**.

Now, there is also some subspace of **b** which can be "reached" by **A**, in the sense that there exists some **x** which is mapped there. This subspace of **b** is called the range of **A**. The dimension of the range is called the *rank* of **A**. If **A** is nonsingular, then its range will be all of the vector space **b**, so its rank is N. If **A** is singular, then the rank will be less than N. In fact, the relevant theorem is " rank plus nullity equals N."

What has this to do with SVD? SVD explicitly constructs orthonormal bases for the nullspace and range of a matrix. Specifically, the columns of

U whose same-numbered elements w_j are *nonzero* are an orthonormal set of basis vectors that span the range; the columns of **V** whose same-numbered elements w_j are *zero* are an orthonormal basis for the nullspace.

Now let's have another look at solving the set of simultaneous linear equations (2.9.6) in the case that **A** is singular. The important question is whether the vector **b** on the right-hand side lies in the range of **A** or not. If it does, then the singular set of equations *does* have a solution **x**; in fact it has more than one solution, since any vector in the nullspace (any column of **V** with a corresponding zero w_j) can be added to **x** in any linear combination.

If we want to single out one particular member of this solution-set of vectors as a representative, we might want to pick the one with the smallest length $|\mathbf{x}|^2$. Here is how to find that vector using SVD: Simply *replace* $1/w_j$ *by zero if* $w_j = 0$. (It is not very often that one gets to set $\infty = 0$!) Then compute (working from right to left)

$$\mathbf{x} = \mathbf{V} \cdot [\mathrm{diag}\ (1/w_j)] \cdot (\mathbf{U}^T \cdot \mathbf{b}) \tag{2.9.7}$$

This will be the solution vector of smallest length; the columns of **V** which are in the nullspace complete the specification of the solution set.

Proof: Consider $|\mathbf{x} + \mathbf{x}'|$, where $\mathbf{x}'$ lies in the nullspace. Then, if $\mathbf{W}^{-1}$ denotes the modified inverse of **W** with some elements zeroed,

$$\begin{aligned} |\mathbf{x} + \mathbf{x}'| &= \left|\mathbf{V} \cdot \mathbf{W}^{-1} \cdot \mathbf{U}^T \cdot \mathbf{b} + \mathbf{x}'\right| \\ &= \left|\mathbf{V} \cdot (\mathbf{W}^{-1} \cdot \mathbf{U}^T \cdot \mathbf{b} + \mathbf{V}^T \cdot \mathbf{x}')\right| \\ &= \left|\mathbf{W}^{-1} \cdot \mathbf{U}^T \cdot \mathbf{b} + \mathbf{V}^T \cdot \mathbf{x}'\right| \end{aligned} \tag{2.9.8}$$

Here the first equality follows from (2.9.7), the second and third from the orthonormality of **V**. If you now examine the two terms which make up the sum on the right-hand side, you will see that the first one has nonzero j components only where $w_j \neq 0$, while the second one, since $\mathbf{x}'$ is in the nullspace, has nonzero j components only where $w_j = 0$. Therefore the minimum length obtains for $\mathbf{x}' = 0$, q.e.d.

If **b** is not in the range of the singular matrix **A**, then the set of equations (2.9.6) has no solution. But here is some good news: If **b** is not in the range of **A**, then equation (2.9.7) can still be used to construct a "solution" vector **x**. This vector **x** will not exactly solve $\mathbf{A} \cdot \mathbf{x} = \mathbf{b}$. But, among all possible vectors **x**, it will do the closest possible job in the least squares sense. In other words (2.9.7) finds

$$\mathbf{x} \quad \text{which minimizes} \quad r \equiv |\mathbf{A} \cdot \mathbf{x} - \mathbf{b}| \tag{2.9.9}$$

The number r is called the *residual* of the solution.

The proof is similar to (2.9.8): Suppose we modify $\mathbf{x}$ by adding some arbitrary $\mathbf{x}'$. Then $\mathbf{A}\cdot\mathbf{x}-\mathbf{b}$ is modified by adding some $\mathbf{b}' \equiv \mathbf{A}\cdot\mathbf{x}'$. Obviously $\mathbf{b}'$ is in the range of $\mathbf{A}$. We then have

$$\begin{aligned}
\left|\mathbf{A}\cdot\mathbf{x}-\mathbf{b}+\mathbf{b}'\right| &= \left|(\mathbf{U}\cdot\mathbf{W}\cdot\mathbf{V}^T)\cdot(\mathbf{V}\cdot\mathbf{W}^{-1}\cdot\mathbf{U}^T\cdot\mathbf{b})-\mathbf{b}+\mathbf{b}'\right| \\
&= \left|(\mathbf{U}\cdot\mathbf{W}\cdot\mathbf{W}^{-1}\cdot\mathbf{U}^T-1)\cdot\mathbf{b}+\mathbf{b}'\right| \\
&= \left|\mathbf{U}\cdot\left[(\mathbf{W}\cdot\mathbf{W}^{-1}-1)\cdot\mathbf{U}^T\cdot\mathbf{b}+\mathbf{U}^T\cdot\mathbf{b}'\right]\right| \\
&= \left|(\mathbf{W}\cdot\mathbf{W}^{-1}-1)\cdot\mathbf{U}^T\cdot\mathbf{b}+\mathbf{U}^T\cdot\mathbf{b}'\right|
\end{aligned} \tag{2.9.10}$$

Now, $(\mathbf{W}\cdot\mathbf{W}^{-1}-1)$ is a diagonal matrix which has nonzero j components only for $w_j=0$, while $\mathbf{U}^T\mathbf{b}'$ has nonzero j components only for $w_j \neq 0$, since $\mathbf{b}'$ lies in the range of $\mathbf{A}$. Therefore the minimum obtains for $\mathbf{b}'=0$, q.e.d.

Figure 2.9.1 summarizes our discussion of SVD thus far.

In the discussion since equation (2.9.6), we have been pretending that a matrix is either singular or else isn't. That is of course true analytically. Numerically, however, the far more common situation is that some of the w_j's are very small but nonzero, so that the matrix is ill-conditioned. In that case, the direct solution methods of *LU* decomposition or Gaussian elimination may actually give a formal solution to the set of equations (that is, a zero pivot may not be encountered); but the solution vector may have wildly large components whose algebraic cancellation, when multiplying by the matrix $\mathbf{A}$, may give a very poor approximation to the right-hand vector $\mathbf{b}$. In such cases, the solution vector $\mathbf{x}$ obtained by *zeroing* the small w_j's and then using equation (2.9.7) is very often better (in the sense of the residual $|\mathbf{A}\cdot\mathbf{x}-\mathbf{b}|$ being smaller) than *both* the direct-method solution *and* the SVD solution where the small w_j's are left nonzero.

It may seem paradoxical that this can be so, since zeroing a singular value corresponds to throwing away one linear combination of the set of equations that we are trying to solve. The resolution of the paradox is that we are throwing away precisely a combination of equations that is so corrupted by roundoff error as to be at best useless; usually it is worse than useless since it "pulls" the solution vector way off towards infinity along some direction that is almost a nullspace vector. In doing this, it compounds the roundoff problem and makes the residual $|\mathbf{A}\cdot\mathbf{x}-\mathbf{b}|$ larger.

SVD cannot be applied blindly, then. You have to exercise some discretion in deciding at what threshold to zero the small w_j's, and/or you have to have some idea what size of computed residual $|\mathbf{A}\cdot\mathbf{x}-\mathbf{b}|$ is acceptable.

As an example, here is a "backsubstitution" routine `SVBKSB` for evaluating equation (2.9.7) and obtaining a solution vector $\mathbf{x}$ from a right-hand side $\mathbf{b}$, given that the SVD of a matrix $\mathbf{A}$ has already been calculated by a call to `SVDCMP`. Note that this routine presumes that *you* have already zeroed the small w_j's. It does not do this for you. If you *haven't* zeroed the small w_j's,

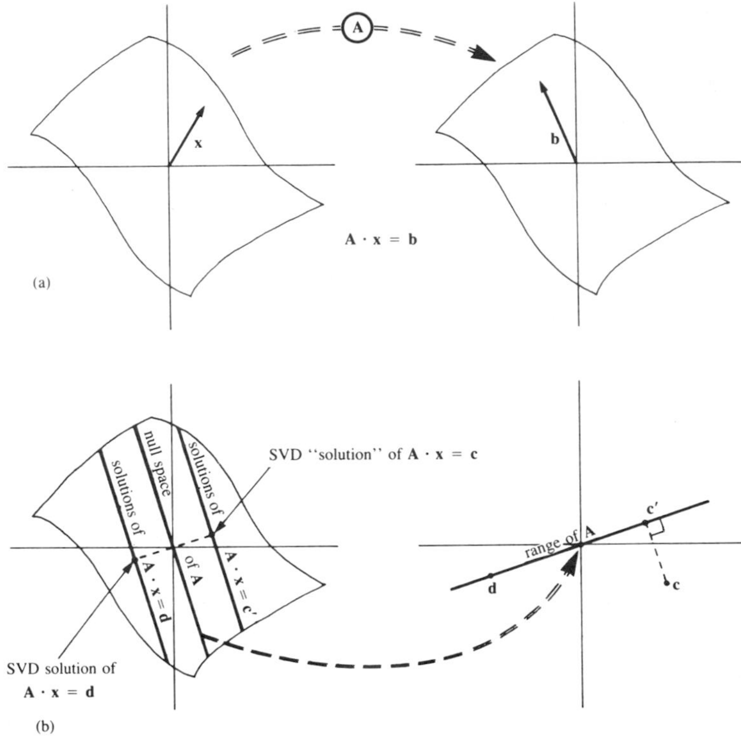

Figure 2.9.1. (a) A nonsingular matrix **A** maps a vector space into one of the same dimension. The vector **x** is mapped into **b**, so that **x** satisfies the equation $\mathbf{A}\cdot\mathbf{x}=\mathbf{b}$. (b) A singular matrix **A** maps a vector space into one of lower dimensionality, here a plane into a line, called the "range" of **A**. The "nullspace" of **A** is mapped to zero. The solutions of $\mathbf{A}\cdot\mathbf{x}=\mathbf{d}$ consist of any one particular solution plus any vector in the nullspace, here forming a line parallel to the nullspace. Singular value decomposition (SVD) selects the particular solution closest to zero, as shown. The point **c** lies outside of the range of **A**, so $\mathbf{A}\cdot\mathbf{x}=\mathbf{c}$ has no solution. SVD finds the least-squares best compromise solution, namely a solution of $\mathbf{A}\cdot\mathbf{x}=\mathbf{c}'$, as shown.

then this routine is just as ill-conditioned as any direct method, and you are misusing SVD.

```
SUBROUTINE SVBKSB(U,W,V,M,N,MP,NP,B,X)
    Solves A · X = B for a vector X, where A is specified by the arrays U, W, V as returned
    by SVDCMP. M and N are the logical dimensions of A, and will be equal for square matrices.
    MP and NP are the physical dimensions of A. B is the input right-hand side. X is the
    output solution vector. No input quantities are destroyed, so the routine may be called
    sequentially with different B's. M must be greater or equal to N; see SVDCMP.
PARAMETER (NMAX=100)                  Maximum anticipated value of N.
DIMENSION U(MP,NP),W(NP),V(NP,NP),B(MP),X(NP),TMP(NMAX)
DO 12 J=1,N                           Calculate U^T B.
    S=0.
    IF(W(J).NE.0.)THEN                     Nonzero result only if w_j is nonzero.
        DO 11 I=1,M
            S=S+U(I,J)*B(I)
11      CONTINUE
        S=S/W(J)                           This is the divide by w_j.
    ENDIF
```

```
    TMP(J)=S
  12 CONTINUE
DO 14 J=1,N                        Matrix multiply by V to get answer.
    S=0.
    DO 13 JJ=1,N
        S=S+V(J,JJ)*TMP(JJ)
      13 CONTINUE
    X(J)=S
  14 CONTINUE
RETURN
END
```

Note that a typical use of SVDCMP and SVBKSB superficially resembles the typical use of LUDCMP and LUBKSB: In both cases, you decompose the right-hand matrix $\mathbf{A}$ just once, and then can use the decomposition either once or many times with different right-hand sides. The crucial difference is the "editing" of the singular values before SVBKSB is called:

```
DIMENSION A(NP,NP),U(NP,NP),W(NP),V(NP,NP),B(NP),X(NP)
...
DO 12 I=1,N                    Copy A into U if you don't want it to be destroyed.
    DO 11 J=1,N
        U(I,J)=A(I,J)
      11 CONTINUE
  12 CONTINUE
CALL SVDCMP(U,N,N,NP,NP,W,V) SVD the square matrix A.
WMAX=0.                        Will be the maximum singular value obtained.
DO 13 J=1,N
    IF(W(J).GT.WMAX)WMAX=W(J)
  13 CONTINUE
WMIN=WMAX*1.0E-6               This is where we set the threshold for singular values allowed to
DO 14 J=1,N                        be nonzero. The constant is typical, but not universal.
    IF(W(J).LT.WMIN)W(J)=0.        You have to experiment with your own application.
  14 CONTINUE
CALL SVBKSB(U,W,V,N,N,NP,NP,B,X)     Now we can backsubstitute.
```

SVD for Fewer Equations than Unknowns

If you have fewer linear equations M than unknowns N, then you are not expecting a unique solution. Usually there will be an $N-M$ dimensional family of solutions. If you want to find this whole solution space, then SVD can readily do the job.

Augment your left-hand side matrix with rows of zeros underneath its M nonzero rows, until it is filled up to be square, $N \times N$. Similarly augment your right-hand side vector with zeros. You now have a singular set of N equations in N unknowns. Apply SVD as described above. You should expect one zero or negligible w_j for each row of zeros that you added, plus additional ones from any degeneracies in your M equations. Be sure that you find this many small w_j's, and zero them before calling SVBKSB, which will give you the particular solution vector $\mathbf{x}$. As before, the columns of $\mathbf{V}$ corresponding to zeroed w_j's are the basis vectors whose linear combinations, added to the particular solution, span the solution space.

SVD for More Equations than Unknowns

This situation will occur in Chapter 14, when we wish to find the least-squares solution to an overdetermined set of linear equations. In tableau, the equations to be solved are

$$\begin{pmatrix} & & \\ & \mathbf{A} & \\ & & \end{pmatrix} \cdot \begin{pmatrix} \mathbf{x} \end{pmatrix} = \begin{pmatrix} \\ \mathbf{b} \\ \\ \end{pmatrix} \tag{2.9.11}$$

The proofs that we gave above for the square case apply without modification to the case of more equations than unknowns. The least-squares solution vector $\mathbf{x}$ is given by (2.9.7), which, with non-square matrices, looks like this,

$$\begin{pmatrix} \mathbf{x} \end{pmatrix} = \begin{pmatrix} & \mathbf{V} & \end{pmatrix} \cdot \begin{pmatrix} \mathrm{diag}(1/w_j) \end{pmatrix} \cdot \begin{pmatrix} & & \mathbf{U}^T & & \end{pmatrix} \cdot \begin{pmatrix} \\ \mathbf{b} \\ \\ \end{pmatrix} \tag{2.9.12}$$

In general, the matrix $\mathbf{W}$ will not be singular, and no w_j's will need to be set to zero. Occasionally, however, there might be column degeneracies in $\mathbf{A}$. In this case you will need to zero some small w_j values after all. The corresponding column in $\mathbf{V}$ gives the linear combination of $\mathbf{x}$'s that is then ill-determined even by the supposedly overdetermined set.

Sometimes, although you do not need to zero any w_j's for *computational* reasons, you may nevertheless want to take note of any that are unusually small: their corresponding columns in $\mathbf{V}$ are linear combinations of $\mathbf{x}$'s which are insensitive to your data. In fact, you may then wish to zero these w_j's, to reduce the number of free parameters in the fit. These matters are discussed more fully in Chapter 14.

Approximation of Matrices

Note that equation (2.9.1) can be rewritten to express any matrix A_{ij} as a sum of outer products of columns of $\mathbf{U}$ and rows of $\mathbf{V}^T$, with the "weighting factors" being the singular values w_j,

$$A_{ij} = \sum_{k=1}^{N} w_k \, U_{ik} V_{jk} \tag{2.9.13}$$

If you ever encounter a situation where *most* of the singular values w_j of a matrix $\mathbf{A}$ are very small, then $\mathbf{A}$ will be well-approximated by only a few terms in the sum (2.9.13). This means that you only have to store a few columns of $\mathbf{U}$ and $\mathbf{V}$ (the same k ones) and you will be able to recover, with good accuracy, the whole matrix. Note also that it is very efficient to multiply such an approximated matrix by a vector $\mathbf{x}$: You just dot $\mathbf{x}$ with each of the stored columns of $\mathbf{V}$, multiply the resulting scalar by the corresponding w_k, and accumulate that multiple of the corresponding column of $\mathbf{U}$. If your matrix is approximated by a small number K of singular values, then this computation of $\mathbf{A} \cdot \mathbf{x}$ takes only about $K(M+N)$ multiplications, instead of MN for the full matrix.

SVD Algorithm

```
SUBROUTINE SVDCMP(A,M,N,MP,NP,W,V)
   Given a matrix A, with logical dimensions M by N and physical dimensions MP by NP, this
   routine computes its singular value decomposition, A = U·W·V^T. The matrix U replaces
   A on output. The diagonal matrix of singular values W is output as a vector W. The matrix
   V (not the transpose V^T) is output as V. M must be greater or equal to N; if it is smaller,
   then A should be filled up to square with zero rows.
PARAMETER (NMAX=100)          Maximum anticipated value of N.
DIMENSION A(MP,NP),W(NP),V(NP,NP),RV1(NMAX)
IF(M.LT.N)PAUSE 'You must augment A with extra zero rows.'
   Householder reduction to bidiagonal form.
G=0.0
SCALE=0.0
ANORM=0.0
DO 25 I=1,N
   L=I+1
   RV1(I)=SCALE*G
   G=0.0
   S=0.0
   SCALE=0.0
   IF (I.LE.M) THEN
      DO 11 K=I,M
         SCALE=SCALE+ABS(A(K,I))
11    CONTINUE
      IF (SCALE.NE.0.0) THEN
         DO 12 K=I,M
            A(K,I)=A(K,I)/SCALE
            S=S+A(K,I)*A(K,I)
12       CONTINUE
         F=A(I,I)
```

```
            G=-SIGN(SQRT(S),F)
            H=F*G-S
            A(I,I)=F-G
            IF (I.NE.N) THEN
                DO 15 J=L,N
                    S=0.0
                    DO 13 K=I,M
                        S=S+A(K,I)*A(K,J)
                      13 CONTINUE
                    F=S/H
                    DO 14 K=I,M
                        A(K,J)=A(K,J)+F*A(K,I)
                      14 CONTINUE
                  15 CONTINUE
            ENDIF
            DO 16 K= I,M
                A(K,I)=SCALE*A(K,I)
              16 CONTINUE
        ENDIF
    ENDIF
    W(I)=SCALE *G
    G=0.0
    S=0.0
    SCALE=0.0
    IF ((I.LE.M).AND.(I.NE.N)) THEN
        DO 17 K=L,N
            SCALE=SCALE+ABS(A(I,K))
          17 CONTINUE
        IF (SCALE.NE.0.0) THEN
            DO 18 K=L,N
                A(I,K)=A(I,K)/SCALE
                S=S+A(I,K)*A(I,K)
              18 CONTINUE
            F=A(I,L)
            G=-SIGN(SQRT(S),F)
            H=F*G-S
            A(I,L)=F-G
            DO 19 K=L,N
                RV1(K)=A(I,K)/H
              19 CONTINUE
            IF (I.NE.M) THEN
                DO 23 J=L,M
                    S=0.0
                    DO 21 K=L,N
                        S=S+A(J,K)*A(I,K)
                      21 CONTINUE
                    DO 22 K=L,N
                        A(J,K)=A(J,K)+S*RV1(K)
                      22 CONTINUE
                  23 CONTINUE
            ENDIF
            DO 24 K=L,N
                A(I,K)=SCALE*A(I,K)
              24 CONTINUE
        ENDIF
    ENDIF
    ANORM=MAX(ANORM,(ABS(W(I))+ABS(RV1(I))))
  25 CONTINUE
    Accumulation of right-hand transformations.
DO 32 I=N,1,-1
    IF (I.LT.N) THEN
        IF (G.NE.0.0) THEN
            DO 26 J=L,N                  Double division to avoid possible underflow:
                V(J,I)=(A(I,J)/A(I,L))/G
```

```
              26 CONTINUE
            DO 29 J=L,N
              S=0.0
              DO 27 K=L,N
                S=S+A(I,K)*V(K,J)
27            CONTINUE
              DO 28 K=L,N
                V(K,J)=V(K,J)+S*V(K,I)
28            CONTINUE
29          CONTINUE
          ENDIF
          DO 31 J=L,N
            V(I,J)=0.0
            V(J,I)=0.0
31        CONTINUE
        ENDIF
        V(I,I)=1.0
        G=RV1(I)
        L=I
32    CONTINUE
```

Accumulation of left-hand transformations.

```
      DO 39 I=N,1,-1
        L=I+1
        G=W(I)
        IF (I.LT.N) THEN
          DO 33 J=L,N
            A(I,J)=0.0
33        CONTINUE
        ENDIF
        IF (G.NE.0.0) THEN
          G=1.0/G
          IF (I.NE.N) THEN
            DO 36 J=L,N
              S=0.0
              DO 34 K=L,M
                S=S+A(K,I)*A(K,J)
34            CONTINUE
              F=(S/A(I,I))*G
              DO 35 K=I,M
                A(K,J)=A(K,J)+F*A(K,I)
35            CONTINUE
36          CONTINUE
          ENDIF
          DO 37 J=I,M
            A(J,I)=A(J,I)*G
37        CONTINUE
        ELSE
          DO 38 J= I,M
            A(J,I)=0.0
38        CONTINUE
        ENDIF
        A(I,I)=A(I,I)+1.0
39    CONTINUE
```

Diagonalization of the bidiagonal form.

```
      DO 49 K=N,1,-1                          Loop over singular values.
        DO 48 ITS=1,30                        Loop over allowed iterations.
          DO 41 L=K,1,-1                      Test for splitting:
            NM=L-1                            Note that RV1(1) is always zero.
            IF ((ABS(RV1(L))+ANORM).EQ.ANORM) GO TO 2
            IF ((ABS(W(NM))+ANORM).EQ.ANORM) GO TO 1
41        CONTINUE
1         C=0.0                               Cancellation of RV1(L), if L> 1 :
          S=1.0
          DO 43 I=L,K
```

```
            F=S*RV1(I)
            IF ((ABS(F)+ANORM).NE.ANORM) THEN
                G=W(I)
                H=SQRT(F*F+G*G)
                W(I)=H
                H=1.0/H
                C= (G*H)
                S=-(F*H)
                DO 42 J=1,M
                    Y=A(J,NM)
                    Z=A(J,I)
                    A(J,NM)=(Y*C)+(Z*S)
                    A(J,I)=-(Y*S)+(Z*C)
42              CONTINUE
            ENDIF
43      CONTINUE
2       Z=W(K)
        IF (L.EQ.K) THEN        Convergence.
            IF (Z.LT.0.0) THEN     Singular value is made nonnegative.
                W(K)=-Z
                DO 44 J=1,N
                    V(J,K)=-V(J,K)
44              CONTINUE
            ENDIF
            GO TO 3
        ENDIF
        IF (ITS.EQ.30) PAUSE 'No convergence in 30 iterations'
        X=W(L)                  Shift from bottom 2-by-2 minor:
        NM=K-1
        Y=W(NM)
        G=RV1(NM)
        H=RV1(K)
        F=((Y-Z)*(Y+Z)+(G-H)*(G+H))/(2.0*H*Y)
        G=SQRT(F*F+1.0)
        F=((X-Z)*(X+Z)+H*((Y/(F+SIGN(G,F)))-H))/X
    Next QR transformation:
        C=1.0
        S=1.0
        DO 47 J=L,NM
            I=J+1
            G=RV1(I)
            Y=W(I)
            H=S*G
            G=C*G
            Z=SQRT(F*F+H*H)
            RV1(J)=Z
            C=F/Z
            S=H/Z
            F= (X*C)+(G*S)
            G=-(X*S)+(G*C)
            H=Y*S
            Y=Y*C
            DO 45 JJ=1,N
                X=V(JJ,J)
                Z=V(JJ,I)
                V(JJ,J)= (X*C)+(Z*S)
                V(JJ,I)=-(X*S)+(Z*C)
45          CONTINUE
            Z=SQRT(F*F+H*H)
            W(J)=Z                 Rotation can be arbitrary if Z=0.
            IF (Z.NE.0.0) THEN
                Z=1.0/Z
                C=F*Z
                S=H*Z
```

```
              ENDIF
              F= (C*G)+(S*Y)
              X=-(S*G)+(C*Y)
              DO 46 JJ=1,M
                  Y=A(JJ,J)
                  Z=A(JJ,I)
                  A(JJ,J)= (Y*C)+(Z*S)
                  A(JJ,I)=-(Y*S)+(Z*C)
46            CONTINUE
47          CONTINUE
          RV1(L)=0.0
          RV1(K)=F
          W(K)=X
48        CONTINUE
3       CONTINUE
49    CONTINUE
      RETURN
      END
```

REFERENCES AND FURTHER READING:

Golub, Gene H., and Van Loan, Charles F. 1983, *Matrix Computations* (Baltimore: Johns Hopkins University Press), §8.3 and Chapter 12.

Wilkinson, J.H., and Reinsch, C. 1971, *Linear Algebra*, vol. II of *Handbook for Automatic Computation* (New York: Springer-Verlag), Chapter I.10 by G.H. Golub and C. Reinsch.

Lawson, Charles L., and Hanson, Richard J. 1974, *Solving Least Squares Problems* (Englewood Cliffs, N.J.: Prentice-Hall), Chapter 18.

Forsythe, George E., Malcolm, Michael A., and Moler, Cleve B. 1977, *Computer Methods for Mathematical Computations* (Englewood Cliffs, N.J.: Prentice-Hall), Chapter 9.

Dongarra, J.J., et al. 1979, *LINPACK User's Guide* (Philadelphia: Society for Industrial and Applied Mathematics), Chapter 11.

Smith, B.T., et al. 1976, *Matrix Eigensystem Routines — EISPACK Guide*, 2nd ed., vol. 6 of Lecture Notes in Computer Science (New York: Springer-Verlag).

Stoer, J., and Bulirsch, R. 1980, *Introduction to Numerical Analysis* (New York: Springer-Verlag), §6.7.

2.10 Sparse Linear Systems

A system of linear equations is called *sparse* if only a relatively small number of its matrix elements a_{ij} are nonzero. It is wasteful to use general methods of linear algebra on such problems, because most of the $O(N^3)$ arithmetic operations devoted to solving the set of equations or inverting the matrix involve zero operands. Furthermore, you might wish to work problems so large as to tax your available memory space, and it is wasteful to reserve storage for unfruitful zero elements. Note that there are two distinct (and not always compatible) goals for any sparse matrix method: saving time and/or saving space.

We have already considered one archetypal sparse form in §2.6, the tridiagonal matrix. There we saw that it was possible to save both time (order N instead of N^3) and space (order N instead of N^2). The method of solution was not different in principle from the general method of LU decomposition; it was just applied cleverly, and with due attention to the bookkeeping of zero elements. Most practical schemes for dealing with sparse problems have this same character. They are fundamentally decomposition schemes, or else elimination schemes akin to Gauss-Jordan, but carefully optimized so as to minimize the number of so-called *fill-ins*, initially zero elements which must become nonzero during the solution process, and for which storage must be reserved.

Direct methods for solving sparse equations, then, depend crucially on the precise pattern of sparsity of the matrix. Patterns which occur frequently, or which are useful as way-stations in the reduction of more general forms, already have special names and special methods of solution. We do not have space here for any detailed review of these. References listed at the end of this section will furnish you with an "in" to the specialized literature, and the following list of buzz words (and Figure 2.10.1) will at least let you hold your own at cocktail parties:

- tridiagonal
- band diagonal (or banded) with bandwidth M
- band triangular
- block diagonal
- block tridiagonal
- block triangular
- cyclic banded
- singly- (or doubly-) bordered block diagonal
- singly- (or doubly-) bordered block triangular
- singly- (or doubly-) bordered band diagonal
- singly- (or doubly-) bordered band triangular
- other (!)

You should also be aware of some of the special sparse forms that occur in the solution of partial differential equations in two or more dimensions. See Chapter 17.

If your particular pattern of sparsity is not a simple one, then you may wish to try an *analyze/factorize/operate* package, which automates the procedure of figuring out how fill-ins are to be minimized. The *analyze* stage is done once only for each pattern of sparsity. The *factorize* stage is done once for each particular matrix that fits the pattern. The *operate* stage is performed once for each right-hand side to be used with the particular matrix. Consult Jacobs for references on this. The NAG library has an analyze/factorize/operate capability. A substantial collection of routines for sparse matrix calculation is also available from IMSL as the *Yale Sparse Matrix Package*.

You should be aware that the special order of interchanges and eliminations, prescribed by a sparse matrix method so as to minimize fill-ins and arithmetic operations, generally acts to decrease the method's numerical stability as compared to, e.g., regular LU decomposition with pivoting. Scaling

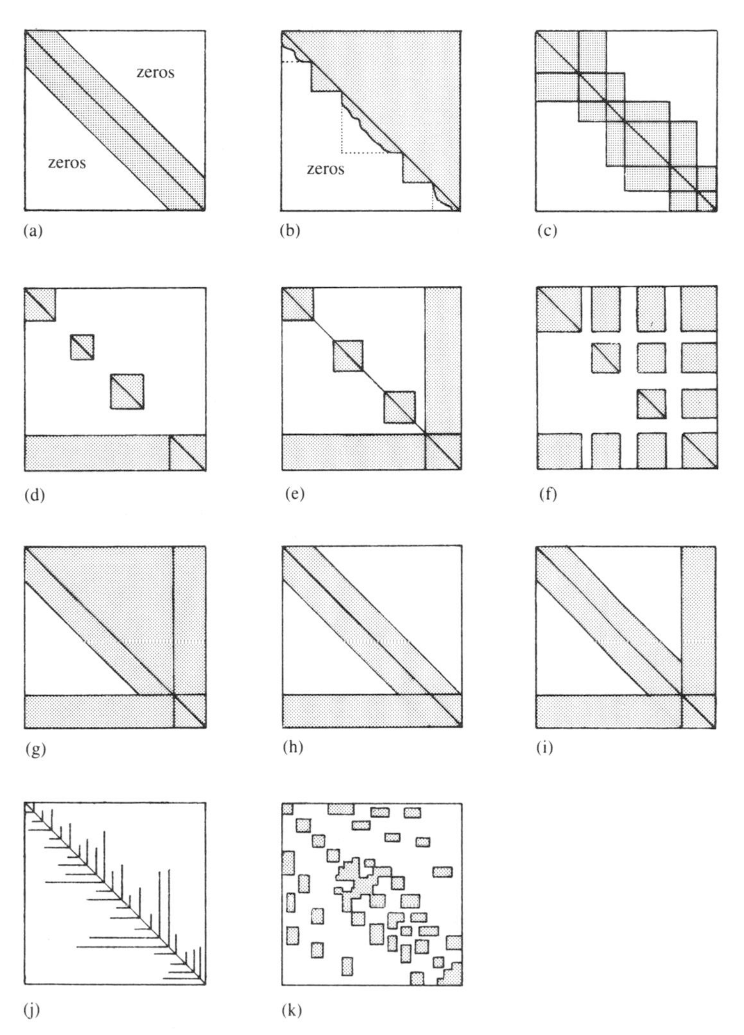

Figure 2.10.1. Some standard forms for sparse matrices. (a) Band diagonal; (b) block triangular; (c) block tridiagonal; (d) singly-bordered block diagonal; (e) doubly-bordered block diagonal; (f) singly-bordered block triangular; (g) bordered band-triangular; (h) and (i) singly- and doubly-bordered band diagonal; (j) and (k) other! (after Tewarson).

your problem so as to make its nonzero matrix elements have comparable magnitudes (if you can do it) will sometimes ameliorate this problem.

In the remainder of this section, we present a couple of ideas which are applicable to some general classes of sparse matrices, and which do not depend on details of the pattern of sparsity.

Sherman-Morrison and Woodbury Formulas

Suppose that you have already obtained, by herculean effort, the inverse matrix $\mathbf{A}^{-1}$ of a square matrix $\mathbf{A}$. Now you want to make a "small" change

in $\mathbf{A}$, for example change one element a_{ij}, or a few elements, or one row, or one column. Is there any way of calculating the corresponding change in $\mathbf{A}^{-1}$ without repeating your difficult labors? Yes, if your change is of the form

$$\mathbf{A} \rightarrow (\mathbf{A} + \mathbf{u} \otimes \mathbf{v}) \tag{2.10.1}$$

for some vectors $\mathbf{u}$ and $\mathbf{v}$. If $\mathbf{u}$ is a unit vector $\mathbf{e}_i$, then (2.10.1) adds the components of $\mathbf{v}$ to the ith row. (Recall that $\mathbf{u} \otimes \mathbf{v}$ is a matrix whose i, j^{th} element is the product of the ith component of $\mathbf{u}$ and the jth component of $\mathbf{v}$.) If $\mathbf{v}$ is a unit vector $\mathbf{e}_j$, then (2.10.1) adds the components of $\mathbf{u}$ to the jth column. If both $\mathbf{u}$ and $\mathbf{v}$ are proportional to unit vectors $\mathbf{e}_i$ and $\mathbf{e}_j$ respectively, then a term is added only to the element a_{ij}.

The *Sherman-Morrison* formula gives the inverse $(\mathbf{A} + \mathbf{u} \otimes \mathbf{v})^{-1}$, and is derived briefly as follows:

$$\begin{aligned}
(\mathbf{A} + \mathbf{u} \otimes \mathbf{v})^{-1} &= (\mathbf{1} + \mathbf{A}^{-1} \cdot \mathbf{u} \otimes \mathbf{v})^{-1} \cdot \mathbf{A}^{-1} \\
&= (\mathbf{1} - \mathbf{A}^{-1} \cdot \mathbf{u} \otimes \mathbf{v} + \mathbf{A}^{-1} \cdot \mathbf{u} \otimes \mathbf{v} \cdot \mathbf{A}^{-1} \cdot \mathbf{u} \otimes \mathbf{v} - \ldots) \cdot \mathbf{A}^{-1} \\
&= \mathbf{A}^{-1} - \mathbf{A}^{-1} \cdot \mathbf{u} \otimes \mathbf{v} \cdot \mathbf{A}^{-1} (1 - \lambda + \lambda^2 - \ldots) \\
&= \mathbf{A}^{-1} - \frac{(\mathbf{A}^{-1} \cdot \mathbf{u}) \otimes (\mathbf{v} \cdot \mathbf{A}^{-1})}{1 + \lambda}
\end{aligned} \tag{2.10.2}$$

where

$$\lambda \equiv \mathbf{v} \cdot \mathbf{A}^{-1} \cdot \mathbf{u} \tag{2.10.3}$$

The second line of (2.10.2) is a formal power series expansion. In the third line, the associativity of outer and inner products is used to factor out the scalars λ.

The use of (2.10.2) is this: Given $\mathbf{A}^{-1}$ and the vectors $\mathbf{u}$ and $\mathbf{v}$, we need only perform two matrix multiplications and a vector dot product,

$$\mathbf{z} \equiv \mathbf{A}^{-1} \cdot \mathbf{u} \qquad \mathbf{w} \equiv (\mathbf{A}^{-1})^T \cdot \mathbf{v} \qquad \lambda = \mathbf{v} \cdot \mathbf{z} \tag{2.10.4}$$

to get the desired change in the inverse

$$\mathbf{A}^{-1} \quad \rightarrow \quad \mathbf{A}^{-1} - \frac{\mathbf{z} \otimes \mathbf{w}}{1 + \lambda} \tag{2.10.5}$$

The whole procedure requires only $3N^2$ multiplies and a like number of adds (an even smaller number if $\mathbf{u}$ or $\mathbf{v}$ is a unit vector).

The Sherman-Morrison formula can be directly applied to a class of sparse problems. If you already have a fast way of calculating the inverse of $\mathbf{A}$ (e.g.,

a tridiagonal matrix, or some other standard sparse form), then (2.10.4)–(2.10.5) allow you to build up to your related but more complicated form, adding for example a row or column at a time. Notice that you can apply the Sherman-Morrison formula more than once successively, using at each stage the most recent update of $\mathbf{A}^{-1}$ (equation 2.10.5). Of course, if you have to modify *every* row, then you are back to an N^3 method. The constant in front of the N^3 is only a few times worse than the better direct methods, but you have deprived yourself of the stabilizing advantages of pivoting — so be careful.

For some other sparse problems, the Sherman-Morrison formula cannot be directly applied for the simple reason that storage of the whole inverse matrix $\mathbf{A}^{-1}$ is not feasible. If you want to add only a single correction of the form $\mathbf{u} \otimes \mathbf{v}$, and solve the linear system

$$(\mathbf{A} + \mathbf{u} \otimes \mathbf{v}) \cdot \mathbf{x} = \mathbf{b} \tag{2.10.6}$$

then you proceed as follows. Using the fast method that is presumed available for the matrix $\mathbf{A}$, solve the two auxiliary problems

$$\mathbf{A} \cdot \mathbf{y} = \mathbf{b} \qquad \mathbf{A} \cdot \mathbf{z} = \mathbf{u} \tag{2.10.7}$$

for the vectors $\mathbf{y}$ and $\mathbf{z}$. In terms of these,

$$\mathbf{x} = \mathbf{y} - \left[\frac{\mathbf{v} \cdot \mathbf{y}}{1 + (\mathbf{v} \cdot \mathbf{z})} \right] \mathbf{z} \tag{2.10.8}$$

as we see by multiplying (2.10.2) on the right by $\mathbf{b}$.

If you want to add more than a single correction term, then you cannot use (2.10.8) repeatedly, since without storing a new $\mathbf{A}^{-1}$ you will not be able to solve the auxiliary problems (2.10.7) efficiently after the first step. Instead, you need the *Woodbury formula*, which is the block-matrix version of the Sherman-Morrison formula,

$$\begin{aligned} &(\mathbf{A} + \mathbf{U} \cdot \mathbf{V}^T)^{-1} \\ &\quad = \mathbf{A}^{-1} - \left[\mathbf{A}^{-1} \cdot \mathbf{U} \cdot (\mathbf{1} + \mathbf{V}^T \cdot \mathbf{A}^{-1} \cdot \mathbf{U})^{-1} \cdot \mathbf{V}^T \cdot \mathbf{A}^{-1} \right] \end{aligned} \tag{2.10.9}$$

Here A is, as usual, an $N \times N$ matrix, while $\mathbf{U}$ and $\mathbf{V}$ are $N \times P$ matrices with $P < N$ and usually $P \ll N$. The inner piece of the correction term may

become clearer if written as the tableau,

$$\left[\begin{array}{c} \\ \\ \mathbf{U} \\ \\ \\ \end{array}\right] \cdot \left[\mathbf{1} + \mathbf{V}^T \cdot \mathbf{A}^{-1} \cdot \mathbf{U}\right]^{-1} \cdot \left[\begin{array}{ccc} & \mathbf{V}^T & \end{array}\right] \tag{2.10.10}$$

where you can see that the matrix whose inverse is needed is only $P \times P$ rather than $N \times N$.

The relation between the Woodbury formula and successive applications of the Sherman-Morrison formula is now clarified by noting that, if $\mathbf{U}$ is the matrix formed by columns out of the P vectors $\mathbf{u}_1, \ldots, \mathbf{u}_P$, and $\mathbf{V}$ is the matrix formed by columns out of the P vectors $\mathbf{v}_1, \ldots, \mathbf{v}_P$,

$$\mathbf{U} \equiv \left[\begin{array}{c} \\ \mathbf{u}_1 \\ \\ \end{array}\right] \cdots \left[\begin{array}{c} \\ \mathbf{u}_P \\ \\ \end{array}\right] \qquad \mathbf{V} \equiv \left[\begin{array}{c} \\ \mathbf{v}_1 \\ \\ \end{array}\right] \cdots \left[\begin{array}{c} \\ \mathbf{v}_P \\ \\ \end{array}\right] \tag{2.10.11}$$

then two ways of expressing the same correction to $\mathbf{A}$ are

$$\left(\mathbf{A} + \sum_{k=1}^{P} \mathbf{u}_k \otimes \mathbf{v}_k\right) = \left(\mathbf{A} + \mathbf{U} \cdot \mathbf{V}^T\right) \tag{2.10.12}$$

(Note that the subscripts on $\mathbf{u}$ and $\mathbf{v}$ do *not* denote components, but rather distinguish the different column vectors.)

Equation (2.10.12) reveals that, if you have $\mathbf{A}^{-1}$ in storage, then you can either make the P corrections in one fell swoop by using (2.10.9), inverting a $P \times P$ matrix, or else make them by applying (2.10.5) P successive times.

If you don't have storage for $\mathbf{A}^{-1}$, then you *must* use (2.10.9) in the following way: To solve the linear equation

$$\left(\mathbf{A} + \sum_{k=1}^{P} \mathbf{u}_k \otimes \mathbf{v}_k\right) \cdot \mathbf{x} = \mathbf{b} \tag{2.10.13}$$

first solve the P auxiliary problems

$$\begin{aligned}
\mathbf{A} \cdot \mathbf{z}_1 &= \mathbf{u}_1 \\
\mathbf{A} \cdot \mathbf{z}_2 &= \mathbf{u}_2 \\
&\cdots \\
\mathbf{A} \cdot \mathbf{z}_P &= \mathbf{u}_P
\end{aligned} \tag{2.10.14}$$

and construct the matrix $\mathbf{Z}$ by columns from the $\mathbf{z}$'s obtained,

$$\mathbf{Z} \equiv \begin{bmatrix} \begin{bmatrix} \\ \mathbf{z}_1 \\ \\ \end{bmatrix} \cdots \begin{bmatrix} \\ \mathbf{z}_P \\ \\ \end{bmatrix} \end{bmatrix} \tag{2.10.15}$$

Next, do the $P \times P$ matrix inversion

$$\mathbf{H} \equiv (\mathbf{1} + \mathbf{V}^T \cdot \mathbf{Z})^{-1} \tag{2.10.16}$$

Finally, solve the one further auxiliary problem

$$\mathbf{A} \cdot \mathbf{y} = \mathbf{b} \tag{2.10.17}$$

In terms of these quantities, the solution is given by

$$\mathbf{x} = \mathbf{y} - \mathbf{Z} \cdot \left[\mathbf{H} \cdot (\mathbf{V}^T \cdot \mathbf{y}) \right] \tag{2.10.18}$$

Conjugate Gradient Method for a Sparse System

Later in this book, in Chapter 10, we will learn efficient iterative methods that converge to the minimum of a function f of a vector variable $\mathbf{x}$. One method in particular, the *conjugate gradient method* demands only that we are able to make two subsidiary calculations, (i) calculate the gradient of the function $\nabla f(\mathbf{x})$ at an arbitrary point, and (ii) minimize f along a specified ray, that is, find the value λ that minimizes the expression $f(\mathbf{x} + \lambda \mathbf{u})$ for specified $\mathbf{x}$ and $\mathbf{u}$. For further explanation and derivation of the conjugate-gradient method, consult §10.6.

Now let us consider the function

$$f(\mathbf{x}) \equiv \frac{1}{2} |\mathbf{A} \cdot \mathbf{x} - \mathbf{b}|^2 \tag{2.10.19}$$

Evidently this function has only a single minimum, at a value $\mathbf{x}$ that satisfies the linear set of equations $\mathbf{A}\cdot\mathbf{x}=\mathbf{b}$. A conjugate-gradient minimization will therefore solve that set of equations.

With a short calculation, you will be able to verify that the required two subsidiary calculations can be done as follows:

$$\nabla f(\mathbf{x})=\mathbf{A}^T\cdot(\mathbf{A}\cdot\mathbf{x}-\mathbf{b}) \tag{2.10.20}$$

$$\lambda=\frac{-\mathbf{u}\cdot\nabla f}{|\mathbf{A}\cdot\mathbf{u}|^2} \tag{2.10.21}$$

What has this to do with sparse matrices? Equations (2.10.20) and (2.10.21) make only two kinds of references to the matrix $\mathbf{A}$, namely multiplying the matrix by a vector and multiplying its transpose by a vector. If the matrix is sparse, then these multiplications require not the usual N^2 operations, but a smaller number equal to the number of nonzero components. You, the "owner" of the matrix $\mathbf{A}$, can be asked to provide subroutines which perform these sparse matrix multiplications as efficiently as possible. We, the "grand strategists" can supply a general routine which solves the set of linear equations using your subroutines.

This scheme is about the closest thing there is to a "general" sparse matrix routine. In truth, it is not quite as general as it appears, for two related reasons. First, the method is iterative and there are no advance guarantees as to how many iterations it will take to converge (e.g.) to your machine accuracy. It usually takes "on the order of" N iterations, each requiring three of your sparse matrix multiplies; but the constant implicit in the phrase "on the order of" is highly variable. Second, the function f is quadratic in $\mathbf{A}$. This means that the condition-number of the matrix that actually occurs in the function f, $\mathbf{A}^T\cdot\mathbf{A}$, is the square of the condition-number of $\mathbf{A}$ (see §2.9 for definition of condition-number). A large condition-number both increases the number of iterations required, and limits the accuracy to which a solution can be obtained.

If you happen to have a sparse matrix $\mathbf{A}$ that is positive definite, then you can eliminate the second of these problems. See the note at the end of §10.6 to find out how to do this. Otherwise, you can give the following routine a try. You may find that it works splendidly – or you may find it to be a bomb – it all depends on your $\mathbf{A}$. (If you want to understand how the routine works, consult §10.6, and compare with the routine FRPRMN.)

```
SUBROUTINE SPARSE(B,N,ASUB,ATSUB,X,RSQ)
```

Solves the linear system $\mathbf{A}\cdot\mathbf{x}=\mathbf{b}$ for the vector X of length N, given the right-hand vector B, and given two subroutines, ASUB(XIN,XOUT) and ATSUB(XIN,XOUT), which respectively calculate $\mathbf{A}\cdot\mathbf{x}$ and $\mathbf{A}^T\cdot\mathbf{x}$ for $\mathbf{x}$ given as their first arguments, returning the result in their second arguments. These subroutines should take every advantage of the sparseness of the matrix $\mathbf{A}$. On input, X should be set to a first guess of the desired solution (all zero components is fine). On output, X is the solution vector, and RSQ is the sum of the squares

```
    of the components of the residual vector A · x − b. If this is not small, then the matrix is
    numerically singular and the solution represents a least-squares best approximation.
  PARAMETER (NMAX=500,EPS=1.E-6)
    Maximum anticipated N, and r.m.s. accuracy desired.
  DIMENSION B(N),X(N),G(NMAX),H(NMAX),XI(NMAX),XJ(NMAX)
  EPS2=N*EPS**2                 Criterion for sum-squared residuals.
  IRST=0                        Number of restarts attempted internally.
1 IRST=IRST+1
  CALL ASUB(X,XI)               Evaluate the starting gradient,
  RP=0.
  BSQ=0.
  DO 11 J=1,N
    BSQ=BSQ+B(J)**2             and the magnitude of the right side.
    XI(J)=XI(J)-B(J)
    RP=RP+XI(J)**2
11 CONTINUE
  CALL ATSUB(XI,G)
  DO 12 J=1,N
    G(J)=-G(J)
    H(J)=G(J)
12 CONTINUE
  DO 19 ITER=1,10*N             Main iteration loop.
    CALL ASUB(H,XI)
    ANUM=0.
    ADEN=0.
    DO 13 J=1,N
      ANUM=ANUM+G(J)*H(J)
      ADEN=ADEN+XI(J)**2
13  CONTINUE
    IF(ADEN.EQ.0.)PAUSE 'very singular matrix'
    ANUM=ANUM/ADEN              Equation (2.10.21).
    DO 14 J=1,N
      XI(J)=X(J)
      X(J)=X(J)+ANUM*H(J)
14  CONTINUE
    CALL ASUB(X,XJ)
    RSQ=0.
    DO 15 J=1,N
      XJ(J)=XJ(J)-B(J)
      RSQ=RSQ+XJ(J)**2
15  CONTINUE
    IF(RSQ.EQ.RP.OR.RSQ.LE.BSQ*EPS2)RETURN   Converged. Normal return.
    IF(RSQ.GT.RP)THEN           Not improving. Do a restart.
      DO 16 J=1,N
        X(J)=XI(J)
16    CONTINUE
      IF(IRST.GE.3)RETURN       Return if too many restarts. This is the normal return when we run
      GO TO 1                       into roundoff error before satisfying the return above.
    ENDIF
    RP=RSQ
    CALL ATSUB(XJ,XI)           Compute gradient for next iteration.
    GG=0.
    DGG=0.
    DO 17 J=1,N
      GG=GG+G(J)**2
      DGG=DGG+(XI(J)+G(J))*XI(J)
17  CONTINUE
    IF(GG.EQ.0.)RETURN          A rare, but normal, return.
    GAM=DGG/GG
    DO 18 J=1,N
      G(J)=-XI(J)
      H(J)=G(J)+GAM*H(J)
18  CONTINUE
19 CONTINUE
```

```
PAUSE 'too many iterations'
RETURN
END
```

So that the specifications for the routines ASUB and ATSUB are clear, we list a couple of dummy versions *which, N.B., do not take any advantage of sparseness!*

```
SUBROUTINE ASUB(X,V)
PARAMETER (N=?)
DIMENSION X(N),V(N)
COMMON /MAT/ A(N,N)              The matrix is stored somewhere.
DO 12 I=1,N
   V(I)=0.
   DO 11 J=1,N
      V(I)=V(I)+A(I,J)*X(J)
11    CONTINUE
12 CONTINUE
RETURN
END
```

```
SUBROUTINE ATSUB(X,V)
PARAMETER (N=?)
DIMENSION X(N),V(N)
COMMON /MAT/ A(N,N)
DO 12 I=1,N
   V(I)=0.
   DO 11 J=1,N
      V(I)=V(I)+A(J,I)*X(J)
11    CONTINUE
12 CONTINUE
RETURN
END
```

REFERENCES AND FURTHER READING:

Golub, Gene H., and Van Loan, Charles F. 1983, *Matrix Computations* (Baltimore: Johns Hopkins University Press), Chapters 5 and 10.

Jacobs, David A.H., ed. 1977, *The State of the Art in Numerical Analysis* (London: Academic Press), Chapter I.3 (by J.K. Reid).

Tewarson, R.P. 1973, *Sparse Matrices* (New York: Academic Press).

Bunch, J.R., and Rose, D.J. (eds.) 1976, *Sparse Matrix Computations* (New York: Academic Press).

Duff, I.S., and Stewart, G.W. (eds.) 1979, *Sparse Matrix Proceedings 1978* (Philadelphia: S.I.A.M.).

Stoer, J., and Bulirsch, R. 1980, *Introduction to Numerical Analysis* (New York: Springer-Verlag), Chapter 8.

IMSL Library Reference Manual, 1980, ed. 8 (IMSL Inc., 7500 Bellaire Boulevard, Houston TX 77036).

NAG Fortran Library Manual Mark 8, 1980 (NAG Central Office, 7 Banbury Road, Oxford OX26NN, U.K.).

2.11 Is Matrix Inversion an N^3 Process?

We close this chapter with a little entertainment, a bit of algorithmic prestidigitation which probes more deeply into the subject of matrix inversion. We start with a seemingly simple question:

How many individual multiplications does it take to perform the matrix multiplication of two 2×2 matrices,

$$\begin{pmatrix} a_{11} & a_{12} \\ a_{21} & a_{22} \end{pmatrix} \cdot \begin{pmatrix} b_{11} & b_{12} \\ b_{21} & b_{22} \end{pmatrix} = \begin{pmatrix} c_{11} & c_{12} \\ c_{21} & c_{22} \end{pmatrix} \quad ? \tag{2.11.1}$$

Eight, right? Here they are written explicitly:

$$\begin{aligned} c_{11} &= a_{11} \times b_{11} + a_{12} \times b_{21} \\ c_{12} &= a_{11} \times b_{12} + a_{12} \times b_{22} \\ c_{21} &= a_{21} \times b_{11} + a_{22} \times b_{21} \\ c_{22} &= a_{21} \times b_{12} + a_{22} \times b_{22} \end{aligned} \tag{2.11.2}$$

Do you think that one can write formulas for the c's that involve only *seven* multiplications? (Try it yourself, before reading on.)

Such a set of formulas was, in fact, discovered by Strassen. The formulas are:

$$\begin{aligned} Q_1 &\equiv (a_{11} + a_{22}) \times (b_{11} + b_{22}) \\ Q_2 &\equiv (a_{21} + a_{22}) \times b_{11} \\ Q_3 &\equiv a_{11} \times (b_{12} - b_{22}) \\ Q_4 &\equiv a_{22} \times (-b_{11} + b_{21}) \\ Q_5 &\equiv (a_{11} + a_{12}) \times b_{22} \\ Q_6 &\equiv (-a_{11} + a_{21}) \times (b_{11} + b_{12}) \\ Q_7 &\equiv (a_{12} - a_{22}) \times (b_{21} + b_{22}) \end{aligned} \tag{2.11.3}$$

in terms of which

$$\begin{aligned} c_{11} &= Q_1 + Q_4 - Q_5 + Q_7 \\ c_{21} &= Q_2 + Q_4 \\ c_{12} &= Q_3 + Q_5 \\ c_{22} &= Q_1 + Q_3 - Q_2 + Q_6 \end{aligned} \tag{2.11.4}$$

What's the use of this? There is one fewer multiply than in equation (2.11.2), but *many more* additions and subtractions. It is not clear that anything has been gained. But notice that in (2.11.3) the a's and b's are never commuted. Therefore (2.11.3) and (2.11.4) are valid when the a's and b's are themselves matrices. The problem of multiplying two very large matrices (of order $N = 2^m$ for some integer m) can now be broken down recursively by partitioning the matrices into quarters, sixteenths, etc. And note the key point: The savings is not just a factor "7/8"; it is that factor at *each* hierarchical level of the recursion. In total it reduces the process of matrix multiplication to order $N^{\log_2 7}$ instead of N^3.

What about all the extra additions in (2.11.3) – (2.11.4)? Don't they outweigh the advantage of the fewer multiplications? For large N, it turns out that there are six times as many additions as multiplications implied by (2.11.3) – (2.11.4). But, if N is very large, this constant factor is no match for the change in the *exponent* from N^3 to $N^{\log_2 7}$.

With this "fast" matrix multiplication, Strassen also obtained a surprising result for matrix inversion. Suppose that the matrices

$$\begin{pmatrix} a_{11} & a_{12} \\ a_{21} & a_{22} \end{pmatrix} \quad \text{and} \quad \begin{pmatrix} c_{11} & c_{12} \\ c_{21} & c_{22} \end{pmatrix} \tag{2.11.5}$$

are inverses of each other. Then the c's can be obtained from the a's by the following operations:

$$\begin{aligned}
R_1 &= \text{Inverse}(a_{11}) \\
R_2 &= a_{21} \times R_1 \\
R_3 &= R_1 \times a_{12} \\
R_4 &= a_{21} \times R_3 \\
R_5 &= R_4 - a_{22} \\
R_6 &= \text{Inverse}(R_5) \\
c_{12} &= R_3 \times R_6 \\
c_{21} &= R_6 \times R_2 \\
R_7 &= R_3 \times c_{21} \\
c_{11} &= R_1 - R_7 \\
c_{22} &= -R_6
\end{aligned} \tag{2.11.6}$$

In (2.11.6) the "inverse" operator occurs just twice. It is to be interpreted as the reciprocal if the a's and c's are scalars, but as matrix inversion if the a's and c's are themselves submatrices. Imagine doing the inversion of a very large matrix, of order $N = 2^m$, recursively by partitions in half. At each

step, halving the order *doubles* the number of inverse operations. But this means that there are only N divisions in all! So divisions don't dominate in the recursive use of (2.11.6). Equation (2.11.6) is dominated, in fact, by its 6 multiplications. Since these can be done by an $N^{\log_2 7}$ algorithm, so can the matrix inversion!

This is fun, but let's look at practicalities: If you estimate how large N has to be before the difference between exponent 3 and exponent $\log_2 7 = 2.807$ is substantial enough to outweigh the bookkeeping overhead, arising from the complicated nature of the recursive Strassen algorithm, you will find that LU decomposition is in no immediate danger of becoming obsolete.

If, on the other hand, you like this kind of fun, then try these: (1) Can you multiply the complex numbers $(a + ib)$ and $(c + id)$ in only *three* real multiplications? [Answer:

$$\begin{aligned} ac - bd &= ac - bd \\ ad + bc &= (a + b)(c + d) - ac - bd \end{aligned} \tag{2.11.7}$$

which requires only the three products ac, bd, $(a + b)(c + d)$.] (2) Can you evaluate a general fourth-degree polynomial in x for many different values of x with only *three* multiplications per evaluation? [Answer: see §5.3.]

REFERENCES AND FURTHER READING:

Strassen, Volker 1969, "Gaussian Elimination is not Optimal," *Numerische Mathematik*, vol. 13, p.354.

Winograd, S. 1971, *Linear Algebra and Its Applications*, vol. 4, pp. 381-388.

Pan, V. Ya. 1980, *S.I.A.M. Journal Comput.*, vol. 9, pp. 321-342.

Pan, V. 1984, *S.I.A.M. Review*, vol. 26, pp. 393-415. [More recent results which show that an exponent of 2.496 can be achieved – theoretically!]

Chapter 3. Interpolation and Extrapolation

3.0 Introduction

We sometimes know the value of a function $f(x)$ at a set of points $x_1, x_2, ..., x_N$ (say, with $x_1 < ... < x_N$), but we don't have an analytic expression for $f(x)$ that lets us calculate its value at an arbitrary point. For example, the $f(x_i)$'s might result from some physical measurement or from long numerical calculation that cannot be cast into a simple functional form. Often the x_i's are equally spaced, but not necessarily.

The task now is to estimate $f(x)$ for arbitrary x by, in some sense, drawing a smooth curve through (and perhaps beyond) the x_i. If the desired x is in between the largest and smallest of the x_i's, the problem is called *interpolation*; if x is outside that range, it is called *extrapolation*, which is considerably more hazardous (as many former stock-market analysts can attest).

Interpolation and extrapolation schemes must model the function, in between or beyond the known points, by some plausible functional form. The form should be sufficiently general so as to be able to approximate large classes of functions which might arise in practice. By far most common among the functional forms used are polynomials (§3.1). Rational functions (quotients of polynomials) also turn out to be extremely useful (§3.2). Trigonometric functions, sines and cosines, give rise to *trigonometric interpolation* and related Fourier methods, which we defer to Chapter 12.

There is an extensive mathematical literature devoted to theorems about what sort of functions can be well approximated by which interpolating functions. These theorems are, alas, almost completely useless in day-to-day work: if we know enough about our function to apply a theorem of any power, we are usually not in the pitiful state of having to interpolate on a table of its values!

Interpolation is related to, but distinct from, *function approximation*. That task consists of finding an approximate (but easily computable) function to use in place of a more complicated one. In the case of interpolation, you are given the function f at points *not of your own choosing*. For the case of function approximation, you are allowed to compute the function f at *any* desired points for the purpose of developing your approximation. We deal with function approximation in Chapter 5.

One can easily find pathological functions that make a mockery of any interpolation scheme. Consider, for example, the function

$$f(x) = 3x^2 + \frac{1}{\pi^4} \ln \left[(\pi - x)^2\right] + 1 \qquad (3.0.1)$$

which is well-behaved everywhere except at $x = \pi$, very mildly singular at $x = \pi$, and otherwise takes on all positive and negative values. Any interpolation based on the values $x = 3.13, 3.14, 3.15, 3.16$, will assuredly get a very wrong answer for the value $x = 3.1416$, even though a graph plotting those five points looks really quite smooth! (Try it on your calculator.)

Because pathologies can lurk anywhere, it is highly desirable that an interpolation and extrapolation routine should return an estimate of its own error. Such an error estimate can never be foolproof, of course. We could have a function that, for reasons known only to its maker, takes off wildly and unexpectedly between two tabulated points. Interpolation always presumes some degree of smoothness for the function interpolated, but within this framework of presumption, deviations from smoothness can be detected.

Conceptually, the interpolation process has two stages: (1) Fit an interpolating function to the data points provided. (2) Evaluate that interpolating function at the target point x.

However, this two-stage method is generally not the best way to proceed in practice. Typically it is computationally less efficient, and more susceptible to roundoff error, than methods which construct a functional estimate $f(x)$ directly from the N tabulated values every time one is desired. Most practical schemes start at a nearby point $f(x_i)$, then add a sequence of (hopefully) decreasing corrections, as information from other $f(x_i)$'s is incorporated. The procedure typically takes $O(N^2)$ operations. If everything is well behaved, the last correction will be the smallest, and it can be used as an informal (though not rigorous) bound on the error.

In the case of polynomial interpolation, it sometimes does happen that the coefficients of the interpolating polynomial are of interest, even though their use in *evaluating* the interpolating function should be frowned on. We deal with this eventuality in §3.5.

Local interpolation, using a finite number of "nearest-neighbor" points, gives interpolated values $f(x)$ that do not, in general, have continuous first or higher derivatives. That happens because, as x crosses the tabulated values x_i, the interpolation scheme switches which tabulated points are the "local" ones. (If such a switch is allowed to occur anywhere *else*, then there will be a discontinuity in the interpolated function itself at that point. Bad practice!)

In situations where continuity of derivatives is a concern, one must use the "stiffer" interpolation provided by a so-called *spline* function. A spline is a polynomial between each pair of table points, but one whose coefficients are determined "slightly" nonlocally. The nonlocality is designed to guarantee global smoothness in the interpolated function up to some order of derivative. Cubic splines (§3.3) are the most popular. They produce an interpolated function that is continuous through the second derivative. Splines tend to be

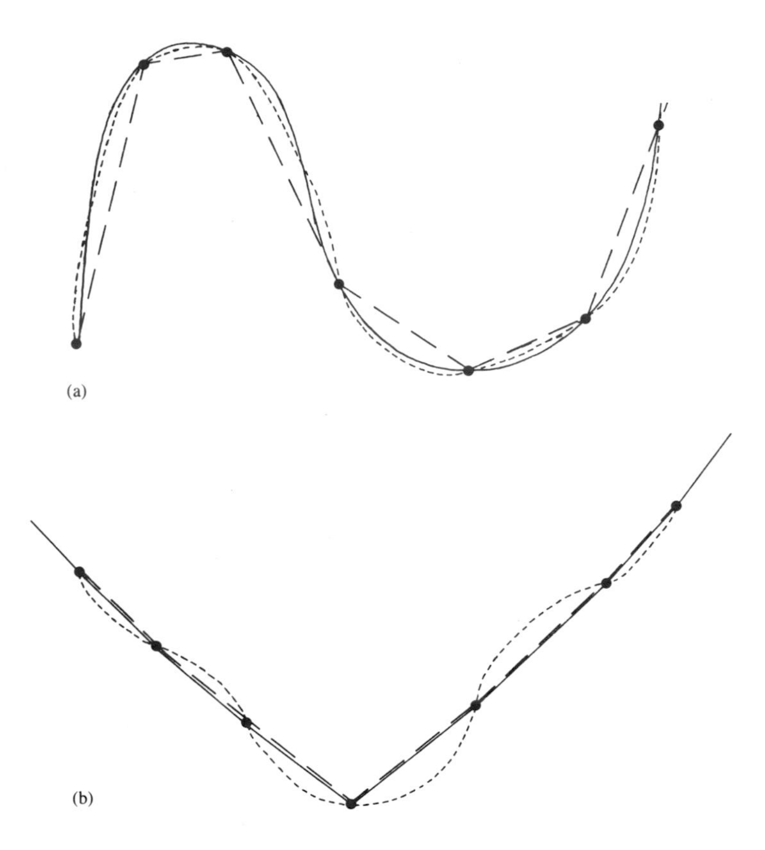

Figure 3.0.1. (a) A smooth function (solid line) is more accurately interpolated by a high-order polynomial (shown schematically as dotted line) than by a low-order polynomial (shown as a piecewise linear dashed line). (b) A function with sharp corners or rapidly changing higher derivatives is *less* accurately approximated by a high-order polynomial (dotted line), which is too "stiff", than by a low-order polynomial (dashed lines). Even some smooth functions, such as exponentials or rational functions, can be badly approximated by high-order polynomials.

stabler than polynomials, with less possibility of wild oscillation between the tabulated points.

The number of points (minus one) used in an interpolation scheme is called the *order* of the interpolation. Increasing the order does not necessarily increase the accuracy, especially in polynomial interpolation. If the added points are distant from the point of interest x, the resulting higher order polynomial, with its additional constrained points, tends to oscillate wildly between the tabulated values. This oscillation may have no relation at all to the behavior of the "true" function (see Figure 3.0.1). Of course, adding points *close* to the desired point usually does help, but a finer mesh implies a larger table of values, not always available.

Unless there is solid evidence that the interpolating function is close in form to the true function f, it is a good idea to be cautious about high-order interpolation. We enthusiastically endorse interpolations with 3 or 4 points,

we are perhaps tolerant of 5 or 6; but we rarely go higher than that unless there is quite rigorous monitoring of estimated errors.

When your table of values contains many more points than the desirable order of interpolation, you must begin each interpolation with a search for the right "local" place in the table. While not strictly a part of the subject of interpolation, this task is important enough (and often enough botched) that we devote §3.4 to its discussion.

The routines given for interpolation are also routines for extrapolation. An important application, in Chapter 15, is their use in the integration of ordinary differential equations. There, considerable care *is* taken on the monitoring of errors. Otherwise, the dangers of extrapolation cannot be overemphasized: An interpolating function, which is perforce an extrapolating function, will typically go berserk when the argument x is outside the range of tabulated values by more than the typical spacing of tabulated points.

Interpolation can be done in more than one dimension, e.g. for a function $f(x, y, z)$. Multidimensional interpolation is generally accomplished by a sequence of one-dimensional interpolations. We discuss this in §3.6.

REFERENCES AND FURTHER READING:

Abramowitz, Milton, and Stegun, Irene A. 1964, *Handbook of Mathematical Functions*, Applied Mathematics Series, vol. 55 (Washington: National Bureau of Standards; reprinted 1968 by Dover Publications, New York), §25.2.

Stoer, J., and Bulirsch, R. 1980, *Introduction to Numerical Analysis* (New York: Springer-Verlag), Chapter 2.

Acton, Forman S. 1970, *Numerical Methods That Work* (New York: Harper and Row), Chapter 3.

Johnson, Lee W., and Riess, R. Dean. 1982, *Numerical Analysis*, 2nd ed. (Reading, Mass.: Addison-Wesley), Chapter 5.

Ralston, Anthony, and Rabinowitz, Philip. 1978, *A First Course in Numerical Analysis*, 2nd ed. (New York: McGraw-Hill), Chapter 3.

Isaacson, Eugene, and Keller, Herbert B. 1966, *Analysis of Numerical Methods* (New York: Wiley), Chapter 6.

3.1 Polynomial Interpolation and Extrapolation

Through any two points there is a unique line. Through any three points, a unique quadratic. Et cetera. The interpolating polynomial of degree $N-1$ through the N points $y_1 = f(x_1), y_2 = f(x_2), \dots y_N = f(x_N)$ is given explicitly by Lagrange's classical formula,

$$P(x) = \frac{(x-x_2)(x-x_3)...(x-x_N)}{(x_1-x_2)(x_1-x_3)...(x_1-x_N)}y_1 + \frac{(x-x_1)(x-x_3)...(x-x_N)}{(x_2-x_1)(x_2-x_3)...(x_2-x_N)}y_2 + \cdots + \frac{(x-x_1)(x-x_2)...(x-x_{N-1})}{(x_N-x_1)(x_N-x_2)...(x_N-x_{N-1})}y_N \tag{3.1.1}$$

There are N terms, each a polynomial of degree $N-1$ and each constructed to be zero for all of the x_i except one, which is constructed to be y_i.

It is not terribly wrong to implement the Lagrange formula straightforwardly, but it is not terribly right either. The resulting algorithm gives no error estimate, and it is also somewhat awkward to program. A much better algorithm (for constructing the same – unique – interpolating polynomial) is *Neville's algorithm*, closely related to and sometimes confused with *Aitken's algorithm*, the latter now considered obsolete.

Let P_1 be the value at x of the unique polynomial of degree zero (i.e. a constant) passing through the point (x_1, y_1); so $P_1 = y_1$. Likewise define $P_2, P_3, \ldots P_N$. Now let P_{12} be the value at x of the unique polynomial of degree one passing through both (x_1, y_1) and (x_2, y_2). Likewise $P_{23}, P_{34}, \ldots P_{(N-1)N}$. Similarly, for higher order polynomials, up to $P_{123\ldots N}$, which is the value of the unique interpolating polynomial through all N points, i.e. the desired answer. The various P's form a "tableau" with "ancestors" on the left leading to a single "descendant" at the extreme right. For example, with $N = 4$,

$$\begin{array}{lllll}
x_1: & y_1 = P_1 & & & \\
 & & P_{12} & & \\
x_2: & y_2 = P_2 & & P_{123} & \\
 & & P_{23} & & P_{1234} \\
x_3: & y_3 = P_3 & & P_{234} & \\
 & & P_{34} & & \\
x_4: & y_4 = P_4 & & &
\end{array} \tag{3.1.2}$$

Neville's algorithm is a recursive way of filling in the numbers in the tableau a column at a time, from left to right. It is based on the relationship between a "daughter" P and its two "parents,"

$$P_{i(i+1)\ldots(i+m)} = \frac{(x - x_{i+m})P_{i(i+1)\ldots(i+m-1)} + (x_i - x)P_{(i+1)(i+2)\ldots(i+m)}}{x_i - x_{i+m}} \tag{3.1.3}$$

This recurrence holds because the two parents already agree at points $x_{i+1} \ldots x_{i+m-1}$.

An improvement on the recurrence (3.1.3) is to keep track of the small *differences* between parents and daughters, namely to define (for $m = 1, 2, \ldots N-1$),

$$\begin{aligned}
C_{m,i} &\equiv P_{i\ldots(i+m)} - P_{i\ldots(i+m-1)} \\
D_{m,i} &\equiv P_{i\ldots(i+m)} - P_{(i+1)\ldots(i+m)}.
\end{aligned} \tag{3.1.4}$$

Then one can easily derive from (3.1.2) the relations

$$D_{m+1,i} = \frac{(x_{i+m+1} - x)(C_{m,i+1} - D_{m,i})}{x_i - x_{i+m+1}}$$
$$C_{m+1,i} = \frac{(x_i - x)(C_{m,i+1} - D_{m,i})}{x_i - x_{i+m+1}} \tag{3.1.5}$$

At each level m, the C's and D's are the corrections which make the interpolation one order higher. The final answer $P_{1\ldots N}$ is equal to the sum of *any* y_i plus a set of C's and/or D's which form a path through the family tree to the rightmost daughter.

Here is a routine for polynomial interpolation or extrapolation:

```
SUBROUTINE POLINT(XA,YA,N,X,Y,DY)
    Given arrays XA and YA, each of length N, and given a value X, this routine returns a
    value Y, and an error estimate DY. If P(x) is the polynomial of degree N − 1 such that
    P(XA_i) = YA_i, i = 1,...,N, then the returned value Y = P(X).
PARAMETER (NMAX=10)              Change NMAX as desired to be the largest anticipated value of N.
DIMENSION XA(N),YA(N),C(NMAX),D(NMAX)
NS=1
DIF=ABS(X-XA(1))
DO 11 I=1,N                      Here we find the index NS of the closest table entry,
    DIFT=ABS(X-XA(I))
    IF (DIFT.LT.DIF) THEN
        NS=I
        DIF=DIFT
    ENDIF
    C(I)=YA(I)                   and initialize the tableau of C's and D's.
    D(I)=YA(I)
11 CONTINUE
Y=YA(NS)                         This is the initial approximation to Y.
NS=NS-1
DO 13 M=1,N-1                    For each column of the tableau,
    DO 12 I=1,N-M                we loop over the current C's and D's and update them.
        HO=XA(I)-X
        HP=XA(I+M)-X
        W=C(I+1)-D(I)
        DEN=HO-HP
        IF(DEN.EQ.O.)PAUSE       This error can occur only if two input XA's are (to within roundoff)
                                     identical.
        DEN=W/DEN
        D(I)=HP*DEN              Here the C's and D's are updated.
        C(I)=HO*DEN
    12 CONTINUE
    IF (2*NS.LT.N-M)THEN         After each column in the tableau is completed, we decide which cor-
        DY=C(NS+1)                   rection, C or D, we want to add to our accumulating value of
    ELSE                             Y, i.e. which path to take through the tableau—forking up or
        DY=D(NS)                     down. We do this in such a way as to take the most "straight
        NS=NS-1                      line" route through the tableau to its apex, updating NS ac-
    ENDIF                            cordingly to keep track of where we are. This route keeps the
    Y=Y+DY                           partial approximations centered (insofar as possible) on the
13 CONTINUE                          target X. The last DY added is thus the error indication.
RETURN
END
```

REFERENCES AND FURTHER READING:

Abramowitz, Milton, and Stegun, Irene A. 1964, *Handbook of Mathematical Functions*, Applied Mathematics Series, vol. 55 (Washington: National Bureau of Standards; reprinted 1968 by Dover Publications, New York), §25.2.

Stoer, J., and Bulirsch, R. 1980, *Introduction to Numerical Analysis* (New York: Springer-Verlag), §2.1.

Gear, C. William. 1971, *Numerical Initial Value Problems in Ordinary Differential Equations* (Englewood Cliffs, N.J.: Prentice-Hall), §6.1.

3.2 Rational Function Interpolation and Extrapolation

Some functions are not well approximated by polynomials, but *are* well approximated by rational functions, that is quotients of polynomials. We denote by $R_{i(i+1)\dots(i+m)}$ a rational function passing through the $m+1$ points $(x_i, y_i)\dots(x_{i+m}, y_{i+m})$. More explicitly, suppose

$$R_{i(i+1)\dots(i+m)} = \frac{P_\mu(x)}{Q_\nu(x)} = \frac{p_0 + p_1 x + \cdots + p_\mu x^\mu}{q_0 + q_1 x + \cdots + q_\nu x^\nu} \tag{3.2.1}$$

Since there are $\mu+\nu+1$ unknown p's and q's (q_0 being arbitrary), we must have

$$m + 1 = \mu + \nu + 1 \tag{3.2.2}$$

In specifying a rational function interpolating function, you must give the desired order of both the numerator and the denominator.

Rational functions are superior to polynomials, roughly speaking, because of their ability to model functions with poles, that is, zeros of the denominator of equation (3.2.1). These poles might occur for real values of x, if the function to be interpolated itself has poles. More often, the function $f(x)$ is finite for all finite *real* x, but has an analytic continuation with poles in the complex x-plane. Such poles can themselves ruin a polynomial approximation, even one restricted to real values of x, just as they can ruin the convergence of an infinite power series in x. If you draw a circle in the complex plane around your m tabulated points, then you should not expect polynomial interpolation to be good unless the nearest pole is rather far outside the circle. A rational function approximation, by contrast, will stay "good" as long as it has enough powers of x in its denominator to account for (cancel) any nearby poles.

For the interpolation problem, a rational function is constructed so as to go through a chosen set of tabulated functional values. However, we should also mention in passing that rational function approximations can be used in analytic work. One sometimes constructs a rational function approximation by the criterion that the rational function of equation (3.2.1) itself have a

power series expansion that agrees with the first $m+1$ terms of the power series expansion of the desired function $f(x)$. This is called *Padé approximation.* It can be quite a powerful technique for turning local information, the derivatives at a point which determine the power series coefficients, into a global capability for evaluating the function, approximately. The approximation is often found to be *remarkably* good.

Bulirsch and Stoer found an algorithm of the Neville type which performs rational function extrapolation on tabulated data. A tableau like that of equation (3.1.2) is constructed column by column, leading to a result and an error estimate. The Bulirsch-Stoer algorithm produces the so-called *diagonal* rational function, with the degrees of numerator and denominator equal (if $m+1$ is even) or with the degree of the denominator larger by one (if $m+1$ is odd, cf. equation 3.2.2 above). For the derivation of the algorithm, refer to Stoer and Bulirsch. The algorithm is summarized by a recurrence relation exactly analogous to equation (3.1.3) for polynomial approximation:

$$R_{i(i+1)\dots(i+m)} = R_{(i+1)\dots(i+m)} + \frac{R_{(i+1)\dots(i+m)} - R_{i\dots(i+m-1)}}{\left(\frac{x-x_i}{x-x_{i+m}}\right)\left(1 - \frac{R_{(i+1)\dots(i+m)} - R_{i\dots(i+m-1)}}{R_{(i+1)\dots(i+m)} - R_{(i+1)\dots(i+m-1)}}\right) - 1} \tag{3.2.3}$$

This recurrence generates the rational functions through $m+1$ points from the ones through m and (second term in the denominator of the denominator of equation 3.2.3) $m-1$ points. It is started with

$$R_i = y_i \tag{3.2.4}$$

and with

$$R \equiv [R_{i(i+1)\dots(i+m)} \quad \text{with} \quad m = -1] = 0 \tag{3.2.5}$$

Now, exactly as in equations (3.1.4) and (3.1.5) above, we can convert the recurrence (3.2.3) to one involving only the small differences

$$\begin{aligned} C_{m,i} &\equiv R_{i\dots(i+m)} - R_{i\dots(i+m-1)} \\ D_{m,i} &\equiv R_{i\dots(i+m)} - R_{(i+1)\dots(i+m)} \end{aligned} \tag{3.2.6}$$

Note that these satisfy the relation

$$C_{m+1,i} - D_{m+1,i} = C_{m,i+1} - D_{m,i} \tag{3.2.7}$$

which is useful in proving the recurrences

$$D_{m+1,i} = \frac{C_{m,i+1}(C_{m,i+1} - D_{m,i})}{\left(\frac{x-x_i}{x-x_{i+m+1}}\right) D_{m,i} - C_{m,i+1}}$$

$$C_{m+1,i} = \frac{\left(\frac{x-x_i}{x-x_{i+m+1}}\right) D_{m,i}(C_{m,i+1} - D_{m,i})}{\left(\frac{x-x_i}{x-x_{i+m+1}}\right) D_{m,i} - C_{m,i+1}} \tag{3.2.8}$$

This recurrence is implemented in the following subroutine, whose use is analogous in every way to POLINT in §3.1.

```
SUBROUTINE RATINT(XA,YA,N,X,Y,DY)
    Given arrays XA and YA, each of length N, and given a value of X, this routine returns a
    value of Y and an accuracy estimate DY. The value returned is that of the diagonal rational
    function, evaluated at X, which passes through the N points (XA_i,YA_i), i = 1...N.
PARAMETER (NMAX=10,TINY=1.E-25)      Largest expected value of N, and a small number.
DIMENSION XA(N),YA(N),C(NMAX),D(NMAX)
NS=1
HH=ABS(X-XA(1))
DO 11 I=1,N
    H=ABS(X-XA(I))
    IF (H.EQ.0.)THEN
        Y=YA(I)
        DY=0.0
        RETURN
    ELSE IF (H.LT.HH) THEN
        NS=I
        HH=H
    ENDIF
    C(I)=YA(I)
    D(I)=YA(I)+TINY          The TINY part is needed to prevent a rare zero-over-zero condition.
11  CONTINUE
Y=YA(NS)
NS=NS-1
DO 13 M=1,N-1
    DO 12 I=1,N-M
        W=C(I+1)-D(I)
        H=XA(I+M)-X          H will never be zero, since this was tested in the initializing loop.
        T=(XA(I)-X)*D(I)/H
        DD=T-C(I+1)
        IF(DD.EQ.0.)PAUSE    This error condition indicates that the interpolating function has a
        DD=W/DD                  pole at the requested value of X.
        D(I)=C(I+1)*DD
        C(I)=T*DD
12      CONTINUE
    IF (2*NS.LT.N-M)THEN
        DY=C(NS+1)
    ELSE
        DY=D(NS)
        NS=NS-1
    ENDIF
    Y=Y+DY
13  CONTINUE
RETURN
END
```

REFERENCES AND FURTHER READING:

Stoer, J., and Bulirsch, R. 1980, *Introduction to Numerical Analysis* (New York: Springer-Verlag), §2.2.

Gear, C. William. 1971, *Numerical Initial Value Problems in Ordinary Differential Equations* (Englewood Cliffs, N.J.: Prentice-Hall), §6.2.

3.3 Cubic Spline Interpolation

Given a tabulated function $y_i = y(x_i)$, $i = 1...N$, focus attention on one particular interval, between x_j and x_{j+1}. Linear interpolation in that interval gives the interpolation formula

$$y = Ay_j + By_{j+1} \tag{3.3.1}$$

where

$$A \equiv \frac{x_{j+1} - x}{x_{j+1} - x_j} \qquad B \equiv 1 - A = \frac{x - x_j}{x_{j+1} - x_j} \tag{3.3.2}$$

Equations (3.3.1) and (3.3.2) are a special case of the general Lagrange interpolation formula (3.1.1).

Now suppose that, in addition to the tabulated values of y_i, we also have tabulated values for another function denoted y'', that is, a set of numbers y''_i. We will see in a moment that y'' is supposed to be the second derivative of the function y, but pretend for now that you don't know this.

We next decide, inscrutably for now, to use the values y''_j and y''_{j+1} as linear coefficients of two linearly independent *cubic* polynomial terms which will not spoil the agreement with the tabulated functional values y_j and y_{j+1} at the endpoints x_j and x_{j+1}, *for any choice of* y''_j *and* y''_{j+1}. A little thought shows that there is only one way to arrange this, namely replacing (3.3.1) by

$$y = Ay_j + By_{j+1} + Cy''_j + Dy''_{j+1} \tag{3.3.3}$$

where A and B are defined in (3.3.2) and

$$C \equiv \frac{1}{6}(A^3 - A)(x_{j+1} - x_j)^2 \qquad D \equiv \frac{1}{6}(B^3 - B)(x_{j+1} - x_j)^2 \tag{3.3.4}$$

Notice that the dependence on the independent variable x in equations (3.3.3) and (3.3.4) is entirely through the linear x-dependence of A and B, and (through A and B) the cubic x-dependence of C and D. The reason that (3.3.4) is unique (up to the choice of multiplicative constants in the definition

of C and D) is that (i) it is a cubic polynomial in x, (ii) it contains four adjustable linear coefficients, $y_j, y_{j+1}, y''_j, y''_{j+1}$, (iii) four is the correct number of linear coefficients necessary to define a general cubic polynomial, and (iv) four is also the sum of the number of constraints (2, the endpoint values) plus free parameters (2, the numerical values of y''_j and y''_{j+1}).

Now we can see that y'' is in fact the second derivative of the interpolating polynomial. We take derivatives of equation (3.3.3) with respect to x, using the definitions of A, B, C, D to compute $dA/dx, dB/dx, dC/dx$, and dD/dx. The result is

$$\frac{dy}{dx} = \frac{y_{j+1} - y_j}{x_{j+1} - x_j} - \frac{3A^2 - 1}{6}(x_{j+1} - x_j)y''_j + \frac{3B^2 - 1}{6}(x_{j+1} - x_j)y''_{j+1} \quad (3.3.5)$$

for the first derivative, and

$$\frac{d^2y}{dx^2} = Ay''_j + By''_{j+1} \quad (3.3.6)$$

for the second derivative. Since $A = 1$ at x_j, $A = 0$ at x_{j+1}, while B is just the other way around, (3.3.6) shows that y'' is just a tabulated second derivative, and also that the second derivative will be continuous across (e.g.) the boundary between the two intervals (x_{j-1}, x_j) and (x_j, x_{j+1}).

Thus far, we could put in *any* numbers we choose for the y''_i's. However, for a "random" choice of numbers, the value of the *first* derivative, computed from equation (3.3.5), would *not* be continuous across the boundary between two intervals. The key idea of a cubic spline is to require this continuity and to use it to get equations for the numbers y''_i.

The required equations are obtained by setting equation (3.3.5) evaluated for $x = x_j$ in the interval (x_{j-1}, x_j) equal to the same equation evaluated for $x = x_j$ but in the interval (x_j, x_{j+1}). With some rearrangement, this gives (for $j = 2, \ldots N - 1$)

$$\frac{x_j - x_{j-1}}{6}y''_{j-1} + \frac{x_{j+1} - x_{j-1}}{3}y''_j + \frac{x_{j+1} - x_j}{6}y''_{j+1} = \frac{y_{j+1} - y_j}{x_{j+1} - x_j} - \frac{y_j - y_{j-1}}{x_j - x_{j-1}} \quad (3.3.7)$$

These are $N-2$ linear equations in the N unknowns $y''_i, i = 1, \ldots N$. Therefore there is a two parameter family of possible solutions.

For a unique solution, we need to specify two further conditions, typically taken as boundary conditions at x_1 and x_N. The most common ways of doing this are either

- set one or both of y_1'' and y_N'' equal to zero, giving the so-called *natural cubic spline*, which has zero second derivative on one or both of its boundaries, or
- set either of y_1'' and y_N'' to values calculated from equation (3.3.5) so as to make the first derivative of the interpolating function have a specified value on either or both boundaries.

One reason that cubic splines are especially practical is that the set of equations (3.3.7), along with the two additional boundary conditions, are not only linear, but also *tridiagonal.* Each y_j'' is coupled only to its nearest neighbors at $j \pm 1$. Therefore, the equations can be solved in $O(N)$ operations by the tridiagonal algorithm (§2.6). That algorithm is concise enough to build right into the spline calculational routine. This makes the routine not completely transparent as an implementation of (3.3.7), so we encourage you to study it carefully, comparing with TRIDAG (§2.6).

```
SUBROUTINE SPLINE(X,Y,N,YP1,YPN,Y2)
```

Given arrays X and Y of length N containing a tabulated function, i.e. $Y_i = f(X_i)$, with $X_1 < X_2 < \ldots < X_N$, and given values YP1 and YPN for the first derivative of the interpolating function at points 1 and N, respectively, this routine returns an array Y2 of length N which contains the second derivatives of the interpolating function at the tabulated points X_i. If YP1 and/or YPN are equal to 1×10^{30} or larger, the routine is signalled to set the corresponding boundary condition for a natural spline, with zero second derivative on that boundary.

```
PARAMETER (NMAX=100)            Change NMAX as desired to be the largest anticipated value of N.
DIMENSION X(N),Y(N),Y2(N),U(NMAX)
IF (YP1.GT..99E30) THEN         The lower boundary condition is set either to be "natural"
    Y2(1)=0.
    U(1)=0.
ELSE                            or else to have a specified first derivative.
    Y2(1)=-0.5
    U(1)=(3./(X(2)-X(1)))*((Y(2)-Y(1))/(X(2)-X(1))-YP1)
ENDIF
DO 11 I=2,N-1                   This is the decomposition loop of the tridiagonal algorithm. Y2 and
    SIG=(X(I)-X(I-1))/(X(I+1)-X(I-1))   U are used for temporary storage of the decomposed
    P=SIG*Y2(I-1)+2.            factors.
    Y2(I)=(SIG-1.)/P
    U(I)=(6.*((Y(I+1)-Y(I))/(X(I+1)-X(I))-(Y(I)-Y(I-1))
*       /(X(I)-X(I-1)))/(X(I+1)-X(I-1))-SIG*U(I-1))/P
11  CONTINUE
IF (YPN.GT..99E30) THEN         The upper boundary condition is set either to be "natural"
    QN=0.
    UN=0.
ELSE                            or else to have a specified first derivative.
    QN=0.5
    UN=(3./(X(N)-X(N-1)))*(YPN-(Y(N)-Y(N-1))/(X(N)-X(N-1)))
ENDIF
Y2(N)=(UN-QN*U(N-1))/(QN*Y2(N-1)+1.)
DO 12 K=N-1,1,-1                This is the backsubstitution loop of the tridiagonal algorithm.
    Y2(K)=Y2(K)*Y2(K+1)+U(K)
12  CONTINUE
RETURN
END
```

It is important to understand that the program SPLINE is called only *once* to process an entire tabulated function in arrays X_i and Y_i. Once this has been done, values of the interpolated function for any value of x are obtained by calls (as many as desired) to a separate routine SPLINT (for "*spl*ine *int*erpolation"):

```
      SUBROUTINE SPLINT(XA,YA,Y2A,N,X,Y)
         Given the arrays XA and YA of length N, which tabulate a function (with the XAi's in order),
         and given the array Y2A, which is the output from SPLINE above, and given a value of X,
         this routine returns a cubic-spline interpolated value Y.
      DIMENSION XA(N),YA(N),Y2A(N)
      KLO=1                           We will find the right place in the table by means of bisection. This is
      KHI=N                                optimal if sequential calls to this routine are at random values
1      IF (KHI-KLO.GT.1) THEN               of X. If sequential calls are in order, and closely spaced, one
        K=(KHI+KLO)/2                      would do better to store previous values of KLO and KHI and
        IF(XA(K).GT.X)THEN                 test if they remain appropriate on the next call.
          KHI=K
        ELSE
          KLO=K
        ENDIF
      GOTO 1
      ENDIF                           KLO and KHI now bracket the input value of X.
      H=XA(KHI)-XA(KLO)
      IF (H.EQ.0.)  PAUSE 'Bad XA input.'      The XA's must be distinct.
      A=(XA(KHI)-X)/H                 Cubic spline polynomial is now evaluated.
      B=(X-XA(KLO))/H
      Y=A*YA(KLO)+B*YA(KHI)+
     *      ((A**3-A)*Y2A(KLO)+(B**3-B)*Y2A(KHI))*(H**2)/6.
      RETURN
      END
```

REFERENCES AND FURTHER READING:

Forsythe, George E., Malcolm, Michael A., and Moler, Cleve B. 1977, *Computer Methods for Mathematical Computations* (Englewood Cliffs, N.J.: Prentice-Hall), §§4.4-4.5.

Stoer, J., and Bulirsch, R. 1980, *Introduction to Numerical Analysis* (New York: Springer-Verlag), §2.4.

Ralston, Anthony, and Rabinowitz, Philip. 1978, *A First Course in Numerical Analysis*, 2nd ed. (New York: McGraw-Hill), §3.8.

3.4 How to Search an Ordered Table

Suppose that you have decided to use some particular interpolation scheme, such as fourth-order polynomial interpolation, to compute a function $f(x)$ from a set of tabulated x_i's and f_i's. Then you will need a fast way of finding your place in the table of x_i's, given some particular value x at which the function evaluation is desired. This problem is not properly

one of numerical analysis, but it occurs so often in practice that it would be negligent of us to ignore it.

Formally, the problem is this: Given an array of abscissas XX(J), J=1, 2, ...,N, with the elements either monotonically increasing or monotonically decreasing, and given a number X, find an integer J such that X lies between XX(J) and XX(J+1). For this task, let us define fictitious array elements X(0) and X(N+1) equal to plus or minus infinity (in whichever order is consistent with the monotonicity of the table). Then J will always be between 0 and N, inclusive; a returned value of 0 indicates "off-scale" at one end of the table, N indicates off-scale at the other end.

In most cases, when all is said and done, it is hard to do better than *bisection*, which will find the right place in the table in about $\log_2 N$ tries. We already did use bisection in the spline evaluation routine SPLINT of the preceding section, so you might glance back at that. Standing by itself, a bisection routine looks like this:

```
      SUBROUTINE LOCATE(XX,N,X,J)
         Given an array XX of length N, and given a value X, returns a value J such that X is between
         XX(J) and XX(J+1). XX must be monotonic, either increasing or decreasing. J=0 or J=N
         is returned to indicate that X is out of range.
      DIMENSION XX(N)
      JL=0                               Initialize lower
      JU=N+1                             and upper limits.
10    IF(JU-JL.GT.1)THEN                 If we are not yet done,
        JM=(JU+JL)/2                     compute a midpoint,
        IF((XX(N).GT.XX(1)).EQV.(X.GT.XX(JM)))THEN
          JL=JM                          and replace either the lower limit
        ELSE
          JU=JM                          or the upper limit, as appropriate.
        ENDIF
      GO TO 10                           Repeat until
      ENDIF                              the test condition 10 is satisfied.
      J=JL                               Then set the output
      RETURN                             and return
      END
```

Note the use of the logical equality relation .EQV., which is true when its two logical operands are either both true or both false. This relation allows the routine to work for both monotonically increasing and monotonically decreasing orders of XX. If you encounter a programming language which does not have this operation, then an expression like L1.EQV.L2 must be replaced by the more awkward (and slower) (L1.AND.L2).OR.(.NOT.(L1.OR.L2)). In Pascal one simply tests for ordinary equality of L1 and L2, since they are both of type boolean.

Search with Correlated Values

Sometimes you will be in the situation of searching a table many times, and with nearly identical abscissas on consecutive searches. For example, you may be generating a function that is used on the right-hand side of a differential equation: Most differential-equation integrators, as we shall see in

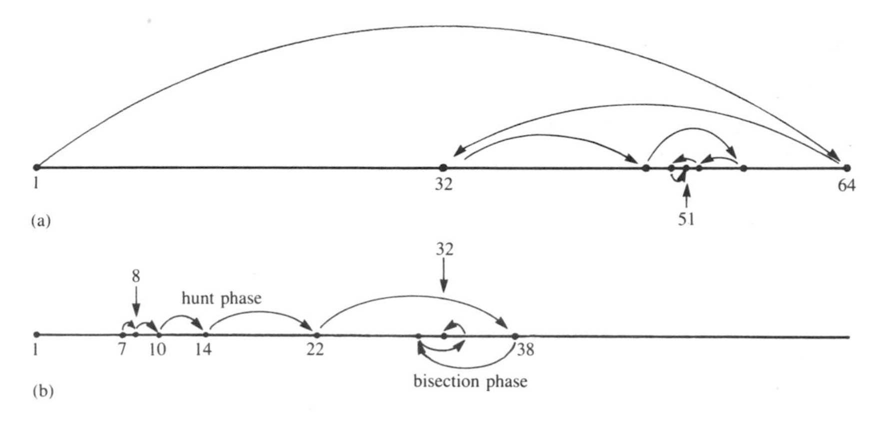

Figure 3.4.1. (a) The routine **LOCATE** finds a table entry by bisection. Shown here is the sequence of steps that converge to element 51 in a table of length 64. (b) The routine **HUNT** searches from a previous known position in the table by increasing steps, then converges by bisection. Shown here is a particularly unfavorable example, converging to element 32 from element 7. A favorable example would be convergence to an element near 7, such as 9, which would require just three "hops."

Chapter 15, call for right-hand side evaluations at points that hop back and forth a bit, but whose trend moves slowly in the direction of the integration.

In such cases it is wasteful to do a full bisection, *ab initio*, on each call. The following routine instead starts with a guessed position in the table. It first "hunts", either up or down, in increments of 1, then 2, then 4, etc., until the desired value is bracketed. Second, it then bisects in the bracketed interval. At worst, this routine is about a factor of 2 slower than LOCATE above (if the hunt phase expands to include the whole table). At best, it can be a factor of $\log_2 N$ faster than LOCATE, if the desired point is usually quite close to the input guess. Figure 3.4.1 compares the two routines.

```
      SUBROUTINE HUNT(XX,N,X,JLO)
         Given an array XX of length N, and given a value X, returns a value JLO such that X is
         between XX(JLO) and XX(JLO+1). XX must be monotonic, either increasing or decreasing.
         JLO=0 or JLO=N is returned to indicate that X is out of range. JLO on input is taken as
         the initial guess for JLO on output.
      DIMENSION XX(N)
      LOGICAL ASCND
      ASCND=XX(N).GT.XX(1)               True if ascending order of table, false otherwise.
      IF(JLO.LE.0.OR.JLO.GT.N)THEN          Input guess not useful. Go immediately to bisection.
        JLO=0
        JHI=N+1
        GO TO 3
      ENDIF
      INC=1                              Set the hunting increment.
      IF(X.GE.XX(JLO).EQV.ASCND)THEN          Hunt up:
1       JHI=JLO+INC
        IF(JHI.GT.N)THEN                 Done hunting, since off end of table.
          JHI=N+1
        ELSE IF(X.GE.XX(JHI).EQV.ASCND)THEN        Not done hunting,
          JLO=JHI
          INC=INC+INC                    so double the increment
          GO TO 1                        and try again.
        ENDIF                            Done hunting, value bracketed.
```

```
      ELSE                            Hunt down:
          JHI=JLO
2         JLO=JHI-INC
          IF(JLO.LT.1)THEN            Done hunting, since off end of table.
              JLO=0
          ELSE IF(X.LT.XX(JLO).EQV.ASCND)THEN      Not done hunting,
              JHI=JLO
              INC=INC+INC             so double the increment
              GO TO 2                 and try again.
          ENDIF                       Done hunting, value bracketed.
      ENDIF
                                      Hunt is done, so begin the final bisection phase:
3     IF(JHI-JLO.EQ.1)RETURN
      JM=(JHI+JLO)/2
      IF(X.GT.XX(JM).EQV.ASCND)THEN
          JLO=JM
      ELSE
          JHI=JM
      ENDIF
      GO TO 3
      END
```

REFERENCES AND FURTHER READING:

Knuth, Donald E. 1973, *Sorting and Searching*, vol. 3 of *The Art of Computer Programming* (Reading, Mass.: Addison-Wesley), §6.2.1.

3.5 Coefficients of the Interpolating Polynomial

Occasionally you may wish to know not the value of the interpolating polynomial that passes through a (small!) number of points, but the coefficients of that polynomial. A valid use of the coefficients might be, for example, to compute simultaneous interpolated values of the function and of several of its derivatives (see §5.3), or to convolve a segment of the tabulated function with some other function, where the moments of that other function (i.e. its convolution with powers of x) are known analytically.

However, please be certain that the coefficients are what you need. Generally the coefficients of the interpolating polynomial can be determined much less accurately than its value at a desired abscissa. Therefore it is not generally a good idea to determine the coefficients only for use in calculating interpolating values. Values thus calculated will not pass exactly through the tabulated points, for example, while values computed by the routines in §3.1–§3.3 will pass exactly through such points.

Also, you should not mistake the interpolating polynomial (and its coefficients) for its cousin, the *best fit* polynomial through a data set. Fitting is a *smoothing* process, since the number of fitted coefficients is typically much less than the number of data points. Therefore, fitted coefficients can be accurately and stably determined even in the presence of statistical errors in the tabulated values. Interpolation, where the number of coefficients and number

of tabulated points are equal, takes the tabulated values as perfect. If they in fact contain statistical errors, these can be magnified into oscillations of the interpolating polynomial in between the tabulated points. The fitting of data to a model is the subject of Chapter 14.

As before, we take the tabulated points to be $y_i \equiv y(x_i)$. If the interpolating polynomial is written as

$$y = c_1 + c_2 x + c_3 x^2 + \cdots + c_N x^{N-1} \tag{3.5.1}$$

then the c_i's are required to satisfy the linear equation

$$\begin{bmatrix} 1 & x_1 & x_1^2 & \cdots & x_1^{N-1} \\ 1 & x_2 & x_2^2 & \cdots & x_2^{N-1} \\ \vdots & \vdots & \vdots & & \vdots \\ 1 & x_N & x_N^2 & \cdots & x_N^{N-1} \end{bmatrix} \cdot \begin{bmatrix} c_1 \\ c_2 \\ \vdots \\ c_N \end{bmatrix} = \begin{bmatrix} y_1 \\ y_2 \\ \vdots \\ y_N \end{bmatrix} \tag{3.5.2}$$

This is a *Vandermonde matrix*, as described in §2.8. One could in principle solve equation (3.5.2) by standard techniques for linear equations generally (§2.3); however the special method that was derived in §2.8 is more efficient by a large factor, of order N, so it is much better.

Remember that Vandermonde systems can be quite ill-conditioned. In such a case, *no* numerical method is going to give a very accurate answer. Such cases do not, please note, imply any difficulty in finding interpolated *values* by the methods of §3.1, but only difficulty in finding *coefficients.*

Like the routine in §2.8, the following is due to G. Rybicki.

```
SUBROUTINE POLCOE(X,Y,N,COF)
    Given arrays X and Y of length N containing a tabulated function Y_i = f(X_i), this routine
    returns an array of coefficients COF, also of length N, such that Y_i = Σ_j COF_j X_i^(j-1)
PARAMETER (NMAX=15)            Change NMAX as desired to be the largest anticipated value of N.
DIMENSION X(N),Y(N),COF(N),S(NMAX)
DO 11 I=1,N
    S(I)=0.
    COF(I)=0.
11  CONTINUE
S(N)=-X(1)
DO 13 I=2,N                    Coefficients S_i of the master polynomial P(x) are found by recur-
    DO 12 J=N+1-I,N-1              rence.
        S(J)=S(J)-X(I)*S(J+1)
12      CONTINUE
    S(N)=S(N)-X(I)
13  CONTINUE
DO 16 J=1,N
    PHI=N
    DO 14 K=N-1,1,-1           The quantity PHI = Π_{j≠k}(x_j - x_k) is found as a derivative of P(x_j).
        PHI=K*S(K+1)+X(J)*PHI
14      CONTINUE
    FF=Y(J)/PHI
    B=1.                       Coefficients of polynomials in each term of the Lagrange formula are
    DO 15 K=N,1,-1                 found by synthetic division of P(x) by (x - x_j). The solution
        COF(K)=COF(K)+B*FF         c_k is accumulated.
        B=S(K)+X(J)*B
```

```
      15 CONTINUE
   16 CONTINUE
RETURN
END
```

Another Method

Another technique is to make use of the function value interpolation routine already given (POLINT §3.1). If we interpolate (or extrapolate) to find the value of the interpolating polynomial at $x = 0$, then this value will evidently be c_1. Now we can subtract c_1 from the y_i's and divide each by its corresponding x_i. Throwing out one point (the one with smallest x_i is a good candidate), we can repeat the procedure to find c_2, and so on.

It is not instantly obvious that this procedure is stable, but we have generally found it to be somewhat *more* stable than the routine immediately preceding. This method is of order N^3, while the preceding one was of order N^2. You will find, however, that neither works very well for large N, because of the intrinsic ill-condition of the Vandermonde problem. In single precision, N up to 8 or 10 is satisfactory; about double this in double precision.

```
SUBROUTINE POLCOF(XA,YA,N,COF)
    Given arrays XA and YA of length N containing a tabulated function YA_i = f(XA_i), this
    routine returns an array of coefficients COF, also of length N, such that YA_i = Σ_j COF_j XA_i^(j-1)
PARAMETER (NMAX=15)          Change NMAX as desired to be the largest anticipated value of N.
DIMENSION XA(N),YA(N),COF(N),X(NMAX),Y(NMAX)
DO 11 J=1,N
    X(J)=XA(J)
    Y(J)=YA(J)
  11 CONTINUE
DO 14 J=1,N
    CALL POLINT(X,Y,N+1-J,0.,COF(J),DY)  This is the polynomial interpolation routine of §3.1.
    XMIN=1.E38                     We extrapolate to x = 0.
    K=0
    DO 12 I=1,N+1-J            Find the remaining X_i of smallest absolute value,
        IF (ABS(X(I)).LT.XMIN)THEN
            XMIN=ABS(X(I))
            K=I
        ENDIF
        IF(X(I).NE.0.)Y(I)=(Y(I)-COF(J))/X(I)     (meanwhile reducing all the terms)
      12 CONTINUE
    IF (K.LT.N+1-J) THEN       and eliminate it.
        DO 13 I=K+1,N+1-J
            Y(I-1)=Y(I)
            X(I-1)=X(I)
          13 CONTINUE
    ENDIF
  14 CONTINUE
RETURN
END
```

If the point $x = 0$ is not in (or at least close to) the range of the tabulated x_i's, then the coefficients of the interpolating polynomial will in general become very large. However the real "information content" of the coefficients is in small differences from the "translation-induced" large values. This is one cause of ill-conditioning, resulting in loss of significance and poorly determined coefficients. You should consider redefining the origin of the problem, to put $x = 0$ in a sensible place.

Another pathology is that, if too high a degree interpolation is attempted on a smooth function, the interpolating polynomial will attempt to use its high-degree coefficients, in combinations with large and almost precisely canceling combinations, to match the tabulated values down to the last possible epsilon of accuracy. This effect is the same as the intrinsic tendency of the interpolating polynomial values to oscillate (wildly) between its constrained points, and would be present even if the machine's floating precision were infinitely good. The above routines POLCOE and POLCOF have slightly different sensitivities to the pathologies that can occur.

Are you still quite certain that the *coefficients* are what you want?

REFERENCES AND FURTHER READING:

Isaacson, Eugene, and Keller, Herbert B. 1966, *Analysis of Numerical Methods* (New York: Wiley), §5.2.

3.6 Interpolation in Two or More Dimensions

In multidimensional interpolation, we seek an estimate of $y(x_1, x_2, \ldots, x_n)$ from an n-dimensional grid of tabulated values y and n one-dimensional vectors giving the tabulated values of each of the independent variables $x_1, x_2, \ldots,$ x_n. We will not here consider the problem of interpolating on a mesh that is not Cartesian, i.e. has tabulated function values at "random" points in n-dimensional space rather than at the vertices of a rectangular array. For clarity, we will consider explicitly only the case of two dimensions, the cases of three or more dimensions being analogous in every way.

In two dimensions, we imagine that we are given an array of functional values YA(J,K), where J varies from 1 to M, and K varies from 1 to N. We are also given an array X1A of length M, and an array X2A of length N. The relation of these input quantities to an underlying function $y(x_1, x_2)$ is

$$\texttt{YA(J,K)} = y(\texttt{X1A(J)}, \texttt{X2A(K)}) \tag{3.6.1}$$

We want to estimate, by interpolation, the function y at some untabulated point (x_1, x_2).

An important concept is that of the *grid square* in which the point (x_1, x_2) falls, that is, the four tabulated points which surround the desired interior

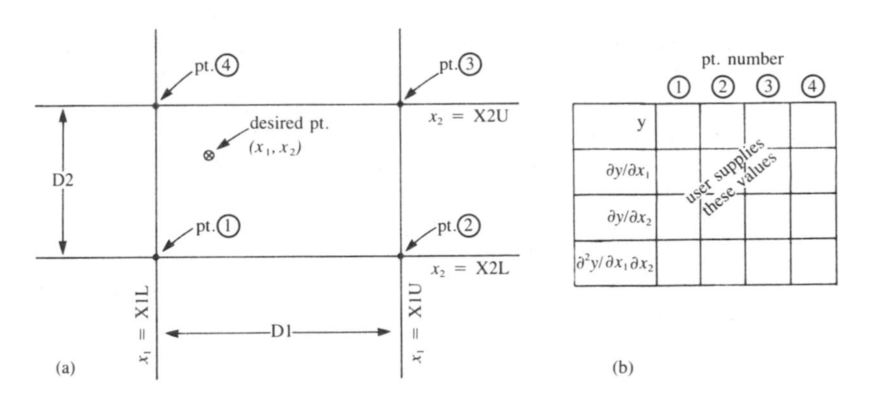

Figure 3.6.1. (a) Labeling of points used in the two-dimensional interpolation routines **BCUINT** and **BCUCOF**. (b) For each of the four points in (a), the user supplies one function value, two first derivatives, and one cross-derivative, a total of 16 numbers.

point. For convenience, we will number these points from 1 to 4, counterclockwise starting from the lower left (see Figure 3.6.1). More precisely, if

$$\begin{aligned} \texttt{X1A(J)} \leq x_1 \leq \texttt{X1A(J+1)} \\ \texttt{X2A(K)} \leq x_2 \leq \texttt{X2A(K+1)} \end{aligned} \tag{3.6.2}$$

defines J and K, then

$$\begin{aligned} y_1 &\equiv \texttt{YA(J,K)} \\ y_2 &\equiv \texttt{YA(J+1,K)} \\ y_3 &\equiv \texttt{YA(J+1,K+1)} \\ y_4 &\equiv \texttt{YA(J,K+1)} \end{aligned} \tag{3.6.3}$$

The simplest interpolation in two dimensions is *bilinear interpolation* on the grid square. Its formulas are:

$$\begin{aligned} t &\equiv (x_1 - \texttt{X1A(J)})/(\texttt{X1A(J+1)} - \texttt{X1A(J)}) \\ u &\equiv (x_2 - \texttt{X2A(K)})/(\texttt{X2A(K+1)} - \texttt{X2A(K)}) \end{aligned} \tag{3.6.4}$$

(so that t and u each lie between 0 and 1), and

$$y(x_1, x_2) = (1-t)(1-u)y_1 + t(1-u)y_2 + tuy_3 + (1-t)uy_4 \tag{3.6.5}$$

Bilinear interpolation is frequently "close enough for government work." As the interpolating point wanders from grid square to grid square, the interpolated function value changes continuously. However, the gradient of the interpolated function changes discontinuously at the boundaries of each grid square.

There are two distinctly different directions that one can take in going beyond bilinear interpolation to higher order methods: One can use higher order to obtain increased accuracy for the interpolated function (for sufficiently smooth functions!), without necessarily trying to fix up the continuity of the gradient and higher derivatives. Or, one can make use of higher order to enforce smoothness of some of these derivatives as the interpolating point crosses grid-square boundaries. We will now consider each of these two directions in turn.

Higher Order for Accuracy

The basic idea is to break up the problem into a succession of one-dimensional interpolations. If we want to do M-1 order interpolation in the x_1 direction, and N-1 order in the x_2 direction, we first locate an M×N sub-block of the tabulated function array that contains our desired point (x_1, x_2). We then do M one-dimensional interpolations in the x_2 direction, i.e. on the rows of the sub-block, to get function values at the points $(\text{X1A(J)}, x_2)$, $\text{J} = 1, \ldots, \text{M}$. Finally, we do a last interpolation in the x_1 direction to get the answer. Using the polynomial interpolation routine POLINT of §3.1, and for a sub-block which is presumed to be already located (and copied into an M by N array YA), the procedure looks like this:

```
SUBROUTINE POLIN2(X1A,X2A,YA,M,N,X1,X2,Y,DY)
    Given arrays X1A (length M) and X2A (length N) of independent variables, and an M by
    N array of function values YA, tabulated at the grid points defined by X1A and X2A; and
    given values X1 and X2 of the independent variables; this routine returns an interpolated
    function value Y, and an accuracy indication DY (based only on the interpolation in the
    X1 direction, however).
PARAMETER (NMAX=20,MMAX=20)  Maximum expected values of N and M, change as desired.
DIMENSION X1A(M),X2A(N),YA(M,N),YNTMP(NMAX),YMTMP(MMAX)
DO 12 J=1,M                      Loop over rows.
    DO 11 K=1,N                  Copy the row into temporary storage.
        YNTMP(K)=YA(J,K)
11  CONTINUE
    CALL POLINT(X2A,YNTMP,N,X2,YMTMP(J),DY)  Interpolate answer into temporary storage.
12 CONTINUE
CALL POLINT(X1A,YMTMP,M,X1,Y,DY) Do the final interpolation.
RETURN
END
```

Higher Order for Smoothness: Bicubic Interpolation

We will give two methods that are in common use, and which are themselves not unrelated. The first is usually called *bicubic interpolation.*

Bicubic interpolation requires the user to specify at each gridpoint not just the function $y(x_1, x_2)$, but also the gradients $\partial y/\partial x_1 \equiv y_{,1}$, $\partial y/\partial x_2 \equiv y_{,2}$ and the cross derivative $\partial^2 y/\partial x_1 \partial x_2 \equiv y_{,12}$. Then an interpolating function that is *cubic* in the scaled coordinates t and u (equation 3.6.4) can be found, with the following properties: (i) the values of the function and the specified derivatives are reproduced exactly on the gridpoints, and (ii) the values of the function and the specified derivatives change continuously as the interpolating point crosses from one grid square to another.

It is important to understand that nothing in the equations of bicubic interpolation requires you to specify the extra derivatives *correctly*! The smoothness properties are tautologically "forced," and have nothing to do with the "accuracy" of the specified derivatives. It is a separate problem for you to decide how to obtain the values that are specified. The better you do, the more *accurate* the interpolation will be. But it will be *smooth* no matter what you do.

Best of all is to know the derivatives analytically, or to be able to compute them accurately by numerical means, at the grid points. Next best is to determine them by numerical differencing from the functional values already tabulated on the grid. The relevant code would be something like this (using centered differencing):

```
Y1A(J,K)=(YA(J+1,K)-YA(J-1,K))/(X1A(J+1)-X1A(J-1))
Y2A(J,K)=(YA(J,K+1)-YA(J,K-1))/(X2A(K+1)-X2A(K-1))
Y12A(J,K)=(YA(J+1,K+1)-YA(J+1,K-1)-YA(J-1,K+1)+YA(J-1,K-1))
    /((X1A(J+1)-X1A(J-1))*(X2A(K+1)-X2A(K-1)))
```

To do a bicubic interpolation within a grid square, given the function Y and the derivatives Y1, Y2, Y12 at each of the four corners of the square, there are two steps: First obtain the sixteen quantities c_{ij}, $i, j = 1, \ldots, 4$ using the routine BCUCOF below. (The formulas that obtain the c's from the function and derivative values are just a complicated linear transformation, with coefficients which, having been determined once in the mists of numerical history, can be tabulated and forgotten.) Next, substitute the c's into any or all of the following bicubic formulas for function and derivatives, as desired:

$$
\begin{aligned}
y(x_1, x_2) &= \sum_{i=1}^{4} \sum_{j=1}^{4} c_{ij} t^{i-1} u^{j-1} \\
y_{,1}(x_1, x_2) &= \sum_{i=1}^{4} \sum_{j=1}^{4} (i-1) c_{ij} t^{i-2} u^{j-1}
\end{aligned}
\tag{3.6.6}
$$

$$y_{,2}(x_1, x_2) = \sum_{i=1}^{4}\sum_{j=1}^{4}(j-1)c_{ij}t^{i-1}u^{j-2}$$

$$y_{,12}(x_1, x_2) = \sum_{i=1}^{4}\sum_{j=1}^{4}(i-1)(j-1)c_{ij}t^{i-2}u^{j-2}$$

where t and u are again given by equation (3.6.4).

```
      SUBROUTINE BCUCOF(Y,Y1,Y2,Y12,D1,D2,C)
         Given arrays Y,Y1,Y2, and Y12, each of length 4, containing the function, gradients, and
         cross derivative at the four grid points of a rectangular grid cell (numbered counterclock-
         wise from the lower left), and given D1 and D2, the length of the grid cell in the 1- and
         2-directions, this routine returns the table C that is used by routine BCUINT for bicubic
         interpolation.
      DIMENSION C(4,4),Y(4),Y1(4),Y2(4),Y12(4),CL(16),X(16),WT(16,16)
      DATA WT/1,0,-3,2,4*0,-3,0,9,-6,2,0,-6,4,8*0,3,0,-9,6,-2,0,6,-4
     *  ,10*0,9,-6,2*0,-6,4,2*0,3,-2,6*0,-9,6,2*0,6,-4
     *  ,4*0,1,0,-3,2,-2,0,6,-4,1,0,-3,2,8*0,-1,0,3,-2,1,0,-3,2
     *  ,10*0,-3,2,2*0,3,-2,6*0,3,-2,2*0,-6,4,2*0,3,-2
     *  ,0,1,-2,1,5*0,-3,6,-3,0,2,-4,2,9*0,3,-6,3,0,-2,4,-2
     *  ,10*0,-3,3,2*0,2,-2,2*0,-1,1,6*0,3,-3,2*0,-2,2
     *  ,5*0,1,-2,1,0,-2,4,-2,0,1,-2,1,9*0,-1,2,-1,0,1,-2,1
     *  ,10*0,1,-1,2*0,-1,1,6*0,-1,1,2*0,2,-2,2*0,-1,1/
      D1D2=D1*D2
      DO 11 I=1,4                              Pack a temporary vector X.
        X(I)=Y(I)
        X(I+4)=Y1(I)*D1
        X(I+8)=Y2(I)*D2
        X(I+12)=Y12(I)*D1D2
11    CONTINUE
      DO 13 I=1,16                             Matrix multiply by the stored table.
        XX=0.
        DO 12 K=1,16
          XX=XX+WT(I,K)*X(K)
12      CONTINUE
        CL(I)=XX
13    CONTINUE
      L=0
      DO 15 I=1,4                              Unpack the result into the output table.
        DO 14 J=1,4
          L=L+1
          C(I,J)=CL(L)
14      CONTINUE
15    CONTINUE
      RETURN
      END
```

The implementation of equation (3.6.6), which performs a bicubic interpolation, returns the interpolated function value and the two gradient values, and uses the above routine BCUCOF, is simply:

```
      SUBROUTINE BCUINT(Y,Y1,Y2,Y12,X1L,X1U,X2L,X2U,X1,X2,ANSY,ANSY1,ANSY2)
         Bicubic interpolation within a grid square. Input quantities are Y,Y1,Y2,Y12 (as described
         in BCUCOF); X1L and X1U, the lower and upper coordinates of the grid square in the 1-
         direction; X2L and X2U likewise for the 2-direction; and X1,X2, the coordinates of the
```

```
    desired point for the interpolation. The interpolated function value is returned as ANSY,
    and the interpolated gradient values as ANSY1 and ANSY2. This routine calls BCUCOF.
DIMENSION Y(4),Y1(4),Y2(4),Y12(4),C(4,4)
CALL BCUCOF(Y,Y1,Y2,Y12,X1U-X1L,X2U-X2L,C)    Get the c's.
IF(X1U.EQ.X1L.OR.X2U.EQ.X2L)PAUSE 'bad input'
T=(X1-X1L)/(X1U-X1L)                        Equation 3.6.4.
U=(X2-X2L)/(X2U-X2L)
ANSY=0.
ANSY2=0.
ANSY1=0.
DO 11 I=4,1,-1                              Equation 3.6.6.
    ANSY=T*ANSY+((C(I,4)*U+C(I,3))*U+C(I,2))*U+C(I,1)
    ANSY2=T*ANSY2+(3.*C(I,4)*U+2.*C(I,3))*U+C(I,2)
    ANSY1=U*ANSY1+(3.*C(4,I)*T+2.*C(3,I))*T+C(2,I)
11 CONTINUE
ANSY1=ANSY1/(X1U-X1L)
ANSY2=ANSY2/(X2U-X2L)
RETURN
END
```

Higher Order for Smoothness: Bicubic Spline

The other common technique for obtaining smoothness in two-dimensional interpolation is the *bicubic spline.* Actually, this is equivalent to a special case of bicubic interpolation: the interpolating function is of the same functional form as equation (3.6.6); the values of the derivatives at the gridpoints are, however, determined "globally" by one-dimensional splines. However, bicubic splines are usually implemented in a form that looks rather different from the above bicubic interpolation routines, instead looking much closer in form to the routine POLIN2 above: To interpolate one functional value, one performs M one-dimensional splines across the rows of the table, followed by one additional one-dimensional spline down the newly created column. It is a matter of taste (and tradeoff between time and memory) as to how much of this process one wants to precompute and store. Instead of precomputing and storing all the derivative information (as in bicubic interpolation), spline users typically precompute and store only one auxiliary table, of second derivatives in one direction only. Then one need only do spline *evaluations* (not constructions) for the M row splines; one must still do a construction *and* an evaluation for the final column spline. (Recall that a spline construction is a process of order N, while a spline evaluation is only of order $\log N$ — and that is just to find the place in the table!)

Here is a routine to precompute the auxiliary second-derivative table:

```
SUBROUTINE SPLIE2(X1A,X2A,YA,M,N,Y2A)
    Given an M by N tabulated function YA, and tabulated independent variables X1A (M values)
    and X2A (N values), this routine constructs one-dimensional natural cubic splines of the
    rows of YA and returns the second-derivatives in the array Y2A.
PARAMETER (NN=100)
DIMENSION X1A(M),X2A(N),YA(M,N),Y2A(M,N),YTMP(NN),Y2TMP(NN)
DO 13 J=1,M
    DO 11 K=1,N
        YTMP(K)=YA(J,K)
11  CONTINUE
    CALL SPLINE(X2A,YTMP,N,1.E30,1.E30,Y2TMP)    Values 1 × 10^30 signal a natural spline.
    DO 12 K=1,N
```

```
        Y2A(J,K)=Y2TMP(K)
      12 CONTINUE
  13 CONTINUE
RETURN
END
```

After the above routine has been executed once, any number of bicubic spline interpolations can be performed by successive calls of the following routine:

```
SUBROUTINE SPLIN2(X1A,X2A,YA,Y2A,M,N,X1,X2,Y)
    Given X1A, X2A, YA, M, N as described in SPLIE2 and Y2A as produced by that routine;
    and given a desired interpolating point X1, X2; this routine returns an interpolated function
    value Y by bicubic spline interpolation.
PARAMETER (NN=100)
DIMENSION X1A(M),X2A(N),YA(M,N),Y2A(M,N),YTMP(NN),Y2TMP(NN),YYTMP(NN)
DO 12 J=1,M                          Perform M evaluations of the row splines constructed by SPLIE2, using
    DO 11 K=1,N                          the one-dimensional spline evaluator SPLINT.
        YTMP(K)=YA(J,K)
        Y2TMP(K)=Y2A(J,K)
      11 CONTINUE
    CALL SPLINT(X2A,YTMP,Y2TMP,N,X2,YYTMP(J))
  12 CONTINUE
CALL SPLINE(X1A,YYTMP,M,1.E30,1.E30,Y2TMP)  Construct the one-dimensional column spline
CALL SPLINT(X1A,YYTMP,Y2TMP,M,X1,Y)          and evaluate it.
RETURN
END
```

REFERENCES AND FURTHER READING:

Abramowitz, Milton, and Stegun, Irene A. 1964, *Handbook of Mathematical Functions*, Applied Mathematics Series, vol. 55 (Washington: National Bureau of Standards; reprinted 1968 by Dover Publications, New York), §25.2.

Kinahan, B.F., and Harm, R. 1975, *Astrophysical Journal*, vol. 200, p.330.

Johnson, Lee W., and Riess, R. Dean. 1982, *Numerical Analysis*, 2nd ed. (Reading, Mass.: Addison-Wesley), §5.2.7.

Dahlquist, Germund, and Bjorck, Ake. 1974, *Numerical Methods* (Englewood Cliffs, N.J.: Prentice-Hall), §7.7.

Chapter 4. Integration of Functions

4.0 Introduction

Numerical integration, which is also called *quadrature*, has a history extending back to the invention of calculus and before. The fact that integrals of elementary functions could not, in general, be computed analytically, while derivatives *could* be, served to give the field a certain panache, and to set it a cut above the arithmetic drudgery of numerical analysis during the whole of the 18th and 19th Centuries.

With the invention of automatic computing, quadrature became just one numerical task among many, and not a very interesting one at that. Automatic computing, even the most primitive sort involving desk calculators and rooms full of "computers" (that were, until the 1950's, people rather than machines), opened to feasibility the much richer field of numerical integration of differential equations. Quadrature is merely the simplest special case: The evaluation of the integral

$$I = \int_a^b f(x)dx \tag{4.0.1}$$

is precisely equivalent to solving for the value $I \equiv y(b)$ the differential equation

$$\frac{dy}{dx} = f(x) \tag{4.0.2}$$

with the boundary condition

$$y(a) = 0 \tag{4.0.3}$$

Chapter 15 of this book deals with the numerical integration of differential equations. In that chapter, much emphasis is given to the concept of "variable" or "adaptive" choices of step-size. We will not, therefore, develop that material here. If the function that you propose to integrate is sharply concentrated in one or more peaks, or if its shape is not readily characterized

by a single length-scale, then it is likely that you should cast the problem in the form of (4.0.2) – (4.0.3) and use the methods of Chapter 15.

The quadrature methods in this chapter are based, in one way or another, on the obvious device of adding up the value of the integrand at a sequence of abscissas within the range of integration. The game is to obtain the integral as accurately as possible with the smallest number of function evaluations of the integrand. Just as in the case of interpolation (Chapter 3), one has the freedom to choose methods of various *orders*, with higher order sometimes, but not always, giving higher accuracy. "Romberg integration," which is discussed in §4.3 is a general formalism for making use of integration methods of a variety of different orders, and we recommend it highly.

Apart from the methods of this chapter and of Chapter 15, there are yet other methods for obtaining integrals. One important class is based on function approximation. We discuss explicitly the integration of Chebyshev-approximated functions in §5.7. Although not explicitly discussed here, you ought to be able to figure out how to do *cubic spline quadrature* using the output of the routine `SPLINE` in §3.3. (Hint: Integrate equation 3.3.3 over x analytically. See Forsythe *et al.*)

Multidimensional integrals are another whole multidimensional bag of worms. Section 4.6 is an introductory discussion in this chapter; the important technique of *Monte-Carlo integration* is treated in Chapter 7.

REFERENCES AND FURTHER READING:

Carnahan, Brice, Luther, H.A., and Wilkes, James O. 1969, *Applied Numerical Methods* (New York: Wiley), Chapter 2.

Isaacson, Eugene, and Keller, Herbert B. 1966, *Analysis of Numerical Methods* (New York: Wiley), Chapter 7.

Acton, Forman S. 1970, *Numerical Methods That Work* (New York: Harper and Row), Chapter 4.

Stoer, J., and Bulirsch, R. 1980, *Introduction to Numerical Analysis* (New York: Springer-Verlag), Chapter 3.

Ralston, Anthony, and Rabinowitz, Philip. 1978, *A First Course in Numerical Analysis*, 2nd ed. (New York: McGraw-Hill), Chapter 4.

Dahlquist, Germund, and Bjorck, Ake. 1974, *Numerical Methods* (Englewood Cliffs, N.J.: Prentice-Hall), §7.4.

Forsythe, George E., Malcolm, Michael A., and Moler, Cleve B. 1977, *Computer Methods for Mathematical Computations* (Englewood Cliffs, N.J.: Prentice-Hall), §5.2, p.89.

4.1 Classical Formulas for Equally-Spaced Abscissas

Where would any book on numerical analysis be without Mr. Simpson and his "rule"? The classical formulas for integrating a function whose value is known at equally-spaced steps have a certain elegance about them, and they are redolent with historical association. Through them, the modern

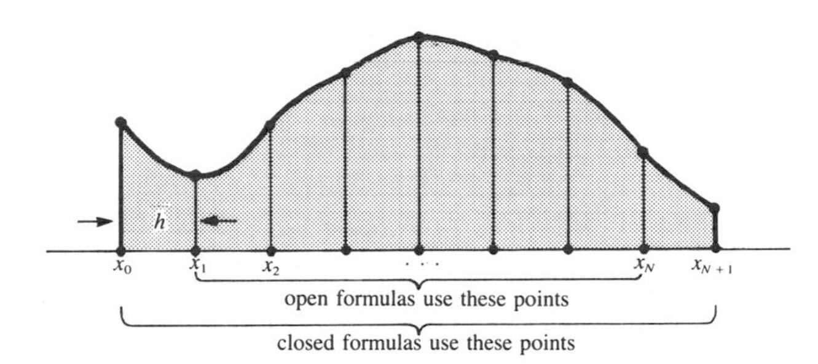

Figure 4.1.1. Quadrature formulas with equally spaced abscissas compute the integral of a function between x_0 and x_{N+1}. Closed formulas evaluate the function on the boundary points, while open formulas refrain from doing so (useful if the evaluation algorithm breaks down on the boundary points).

numerical analyst communes with the spirits of his or her predecessors back across the centuries, as far as the time of Newton, if not farther. Alas, times *do* change; with the exception of two of the most modest formulas ("extended trapezoidal rule," equation 4.1.11, and "extended midpoint rule," equation 4.1.19, see §4.2), the classical formulas are almost entirely useless. They are museum pieces, but beautiful ones.

Some notation: We have a sequence of abscissas, denoted $x_0, x_1, \ldots, x_N, x_{N+1}$ which are spaced apart by a constant step h,

$$x_i = x_0 + ih \qquad i = 0, 1, \ldots, N+1 \tag{4.1.1}$$

A function $f(x)$ has known values at the x_i's,

$$f(x_i) \equiv f_i. \tag{4.1.2}$$

We want to integrate the function $f(x)$ between a lower limit a and an upper limit b, where a and b are each equal to one or the other of the x_i's. An integration formula that uses the value of the function at the endpoints, $f(a)$ or $f(b)$, is called a *closed* formula. Occasionally, we want to integrate a function whose value at one or both endpoints is difficult to compute (e.g., the computation of f goes to a limit of zero over zero there, or worse yet has an integrable singularity there). In this case we want an *open* formula, which estimates the integral using only x_i's strictly *between* a and b (see Figure 4.1.1).

The basic building blocks of the classical formulas are rules for integrating a function over a small number of intervals. As that number increases, we can find rules that are exact for polynomials of increasingly high order. (Keep in mind that higher order does not always imply higher accuracy in real cases.) A sequence of such closed formulas is now given.

Closed Newton-Cotes Formulas

Trapezoidal rule:

$$\int_{x_1}^{x_2} f(x)dx = h[\frac{1}{2}f_1 + \frac{1}{2}f_2] \quad + O(h^3 f'') \tag{4.1.3}$$

Here the error term $O(\ )$ signifies that the true answer differs from the estimate by an amount that is the product of some numerical coefficient times h^3 times the value of the functions's second derivative somewhere in the interval of integration. The coefficient is knowable, and it can be found in all the standard references on this subject. The point at which the second derivative is to be evaluated is, however, unknowable. If we knew it, we could evaluate the function there and have a higher-order method! Since the product of a knowable and an unknowable is unknowable, we will streamline our formulas and write only $O(\ )$, instead of the coefficient.

Equation (4.1.3) is a two-point formula (x_1 and x_2). It is exact for polynomials up to and including degree 1, i.e. $f(x) = x$. One anticipates that there is a three-point formula exact up to polynomials of degree 2. This is true; moreover, by a cancellation of coefficients due to left-right symmetry of the formula, the three-point formula is exact for polynomials up to and including degree 3, i.e. $f(x) = x^3$:

Simpson's rule:

$$\int_{x_1}^{x_3} f(x)dx = h[\frac{1}{3}f_1 + \frac{4}{3}f_2 + \frac{1}{3}f_3] \quad + O(h^5 f^{(4)}) \tag{4.1.4}$$

Here $f^{(4)}$ means the fourth derivative of the function f evaluated at an unknown place in the interval. Note also that the formula gives the integral over an interval of size $2h$, so the coefficients add up to 2.

There is no lucky cancellation in the four-point formula, so it is also exact for polynomials up to and including degree 3.

Simpson's $\frac{3}{8}$ rule:

$$\int_{x_1}^{x_4} f(x)dx = h[\frac{3}{8}f_1 + \frac{9}{8}f_2 + \frac{9}{8}f_3 + \frac{3}{8}f_4] \quad + O(h^5 f^{(4)}) \tag{4.1.5}$$

The five-point formula again benefits from a cancellation:

Bode's rule:

$$\int_{x_1}^{x_5} f(x)dx = h[\frac{14}{45}f_1 + \frac{64}{45}f_2 + \frac{24}{45}f_3 + \frac{64}{45}f_4 + \frac{14}{45}f_5] \quad + O(h^7 f^{(6)}) \qquad (4.1.6)$$

This is exact for polynomials up to and including degree 5.

At this point the formulas stop being named after famous personages, so we will not go any further. Consult Abramowitz and Stegun for additional formulas in the sequence.

Extrapolative Formulas for a Single Interval

We are going to depart from historical practice for a moment. Most texts would give, at this point, a sequence of "Newton-Cotes Formulas of Open Type." Here is an example:

$$\int_{x_0}^{x_5} f(x)dx = h[\frac{55}{24}f_1 + \frac{5}{24}f_2 + \frac{5}{24}f_3 + \frac{55}{24}f_4] \quad + O(h^5 f^{(4)})$$

Notice that the integral from $a = x_0$ to $b = x_5$ is estimated, using only the interior points x_1, x_2, x_3, x_4. In our opinion, formulas of this type are not useful for the reasons that (i) they cannot usefully be strung together to get "extended" rules, as we are about to do with the closed formulas, and (ii) for all other possible uses they are dominated by the Gaussian integration formulas which we will introduce in §4.3.

Instead of the Newton-Cotes open formulas, let us set out the formulas for estimating the integral in the single interval from x_0 to x_1, using values of the function f at $x_1, x_2, \ldots$. These will be useful building blocks for the "extended" open formulas.

$$\int_{x_0}^{x_1} f(x)dx = h[f_1] \quad + O(h^2 f') \qquad (4.1.7)$$

$$\int_{x_0}^{x_1} f(x)dx = h[\frac{3}{2}f_1 - \frac{1}{2}f_2] \quad + O(h^3 f'') \qquad (4.1.8)$$

$$\int_{x_0}^{x_1} f(x)dx = h[\frac{23}{12}f_1 - \frac{16}{12}f_2 + \frac{5}{12}f_3] \quad + O(h^4 f^{(3)}) \qquad (4.1.9)$$

$$\int_{x_0}^{x_1} f(x)dx = h[\frac{55}{24}f_1 - \frac{59}{24}f_2 + \frac{37}{24}f_3 - \frac{9}{24}f_4] \quad + O(h^5 f^{(4)}) \qquad (4.1.10)$$

Perhaps a word here would be in order about how formulas like the above can be derived. There are elegant ways, but the most straightforward is to

write down the basic form of the formula, replacing the numerical coefficients with unknowns, say p, q, r, s. Without loss of generality take $x_0 = 0$ and $x_1 = 1$, so $h = 1$. Substitute in turn for $f(x)$ (and for f_1, f_2, f_3, f_4) the functions $f(x) = 1$, $f(x) = x$, $f(x) = x^2$, and $f(x) = x^3$. Doing the integral in each case reduces the right-hand side to a number, and the left-hand side to a linear equation for the unknowns p, q, r, s. Solving the four equations produced in this way gives the coefficients.

Extended Formulas (Closed)

If we use equation (4.1.3) $N - 1$ times, to do the integration in the intervals $(x_1, x_2), (x_2, x_3), \ldots, (x_{N-1}, x_N)$, and then add the results, we obtain an "extended" or "composite" formula for the integral from x_1 to x_N.

Extended trapezoidal rule:

$$\begin{aligned}\int_{x_1}^{x_N} f(x)dx = h[\frac{1}{2}f_1 + f_2 + f_3 + \\ \cdots + f_{N-1} + \frac{1}{2}f_N] \quad + O\left(\frac{(b-a)^3 f''}{N^2}\right)\end{aligned} \tag{4.1.11}$$

Here we have written the error estimate in terms of the interval $b - a$ and the number of points N instead of in terms of h. This is clearer, since one is usually holding a and b fixed and wanting to know (e.g.) how much the error will be decreased by taking twice as many steps (in this case, it is by a factor of 4). In subsequent equations we will show *only* the scaling of the error term with the number of steps.

For reasons which will not become clear until §4.2, equation (4.1.11) is in fact the most important equation in this section, the basis for most practical quadrature schemes.

The *extended formula of order* $1/N^3$ is:

$$\begin{aligned}\int_{x_1}^{x_N} f(x)dx = h[\frac{5}{12}f_1 + \frac{13}{12}f_2 + f_3 + f_4 + \\ \cdots + f_{N-2} + \frac{13}{12}f_{N-1} + \frac{5}{12}f_N] \quad + O\left(\frac{1}{N^3}\right)\end{aligned} \tag{4.1.12}$$

(We will see in a moment where this comes from.)

If we apply equation (4.1.4) to successive, non-overlapping *pairs* of intervals, we get the *extended Simpson's rule:*

$$\int_{x_1}^{x_N} f(x)dx = h\left[\frac{1}{3}f_1 + \frac{4}{3}f_2 + \frac{2}{3}f_3 + \frac{4}{3}f_4 + \cdots + \frac{2}{3}f_{N-2} + \frac{4}{3}f_{N-1} + \frac{1}{3}f_N\right] \quad + O\left(\frac{1}{N^4}\right) \tag{4.1.13}$$

Notice that the 2/3, 4/3 alternation continues throughout the interior of the evaluation. Many people believe that the wobbling alternation somehow contains information about the integral of their function which is not apparent to mortal eyes. In fact, the alternation is an artifact of using only the building block (4.1.4) and not (4.1.5), which is of the same order. If we take one three-interval step using (4.1.5) and then subsequent two-interval steps of (4.1.4), the alternation of 2/3, 4/3 will be exactly out of phase with that in (4.1.13). Averaging the result with (4.1.13) gives the *alternative extended Simpson's rule*:

$$\int_{x_1}^{x_N} f(x)dx = h\left[\frac{17}{48}f_1 + \frac{59}{48}f_2 + \frac{43}{48}f_3 + \frac{49}{48}f_4 + f_5 + f_6 + \cdots + f_{N-4} + \frac{49}{48}f_{N-3} + \frac{43}{48}f_{N-2} + \frac{59}{48}f_{N-1} + \frac{17}{48}f_N\right] + O\left(\frac{1}{N^4}\right) \tag{4.1.14}$$

This is a somewhat ugly formula because of the peculiar rational coefficients, but your computer won't care about that.

We can now tell you where equation (4.1.12) came from. It is Simpson's extended rule averaged with the sequence: one step of trapezoidal (4.1.3) followed by Simpson's extended rule. The trapezoidal step is *two* orders lower than Simpson's rule; however, its contribution to the integral goes down as an additional power of N (since it is used only twice, not N times). This makes the resulting formula of degree *one* less than Simpson. It is likewise true that, in constructing (4.1.14), we could have used a starting step of one degree *less* than Simpson, instead of using (4.1.5). However there *is* no step of degree one less than Simpson!

Extended Formulas (Open and Semi-open)

We can construct open and semi-open extended formulas by adding the closed formulas (4.1.11) - (4.1.14), evaluated for the second and subsequent steps, to the extrapolative open formulas for the first step, (4.1.7) - (4.1.10). As discussed immediately above, it is consistent to use an end step that is of

one order lower than the (repeated) interior step. The resulting formulas for an interval open at both ends are as follows:

Equations (4.1.7) and (4.1.11) give

$$\int_{x_1}^{x_N} f(x)dx = h[\frac{3}{2}f_2 + f_3 + f_4 + \cdots + f_{N-2} + \frac{3}{2}f_{N-1}] \quad + O\left(\frac{1}{N^2}\right) \tag{4.1.15}$$

Equations (4.1.8) and (4.1.12) give

$$\begin{aligned} \int_{x_1}^{x_N} f(x)dx = h[\frac{23}{12}f_2 + \frac{7}{12}f_3 + f_4 + f_5 + \\ \cdots + f_{N-3} + \frac{7}{12}f_{N-2} + \frac{23}{12}f_{N-1}] \\ + O\left(\frac{1}{N^3}\right) \end{aligned} \tag{4.1.16}$$

Equations (4.1.9) and (4.1.13) give

$$\begin{aligned} \int_{x_1}^{x_N} f(x)dx = h[\frac{27}{12}f_2 + 0 + \frac{13}{12}f_4 + \frac{4}{3}f_5 + \\ \cdots + \frac{4}{3}f_{N-4} + \frac{13}{12}f_{N-3} + 0 + \frac{27}{12}f_{N-1}] \\ + O\left(\frac{1}{N^4}\right) \end{aligned} \tag{4.1.17}$$

The interior points alternate 4/3 and 2/3. If we want to avoid this alternation, we can combine equations (4.1.9) and (4.1.14), giving

$$\begin{aligned} \int_{x_1}^{x_N} f(x)dx = h[\frac{109}{48}f_2 - \frac{5}{48}f_3 + \frac{63}{48}f_4 + \frac{49}{48}f_5 + f_6 + f_7 + \\ \cdots + f_{N-5} + \frac{49}{48}f_{N-4} + \frac{63}{48}f_{N-3} - \frac{5}{48}f_{N-2} + \frac{109}{48}f_{N-1}] \\ + O\left(\frac{1}{N^4}\right) \end{aligned} \tag{4.1.18}$$

We should mention in passing another extended open formula, for use where the limits of integration are located halfway between tabulated abscissas. This one is known as the *extended midpoint rule*, and is accurate to the

same order as (4.1.15):

$$\int_{x_1}^{x_N} f(x)dx = h[f_{3/2} + f_{5/2} + f_{7/2} + \cdots + f_{N-3/2} + f_{N-1/2}] \quad + O\left(\frac{1}{N^2}\right) \tag{4.1.19}$$

There are also formulas of higher order for this situation, but we will refrain from giving them.

The *semi-open formulas* are just the obvious combinations of equations (4.1.11) through (4.1.14) with (4.1.15) through (4.1.18) respectively. At the closed end of the integration, use the weights from the former equations; at the open end use the weights from the latter equations. One example should give the idea, the formula with error term decreasing as $1/N^3$ which is closed on the right and open on the left:

$$\int_{x_1}^{x_N} f(x)dx = h[\frac{23}{12}f_2 + \frac{7}{12}f_3 + f_4 + f_5 + \cdots + f_{N-2} + \frac{13}{12}f_{N-1} + \frac{5}{12}f_N] \quad + O\left(\frac{1}{N^3}\right) \tag{4.1.20}$$

REFERENCES AND FURTHER READING:

Abramowitz, Milton, and Stegun, Irene A. 1964, *Handbook of Mathematical Functions*, Applied Mathematics Series, vol. 55 (Washington: National Bureau of Standards; reprinted 1968 by Dover Publications, New York), §25.4.

Isaacson, Eugene, and Keller, Herbert B. 1966, *Analysis of Numerical Methods* (New York: Wiley), §7.1.

4.2 Elementary Algorithms

Our starting point is equation (4.1.11), the extended trapezoidal rule. There are two facts about the trapezoidal rule which make it the starting point for a variety of algorithms. One fact is rather obvious, while the second is rather "deep."

The obvious fact is that, for a fixed function $f(x)$ to be integrated between fixed limits a and b, one can double the number of intervals in the extended trapezoidal rule without losing the benefit of previous work. The coarsest implementation of the trapezoidal rule is to average the function at its endpoints a and b. The first stage of refinement is to add to this average the value of the function at the halfway point. The second stage of refinement is to add the values at the 1/4 and 3/4 points. And so on (see Figure 4.2.1).

Without further ado we can write a routine with this kind of logic to it:

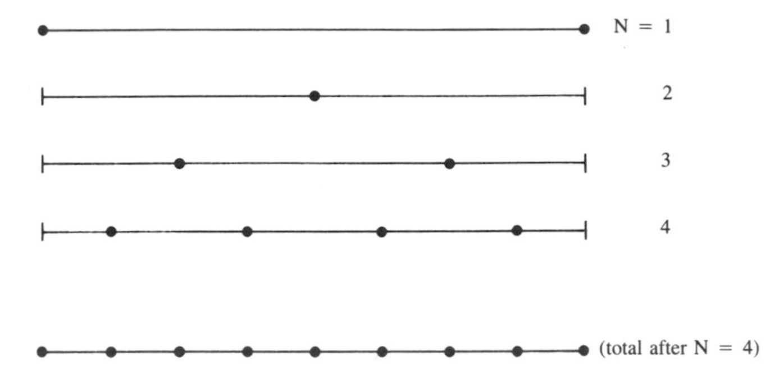

Figure 4.2.1. Sequential calls to the routine **TRAPZD** incorporate the information from previous calls and evaluate the integrand only at those new points necessary to refine the grid. The bottom line shows the totality of function evaluations after the fourth call. The routine **QSIMP**, by weighting the intermediate results, transforms the trapezoid rule into Simpson's rule with essentially no additional overhead.

```
SUBROUTINE TRAPZD(FUNC,A,B,S,N)
```

This routine computes the N'th stage of refinement of an extended trapezoidal rule. **FUNC** is input as the name of the function to be integrated between limits **A** and **B**, also input. When called with N=1, the routine returns as **S** the crudest estimate of $\int_a^b f(x)dx$. Subsequent calls with N=2,3,... (in that sequential order) will improve the accuracy of **S** by adding 2^{N-2} additional interior points. **S** should not be modified between sequential calls.

```
IF (N.EQ.1) THEN
    S=0.5*(B-A)*(FUNC(A)+FUNC(B))
    IT=1                         IT is the number of points to be added on the next call.
ELSE
    TNM=IT
    DEL=(B-A)/TNM                This is the spacing of the points to be added.
    X=A+0.5*DEL
    SUM=0.
    DO 11 J=1,IT
        SUM=SUM+FUNC(X)
        X=X+DEL
11  CONTINUE
    S=0.5*(S+(B-A)*SUM/TNM)      This replaces S by its refined value.
    IT=2*IT
ENDIF
RETURN
END
```

The above routine (TRAPZD) is a work horse which can be harnessed in several ways. The simplest and crudest is to integrate a function by the extended trapezoidal rule where you know in advance (we can't imagine how!) the number of steps you want. If you want $2^M + 1$, you can accomplish this by the fragment

```
DO J=1,M+1
    CALL TRAPZD(FUNC,A,B,S,J)
ENDDO
```

with the answer returned in S.

Much better, of course, is to refine the trapezoidal rule until some specified degree of accuracy has been achieved:

```
SUBROUTINE QTRAP(FUNC,A,B,S)
   Returns as S the integral of the function FUNC from A to B. The parameters EPS can be
   set to the desired fractional accuracy and JMAX so that 2^(JMAX-1) is the maximum allowed
   number of steps. Integration is performed by the trapezoidal rule.
PARAMETER (EPS=1.E-6, JMAX=20)
OLDS=-1.E30                     Any number that is unlikely to be the average of the function at its
DO 11 J=1,JMAX                          endpoints will do here.
    CALL TRAPZD(FUNC,A,B,S,J)
    IF (ABS(S-OLDS).LT.EPS*ABS(OLDS)) RETURN
    OLDS=S
 11 CONTINUE
PAUSE 'Too many steps.'
END
```

Unsophisticated as it is, routine QTRAP is in fact a fairly robust way of doing integrals of functions that are not very smooth. Increased sophistication will usually translate into a higher order method whose efficiency will be greater only for sufficiently smooth integrands. QTRAP is the method of choice, e.g., for an integrand which is a function of a variable that is linearly interpolated between measured data points. Be sure that you do not require too stringent an EPS, however: if QTRAP takes too many steps in trying to achieve your required accuracy, accumulated roundoff errors may start increasing, and the routine may never converge. The value 10^{-6} used above is just on the edge of trouble for most 32-bit machines; it is achievable when the convergence is moderately rapid, but not otherwise.

We come now to the "deep" fact about the extended trapezoidal rule, equation (4.1.11). It is this: the error of the approximation, which begins with a term of order $1/N^2$ is in fact *entirely even* when expressed in powers of $1/N$. This follows directly from the *Euler-Maclaurin Summation Formula*,

$$\int_{x_1}^{x_N} f(x)dx = h[\frac{1}{2}f_1 + f_2 + f_3 + \cdots + f_{N-1} + \frac{1}{2}f_N] \\ - \frac{B_2 h^2}{2!}(f'_N - f'_1) - \cdots - \frac{B_{2k}h^{2k}}{(2k)!}(f_N^{(2k-1)} - f_1^{(2k-1)}) - \cdots \tag{4.2.1}$$

Here B_{2k} is a *Bernoulli number*, defined by the generating function

$$\frac{t}{e^t - 1} = \sum_{n=0}^{\infty} B_n \frac{t^n}{n!} \tag{4.2.2}$$

with the first few even values (odd values vanish except for $B_1 = -1/2$)

$$B_0 = 1 \quad B_2 = \frac{1}{6} \quad B_4 = -\frac{1}{30} \quad B_6 = \frac{1}{42} \\ B_8 = -\frac{1}{30} \quad B_{10} = \frac{5}{66} \quad B_{12} = -\frac{691}{2730} \tag{4.2.3}$$

Equation (4.2.1) is not a convergent expansion, but rather only an asymptotic expansion whose error when truncated at any point is always less than twice the magnitude of the first neglected term. The reason that it is not convergent is that the Bernoulli numbers become very large, e.g.

$$B_{50} = \frac{495057205241079648212477525}{66}$$

The key point is that only even powers of h occur in the error series of (4.2.1). This fact is not, in general, shared by the higher order quadrature rules in §4.1. For example, equation (4.1.13) has an error series beginning with $O(1/N^3)$, but continuing with all subsequent powers of N: $1/N^4$, $1/N^5$, etc.

Suppose we evaluate (4.1.11) with N steps, getting a result S_N, and then again with $2N$ steps, getting a result S_{2N}. (This is done by any two consecutive calls of TRAPZD.) The leading error term in the second evaluation will be 1/4 the size of the error in the first evaluation. Therefore the combination

$$S = \frac{4}{3}S_{2N} - \frac{1}{3}S_N \tag{4.2.4}$$

will cancel out the leading order error term. But there *is* no error term of order $1/N^3$, by (4.2.1). The surviving error is of order $1/N^4$, the same as Simpson's rule. In fact, it should not take long for you to see that (4.2.4) is *exactly* Simpson's rule (4.1.13), alternating 2/3's, 4/3's and all. This is the preferred method for evaluating that rule, and we can write it as a routine exactly analogous to QTRAP above:

```
SUBROUTINE QSIMP(FUNC,A,B,S)
    Returns as S the integral of the function FUNC from A to B. The parameters EPS can be
    set to the desired fractional accuracy and JMAX so that 2^(JMAX-1) is the maximum allowed
    number of steps. Integration is performed by Simpson's rule.
PARAMETER (EPS=1.E-6, JMAX=20)
OST=-1.E30
OS= -1.E30
DO 11 J=1,JMAX
    CALL TRAPZD(FUNC,A,B,ST,J)
    S=(4.*ST-OST)/3.            Compare equation (4.2.4), above.
    IF (ABS(S-OS).LT.EPS*ABS(OS)) RETURN
    OS=S
    OST=ST
11  CONTINUE
PAUSE 'Too many steps.'
END
```

The routine QSIMP will in general be more efficient than QTRAP (i.e., require fewer function evaluations) when the function to be integrated has a finite 4th derivative (i.e., a continuous 3rd derivative). The combination of QSIMP and its necessary work-horse TRAPZD is a good one for light-duty work where you have to type the routines into an unfamiliar machine.

REFERENCES AND FURTHER READING:

Stoer, J., and Bulirsch, R. 1980, *Introduction to Numerical Analysis* (New York: Springer-Verlag), §3.3.

Dahlquist, Germund, and Bjorck, Ake. 1974, *Numerical Methods* (Englewood Cliffs, N.J.: Prentice-Hall), §§7.4.1-7.4.2.

Forsythe, George E., Malcolm, Michael A., and Moler, Cleve B. 1977, *Computer Methods for Mathematical Computations* (Englewood Cliffs, N.J.: Prentice-Hall), §5.3.

4.3 Romberg Integration

We can view Romberg's method as the natural generalization of the routine QSIMP in the last section to integration schemes which are of higher order than Simpson's rule. The basic idea is to use the results from k successive refinements of the extended trapezoidal rule (implemented in TRAPZD) to remove all terms in the error series up to but not including $O(N^{2k})$. QSIMP is the case of $k = 2$. This is one example of a very general idea which goes by the name of *Richardson's deferred approach to the limit*: perform some numerical algorithm for various values of a parameter h, and then extrapolate the result to the continuum limit $h = 0$.

Equation (4.2.4), which subtracts off the leading error term, is a special case of polynomial extrapolation. In the more general Romberg case, we can use Neville's algorithm (see §3.1) to extrapolate the successive refinements to zero step-size. Neville's algorithm can in fact be coded very concisely within a Romberg integration routine. For clarity of the program, however, it seems better to do the extrapolation by subroutine call to POLINT, already given in §3.1.

```
SUBROUTINE QROMB(FUNC,A,B,SS)
    Returns as S the integral of the function FUNC from A to B. Integration is performed by
    Romberg's method of order 2K, where, e.g., K=2 is Simpson's rule.
PARAMETER (EPS=1.E-6, JMAX=20, JMAXP=JMAX+1, K=5, KM=K-1)
    Here EPS is the fractional accuracy desired, as determined by the extrapolation error
    estimate; JMAX limits the total number of steps; K is the number of points used in the
    extrapolation.
DIMENSION S(JMAXP),H(JMAXP)   These store the successive trapezoidal approximations and their rel-
H(1)=1.                           ative step-sizes.
DO 11 J=1,JMAX
    CALL TRAPZD(FUNC,A,B,S(J),J)
    IF (J.GE.K) THEN
        CALL POLINT(H(J-KM),S(J-KM),K,0.,SS,DSS)
        IF (ABS(DSS).LT.EPS*ABS(SS)) RETURN
    ENDIF
```

```
    S(J+1)=S(J)
    H(J+1)=0.25*H(J)          This is a key step: The factor is 0.25 even though the step-size
11  CONTINUE                      is decreased by only 0.5. This makes the extrapolation a
PAUSE 'Too many steps.'           polynomial in h^2 as allowed by equation (4.2.1), not just a
END                               polynomial in h.
```

The routine QROMB, along with its required TRAPZD and POLINT, is quite powerful for sufficiently smooth (e.g., analytic) integrands, integrated over intervals which contain no singularities, and where the endpoints are also nonsingular. QROMB, in such circumstances, takes many, *many* fewer function evaluations than either of the routines in §4.2. For example, the integral

$$\int_0^2 x^4 \log(x + \sqrt{x^2 + 1})dx$$

converges (with parameters as shown above) on the very first extrapolation, after just 5 calls to TRAPZD, while QSIMP requires 8 calls (8 times as many evaluations of the integrand) and QTRAP requires 13 calls (making 256 times as many evaluations of the integrand).

REFERENCES AND FURTHER READING:

Stoer, J., and Bulirsch, R. 1980, *Introduction to Numerical Analysis* (New York: Springer-Verlag), §§3.4-3.5.

Dahlquist, Germund, and Bjorck, Ake. 1974, *Numerical Methods* (Englewood Cliffs, N.J.: Prentice-Hall), §§7.4.1-7.4.2.

Ralston, Anthony, and Rabinowitz, Philip. 1978, *A First Course in Numerical Analysis*, 2nd ed. (New York: McGraw-Hill), §4.10-2.

4.4 Improper Integrals

For our present purposes, an integral will be "improper" if it has any of the following problems:

- its integrand goes to a finite limiting value at finite upper and lower limits, but cannot be evaluated *right on* one of those limits [e.g., $\sin(x)/x$ at $x = 0$]
- its upper limit is ∞ , or its lower limit is $-\infty$
- it has an integrable singularity at either limit (e.g., $x^{-1/2}$ at $x = 0$)
- it has an integrable singularity at a known place between its upper and lower limits

- it has an integrable singularity at an unknown place between its upper and lower limits

If an integral is infinite (e.g. $\int_1^\infty x^{-1}dx$), or does not exist in a limiting sense [e.g. $\int_{-\infty}^{\infty} \cos(x)dx$], we do not call it improper; we call it impossible. No amount of clever algorithmics will return a meaningful answer to an ill-posed problem.

In this section we will generalize the techniques of the preceding two sections to cover the first four problems on the above list. The fifth problem, singularity at unknown location, can really only be handled by the use of a variable-step-size differential equation integration routine, as will be given in Chapter 15.

We need a work-horse like the extended trapezoidal rule (equation 4.1.11), but one which is an *open* formula in the sense of §4.1., i.e. does not require the integrand to be evaluated at the endpoints. Equation (4.1.19), the extended midpoint rule, is the best choice. The reason is that (4.1.19) shares with (4.1.11) the "deep" property of having an error series that is entirely even in h. Indeed there is a formula, not as well known as it ought to be, called the *Second Euler-Maclaurin summation formula*,

$$\begin{aligned}\int_{x_1}^{x_N} f(x)dx = h[f_{3/2} + f_{5/2} + f_{7/2} + \cdots + f_{N-3/2} + f_{N-1/2}]\\ - \frac{B_2 h^2}{4}(f'_N - f'_1) - \cdots \\ - \frac{B_{2k} h^{2k}}{(2k)!}(1 - 2^{-2k+1})(f_N^{(2k-1)} - f_1^{(2k-1)})\end{aligned} \tag{4.4.1}$$

This equation can be derived by writing out (4.2.1) with step-size h, then writing it out again with step-size $h/2$, then subtracting twice the second from the first.

It is not possible to double the number of steps in the extended midpoint rule and still have the benefit of previous function evaluations (try it!). However, it is possible to ***triple*** the number of steps and do so. Shall we do this, or double and accept the loss? On the average, tripling does a factor $\sqrt{3}$ of unnecessary work, since the "right" number of steps for a desired accuracy criterion may in fact fall anywhere in the logarithmic interval implied by tripling. For doubling, the factor is only $\sqrt{2}$, but we lose an extra factor of 2 in being unable to use all the previous evaluations. Since $1.732 < 2 \times 1.414$, it is better to triple.

Here is the resulting routine, which is directly comparable to TRAPZD.

```
SUBROUTINE MIDPNT(FUNC,A,B,S,N)
    This routine computes the N'th stage of refinement of an extended midpoint rule. FUNC is
    input as the name of the function to be integrated between limits A and B, also input. When
    called with N=1, the routine returns as S the crudest estimate of ∫_a^b f(x)dx. Subsequent
    calls with N=2,3,... (in that sequential order) will improve the accuracy of S by adding
    (2/3) × 3^(N-1) additional interior points. S should not be modified between sequential calls.
IF (N.EQ.1) THEN
  S=(B-A)*FUNC(0.5*(A+B))
```

```
    IT=1                          2×IT points will be added on the next refinement.
ELSE
    TNM=IT
    DEL=(B-A)/(3.*TNM)
    DDEL=DEL+DEL                  The added points alternate in spacing between DEL and DDEL.
    X=A+0.5*DEL
    SUM=0.
    DO 11 J=1,IT
        SUM=SUM+FUNC(X)
        X=X+DDEL
        SUM=SUM+FUNC(X)
        X=X+DEL
11  CONTINUE
    S=(S+(B-A)*SUM/TNM)/3.        The new sum is combined with the old integral to give a refined
    IT=3*IT                           integral.
ENDIF
RETURN
END
```

The routine MIDPNT can exactly replace TRAPZD in a driver routine like QTRAP (§4.2); one simply changes CALL TRAPZD to CALL MIDPNT, and perhaps also decreases the parameter JMAX since $3^{\text{JMAX}-1}$ (from step tripling) is a much larger number than $2^{\text{JMAX}-1}$ (step doubling).

The open formula implementation analogous to Simpson's rule (QSIMP in §4.2) substitutes MIDPNT for TRAPZD and decreases JMAX as above, but now also changes the extrapolation step to be

```
    S=(9.*ST-OST)/8.
```

since, when the number of steps is tripled, the error decreases to $1/9^{th}$ its size, not $1/4^{th}$ as with step doubling.

Either the modified QTRAP or the modified QSIMP will fix the first problem on the list at the beginning of this section. Yet more sophisticated is to generalize Romberg integration in like manner:

```
SUBROUTINE QROMO(FUNC,A,B,SS,CHOOSE)
    Romberg integration on an open interval. Returns as SS the integral of the function FUNC
    from A to B, using any specified integrating subroutine CHOOSE and Romberg's method.
    Normally CHOOSE will be an open formula, not evaluating the function at the endpoints. It
    is assumed that CHOOSE triples the number of steps on each call, and that its error series
    contains only even powers of the number of steps. The routines MIDPNT, MIDINF, MIDSQL,
    MIDSQU, are possible choices for CHOOSE.
PARAMETER (EPS=1.E-6, JMAX=14, JMAXP=JMAX+1, K=5, KM=K-1)
    The parameters have the same meaning as in QROMB.
DIMENSION S(JMAXP),H(JMAXP)
H(1)=1.
DO 11 J=1,JMAX
    CALL CHOOSE(FUNC,A,B,S(J),J)
    IF (J.GE.K) THEN
        CALL POLINT(H(J-KM),S(J-KM),K,0.,SS,DSS)
        IF (ABS(DSS).LT.EPS*ABS(SS)) RETURN
    ENDIF
    S(J+1)=S(J)
    H(J+1)=H(J)/9.                This is where the assumption of step tripling and an even error series
11  CONTINUE                          is used.
PAUSE 'Too many steps.'
END
```

The differences between QROMO and QROMB (§4.3) are so slight that it is perhaps gratuitous to list QROMO in full. It, however, is an excellent driver routine for solving all the other problems of improper integrals in our first list (except the intractable fifth), as we shall now see.

The basic trick for improper integrals is to make a change of variables to eliminate the singularity, or to map an infinite range of integration to a finite one. For example, the identity

$$\int_a^b f(x)dx = \int_{1/b}^{1/a} \frac{1}{t^2} f\left(\frac{1}{t}\right) dt \qquad ab > 0 \tag{4.4.2}$$

can be used with *either* $b \to \infty$ and a positive, *or* with $a \to -\infty$ and b negative, and works for any function which decreases towards infinity at least as fast as $1/x^2$.

You can make the change of variable implied by (4.4.2) either analytically and then use (e.g.) QROMO and MIDPNT to do the numerical evaluation, *or* you can let the numerical algorithm make the change of variable for you. We prefer the latter method as being more transparent to the user. To implement equation (4.4.2) we simply write a modified version of MIDPNT, called MIDINF, which allows b to be infinite (or, more precisely, a very large number on your particular machine, such as 1×10^{30}), or a to be negative infinite.

```
SUBROUTINE MIDINF(FUNK,AA,BB,S,N)
    This routine is an exact replacement for MIDPNT, i.e. returns as S the N^th stage of
    refinement of the integral of FUNK from AA to BB, except that the function is evaluated at
    evenly spaced points in 1/x rather than in x. This allows the upper limit BB to be as large
    and positive as the computer allows, or the lower limit AA to be as large and negative,
    but not both. AA and BB must have the same sign.
FUNC(X)=FUNK(1./X)/X**2          This is a statement function which effects the change of variable.
B=1./AA                                These two statements change the limits of integration ac-
A=1./BB                                cordingly.
IF (N.EQ.1) THEN                 From this point on, the routine is exactly identical to MIDPNT.
    S=(B-A)*FUNC(0.5*(A+B))
    IT=1
ELSE
    TNM=IT
    DEL=(B-A)/(3.*TNM)
    DDEL=DEL+DEL
    X=A+0.5*DEL
    SUM=0.
    DO 11 J=1,IT
        SUM=SUM+FUNC(X)
        X=X+DDEL
        SUM=SUM+FUNC(X)
        X=X+DEL
 11 CONTINUE
    S=(S+(B-A)*SUM/TNM)/3.
    IT=3*IT
ENDIF
RETURN
END
```

If you need to integrate from a negative lower limit to positive infinity, you do this by breaking the integral into two pieces at some positive value, for example,

```
CALL QROMO(FUNK,-5.,2.,S1,MIDPNT)
CALL QROMO(FUNK,2.,1.E30,S2,MIDINF)
ANSWER=S1+S2
```

Where should you choose the breakpoint? At a sufficiently large positive value so that the function FUNK is at least beginning to approach its asymptotic decrease to zero value at infinity. The polynomial extrapolation implicit in the second call to QROMO deals with a polynomial in $1/x$, not in x.

To deal with an integral that has an integrable power-law singularity at its lower limit, one also makes a change of variable. If the integrand diverges as $(x-a)^\gamma$, $0 \le \gamma < 1$, near $x = a$, use the identity

$$\int_a^b f(x)dx = \frac{1}{1-\gamma}\int_0^{(b-a)^{1-\gamma}} t^{\frac{\gamma}{1-\gamma}} f(t^{\frac{1}{1-\gamma}} + a)dt \qquad (b > a) \tag{4.4.3}$$

If the singularity is at the upper limit, use the identity

$$\int_a^b f(x)dx = \frac{1}{1-\gamma}\int_0^{(b-a)^{1-\gamma}} t^{\frac{\gamma}{1-\gamma}} f(b - t^{\frac{1}{1-\gamma}})dt \qquad (b > a) \tag{4.4.4}$$

If there is a singularity at both limits, divide the integral at an interior breakpoint as in the example above.

Equations (4.4.3) and (4.4.4) are particularly simple in the case of inverse square-root singularities, a case that occurs frequently in practice:

$$\int_a^b f(x)dx = \int_0^{\sqrt{b-a}} 2tf(a+t^2)dt \qquad (b > a) \tag{4.4.5}$$

for a singularity at a, an

$$\int_a^b f(x)dx = \int_0^{\sqrt{b-a}} 2tf(b-t^2)dt \qquad (b > a) \tag{4.4.6}$$

for a singularity at b. Once again, we can implement these changes of variable transparently to the user by defining substitute routines for MIDPNT which make the change of variable automatically:

```
      SUBROUTINE MIDSQL(FUNK,AA,BB,S,N)
```
This routine is an exact replacement for MIDPNT, except that it allows for an inverse square-root singularity in the integrand at the lower limit AA.
```
      FUNC(X)=2.*X*FUNK(AA+X**2)
      B=SQRT(BB-AA)
      A=0.
      IF (N.EQ.1) THEN
```
The rest of the routine is exactly like MIDPNT and is omitted.

Exactly similarly,

```
      SUBROUTINE MIDSQU(FUNK,AA,BB,S,N)
```
This routine is an exact replacement for MIDPNT, except that it allows for an inverse square-root singularity in the integrand at the upper limit BB.
```
      FUNC(X)=2.*X*FUNK(BB-X**2)
      B=SQRT(BB-AA)
      A=0.
      IF (N.EQ.1) THEN
```
The rest of the routine is exactly like MIDPNT and is omitted.

One last example should suffice to show how these formulas are derived in general. Suppose the upper limit of integration is infinite, and the integrand falls off exponentially. Then we want a change of variable that maps $e^{-x}dx$ into $(\pm)dt$ (with the sign chosen to keep the upper limit of the new variable larger than the lower limit). Doing the integration gives by inspection

$$t = e^{-x} \qquad \text{or} \qquad x = -\log t \tag{4.4.7}$$

so that

$$\int_{x=a}^{x=\infty} f(x)dx = \int_{t=0}^{t=e^{-a}} f(-\log t)\frac{dt}{t} \quad . \tag{4.4.8}$$

The user-transparent implementation would be

```
      SUBROUTINE MIDEXP(FUNK,AA,BB,S,N)
```
This routine is an exact replacement for MIDPNT, except that BB is assumed to be infinite (value passed not actually used). It is assumed that the function FUNK decreases exponentially rapidly at infinity.
```
      FUNC(X)=FUNK(-ALOG(X))/X
      B=EXP(-AA)
      A=0.
      IF (N.EQ.1) THEN
```
The rest of the routine is exactly like MIDPNT and is omitted.

REFERENCES AND FURTHER READING:

Acton, Forman S. 1970, *Numerical Methods That Work* (New York: Harper and Row), Chapter 4.

Dahlquist, Germund, and Bjorck, Ake. 1974, *Numerical Methods* (Englewood Cliffs, N.J.: Prentice-Hall), §7.4.3, p.294.

Stoer, J., and Bulirsch, R. 1980, *Introduction to Numerical Analysis* (New York: Springer-Verlag), §3.7, p.152.

4.5 Gaussian Quadratures

In the formulas of §4.1, the integral of a function was approximated by the sum of its functional values at a set of equally spaced points, multiplied by certain aptly chosen weighting coefficients. We saw that as we allowed ourselves more freedom in choosing the coefficients, we could achieve integration formulas of higher and higher order. The idea of *Gaussian quadratures* is to give ourselves the freedom to choose not only the weighting coefficients, but also the location of the abscissas at which the function is to be evaluated: they will no longer be equally spaced. Thus, we will have *twice* the number of degrees of freedom at our disposal; it will turn out that we can achieve Gaussian quadrature formulas whose order is, essentially, twice that of the Newton-Cotes formula with the same number of function evaluations.

Does this sound too good to be true? Well, in a sense it is. The catch is a familiar one, which cannot be overemphasized: high order is not the same as high accuracy. High order translates to high accuracy only when the integrand is very smooth, in the sense of being "well-approximated by a polynomial."

There is, however, one additional feature of Gaussian quadrature formulas which adds to their usefulness: We can arrange the choice of weights and abscissas to make the integral exact for a class of integrands "polynomials times some known function $W(x)$" rather than for the usual class of integrands "polynomials." The function $W(x)$ can then be chosen to remove integrable singularities from the desired integral. Given $W(x)$, in other words, and given an integer N, we can find a set of weights w_i and abscissas x_i such that the approximation

$$\int_a^b W(x)f(x)dx \approx \sum_{i=1}^{N} w_i f(x_i) \tag{4.5.1}$$

is exact if $f(x)$ is a polynomial. For example, to do the integral

$$\int_{-1}^{1} \frac{\exp(-\cos^2 x)}{\sqrt{1-x^2}} dx \tag{4.5.2}$$

(not a very natural looking integral, it must be admitted), we might well be interested in a Gaussian quadrature formula based on the choice

$$W(x) = \frac{1}{\sqrt{1-x^2}} \tag{4.5.3}$$

in the interval $(-1, 1)$. (This particular choice is called *Gauss-Chebyshev integration*, for reasons that will become clear shortly.)

Notice that the integration formula (4.5.1) can also be written with the weight function $W(x)$ not overtly visible: Define $g(x) \equiv W(x)f(x)$ and $v_i \equiv w_i/W(x_i)$. Then (4.5.1) becomes

$$\int_a^b g(x)dx \approx \sum_{i=1}^{N} v_i g(x_i) \tag{4.5.4}$$

Where did the function $W(x)$ go? It is lurking there, ready to give high-order accuracy to integrands of the form polynomials times $W(x)$, and ready to *deny* high-order accuracy to integrands that are otherwise perfectly smooth and well-behaved. When you find tabulations of the weights and abscissas for a given $W(x)$, you have to determine carefully whether they are to be used with a formula in the form of (4.5.1), or like (4.5.4).

Here is an example of a quadrature routine which contains the tabulated abscissas and weights for the case $W(x) = 1$ and $N = 10$. Since the weights and abscissas are, in this case, symmetric around the midpoint of the range of integration, there are actually only five distinct values of each:

```
SUBROUTINE QGAUS(FUNC,A,B,SS)
   Returns as SS the integral of the function FUNC between A and B, by ten-point Gauss-
   Legendre integration: the function is evaluated exactly ten times at interior points in the
   range of integration.
DIMENSION X(5),W(5)              The abscissas and weights.
DATA X/.1488743389,.4333953941,.6794095682,.8650633666,.9739065285/
DATA W/.2955242247,.2692667193,.2190863625,.1494513491,.0666713443/
XM=0.5*(B+A)
XR=0.5*(B-A)
SS=0                             Will be twice the average value of the function, since the ten weights
DO 11 J=1,5                          (five numbers above each used twice) sum to 2.
   DX=XR*X(J)
   SS=SS+W(J)*(FUNC(XM+DX)+FUNC(XM-DX))
11 CONTINUE
SS=XR*SS                         Scale the answer to the range of integration.
RETURN
END
```

The above routine illustrates that one can use Gaussian quadratures without necessarily understanding the theory behind them: one just locates tabulated weights and abscissas in a book (e.g. Abramowitz and Stegun). However, the theory is very pretty, and it will come in handy if you ever need to construct your own tabulation of weights and abscissas for an unusual choice of $W(x)$. We will therefore give, without any proofs, some useful results that will enable you to do this.

Finally, we will include a routine for computing the weights and abscissas for the most common and useful case, namely the simple choice $W(x) \equiv 1$. This special case is most accurately termed *Gauss-Legendre integration*, but it is often (confusingly) called simply *Gaussian integration*. These weights and abscissas were, e.g. used above in QGAUS.

The theory behind Gaussian quadratures is closely tied to that of orthogonal polynomials. Let us fix the interval of interest to (a, b). We can define the "scalar product of two functions f and g over a weight function W" as

$$\langle f|g\rangle \equiv \int_a^b W(x)f(x)g(x)dx \tag{4.5.5}$$

The scalar product is a number, not a function of x. Two functions are said to be *orthogonal* if their scalar product is zero. A function is said to be *normalized* if its scalar product with itself is unity. A set of functions that are all mutually orthogonal and also all individually normalized is called an *orthonormal* set.

We can find a set of polynomials (i) that includes exactly one polynomial of order j, called $p_j(x)$, for each $j = 0, 1, 2, \ldots$, and (ii) all of which are mutually orthogonal over the specified weight function $W(x)$. A constructive procedure for finding such a set is the recurrence relation

$$\begin{aligned} p_0(x) &\equiv 1 \\ p_{i+1}(x) &= \left[x - \frac{\langle xp_i|p_i\rangle}{\langle p_i|p_i\rangle}\right] p_i(x) - \left[\frac{\langle p_i|p_i\rangle}{\langle p_{i-1}|p_{i-1}\rangle}\right] p_{i-1}(x) \end{aligned} \tag{4.5.6}$$

plus the special rule that the second term of (4.5.6) be omitted for the case of $i = 0$ (there is no p_{-1}).

The polynomials defined by (4.5.6) are *monic*, i.e. the coefficient of their leading term [x^j for $p_j(x)$] is unity. If we divide each $p_j(x)$ by the constant $[\langle p_j|p_j\rangle]^{1/2}$ we can render the set of polynomials orthonormal. One also encounters orthogonal polynomials with various other normalizations.

The polynomial $p_j(x)$ can be shown to have exactly j distinct roots in the interval (a, b). Moreover, it can be shown that the roots of $p_j(x)$ "interleave" the $j - 1$ roots of $p_{j-1}(x)$, i.e. there is exactly one root of the former in between each two adjacent roots of the latter. This fact comes in handy if you need to find all the roots: you can start with the one root of $p_1(x)$ and then, in turn, bracket the roots of each higher j, pinning them down at each stage more precisely by Newton's rule or some other root-finding scheme (see Chapter 9).

Why would you ever want to find all the roots of an orthogonal polynomial $p_j(x)$? Because the abscissas of the N-point Gaussian quadrature formulas (4.5.1) and (4.5.4) with weighting function $W(x)$ in the interval (a, b) are precisely the roots of the orthogonal polynomial $p_N(x)$ for the same interval and weighting function. This is the fundamental theorem of Gaussian quadratures, and lets you find the abscissas for any particular case.

Once you know the abscissas $x_1 \dots x_N$, you need to find the weights w_i, $i = 1 \dots N$. One way to do this is to solve the set of linear equations

$$\begin{bmatrix} p_0(x_1) & \dots & p_0(x_N) \\ p_1(x_1) & \dots & p_1(x_N) \\ \vdots & & \vdots \\ p_{N-1}(x_1) & \dots & p_{N-1}(x_N) \end{bmatrix} \begin{bmatrix} w_1 \\ w_2 \\ \vdots \\ w_N \end{bmatrix} = \begin{bmatrix} \int_a^b W(x)p_0(x)dx \\ 0 \\ \vdots \\ 0 \end{bmatrix} \tag{4.5.7}$$

Equation (4.5.7) simply solves for those weights such that the quadrature (4.5.1) gives the correct answer for the integral of the first $N-1$ orthogonal polynomials. It can be shown that, with those weights, the integral of the *next* N polynomials is also exact, so that the quadrature is exact for all polynomials of degree $2N-1$ or less.

There is another, trickier, way to find the weights w_i, based on some deeper theorems. Let us define a new sequence of polynomials $\phi_j(x)$ by the following recurrence:

$$\begin{aligned} \phi_0(x) &\equiv 0 \\ \phi_1(x) &\equiv p_1' \int_a^b W(x)dx \\ \phi_{i+1} &= \left[x - \frac{\langle xp_i|p_i\rangle}{\langle p_i|p_i\rangle} \right] \phi_i(x) - \left[\frac{\langle p_i|p_i\rangle}{\langle p_{i-1}|p_{i-1}\rangle} \right] \phi_{i-1}(x) \end{aligned} \tag{4.5.8}$$

Here p_1' is the derivative of $p_1(x)$ with respect to x, a constant since p_1 is linear. Notice that (4.5.6) says that the ϕ_j's obey *exactly the same recurrence relation* as the p_j's, but with different starting values. Also notice that ϕ_j is of degree one lower than the corresponding p_j. If you know the coefficients of the recurrence relation that generates the p_j's (that is, you know the various scalar products of p_i's that appear in [4.5.6]), then you can easily evaluate $\phi_j(x)$ from (4.5.8) for any desired x and j, by recurrence. Now the wonderful result for the weights of the N-point Gaussian quadrature is

$$w_i = \frac{\phi_N(x_i)}{p_N'(x_i)} \qquad i = 1, \dots, N \tag{4.5.9}$$

Since any constant factor common to both p_N and ϕ_N will cancel in (4.5.9), that equation also holds when the recurrence for orthonormal polynomials, or any other normalization, is used instead of the recurrence for the corresponding monic polynomials, (4.5.6) or (4.5.8).

Here are some of the intervals, weighting functions, and recurrence relations which generate commonly used orthogonal polynomials and their corre-

sponding Gaussian quadrature formulas:

(a,b)	$W(x)$	Recurrence	Gauss–
$(-1,1)$	1	$(i+1)P_{i+1} = (2i+1)xP_i - iP_{i-1}$	Legendre
$(-1,1)$	$(1-x^2)^{-1/2}$	$T_{i+1} = 2xT_i - T_{i-1}$	Chebyshev
$(0,\infty)$	$x^c e^{-x}$	$(i+1)L^c_{i+1} = (-x+2i+c+1)L^c_i - (i+c)L^c_{i-1}$	Laguerre $c = 0, 1, \ldots$
$(-\infty,\infty)$	e^{-x^2}	$H_{i+1} = 2xH_i - iH_{i-1}$	Hermite

Tabulations of the abscissas and weights for these, and other, possibilities can be found in standard references.

Here is an example of a routine for calculating one set of abscissas and weights, those of Gauss-Legendre. The routine, due to G. Rybicki, has three departures from the general methods for arbitrary weight functions that we have just discussed: (i) Instead of bracketing the roots by their "interleaving" property, it uses a clever approximation to jump directly to the neighborhood of the desired root, where it converges by Newton's method (to be discussed in §9.4). (ii) Instead of using equation (4.5.9) to find the weights, it uses a special formula that holds for the Gauss-Legendre case,

$$w_i = \frac{2}{(1-x_i^2)[P'_N(x_i)]^2} \tag{4.5.10}$$

(iii) The routine scales the range of integration from (x_1, x_2) to $(-1, 1)$, and provides abscissas x_i and weights w_i for the Gaussian formula

$$\int_{x_1}^{x_2} f(x)dx = \sum_{i=1}^{N} w_i f(x_i) \tag{4.5.11}$$

```
      SUBROUTINE GAULEG(X1,X2,X,W,N)
         Given the lower and upper limits of integration X1 and X2, and given N, this routine returns
         arrays X and W of length N, containing the abscissas and weights of the Gauss-Legendre
         N-point quadrature formula.
      IMPLICIT REAL*8 (A-H,O-Z)      High precision is a good idea for this routine.
      REAL*4 X1,X2,X(N),W(N)
      PARAMETER (EPS=3.D-14)         Increase if you don't have this floating precision.
      M=(N+1)/2                      The roots are symmetric in the interval, so we only have to find half
      XM=0.5D0*(X2+X1)                   of them.
      XL=0.5D0*(X2-X1)
      DO 12 I=1,M                    Loop over the desired roots.
         Z=COS(3.141592654D0*(I-.25D0)/(N+.5D0))
         Starting with the above approximation to the Ith root, we enter the main loop of refinement
         by Newton's method.
1        CONTINUE
            P1=1.D0
            P2=0.D0
            DO 11 J=1,N              Loop up the recurrence relation to get the Legendre polynomial eval-
                                         uated at Z.
```

```
          P3=P2
          P2=P1
          P1=((2.D0*J-1.D0)*Z*P2-(J-1.D0)*P3)/J
11      CONTINUE
```

P1 is now the desired Legendre polynomial. We next compute PP, its derivative, by a standard relation involving also P2, the polynomial of one lower order.

```
        PP=N*(Z*P1-P2)/(Z*Z-1.D0)
        Z1=Z
        Z=Z1-P1/PP                    Newton's method.
      IF(ABS(Z-Z1).GT.EPS)GO TO 1
      X(I)=XM-XL*Z                    Scale the root to the desired interval,
      X(N+1-I)=XM+XL*Z                and put in its symmetric counterpart.
      W(I)=2.D0*XL/((1.D0-Z*Z)*PP*PP)       Compute the weight
      W(N+1-I)=W(I)                   and its symmetric counterpart.
12    CONTINUE
      RETURN
      END
```

REFERENCES AND FURTHER READING:

Abramowitz, Milton, and Stegun, Irene A. 1964, *Handbook of Mathematical Functions*, Applied Mathematics Series, vol. 55 (Washington: National Bureau of Standards; reprinted 1968 by Dover Publications, New York), §25.4.

Stoer, J., and Bulirsch, R. 1980, *Introduction to Numerical Analysis* (New York: Springer-Verlag), §3.6.

Johnson, Lee W., and Riess, R. Dean. 1982, *Numerical Analysis*, 2nd ed. (Reading, Mass.: Addison-Wesley), §6.5.

Carnahan, Brice, Luther, H.A., and Wilkes, James O. 1969, *Applied Numerical Methods* (New York: Wiley), §§2.9-2.10.

Ralston, Anthony, and Rabinowitz, Philip. 1978, *A First Course in Numerical Analysis*, 2nd ed. (New York: McGraw-Hill), §§4.4-4.8.

4.6 Multidimensional Integrals

Integrals of functions of several variables, over regions with dimension greater than one, are *not easy*. There are two reasons for this. First, the number of function evaluations needed to sample an N-dimensional space increases as the Nth power of the number needed to do a one-dimensional integral. If you need 30 function evaluations to do a one-dimensional integral crudely, then you will likely need on the order of 30000 evaluations to reach the same crude level for a 3-dimensional integral. Second, the region of integration in N-dimensional space is defined by an $N-1$ dimensional boundary which can itself be terribly complicated: it need not be convex or simply connected, for example. By contrast, the boundary of a one-dimensional integral consists of two numbers, its upper and lower limits.

The first question to be asked, when faced with a multidimensional integral, is, "can it be reduced analytically to a lower dimensionality?" For

example, so-called *iterated integrals* of a function of one variable $f(t)$ can be reduced to one-dimensional integrals by the formula

$$\begin{aligned}\int_0^x dt_n \int_0^{t_n} dt_{n-1} \cdots \int_0^{t_3} dt_2 \int_0^{t_2} f(t_1)dt_1 \\ = \frac{1}{(n-1)!}\int_0^x (x-t)^{n-1} f(t)dt\end{aligned} \tag{4.6.1}$$

Alternatively, the function may have some special symmetry in the way it depends on its independent variables. If the boundary also has this symmetry, then the dimension can be reduced. In three dimensions, for example, the integration of a spherically-symmetric function over a spherical region reduces, in polar coordinates, to a one-dimensional integral.

The next questions to be asked will guide your choice between two entirely different approaches to doing the problem. The questions are: Is the shape of the boundary of the region of integration simple or complicated? Inside the region, is the integrand smooth and simple, or complicated, or locally strongly peaked? Does the problem require high accuracy, or does it require an answer accurate only to a percent, or a few percent?

If your answers are that the boundary is complicated, the integrand is *not* strongly peaked in very small regions, and relatively low accuracy is tolerable, then your problem is a good candidate for *Monte Carlo integration.* This method is also very straightforward to program, especially in its cruder forms. One needs only to know a region with simple boundaries that *includes* the complicated region of integration, plus a method of determining whether a random point is inside or outside the region of integration. Monte Carlo integration evaluates the function at a random sample of points, and estimates its integral based on that random sample. We will discuss it in more detail, and with more sophistication, in Chapter 7.

If the boundary is simple, and the function is very smooth, then the remaining approach, breaking up the problem into repeated one-dimensional integrals, will be effective and relatively fast. If you require high accuracy, this approach is in any case the *only* one available to you, since Monte Carlo methods are by nature asymptotically slow to converge.

For low accuracy, use repeated one-dimensional integration when the integrand is slowly varying and smooth in the region of integration, Monte Carlo when the integrand is oscillatory or discontinuous, but not strongly peaked in small regions.

If the integrand *is* strongly peaked in small regions, and you know where those regions are, break the integral up into several regions so that the integrand is smooth in each, and do each separately. If you don't know where the strongly peaked regions are, you might as well (at the level of sophistication of this book) quit: It is hopeless to expect an integration routine to search out unknown pockets of large contribution in a huge N-dimensional space.

If, on the basis of the above guidelines, you decide to pursue the repeated one-dimensional integration approach, here is how it works. For definiteness,

we will consider the case of a three-dimensional integral in x, y, z-space. Two dimensions, or more than three dimensions are entirely analogous.

The first step is to specify the region of integration by (i) its lower and upper limits in x, which we will denote x_1 and x_2; (ii) its lower and upper limits in y at a specified value of x, denoted $y_1(x)$ and $y_2(x)$; and (iii) its lower and upper limits in z at specified x and y, denoted $z_1(x,y)$ and $z_2(x,y)$. In other words, find the numbers x_1 and x_2, and the functions $y_1(x), y_2(x), z_1(x,y)$, and $z_2(x,y)$ such that

$$\begin{aligned} I &\equiv \int\int\int dx dy dz f(x,y,z) \\ &= \int_{x_1}^{x_2} dx \int_{y_1(x)}^{y_2(x)} dy \int_{z_1(x,y)}^{z_2(x,y)} dz\ f(x,y,z) \end{aligned} \tag{4.6.2}$$

For example, a two-dimensional integral over a circle of radius one centered on the origin becomes

$$\int_{-1}^{1} dx \int_{-\sqrt{1-x^2}}^{\sqrt{1-x^2}} dy\ f(x,y) \tag{4.6.3}$$

Now we can define a function $G(x,y)$ that does the innermost integral,

$$G(x,y) \equiv \int_{z_1(x,y)}^{z_2(x,y)} f(x,y,z)dz \tag{4.6.4}$$

and a function $H(x)$ that does the integral of $G(x,y)$,

$$H(x) \equiv \int_{y_1(x)}^{y_2(x)} G(x,y)dy \tag{4.6.5}$$

and finally our answer as an integral over $H(x)$

$$I = \int_{x_1}^{x_2} H(x)dx \tag{4.6.6}$$

To implement equations (4.6.4) – (4.6.6) in a program, one needs three separate copies of a basic one-dimensional integration routine (and of any subroutines called by it), one each for the x, y, and z integrations. If you try to make do with only one copy, then it will call itself recursively, since (e.g.) the function evaluations of H for the x integration will themselves call the integration routine to do the y integration (see Figure 4.6.1). In our example, let us suppose that we plan to use the one-dimensional integrator `QGAUS` of

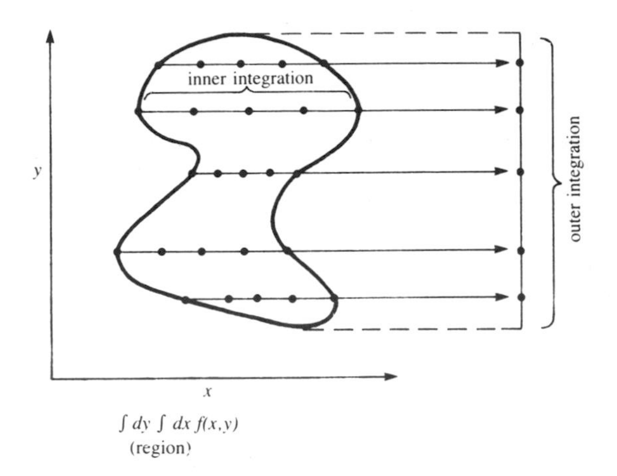

Figure 4.6.1. Function evaluations for a two-dimensional integral over an irregular region, shown schematically. The outer integration routine, in y, requests values of the inner, x, integral at locations along the y axis of its own choosing. The inner integration routine then evaluates the function at x locations suitable to *it*. This is more accurate in general than, e.g., evaluating the function on a Cartesian mesh of points.

§4.5. Then we make three identical copies and call them QGAUSX, QGAUSY, and QGAUSZ. The basic program for three-dimensional integration then is as follows:

```
SUBROUTINE QUAD3D(X1,X2,SS)
   Returns as SS the integral of a user-supplied function FUNC over a three-dimensional region
   specified by the limits X1, X2, and by the user-supplied functions Y1, Y2, Z1, and Z2,
   as defined in (4.6.2).
EXTERNAL H
CALL QGAUSX(H,X1,X2,SS)
RETURN
END


FUNCTION F(ZZ)                       Called by QGAUSZ. Calls FUNC.
COMMON /XYZ/ X,Y,Z
Z=ZZ
F=FUNC(X,Y,Z)
RETURN
END

FUNCTION G(YY)                       Called by QGAUSY. Calls QGAUSZ.
EXTERNAL F
COMMON /XYZ/ X,Y,Z
Y=YY
CALL QGAUSZ(F,Z1(X,Y),Z2(X,Y),SS)
G=SS
RETURN
END

FUNCTION H(XX)                       Called by QGAUSX. Calls QGAUSY.
EXTERNAL G
COMMON /XYZ/ X,Y,Z
X=XX
CALL QGAUSY(G,Y1(X),Y2(X),SS)
H=SS
```

```
RETURN
END
```

The necessary user-supplied functions have the following calling sequences:

```
FUNCTION FUNC(X,Y,Z)          The 3-dimensional function to be integrated

FUNCTION Y1(X)
FUNCTION Y2(X)
FUNCTION Z1(X,Y)
FUNCTION Z2(X,Y)
```

REFERENCES AND FURTHER READING:

Dahlquist, Germund, and Bjorck, Ake. 1974, *Numerical Methods* (Englewood Cliffs, N.J.: Prentice-Hall), §7.7, p.318.

Johnson, Lee W., and Riess, R. Dean. 1982, *Numerical Analysis*, 2nd ed. (Reading, Mass.: Addison-Wesley), §6.2.5, p.307.

Abramowitz, Milton, and Stegun, Irene A. 1964, *Handbook of Mathematical Functions*, Applied Mathematics Series, vol. 55 (Washington: National Bureau of Standards; reprinted 1968 by Dover Publications, New York), equations 25.4.58 ff.

Chapter 5. Evaluation of Functions

5.0 Introduction

The purpose of this chapter is to acquaint you with a selection of the techniques which are frequently used in evaluating functions. In Chapter 6, we will apply and illustrate these techniques by giving routines for a variety of specific functions. The purposes of this chapter and the next are thus mostly in harmony, but there is nevertheless some tension between them: Routines that are clearest and most illustrative of the general techniques of this chapter are not always the methods of choice for a particular special function. By comparing this chapter to the next one, you should get some idea of the balance between "general" and "special" methods that occurs in practice.

Insofar as that balance favors general methods, this chapter should give you ideas about how to write your own routines for the evaluation of a function which, while "special" to you, is not so special as to be included in the Chapter 6 or the standard program libraries.

REFERENCES AND FURTHER READING:

Fike, C.T. 1968, *Computer Evaluation of Mathematical Functions* (Englewood Cliffs, N.J.: Prentice-Hall).

Lanczos, Cornelius. 1956, *Applied Analysis* (Englewood Cliffs, N.J.: Prentice-Hall), Chapter 7.

5.1 Series and Their Convergence

Everybody knows that an analytic function can be expanded in the neighborhood of a point x_0 in a power series,

$$f(x) = \sum_{k=0}^{\infty} a_k (x - x_0)^k \tag{5.1.1}$$

Such series are straightforward to evaluate. You don't, of course, evaluate the kth power of $x - x_0$ *ab initio* for each term; rather you keep the $k - 1^{st}$ power and update it with a multiply. Similarly, the form of the coefficients a is often such as to make use of previous work: terms like $k!$ or $(2k)!$ can be updated in a multiply or two.

How do you know when you have summed enough terms? In practice, the terms had better be getting small fast, otherwise the series is not a good technique to use in the first place. While not mathematically rigorous in all cases, standard practice is to quit when the term you have just added is smaller in magnitude than some small ϵ times the magnitude of the sum thus far accumulated. (But watch out if isolated instances of $a_k = 0$ are possible!).

A weakness of a power series representation is that it is guaranteed *not* to converge farther than that distance from x_0 at which a singularity is encountered *in the complex plane.* This catastrophe is not usually unexpected: When you find a power series in a book (or when you work one out yourself), you will generally also know the radius of convergence. An insidious problem occurs with series that converge everywhere (in the mathematical sense), but almost nowhere fast enough to be useful in a numerical method. Two familiar examples are the sine function and the Bessel function of the first kind,

$$\sin(x) = \sum_{k=0}^{\infty} \frac{(-1)^k}{(2k+1)!} x^{(2k+1)} \tag{5.1.2}$$

$$J_n(x) = \left(\frac{x}{2}\right)^n \sum_{k=0}^{\infty} \frac{(-\frac{1}{4}x^2)^k}{k!(k+n)!} \tag{5.1.3}$$

Both of these series converge for all x. But both don't even start to converge until $k \gg |x|$; before this, their terms are increasing. This makes these series rather useless for large x.

Accelerating the Convergence of Series

There are several tricks for accelerating the rate of convergence of a series (or, equivalently, of a sequence of partial sums). These tricks will *not*

generally help in cases like (5.1.2) or (5.1.3) while the size of the terms is still increasing. For series with terms of decreasing magnitude, however, some accelerating methods can be startlingly good. *Aitken's δ^2-process* is simply a formula for extrapolating the partial sums of a series whose convergence is approximately geometric. If S_{n-1}, S_n, S_{n+1} are three successive partial sums, then an improved estimate is

$$S'_n \equiv S_{n+1} - \frac{(S_{n+1} - S_n)^2}{S_{n+1} - 2S_n + S_{n-1}} \tag{5.1.4}$$

You can also use (5.1.4) with $n+1$ and $n-1$ replaced by $n+p$ and $n-p$ respectively, for any integer p. If you form the sequence of S'_i's, you can apply (5.1.4) a second time to *that* sequence, and so on. (In practice, this iteration will only rarely do much for you after the first stage.) Note that equation (5.1.4) should be computed as written; there exist algebraically equivalent forms that are much more susceptible to roundoff error.

For *alternating series* (where the terms in the sum alternate in sign), *Euler's transformation* can be a powerful tool. Generally it is advisable to do a small number $n-1$ of terms directly, then apply the transformation to the rest of the series beginning with the nth term. The formula (for n even) is

$$\sum_{s=0}^{\infty}(-)^s u_s = u_0 - u_1 + u_2 \ldots - u_{n-1} + \sum_{s=0}^{\infty}\frac{(-)^s}{2^{s+1}}[\Delta^s u_n] \tag{5.1.5}$$

Here Δ is the *forward difference operator*, i.e.

$$\begin{aligned} \Delta u_n &\equiv u_{n+1} - u_n \\ \Delta^2 u_n &\equiv u_{n+2} - 2u_{n+1} + u_n \\ \Delta^3 u_n &\equiv u_{n+3} - 3u_{n+2} + 3u_{n+1} - u_n \qquad \text{etc.} \end{aligned} \tag{5.1.6}$$

Of course you don't actually do the infinite sum on the right hand side of (5.1.5), but only the first, say, p terms, thus requiring the first p differences (5.1.6) obtained from the terms starting at u_n.

Euler's transformation can be applied not only to convergent series. In some cases it will produce accurate answers from the first terms of a series that is formally divergent. It is widely used in the summation of asymptotic series. In this case it is generally wise not to sum farther than where the terms start increasing in magnitude; and you should devise some independent numerical check that the results are meaningful.

There is an elegant and subtle implementation of Euler's transformation due to van Wijngaarden: It incorporates the terms of the original alternating series one at a time, in order. For each incorporation it *either* increases p by 1,

equivalent to computing one further difference (5.1.6); or else *retroactively* increases n by 1, without having to redo all the difference calculations based on the old n value! The decision as to which to increase, n or p, is taken in such a way as to make the convergence most rapid. Van Wijngaarden's technique requires only one vector of saved partial differences. Here is the algorithm:

```
SUBROUTINE EULSUM(SUM,TERM,JTERM,WKSP)
    Incorporates into SUM the JTERM^th term, with value TERM, of an alternating series. SUM is
    input as the previous partial sum, and is output as the new partial sum. The first call
    to this routine, with the first TERM in the series, should be with JTERM=1. On the second
    call, TERM should be set to the second term of the series, with sign opposite to that of
    the first call, and JTERM should be 2. And so on.
DIMENSION WKSP(JTERM)              Workspace, provided by the calling program.
IF(JTERM.EQ.1)THEN                 Initialize:
    NTERM=1                        Number of saved differences in WKSP.
    WKSP(1)=TERM
    SUM=0.5*TERM                   Return first estimate.
ELSE
    TMP=WKSP(1)
    WKSP(1)=TERM
    DO 11 J=1,NTERM-1                  Update saved quantities by van Wijngaarden's algorithm.
        DUM=WKSP(J+1)
        WKSP(J+1)=0.5*(WKSP(J)+TMP)
        TMP=DUM
11  CONTINUE
    WKSP(NTERM+1)=0.5*(WKSP(NTERM)+TMP)
    IF(ABS(WKSP(NTERM+1)).LE.ABS(WKSP(NTERM)))THEN  Favorable to increase p,
        SUM=SUM+0.5*WKSP(NTERM+1)
        NTERM=NTERM+1                               and the table becomes longer.
    ELSE                                            Favorable to increase n,
        SUM=SUM+WKSP(NTERM+1)                       the table doesn't become longer.
    ENDIF
ENDIF
RETURN
END
```

The powerful Euler technique is not directly applicable to a series of positive terms. Occasionally it is useful to convert a series of positive terms into an alternating series, just so that the Euler transformation can be used! Van Wijngaarden has given a transformation for accomplishing this:

$$\sum_{r=1}^{\infty} v_r = \sum_{r=1}^{\infty} (-)^{r-1} w_r \tag{5.1.7}$$

where

$$w_r \equiv v_r + 2v_{2r} + 4v_{4r} + 8v_{8r} + \cdots \tag{5.1.8}$$

Equations (5.1.7) and (5.1.8) replace a simple sum by a two-dimensional sum, each term in (5.1.7) being itself an infinite sum (5.1.8). This may seem a strange way to save on work! Since, however, the indices in (5.1.8) increase tremendously rapidly, as powers of 2, it often requires only a few terms to converge (5.1.8) to extraordinary accuracy. You do, however, need to be able

to compute the v_r's efficiently for "random" values r. The standard "updating" tricks for sequential r's, mentioned above following equation (5.1.1), can't be used.

Sometimes you will want to compute a function from a series representation even when the computation is *not* efficient. For example, you may be using the values obtained to fit the function to an approximating form that you will use subsequently (cf. §5.6). If you are summing very large numbers of slowly convergent terms, pay attention to roundoff errors! In floating-point representation it is more accurate to sum a list of numbers in the order starting with the smallest one, rather than starting with the largest one. It is even better to group terms pairwise, then in pairs of pairs, etc., so that all additions involve operands of comparable magnitude.

REFERENCES AND FURTHER READING:

Goodwin, E.T. (ed.) 1961, *Modern Computing Methods*, 2nd ed. (New York: Philosophical Library), Chapter 13 [van Wijngaarden's transformations].

Dahlquist, Germund, and Bjorck, Ake. 1974, *Numerical Methods* (Englewood Cliffs, N.J.: Prentice-Hall), Chapter 3.

Abramowitz, Milton, and Stegun, Irene A. 1964, *Handbook of Mathematical Functions*, Applied Mathematics Series, vol. 55 (Washington: National Bureau of Standards; reprinted 1968 by Dover Publications, New York), §3.6.

5.2 Evaluation of Continued Fractions

A continued fraction looks like this:

$$f(x) = b_0 + \cfrac{a_1}{b_1 + \cfrac{a_2}{b_2 + \cfrac{a_3}{b_3 + \cfrac{a_4}{b_4 + \cfrac{a_5}{b_5 + \cdots}}}}} \tag{5.2.1}$$

Printers prefer to write this as

$$f(x) = b_0 + \frac{a_1}{b_1+}\,\frac{a_2}{b_2+}\,\frac{a_3}{b_3+}\,\frac{a_4}{b_4+}\,\frac{a_5}{b_5+}\cdots \tag{5.2.2}$$

In either (5.2.1) or (5.2.2), the a's and b's can themselves be functions of x, usually linear or quadratic monomials at worst (i.e. constants times x or times x^2). For example, the continued fraction representation of the tangent function is

$$\tan(x) = \frac{x}{1-}\,\frac{x^2}{3-}\,\frac{x^2}{5-}\,\frac{x^2}{7-}\cdots \tag{5.2.3}$$

Continued fractions frequently converge much more rapidly than power series expansions, and in a much larger domain (not necessarily including the domain of convergence of the series, however). Sometimes the continued fraction converges best where the series does worst, although this is not a general rule.

How do you tell how far to go with a continued fraction? Unlike a series, you can't just evaluate equation (5.2.1) from left to right, stopping when the change is small. Written in the form of (5.2.1), the only way to evaluate the continued fraction is from right to left, first (blindly!) guessing how far out to start. This is not the right way.

The right way is to use a result which relates continued fractions to rational approximations, and which gives a means of evaluating (5.2.1) or (5.2.2) from left to right. Let f_n denote the result of evaluating (5.2.2) with coefficients through a_n and b_n. Then

$$f_n = \frac{A_n}{B_n} \tag{5.2.4}$$

where A_n and B_n are given by the following recurrence:

$$\begin{aligned} &A_{-1} \equiv 1 \qquad B_{-1} \equiv 0 \\ &A_0 \equiv b_0 \qquad B_0 \equiv 1 \\ A_j = b_j A_{j-1} + a_j A_{j-2} \qquad &B_j = b_j B_{j-1} + a_j B_{j-2} \qquad j = 1, 2, \ldots, n \end{aligned} \tag{5.2.5}$$

(You can easily prove this by induction if you like.)

In practice, the recurrence (5.2.5) frequently generates very large or very small values for the partial numerators and denominators A_j and B_j. There is thus the danger of overflow or underflow of the floating-point representation. However, the recurrence (5.2.5) is linear in the A's and B's. At any point you can rescale the currently-saved two levels of the recurrence, e.g. divide A_j, B_j, A_{j-1}, and B_{j-1} all by B_j. This incidentally makes $A_j = f_j$ and is convenient for testing whether you have gone far enough: see if f_j and f_{j-1} from the last iteration are as close as you would like them to be. (If B_j happens to be zero, which can happen, just skip the renormalization for this cycle. A fancier level of optimization is to renormalize only when an overflow is imminent, saving the unnecessary divides.)

There are standard techniques, including the important *quotient-difference algorithm*, for going back and forth between continued fraction approximations, power series approximations, and rational function approximations. Consult Acton for an introduction to this subject, and Fike for further details and references.

We will make frequent use of continued fractions in the next chapter.

REFERENCES AND FURTHER READING:

Abramowitz, Milton, and Stegun, Irene A. 1964, *Handbook of Mathematical Functions*, Applied Mathematics Series, vol. 55 (Washington: National Bureau of Standards; reprinted 1968 by Dover Publications, New York), §3.10.

Acton, Forman S. 1970, *Numerical Methods That Work* (New York: Harper and Row), Chapter 11.

Fike, C.T. 1968, *Computer Evaluation of Mathematical Functions* (Englewood Cliffs, N.J.: Prentice-Hall), §§8.2, 10.4, and 10.5.

5.3 Polynomials and Rational Functions

A polynomial of degree $N-1$ is represented numerically as a stored array of coefficients, C(J) with J= $1,\ldots,N$. We will always take C(1) to be the constant term in the polynomial, C(N) the coefficient of x^{N-1}; but of course other conventions are possible. There are two kinds of manipulations that you can do with a polynomial: *numerical* manipulations (such as evaluation), where you are given the numerical value of its argument, or *algebraic* manipulations, where you want to transform the coefficient array in some way without choosing any particular argument. Let's start with the numerical.

We assume that you know enough *never* to evaluate a polynomial this way:

```
P=C(1)+C(2)*X+C(3)*X**2+C(4)*X**3+C(5)*X**4
```

Come the (computer) revolution, all persons found guilty of such criminal behavior will be summarily executed, and their programs won't be! It is a matter of taste, however, whether to write

```
P=C(1)+X*(C(2)+X*(C(3)+X*(C(4)+X*C(5))))
```

or

```
P=(((C(5)*X+C(4))*X+C(3))*X+C(2))*X+C(1)
```

If the number of coefficients is a large number N, one writes

```
P=C(N)
DO 11 J=N-1,1,-1
    P=P*X+C(J)
11 CONTINUE
```

Another useful trick is for evaluating a polynomial $P(x)$ and its derivative $dP(x)/dx$ simultaneously:

```
P=C(N)
DP=0.
DO 11 J=N-1,1,-1
    DP=DP*X+P
    P=P*X+C(J)
11 CONTINUE
```

which returns the polynomial as P and its derivative as DP.

The above trick, which is basically *synthetic division* (see Acton, or Mathews and Walker), generalizes to the evaluation of the polynomial and ND-1 of its derivatives simultaneously:

```
SUBROUTINE DDPOLY(C,NC,X,PD,ND)
    Given the NC coefficients of a polynomial of degree NC-1 as an array C with C(1) being
    the constant term, and given a value X, and given a value ND>1, this routine returns the
    polynomial evaluated at X as PD(1) and ND-1 derivatives as PD(2)...PD(ND).
DIMENSION C(NC),PD(ND)
PD(1)=C(NC)
DO 11 J=2,ND
    PD(J)=0.
11  CONTINUE
DO 13 I=NC-1,1,-1
    NND=MIN(ND,NC+1-I)
    DO 12 J=NND,2,-1
        PD(J)=PD(J)*X+PD(J-1)
12      CONTINUE
    PD(1)=PD(1)*X+C(I)
13  CONTINUE
CONST=2.                           After the first derivative, factorial constants come in.
DO 14 I=3,ND
    PD(I)=CONST*PD(I)
    CONST=CONST*I
14  CONTINUE
RETURN
END
```

As a curiosity, you might be interested to know that polynomials of degree $n > 3$ can be evaluated in *fewer* than n multiplications, at least if you are willing to precompute some auxiliary coefficients and, in some cases, do an extra addition. For example, the polynomial

$$P(x) = a_0 + a_1 x + a_2 x^2 + a_3 x^3 + a_4 x^4 \tag{5.3.1}$$

where $a_4 > 0$, can be evaluated with 3 multiplications and 5 additions as follows:

$$P(x) = [(Ax + B)^2 + Ax + C][(Ax + B)^2 + D] + E \tag{5.3.2}$$

where A, B, C, D, and E are to be precomputed by

$$\begin{aligned} A &= (a_4)^{1/4} \\ B &= \frac{a_3 - A^3}{4A^3} \\ D &= 3B^2 + 8B^3 + \frac{a_1 A - 2a_2 B}{A^2} \\ C &= \frac{a_2}{A^2} - 2B - 6B^2 - D \\ E &= a_0 - B^4 - B^2(C + D) - CD \end{aligned} \tag{5.3.3}$$

Fifth degree polynomials can be evaluated in 4 multiplies and 5 adds; sixth degree polynomials can be evaluated in 4 multiplies and 7 adds; if any of this strikes you as interesting, consult Knuth or Winograd below. The subject has something of the same entertaining, if impractical, flavor as that of fast matrix multiplication, discussed in §2.11.

Turn now to algebraic manipulations. You multiply a polynomial of degree N-1 (array of length N) by a monomial factor $x -$ A by a bit of code like the following,

```
C(N+1)=C(N)
DO J=N,2,-1
    C(J)=C(J-1)-C(J)*A
ENDDO
C(1)=-C(1)*A
```

Likewise, you divide a polynomial of degree N-1 by a monomial factor $x -$ A (synthetic division again) using

```
REM=C(N)
C(N)=0.
DO I=N-1,1,-1
    SWAP=C(I)
    C(I)=REM
    REM=SWAP+REM*A
ENDDO
```

which leaves you with a new polynomial array and a numerical remainder REM.

Multiplication of two general polynomials involves straightforward summing of the products, each involving one coefficient from each polynomial. Division of two general polynomials, while it can be done awkwardly in the fashion taught using pencil and paper, is susceptible to a good deal of streamlining. Witness the following routine based on the algorithm in Knuth.

```
SUBROUTINE POLDIV(U,N,V,NV,Q,R)
   Given the N coefficients of a polynomial in U, and the NV coefficients of another polynomial
   in V, divide the polynomial U by the polynomial V ("U"/"V") giving a quotient polynomial
   whose coefficients are returned in Q, and a remainder polynomial whose coefficients are
   returned in R. The arrays Q and R are dimensioned with lengths N, but the elements R(NV)
   and Q(N-NV+2)...Q(N) will be returned as zero.
DIMENSION U(N),V(NV),Q(N),R(N)
DO 11 J=1,N
    R(J)=U(J)
    Q(J)=0.
 11 CONTINUE
DO 13 K=N-NV,0,-1
    Q(K+1)=R(NV+K)/V(NV)
    DO 12 J=NV+K-1,K+1,-1
        R(J)=R(J)-Q(K+1)*V(J-K)
     12 CONTINUE
 13 CONTINUE
R(NV)=0.
RETURN
END
```

Rational Functions

You evaluate a rational function like

$$R(x)=\frac{P_\mu(x)}{Q_\nu(x)}=\frac{p_0+p_1x+\cdots+p_\mu x^\mu}{q_0+q_1x+\cdots+q_\nu x^\nu} \tag{5.3.4}$$

in the obvious way, namely as two separate polynomials followed by a divide.

REFERENCES AND FURTHER READING:

Knuth, Donald E. 1981, *Seminumerical Algorithms*, 2nd ed., vol. 2 of *The Art of Computer Programming* (Reading, Mass.: Addison-Wesley), §4.6.

Fike, C.T. 1968, *Computer Evaluation of Mathematical Functions* (Englewood Cliffs, N.J.: Prentice-Hall), Chapter 4.

Acton, Forman S. 1970, *Numerical Methods That Work* (New York: Harper and Row), pp. 183, 190.

Mathews, Jon, and Walker, R.L. 1970, *Mathematical Methods of Physics*, 2nd ed. (Reading, Mass.: W.A. Benjamin/Addison-Wesley), pp. 361–363.

Winograd, S. 1970, *Commun. on Pure and Appl. Math.*, vol. 23, pp. 165–179.

5.4 Recurrence Relations and Clenshaw's Recurrence Formula

Many useful functions satisfy recurrence relations, e.g.,

$$(n+1)P_{n+1}(x) = (2n+1)xP_n(x) - nP_{n-1}(x) \tag{5.4.1}$$

$$J_{n+1}(x) = \frac{2n}{x}J_n(x) - J_{n-1}(x) \tag{5.4.2}$$

$$nE_{n+1}(x) = e^{-x} - xE_n(x) \tag{5.4.3}$$

$$\cos(n\theta) = 2\cos(\theta)\cos([n-1]\theta) - \cos([n-2]\theta) \tag{5.4.4}$$

$$\sin(n\theta) = 2\cos(\theta)\sin([n-1]\theta) - \sin([n-2]\theta) \tag{5.4.5}$$

where the first three functions are Legendre polynomials, Bessel functions of the first kind, and exponential integrals, respectively. (For notation see Abramowitz and Stegun.) These relations are useful for extending computational methods from two successive values of n to other values, either larger or smaller.

You must be aware, however, that recurrence relations are not necessarily *stable* against roundoff error in the direction that you propose to go (either increasing n or decreasing n). A three term linear recurrence relation has two linearly independent solutions. Only one of these corresponds to the sequence of functions that you are trying to generate. The other one *may* be exponentially growing in the direction that you want to go, or exponentially damped, or exponentially neutral (growing or dying as some power law, for example). If it is exponentially growing, then the recurrence relation is of little or no practical use in that direction. This is the case, e.g., for (5.4.2) in the direction of increasing n, when $x < n$. You cannot generate Bessel functions of high n by forward recurrence on (5.4.2).

Abramowitz and Stegun (in their Introduction) give a list of recurrences that are stable in the increasing or decreasing directions. That list does not contain all possible formulas, of course. Given a recurrence relation for some function $X_n(x)$ you can test it yourself with about five minutes of (human) labor: For a fixed x in your range of interest, start the recurrence not with true values of $X_j(x)$ and $X_{j+1}(x)$, but (first) with the values 1 and 0 respectively,

and then (second) with 0 and 1 respectively. Generate 10 or 20 terms of the recursive sequences in the direction that you want to go (increasing or decreasing from j), for each of the two starting conditions. Look at the difference between the corresponding members of the two sequences. If the differences stay of order unity (absolute value less than 10, say), then the recurrence is stable. If they increase slowly, then the recurrence may be mildly unstable but quite tolerably so. If they increase catastrophically, then there is an exponentially growing solution of the recurrence. If you know that the function that you want actually corresponds to the growing solution, then you can keep the recurrence formula anyway e.g. the case of the Bessel function $Y_n(x)$ for increasing n, see §6.4; if you don't know which solution your function corresponds to, you must at this point reject the recurrence formula. Notice that you can do this test *before* you go to the trouble of finding a numerical method for computing the two starting functions $X_j(x)$ and $X_{j+1}(x)$: stability is a property of the recurrence, not of the starting values.

An alternative heuristic procedure for testing stability is to replace the recurrence relation by a similar one that is linear with constant coefficients. For example, the relation (5.4.2) becomes

$$y_{n+1} = 2\gamma y_n - y_{n-1}$$

where $\gamma \equiv n/x$ is treated as a constant. You solve such recurrence relations by trying solutions of the form $y_n = a^n$. Substituting into the above recurrence gives

$$a^2 - 2\gamma a + 1 = 0 \qquad \text{or} \qquad a = \gamma \pm \sqrt{\gamma^2 - 1}$$

The recurrence is stable if $|a| \leq 1$ for all solutions a. This holds (as you can verify) if $|\gamma| \leq 1$ or $n \leq x$. The recurrence (5.4.2) thus cannot be used, starting with $J_0(x)$ and $J_1(x)$, to compute $J_n(x)$ for large n.

Every cloud has a silver lining: if a recurrence relation is catastrophically unstable in one direction, then that (undesired) solution will decrease very rapidly in the reverse direction. This means that you can start with *any* seed values for the consecutive X_j and X_{j+1} and (when you have gone enough steps in the stable direction) you will converge to the sequence of functions that you want, times an unknown normalization factor. If there is some other way to normalize the sequence (e.g., by a formula for the sum of the X_n's), then this can be a practical means of function evaluation. The method is called *Miller's algorithm.* An example often given uses equation (5.4.2) in just this way, along with the normalization formula

$$1 = J_0(x) + 2J_2(x) + 2J_4(x) + 2J_6(x) + \cdots \tag{5.4.6}$$

(see Acton, or Abramowitz and Stegun).

Clenshaw's recurrence formula is an elegant and efficient way to evaluate a sum of coefficients times functions that obey a recurrence formula, e.g.,

$$f(\theta) = \sum_{k=0}^{N} c_k \cos(k\theta) \qquad \text{or} \qquad f(x) = \sum_{k=0}^{N} c_k P_k(x)$$

Here is how it works: Suppose that the desired sum is

$$f(x) = \sum_{k=0}^{N} c_k F_k(x) \tag{5.4.7}$$

and that F_k obeys the recurrence relation

$$F_{n+1}(x) = \alpha(n,x)F_n(x) + \beta(n,x)F_{n-1}(x) \tag{5.4.8}$$

for some functions $\alpha(n,x)$ and $\beta(n,x)$. Now define the quantities y_k ($k = N, N-1, \ldots, 1$) by the following recurrence:

$$\begin{aligned} &y_{N+2} = y_{N+1} = 0 \\ &\quad y_k = \alpha(k,x)y_{k+1} + \beta(k+1,x)y_{k+2} + c_k \quad (k = N, N-1, \ldots, 1) \end{aligned} \tag{5.4.9}$$

If you solve equation (5.4.9) for c_k on the left, and then write out explicitly the sum (5.4.7), it will look (in part) like this:

$$\begin{aligned} f(x) = \cdots & \\ &+ [y_8 - \alpha(8,x)y_9 - \beta(9,x)y_{10}]F_8(x) \\ &+ [y_7 - \alpha(7,x)y_8 - \beta(8,x)y_9]F_7(x) \\ &+ [y_6 - \alpha(6,x)y_7 - \beta(7,x)y_8]F_6(x) \\ &+ [y_5 - \alpha(5,x)y_6 - \beta(6,x)y_7]F_5(x) \\ &+ \cdots \\ &+ [y_2 - \alpha(2,x)y_3 - \beta(3,x)y_4]F_2(x) \\ &+ [y_1 - \alpha(1,x)y_2 - \beta(2,x)y_3]F_1(x) \\ &+ [c_0 + \beta(1,x)y_2 - \beta(1,x)y_2]F_0(x) \end{aligned} \tag{5.4.10}$$

Notice that we have added and subtracted $\beta(1,x)y_2$ in the last line. If you examine the terms containing a factor of y_8 in (5.4.10), you will find that they

sum to zero as a consequence of the recurrence relation (5.4.8); similarly all the other y_k's down through y_2. The only surviving terms in (5.4.10) are

$$f(x) = \beta(1,x)F_0(x)y_2 + F_1(x)y_1 + F_0(x)c_0 \tag{5.4.11}$$

Equations (5.4.9) and (5.4.11) are *Clenshaw's recurrence formula* for doing the sum (5.4.7): You make one pass down through the y_k's using (5.4.9); when you have reached y_2 and y_1 you apply (5.4.11) to get the desired answer.

Clenshaw's recurrence as written above incorporates the coefficients c_k in a downward order, with k decreasing. At each stage, the effect of all previous c_k's is "remembered" as two coefficients which multiply the functions F_{k+1} and F_k (ultimately F_0 and F_1). If the functions F_k are small when k is large, *and* if the coefficients c_k are small when k is *small*, then the sum can be dominated by small F_k's. In this case the remembered coefficients will involve a delicate cancellation and there can be a catastrophic loss of significance. An example would be to sum the trivial series

$$J_{15}(1) = 0 \times J_0(1) + 0 \times J_1(1) + \ldots + 0 \times J_{14}(1) + 1 \times J_{15}(1) \tag{5.4.12}$$

Here J_{15}, which is tiny, ends up represented as a cancelling linear combination of J_0 and J_1, which are of order unity.

The solution in such cases is to use an alternative Clenshaw recurrence that incorporates c_k's in an upward direction. The relevant equations are

$$\begin{aligned} &y_{-2} = y_{-1} = 0 \\ &y_k = \frac{1}{\beta(k+1,x)}[y_{k-2} - \alpha(k,x)y_{k-1} - c_k] \quad (k = 0,1,\ldots,N-1) \end{aligned} \tag{5.4.13}$$

$$f(x) = c_N F_N(x) - \beta(N,x)F_{N-1}(x)y_{N-1} - F_N(x)y_{N-2} \tag{5.4.14}$$

The rare case where equations (5.4.13) – (5.4.14) should be used instead of equations (5.4.9) and (5.4.11) can be detected automatically by testing whether the operands in the first sum in (5.4.11) are opposite in sign and nearly equal in magnitude. Other than in this special case, Clenshaw's recurrence is always stable, independent of whether the recurrence for the functions F_k is stable in the upward or downward direction.

REFERENCES AND FURTHER READING:

Abramowitz, Milton, and Stegun, Irene A. 1964, *Handbook of Mathematical Functions*, Applied Mathematics Series, vol. 55 (Washington: National Bureau of Standards; reprinted 1968 by Dover Publications, New York), pp.xiii, 697.

Acton, Forman S. 1970, *Numerical Methods That Work* (New York: Harper and Row), pp.20 ff.

Dahlquist, Germund, and Bjorck, Ake. 1974, *Numerical Methods* (Englewood Cliffs, N.J.: Prentice-Hall), §4.4.3, p.111.

Goodwin, E.T. (ed.) 1961, *Modern Computing Methods*, 2nd ed. (New York: Philosophical Library), p.76.

Clenshaw, C.W., 1962, "Chebyshev Series for Mathematical Functions," in *Mathematical Tables*, vol. 5, National Physical Laboratory, (London: H.M. Stationery Office).

5.5 Quadratic and Cubic Equations

The roots of simple algebraic equations can be viewed as being functions of the equations' coefficients. We are taught these functions in elementary algebra. Yet, surprisingly many people don't know the right way to solve a quadratic equation with two real roots, or to obtain the real roots of a cubic equation.

There are two ways to write the solution of the quadratic equation

$$ax^2 + bx + c = 0 \tag{5.5.1}$$

with real coefficients a, b, c, namely

$$x = \frac{-b \pm \sqrt{b^2 - 4ac}}{2a} \tag{5.5.2}$$

and

$$x = \frac{2c}{-b \pm \sqrt{b^2 - 4ac}} \tag{5.5.3}$$

If you use *either* (5.5.2) *or* (5.5.3) to get the two roots, you are asking for trouble: if either a or c (or both) are small, then one of the roots will involve the subtraction of b from a very nearly equal quantity (the discriminant); you will get that root very inaccurately. The correct way to compute the roots is

$$q \equiv -\frac{1}{2}\left[b + \text{sgn}(b)\sqrt{b^2 - 4ac}\right] \tag{5.5.4}$$

Then the two roots are

$$x_1 = \frac{q}{a} \qquad \text{and} \qquad x_2 = \frac{c}{q} \tag{5.5.5}$$

For the cubic equation

$$x^3 + a_1 x^2 + a_2 x + a_3 = 0 \tag{5.5.6}$$

with real coefficients a_1, a_2, a_3, first compute

$$Q \equiv \frac{a_1^2 - 3a_2}{9} \qquad \text{and} \qquad R \equiv \frac{2a_1^3 - 9a_1 a_2 + 27a_3}{54} \tag{5.5.7}$$

Next, check that $Q^3 - R^2 \geq 0$. If this holds, then your cubic equation has three real roots. Find them by computing

$$\theta = \arccos(R/\sqrt{Q^3}) \tag{5.5.8}$$

in terms of which the three roots are

$$\begin{aligned} x_1 &= -2\sqrt{Q}\cos\left(\frac{\theta}{3}\right) - \frac{a_1}{3} \\ x_2 &= -2\sqrt{Q}\cos\left(\frac{\theta + 2\pi}{3}\right) - \frac{a_1}{3} \\ x_3 &= -2\sqrt{Q}\cos\left(\frac{\theta + 4\pi}{3}\right) - \frac{a_1}{3} \end{aligned} \tag{5.5.9}$$

(This equation first appears in Chapter VI of François Viète's treatise "De emendatione," published in 1615!)

If, on the other hand, $R^2 - Q^3 > 0$, then your cubic has only one real root, given by

$$x_1 = -\text{sgn}(R)\left[(\sqrt{R^2 - Q^3} + |R|)^{1/3} + \frac{Q}{(\sqrt{R^2 - Q^3} + |R|)^{1/3}}\right] - \frac{a_1}{3} \tag{5.5.10}$$

If you need to solve many cubic equations with only slightly different coefficients, it is more efficient to use Newton's method (§9.4).

REFERENCES AND FURTHER READING:

Handbook of Tables for Mathematics, 3rd ed., Robert C. Weast, ed. (Cleveland: The Chemical Rubber Co.), pp. 130-133.

5.6 Chebyshev Approximation

The Chebyshev polynomial of degree n is denoted $T_n(x)$, and is given by the explicit formula

$$T_n(x) = \cos(n \arccos x) \tag{5.6.1}$$

This may look trigonometric at first glance (and there is in fact a close relation between the Chebyshev polynomials and the discrete Fourier transform); however (5.6.1) can be combined with trigonometric identities to yield explicit expressions for $T_n(x)$ (see Figure 5.6.1),

$$\begin{aligned}
T_0(x) &= 1 \\
T_1(x) &= x \\
T_2(x) &= 2x^2 - 1 \\
T_3(x) &= 4x^3 - 3x \\
T_4(x) &= 8x^4 - 8x^2 + 1 \\
&\cdots \\
T_{n+1}(x) &= 2xT_n(x) - T_{n-1}(x) \quad n \geq 1.
\end{aligned} \tag{5.6.2}$$

The Chebyshev polynomials are orthogonal in the interval $[-1,1]$ over a weight $(1-x^2)^{-1/2}$. In particular,

$$\int_{-1}^{1} \frac{T_i(x)T_j(x)}{\sqrt{1-x^2}} dx = \begin{cases} 0 & i \neq j \\ \pi/2 & i = j \neq 0 \\ \pi & i = j = 0 \end{cases} \tag{5.6.3}$$

The polynomial $T_n(x)$ has n zeros in the interval $[-1,1]$, and they are located at the points

$$x = \cos\left(\frac{\pi(k-\frac{1}{2})}{n}\right) \qquad k = 1, 2, \ldots, n \tag{5.6.4}$$

In this same interval there are $n+1$ extrema (maxima and minima), located at

$$x = \cos\left(\frac{\pi k}{n}\right) \qquad k = 0, 1, \ldots, n \tag{5.6.5}$$

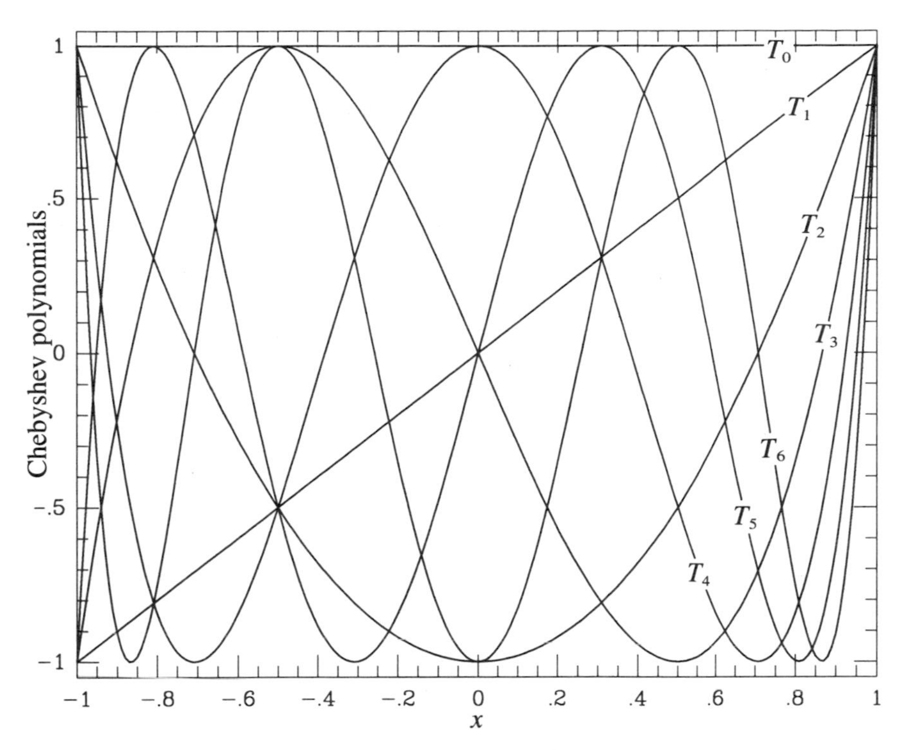

Figure 5.6.1. Chebyshev polynomials $T_0(x)$ through $T_6(x)$. Note that T_j has j roots in the interval $(-1,1)$ and that all the polynomials are bounded between ± 1.

At all of the maxima $T_n(x) = 1$, while at all of the minima $T_n(x) = -1$; it is precisely this property that makes the Chebyshev polynomials so useful in polynomial approximation of functions.

The Chebyshev polynomials satisfy a discrete orthogonality relation as well as the continuous one (5.6.3): If x_k $(k = 1, \ldots, m)$ are the m zeros of $T_m(x)$ given by (5.6.4), then

$$\sum_{k=1}^{m} T_i(x_k)T_j(x_k) = \begin{cases} 0 & i \neq j \\ m/2 & i = j \neq 0 \\ m & i = j = 0 \end{cases} \tag{5.6.6}$$

It is not too difficult to combine equations (5.6.1), (5.6.4), and (5.6.6) to prove the following theorem: If $f(x)$ is an arbitrary function in the interval $[-1,1]$, and if N coefficients $c_j, j = 1, \ldots, N$, are defined by

$$\begin{aligned} c_j &= \frac{2}{N} \sum_{k=1}^{N} f(x_k)T_{j-1}(x_k) \\ &= \frac{2}{N} \sum_{k=1}^{N} f\left[\cos\left(\frac{\pi(k-\frac{1}{2})}{N}\right)\right] \cos\left(\frac{\pi(j-1)(k-\frac{1}{2})}{N}\right) \end{aligned} \tag{5.6.7}$$

then the approximation formula

$$f(x) \approx \left[\sum_{k=1}^{N} c_k T_{k-1}(x)\right] - \frac{1}{2}c_1 \tag{5.6.8}$$

is *exact* for x equal to all of the N zeros of $T_N(x)$.

For a fixed N, equation (5.6.8) is a polynomial in x which approximates the function $f(x)$ in the interval $[-1, 1]$ (where all the zeros of $T_N(x)$ are located). Why is this particular approximating polynomial better than any other one, exact on some other set of N points? The answer is *not* that (5.6.8) is necessarily more accurate than some other approximating polynomial of the same order N (for some specified definition of "accurate"), but rather that (5.6.8) can be truncated to a polynomial of *lower* degree $m \ll N$ in a very graceful way, one which *does* yield the "most accurate" approximation of degree m (in a sense which can be made precise). Suppose N is so large that (5.6.8) is virtually a perfect approximation of $f(x)$. Now consider the truncated approximation

$$f(x) \approx \left[\sum_{k=1}^{m} c_k T_{k-1}(x)\right] - \frac{1}{2}c_1 \tag{5.6.9}$$

with the same c_j's, computed from (5.6.7). Since the $T_k(x)$'s are all bounded between ± 1, the difference between (5.6.9) and (5.6.8) can be no larger than the sum of the neglected c_k's ($k = m+1, \ldots, N$). In fact, if the c_k's are rapidly decreasing (which is the typical case), then the error is dominated by $c_{m+1}T_m(x)$, an oscillatory function with $m+1$ equal extrema distributed smoothly over the interval $[-1, 1]$. This smooth spreading out of the error is a very important property: The Chebyshev approximation (5.6.9) is very nearly the same polynomial as that holy grail of approximating polynomials the *minimax polynomial*, which (among all polynomials of the same degree) has the smallest maximum deviation from the true function $f(x)$. The minimax polynomial is very difficult to find; the Chebyshev approximating polynomial is almost identical and is very easy to compute!

So, given some (perhaps difficult) means of computing the function $f(x)$, we now need algorithms for implementing (5.6.7) and (after inspection of the resulting c_k's and choice of a truncating value m) evaluating (5.6.9). The latter equation then becomes an easy way of computing $f(x)$ for all subsequent time.

The first of these tasks is straightforward. A generalization of equation (5.6.7) that is here implemented is to allow the range of approximation to be between two arbitrary limits a and b, instead of just -1 to 1. This is effected by a change of variable

$$y \equiv \frac{x - \frac{1}{2}(b+a)}{\frac{1}{2}(b-a)} \tag{5.6.10}$$

and by the approximation of $f(x)$ by a Chebyshev polynomial in y.

```
SUBROUTINE CHEBFT(A,B,C,N,FUNC)
   Chebyshev fit: Given a function FUNC, lower and upper limits of the interval [A,B], and
   a maximum degree N, this routine computes the N coefficients C_k such that FUNC(x) ≈
   [Σ_{k=1}^N C_k T_{k-1}(y)] - c_1/2, where y and x are related by (5.6.10). This routine is to be
   used with moderately large N (e.g. 30 or 50), the array of C's subsequently to be truncated
   at the smaller value m such that C_{m+1} and subsequent elements are negligible.
PARAMETER (NMAX=50, PI=3.141592653589793D0)
REAL*8 SUM
DIMENSION C(N),F(NMAX)
BMA=0.5*(B-A)
BPA=0.5*(B+A)
DO 11 K=1,N                      We evaluate the function at the N points required by (5.6.7).
    Y=COS(PI*(K-0.5)/N)
    F(K)=FUNC(Y*BMA+BPA)
11  CONTINUE
FAC=2./N
DO 13 J=1,N
    SUM=0.D0                     We will accumulate the sum in double precision, a nicety which you
    DO 12 K=1,N                      can ignore.
        SUM=SUM+F(K)*COS((PI*(J-1))*((K-0.5D0)/N))
12      CONTINUE
    C(J)=FAC*SUM
13  CONTINUE
RETURN
END
```

Now that we have the Chebyshev coefficients how do we evaluate the approximation? One could use the recurrence relation of equation (5.6.2) to generate values for $T_k(x)$ from $T_0 = 1, T_1 = x$, while also accumulating the sum of (5.6.9). It is better to use Clenshaw's recurrence formula (§5.4), effecting the two processes simultaneously. Applied to the Chebyshev series (5.6.9), the recurrence is

$$\begin{aligned} d_{m+2} &\equiv d_{m+1} \equiv 0 \\ d_j &= 2x d_{j+1} - d_{j+2} + c_j \qquad j = m, m-1, \ldots, 2 \\ f(x) \equiv d_0 &= x d_2 - d_3 + \frac{1}{2} c_1 \end{aligned} \tag{5.6.11}$$

```
FUNCTION CHEBEV(A,B,C,M,X)
   Chebyshev evaluation: All arguments are input. C is an array of Chebyshev coefficients,
   of length M, the first M elements of C output from CHEBFT (which must have been called
   with the same A and B). The Chebyshev polynomial is evaluated at a point Y determined
   from X, A, and B, and the result is returned as the function value.
DIMENSION C(M)
IF ((X-A)*(X-B).GT.0.)  PAUSE 'X not in range.'
D=0.
DD=0.
Y=(2.*X-A-B)/(B-A)              Change of variable.
Y2=2.*Y
DO 11 J=M,2,-1                  Clenshaw's recurrence.
    SV=D
    D=Y2*D-DD+C(J)
```

```
      DD=SV
   11 CONTINUE
      CHEBEV=Y*D-DD+0.5*C(1)          Last step is different.
      RETURN
      END
```

REFERENCES AND FURTHER READING:

Goodwin, E.T. (ed.) 1961, *Modern Computing Methods*, 2nd ed. (New York: Philosophical Library), Chapter 8.

Dahlquist, Germund, and Bjorck, Ake. 1974, *Numerical Methods* (Englewood Cliffs, N.J.: Prentice-Hall), §4.4.1, p.104.

Johnson, Lee W., and Riess, R. Dean. 1982, *Numerical Analysis*, 2nd ed. (Reading, Mass.: Addison-Wesley), §6.5.2, p.334.

Carnahan, Brice, Luther, H.A., and Wilkes, James O. 1969, *Applied Numerical Methods* (New York: Wiley), §1.10, p.39.

5.7 Derivatives or Integrals of a Chebyshev-approximated Function

If you have obtained the Chebyshev coefficients that approximate a function in a certain range (e.g. from CHEBFT in §5.6), then it is a simple matter to transform them to Chebyshev coefficients corresponding to the derivative or integral of the function. Having done this, you can evaluate the derivative or integral just as if it were a function that you had Chebyshev-fitted *ab initio*.

The relevant formulas are these: If c_i, $i = 1, \ldots, m$ are the coefficients that approximate a function f in equation (5.6.9), C_i are the coefficients that approximate the indefinite integral of f, and c'_i are the coefficients that approximate the derivative of f, then

$$C_i = \frac{c_{i-1} - c_{i+1}}{2(i-1)} \qquad (i > 1) \tag{5.7.1}$$

$$c'_{i-1} = c'_{i+1} + 2(i-1)c_i \qquad (i = m-1, m-2, \ldots, 2) \tag{5.7.2}$$

Equation (5.7.1) is augmented by an arbitrary choice of C_1, corresponding to an arbitrary constant of integration. Equation (5.7.2), which is a recurrence, is started with the values $c'_m = c'_{m+1} = 0$, corresponding to no information about the $m + 1^{st}$ Chebyshev coefficient of the original function f.

Here are routines for implementing equations (5.7.1) and (5.7.2).

```
SUBROUTINE CHINT(A,B,C,CINT,N)
```

Given **A,B,C**, as output from routine **CHEBFT** §5.6, and given N, the desired degree of approximation (length of **C** to be used), this routine returns the array **CINT**, the Chebyshev coefficients of the integral of the function whose coefficients are **C**. The constant of integration is set so that the integral vanishes at **A**.

```
DIMENSION C(N),CINT(N)
CON=0.25*(B-A)                      Factor which normalizes to the interval B − A.
SUM=0.                              Accumulates the constant of integration.
FAC=1.                              Will equal ±1.
DO 11 J=2,N-1
    CINT(J)=CON*(C(J-1)-C(J+1))/(J-1)     Equation 5.7.1.
    SUM=SUM+FAC*CINT(J)
    FAC=-FAC
11  CONTINUE
CINT(N)=CON*C(N-1)/(N-1)            Special case of 5.7.1 for N.
SUM=SUM+FAC*CINT(N)
CINT(1)=2.*SUM                      Set the constant of integration.
RETURN
END
```

```
SUBROUTINE CHDER(A,B,C,CDER,N)
```

Given **A,B,C**, as output from routine **CHEBFT** §5.6, and given N, the desired degree of approximation (length of **C** to be used), this routine returns the array **CDER**, the Chebyshev coefficients of the derivative of the function whose coefficients are **C**.

```
DIMENSION C(N),CDER(N)
CDER(N)=0.                          N and N-1 are special cases.
CDER(N-1)=2*(N-1)*C(N)
IF(N.GE.3)THEN
    DO 11 J=N-2,1,-1
        CDER(J)=CDER(J+2)+2*J*C(J+1)      Equation 5.7.2.
11      CONTINUE
ENDIF
CON=2./(B-A)
DO 12 J=1,N                         Normalize to the interval B − A.
    CDER(J)=CDER(J)*CON
12  CONTINUE
RETURN
END
```

REFERENCES AND FURTHER READING:

Goodwin, E.T. (ed.) 1961, *Modern Computing Methods*, 2nd ed. (New York: Philosophical Library), pp.78-79.

5.8 Polynomial Approximation from Chebyshev Coefficients

You may well ask after reading the preceding two sections, "Must I store and evaluate my Chebyshev approximation as an array of Chebyshev coefficients for a transformed variable y? Can't I convert the c_k's into actual polynomial coefficients in the original variable x and have an approximation of the following form?"

$$f(x) \approx \sum_{k=1}^{m} g_k x^{k-1} \tag{5.8.1}$$

Yes, you can do this (and we will give you the algorithm to do it), but we caution you against it: Evaluating equation (5.8.1), where the coefficient g's reflect an underlying Chebyshev approximation, usually requires more significant figures than evaluation of the Chebyshev sum directly (as by CHEBEV). This is because the Chebyshev polynomials themselves exhibit a rather delicate cancellation: The leading coefficient of $T_n(x)$, for example is 2^{n-1}; other coefficients of $T_n(x)$ are even bigger; yet they all manage to combine into a polynomial which lies between ± 1. *Only* when m is no larger than 7 or 8 should you contemplate writing a Chebyshev fit as a direct polynomial, and even in those cases you should be willing to tolerate two or so significant figures less accuracy than the roundoff limit of your machine.

You get the g's in equation (5.8.1) from the c's output from CHEBFT (suitably truncated at a modest value of m) by calling in sequence the following two procedures:

```
SUBROUTINE CHEBPC(C,D,N)
```

Chebyshev polynomial coefficients. Given a coefficient array C of length N, this routine generates a coefficient array D of length N such that $\sum_{k=1}^{N} D_k y^{k-1} = \sum_{k=1}^{N} C_k T_{k-1}(y)$. The method is Clenshaw's recurrence (5.6.11), but now applied algebraically rather than arithmetically.

```
PARAMETER (NMAX=50)
DIMENSION C(N),D(N),DD(NMAX)
DO 11 J=1,N
   D(J)=0.
   DD(J)=0.
11 CONTINUE
D(1)=C(N)
DO 13 J=N-1,2,-1
   DO 12 K=N-J+1,2,-1
      SV=D(K)
      D(K)=2.*D(K-1)-DD(K)
      DD(K)=SV
12 CONTINUE
   SV=D(1)
   D(1)=-DD(1)+C(J)
   DD(1)=SV
13 CONTINUE
DO 14 J=N,2,-1
   D(J)=D(J-1)-DD(J)
14 CONTINUE
```

```
      D(1)=-DD(1)+0.5*C(1)
      RETURN
      END
```

```
      SUBROUTINE PCSHFT(A,B,D,N)
```

Polynomial coefficient shift. Given a coefficient array D of length N, this routine generates a coefficient array g of length N such that $\sum_{k=1}^{N} D_k y^{k-1} = \sum_{k=1}^{N} g_k x^{k-1}$, where x and y are related by (5.6.10), i.e. the interval $-1 < y < 1$ is mapped to the interval $A < x < B$. The array g is returned in D.

```
      DIMENSION D(N)
      CONST=2./(B-A)
      FAC=CONST
      DO 11 J=2,N                 First we rescale by the factor CONST...
        D(J)=D(J)*FAC
        FAC=FAC*CONST
11    CONTINUE
      CONST=0.5*(A+B)             ...which is then redefined as the desired shift.
      DO 13 J=1,N-1               We accomplish the shift by synthetic division. Synthetic division is a
        DO 12 K=N-1,J,-1              miracle of high-school algebra. If you never learned it, go do
          D(K)=D(K)-CONST*D(K+1)      so. You won't be sorry.
12      CONTINUE
13    CONTINUE
      RETURN
      END
```

REFERENCES AND FURTHER READING:

Carnahan, Brice, Luther, H.A., and Wilkes, James O. 1969, *Applied Numerical Methods* (New York: Wiley), Example 1.3. [Unfortunately these authors do not seem to know about either Clenshaw or synthetic division, so their algorithms are quite unwieldy.]

Acton, Forman S. 1970, *Numerical Methods That Work* (New York: Harper and Row), pp.59,182-183 [synthetic division].

Chapter 6. Special Functions

6.0 Introduction

There is nothing particularly special about a *special function*, except that some person in authority or textbook writer (not the same thing!) has decided to bestow the moniker. Special functions are also called *higher transcendental functions* (higher than what?) or *functions of mathematical physics* (but they occur in other fields also) or *functions which satisfy certain frequently-occurring second-order differential equations* (but not all special functions do). One might simply call them "useful functions" and let it go at that; it is surely only a matter of taste which functions we have chosen to include in this chapter.

Good commercially available program libraries, such as NAG or IMSL, contain routines for a number of special functions. These routines are intended for users who will have no idea what goes on inside them. Such state of the art "black boxes" are often very messy things, full of branches to completely different methods depending on the value of the calling arguments. Black boxes have, or should have, careful control of accuracy, to some stated uniform precision in all regimes.

We will not be quite so fastidious in our examples, in part because we want to illustrate techniques from Chapter 5, and in part because we *want* you to understand what goes on in the routines presented. Some of our routines have an accuracy parameter which can be made as small as desired, while others (especially those involving polynomial fits) give only a certain accuracy, one which we believe serviceable (typically six significant figures or so). We do *not* certify that the routines are perfect black boxes. We do hope that, if you ever encounter trouble in a routine, you will be able to diagnose and correct the problem on the basis of the information that we have given.

In short, the special function routines of this chapter are meant to be used — we use them all the time — but only to the extent that the user is prepared to understand their inner workings.

REFERENCES AND FURTHER READING:

Abramowitz, Milton, and Stegun, Irene A. 1964, *Handbook of Mathematical Functions*, Applied Mathematics Series, vol. 55 (Washington: National Bureau of Standards; reprinted 1968 by Dover Publications, New York)[full of useful numerical approximations to a great variety of functions].

IMSL Library Reference Manual, 1980, ed. 8 (IMSL Inc., 7500 Bellaire Boulevard, Houston TX 77036), Chapter M.

NAG Fortran Library Manual Mark 8, 1980 (NAG Central Office, 7 Banbury Road, Oxford OX26NN, U.K.), Chapter S.

Hart, John F., et al. 1968, *Computer Approximations* (New York: Wiley).

Hastings, Cecil. 1955, *Approximations for Digital Computers* (Princeton: Princeton University Press).

Luke, Yudell L. 1975, *Mathematical Functions and Their Approximations* (New York: Academic Press).

6.1 Gamma Function, Beta Function, Factorials, Binomial Coefficients

The gamma function is defined by the integral

$$\Gamma(z) = \int_0^\infty t^{z-1} e^{-t} dt \tag{6.1.1}$$

When the argument z is an integer, the gamma function is just the familiar factorial function, but offset by one,

$$n! = \Gamma(n+1) \tag{6.1.2}$$

The gamma function satisfies the recurrence relation

$$\Gamma(z+1) = z\Gamma(z) \tag{6.1.3}$$

If the function is known for arguments $z > 1$ or, more generally, in the half complex plane $\mathrm{Re}(z) > 1$ it can be obtained for $z < 1$ or $\mathrm{Re}\,(z) < 1$ by the reflection formula

$$\Gamma(1-z) = \frac{\pi}{\Gamma(z)\sin(\pi z)} = \frac{\pi z}{\Gamma(1+z)\sin(\pi z)} \tag{6.1.4}$$

Notice that $\Gamma(z)$ has a pole at $z = 0$, and at all negative integer values of z.

There are a variety of methods in use for calculating the function $\Gamma(z)$ numerically, but none is quite as neat as the approximation derived by Lanczos. This scheme is entirely specific to the gamma function, seemingly plucked from thin air. We will not attempt to derive the approximation, but only state

the resulting formula: For certain integer choices of γ and N, and for certain coefficients $c_1, c_2, \ldots, c_N$, the gamma function is given by

$$\Gamma(z+1) = (z+\gamma+\tfrac{1}{2})^{z+\frac{1}{2}} e^{-(z+\gamma+\frac{1}{2})} \times \sqrt{2\pi}\left[c_0 + \frac{c_1}{z+1} + \frac{c_2}{z+2} + \cdots + \frac{c_N}{z+N} + \epsilon\right] \quad (z>0) \tag{6.1.5}$$

You can see that this is a sort of take-off on Stirling's approximation, but with a series of corrections which take into account the first few poles in the left complex plane. The constant c_0 is very nearly equal to 1. The error term is parametrized by ϵ. For $\gamma = 5$, $N = 6$, and a certain set of c's, the error is smaller than $|\epsilon| < 2 \times 10^{-10}$. Impressed? If not, then perhaps you will be impressed by the fact that (with these same parameters) the formula (6.1.5) and bound on ϵ apply for the *complex* gamma function, *everywhere in the half complex plane Re* $z > 0$.

It is better to implement $\ln\Gamma(x)$ than $\Gamma(x)$, since the latter will overflow many computers' floating point representation at quite modest values of x. Often the gamma function is used in calculations where the large values of $\Gamma(x)$ are divided by other large numbers with the result being a perfectly ordinary value. Such operations would normally be coded as subtraction of logarithms. With (6.1.5) in hand, we can compute the logarithm of the gamma function with two calls to a logarithm and 25 or so arithmetic operations. This makes it not much more difficult than other built-in functions that we take for granted, such as $\sin(x)$ or e^x:

```
      FUNCTION GAMMLN(XX)
C     Returns the value ln[Γ(XX)] for XX > 0. Full accuracy is obtained for XX > 1. For
C     0 < XX < 1, the reflection formula (6.1.4) can be used first.
      REAL*8 COF(6),STP,HALF,ONE,FPF,X,TMP,SER
C     Internal arithmetic will be done in double precision, a nicety that you can omit if five-
C     figure accuracy is good enough.
      DATA COF,STP/76.18009173D0,-86.50532033D0,24.01409822D0,
     *    -1.231739516D0,.120858003D-2,-.536382D-5,2.50662827465D0/
      DATA HALF,ONE,FPF/0.5D0,1.0D0,5.5D0/
      X=XX-ONE
      TMP=X+FPF
      TMP=(X+HALF)*LOG(TMP)-TMP
      SER=ONE
      DO 11 J=1,6
        X=X+ONE
        SER=SER+COF(J)/X
11    CONTINUE
      GAMMLN=TMP+LOG(STP*SER)
      RETURN
      END
```

How shall we write a routine for the factorial function $n!$? Generally the factorial function will be called for small integer values (for large values it will overflow anyway!), and in most applications the same integer value will be called for many times. It is a profligate waste of computer time to call EXP(GAMMLN(N+1.)) for each required factorial. Better to go back to basics, holding GAMMLN in reserve for unlikely calls:

```
FUNCTION FACTRL(N)
   Returns the value N! as a floating point number.
DIMENSION A(33)                  Table to be filled in only as required
DATA NTOP,A(1)/0,1./             Table initialized with 0! only.
IF (N.LT.0) THEN
    PAUSE 'negative factorial'
ELSE IF (N.LE.NTOP) THEN         Already in table.
    FACTRL=A(N+1)
ELSE IF (N.LE.32) THEN           Fill in table up to desired value.
    DO 11 J=NTOP+1,N
        A(J+1)=J*A(J)
11  CONTINUE
    NTOP=N
    FACTRL=A(N+1)
ELSE                             Larger value than size of table is required. Actually, this big a value is
    FACTRL=EXP(GAMMLN(N+1.))         going to overflow on many computers, but no harm in trying.
ENDIF
RETURN
END
```

A useful point is that FACTRL will be *exact* for the smaller values of N, since floating point multiplies on small integers are exact on all computers. This exactness will not hold if we turn to the logarithm of the factorials. For binomial coefficients, however, we must do exactly this, since the individual factorials in a binomial coefficient will overflow long before the coefficient itself will.

The binomial coefficient is defined by

$$\binom{n}{k} = \frac{n!}{k!(n-k)!} \qquad 0 \le k \le n \tag{6.1.6}$$

```
FUNCTION BICO(N,K)
   Returns the binomial coefficient (n k) as a floating-point number.
BICO=ANINT(EXP(FACTLN(N)-FACTLN(K)-FACTLN(N-K)))
RETURN                           The nearest-integer function cleans up roundoff error for smaller val-
END                                  ues of N and K.
```

which uses

```
FUNCTION FACTLN(N)
   Returns ln(N!).
DIMENSION A(100)
DATA A/100*-1./                   Initialize the table to negative values.
IF (N.LT.0) PAUSE 'negative factorial'
IF (N.LE.99) THEN                 In range of the table.
    IF (A(N+1).LT.0.)  A(N+1)=GAMMLN(N+1.)  If not already in the table, put it in.
    FACTLN=A(N+1)
ELSE
    FACTLN=GAMMLN(N+1.)           Out of range of the table.
ENDIF
RETURN
END
```

Finally, turning away from the combinatorial functions with integer valued arguments, we come to the beta function,

$$B(z,w) = B(w,z) = \int_0^1 t^{z-1}(1-t)^{w-1}dt \tag{6.1.7}$$

which is related to the gamma function by

$$B(z,w) = \frac{\Gamma(z)\Gamma(w)}{\Gamma(z+w)} \tag{6.1.8}$$

hence

```
FUNCTION BETA(Z,W)
   Returns the value of the beta function B(z, w).
BETA=EXP(GAMMLN(Z)+GAMMLN(W)-GAMMLN(Z+W))
RETURN
END
```

REFERENCES AND FURTHER READING:

Abramowitz, Milton, and Stegun, Irene A. 1964, *Handbook of Mathematical Functions*, Applied Mathematics Series, vol. 55 (Washington: National Bureau of Standards; reprinted 1968 by Dover Publications, New York), Chapter 6.

Lanczos, C., 1964, *Journal S.I.A.M. Numerical Analysis*, ser. B, vol. 1, p.86.

6.2 Incomplete Gamma Function, Error Function, Chi-Square Probability Function, Cumulative Poisson Function

The incomplete gamma function is defined by

$$P(a,x) \equiv \frac{\gamma(a,x)}{\Gamma(a)} \equiv \frac{1}{\Gamma(a)} \int_0^x e^{-t} t^{a-1} dt \qquad (a > 0) \tag{6.2.1}$$

It has the limiting values

$$P(a,0) = 0 \qquad \text{and} \qquad P(a,\infty) = 1 \tag{6.2.2}$$

The incomplete gamma function $P(a,x)$ is monotonic and (for a greater than one or so) rises from "near-zero" to "near-unity" in a range of x centered on about $a-1$, and of width about $\sqrt{a}$ (see Figure 6.2.1).

The complement of $P(a,x)$ is also confusingly called an incomplete gamma function,

$$Q(a,x) \equiv 1 - P(a,x) \equiv \frac{\Gamma(a,x)}{\Gamma(a)} \equiv \frac{1}{\Gamma(a)} \int_x^\infty e^{-t} t^{a-1} dt \qquad (a > 0) \tag{6.2.3}$$

It has the limiting values

$$Q(a,0) = 1 \qquad \text{and} \qquad Q(a,\infty) = 0 \tag{6.2.4}$$

The notations $P(a,x), \gamma(a,x)$, and $\Gamma(a,x)$ are standard; the notation $Q(a,x)$ is specific to this book.

There is a series development for $\gamma(a,x)$ as follows:

$$\gamma(a,x) = e^{-x} x^a \sum_{n=0}^{\infty} \frac{\Gamma(a)}{\Gamma(a+1+n)} x^n \tag{6.2.5}$$

One does not actually need to compute a new $\Gamma(a+1+n)$ for each n; one rather uses equation (6.1.3) and the previous coefficient.

A continued fraction development for $\Gamma(a,x)$ is

$$\Gamma(a,x) = e^{-x} x^a \left(\frac{1}{x+} \, \frac{1-a}{1+} \, \frac{1}{x+} \, \frac{2-a}{1+} \, \frac{2}{x+} \cdots \right) \qquad (x > 0) \tag{6.2.6}$$

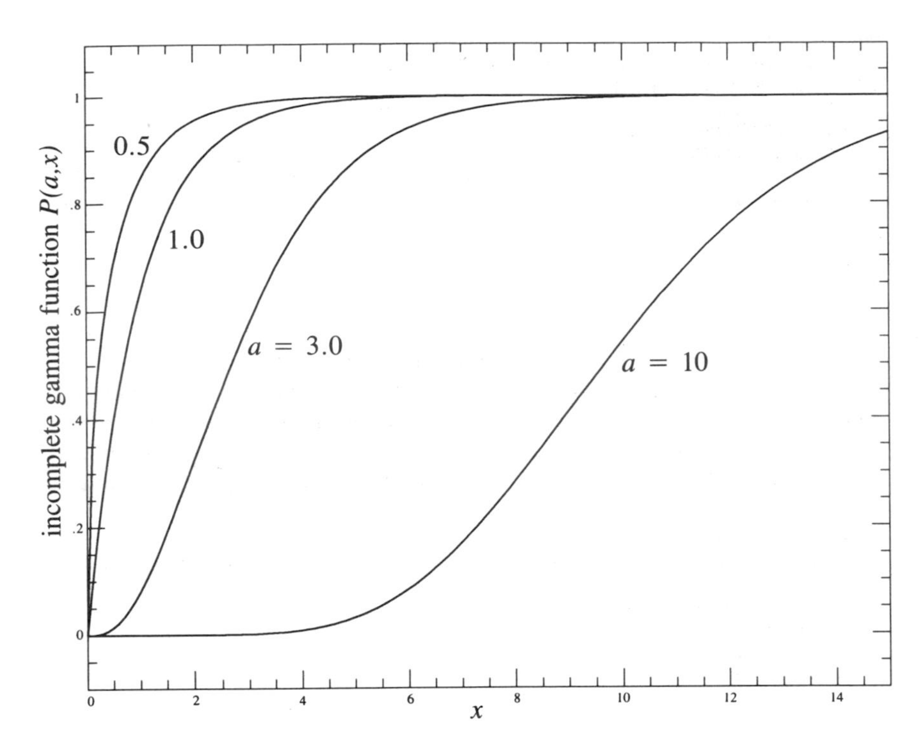

Figure 6.2.1. The incomplete gamma function $P(a, x)$ for four values of a.

It turns out that (6.2.5) converges rapidly for x less than about $a+1$, while (6.2.6) converges rapidly for x greater than about $a + 1$. In these respective regimes each requires at most a few times $\sqrt{a}$ terms to converge, and this many only near $x = a$, where the incomplete gamma functions are varying most rapidly. Thus (6.2.5) and (6.2.6) together allow evaluation of the function for all positive a and x. An extra dividend is that we never need compute a function value near zero by subtracting two nearly equal numbers. The higher-level calls for $P(a, x)$ and $Q(a, x)$ are

```
FUNCTION GAMMP(A,X)
    Returns the incomplete gamma function P(a,x)
IF(X.LT.0..OR.A.LE.0.)PAUSE
IF(X.LT.A+1.)THEN                 Use the series representation.
    CALL GSER(GAMSER,A,X,GLN)
    GAMMP=GAMSER
ELSE                              Use the continued fraction representation
    CALL GCF(GAMMCF,A,X,GLN)
    GAMMP=1.-GAMMCF               and take its complement.
ENDIF
RETURN
END
```

```
      FUNCTION GAMMQ(A,X)
C         Returns the incomplete gamma function Q(a,x) ≡ 1 - P(a,x).
      IF(X.LT.0..OR.A.LE.0.)PAUSE
      IF(X.LT.A+1.)THEN                   Use the series representation
          CALL GSER(GAMSER,A,X,GLN)
          GAMMQ=1.-GAMSER                 and take its complement.
      ELSE                                Use the continued fraction representation.
          CALL GCF(GAMMCF,A,X,GLN)
          GAMMQ=GAMMCF
      ENDIF
      RETURN
      END
```

The argument GLN is returned by both the series and continued fraction procedures containing the value $\ln\Gamma(a)$; the reason for this is so that it is available to you if you want to modify the above two procedures to give $\gamma(a,x)$ and $\Gamma(a,x)$, in addition to $P(a,x)$ and $Q(a,x)$ (cf. equations 6.2.1 and 6.2.3).

The procedures GSER and GCF which implement (6.2.5) and (6.2.6) are

```
      SUBROUTINE GSER(GAMSER,A,X,GLN)
C         Returns the incomplete gamma function P(a,x) evaluated by its series representation as
C         GAMSER. Also returns ln Γ(a) as GLN.
      PARAMETER (ITMAX=100,EPS=3.E-7)
      GLN=GAMMLN(A)
      IF(X.LE.0.)THEN
          IF(X.LT.0.)PAUSE
          GAMSER=0.
          RETURN
      ENDIF
      AP=A
      SUM=1./A
      DEL=SUM
      DO 11 N=1,ITMAX
          AP=AP+1.
          DEL=DEL*X/AP
          SUM=SUM+DEL
          IF(ABS(DEL).LT.ABS(SUM)*EPS)GO TO 1
11    CONTINUE
      PAUSE 'A too large, ITMAX too small'
1     GAMSER=SUM*EXP(-X+A*LOG(X)-GLN)
      RETURN
      END
```

```
      SUBROUTINE GCF(GAMMCF,A,X,GLN)
C         Returns the incomplete gamma function Q(a,x) evaluated by its continued fraction rep-
C         resentation as GAMMCF. Also returns Γ(a) as GLN.
      PARAMETER (ITMAX=100,EPS=3.E-7)
      GLN=GAMMLN(A)
      GOLD=0.                This is the previous value, tested against for convergence.
      A0=1.
      A1=X                   We are here setting up the A's and B's of equation (5.2.4) for eval-
      B0=0.                      uating the continued fraction.
      B1=1.
      FAC=1.                 FAC is the renormalization factor for preventing overflow of the partial
      DO 11 N=1,ITMAX            numerators and denominators.
          AN=FLOAT(N)
          ANA=AN-A
          A0=(A1+A0*ANA)*FAC     One step of the recurrence (5.2.5).
```

```
      BO=(B1+BO*ANA)*FAC
      ANF=AN*FAC
      A1=X*AO+ANF*A1              The next step of the recurrence (5.2.5).
      B1=X*BO+ANF*B1
      IF(A1.NE.O.)THEN            Shall we renormalize?
        FAC=1./A1                 Yes. Set FAC so it happens.
        G=B1*FAC                  New value of answer.
        IF(ABS((G-GOLD)/G).LT.EPS)GO TO 1     Converged? If so, exit.
        GOLD=G                          If not, save value.
      ENDIF
11  CONTINUE
    PAUSE 'A too large, ITMAX too small'
1   GAMMCF=EXP(-X+A*ALOG(X)-GLN)*G        Put factors in front.
    RETURN
    END
```

Error Function

The error function and complementary error function are special cases of the incomplete gamma function, and are obtained moderately efficiently by the above procedures. Their definitions are

$$\operatorname{erf}(x) = \frac{2}{\sqrt{\pi}} \int_0^x e^{-t^2} dt \tag{6.2.7}$$

and

$$\operatorname{erfc}(x) \equiv 1 - \operatorname{erf}(x) = \frac{2}{\sqrt{\pi}} \int_x^\infty e^{-t^2} dt \tag{6.2.8}$$

The functions have the following limiting values and symmetries:

$$\operatorname{erf}(0) = 0 \qquad \operatorname{erf}(\infty) = 1 \qquad \operatorname{erf}(-x) = -\operatorname{erf}(x) \tag{6.2.9}$$

$$\operatorname{erfc}(0) = 1 \qquad \operatorname{erfc}(\infty) = 0 \qquad \operatorname{erfc}(-x) = 2 - \operatorname{erfc}(x) \tag{6.2.10}$$

They are related to the incomplete gamma functions by

$$\operatorname{erf}(x) = P(\frac{1}{2}, x^2) \qquad (x \geq 0) \tag{6.2.11}$$

and

$$\operatorname{erfc}(x) = Q(\frac{1}{2}, x^2) \qquad (x \geq 0) \tag{6.2.12}$$

Hence we have

```
FUNCTION ERF(X)
    Returns the error function erf(X).
IF(X.LT.0.)THEN
    ERF=-GAMMP(.5,X**2)
ELSE
    ERF=GAMMP(.5,X**2)
ENDIF
RETURN
END
```

```
FUNCTION ERFC(X)
    Returns the complementary error function erfc(X).
IF(X.LT.0.)THEN
    ERFC=1.+GAMMP(.5,X**2)
ELSE
    ERFC=GAMMQ(.5,X**2)
ENDIF
RETURN
END
```

If you care to do so, you can easily remedy the minor inefficiency in ERF and ERFC, namely that $\Gamma(0.5) = \sqrt{\pi}$ is computed unnecessarily when GAMMP or GAMMQ is called. Before you do that, however, you might wish to consider the following routine, based on Chebyshev fitting to an inspired guess as to the functional form:

```
      FUNCTION ERFCC(X)
          Returns the complementary error function erfc(X) with fractional error everywhere less
          than 1.2 × 10^-7.
      Z=ABS(X)
      T=1./(1.+0.5*Z)
      ERFCC=T*EXP(-Z*Z-1.26551223+T*(1.00002368+T*(.37409196+
*         T*(.09678418+T*(-.18628806+T*(.27886807+T*(-1.13520398+
*         T*(1.48851587+T*(-.82215223+T*.17087277)))))))))
      IF (X.LT.0.)  ERFCC=2.-ERFCC
      RETURN
      END
```

There are also some functions of *two* variables which are special cases of the incomplete gamma function:

Cumulative Poisson Probability Function

$P_x(< k)$, for positive x and integer $k \geq 1$, denotes the *cumulative Poisson probability* function. It is defined as the probability that the number of Poisson random events occurring will be between 0 and $k-1$ *inclusive*, if the expected mean number is x. It has the limiting values

$$P_x(< 1) = e^{-x} \qquad P_x(\infty) = 1 \tag{6.2.13}$$

Its relation to the incomplete gamma function is simply

$$P_x(< k) = Q(k, x) = \texttt{GAMMQ}\ (k, x) \tag{6.2.14}$$

Chi-Square Probability Function

$P(\chi^2|\nu)$ is defined as the probability that the observed chi-square for a correct model should be less than a value χ^2. (We will discuss the use of this function in Chapter 14.) Its complement $Q(\chi^2|\nu)$ is the probability that the observed chi-square will exceed the value χ^2 by chance *even* for a correct model. In both cases ν is an integer, the number of degrees of freedom. The functions have the limiting values

$$P(0|\nu) = 0 \qquad P(\infty|\nu) = 1 \tag{6.2.15}$$

$$Q(0|\nu) = 1 \qquad Q(\infty|\nu) = 0 \tag{6.2.16}$$

and the following relation to the incomplete gamma functions,

$$P(\chi^2|\nu) = P(\frac{\nu}{2}, \frac{\chi^2}{2}) = \texttt{GAMMP}\ (\frac{\nu}{2}, \frac{\chi^2}{2}) \tag{6.2.17}$$

$$Q(\chi^2|\nu) = Q(\frac{\nu}{2}, \frac{\chi^2}{2}) = \texttt{GAMMQ}\ (\frac{\nu}{2}, \frac{\chi^2}{2}) \tag{6.2.18}$$

REFERENCES AND FURTHER READING:

Abramowitz, Milton, and Stegun, Irene A. 1964, *Handbook of Mathematical Functions*, Applied Mathematics Series, vol. 55 (Washington: National Bureau of Standards; reprinted 1968 by Dover Publications, New York), Chapters 6,7, and 26.

Pearson, K. (ed.) 1951, *Tables of the Incomplete Gamma Function* (Cambridge: Cambridge University Press).

6.3 Incomplete Beta Function, Student's Distribution, F-Distribution, Cumulative Binomial Distribution

The incomplete beta function is defined by

$$I_x(a,b) \equiv \frac{B_x(a,b)}{B(a,b)} \equiv \frac{1}{B(a,b)} \int_0^x t^{a-1}(1-t)^{b-1}dt \qquad (a,b>0) \tag{6.3.1}$$

It has the limiting values

$$I_0(a,b) = 0 \qquad I_1(a,b) = 1 \tag{6.3.2}$$

and the symmetry relation

$$I_x(a,b) = 1 - I_{1-x}(b,a) \tag{6.3.3}$$

If a and b are both rather greater than one, then $I_x(a,b)$ rises from "near-zero" to "near-unity" quite sharply at about $x = a/(a+b)$. Figure 6.3.1 plots the function for several pairs (a,b).

The incomplete beta function has a series expansion

$$I_x(a,b) = \frac{x^a(1-x)^b}{aB(a,b)} \left[1 + \sum_{n=0}^{\infty} \frac{B(a+1,n+1)}{B(a+b,n+1)} x^{n+1}\right], \tag{6.3.4}$$

but this does not prove to be very useful in its numerical evaluation. (Note, however, that the beta functions in the coefficients can be evaluated for each value of n with just the previous value and a few multiplies, using equations 6.1.8 and 6.1.3.)

The continued fraction representation proves to be much more useful,

$$I_x(a,b) = \frac{x^a(1-x)^b}{aB(a,b)} \left[\frac{1}{1+}\frac{d_1}{1+}\frac{d_2}{1+}\cdots\right] \tag{6.3.5}$$

where

$$\begin{aligned} d_{2m+1} &= -\frac{(a+m)(a+b+m)x}{(a+2m)(a+2m+1)} \\ d_{2m} &= \frac{m(b-m)x}{(a+2m-1)(a+2m)} \end{aligned} \tag{6.3.6}$$

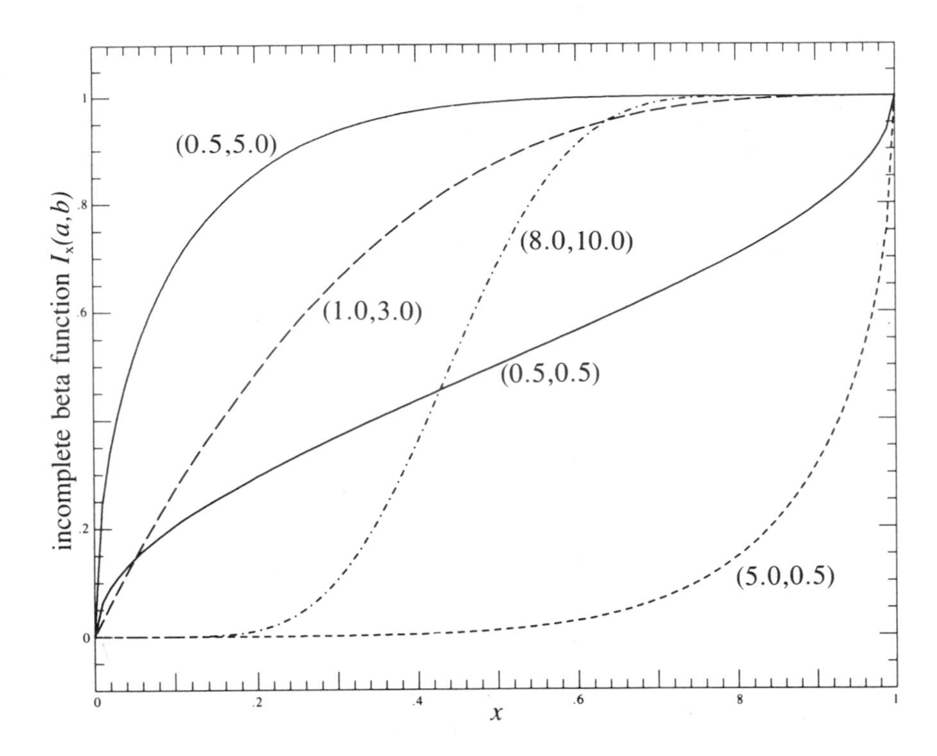

Figure 6.3.1. The incomplete beta function $I_x(a,b)$ for five different pairs of (a,b). Notice that the pairs $(0.5, 5.0)$ and $(5.0, 0.5)$ are related by reflection symmetry around the diagonal (cf. equation 6.3.3).

This continued fraction converges rapidly for $x < (a+1)/(a+b+2)$, taking in the worst case $O(\sqrt{\max(a,b)})$ iterations. But for $x > (a+1)/(a+b+2)$ we can just use the symmetry relation (6.3.3) to obtain an equivalent computation where the continued fraction will also converge rapidly. Hence we have

```
      FUNCTION BETAI(A,B,X)
C        Returns the incomplete beta function I_X(A,B).
      IF(X.LT.0..OR.X.GT.1.)PAUSE 'bad argument X in BETAI'
      IF(X.EQ.0..OR.X.EQ.1.)THEN
        BT=0.
      ELSE                                 Factors in front of the continued fraction.
        BT=EXP(GAMMLN(A+B)-GAMMLN(A)-GAMMLN(B)
     *      +A*ALOG(X)+B*ALOG(1.-X))
      ENDIF
      IF(X.LT.(A+1.)/(A+B+2.))THEN Use continued fraction directly.
        BETAI=BT*BETACF(A,B,X)/A
        RETURN
      ELSE
        BETAI=1.-BT*BETACF(B,A,1.-X)/B  Use continued fraction after making the symmetry trans-
        RETURN                              formation.
      ENDIF
      END
```

which calls the continued fraction evaluation routine

```
FUNCTION BETACF(A,B,X)
   Continued fraction for incomplete beta function, used by BETAI.
PARAMETER (ITMAX=100,EPS=3.E-7)
AM=1.
BM=1.
AZ=1.
QAB=A+B                      These Q's will be used in factors which occur in the coefficients
QAP=A+1.                        (6.3.6).
QAM=A-1.
BZ=1.-QAB*X/QAP
DO 11 M=1,ITMAX              Continued fraction evaluation by the recurrence method (5.2.5).
    EM=M
    TEM=EM+EM
    D=EM*(B-M)*X/((QAM+TEM)*(A+TEM))
    AP=AZ+D*AM               One step (the even one) of the recurrence.
    BP=BZ+D*BM
    D=-(A+EM)*(QAB+EM)*X/((A+TEM)*(QAP+TEM))
    APP=AP+D*AZ              Next step of the recurrence (the odd one).
    BPP=BP+D*BZ
    AOLD=AZ                  Save the old answer.
    AM=AP/BPP                Renormalize to prevent overflows.
    BM=BP/BPP
    AZ=APP/BPP
    BZ=1.
    IF(ABS(AZ-AOLD).LT.EPS*ABS(AZ))GO TO 1      Are we done?
11  CONTINUE
PAUSE 'A or B too big, or ITMAX too small'
1   BETACF=AZ
RETURN
END
```

Student's Distribution Probability Function

Student's distribution, denoted $A(t|\nu)$, is useful in several statistical contexts, notably in the test of whether two observed distributions have the same mean. $A(t|\nu)$ is the probability, for ν degrees of freedom, that a certain statistic t (measuring the observed difference of means) would be smaller than the observed value if the means were in fact the same. (See Chapter 14 for further details.) Two means are significantly different if, e.g., $A(t|\nu) > 0.99$. In other words, $1 - A(t|\nu)$ is the significance level at which the hypothesis that the means are equal is disproved.

The mathematical definition of the function is

$$A(t|\nu) = \frac{1}{\nu^{1/2} B(\frac{1}{2}, \frac{\nu}{2})} \int_{-t}^{t} (1 + \frac{x^2}{\nu})^{-\frac{\nu+1}{2}} dx \tag{6.3.7}$$

Limiting values are

$$A(0|\nu) = 0 \qquad A(\infty|\nu) = 1 \tag{6.3.8}$$

$A(t|\nu)$ is related to the incomplete beta function $I_x(a,b)$ by

$$A(t|\nu) = 1 - I_{\frac{\nu}{\nu+t^2}}(\frac{\nu}{2}, \frac{1}{2}) \qquad (6.3.9)$$

So, you can use (6.3.9) and the above routine BETAI to evaluate the function.

F-Distribution Probability Function

This function occurs in the statistical test of whether two observed samples have the same variance. A certain statistic F, essentially the ratio of the observed dispersion of the first sample to that of the second one, is calculated. (For further details, see Chapter 14.) The probability that F would be as *large* as it is if the first sample's underlying distribution actually has *smaller* variance than the second's is denoted $Q(F|\nu_1, \nu_2)$, where ν_1 and ν_2 are the number of degrees of freedom in the first and second samples respectively. In other words, $Q(F|\nu_1, \nu_2)$ is the significance level at which the hypothesis "1 has smaller variance than 2" can be rejected. A small numerical value implies a very significant rejection, in turn implying high confidence in the hypothesis "1 has variance greater or equal to 2."

$Q(F|\nu_1, \nu_2)$ has the limiting values

$$Q(0|\nu_1, \nu_2) = 1 \qquad Q(\infty|\nu_1, \nu_2) = 0 \qquad (6.3.10)$$

Its relation to the incomplete beta function $I_x(a,b)$ as evaluated by BETAI above is

$$Q(F|\nu_1, \nu_2) = I_{\frac{\nu_2}{\nu_2+\nu_1 F}}(\frac{\nu_2}{2}, \frac{\nu_1}{2}) \qquad (6.3.11)$$

Cumulative Binomial Probability Distribution

Suppose an event occurs with probability p per trial. Then the probability P of its occurring k *or more* times in n trials is termed a *cumulative binomial probability*, and is related to the incomplete beta function $I_x(a,b)$ as follows:

$$P \equiv \sum_{j=k}^{n} \binom{n}{j} p^j (1-p)^{n-j} = I_p(k, n-k+1) \qquad (6.3.12)$$

For n larger than a dozen or so, BETAI is a much better way to evaluate the sum in (6.3.12) than would be the straightforward sum with concurrent computation of the binomial coefficients. (For n smaller than a dozen, either method is acceptable.)

REFERENCES AND FURTHER READING:

Abramowitz, Milton, and Stegun, Irene A. 1964, *Handbook of Mathematical Functions*, Applied Mathematics Series, vol. 55 (Washington: National Bureau of Standards; reprinted 1968 by Dover Publications, New York), Chapters 6 and 26.

Pearson, E. and Johnson, N. 1968, *Tables of the Incomplete Beta Function* (Cambridge: Cambridge University Press).

6.4 Bessel Functions of Integer Order

For any real ν, the Bessel function $J_\nu(z)$ can be defined by the series representation

$$J_\nu(z) = \left(\frac{1}{2}z\right)^\nu \sum_{k=0}^{\infty} \frac{(-\frac{1}{4}z^2)^k}{k!\Gamma(\nu+k+1)} \tag{6.4.1}$$

The series converges for all z, but it is not computationally very useful for $z \gg 1$.

For ν *not* an integer the Bessel function $Y_\nu(z)$ is given by

$$Y_\nu(z) = \frac{J_\nu(z)\cos(\nu\pi) - J_{-\nu}(z)}{\sin(\nu\pi)} \tag{6.4.2}$$

The right-hand side goes to the correct limiting value $Y_n(z)$ as ν goes to some integer n, but this is not computationally useful.

For arguments $x < \nu$, both Bessel functions look qualitatively like simple power laws, with the asymptotic forms for $0 < x \ll \nu$

$$\begin{aligned} J_\nu(x) &\sim \frac{1}{\Gamma(\nu+1)}(\frac{1}{2}x)^\nu \qquad \nu \ge 0 \\ Y_0(x) &\sim \frac{2}{\pi}\ln(x) \\ Y_\nu(x) &\sim -\frac{\Gamma(\nu)}{\pi}(\frac{1}{2}x)^{-\nu} \qquad \nu > 0 \end{aligned} \tag{6.4.3}$$

For $x > \nu$, both Bessel functions look qualitatively like sine or cosine waves whose amplitude decays as $x^{-1/2}$. The asymptotic forms for $x \gg \nu$ are

$$\begin{aligned} J_\nu(x) &\sim \sqrt{\frac{2}{\pi x}}\cos(x - \frac{1}{2}\nu\pi - \frac{1}{4}\pi) \\ Y_\nu(x) &\sim \sqrt{\frac{2}{\pi x}}\sin(x - \frac{1}{2}\nu\pi - \frac{1}{4}\pi) \end{aligned} \tag{6.4.4}$$

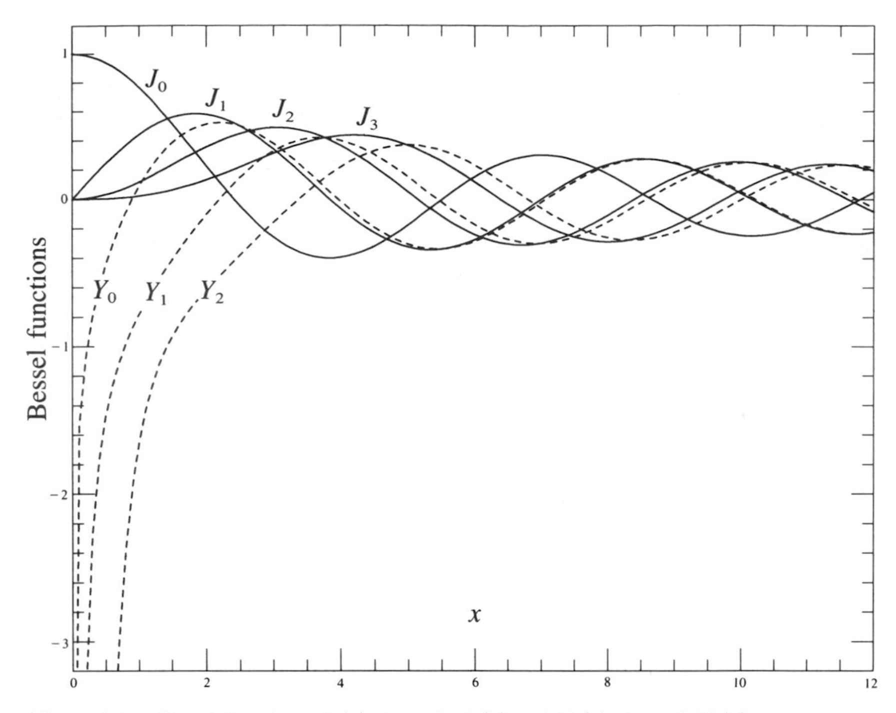

Figure 6.4.1. Bessel functions $J_0(x)$ through $J_3(x)$ and $Y_0(x)$ through $Y_2(x)$.

In the transition region where $x \sim \nu$, the typical amplitudes of the Bessel functions are on the order

$$
\begin{aligned}
J_\nu(\nu) &\sim \frac{2^{1/3}}{3^{2/3}\Gamma(\frac{2}{3})}\frac{1}{\nu^{1/3}} \sim \frac{0.4473}{\nu^{1/3}} \\
Y_\nu(\nu) &\sim -\frac{2^{1/3}}{3^{1/6}\Gamma(\frac{2}{3})}\frac{1}{\nu^{1/3}} \sim -\frac{0.7748}{\nu^{1/3}}
\end{aligned}
\qquad (6.4.5)
$$

which holds asymptotically for large ν. Figure 6.4.1 plots the first few Bessel functions of each kind.

The Bessel functions satisfy the recurrence relations

$$
J_{n+1}(x) = \frac{2n}{x}J_n(x) - J_{n-1}(x) \qquad (6.4.6)
$$

and

$$
Y_{n+1}(x) = \frac{2n}{x}Y_n(x) - Y_{n-1}(x) \qquad (6.4.7)
$$

As already mentioned in §5.4, only the second of these (6.4.7) is stable in the direction of increasing n for $x < n$. The reason that (6.4.6) is unstable in the direction of increasing n is simply that it is *the same recurrence* as (6.4.7): a small amount of "polluting" Y_n introduced by roundoff error will quickly come to swamp the desired J_n, according to equation (6.4.3).

A practical strategy for computing the Bessel functions of integer order divides into two tasks: first, how to compute J_0, J_1, Y_0, and Y_1, and second, how to use the recurrence relations stably to find other J's and Y's. We treat the first task first:

For x between zero and some arbitrary value (we will use the value 8), approximate $J_0(x)$ and $J_1(x)$ by rational functions in x. Likewise approximate by rational functions the "regular part" of $Y_0(x)$ and $Y_1(x)$, defined as

$$Y_0(x) - \frac{2}{\pi} J_0(x) \ln(x) \qquad \text{and} \qquad Y_1(x) - \frac{2}{\pi}\left[J_1(x)\ln(x) - \frac{1}{x}\right] \tag{6.4.8}$$

For $8 < x < \infty$, use the approximating forms ($n = 0, 1$)

$$J_n(x) = \sqrt{\frac{2}{\pi x}}\left[P_n\left(\frac{8}{x}\right)\cos(X_n) - Q_n\left(\frac{8}{x}\right)\sin(X_n)\right] \tag{6.4.9}$$

$$Y_n(x) = \sqrt{\frac{2}{\pi x}}\left[P_n\left(\frac{8}{x}\right)\sin(X_n) + Q_n\left(\frac{8}{x}\right)\cos(X_n)\right] \tag{6.4.10}$$

where

$$X_n \equiv x - \frac{2n+1}{4}\pi \tag{6.4.11}$$

and where P_0, P_1, Q_0, and Q_1 are each polynomials in their arguments, for $0 < 8/x < 1$. The P's are even polynomials, the Q's odd.

Coefficients of the various rational functions and polynomials are given by Hart (1968), for various levels of desired accuracy. A straightforward implementation is

```
      FUNCTION BESSJ0(X)
         Returns the Bessel function J0(X) for any real X.
      REAL*8 Y,P1,P2,P3,P4,P5,Q1,Q2,Q3,Q4,Q5,R1,R2,R3,R4,R5,R6,
     *    S1,S2,S3,S4,S5,S6          We'll accumulate polynomials in double precision.
      DATA P1,P2,P3,P4,P5/1.D0,-.1098628627D-2,.2734510407D-4,
     *    -.2073370639D-5,.2093887211D-6/, Q1,Q2,Q3,Q4,Q5/-.1562499995D-1,
     *    .1430488765D-3,-.6911147651D-5,.7621095161D-6,-.934945152D-7/
      DATA R1,R2,R3,R4,R5,R6/57568490574.D0,-13362590354.D0,651619640.7D0,
     *    -11214424.18D0,77392.33017D0,-184.9052456D0/,
     *    S1,S2,S3,S4,S5,S6/57568490411.D0,1029532985.D0,
     *    9494680.718D0,59272.64853D0,267.8532712D0,1.D0/
```

```
      IF(ABS(X).LT.8.)THEN          Direct rational function fit.
        Y=X**2
        BESSJ0=(R1+Y*(R2+Y*(R3+Y*(R4+Y*(R5+Y*R6)))))
     *      /(S1+Y*(S2+Y*(S3+Y*(S4+Y*(S5+Y*S6)))))
      ELSE                          Fitting function (6.4.9).
        AX=ABS(X)
        Z=8./AX
        Y=Z**2
        XX=AX-.785398164
        BESSJ0=SQRT(.636619772/AX)*(COS(XX)*(P1+Y*(P2+Y*(P3+Y*(P4+Y
     *      *P5))))-Z*SIN(XX)*(Q1+Y*(Q2+Y*(Q3+Y*(Q4+Y*Q5)))))
      ENDIF
      RETURN
      END
```

```
      FUNCTION BESSY0(X)
        Returns the Bessel function Y0(X) for positive X.
      REAL*8 Y,P1,P2,P3,P4,P5,Q1,Q2,Q3,Q4,Q5,R1,R2,R3,R4,R5,R6,
     *    S1,S2,S3,S4,S5,S6         We'll accumulate polynomials in double precision.
      DATA P1,P2,P3,P4,P5/1.D0,-.1098628627D-2,.2734510407D-4,
     *    -.2073370639D-5,.2093887211D-6/, Q1,Q2,Q3,Q4,Q5/-.1562499995D-1,
     *    .1430488765D-3,-.6911147651D-5,.7621095161D-6,-.934945152D-7/
      DATA R1,R2,R3,R4,R5,R6/-2957821389.D0,7062834065.D0,-512359803.6D0,
     *    10879881.29D0,-86327.92757D0,228.4622733D0/,
     *    S1,S2,S3,S4,S5,S6/40076544269.D0,745249964.8D0,
     *    7189466.438D0,47447.26470D0,226.1030244D0,1.D0/
      IF(X.LT.8.)THEN               Rational function approximation of (6.4.8).
        Y=X**2
        BESSY0=(R1+Y*(R2+Y*(R3+Y*(R4+Y*(R5+Y*R6)))))/(S1+Y*(S2+Y
     *      *(S3+Y*(S4+Y*(S5+Y*S6)))))+.636619772*BESSJ0(X)*LOG(X)
      ELSE                          Fitting function 6.4.10.
        Z=8./X
        Y=Z**2
        XX=X-.785398164
        BESSY0=SQRT(.636619772/X)*(SIN(XX)*(P1+Y*(P2+Y*(P3+Y*(P4+Y*
     *      P5))))+Z*COS(XX)*(Q1+Y*(Q2+Y*(Q3+Y*(Q4+Y*Q5)))))
      ENDIF
      RETURN
      END
```

```
      FUNCTION BESSJ1(X)
        Returns the Bessel function J1(X) for any real X.
      REAL*8 Y,P1,P2,P3,P4,P5,Q1,Q2,Q3,Q4,Q5,R1,R2,R3,R4,R5,R6,
     *    S1,S2,S3,S4,S5,S6         We'll accumulate polynomials in double precision.
      DATA R1,R2,R3,R4,R5,R6/72362614232.D0,-7895059235.D0,242396853.1D0,
     *    -2972611.439D0,15704.48260D0,-30.16036606D0/,
     *    S1,S2,S3,S4,S5,S6/144725228442.D0,2300535178.D0,
     *    18583304.74D0,99447.43394D0,376.9991397D0,1.D0/
      DATA P1,P2,P3,P4,P5/1.D0,.183105D-2,-.3516396496D-4,.2457520174D-5,
     *    -.240337019D-6/, Q1,Q2,Q3,Q4,Q5/.04687499995D0,-.2002690873D-3,
     *    .8449199096D-5,-.88228987D-6,.105787412D-6/
      IF(ABS(X).LT.8.)THEN          Direct rational approximation.
        Y=X**2
        BESSJ1=X*(R1+Y*(R2+Y*(R3+Y*(R4+Y*(R5+Y*R6)))))
     *      /(S1+Y*(S2+Y*(S3+Y*(S4+Y*(S5+Y*S6)))))
      ELSE                          Fitting function (6.4.9).
        AX=ABS(X)
        Z=8./AX
```

```
      Y=Z**2
      XX=AX-2.356194491
      BESSJ1=SQRT(.636619772/AX)*(COS(XX)*(P1+Y*(P2+Y*(P3+Y*(P4+Y
*         *P5))))-Z*SIN(XX)*(Q1+Y*(Q2+Y*(Q3+Y*(Q4+Y*Q5)))))
*         *SIGN(1.,X)
    ENDIF
    RETURN
    END
```

```
    FUNCTION BESSY1(X)
      Returns the Bessel function Y1(X) for positive X.
    REAL*8 Y,P1,P2,P3,P4,P5,Q1,Q2,Q3,Q4,Q5,R1,R2,R3,R4,R5,R6,
*       S1,S2,S3,S4,S5,S6,S7     We'll accumulate polynomials in double precision.
    DATA P1,P2,P3,P4,P5/1.D0,.183105D-2,-.3516396496D-4,.2457520174D-5,
*       -.240337019D-6/, Q1,Q2,Q3,Q4,Q5/.04687499995D0,-.2002690873D-3,
*       .8449199096D-5,-.88228987D-6,.105787412D-6/
    DATA R1,R2,R3,R4,R5,R6/-.4900604943D13,.1275274390D13,-.5153438139D11,
*       .7349264551D9,-.4237922726D7,.8511937935D4/,
*       S1,S2,S3,S4,S5,S6,S7/.2499580570D14,.4244419664D12,
*       .3733650367D10,.2245904002D8,.1020426050D6,.3549632885D3,1.D0/
    IF(X.LT.8.)THEN                   Rational function approximation of (6.4.8).
      Y=X**2
      BESSY1=X*(R1+Y*(R2+Y*(R3+Y*(R4+Y*(R5+Y*R6)))))/(S1+Y*(S2+Y*
*         (S3+Y*(S4+Y*(S5+Y*(S6+Y*S7))))))+.636619772
*         *(BESSJ1(X)*LOG(X)-1./X)
    ELSE                              Fitting function (6.4.10).
      Z=8./X
      Y=Z**2
      XX=X-2.356194491
      BESSY1=SQRT(.636619772/X)*(SIN(XX)*(P1+Y*(P2+Y*(P3+Y*(P4+Y
*         *P5))))+Z*COS(XX)*(Q1+Y*(Q2+Y*(Q3+Y*(Q4+Y*Q5)))))
    ENDIF
    RETURN
    END
```

We now turn to the second task, namely how to use the recurrence formulas (6.4.6) and (6.4.7) to get the Bessel functions $J_n(x)$ and $Y_n(x)$ for $n \geq 2$. The latter of these is straightforward, since its upward recurrence is always stable:

```
    FUNCTION BESSY(N,X)
      Returns the Bessel function YN(X) for positive X and N≥ 2.
    IF(N.LT.2)PAUSE 'bad argument N in BESSY'
    TOX=2./X
    BY=BESSY1(X)                      Starting values for the recurrence.
    BYM=BESSY0(X)
    DO 11 J=1,N-1                     Recurrence (6.4.7).
      BYP=J*TOX*BY-BYM
      BYM=BY
      BY=BYP
11  CONTINUE
    BESSY=BY
    RETURN
    END
```

The cost of this algorithm is the call to BESSY1 and BESSY0 (which generate a call to each of BESSJ1 and BESSJ0), plus $O(\mathrm{N})$ operations in the recurrence.

As for $J_n(x)$, things are a bit more complicated. We can start the recurrence upward on n from J_0 and J_1, but it will remain stable only while n does not exceed x. This is, however, just fine for calls with large x and small n, a case which occurs frequently in practice.

The harder case to provide for is that with $x < n$. The best thing to do here is to use Miller's algorithm (see discussion preceding equation 5.4.6), applying the recurrence *downward* from some arbitrary starting value and making use of the upward-unstable nature of the recurrence to put us *onto* the correct solution. When we finally arrive at J_0 or J_1 we are able to normalize the solution with the sum (5.4.6) accumulated along the way.

The only subtlety is in deciding at how large an n we need start the downward recurrence so as to obtain a desired accuracy by the time we reach the n that we really want. If you play with the asymptotic forms (6.4.3) and (6.4.5), you should be able to convince yourself that the answer is to start larger than the desired n by an additive amount of order $[\text{constant} \times n]^{1/2}$, where the square root of the constant is, very roughly, the number of significant figures of accuracy.

The above considerations lead to the following routine.

```
FUNCTION BESSJ(N,X)
   Returns the Bessel function J_N(X) for any real X and N≥ 2.
PARAMETER (IACC=40,BIGNO=1.E10,BIGNI=1.E-10)
IF(N.LT.2)PAUSE 'bad argument N in BESSJ'
AX=ABS(X)
IF(AX.EQ.0.)THEN
    BESSJ=0.
ELSE IF(AX.GT.FLOAT(N))THEN          Use upwards recurrence from J0 and J1.
    TOX=2./AX
    BJM=BESSJ0(AX)
    BJ=BESSJ1(AX)
    DO 11 J=1,N-1
        BJP=J*TOX*BJ-BJM
        BJM=BJ
        BJ=BJP
  11    CONTINUE
    BESSJ=BJ
ELSE                         Use downwards recurrence from an even value M here computed.
    TOX=2./AX                              Make IACC larger to increase accuracy.
    M=2*((N+INT(SQRT(FLOAT(IACC*N))))/2)
    BESSJ=0.
    JSUM=0                   JSUM will alternate between 0 and 1; when it is 1, we accumulate in
    SUM=0.                       SUM the even terms in (5.4.6).
    BJP=0.
    BJ=1.
    DO 12 J=M,1,-1           The downward recurrence.
        BJM=J*TOX*BJ-BJP
        BJP=BJ
        BJ=BJM
        IF(ABS(BJ).GT.BIGNO)THEN            Renormalize to prevent overflows.
            BJ=BJ*BIGNI
            BJP=BJP*BIGNI
            BESSJ=BESSJ*BIGNI
            SUM=SUM*BIGNI
        ENDIF
        IF(JSUM.NE.0)SUM=SUM+BJ  Accumulate the sum.
```

```
        JSUM=1-JSUM               Change 0 to 1 or vice-versa.
        IF(J.EQ.N)BESSJ=BJP       Save the unnormalized answer.
12    CONTINUE
      SUM=2.*SUM-BJ               Compute (5.4.6)
      BESSJ=BESSJ/SUM             and use it to normalize the answer.
    ENDIF
    IF(X.LT.0..AND.MOD(N,2).EQ.1)BESSJ=-BESSJ
    RETURN
    END
```

REFERENCES AND FURTHER READING:

Abramowitz, Milton, and Stegun, Irene A. 1964, *Handbook of Mathematical Functions*, Applied Mathematics Series, vol. 55 (Washington: National Bureau of Standards; reprinted 1968 by Dover Publications, New York), Chapter 9.

Hart, John F., et al. 1968, *Computer Approximations* (New York: Wiley), §6.8, p.141.

6.5 Modified Bessel Functions of Integer Order

The modified Bessel functions $I_n(x)$ and $K_n(x)$ are equivalent to the usual Bessel functions J_n and Y_n evaluated for purely imaginary arguments. In detail, the relationship is

$$\begin{aligned} I_n(x) &= (-i)^n J_n(ix) \\ K_n(x) &= \frac{\pi}{2} i^{n+1}[J_n(ix) + iY_n(ix)] \end{aligned} \tag{6.5.1}$$

The particular choice of prefactor and of the linear combination of J_n and Y_n to form K_n are simply choices which make the functions real-valued for real arguments x.

For small arguments $x \ll n$, both $I_n(x)$ and $K_n(x)$ become, asympotically, simple powers of their argument

$$\begin{aligned} I_n(x) &\approx \frac{1}{n!}\left(\frac{x}{2}\right)^n \qquad n \geq 0 \\ K_0(x) &\approx -\ln(x) \\ K_n(x) &\approx \frac{(n-1)!}{2}\left(\frac{x}{2}\right)^{-n} \qquad n > 0 \end{aligned} \tag{6.5.2}$$

These expressions are virtually identical to those for $J_n(x)$ and $Y_n(x)$ in this region, except for the factor of $-2/\pi$ difference between $Y_n(x)$ and $K_n(x)$.

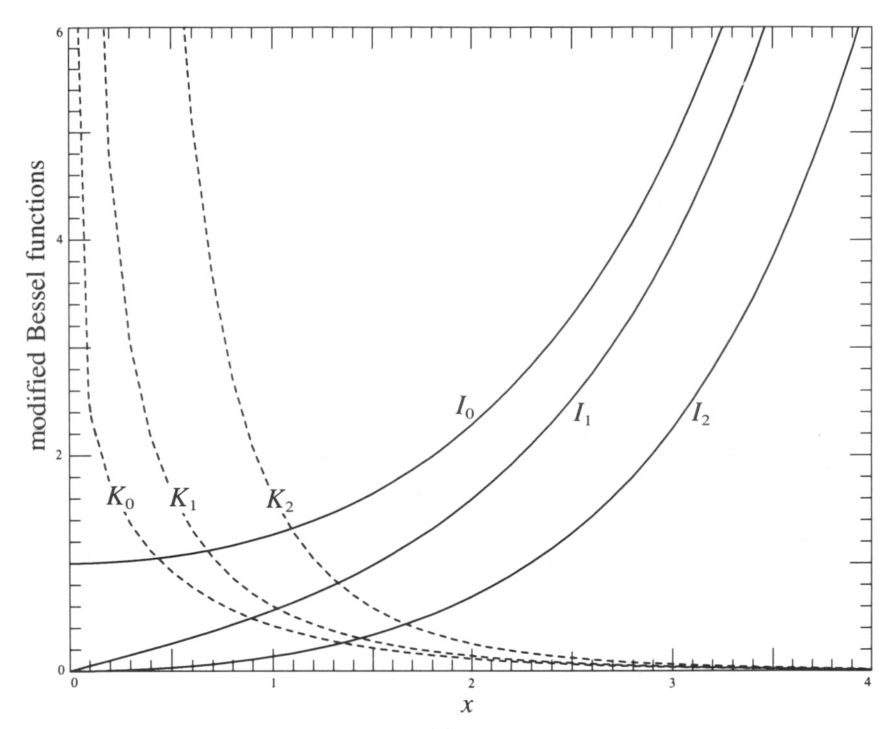

Figure 6.5.1. Modified Bessel functions $I_0(x)$, $I_1(x)$, $I_2(x)$, $K_0(x)$, $K_1(x)$, $K_2(x)$.

In the region $x \gg n$, however, the modified functions have quite different behavior than the Bessel functions,

$$
\begin{aligned}
I_n(x) &\approx \frac{1}{\sqrt{2\pi x}} \exp(x) \\
K_n(x) &\approx \frac{\pi}{\sqrt{2\pi x}} \exp(-x)
\end{aligned}
\tag{6.5.3}
$$

The modified functions evidently have exponential rather than sinusoidal behavior for large arguments (see Figure 6.5.1). The smoothness of the modified Bessel functions, once the exponential factor is removed, makes a simple polynomial approximation of a few terms quite suitable for the functions I_0, I_1, K_0 and K_1. The following routines, based on polynomial coefficients given by Abramowitz and Stegun, evaluate these four functions, and will provide the basis for upward recursion for $n > 1$ when $x > n$.

```
      FUNCTION BESSIO(X)
         Returns the modified Bessel function I0(X) for any real X.
      REAL*8 Y,P1,P2,P3,P4,P5,P6,P7,
     *    Q1,Q2,Q3,Q4,Q5,Q6,Q7,Q8,Q9   Accumulate polynomials in double precision.
      DATA P1,P2,P3,P4,P5,P6,P7/1.0D0,3.5156229D0,3.0899424D0,1.2067492D0,
     *    0.2659732D0,0.360768D-1,0.45813D-2/
      DATA Q1,Q2,Q3,Q4,Q5,Q6,Q7,Q8,Q9/0.39894228D0,0.1328592D-1,
     *    0.225319D-2,-0.157565D-2,0.916281D-2,-0.2057706D-1,
```

```
     *      0.2635537D-1,-0.1647633D-1,0.392377D-2/
      IF (ABS(X).LT.3.75) THEN
          Y=(X/3.75)**2
          BESSI0=P1+Y*(P2+Y*(P3+Y*(P4+Y*(P5+Y*(P6+Y*P7)))))
      ELSE
          AX=ABS(X)
          Y=3.75/AX
          BESSI0=(EXP(AX)/SQRT(AX))*(Q1+Y*(Q2+Y*(Q3+Y*(Q4
     *          +Y*(Q5+Y*(Q6+Y*(Q7+Y*(Q8+Y*Q9))))))))
      ENDIF
      RETURN
      END
```

```
      FUNCTION BESSK0(X)
          Returns the modified Bessel function K0(X) for positive real X.
      REAL*8 Y,P1,P2,P3,P4,P5,P6,P7,
     *     Q1,Q2,Q3,Q4,Q5,Q6,Q7     Accumulate polynomials in double precision.
      DATA P1,P2,P3,P4,P5,P6,P7/-0.57721566D0,0.42278420D0,0.23069756D0,
     *     0.3488590D-1,0.262698D-2,0.10750D-3,0.74D-5/
      DATA Q1,Q2,Q3,Q4,Q5,Q6,Q7/1.25331414D0,-0.7832358D-1,0.2189568D-1,
     *     -0.1062446D-1,0.587872D-2,-0.251540D-2,0.53208D-3/
      IF (X.LE.2.0) THEN                Polynomial fit.
          Y=X*X/4.0
          BESSK0=(-LOG(X/2.0)*BESSI0(X))+(P1+Y*(P2+Y*(P3+
     *            Y*(P4+Y*(P5+Y*(P6+Y*P7))))))
      ELSE
          Y=(2.0/X)
          BESSK0=(EXP(-X)/SQRT(X))*(Q1+Y*(Q2+Y*(Q3+
     *            Y*(Q4+Y*(Q5+Y*(Q6+Y*Q7))))))
      ENDIF
      RETURN
      END
```

```
      FUNCTION BESSI1(X)
          Returns the modified Bessel function I1(X) for any real X.
      REAL*8 Y,P1,P2,P3,P4,P5,P6,P7,
     *     Q1,Q2,Q3,Q4,Q5,Q6,Q7,Q8,Q9  Accumulate polynomials in double precision.
      DATA P1,P2,P3,P4,P5,P6,P7/0.5D0,0.87890594D0,0.51498869D0,
     *     0.15084934D0,0.2658733D-1,0.301532D-2,0.32411D-3/
      DATA Q1,Q2,Q3,Q4,Q5,Q6,Q7,Q8,Q9/0.39894228D0,-0.3988024D-1,
     *     -0.362018D-2,0.163801D-2,-0.1031555D-1,0.2282967D-1,
     *     -0.2895312D-1,0.1787654D-1,-0.420059D-2/
      IF (ABS(X).LT.3.75) THEN        polynomial fit
          Y=(X/3.75)**2
          BESSI1=X*(P1+Y*(P2+Y*(P3+Y*(P4+Y*(P5+Y*(P6+Y*P7))))))
      ELSE
          AX=ABS(X)
          Y=3.75/AX
          BESSI1=(EXP(AX)/SQRT(AX))*(Q1+Y*(Q2+Y*(Q3+Y*(Q4+
     *          Y*(Q5+Y*(Q6+Y*(Q7+Y*(Q8+Y*Q9))))))))
          IF(X.LT.0.)BESSI1=-BESSI1
      ENDIF
      RETURN
      END
```

```
      FUNCTION BESSK1(X)
         Returns the modified Bessel function K1(X) for positive real X.
      REAL*8 Y,P1,P2,P3,P4,P5,P6,P7,
     *    Q1,Q2,Q3,Q4,Q5,Q6,Q7     Accumulate polynomials in double precision.
      DATA P1,P2,P3,P4,P5,P6,P7/1.0D0,0.15443144D0,-0.67278579D0,
     *    -0.18156897D0,-0.1919402D-1,-0.110404D-2,-0.4686D-4/
      DATA Q1,Q2,Q3,Q4,Q5,Q6,Q7/1.25331414D0,0.23498619D0,-0.3655620D-1,
     *    0.1504268D-1,-0.780353D-2,0.325614D-2,-0.68245D-3/
      IF (X.LE.2.0) THEN                 Polynomial fit.
         Y=X*X/4.0
         BESSK1=(LOG(X/2.0)*BESSI1(X))+(1.0/X)*(P1+Y*(P2+
     *       Y*(P3+Y*(P4+Y*(P5+Y*(P6+Y*P7))))))
      ELSE
         Y=2.0/X
         BESSK1=(EXP(-X)/SQRT(X))*(Q1+Y*(Q2+Y*(Q3+
     *       Y*(Q4+Y*(Q5+Y*(Q6+Y*Q7))))))
      ENDIF
      RETURN
      END
```

The recurrence relation for $I_n(x)$ and $K_n(x)$ is the same as that for $J_n(x)$ and $Y_n(x)$ provided that ix is substituted for x. This has the effect of changing a sign in the relation,

$$
\begin{aligned}
I_{n+1}(x) &= -\left(\frac{2n}{x}\right) I_n(x) + I_{n-1}(x) \\
K_{n+1}(x) &= +\left(\frac{2n}{x}\right) K_n(x) + K_{n-1}(x)
\end{aligned}
\tag{6.5.4}
$$

These relations are always *unstable* for upward recurrence. For K_n, itself growing, this presents no problem. For I_n, however, the strategy of downward recursion is therefore required once again, and the starting point for the recursion may be chosen in the same manner as for the routine BESSJ. The only fundamental difference is that the normalization formula for $I_n(x)$ has an alternating minus sign in successive terms, which again arises from the substitution of ix for x in the formula used previously for J_n

$$1 = I_0(x) - 2I_2(x) + 2I_4(x) - 2I_6(x) + \cdots \tag{6.5.5}$$

In fact, we prefer simply to normalize with a call to BESSI0.

With this simple modification, the recursion routines BESSJ and BESSY become the new routines BESSI and BESSK:

```
      FUNCTION BESSK(N,X)
         Returns the modified Bessel function KN(x) for positive X and N≥ 2.
      IF (N.LT.2) PAUSE 'bad argument N in BESSK'
      TOX=2.0/X
      BKM=BESSK0(X)                  Upward recurrence for all X
      BK=BESSK1(X)
      DO 11 J=1,N-1                  ...and here it is.
         BKP=BKM+J*TOX*BK
         BKM=BK
```

```
      BK=BKP
11    CONTINUE
      BESSK=BK
      RETURN
      END

      FUNCTION BESSI(N,X)
C     Returns the modified Bessel function I_N(X) for any real X and N≥ 2.
      PARAMETER(IACC=40,BIGNO=1.0E10,BIGNI=1.0E-10)
      IF (N.LT.2) PAUSE 'bad argument N in BESSI'
      IF (X.EQ.0.)  THEN
          BESSI=0.
      ELSE
          TOX=2.0/ABS(X)
          BIP=0.0
          BI=1.0
          BESSI=0.
          M=2*((N+INT(SQRT(FLOAT(IACC*N)))))       Downward recurrence from an even value M.
          DO 11 J=M,1,-1                           Make IACC larger to increase accuracy.
              BIM=BIP+FLOAT(J)*TOX*BI              The downward recurrence.
              BIP=BI
              BI=BIM
              IF (ABS(BI).GT.BIGNO) THEN           Renormalize to prevent overflows.
                  BESSI=BESSI*BIGNI
                  BI=BI*BIGNI
                  BIP=BIP*BIGNI
              ENDIF
              IF (J.EQ.N) BESSI=BIP
11        CONTINUE
          BESSI=BESSI*BESSI0(X)/BI                 Normalize with BESSI0.
          IF (X.LT.0..AND.MOD(N,2).EQ.1) BESSI=-BESSI
      ENDIF
      RETURN
      END
```

REFERENCES AND FURTHER READING:

Abramowitz, Milton, and Stegun, Irene A. 1964, *Handbook of Mathematical Functions*, Applied Mathematics Series, vol. 55 (Washington: National Bureau of Standards; reprinted 1968 by Dover Publications, New York), §9.8.

Carrier, G.F., Krook, M. and Pearson, C.E. 1966, *Functions of a Complex Variable* (New York: McGraw-Hill), pp. 220ff.

6.6 Spherical Harmonics

Spherical harmonics occur in a large variety of physical problems, for example, whenever a wave equation, or Laplace's equation, is solved by separation of variables in spherical coordinates. The spherical harmonic $Y_{lm}(\theta, \phi)$, $-l \leq m \leq l$, is a function of the two coordinates θ, ϕ on the surface of a sphere.

The spherical harmonics are orthogonal for different l and m, and they are normalized so that their integrated square over the sphere is unity:

$$\int_0^{2\pi} d\phi \int_{-1}^{1} d(\cos\theta) Y_{l'm'}{}^*(\theta, \phi) Y_{lm}(\theta, \phi) = \delta_{l'l}\delta_{m'm} \tag{6.6.1}$$

Here asterisk denotes complex conjugation.

Mathematically, the spherical harmonics are related to *associated Legendre polynomials* by the equation

$$Y_{lm}(\theta,\phi) = \sqrt{\frac{2l+1}{4\pi}\frac{(l-m)!}{(l+m)!}}P_l^m(\cos\theta)e^{im\phi} \tag{6.6.2}$$

By using the relation

$$Y_{l,-m}(\theta,\phi) = (-1)^m Y_{lm}{}^*(\theta,\phi) \tag{6.6.3}$$

we can always relate a spherical harmonic to an associated Legendre polynomial with $m \geq 0$. With $x \equiv \cos\theta$, these are defined in terms of the ordinary Legendre polynomials (cf. §§4.5 and 5.4) by

$$P_l^m(x) = (-1)^m(1-x^2)^{m/2}\frac{d^m}{dx^m}P_l(x) \tag{6.6.4}$$

The first few associated Legendre polynomials, and their corresponding normalized spherical harmonics, are

$P_0^0(x) =$	1	$Y_{00} =$	$\sqrt{\frac{1}{4\pi}}$
$P_1^1(x) =$	$-\ (1-x^2)^{1/2}$	$Y_{11} =$	$-\ \sqrt{\frac{3}{8\pi}}\sin\theta e^{i\phi}$
$P_1^0(x) =$	x	$Y_{10} =$	$\sqrt{\frac{3}{4\pi}}\cos\theta$
$P_2^2(x) =$	$3\ (1-x^2)$	$Y_{22} =$	$\frac{1}{4}\sqrt{\frac{15}{2\pi}}\sin^2\theta e^{2i\phi}$
$P_2^1(x) =$	$-3\ (1-x^2)^{1/2}x$	$Y_{21} =$	$-\ \sqrt{\frac{15}{8\pi}}\sin\theta\cos\theta e^{i\phi}$
$P_2^0(x) =$	$\frac{1}{2}\ (3x^2-1)$	$Y_{20} =$	$\sqrt{\frac{5}{4\pi}}(\frac{3}{2}\cos^2\theta - \frac{1}{2})$

(6.6.5)

There are many bad ways to evaluate associated Legendre polynomials numerically. For example, there are explicit expressions, such as

$$P_l^m(x) = \frac{(-1)^m(l+m)!}{2^m m!(l-m)!}(1-x^2)^{m/2}\left[1 - \frac{(l-m)(m+l+1)}{1!(m+1)}\left(\frac{1-x}{2}\right)\right.$$
$$\left. + \frac{(l-m)(l-m-1)(m+l+1)(m+l+2)}{2!(m+1)(m+2)}\left(\frac{1-x}{2}\right)^2 - \cdots\right] \tag{6.6.6}$$

where the polynomial continues up through the term in $(1-x)^{l-m}$. (See Magnus and Oberhettinger for this and related formulas.) This is not a satisfactory method because evaluation of the polynomial involves delicate cancellations between successive terms, which alternate in sign. For large l, the individual terms in the polynomial become very much larger than their sum, and all accuracy is lost.

In practice, (6.6.6) can only be used in single precision (32-bit) for l up to 6 or 8, and in double precision (64-bit) for l up to 15 or 18, depending on the precision required for the answer. A more robust computational procedure is therefore desirable, as follows:

The associated Legendre functions satisfy numerous recurrence relations, tabulated in the references below. These are recurrences on l alone, on m alone, and on both l and m simultaneously. Most of the recurrences involving m are unstable, and so dangerous for numerical work. The following recurrence on l is, however, stable (compare 5.4.1):

$$(l-m)P_l^m = x(2l-1)P_{l-1}^m - (l+m-1)P_{l-2}^m \tag{6.6.7}$$

It is useful because there is a closed-form expression for the starting value,

$$P_m^m = (-1)^m(2m-1)!!(1-x^2)^{m/2} \tag{6.6.8}$$

(The notation $n!!$ denotes the product of all *odd* integers less than or equal to n.) Using (6.6.7) with $l = m+1$, and setting $P_{m-1}^m = 0$, we find

$$P_{m+1}^m = x(2m+1)P_m^m \tag{6.6.9}$$

Equations (6.6.8) and (6.6.9) provide the two starting values required for (6.6.7) for general l.

The routine that implements this is

```
FUNCTION PLGNDR(L,M,X)
    Computes the associated Legendre polynomial P_l^m(x). Here m and l are integers satisfying
    0 <= m <= l, while x lies in the range -1 <= x <= 1.
IF(M.LT.0.OR.M.GT.L.OR.ABS(X).GT.1.)PAUSE 'bad arguments'
PMM=1.                                Compute P_m^m.
IF(M.GT.0) THEN
    SOMX2=SQRT((1.-X)*(1.+X))
    FACT=1.
    DO 11 I=1,M
        PMM=-PMM*FACT*SOMX2
        FACT=FACT+2.
11  CONTINUE
ENDIF
IF(L.EQ.M) THEN
    PLGNDR=PMM
ELSE                                  Compute P_{m+1}^m.
    PMMP1=X*(2*M+1)*PMM
    IF(L.EQ.M+1) THEN
        PLGNDR=PMMP1
```

```
    ELSE                          Compute P_l^m, l > m+1.
        DO 12 LL=M+2,L
            PLL=(X*(2*LL-1)*PMMP1-(LL+M-1)*PMM)/(LL-M)
            PMM=PMMP1
            PMMP1=PLL
12      CONTINUE
        PLGNDR=PLL
    ENDIF
ENDIF
RETURN
END
```

REFERENCES AND FURTHER READING:

Magnus, Wilhelm, and Oberhettinger, Fritz. 1949, *Formulas and Theorems for the Functions of Mathematical Physics* (New York: Chelsea), pp. 54ff.

Abramowitz, Milton, and Stegun, Irene A. 1964, *Handbook of Mathematical Functions*, Applied Mathematics Series, vol. 55 (Washington: National Bureau of Standards; reprinted 1968 by Dover Publications, New York), Chapter 8.

6.7 Elliptic Integrals and Jacobian Elliptic Functions

Algorithms for computing elliptic integrals are highly specific to the nature of the functions. It would take us too far afield to delve into the theory of these algorithms, so we shall be content to give a "user's guide" and a couple of black-box routines that are established as state-of-the-art.

Elliptic integrals occur in a variety of applications, since they are the functions in terms of which the general integral

$$\int dx \frac{A(x) + B(x)\sqrt{S(x)}}{C(x) + D(x)\sqrt{S(x)}} \tag{6.7.1}$$

can always be expressed, where A, B, C, and D are arbitrary polynomials, and S is a cubic or quartic polynomial.

One of the harder things about using elliptic integrals is the notational thicket that surrounds the subject in the literature. Beware: there are various notational conventions in use. We will follow Bulirsch, whose algorithms we use; this notation is almost, but not quite, consistent with that in Abramowitz and Stegun. (We note the differences.)

A useful starting point is to define the *general elliptic integral of the 2nd kind* as the following function of four variables,

$$el2(y,k_c,a,b) \equiv \int_0^y \frac{(a+bx^2)dx}{(1+x^2)\sqrt{(1+x^2)(1+k_c^2x^2)}} \tag{6.7.2}$$

where $x \geq 0$ and a, b, k_c can each have any real value. With the change of variables

$$x \equiv \tan\phi \qquad k^2 \equiv 1-k_c^2 \tag{6.7.3}$$

the function $el2(y,k_c,a,b)$ can easily be seen also to equal

$$= \int_0^{\tan^{-1}y} \frac{a+b\tan^2\phi}{\sqrt{(1+\tan^2\phi)(1+k_c^2\tan^2\phi)}}d\phi \tag{6.7.4}$$

$$= \int_0^{\tan^{-1}y} \frac{a+(b-a)\sin^2\phi}{\sqrt{1-k^2\sin^2\phi}}d\phi \tag{6.7.5}$$

In the limit $x \to \infty$ or $\phi \to \pi/2$, (6.7.2) becomes a *complete elliptic integral.* It is useful, however, to define a more general form, the *general complete elliptic integral,*

$$\begin{aligned} cel(k_c,p,a,b) &\equiv \int_0^\infty \frac{a+bx^2}{(1+px^2)\sqrt{(1+x^2)(1+k_c^2x^2)}}dx \\ &= \int_0^{\pi/2} \frac{(a\cos^2\phi+b\sin^2\phi)d\phi}{(\cos^2\phi+p\sin^2\phi)\sqrt{\cos^2\phi+k_c^2\sin^2\phi}} \end{aligned} \tag{6.7.6}$$

Below we will give Bulirsch's algorithms for the functions $el2(x,k_c,a,b)$ and $cel(k_c,p,a,b)$. Here is how to use them with other notational conventions:

The *Legendre elliptic integral of the 1st kind* is defined by any of the following expressions

$$\begin{aligned} F(\phi,k) = F(\phi\backslash\alpha) &= \int_0^\phi \frac{d\phi}{\sqrt{1-k^2\sin^2\phi}} = \int_0^y \frac{dy}{\sqrt{(1-y^2)(1-k^2y^2)}} \\ &= \int_0^x \frac{dx}{\sqrt{(1+x^2)(1+k_c^2x^2)}} = el2(x,k_c,1,1) \end{aligned} \tag{6.7.7}$$

where the various quantities are related by

$$y = \sin\phi, \quad x = \tan\phi, \quad k_c^2 = 1 - k^2, \quad k_c = \cos\alpha, \quad k = \sin\alpha \tag{6.7.8}$$

The *complete elliptic integral of the 1st kind* is given by

$$K(k) \equiv F(\frac{\pi}{2}, k) = cel(k_c, 1, 1, 1) \tag{6.7.9}$$

The *Legendre elliptic integral of the 2nd kind* is defined by any of the following expressions

$$E(\phi, k) = E(\phi \backslash \alpha) = \int_0^\phi \sqrt{1 - k^2 \sin^2\phi}\, d\phi = \int_0^y \frac{\sqrt{1 - k^2 y^2}\, dy}{\sqrt{(1 - y^2)}}$$

$$= \int_0^x \frac{\sqrt{1 + k_c^2 x^2} dx}{(1 + x^2)\sqrt{(1 + x^2)}} = el2(x, k_c, 1, k_c^2) \tag{6.7.10}$$

where the various quantities are related by (6.7.8).

The *complete elliptic integral of the 2nd kind* is given by

$$E(k) \equiv E(\frac{\pi}{2}, k) = cel(k_c, 1, 1, k_c^2) \tag{6.7.11}$$

One occasionally sees linear combinations of the elliptic integrals of the first and second kinds, as follows

$$B(\phi, k) = \int_0^\phi \frac{\cos^2\phi}{\sqrt{1 - k^2 \sin^2\phi}} d\phi = el2(x, k_c, 1, 0) \tag{6.7.12}$$

$$D(\phi, k) = \int_0^\phi \frac{\sin^2\phi}{\sqrt{1 - k^2 \sin^2\phi}} d\phi = el2(x, k_c, 0, 1) \tag{6.7.13}$$

Finally we come to the *elliptic integral of the 3rd kind*, which occurs less frequently in practice, but is also harder to compute. It cannot be written in terms of *el*2 defined above. We will refer you to the literature for Bulirsch's rather lengthy algorithm *el*3, in terms of which

$$\begin{aligned}\Pi(\phi,n,k) &= \int_0^{\phi} \frac{d\phi}{(1+n\sin^2\phi)\sqrt{1-k^2\sin^2\phi}} \\ &= \int_0^{y} \frac{dy}{(1+ny^2)\sqrt{(1-y^2)(1-k^2y^2)}} \\ &= \int_0^{x} \frac{1+x^2}{(1+px^2)\sqrt{(1+x^2)(1+k_c^2x^2)}}dx = el3(x,k_c,p)\end{aligned} \tag{6.7.14}$$

where the relations (6.7.8) still hold, and $p \equiv n+1$. (Note, however, that Abramowitz and Stegun define n with the opposite sign.)

For the case of the *complete elliptic integral of the 3rd kind*, however, the function *cel* above *is* sufficiently general:

$$\Pi(\frac{\pi}{2},n,k) = cel(k_c,p,1,1) \tag{6.7.15}$$

where again $p \equiv n+1$.

So here are the routines that are needed, given as black boxes (a violation of our general philosophy on such matters):

```
FUNCTION EL2(X,QQC,AA,BB)
```
Returns the general elliptic integral of the second kind, $el2(x,k_c,a,b)$ with $X = x > 0$, $QQC = k_c$, $AA = a$, and $BB = b$.
```
PARAMETER(PI=3.14159265, CA=.0003, CB=1.E-9)
```
The desired accuracy is the square of CA, while CB should be set to 0.01 times the desired accuracy.
```
      IF(X.EQ.0.)THEN
        EL2=0.
      ELSE IF(QQC.NE.0.)THEN
        QC=QQC
        A=AA
        B=BB
        C=X**2
        D=1.+C
        P=SQRT((1.+QC**2*C)/D)
        D=X/D
        C=D/(2.*P)
        Z=A-B
        EYE=A
        A=0.5*(B+A)
        Y=ABS(1./X)
        F=0.
        L=0
        EM=1.
        QC=ABS(QC)
1       B=EYE*QC+B
        E=EM*QC
```

```
        G=E/P
        D=F*G+D
        F=C
        EYE=A
        P=G+P
        C=0.5*(D/P+C)
        G=EM
        EM=QC+EM
        A=0.5*(B/EM+A)
        Y=-E/Y+Y
        IF(Y.EQ.0.)Y=SQRT(E)*CB
        IF(ABS(G-QC).GT.CA*G)THEN
            QC=SQRT(E)*2.
            L=L+L
            IF(Y.LT.0.)L=L+1
            GO TO 1
        ENDIF
        IF(Y.LT.0.)L=L+1
        E=(ATAN(EM/Y)+PI*L)*A/EM
        IF(X.LT.0.)E=-E
        EL2=E+C*Z
    ELSE
        PAUSE 'failure in EL2'    Argument QQC was zero.
    ENDIF
    RETURN
    END
```

```
    FUNCTION CEL(QQC,PP,AA,BB)
```

Returns the general complete elliptic integral $cel(k_c, p, a, b)$ with QQC $= k_c$, PP $= p$, AA $= a$, and BB $= b$.

```
    PARAMETER (CA=.0003, PIO2=1.5707963268)
```

The desired accuracy is the square of CA.

```
    IF(QQC.EQ.0.)PAUSE 'failure in CEL'
    QC=ABS(QQC)
    A=AA
    B=BB
    P=PP
    E=QC
    EM=1.
    IF(P.GT.0.)THEN
        P=SQRT(P)
        B=B/P
    ELSE
        F=QC*QC
        Q=1.-F
        G=1.-P
        F=F-P
        Q=Q*(B-A*P)
        P=SQRT(F/G)
        A=(A-B)/G
        B=-Q/(G*G*P)+A*P
    ENDIF
1   F=A
    A=A+B/P
    G=E/P
    B=B+F*G
    B=B+B
    P=G+P
    G=EM
    EM=QC+EM
    IF(ABS(G-QC).GT.G*CA)THEN
```

```
    QC=SQRT(E)
    QC=QC+QC
    E=QC*EM
    GO TO 1
ENDIF
CEL=PIO2*(B+A*EM)/(EM*(EM+P))
RETURN
END
```

Jacobian Elliptic Functions

The Jacobian elliptic function sn is defined as follows: instead of considering the elliptic integral

$$u(y,k) \equiv u = F(\phi, k) \tag{6.7.16}$$

consider the *inverse* function

$$y = \sin\phi = \mathrm{sn}(u,k) \tag{6.7.17}$$

Equivalently,

$$u = \int_0^{\mathrm{sn}} \frac{dy}{\sqrt{(1-y^2)(1-k^2y^2)}} \tag{6.7.18}$$

When $k = 0$, sn is just sin. The functions cn and dn are defined by the relations

$$\mathrm{sn}^2 + \mathrm{cn}^2 = 1, \qquad k^2\mathrm{sn}^2 + \mathrm{dn}^2 = 1 \tag{6.7.19}$$

The routine given below actually takes $m_c \equiv k_c^2 = 1 - k^2$ as an input parameter. It also computes all three functions sn, cn and dn since computing all three is no harder than computing any one of them.

```
      SUBROUTINE SNCNDN(UU,EMMC,SN,CN,DN)
```

Returns the Jacobian elliptic functions $\mathrm{sn}(u, k_c)$, $\mathrm{cn}(u, k_c)$ and $\mathrm{dn}(u, k_c)$. Here UU$=u$, while EMMC$=k_c^2$.

```
      PARAMETER (CA=.0003)            The accuracy is the square of CA.
      LOGICAL BO
      DIMENSION EM(13),EN(13)
      EMC=EMMC
      U=UU
      IF(EMC.NE.0.)THEN
          BO=(EMC.LT.0.)
          IF(BO)THEN
              D=1.-EMC
              EMC=-EMC/D
              D=SQRT(D)
              U=D*U
          ENDIF
          A=1.
          DN=1.
          DO 11 I=1,13
              L=I
              EM(I)=A
              EMC=SQRT(EMC)
              EN(I)=EMC
              C=0.5*(A+EMC)
              IF(ABS(A-EMC).LE.CA*A)GO TO 1
              EMC=A*EMC
              A=C
11        CONTINUE
1         U=C*U
          SN=SIN(U)
          CN=COS(U)
          IF(SN.EQ.0.)GO TO 2
          A=CN/SN
          C=A*C
          DO 12 II=L,1,-1
              B=EM(II)
              A=C*A
              C=DN*C
              DN=(EN(II)+A)/(B+A)
              A=C/B
12        CONTINUE
          A=1./SQRT(C**2+1.)
          IF(SN.LT.0.)THEN
              SN=-A
          ELSE
              SN=A
          ENDIF
          CN=C*SN
2         IF(BO)THEN
              A=DN
              DN=CN
              CN=A
              SN=SN/D
          ENDIF
      ELSE
          CN=1./COSH(U)
          DN=CN
          SN=TANH(U)
      ENDIF
      RETURN
      END
```

REFERENCES AND FURTHER READING:

Bulirsch, Roland, 1965, *Numerische Mathematik*, vol. 7, p.78; 1965, *op. cit.*, vol. 7, p.353; 1969, *op. cit.*, vol. 13, p.305.

Abramowitz, Milton, and Stegun, Irene A. 1964, *Handbook of Mathematical Functions*, Applied Mathematics Series, vol. 55 (Washington: National Bureau of Standards; reprinted 1968 by Dover Publications, New York), Chapter 17.

Mathews, Jon, and Walker, R.L. 1970, *Mathematical Methods of Physics*, 2nd ed. (Reading, Mass.: W.A. Benjamin/Addison-Wesley), pp. 78-79.

Chapter 7. Random Numbers

7.0 Introduction

It may seem perverse to use a computer, that most precise and deterministic of all machines conceived by the human mind, to produce "random" numbers. More than perverse, it may seem to be a conceptual impossibility. Any program, after all, will produce output that is entirely predictable, hence not truly "random."

Nevertheless, practical computer "random number generators" are in common use. We will leave it to philosophers of the computer age to resolve the paradox in a deep way (see, e.g., Knuth §3.5 for discussion and references). One sometimes hears computer-generated sequences termed *quasi-random*, while the word *random* is reserved for the output of an intrinsically random physical process, like the elapsed time between clicks of a Geiger counter placed next to a sample of some radioactive element. We will not try to make such fine distinctions.

A working, though imprecise, definition of randomness in the context of computer-generated sequences, is to say that the deterministic program that produces a random sequence should be different from, and — in all measurable respects — statistically uncorrelated with, the computer program that *uses* its output. In other words, any two different random number generators ought to produce statistically the same results when coupled to your particular applications program. If they don't, then at least one of them is not (from your point of view) a good generator.

The above definition may seem circular, comparing, as it does, one generator to another. However, there exists a body of random number generators which mutually do satisfy the definition over a very, very broad class of applications programs. And it is also found empirically that statistically identical results are obtained from random numbers produced by physical processes. So, because such generators are known to exist, we can leave to the philosophers the problem of defining them.

A pragmatic point of view, then, is that randomness is in the eye of the beholder (or applications programmer). What is random enough for one application may not be random enough for another. Still, one is not entirely adrift in a sea of incommensurable applications programs: There is a certain list of statistical tests, some sensible and some merely enshrined by

history, which on the whole will do a very good job of ferreting out any correlations that are likely to be detected by an applications program (in this case, yours). Good random number generators ought to pass all of these tests; or else the user had better at least be aware of any that they fail, so that he or she is able to judge whether they are relevant to the case at hand.

As for references on this subject, there is really only one worth turning to first, and that is Knuth. Only a few of the standard books on numerical methods treat topics relating to random numbers.

REFERENCES AND FURTHER READING:

Knuth, Donald E. 1981, *Seminumerical Algorithms*, 2nd ed., vol. 2 of *The Art of Computer Programming* (Reading, Mass.: Addison-Wesley), Chapter 3.

Dahlquist, Germund, and Bjorck, Ake. 1974, *Numerical Methods* (Englewood Cliffs, N.J.: Prentice-Hall), Chapter 11.

Forsythe, George E., Malcolm, Michael A., and Moler, Cleve B. 1977, *Computer Methods for Mathematical Computations* (Englewood Cliffs, N.J.: Prentice-Hall), Chapter 10.

7.1 Uniform Deviates

Uniform deviates are just random numbers which lie within a specified range (typically 0 to 1), with any one number in the range just as likely as any other. They are, in other words, what you probably think "random numbers" are; however we want to distinguish uniform deviates from other sorts of "random numbers," for example, numbers drawn from a normal (Gaussian) distribution of specified mean and standard deviation. These other sorts of deviates are almost always generated by performing appropriate operations on one or more uniform deviates, as we will see in subsequent sections. So, a reliable source of random uniform deviates, the subject of this section, is an essential building block for any sort of stochastic modeling or Monte Carlo computer work.

System-Supplied Random Number Generators

Your computer very likely has lurking within it a library routine which is called a "random number generator." That routine typically has an unforgettable name like "RAN," and a calling sequence like

```
X=RAN(ISEED)          sets X to the next random number and updates ISEED
```

You initialize ISEED to a (usually) arbitrary value before the first call to RAN. Each initializing value will typically return a different subsequent random sequence, or at least a different subsequence of some one enormously

long sequence. The *same* initializing value of ISEED will always return the *same* random sequence, however.

Now our first, and perhaps most important, lesson in this chapter is: be *very, very* suspicious of a system-supplied RAN which resembles the one just described. If all scientific papers whose results are in doubt because of bad RANs were to disappear from library shelves, there would be a gap on each shelf about as big as your fist. System-supplied RANs are almost always *linear congruential generators*, which generate a sequence of integers $I_1, I_2, I_3, \ldots$, each between 0 and $m-1$ (a large number) by the recurrence relation

$$I_{j+1} = aI_j + c \pmod{m} \tag{7.1.1}$$

Here m is called the *modulus*, and a and c are positive integers called the *multiplier* and the *increment* respectively. The recurrence (7.1.1) will eventually repeat itself, with a period that is obviously no greater than m. If m, a, and c are properly chosen, then the period will be of maximal length, i.e. of length m. In that case, all possible integers between 0 and $m-1$ occur at some point, so any initial "seed" choice of I_0 is as good as any other: the sequence just takes off from that point. The real number between 0 and 1 which is returned is generally I_{j+1}/m, so that it is strictly less than 1, but occasionally (once in m calls) exactly equal to zero. ISEED is returned as I_{j+1} (or some encoding of it), so that it can be used on the next call to generate I_{j+2}, and so on.

The linear congruential method has the advantage of being very fast, requiring only a few operations per call, hence its almost universal use. It has the disadvantage that it is not free of sequential correlation on successive calls. If k random numbers at a time are used to plot points in k dimensional space (with each coordinate between 0 and 1), then the points will not tend to "fill up" the k-dimensional space, but rather will lie on $k-1$-dimensional "planes". There will be *at most* about $m^{1/k}$ such planes. *And if the constants m, a, and c are not very carefully chosen, there will be many fewer than that.* The number m is usually about the wordsize of the machine, e.g. 2^{32}. So, for example, the number of planes on which triples of points lie in three-dimensional space is usually no greater than about the cube root of 2^{32}, about 1600. You might well be focusing attention on a physical process that occurs in a small fraction of the total volume, so that the discreteness of the planes can be very pronounced.

Even worse, you might be using a RAN whose choices of m, a, and c have been botched. One infamous such routine was widespread on IBM computers for many years, and widely copied onto other systems. One of us recalls producing a "random" plot with only 11 planes, and being told by his computer center's programming consultant that he had misused the random number generator: "We guarantee that each number is random individually, but we *don't* guarantee that more than one of them is random."

Correlation in k-space is not the only weakness of linear congruential generators. Such generators often have their low-order (least significant) bits

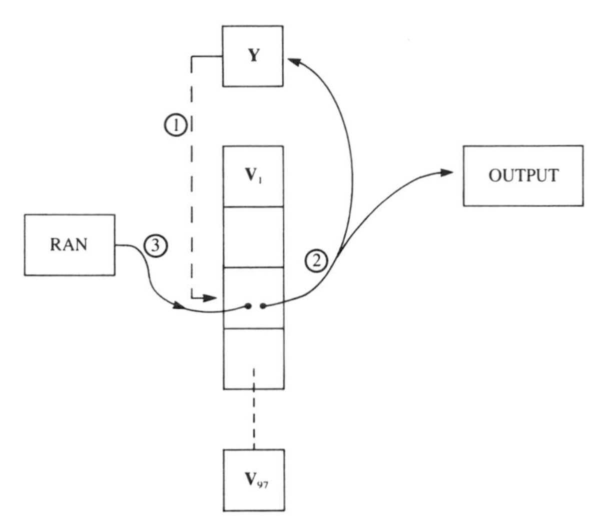

Figure 7.1.1. Shuffling procedure used in **RAN0** to break up sequential correlations in a system-supplied random number generator. Circled numbers indicate the sequence of events: On each call, the random number in **Y** is used to choose a random element in the array **V**. That element becomes the output random number, and also is the next **Y**. Its spot in **V** is refilled from the system-supplied routine.

much less random than their high-order bits. If you want to generate a random integer between 1 and 10, you should always do it by

```
J=1+INT(10.*RAN(ISEED))
```

and never by anything resembling

```
J=1+MOD(INT(1000000.*RAN(ISEED)),10)
```

(which uses lower-order bits). Similarly you should never try to take apart a "RAN" number into several supposedly-random pieces. Instead use separate calls for every piece.

How to Improve a System-Supplied Routine

So, if RAN is so dangerous, what is a body to do? A minimum is to do an additional randomizing shuffle on the numbers generated by RAN. Instead of calling RAN, you can call RAN0, the following routine, based on the algorithm of Bays and Durham as described in Knuth and illustrated in Figure 7.1.1.

```
FUNCTION RANO(IDUM)
    Returns a uniform random deviate between 0.0 and 1.0 using a system-supplied routine
    RAN(ISEED). Set IDUM to any negative value to initialize or reinitialize the sequence.
DIMENSION V(97)                   The exact number 97 is unimportant.
DATA IFF /0/                      As a precaution against misuse, we will always initialize on the first
IF(IDUM.LT.O.OR.IFF.EQ.O)THEN         call, even if IDUM is not set negative.
    IFF=1
    ISEED=ABS(IDUM)
    IDUM=1
    DO 11 J=1,97                  Exercise the system routine, especially important if the system's mul-
        DUM=RAN(ISEED)                tiplier is small.
11  CONTINUE
    DO 12 J=1,97                  Then save 97 values
        V(J)=RAN(ISEED)
12  CONTINUE
    Y=RAN(ISEED)                  and a 98th.
ENDIF
J=1+INT(97.*Y)                    This is where we start if not initializing. Use the previously saved
IF(J.GT.97.OR.J.LT.1)PAUSE            random number Y to get an index J between 1 and 97. Then
Y=V(J)                                use the corresponding V(J) for both the next J and as the
RANO=Y                                output number.
V(J)=RAN(ISEED)                   Finally, refill the table entry with the next random number from RAN.
RETURN
END
```

The sequence returned by RANO will be effectively free of sequential correlation. Unless your system-supplied RAN is truly a botch, the output of RANO ought to be good for all purposes, although we would tend to remain suspicious of its lowest-order (least significant) bits.

Portable Random Number Generators

You may well want to have a "portable" random number generator which can be programmed in a high-level language, and which will generate the same random sequence (from a given seed) on all machines. We can distinguish two classes of use for such portable routines: First, one might want a fully reliable generator, at least as good as RANO given above. For this, and portability too, you are going to have to pay a price: the routines that we are about to give you will run considerably more slowly than RANO. Second, one frequently would like a quick-and-dirty generator to embed in a program, perhaps taking only one or two lines of code, just to *somewhat* randomize things. One might wish to process data from an experiment not always in exactly the same order, for example, so that the first output is more "typical" than might otherwise be the case.

Let us address the second class of use first. All we really need is a list of "good" choices for m, a, and c. Then we can easily embed in our programs

```
JRAN=MOD(JRAN*IA+IC,IM)
RAN=FLOAT(JRAN)/FLOAT(IM)
```

whenever we want a quick-and-dirty uniform deviate, or

```
      JRAN=MOD(JRAN*IA+IC,IM)
      J=JLO+((JHI-JLO+1)*JRAN)/IM
```

whenever we want an integer between JLO and JHI, inclusive. (In both cases JRAN was once initialized to any seed value between 0 and IM-1.)

There is a reason that the above fragments of high-level language cannot be made as good as the system-supplied RAN: we must choose IM (which is a bound on JRAN) and IA small enough so that their product does not produce an integer overflow. The machine-language programmer has no such restriction: she can multiply two integers each as large as the machine's wordsize into a double-width register, then manipulate the high- and low-order words separately to take the MOD. So our "good" values of IM, IA, and IC are restricted by the largest product which does not overflow. (You can also use a machine's floating point arithmetic: in this case the restriction is to the largest product which represents an integer *exactly*, i.e. without losing the low-significance bits. Some computers have only 16-bit integers, but have 8-byte floating formats with up to 48 bits of mantissa; in this case a floating implementation is vastly preferred.)

Be sure to remember that when IM is small, the kth root of it, which is the number of planes in k-space, is even smaller!

With these caveats, some "good" choices for the constants are given in the accompanying table. These constants (i) give a period of maximal length IM, and, more important, (ii) pass Knuth's "spectral test" for dimensions 2, 3, 4, 5, and 6. The increment IC is a prime, close to the value $(\frac{1}{2} - \frac{1}{6}\sqrt{3})$IM; actually almost any value of IC that is relatively prime to IM will do just as well, but there is some "lore" favoring this choice (see Knuth, p. 84).

We now turn to the first use of a portable random number generator, namely to generate good random numbers. We will give three suggested routines.

The first, RAN1, is based on three linear congruential generators from the above table. One generator is used for the most significant part of the output number, the second for the least significant part, and the third to control a shuffling routine (here absolutely essential because of the relatively small values of multiplier m available to us). The shuffling routine used is not the same as that illustrated in RAN0 above, but is another one which has been suggested by Knuth and widely used. RAN1 is, of course, very much better than any of its three generators individually: its period is (for all practical purposes) infinite, and it ought to have no sensible sequential correlations. The choice of which constants from the accompanying table to use is arbitrary, as is the length of the shuffling array.

```
FUNCTION RAN1(IDUM)
   Returns a uniform random deviate between 0.0 and 1.0. Set IDUM to any negative value
   to initialize or reinitialize the sequence.
DIMENSION R(97)
PARAMETER (M1=259200,IA1=7141,IC1=54773,RM1=1./M1)
PARAMETER (M2=134456,IA2=8121,IC2=28411,RM2=1./M2)
PARAMETER (M3=243000,IA3=4561,IC3=51349)
DATA IFF /0/                    As above, initialize on first call even if IDUM is not negative.
IF (IDUM.LT.0.OR.IFF.EQ.0) THEN
```

```
      IFF=1
      IX1=MOD(IC1-IDUM,M1)          Seed the first routine,
      IX1=MOD(IA1*IX1+IC1,M1)
      IX2=MOD(IX1,M2)               and use it to seed the second
      IX1=MOD(IA1*IX1+IC1,M1)
      IX3=MOD(IX1,M3)               and third routines.
      DO 11 J=1,97                  Fill the table with sequential uniform deviates generated by the first
          IX1=MOD(IA1*IX1+IC1,M1)       two routines.
          IX2=MOD(IA2*IX2+IC2,M2)
          R(J)=(FLOAT(IX1)+FLOAT(IX2)*RM2)*RM1      Low- and high-order pieces combined here.
11    CONTINUE
      IDUM=1
ENDIF
IX1=MOD(IA1*IX1+IC1,M1)             Except when initializing, this is where we start. Generate the next
IX2=MOD(IA2*IX2+IC2,M2)                 number for each sequence.
IX3=MOD(IA3*IX3+IC3,M3)
J=1+(97*IX3)/M3                     Use the third sequence to get an integer between 1 and 97.
IF(J.GT.97.OR.J.LT.1)PAUSE
RAN1=R(J)                           Return that table entry,
R(J)=(FLOAT(IX1)+FLOAT(IX2)*RM2)*RM1    and refill it.
RETURN
END
```

For many purposes we *do* care that there be no sequential correlations, but we *don't* care that the discreteness of the random values returned be as fine as is allowed by all significant bits of the wordsize. In this case, one might desire the gain in speed that comes from using only one linear congruential generator, shuffling as was done in RAN0 above. Then the following routine is perfectly adequate:

```
FUNCTION RAN2(IDUM)
    Returns a uniform random deviate between 0.0 and 1.0. Set IDUM to any negative value
    to initialize or reinitialize the sequence.
PARAMETER (M=714025,IA=1366,IC=150889,RM=1./M)
DIMENSION IR(97)
DATA IFF /0/                        As above.
IF(IDUM.LT.0.OR.IFF.EQ.0)THEN
      IFF=1
      IDUM=MOD(IC-IDUM,M)
      DO 11 J=1,97                  Initialize the shuffle table.
          IDUM=MOD(IA*IDUM+IC,M)
          IR(J)=IDUM
11    CONTINUE
      IDUM=MOD(IA*IDUM+IC,M)
      IY=IDUM                       Compare to RAN0, above.
ENDIF
J=1+(97*IY)/M                       Here is where we start except on initialization.
IF(J.GT.97.OR.J.LT.1)PAUSE
IY=IR(J)
RAN2=IY*RM
IDUM=MOD(IA*IDUM+IC,M)
IR(J)=IDUM
RETURN
END
```

Constants for Portable Random Number Generators

overflow at	IM	IA	IC	overflow at	IM	IA	IC
	6075	106	1283		117128	1277	24749
2^{20}					312500	741	66037
	7875	211	1663		121500	2041	25673
2^{21}				2^{28}			
	7875	421	1663		120050	2311	25367
2^{22}					214326	1807	45289
	11979	430	2531		244944	1597	51749
	6655	936	1399		233280	1861	49297
	6075	1366	1283		175000	2661	36979
2^{23}					121500	4081	25673
	53125	171	11213		145800	3661	30809
	11979	859	2531	2^{29}			
	29282	419	6173		139968	3877	29573
	14406	967	3041		214326	3613	45289
2^{24}					714025	1366	150889
	134456	141	28411	2^{30}			
	31104	625	6571		134456	8121	28411
	14000	1541	2957		243000	4561	51349
	12960	1741	2731		259200	7141	54773
	21870	1291	4621	2^{31}			
	139968	205	29573		233280	9301	49297
2^{25}					714025	4096	150889
	81000	421	17117	2^{32}			
	29282	1255	6173		1771875	2416	374441
	134456	281	28411	2^{33}			
2^{26}					510300	17221	107839
	86436	1093	18257		312500	36261	66037
	259200	421	54773	2^{34}			
	116640	1021	24631		217728	84589	45989
	121500	1021	25673	2^{35}			
2^{27}							

The period of RAN2 is again effectively infinite. Its principal limitation is that it returns one of only 714025 possible values, equally spaced as a "comb" in the interval $[0, 1)$.

Finally, we give you Knuth's suggestion for a portable routine, which we have translated to the present conventions as RAN3. This is not based on the linear congruential method at all, but rather on a *subtractive method.* One might hope that its weaknesses, if any, are therefore of a highly different character from the weaknesses, if any, of RAN1 above. If you ever suspect trouble with one routine, it is a good idea to try the other in the same application. RAN3 has one nice feature: if your machine is poor on integer arithmetic (i.e.

is limited to 16-bit integers), substitution of the two "commented" lines for the one directly following them will render the routine entirely floating point.

```
      FUNCTION RAN3(IDUM)
         Returns a uniform random deviate between 0.0 and 1.0. Set IDUM to any negative value
         to initialize or reinitialize the sequence.
C           IMPLICIT REAL*4(M)
C           PARAMETER (MBIG=4000000.,MSEED=1618033.,MZ=0.,FAC=1./MBIG)
      PARAMETER (MBIG=1000000000,MSEED=161803398,MZ=0,FAC=1./MBIG)
         According to Knuth, any large MBIG, and any smaller (but still large) MSEED can be sub-
         stituted for the above values.
      DIMENSION MA(55)              This value is special and should not be modified; see Knuth.
      DATA IFF /0/
      IF(IDUM.LT.0.OR.IFF.EQ.0)THEN   Initialization.
          IFF=1
          MJ=MSEED-IABS(IDUM)       Initialize MA(55) using the seed IDUM and the large number MSEED.
          MJ=MOD(MJ,MBIG)
          MA(55)=MJ
          MK=1
          DO 11 I=1,54              Now initialize the rest of the table,
              II=MOD(21*I,55)       in a slightly random order,
              MA(II)=MK             with numbers that are not especially random.
              MK=MJ-MK
              IF(MK.LT.MZ)MK=MK+MBIG
              MJ=MA(II)
11        CONTINUE
          DO 13 K=1,4               We randomize them by "warming up the generator."
              DO 12 I=1,55
                  MA(I)=MA(I)-MA(1+MOD(I+30,55))
                  IF(MA(I).LT.MZ)MA(I)=MA(I)+MBIG
12            CONTINUE
13        CONTINUE
          INEXT=0                   Prepare indices for our first generated number.
          INEXTP=31                 The constant 31 is special; see Knuth.
          IDUM=1
      ENDIF
      INEXT=INEXT+1                 Here is where we start, except on initialization. Increment INEXT,
      IF(INEXT.EQ.56)INEXT=1             wrapping around 56 to 1.
      INEXTP=INEXTP+1               Ditto for INEXTP.
      IF(INEXTP.EQ.56)INEXTP=1
      MJ=MA(INEXT)-MA(INEXTP)       Now generate a new random number subtractively.
      IF(MJ.LT.MZ)MJ=MJ+MBIG        Be sure that it is in range.
      MA(INEXT)=MJ                  Store it,
      RAN3=MJ*FAC                   and output the derived uniform deviate.
      RETURN
      END
```

REFERENCES AND FURTHER READING:

Knuth, Donald E. 1981, *Seminumerical Algorithms*, 2nd ed., vol. 2 of *The Art of Computer Programming* (Reading, Mass.: Addison-Wesley), §§3.2–3.3.

Forsythe, George E., Malcolm, Michael A., and Moler, Cleve B. 1977, *Computer Methods for Mathematical Computations* (Englewood Cliffs, N.J.: Prentice-Hall), Chapter 10.

7.2 Transformation Method: Exponential and Normal Deviates

In the previous section, we learned how to generate random deviates with a uniform probability distribution, so that the probability of generating a number between x and $x + dx$, denoted $p(x)dx$, is given by

$$p(x)dx = \begin{cases} dx & 0 < x < 1 \\ 0 & \text{otherwise} \end{cases} \tag{7.2.1}$$

The probability distribution $p(x)$ is of course normalized, so that

$$\int_{-\infty}^{\infty} p(x)dx = 1 \tag{7.2.2}$$

Now suppose that we generate a uniform deviate x and then take some prescribed function of it, $y(x)$. The probability distribution of y, denoted $p(y)dy$, is determined by the fundamental transformation law of probabilities, which is simply

$$|p(y)dy| = |p(x)dx| \tag{7.2.3}$$

or

$$p(y) = p(x)\left|\frac{dx}{dy}\right| \tag{7.2.4}$$

Exponential Deviates

As an example, suppose that $y(x) \equiv -\ln(x)$, and that $p(x)$ is as given by equation (7.2.1) for a uniform deviate. Then

$$p(y)dy = \left|\frac{dx}{dy}\right| dy = e^{-y}dy \tag{7.2.5}$$

which is distributed exponentially. This exponential distribution occurs frequently in real problems, usually as the distribution of waiting times between independent Poisson-random events, for example the radioactive decay of nuclei. You can also easily see (from 7.2.4) that the quantity y/λ has the probability distribution $\lambda e^{-\lambda y}$.

So we have

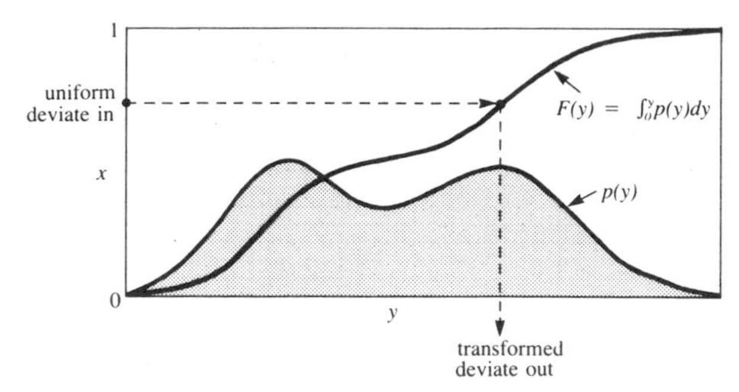

Figure 7.2.1. Transformation method for generating a random deviate y from a known probability distribution $p(y)$. The indefinite integral of $p(y)$ must be known and invertible. A uniform deviate x is chosen between 0 and 1. Its corresponding y on the definite-integral curve is the desired deviate.

```
FUNCTION EXPDEV(IDUM)
   Returns an exponentially distributed, positive, random deviate of unit mean, using
   RAN1(IDUM) as the source of uniform deviates.
EXPDEV=-LOG(RAN1(IDUM))
RETURN
END
```

Let's see what is involved in using the above *transformation method* to generate some arbitrary desired distribution of y's, say one with $p(y) = f(y)$ for some positive function f whose integral is 1. (See Figure 7.2.1.) According to (7.2.4), we need to solve the differential equation

$$\frac{dx}{dy} = f(y) \tag{7.2.6}$$

But the solution of this is just $x = F(y)$, where $F(y)$ is the indefinite integral of $f(y)$. The desired transformation which takes a uniform deviate into one distributed as $f(y)$ is therefore

$$y(x) = F^{-1}(x) \tag{7.2.7}$$

where F^{-1} is the inverse function to F. Whether (7.2.7) is feasible to implement depends on whether the *inverse function of the integral of f(y)* is itself feasible to compute, either analytically or numerically. Sometimes it is, and sometimes it isn't.

Incidentally, (7.2.7) has an immediate geometric interpretation: Since $F(y)$ is the area under the probability curve to the left of y, (7.2.7) is just the prescription: choose a uniform random x, then find the value y that has that fraction x of probability area to its left, and return the value y.

Normal (Gaussian) Deviates

Transformation methods generalize to more than one dimension. If $x_1, x_2, \ldots$ are random deviates with a *joint* probability distribution $p(x_1, x_2, \ldots)$ $dx_1 dx_2 \ldots$, and if $y_1, y_2, \ldots$ are each functions of all the x's (same number of y's as x's), then the joint probability distribution of the y's is

$$p(y_1, y_2, \ldots) dy_1 dy_2 \ldots = p(x_1, x_2, \ldots) \left| \frac{\partial(x_1, x_2, \ldots)}{\partial(y_1, y_2, \ldots)} \right| dy_1 dy_2 \ldots \tag{7.2.8}$$

where $|\partial(\ )/\partial(\ )|$ is the Jacobian determinant of the x's with respect to the y's (or reciprocal of the Jacobian determinant of the y's with respect to the x's).

An important example of the use of (7.2.8) is the *Box-Muller* method for generating random deviates with a normal (Gaussian) distribution,

$$p(y)dy = \frac{1}{\sqrt{2\pi}} e^{-y^2/2} dy \tag{7.2.9}$$

Consider the transformation between two uniform deviates on (0,1), x_1, x_2, and two quantities y_1, y_2,

$$\begin{aligned} y_1 &= \sqrt{-2 \ln x_1} \cos 2\pi x_2 \\ y_2 &= \sqrt{-2 \ln x_1} \sin 2\pi x_2 \end{aligned} \tag{7.2.10}$$

Equivalently we can write

$$\begin{aligned} x_1 &= \exp\left[-\frac{1}{2}(y_1^2 + y_2^2) \right] \\ x_2 &= \frac{1}{2\pi} \arctan \frac{y_2}{y_1} \end{aligned} \tag{7.2.11}$$

Now the Jacobian determinant can readily be calculated (try it!):

$$\frac{\partial(x_1, x_2)}{\partial(y_1, y_2)} = \begin{vmatrix} \frac{\partial x_1}{\partial y_1} & \frac{\partial x_1}{\partial y_2} \\ \frac{\partial x_2}{\partial y_1} & \frac{\partial x_2}{\partial y_2} \end{vmatrix} = -\left[\frac{1}{\sqrt{2\pi}} e^{-y_1^2/2} \right] \left[\frac{1}{\sqrt{2\pi}} e^{-y_2^2/2} \right] \tag{7.2.12}$$

Since this is the product of a function of y_2 alone and a function of y_1 alone, we see that each y is independently distributed according to the normal distribution (7.2.9).

One further trick is useful in applying (7.2.10). Suppose that, instead of picking uniform deviates x_1 and x_2 in the unit square, we instead pick v_1 and

v_2 as the ordinate and abscissa of a random point inside the unit circle around the origin. Then the sum of their squares, $R \equiv v_1^2 + v_2^2$ is a uniform deviate, which can be used for x_1, while the angle that (v_1, v_2) defines with respect to the v_1 axis can serve as the random angle $2\pi x_2$. What's the advantage? It's that the cosine and sine in (7.2.10) can now be written as $v_1/\sqrt{R}$ and $v_2/\sqrt{R}$, obviating the trigonometric function calls!

We thus have

```
      FUNCTION GASDEV(IDUM)
         Returns a normally distributed deviate with zero mean and unit variance, using RAN1(IDUM)
         as the source of uniform deviates.
      DATA ISET/0/
      IF (ISET.EQ.0) THEN              We don't have an extra deviate handy, so
1         V1=2.*RAN1(IDUM)-1.          pick two uniform numbers in the square extending from -1 to +1 in
          V2=2.*RAN1(IDUM)-1.                each direction,
          R=V1**2+V2**2                see if they are in the unit circle,
          IF(R.GE.1.)GO TO 1           and if they are not, try again.
          FAC=SQRT(-2.*LOG(R)/R)       Now make the Box-Muller transformation
          GSET=V1*FAC                  to get two normal deviates. Return one and save the other for next
          GASDEV=V2*FAC                      time.
          ISET=1                       Set flag.
      ELSE                             We have an extra deviate handy,
          GASDEV=GSET                  so return it,
          ISET=0                       and unset the flag.
      ENDIF
      RETURN
      END
```

REFERENCES AND FURTHER READING:

Knuth, Donald E. 1981, *Seminumerical Algorithms*, 2nd ed., vol. 2 of *The Art of Computer Programming* (Reading, Mass.: Addison-Wesley), pp. 116ff.

7.3 Rejection Method: Gamma, Poisson, Binomial Deviates

The *rejection method* is a powerful, general technique for generating random deviates whose distribution function $p(x)dx$ (probability of a value occurring between x and $x + dx$) is known and computable. The rejection method does *not* require that the cumulative distribution function [indefinite integral of $p(x)$] be readily computible, much less the inverse of that function — which was required for the transformation method in the previous section.

The rejection method is based on a simple geometrical argument:

Draw a graph of the probability distribution $p(x)$ that you wish to generate, so that the area under the curve in any range of x corresponds to the desired probability of generating an x in that range. If we had some way of choosing a random point *in two dimensions*, with uniform probability in the

area under your curve, then the x value of that random point would have the desired distribution.

Now, on the same graph, draw any other curve $f(x)$ which has finite (not infinite) area and lies everywhere *above* your original probability distribution. (This is always possible, because your original curve encloses only unit area, by definition of probability.) We will call this $f(x)$ the *comparison function.* Imagine now that you have some way of choosing a random point in two dimensions that is uniform in the area under the comparison function. Whenever that point lies outside the area under the original probability distribution, we will *reject* it and choose another random point. Whenever it lies inside the area under the original probability distribution, we will *accept* it. It should be obvious that the accepted points are uniform in the accepted area, so that their x values have the desired distribution. It should also be obvious that the fraction of points rejected just depends on the ratio of the area of the comparison function to the area of the probability distribution function, not on the details of shape of either function. For example, a comparison function whose area is less than 2 will reject fewer than half the points, even if it approximates the probability function very badly at some values of x, e.g. remains finite in some region where x is zero.

It remains only to suggest how to choose a uniform random point in two dimensions under the comparison function $f(x)$. A variant of the transformation method (§7.2) does nicely: Be sure to have chosen a comparison function whose indefinite integral is known analytically, and is also analytically invertible to give x as a function of "area under the comparison function to the left of x." Now pick a uniform deviate between 0 and A, where A is the total area under $f(x)$, and use it to get a corresponding x. Then pick a uniform deviate between 0 and $f(x)$ as the y value for the two-dimensional point. You should be able to convince yourself that the point (x, y) is uniformly distributed in the area under the comparison function $f(x)$.

An equivalent procedure is to pick the second uniform deviate between zero and one, and accept or reject according to whether it is respectively less than or greater than the ratio $p(x)/f(x)$.

So, to summarize, the rejection method for some given $p(x)$ requires that one find, once and for all, some reasonably good comparison function $f(x)$. Thereafter, each deviate generated requires two uniform random deviates, one evaluation of f (to get the coordinate y), and one evaluation of p (to decide whether to accept or reject the point x, y). Figure 7.3.1 illustrates the procedure. Then, of course, this procedure must be repeated, on the average, A times before the final deviate is obtained.

Gamma Distribution

The gamma distribution of integer order $a > 0$ is the waiting time to the ath event in a Poisson random process of unit mean. For example, when $a = 1$, it is just the exponential distribution of §7.2, the waiting time to the first event.

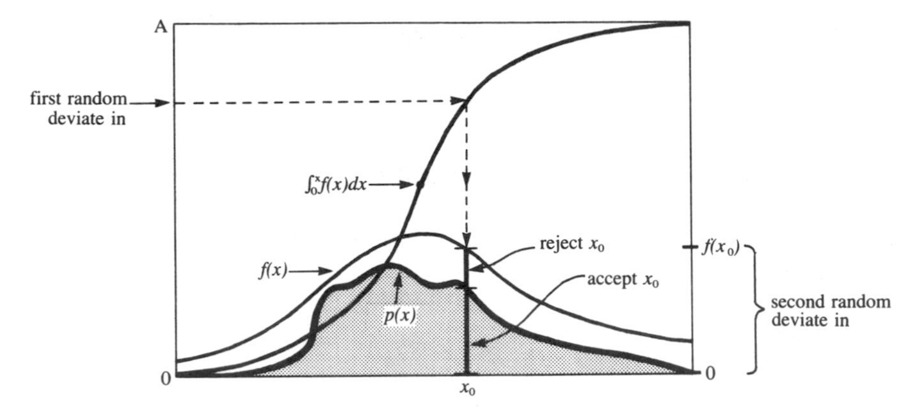

Figure 7.3.1. Rejection method for generating a random deviate x from a known probability distribution $p(x)$ that is everywhere less than some other function $f(x)$. The transformation method is first used to generate a random deviate x of the distribution f (compare Figure 7.2.1). A second uniform deviate is used to decide whether to accept or reject that x. If it is rejected, a new deviate of f is found; and so on. The ratio of accepted to rejected points is the ratio of the area under p to the area between p and f.

A gamma deviate has probability $p_a(x)dx$ of occurring with a value between x and $x + dx$, where

$$p_a(x)dx = \frac{x^{a-1}e^{-x}}{\Gamma(a)}dx \qquad x > 0 \tag{7.3.1}$$

To generate deviates of (7.3.1) for small values of a, it is best to add up a exponentially distributed waiting times, i.e. logarithms of uniform deviates. Since the sum of logarithms is the logarithm of the product, one really has only to generate the product of a uniform deviates, then take the log.

For larger values of a, the distribution (7.3.1) has a typically "bell-shaped" form, with a peak at $x = a$ and a half-width of about $\sqrt{a}$.

We will be interested in several probability distributions with this same qualitative form. A useful comparison function in such cases is derived from the *Lorentzian distribution*

$$p(y)dy = \frac{1}{\pi}\left(\frac{1}{1+y^2}\right)dy \tag{7.3.2}$$

whose inverse indefinite integral is just the tangent function. It follows that the x-coordinate of an area-uniform random point under the comparison function

$$f(x) = \frac{c_0}{1+(x-x_0)^2/a_0^2} \tag{7.3.3}$$

for any constants a_0, c_0, and x_0, can be generated by the prescription

$$x = a_0 \tan(\pi U) + x_0 \tag{7.3.4}$$

where U is a uniform deviate between 0 and 1. Thus, for some specific "bell-shaped" $p(x)$ probability distribution, we need only find constants a_0, c_0, x_0, with the product $a_0 c_0$ (which determines the area) as small as possible, such that (7.3.3) is everywhere greater than $p(x)$.

Ahrens has done this for the gamma distribution, yielding the following algorithm (as described in Knuth):

```
      FUNCTION GAMDEV(IA,IDUM)
          Returns a deviate distributed as a gamma distribution of integer order IA, i.e. a waiting
          time to the IAth event in a Poisson process of unit mean, using RAN1(IDUM) as the source
          of uniform deviates.
      IF(IA.LT.1)PAUSE
      IF(IA.LT.6)THEN                     Use direct method, adding waiting times.
        X=1.
        DO 11 J=1,IA
          X=X*RAN1(IDUM)
11      CONTINUE
        X=-LOG(X)
      ELSE                                Use rejection method.
1         V1=2.*RAN1(IDUM)-1.             These four lines generate the tangent of a random angle, i.e. are
          V2=2.*RAN1(IDUM)-1.                  equivalent to Y = TAN(3.14159265 * RAN1(IDUM)).
        IF(V1**2+V2**2.GT.1.)GO TO 1
          Y=V2/V1
          AM=IA-1
          S=SQRT(2.*AM+1.)
          X=S*Y+AM                        We decide whether to reject X:
        IF(X.LE.0.)GO TO 1                Reject in region of zero probability.
          E=(1.+Y**2)*EXP(AM*LOG(X/AM)-S*Y)    Ratio of probability fn. to comparison fn.
        IF(RAN1(IDUM).GT.E)GO TO 1        Reject on basis of a second uniform deviate.
      ENDIF
      GAMDEV=X
      RETURN
      END
```

Poisson Deviates

The Poisson distribution is conceptually related to the gamma distribution. It gives the probability of a certain integer number m of unit rate Poisson random events occurring in a given interval of time x, while the gamma distribution was the probability of waiting time between x and $x + dx$ to the mth event. Note that m takes on only integer values ≥ 0, so that the Poisson distribution, viewed as a continuous distribution function $p_x(m)dm$, is zero everywhere except where m is an integer ≥ 0. At such places, it is infinite, such that the integrated probability over a region containing the integer is some finite number. The total probability at an integer j is

$$\text{Prob}(j) = \int_{j-\epsilon}^{j+\epsilon} p_x(m)dm = \frac{x^j e^{-x}}{j!} \tag{7.3.5}$$

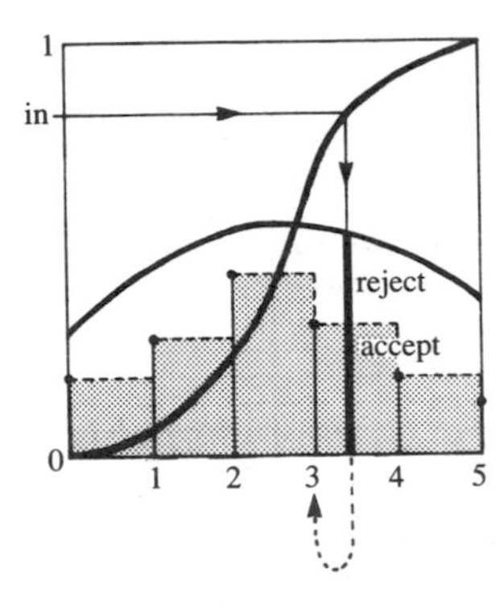

Figure 7.3.2. Rejection method as applied to an integer-valued distribution. The method is performed on the step function shown as a dashed line, yielding a real-valued deviate. This deviate is rounded down to the next lower integer, which is output.

At first sight this might seem an unlikely candidate distribution for the rejection method, since no continuous comparison function can be larger than the infinitely tall, but infinitely narrow, *Dirac delta functions* in $p_x(m)$. However there is a trick that we can do: Spread the finite area in the spike at j uniformly into the interval between j and $j+1$. This defines a continuous distribution $q_x(m)dm$ given by

$$q_x(m)dm = \frac{x^{[m]}e^{-x}}{[m]!} \tag{7.3.6}$$

where $[m]$ represents the largest integer less than m. If we now use the rejection method to generate a (non-integer) deviate from (7.3.6), and then take the integer part of that deviate, it will be as if drawn from the desired distribution (7.3.5). (See Figure 7.3.2.) This trick is general for any integer-valued probability distribution.

For x large enough, the distribution (7.3.6) is qualitatively bell-shaped (albeit with a bell made out of small, square steps), and we can use the same kind of Lorentzian comparison function as was already used above. For small x, we can generate independent exponential deviates (waiting times between events); when the sum of these first exceeds x, then the number of events which would have occurred in waiting time x becomes known and is one less than the number of terms in the sum.

These ideas produce the following routine:

```
FUNCTION POIDEV(XM,IDUM)
    Returns as a floating-point number an integer value that is a random deviate drawn from a
    Poisson distribution of mean XM, using RAN1(IDUM) as a source of uniform random deviates.
PARAMETER (PI=3.141592654)
DATA OLDM /-1./                 Flag for whether XM has changed since last call.
IF (XM.LT.12.)THEN              Use direct method.
    IF (XM.NE.OLDM) THEN
        OLDM=XM
        G=EXP(-XM)              If XM is new, compute the exponential.
    ENDIF
    EM=-1
```

```
          T=1.
2         EM=EM+1.                  Instead of adding exponential deviates it is equivalent to multiply
          T=T*RAN1(IDUM)                uniform deviates. Then we never actually have to take the
          IF (T.GT.G) GO TO 2           log, merely compare to the pre-computed exponential.
      ELSE                          Use rejection method.
          IF (XM.NE.OLDM) THEN      If XM has changed since the last call, then precompute some functions
              OLDM=XM                   which occur below.
              SQ=SQRT(2.*XM)
              ALXM=ALOG(XM)
              G=XM*ALXM-GAMMLN(XM+1.)   The function GAMMLN is the natural log of the gamma func-
          ENDIF                         tion, as given in §6.2.
1         Y=TAN(PI*RAN1(IDUM))      Y is a deviate from a Lorentzian comparison function.
          EM=SQ*Y+XM                EM is Y, shifted and scaled.
          IF (EM.LT.0.)  GO TO 1    Reject if in regime of zero probability.
          EM=INT(EM)                The trick for integer-valued distributions.
          T=0.9*(1.+Y**2)*EXP(EM*ALXM-GAMMLN(EM+1.)-G)   The ratio of the desired distribution to
          IF (RAN1(IDUM).GT.T) GO TO 1  the comparison function; we accept or reject by comparing it
      ENDIF                             to another uniform deviate. The factor 0.9 is chosen so that
      POIDEV=EM                         T never exceeds 1.
      RETURN
      END
```

Binomial Deviates

If an event occurs with probability q, and we make n trials, then the number of times m that it occurs has the binomial distribution,

$$\int_{j-\epsilon}^{j+\epsilon} p_{n,q}(m)dm = \binom{n}{j} q^j (1-q)^{n-j} \tag{7.3.7}$$

The binomial distribution is integer valued, with m taking on possible values from 0 to n. It depends on *two* parameters, n and q, so is correspondingly a bit harder to implement than our previous examples. Nevertheless, the techniques already illustrated are sufficiently powerful to do the job:

```
      FUNCTION BNLDEV(PP,N,IDUM)
```
Returns as a floating-point number an integer value that is a random deviate drawn from a binomial distribution of N trials each of probability PP, using RAN1(IDUM) as a source of uniform random deviates.
```
      PARAMETER (PI=3.141592654)
      DATA NOLD /-1/, POLD /-1./    Arguments from previous calls.
      IF(PP.LE.0.5)THEN             The binomial distribution is invariant under changing PP to 1.-PP, if
          P=PP                          we also change the answer to N minus itself; we'll remember
      ELSE                              to do this below.
          P=1.-PP
      ENDIF
      AM=N*P                        This is the mean of the deviate to be produced.
      IF (N.LT.25)THEN              Use the direct method while N is not too large. This can require up
          BNLDEV=0.                     to 25 calls to RAN1.
          DO 11 J=1,N
              IF(RAN1(IDUM).LT.P)BNLDEV=BNLDEV+1.
11        CONTINUE
      ELSE IF (AM.LT.1.)  THEN      If fewer than one event is expected out of 25 or more trials, then the
          G=EXP(-AM)                    distribution is quite accurately Poisson. Use direct Poisson
          T=1.                          method.
```

```
      DO 12 J=0,N
        T=T*RAN1(IDUM)
        IF (T.LT.G) GO TO 1
12    CONTINUE
      J=N
1     BNLDEV=J
    ELSE                          Use the rejection method.
      IF (N.NE.NOLD) THEN         If N has changed, then compute useful quantities.
        EN=N
        OLDG=GAMMLN(EN+1.)
        NOLD=N
      ENDIF
      IF (P.NE.POLD) THEN         If P has changed, then compute useful quantities.
        PC=1.-P
        PLOG=LOG(P)
        PCLOG=LOG(PC)
        POLD=P
      ENDIF
      SQ=SQRT(2.*AM*PC)           The following code should by now seem familiar: rejection method
2     Y=TAN(PI*RAN1(IDUM))            with a Lorentzian comparison function.
      EM=SQ*Y+AM
      IF (EM.LT.0..OR.EM.GE.EN+1.)  GO TO 2        Reject.
      EM=INT(EM)                  Trick for integer-valued distribution.
      T=1.2*SQ*(1.+Y**2)*EXP(OLDG-GAMMLN(EM+1.)
*       -GAMMLN(EN-EM+1.)+EM*PLOG+(EN-EM)*PCLOG)
      IF (RAN1(IDUM).GT.T) GO TO 2    Reject. This happens about 1.5 times per deviate, on av-
      BNLDEV=EM                         erage.
    ENDIF
    IF (P.NE.PP) BNLDEV=N-BNLDEV      Remember to undo the symmetry transformation.
    RETURN
    END
```

REFERENCES AND FURTHER READING:

Knuth, Donald E. 1981, *Seminumerical Algorithms*, 2nd ed., vol. 2 of *The Art of Computer Programming* (Reading, Mass.: Addison-Wesley), pp. 120ff.

7.4 Generation of Random Bits

This topic is not very useful for programming in high-level languages, but it can be quite useful when you have access to the machine-language level of a machine or when you are in a position to build special-purpose hardware out of readily available chips.

The problem is how to generate single random bits, with 0 and 1 equally probable. Of course you can just generate uniform random deviates between zero and one and use their first bit (i.e. test if they are greater than or less than 0.5). However this takes a lot of arithmetic; there are special purpose applications, such as real-time signal processing, where you want to generate bits very much faster than that.

One method for generating random bits, with two variant implementations, is based on the theory of "primitive polynomials modulo 2." It is

beyond our scope to discuss this theory. Suffice it to say that there are such things, special polynomials among those whose coefficients are zero or one. An example is

$$x^{18} + x^5 + x^2 + x^1 + x^0 \tag{7.4.1}$$

which we can abbreviate by just writing the nonzero powers of x, e.g.,

$$(18, 5, 2, 1, 0)$$

Every primitive polynomial modulo 2 of order n (=18 above) defines a recurrence relation for obtaining a new random bit from the n preceding ones. The recurrence relation is guaranteed to produce a sequence of maximal length, i.e. cycle through all possible sequences of n bits (except all zeros) before it repeats. Therefore one can seed the sequence with any initial bit pattern (except all zeros), and get $2^n - 1$ random bits before the sequence repeats.

Let the bits be numbered from 1 (most recently generated) through n (generated n steps ago), and denoted $a_1, a_2, \ldots, a_n$. We want to give a formula for a new bit a_0. After generating a_0 we will shift all the bits by one, so that the old a_n is finally lost, and the new a_0 becomes a_1. We then apply the formula again, and so on.

"Method I" is the easiest to implement in hardware, requiring only a single shift register n bits long and a few XOR ("exclusive or" or bit addition mod 2) gates. For the primitive polynomial given above, the recurrence formula is

$$a_0 = a_{18}\,.\texttt{XOR}.\,a_5\,.\texttt{XOR}.\,a_2\,.\texttt{XOR}.\,a_1 \tag{7.4.2}$$

The terms that are XOR'd together can be thought of as "taps" on the shift register, XOR'd into the register's input. More generally, there is precisely one term for each nonzero coefficient in the primitive polynomial except the constant (zero bit) term. So the first term will always be a_n for a primitive polynomial of degree n, while the last term might or might not be a_1, depending on whether the primitive polynomial has a term in x^1.

It is rather cumbersome to illustrate the method in FORTRAN. Assume that IAND is a bitwise AND function, NOT is bitwise complement, ISHFT(,1) is leftshift by one bit, IOR is bitwise OR. (These are available, e.g., in VAX-11 FORTRAN.) Then we have:

```
FUNCTION IRBIT1(ISEED)
   Returns as an integer a random bit, based on the 18 low-significance bits in ISEED (which
   is modified for the next call).
LOGICAL NEWBIT                          The accumulated XOR's.
PARAMETER (IB1=1,IB2=2,IB5=16,IB18=131072)         Powers of 2.
NEWBIT=IAND(ISEED,IB18).NE.0 Get bit 18.
IF(IAND(ISEED,IB5).NE.0)NEWBIT=.NOT.NEWBIT  XOR with bit 5.
IF(IAND(ISEED,IB2).NE.0)NEWBIT=.NOT.NEWBIT  XOR with bit 2.
```

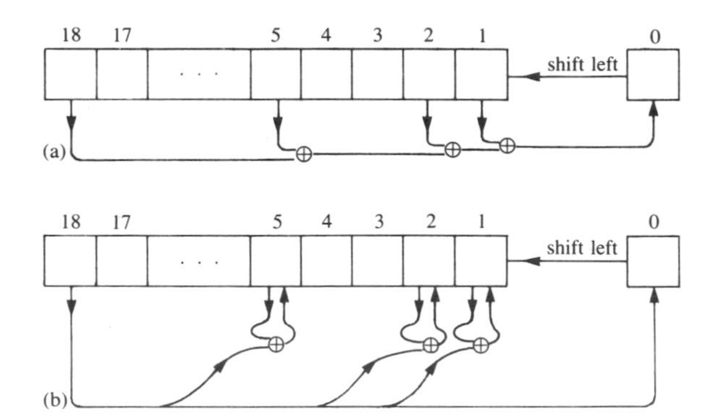

Figure 7.4.1. Two related methods for obtaining random bits from a shift register and a primitive polynomial modulo 2. (a) The contents of selected taps are combined by exclusive-or (addition modulo 2), and the result is shifted in from the right. This method is easiest to implement in hardware. (b) Selected bits are modified by exclusive-or with the leftmost bit, which is then shifted in from the right. This method is easiest to implement in software.

```
IF(IAND(ISEED,IB1).NE.0)NEWBIT=.NOT.NEWBIT  XOR with bit 1.
IRBIT1=0
ISEED=IAND(ISHFT(ISEED,1),NOT(IB1)) Leftshift the seed and put a zero in its bit 1.
IF(NEWBIT)THEN                      But if the XOR calculation gave a 1,
    IRBIT1=1
    ISEED=IOR(ISEED,IB1)            then put that in bit 1 instead.
ENDIF
RETURN
END
```

"Method II" is less suited to direct hardware implementation (though still possible), but is more suited to machine-language implementation. It modifies more than one bit among the saved n bits as each new bit is generated (Figure 7.4.1). It generates the maximal length sequence, but not in the same order as Method I. The prescription for the primitive polynomial (7.4.1) is:

$$\begin{aligned} a_0 &= a_{18} \\ a_5 &= a_5 \,.\mathtt{XOR}.\, a_0 \\ a_2 &= a_2 \,.\mathtt{XOR}.\, a_0 \\ a_1 &= a_1 \,.\mathtt{XOR}.\, a_0 \end{aligned} \tag{7.4.3}$$

In general there will be an XOR to be done for each nonzero term in the primitive polynomial except 0 and n. The nice feature about Method II is that all the XOR's can usually be done as a single masked word XOR (here assumed to be the FORTRAN function IEOR):

Primitive Polynomials Modulo 2

(1, 0)
(2, 1, 0)
(3, 1, 0)
(4, 1, 0)
(5, 2, 0)
(6, 1, 0)
(7, 1, 0)
(8, 4, 3, 2, 0)
(9, 4, 0)
(10, 3, 0)
(11, 2, 0)
(12, 6, 4, 1, 0)
(13, 4, 3, 1, 0)
(14, 5, 3, 1, 0)
(15, 1, 0)
(16, 5, 3, 2, 0)
(17, 3, 0)
(18, 5, 2, 1, 0)
(19, 5, 2, 1, 0)
(20, 3, 0)
(21, 2, 0)
(22, 1, 0)
(23, 5, 0)
(24, 4, 3, 1, 0)
(25, 3, 0)
(26, 6, 2, 1, 0)
(27, 5, 2, 1, 0)
(28, 3, 0)
(29, 2, 0)
(30, 6, 4, 1, 0)
(31, 3, 0)
(32, 7, 5, 3, 2, 1, 0)
(33, 6, 4, 1, 0)
(34, 7, 6, 5, 2, 1, 0)
(35, 2, 0)
(36, 6, 5, 4, 2, 1, 0)
(37, 5, 4, 3, 2, 1, 0)
(38, 6, 5, 1, 0)
(39, 4, 0)
(40, 5, 4 3, 0)
(41, 3, 0)
(42, 5, 4, 3, 2, 1, 0)
(43, 6, 4, 3, 0)
(44, 6, 5, 2, 0)
(45, 4, 3, 1, 0)
(46, 8, 5, 3, 2, 1, 0)
(47, 5, 0)
(48, 7, 5, 4, 2, 1, 0)
(49, 6, 5, 4, 0)
(50, 4, 3, 2, 0)
(51, 6, 3, 1, 0)
(52, 3, 0)
(53, 6, 2, 1, 0)
(54, 6, 5, 4, 3, 2, 0)
(55, 6, 2, 1, 0)
(56, 7, 4, 2, 0)
(57, 5, 3, 2, 0)
(58, 6, 5, 1, 0)
(59, 6, 5, 4, 3, 1, 0)
(60, 1, 0)
(61, 5, 2, 1, 0)
(62, 6, 5, 3, 0)
(63, 1, 0)
(64, 4, 3, 1, 0)
(65, 4, 3, 1, 0)
(66, 8, 6, 5, 3, 2, 0)
(67, 5, 2, 1, 0)
(68, 7, 5, 1, 0)
(69, 6, 5, 2, 0)
(70, 5, 3, 1, 0)
(71, 5, 3, 1, 0)
(72, 6, 4, 3, 2, 1, 0)
(73, 4, 3, 2, 0)
(74, 7, 4, 3, 0)
(75, 6, 3, 1, 0)
(76, 5, 4, 2, 0)
(77, 6, 5, 2, 0)
(78, 7, 2, 1, 0)
(79, 4, 3, 2, 0)
(80, 7, 5, 3, 2, 1, 0)
(81, 4 0)
(82, 8, 7, 6, 4, 1, 0)
(83, 7, 4, 2, 0)
(84, 8, 7, 5, 3, 1, 0)
(85, 8, 2, 1, 0)
(86, 6, 5, 2, 0)
(87, 7, 5, 1, 0)
(88, 8, 5, 4, 3, 1, 0)
(89, 6, 5, 3, 0)
(90, 5, 3, 2, 0)
(91, 7, 6, 5, 3, 2, 0)
(92, 6, 5, 2, 0)
(93, 2, 0)
(94, 6, 5, 1, 0)
(95, 6, 5, 4, 2, 1, 0)
(96, 7, 6, 4, 3, 2, 0)
(97, 6, 0)
(98, 7, 4, 3, 2, 1, 0)
(99, 7, 5, 4, 0)
(100, 8, 7, 2, 0)

```
FUNCTION IRBIT2(ISEED)
    Returns as an integer a random bit, based on the 18 low-significance bits in ISEED (which
    is modified for the next call).
PARAMETER (IB1=1,IB2=2,IB5=16,IB18=131072,MASK=IB1+IB2+IB5)
IF(IAND(ISEED,IB18).NE.0)THEN     Change all masked bits, shift, and put 1 into bit 1.
    ISEED=IOR(ISHFT(IEOR(ISEED,MASK),1),IB1)
    IRBIT2=1
ELSE                              Shift and put 0 into bit 1.
    ISEED=IAND(ISHFT(ISEED,1),NOT(IB1))
    IRBIT2=0
ENDIF
RETURN
END
```

A word of caution is: Don't use sequential bits from these routines as the bits of a large, supposedly random, integer, or as the bits in the mantissa of a supposedly random floating point number. They are not very random for that purpose; see Knuth. Examples of acceptable uses of these random bits are: (i) multiplying a signal randomly by ± 1 at a rapid "chip rate," so as to spread its spectrum uniformly (but recoverably) across some desired bandpass, or (ii) Monte Carlo exploration of a binary tree, where decisions as to whether to branch left or right are to be made randomly.

Now we do not want you to go through life thinking that there is something special about the primitive polynomial of degree 18 used in the above examples. (We chose 18 because 2^{18} is small enough for you to verify our claims directly by numerical experiment.) The accompanying table lists some (random) primitive polynomials mod 2, as tabulated by Watson.

REFERENCES AND FURTHER READING:

Knuth, Donald E. 1981, *Seminumerical Algorithms*, 2nd ed., vol. 2 of *The Art of Computer Programming* (Reading, Mass.: Addison-Wesley), pp. 29ff.

Horowitz, Paul, and Hill, Winfield 1980, *The Art of Electronics* (Cambridge: Cambridge University Press), §§9.34–9.39.

Tausworthe, Robert C. 1965, "Random Numbers Generated by Linear Recurrence Modulo Two," *Mathematics of Computation*, vol. 19, p. 201.

Watson, E.J. 1962, "Primitive Polynomials (Mod 2)," *Mathematics of Computation*, vol. 16, p. 368.

7.5 The Data Encryption Standard

At the time of writing, this section is intended for your cultural edification, not for your day-to-day use, *unless* you know how to build special purpose hardware out of commercially available chips. We expect, however, that the hardware that we describe will be available on many computer systems quite soon.

After reading all the caveats and conditions which accompany the uniform random number generators which were described in §7.1, you might very well ask: "Why doesn't someone just do the job *right*, design a random number generator so good that there are no ifs-ands-or-buts about it?" The answer is that the U.S. Government has done so, but for purposes rather different from those of interest to us in this book.

There is a close connection between random number generation and encryption, the latter being a subject of great interest in the "real world" of governmental and industrial secrecy. Suppose that you have a stream of data ("plaintext") to encrypt. One way to do this, which can be shown to be as mathematically perfect as any other way, is to add the data stream bit-by-bit modulo 2 to a stream of perfectly random numbers. Here, the notion of perfect randomness is a strong one. It requires not only that the random bit stream have no detectable correlations of any sort, but also that it be generated from a sequence so long (or from a universe of sequences so large) that even a very long stretch of its past behavior does not provide enough information, by itself, to predict the future bit stream. This latter condition is required so that the enciphering scheme is resistant to "plaintext attack," where the enemy cryptanalyst has by other means obtained the unciphered data stream corresponding to a part of your encoded message.

Most encryption schemes that are any good are, of course, highly secret. An exception is the *Data Encryption Standard* (DES), issued by the U.S. National Bureau of Standards (NBS). The DES has a somewhat controversial history: Strictly speaking, it derives from IBM's response to a public solicitation put forth by NBS. It has been widely reported, however, that the supersecret National Security Agency (NSA) was intimately involved in the development and acceptance testing of the final algorithm. (The DES document itself uses the guarded language, "NBS, supported by the technical assistance of appropriate Government agencies....") A key controversial question is whether NSA purposely weakened the algorithm, so that it had vulnerabilities significant enough to be exploited by NSA's own multi-billion dollar resources, but not so significant as to be exploitable by anyone else. For our purposes we hardly need know the answer to this: A random number generator whose deviations from randomness can be discerned only by concerted attack with resources comparable to NSA — that random number generator should surely be a contender for the "World's Best" title!

Unfortunately for those of us who work with high-level computer languages, the DES consists almost entirely of bit-level permutations, substitutions, shifts and shufflings. It is designed for hardware implementation on a single large-scale integrated chip; such chips are available as we write, and

should soon be cheap.

The basic DES is a substitution cipher which takes a block of 64 bits of input into a unique block of 64 bits of output, under the control of a 64 bit key, which is known only to the persons intended to read the message. (One point of controversy is that only 56 bits in the key are actually used, the remaining 8 being predictable parity checks.) As long as the key is held fixed, the same 64 bits of input will always produce the same 64 bits of output. The mapping of input to output is, of course, one-to-one and invertible, so that the message can be decrypted. Thus, if we imagine looping through all 2^{64} possible input bit configurations, all possible output configurations will also be produced, but in a highly scrambled order. We can therefore generate random numbers by generating a random bit stream by the methods of §7.4 and running that stream through DES.

The routines which follow are illustrative (and work), but are terribly inefficient: They store one bit per FORTRAN integer word, and do all bit manipulations as explicit operations on full words.

Random Number Generator Which Calls DES

Here is a routine which can be substituted for any of the RANs of §7.1. It presumes the existence of a subroutine called DES, whose arguments are 64 bits of input, 64 bits of key, 64 bits of output, an encrypt/decrypt flag (input to the routine), and a flag telling that the key has been changed since the previous call (input to the routine). If you have the DES available on your system as a system call, you should be able to rewrite this routine to take advantage of it (presumably using bits instead of words!).

```
FUNCTION RAN4(IDUM)
   Returns a uniform random deviate between 0.0 and 1.0 using DES. Set IDUM negative to
   initialize. There are IM possible initializations. This routine is extremely slow and should
   be used for demonstration purposes only.
PARAMETER (IM=11979,IA=430,IC=2531,NACC=24)
   The first three parameters are used in initializing, cf. §7.1. NACC is the number of bits
   floating point precision desired on the random deviate.
DIMENSION INP(64),JOT(64),KEY(64),POW(65)
DATA IFF/0/
IF(IDUM.LT.0.OR.IFF.EQ.0)THEN           Initialize:
    IFF=1
    IDUM=MOD(IDUM,IM)
    IF(IDUM.LT.0)IDUM=IDUM+IM
    POW(1)=0.5
    DO 11 J=1,64                       Set both the 64 bits of key and also the starting configuration of the
        IDUM=MOD(IDUM*IA+IC,IM)            64-bit input array.
        KEY(J)=(2*IDUM)/IM                 Highest order bit.
        INP(J)=MOD((4*IDUM)/IM,2)          Next highest order bit.
        POW(J+1)=0.5*POW(J)                Inverse powers of 2 in floating point.
11  CONTINUE
    NEWKEY=1                           Set this flag.
ENDIF
ISAV=INP(64)                           Start here except on initialization.
IF(ISAV.NE.0)THEN                      Generate the next input bit configuration; cf. §7.4.
    INP(4)=1-INP(4)                    I.e., change 1 to 0 and 0 to 1.
    INP(3)=1-INP(3)
```

```
    INP(1)=1-INP(1)
ENDIF
DO 12 J=64,2,-1
    INP(J)=INP(J-1)
  12 CONTINUE
INP(1)=ISAV                          Input bit configuration now ready.
CALL DES(INP,KEY,NEWKEY,0,JOT)       Here is the real business.
RAN4=0.0                             It remains only to make a floating number out of random bits.
DO 13 J=1,NACC
    IF(JOT(J).NE.0)RAN4=RAN4+POW(J)
  13 CONTINUE
RETURN
END
```

Details of the Data Encryption Standard

Here we give a FORTRAN implementation of the Data Encryption Standard for illustrative purposes, terribly inefficient to use in this high-level form. (Technically this is *not* an implementation of DES since the standard itself states that *no* implementation in software is acceptably secure! We, however, are not concerned with security but only with algorithmics.)

The DES has three distinguishable components. First there is the *key schedule.* This is simply a certain prescription for taking the 56 active bits in the key and shuffling them into 16 different configurations, each 48 bits long. The shuffling procedure is complicated, but straightforward: bits are never combined with other bits, or modified, but only moved around from place to place. In other words, the key schedule makes 16 "sub-master keys" out of a "master key." This need be done only whenever the key is changed, not every time an input block is enciphered.

Second, as the heart and soul of the DES, there is the *cipher function.* This is a fixed, highly non-linear function which combines 32 bits of input (a half-word of the total block being encrypted) with 48 bits of sub-master key to produce 32 bits of highly random output. The procedure in detail is to expand the 32 input bits to 48 bits by a permutation that repeats some bits twice, then to add the expanded input to the 48-bit submaster key (bit by bit modulo 2). Then, most important, the 48 bits are reduced down to 32 bits of output by a table look-up which maps every 6 sequential bits into 4 bits. This table, which is called the *S-box,* is the soul of the algorithm. It is a point of controversy that the theory behind the design of the S-box has never been fully revealed. Finally in the cipher function, a bitwise permutation of the 32 output bits is performed.

It is important to understand that the cipher function is not particularly invertible: to decode a message, it is *not* necessary to recover the input to the cipher function from its output and a knowledge of its key. In fact, to be resistant to plaintext attack, the cipher function should be highly non-invertible. The DES gets its invertibility (which, by the way, we don't need for the purpose of generating random numbers) from its overall encoding strategy, the third component of DES that we now describe:

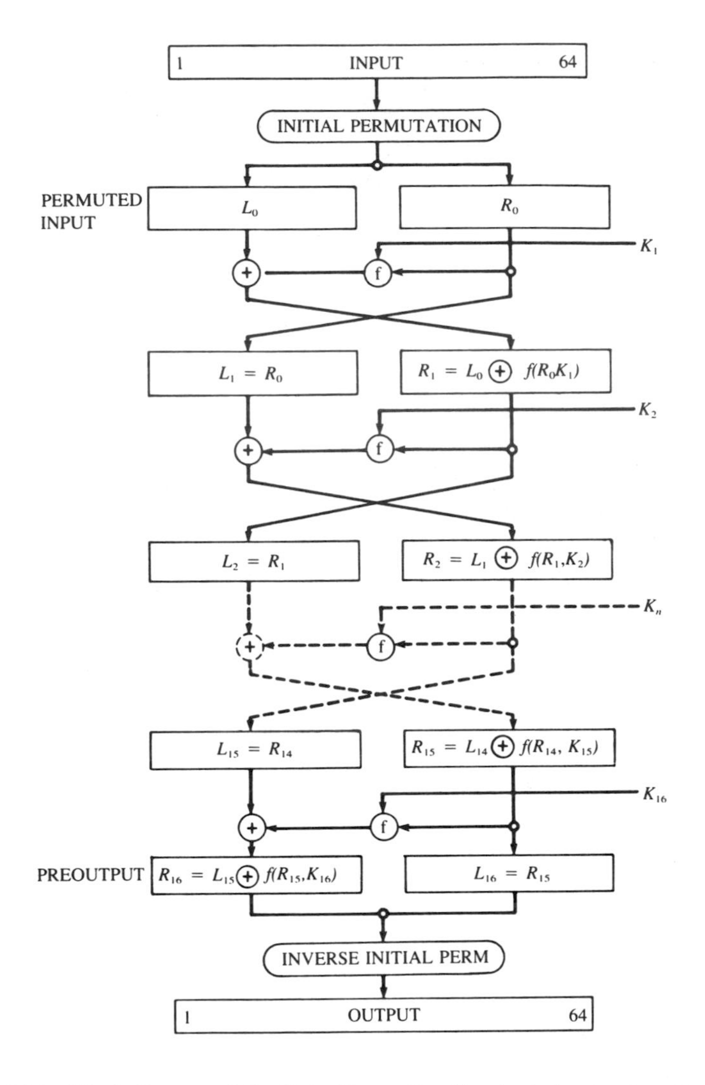

Figure 7.5.1. Block diagram of the Data Encryption Standard showing how the 16 sub-master keys $K_1 \dots K_{16}$ are combined with 32-bit half-words by means of a highly nonlinear cipher function f.

The original 64 bits of input are put through a fixed initial permutation and then divided into two half-words of 32 bits each. Now 16 stages of the following procedure are performed: One half-word is passed on to the next stage, untouched. That same half-word is also sent, with one of the 16 sub-master keys, to the cipher function, producing a 32 bit "random" configuration. The half-word which has not yet been used is now encrypted by adding it (bitwise modulo 2) to this random configuration and then passed on to the next stage. Figure 7.5.1 should make the process clear.

Between each stage, the half-words are exchanged, so that the one passed on unchanged is not the one that was passed on unchanged in the previous

transition. After all 16 stages are complete, the two half-words are recombined through a specified bitwise permutation into the output 64-bit block.

The reason that the whole DES procedure is invertible is that (after the final, fixed permutation is undone) one half-word of the output is precisely the bit configuration which (with a sub-master key) was used to encode the other half. Thus, by using the sub-master keys in the reverse order, one can "peel away" the 16 stages of encryption one after the other. That is why one half-word is always passed through unchanged: to provide the means of decrypting the other half-word!

Referring to the above description, you should find the following routines comprehensible:

```
      SUBROUTINE DES(INPUT,KEY,NEWKEY,ISW,JOTPUT)
         Data Encryption Standard. Encrypts 64 bits, stored one bit per word, in array INPUT into
         JOTPUT using KEY. Set NEWKEY=1 when the key is new. Set ISW=0 for encryption, =1 for
         decryption. Timing: about 24 ms per call on a VAX11/780.
      DIMENSION INPUT(64),KEY(64),JOTPUT(64),ITMP(64),IP(64),IPM(64)
     *    ,ICF(32),KNS(48,16)
      DATA IP/58,50,42,34,26,18,10,2,60,52,44,36,28,20,12,4,62,54,46
     *    ,38,30,22,14,6,64,56,48,40,32,24,16,8,57,49,41,33,25,17,9,1,59,51
     *    ,43,35,27,19,11,3,61,53,45,37,29,21,13,5,63,55,47,39,31,23,15,7/
      DATA IPM/40,8,48,16,56,24,64,32,39,7,47,15,55,23,63,31,38,6,46,14
     *    ,54,22,62,30,37,5,45,13,53,21,61,29,36,4,44,12,52,20,60,28,35,3
     *    ,43,11,51,19,59,27,34,2,42,10,50,18,58,26,33,1,41,9,49,17,57,25/
      IF(NEWKEY.NE.0)THEN            Get the 16 sub-master keys from the master key.
         NEWKEY=0
         DO 11 I=1,16
            CALL KS(KEY,I,KNS(1,I))
11       CONTINUE
      ENDIF
      DO 12 J=1,64                   The initial permutation.
         ITMP(J)=INPUT(IP(J))
12    CONTINUE
      DO 14 I=1,16                   The 16 stages of encryption.
         II=I
         IF(ISW.EQ.1)II=17-I         Use the sub-master keys in reverse order for decryption.
         CALL CYFUN(ITMP(33),KNS(1,II),ICF)   Get cipher function.
         DO 13 J=1,32                Pass one half-word through unchanged, while encrypting the other
            IC=ICF(J)+ITMP(J)           half-word and exchanging the two half-words output.
            ITMP(J)=ITMP(J+32)
            ITMP(J+32)=IAND(IC,1)
         If you don't have the IAND function, instead use:
C           ITMP(J+32)=MOD(MOD(IC,2)+2,2)
13       CONTINUE
14    CONTINUE                       Done with the 16 stages.
      DO 15 J=1,32                   A final exchange of the two half-words is required.
         IC=ITMP(J)
         ITMP(J)=ITMP(J+32)
         ITMP(J+32)=IC
15    CONTINUE
      DO 16 J=1,64                   Final output permutation.
         JOTPUT(J)=ITMP(IPM(J))
16    CONTINUE
      RETURN
      END
```

```
      SUBROUTINE KS(KEY,N,KN)
         Key schedule calculation, returns KN given KEY and N=1,2,...,16; must be called with N
         in that order.
      DIMENSION KEY(64),KN(48),ICD(56),IPC1(56),IPC2(48)
      DATA IPC1/57,49,41,33,25,17,9,1,58,50,42,34,26,18
     *    ,10,2,59,51,43,35,27,19,11,3,60,52,44,36,63,55,47,39,31,23,15
     *    ,7,62,54,46,38,30,22,14,6,61,53,45,37,29,21,13,5,28,20,12,4/
      DATA IPC2/14,17,11,24,1,5,3,28,15,6,21,10
     *    ,23,19,12,4,26,8,16,7,27,20,13,2,41,52,31,37,47,55
     *    ,30,40,51,45,33,48,44,49,39,56,34,53,46,42,50,36,29,32/
      IF(N.EQ.1)THEN                  Initial selection and permutation.
        DO 11 J=1,56
          ICD(J)=KEY(IPC1(J))
11      CONTINUE
      ENDIF
      IT=2                            For most values of N perform two shifts,
      IF(N.EQ.1.OR.N.EQ.2.OR.N.EQ.9.OR.N.EQ.16)IT=1   but for these perform only one.
      DO 13 I=1,IT                    Circular left-shifts of the two halves of the array ICD.
        IC=ICD(1)
        ID=ICD(29)
        DO 12 J=1,27
          ICD(J)=ICD(J+1)
          ICD(J+28)=ICD(J+29)
12      CONTINUE
        ICD(28)=IC
        ICD(56)=ID
13    CONTINUE                        Done with the shifts.
      DO 14 J=1,48                    The sub-master key is a selection of bits from the shifted ICD.
        KN(J)=ICD(IPC2(J))
14    CONTINUE
      RETURN
      END

      SUBROUTINE CYFUN(IR,K,IOUT)
         Returns the cipher function of IR and K in IOUT.
      DIMENSION IR(32),K(48),IOUT(32),IE(48),IET(48),IP(32)
     *    ,ITMP(32),IS(16,4,8),IBIN(4,16)
      DATA IET/32,1,2,3,4,5,4,5,6,7,8,9,8,9,10,11,12,13,12,13
     *    ,14,15,16,17,16,17,18,19,20,21,20,21,22,23,24
     *    ,25,24,25,26,27,28,29,28,29,30,31,32,1/
      DATA IP/16,7,20,21,29,12,28,17,1,15,23,26,5,18,31,10
     *    ,2,8,24,14,32,27,3,9,19,13,30,6,22,11,4,25/
         Here follows the S-Box, in full glory. Alternatively, you might read IS from a data file.
      DATA IS/14,4,13,1,2,15,11,8,3,10,6,12,5,9,0,7
     *    ,0,15,7,4,14,2,13,1,10,6,12,11,9,5,3,8
     *    ,4,1,14,8,13,6,2,11,15,12,9,7,3,10,5,0
     *    ,15,12,8,2,4,9,1,7,5,11,3,14,10,0,6,13
     *    ,15,1,8,14,6,11,3,4,9,7,2,13,12,0,5,10
     *    ,3,13,4,7,15,2,8,14,12,0,1,10,6,9,11,5
     *    ,0,14,7,11,10,4,13,1,5,8,12,6,9,3,2,15
     *    ,13,8,10,1,3,15,4,2,11,6,7,12,0,5,14,9
     *    ,10,0,9,14,6,3,15,5,1,13,12,7,11,4,2,8
     *    ,13,7,0,9,3,4,6,10,2,8,5,14,12,11,15,1
     *    ,13,6,4,9,8,15,3,0,11,1,2,12,5,10,14,7
     *    ,1,10,13,0,6,9,8,7,4,15,14,3,11,5,2,12
     *    ,7,13,14,3,0,6,9,10,1,2,8,5,11,12,4,15
     *    ,13,8,11,5,6,15,0,3,4,7,2,12,1,10,14,9
     *    ,10,6,9,0,12,11,7,13,15,1,3,14,5,2,8,4
     *    ,3,15,0,6,10,1,13,8,9,4,5,11,12,7,2,14
     *    ,2,12,4,1,7,10,11,6,8,5,3,15,13,0,14,9
     *    ,14,11,2,12,4,7,13,1,5,0,15,10,3,9,8,6
     *    ,4,2,1,11,10,13,7,8,15,9,12,5,6,3,0,14
     *    ,11,8,12,7,1,14,2,13,6,15,0,9,10,4,5,3
     *    ,12,1,10,15,9,2,6,8,0,13,3,4,14,7,5,11
```

```
*     ,10,15,4,2,7,12,9,5,6,1,13,14,0,11,3,8
*     ,9,14,15,5,2,8,12,3,7,0,4,10,1,13,11,6
*     ,4,3,2,12,9,5,15,10,11,14,1,7,6,0,8,13
*     ,4,11,2,14,15,0,8,13,3,12,9,7,5,10,6,1
*     ,13,0,11,7,4,9,1,10,14,3,5,12,2,15,8,6
*     ,1,4,11,13,12,3,7,14,10,15,6,8,0,5,9,2
*     ,6,11,13,8,1,4,10,7,9,5,0,15,14,2,3,12
*     ,13,2,8,4,6,15,11,1,10,9,3,14,5,0,12,7
*     ,1,15,13,8,10,3,7,4,12,5,6,11,0,14,9,2
*     ,7,11,4,1,9,12,14,2,0,6,10,13,15,3,5,8
*     ,2,1,14,7,4,10,8,13,15,12,9,0,3,5,6,11/
         Next follows the table of bits in the integers 0 to 15:
      DATA IBIN/0,0,0,0,0,0,0,1,0,0,1,0,0,0,1,1
*     ,0,1,0,0,0,1,0,1,0,1,1,0,0,1,1,1,1,0,0,0,1,0,0,1
*     ,1,0,1,0,1,0,1,1,1,1,0,0,1,1,0,1,1,1,1,0,1,1,1,1/
      DO 15 J=1,48                      Expand IR to 48 bits and combine it with K.
        IE(J)=IAND(IR(IET(J))+K(J),1)
        If you don't have the IAND function, instead use:
C         IE(J)=MOD(MOD(IR(IET(J))+K(J),2)+2,2)
15    CONTINUE
      DO 17 JJ=1,8                      Loop over 8 groups of 6 bits.
        J=6*JJ-5
        IROW=IOR(IE(J+5),ISHFT(IE(J),1))      Find place in the S-box table.
        ICOL=IOR(IE(J+4),ISHFT(IOR(IE(J+3),ISHFT(IOR(IE(J+2),
*            ISHFT(IE(J+1),1)),1)),1))
        If you don't have the above bit functions, instead use:
C         IROW=2*IE(J)+IE(J+5)
C         ICOL=8*IE(J+1)+4*IE(J+2)+2*IE(J+3)+IE(J+4)
        ISS=IS(ICOL+1,IROW+1,JJ)         Look up the number in the S-box table
        KK=4*(JJ-1)
        DO 16 KI=1,4                     and plug its bits into the output.
          ITMP(KK+KI)=IBIN(KI,ISS+1)
16      CONTINUE
17    CONTINUE
      DO 18 J=1,32                      Final permutation.
        IOUT(J)=ITMP(IP(J))
18    CONTINUE
      RETURN
      END
```

REFERENCES AND FURTHER READING:

Data Encryption Standard, 1977 January 15, Federal Information Processing Standards Publication, number 46 (Washington: U.S. Department of Commerce, National Bureau of Standards).

Guidelines for Implementing and Using the NBS Data Encryption Standard, 1981 April 1, Federal Information Processing Standards Publication, number 74 (Washington: U.S. Department of Commerce, National Bureau of Standards).

Validating the Correctness of Hardware Implementations of the NBS Data Encryption Standard, 1980, NBS Special Publication 500-20 (Washington: U.S. Department of Commerce, National Bureau of Standards).

7.6 Monte Carlo Integration

Suppose that we pick N points, uniformly randomly in a multidimensional volume V. Call them $x_1, \ldots, x_N$. Then the basic theorem of Monte Carlo integration estimates the integral of a function f over the multidimensional volume,

$$\int f\,dV \approx V\,\langle f\rangle\; \pm V\sqrt{\frac{\langle f^2\rangle - \langle f\rangle^2}{N}} \tag{7.6.1}$$

Here the angle brackets denote taking the arithmetic mean over the N sample points,

$$\langle f\rangle \equiv \frac{1}{N}\sum_{i=1}^{N} f(x_i) \qquad \langle f^2\rangle \equiv \frac{1}{N}\sum_{i=1}^{N} f^2(x_i) \tag{7.6.2}$$

The "plus-or-minus" term in (7.6.1) is a one standard deviation error estimate for the integral, not a rigorous bound; further, there is no guarantee that the error is distributed as a Gaussian, so the error term should be taken only as a rough indication of probable error.

Suppose that you want to integrate a function g over a region W which is not easy to sample randomly. For example, W might have a very complicated shape. No problem. Just find a region V which *includes* W and which *can* easily be sampled (Figure 7.6.1), and then define f to be equal to g for points in W and equal to zero for points outside of W (but still inside the sampled V). You want to try to make V enclose W as closely as possible, because the zero values of f will increase the error estimate term of (7.6.1). And well they should: points chosen outside of W have no information content, so the effective value of N, the number of points, is reduced. The error estimate in (7.6.1) takes this into account.

It is not feasible to give a general purpose routine for Monte Carlo integration, but a worked example ought to show you how it is done. Suppose that we want to find the weight and the position of the center of mass of an object of complicated shape, namely the intersection of a torus with the edge of a large box. In particular let the object be defined by the three simultaneous conditions

$$z^2 + \left(\sqrt{x^2+y^2} - 3\right)^2 \le 1 \tag{7.6.3}$$

(torus centered on the origin with major radius =4, minor radius =2,)

$$x \ge 1 \qquad y \ge -3 \tag{7.6.4}$$

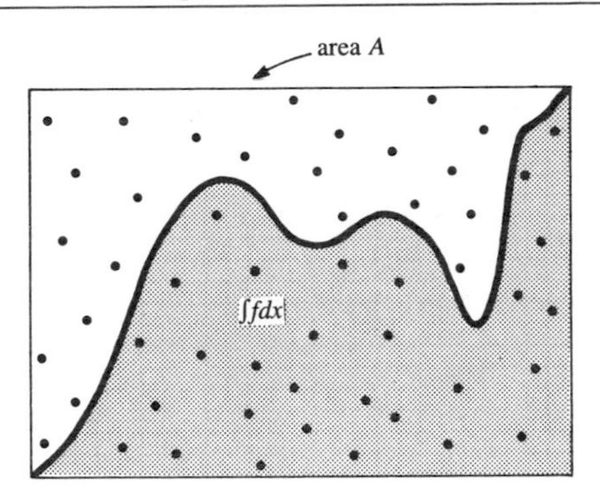

Figure 7.6.1. Monte Carlo integration. Random points are chosen within the area A. The integral of the function f is estimated as the area of A multiplied by the fraction of random points that fall below the curve f. Refinements on this procedure can improve the accuracy of the method; see text.

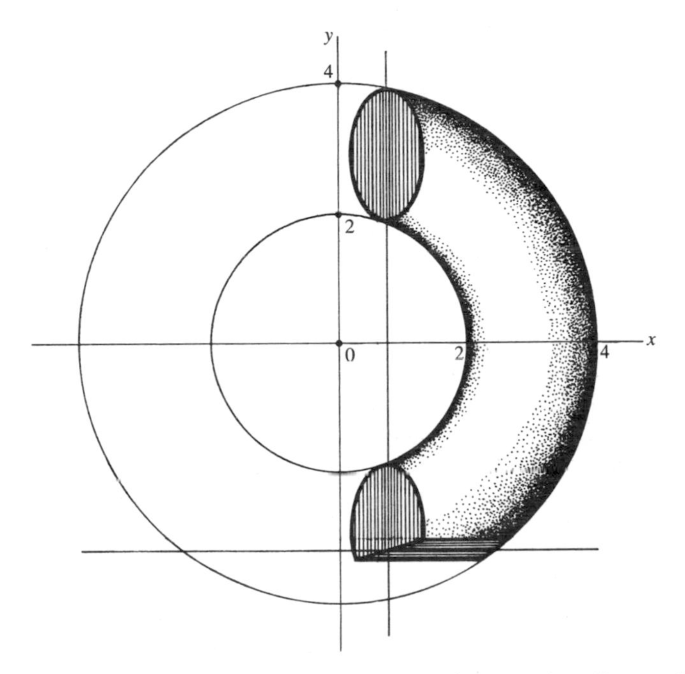

Figure 7.6.2. Example of Monte Carlo integration (see text). The region of interest is a piece of a torus, bounded by the intersection of two planes. The limits of integration of the region cannot easily be written in analytically closed form, so Monte Carlo is a useful technique.

(two faces of the box, see Figure 7.6.2). Suppose for the moment that the object has a constant density ρ.

We want to estimate the following integrals over the interior of the complicated object:

$$\int \rho \, dxdydz \qquad \int x\rho \, dxdydz \qquad \int y\rho \, dxdydz \qquad \int z\rho \, dxdydz \tag{7.6.5}$$

The coordinates of the center of mass will be the ratio of the latter three integrals (linear moments) to the first one (the weight).

In the following fragment, the region V, enclosing the piece-of-torus W, is the rectangular box extending from 1 to 4 in x, -3 to 4 in y, and -1 to 1 in z.

```
N=                                    Set to the number of sample points desired.
DEN=                                  Set to the constant value of the density.
SW=0.                                 Zero the various sums to be accumulated.
SWX=0.
SWY=0.
SWZ=0.
VARW=0.
VARX=0.
VARY=0.
VARZ=0.
VOL=3.*7.*2.                          Volume of the sampled region.
DO 11 J=1,N
    X=1.+3.*RAN2(IDUM)                Pick a point randomly in the sampled region.
    Y=-3.+7.*RAN2(IDUM)
    Z=-1.+2.*RAN2(IDUM)
    IF (Z**2+(SQRT(X**2+Y**2)-3.)**2.LE.1.)THEN Is it in the torus?
        SW=SW+DEN                     If so, add to the various cumulants.
        SWX=SWX+X*DEN
        SWY=SWY+Y*DEN
        SWZ=SWZ+Z*DEN
        VARW=VARW+DEN**2
        VARX=VARX+(X*DEN)**2
        VARY=VARY+(Y*DEN)**2
        VARZ=VARZ+(Z*DEN)**2
    ENDIF
11 CONTINUE
W=VOL*SW/N                            The values of the integrals (7.6.5),
X=VOL*SWX/N
Y=VOL*SWY/N
Z=VOL*SWZ/N
DW=VOL*SQRT((VARW/N-(SW/N)**2)/N)           and their corresponding error estimates.
DX=VOL*SQRT((VARX/N-(SWX/N)**2)/N)
DY=VOL*SQRT((VARY/N-(SWY/N)**2)/N)
DZ=VOL*SQRT((VARZ/N-(SWZ/N)**2)/N)
```

Next, suppose that we want to evaluate the same integrals, but for a piece-of-torus whose density is a strong function of z, in fact varying according to

$$\rho(x, y, z) = e^{5z} \tag{7.6.6}$$

One way to do this is to put the statement

```
DEN=EXP(5.*Z)
```

inside the `IF...THEN` block, just before `DEN` is first used. This will work, but it is not the optimal way to proceed. Since (7.6.6) falls so rapidly to zero as z decreases (down to its lower limit -1), most sampled points contribute almost nothing to the sum of the weight or moments. These points are effectively wasted, almost as badly as those that fall outside of the region W. The right

way to do the problem is to make use of a change of variable, exactly as was done in developing the transformation methods of §7.2. Let

$$ds = e^{5z}dz \qquad \text{so that} \qquad s = \frac{1}{5}e^{5z}, \quad z = \frac{1}{5}\ln(5s) \tag{7.6.7}$$

Then $\rho dz = ds$, and the limits $-1 < z < 1$ become $.00135 < s < 29.682$. The program fragment now looks like this

```
N=                                  Set to the number of sample points desired.
SW=0.
SWX=0.
SWY=0.
SWZ=0.
VARW=0.
VARX=0.
VARY=0.
VARZ=0.
SS=(0.2*(EXP(5.)-EXP(-5.)))         Interval of s to be random sampled.
VOL=3.*7.*SS                        Volume in x,y,s-space.
DO 11 J=1,N
    X=1.+3.*RAN2(IDUM)
    Y=-3.+7.*RAN2(IDUM)
    S=.00135+SS*RAN2(IDUM)          Pick a point in s.
    Z=0.2*LOG(5.*S)                 Equation (7.6.7).
    IF (Z**2+(SQRT(X**2+Y**2)-3.)**2.LT.1.)THEN
        SW=SW+1.                    Density is 1, since absorbed into definition of s.
        SWX=SWX+X
        SWY=SWY+Y
        SWZ=SWZ+Z
        VARW=VARW+1.
        VARX=VARX+X**2
        VARY=VARY+Y**2
        VARZ=VARZ+Z**2
    ENDIF
11  CONTINUE
W=VOL*SW/N                          The values of the integrals (7.6.5),
X=VOL*SWX/N
Y=VOL*SWY/N
Z=VOL*SWZ/N
DW=VOL*SQRT((VARW/N-(SW/N)**2)/N)   and their corresponding error estimates.
DX=VOL*SQRT((VARX/N-(SWX/N)**2)/N)
DY=VOL*SQRT((VARY/N-(SWY/N)**2)/N)
DZ=VOL*SQRT((VARZ/N-(SWZ/N)**2)/N)
```

If you think for a minute, you will realize that equation (7.6.7) was useful only because the part of the integrand that we wanted to eliminate (e^{5z}) was both integrable analytically, and had an integral that could be analytically inverted. (Compare §7.2!) In general these properties will not hold. Question: What then? Answer: Pull out of the integrand the "best" factor that *can* be integrated and inverted. The criterion for "best" is to try to reduce the remaining integrand to a function that is as close as possible to constant.

The limiting case is instructive: If you manage to make the integrand f *exactly* constant, and if the region V, of known volume, *exactly* encloses the desired region W, then the average of f that you compute will be exactly its constant value, and the error estimate in equation (7.6.1) will exactly vanish.

You will, in fact, have done the integral exactly, and the Monte Carlo numerical evaluations are superfluous. So, backing off from the extreme limiting case, *to the extent* that you are able to make f approximately constant by change of variable, and *to the extent* that you can sample a region only slightly larger than W, you will increase the accuracy of the Monte Carlo integral. This technique is generically called *reduction of variance* in the literature.

The fundamental disadvantage of Monte Carlo integration is just that its accuracy increases only as the square root of N, the number of sampled points. If your accuracy requirements are modest, or if your computer budget is large, then the technique is highly recommended as one of great generality.

REFERENCES AND FURTHER READING:

Hammersley, J.M., and Handscomb, D.C. 1964, *Monte Carlo Methods* (London: Methuen).

Shreider, Yu. A., ed. 1966, *The Monte Carlo Method* (Oxford: Pergamon).

Chapter 8. Sorting

8.0 Introduction

This chapter doesn't quite belong in a book on *numerical* methods. However, some practical knowledge of techniques for sorting is an indispensable part of any good programmer's repertory of expertise. We do not want you to consider yourself expert in numerical techniques while remaining ignorant of so basic a subject as sorting.

In conjunction with numerical work, sorting is frequently necessary when data (either experimental or numerically generated) are being handled. One has tables or lists of numbers, representing one or more independent (or "control") variables, and one or more dependent (or "measured") variables. One may wish to arrange these data, in various circumstances, in order by one or another of these variables. Alternatively, one may simply wish to identify the "median" value, or the "upper quartile" value of one of the lists of values. This task also requires that a list be sorted.

Here, more specifically, are the tasks that this chapter will deal with:

- Sort, i.e. rearrange, an array of numbers into numerical order.
- Rearrange an array into numerical order while performing the corresponding rearrangement of one or more additional arrays, so that the correspondence between elements in all arrays is maintained.
- Given an array, prepare an *index table* for it, i.e., a table of pointers telling which number array element comes first in numerical order, which second, and so on.
- Given an array, prepare a *rank table* for it, i.e., a table telling what is the numerical rank of the first array element, the second array element, and so on.

For the basic task of sorting N elements, the best algorithms require on the order of several times $N \log_2 N$ operations. The algorithm inventor tries to reduce the constant in front of this estimate to as small a value as possible. Two of the best algorithms are *Heapsort* (§8.2), invented by J.W.J. Williams, and *Quicksort* (§8.4), invented by the inimitable C.A.R. Hoare.

For large N (say > 1000), Quicksort is faster, on most machines, by a factor of 1.5 or 2; it requires a bit of extra memory, however, and is a moderately complicated program. Heapsort is a true "sort in place," and is

somewhat more compact to program and therefore a bit easier to modify for special purposes. On balance, we prefer Heapsort, but we give both routines for you to use.

For small N one does better to use an algorithm whose operation count goes as a higher, i.e. poorer, power of N, if the constant in front is small enough. For $N < 50$, roughly, the method of *straight insertion* (§8.1) is concise and fast enough. We include it with some trepidation: it is an N^2 algorithm, whose potential for misuse (by using it for too large an N) is great. The resultant waste of computer time is so awesome, that we were tempted not to include any N^2 routine at all. We *will* draw the line, however, at the inefficient N^2 algorithm *bubble sort*. If you know what bubble sort is, wipe it from your mind; if you don't know, make a point of never finding out!

For $50 < N < 1000$, roughly, *Shell's method* (§8.1), only slightly more complicated to program than straight insertion, is usually the method of choice. This method goes as $N^{3/2}$ in the worst case, but is usually faster.

Knuth is the preeminent source for all further information on the subject of sorting, and for detailed references to the literature.

REFERENCES AND FURTHER READING:

Knuth, Donald E. 1973, *Sorting and Searching*, vol. 3 of *The Art of Computer Programming* (Reading, Mass.: Addison-Wesley).

8.1 Straight Insertion and Shell's Method

Straight insertion is an N^2 routine, and should only be used for small N, say < 50.

The technique is exactly the one used by experienced card players to sort their cards: Pick out the second card and put it in order with respect to the first; then pick out the third card and insert it into the sequence among the first two; and so on until the last card has been picked out and inserted.

```
      SUBROUTINE PIKSRT(N,ARR)
         Sorts an array ARR of length N into ascending numerical order, by straight insertion. N is
         input; ARR is replaced on output by its sorted rearrangement.
      DIMENSION ARR(N)
      DO 12 J=2,N                         Pick out each element in turn.
        A=ARR(J)
        DO 11 I=J-1,1,-1                  Look for the place to insert it.
          IF(ARR(I).LE.A)GO TO 10
          ARR(I+1)=ARR(I)
11      CONTINUE
        I=0
10      ARR(I+1)=A                        Insert it.
12    CONTINUE
      RETURN
      END
```

What if you also want to rearrange an array BRR at the same time as you sort ARR? Simply move an element of BRR whenever you move an element of ARR:

```
      SUBROUTINE PIKSR2(N,ARR,BRR)
C        Sorts an array ARR of length N into ascending numerical order, by straight insertion, while
C        making the corresponding rearrangement of the array BRR.
      DIMENSION ARR(N),BRR(N)
      DO 12 J=2,N                       Pick out each element in turn.
        A=ARR(J)
        B=BRR(J)
        DO 11 I=J-1,1,-1                Look for the place to insert it.
          IF(ARR(I).LE.A)GO TO 10
          ARR(I+1)=ARR(I)
          BRR(I+1)=BRR(I)
11      CONTINUE
        I=0
10      ARR(I+1)=A                      Insert it.
        BRR(I+1)=B
12    CONTINUE
      RETURN
      END
```

For the case of rearranging a larger number of arrays by sorting on one of them, see §8.3.

Shell's Method

This is actually a variant on straight insertion, but a very powerful variant indeed. The rough idea, e.g. for the case of sorting 16 numbers $n_1 \ldots n_{16}$, is this: First sort, by straight insertion, each of the 8 groups of 2 (n_1, n_9), (n_2, n_{10}), ..., (n_8, n_{16}). Next, sort each of the 4 groups of 4 (n_1, n_5, n_9, n_{13}), ..., $(n_4, n_8, n_{12}, n_{16})$. Next sort the 2 groups of 8 records, beginning with $(n_1, n_3, n_5, n_7, n_9, n_{11}, n_{13}, n_{15})$. Finally, sort the whole list of 16 numbers.

Of course, only the *last* sort is *necessary* for putting the numbers into order. So what is the purpose of the previous partial sorts? The answer is that the previous sorts allow numbers efficiently to filter up or down to positions close to their final resting places. Therefore, the straight insertion passes on the final sort rarely have to go past more than a "few" elements before finding the right place. (Think of sorting a hand of cards which are already almost in order.)

It can be shown (see Knuth) that the number of operations required in all is of order $N^{3/2}$ for the worst possible ordering of the original data. For "randomly" ordered data, the operations count goes approximately as $N^{1.27}$, at least for $N < 60000$. The program follows:

```
      SUBROUTINE SHELL(N,ARR)
C     Sorts an array ARR of length N into ascending numerical order, by the Shell-Mezgar algo-
C     rithm (diminishing increment sort). N is input; ARR is replaced on output by its sorted
C     rearrangement.
      PARAMETER (ALN2I=1./0.69314718, TINY=1.E-5)
      DIMENSION ARR(N)
      LOGNB2=INT(ALOG(FLOAT(N))*ALN2I+TINY)
      M=N
      DO 12 NN=1,LOGNB2                Loop over the partial sorts.
        M=M/2
        K=N-M
        DO 11 J=1,K                    Outer loop of straight insertion.
          I=J
3         CONTINUE                     Inner loop of straight insertion.
          L=I+M
          IF(ARR(L).LT.ARR(I)) THEN
            T=ARR(I)
            ARR(I)=ARR(L)
            ARR(L)=T
            I=I-M
            IF(I.GE.1)GO TO 3
          ENDIF
11      CONTINUE
12    CONTINUE
      RETURN
      END
```

REFERENCES AND FURTHER READING:

Knuth, Donald E. 1973, *Sorting and Searching*, vol. 3 of *The Art of Computer Programming* (Reading, Mass.: Addison-Wesley), §5.2.1.

8.2 Heapsort

Heapsort is our favorite sorting routine. It can be recommended wholeheartedly for a variety of sorting applications. It is a true "in-place" sort, requiring no auxiliary storage. It is an $N \log_2 N$ process, not only on average, but also for the worst-case order of input data. In fact, its worst case is only 20 percent or so worse than its average running time.

It is beyond our scope to give a complete exposition on the theory of Heapsort. We will mention the general principles, then let you refer to Knuth, or analyze the program yourself, if you want to understand the details.

A set of N numbers a_i, $i = 1, \ldots, N$, is said to form a "heap" if it satisfies the relation

$$a_{j/2} \geq a_j \quad \text{for} \quad 1 \leq j/2 < j \leq N \tag{8.2.1}$$

Here the division in $j/2$ means "integer divide," i.e. is an exact integer or else is rounded down to the closest integer. Definition (8.2.1) will make sense if

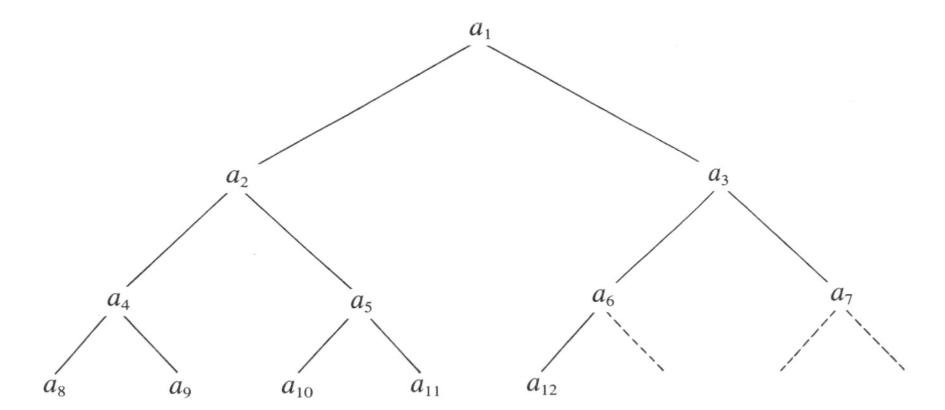

Figure 8.2.1. Ordering implied by a "heap," here of 12 elements. Elements connected by an upward path are sorted with respect to one another, but there is not necessarily any ordering among elements related only "laterally."

you think of the numbers a_i as being arranged in a binary tree, with the top, "boss," node being a_1, the two "underling" nodes being a_2 and a_3, *their* four underling nodes being a_4 through a_7, etc. (See Figure 8.2.1.) In this form, a heap has every "supervisor" greater than or equal to its two "supervisees," down through the levels of the hierarchy.

If you have managed to rearrange your array into an order that forms a heap, then sorting it is very easy: You pull off the "top of the heap," which will be the largest element yet unsorted. Then you "promote" to the top of the heap its largest underling. Then you promote *its* largest underling, and so on. The process is like what happens (or is supposed to happen) in a large corporation when the chairman of the board retires. You then repeat the whole process by retiring the new chairman of the board. Evidently the whole thing is an $N \log_2 N$ process, since each retiring chairman leads to $\log_2 N$ promotions of underlings.

Well, how do you arrange the array into a heap in the first place? The answer is again a "sift-up" process like corporate promotion. Imagine that the corporation starts out with $N/2$ employees on the production line, but with no supervisors. Now a supervisor is hired to supervise two workers. If he is less capable than both his workers, one of them is promoted in his place, and he joins the production line. After supervisors are hired, then supervisors of supervisors are hired, and so on up the corporate ladder. Each employee is brought in at the top of the tree, but then immediately sifted down, with more capable workers promoted until their proper corporate level has been reached.

In the Heapsort implementation, the same "sift-down" code can be used for the initial creation of the heap and for the subsequent retirement-and-promotion phase. One execution of the Heapsort subroutine represents the entire life-cycle of a giant corporation: $N/2$ workers are hired; $N/2$ potential supervisors are hired; there is a sifting up in the ranks, a sort of Parkinson's Law; finally, in due course, each of the original employees gets promoted to chairman of the board.

```
      SUBROUTINE SORT(N,RA)
C     Sorts an array RA of length N into ascending numerical order using the Heapsort algorithm.
C     N is input; RA is replaced on output by its sorted rearrangement.
      DIMENSION RA(N)
      L=N/2+1
      IR=N
C     The index L will be decremented from its initial value down to 1 during the "hiring" (heap
C     creation) phase. Once it reaches 1, the index IR will be decremented from its initial value
C     down to 1 during the "retirement-and-promotion" (heap selection) phase.
10    CONTINUE
        IF(L.GT.1)THEN                Still in hiring phase.
          L=L-1
          RRA=RA(L)
        ELSE                          In retirement-and-promotion phase.
          RRA=RA(IR)                  Clear a space at end of array.
          RA(IR)=RA(1)                Retire the top of the heap into it.
          IR=IR-1                     Decrease the size of the corporation.
          IF(IR.EQ.1)THEN             Done with the last promotion.
            RA(1)=RRA                 The least competent worker of all!
            RETURN
          ENDIF
        ENDIF
        I=L                           Whether we are in the hiring phase or promotion phase, we here set
        J=L+L                               up to sift down element RRA to its proper level.
20      IF(J.LE.IR)THEN                   "Do while J.LE.IR:"
          IF(J.LT.IR)THEN
            IF(RA(J).LT.RA(J+1))J=J+1              Compare to the better underling.
          ENDIF
          IF(RRA.LT.RA(J))THEN            Demote RRA.
            RA(I)=RA(J)
            I=J
            J=J+J
          ELSE                            This is RRA's level. Set J to terminate the sift-down.
            J=IR+1
          ENDIF
        GO TO 20
        ENDIF
        RA(I)=RRA                         Put RRA into its slot.
      GO TO 10
      END
```

As usual you can move any other arrays around at the same time as you sort RRA. At the risk of repetitiousness:

```
      SUBROUTINE SORT2(N,RA,RB)
C     Sorts an array RA of length N into ascending numerical order using the Heapsort algorithm,
C     while making the corresponding rearrangement of the array RB.
      DIMENSION RA(N),RB(N)
      L=N/2+1
      IR=N
10    CONTINUE
        IF(L.GT.1)THEN
          L=L-1
          RRA=RA(L)
          RRB=RB(L)
        ELSE
          RRA=RA(IR)
          RRB=RB(IR)
          RA(IR)=RA(1)
          RB(IR)=RB(1)
          IR=IR-1
```

```
          IF(IR.EQ.1)THEN
              RA(1)=RRA
              RB(1)=RRB
              RETURN
          ENDIF
      ENDIF
      I=L
      J=L+L
20    IF(J.LE.IR)THEN
          IF(J.LT.IR)THEN
              IF(RA(J).LT.RA(J+1))J=J+1
          ENDIF
          IF(RRA.LT.RA(J))THEN
              RA(I)=RA(J)
              RB(I)=RB(J)
              I=J
              J=J+J
          ELSE
              J=IR+1
          ENDIF
      GO TO 20
      ENDIF
      RA(I)=RRA
      RB(I)=RRB
  GO TO 10
  END
```

You could, in principle, rearrange any number of additional arrays along with RRB, but this becomes wasteful as the number of such arrays becomes large. The preferred technique is to make use of an index table, as described in the next section.

8.3 Indexing and Ranking

As an alternative to sorting a large number N of records into the order of their *keys* K_i, $i = 1, \ldots, N$, one can instead construct an *index table* I_j, $j = 1, \ldots, N$, such that the smallest K_i has $i = I_1$, the second smallest has $i = I_2$, and so on up to the largest K_i with $i = I_N$. In other words, the array

$$K_{I_j} \quad j = 1, 2, \ldots, N \tag{8.3.1}$$

is in sorted order when indexed by j.

When an index table is available, one never need move records from their original order. Further, different index tables can be made from the same set of records, indexing them to different keys. For example, one might wish to loop over measured data points sometimes in the order of one control variable, sometimes in the order of another.

The algorithm for constructing an index table is straightforward: Initialize the index array with the integers from 1 to N, then move the elements

around *as if* one were sorting the keys. The integer that initially numbered the smallest key thus ends up in the number one position, and so on.

```
      SUBROUTINE INDEXX(N,ARRIN,INDX)
C        Indexes an array ARRIN of length N, i.e. outputs the array INDX such that ARRIN(INDX(J))
C        is in ascending order for J= 1,2,...,N. The input quantities N and ARRIN are not changed.
      DIMENSION ARRIN(N),INDX(N)
      DO 11 J=1,N                       Initialize the index array with consecutive integers.
        INDX(J)=J
11    CONTINUE
      L=N/2+1                           From here on, we just have Heapsort, but with indirect indexing
      IR=N                                  through INDX in all references to ARRIN.
10    CONTINUE
        IF(L.GT.1)THEN
          L=L-1
          INDXT=INDX(L)
          Q=ARRIN(INDXT)
        ELSE
          INDXT=INDX(IR)
          Q=ARRIN(INDXT)
          INDX(IR)=INDX(1)
          IR=IR-1
          IF(IR.EQ.1)THEN
            INDX(1)=INDXT
            RETURN
          ENDIF
        ENDIF
        I=L
        J=L+L
20      IF(J.LE.IR)THEN
          IF(J.LT.IR)THEN
            IF(ARRIN(INDX(J)).LT.ARRIN(INDX(J+1)))J=J+1
          ENDIF
          IF(Q.LT.ARRIN(INDX(J)))THEN
            INDX(I)=INDX(J)
            I=J
            J=J+J
          ELSE
            J=IR+1
          ENDIF
        GO TO 20
        ENDIF
        INDX(I)=INDXT
      GO TO 10
      END
```

If you want to sort an array while making the corresponding rearrangement of several or many other arrays, you should first make an index table, then use it to rearrange each array in turn. This requires two arrays of working space: one to hold the index, and another into which an array is temporarily moved, and from which it is redeposited back on itself in the rearranged order. For 3 arrays, the procedure looks like this:

```
      SUBROUTINE SORT3(N,RA,RB,RC,WKSP,IWKSP)
C        Sorts an array RA of length N into ascending numerical order while making the correspond-
C        ing rearrangements of the arrays RB and RC. An index table is constructed via the routine
C        INDEXX.
      DIMENSION RA(N),RB(N),RC(N),WKSP(N),IWKSP(N)
      CALL INDEXX(N,RA,IWKSP)           Make the index table.
```

```
DO 11 J=1,N                    Save the array RA.
    WKSP(J)=RA(J)
  11 CONTINUE
DO 12 J=1,N                    Copy it back in the rearranged order.
    RA(J)=WKSP(IWKSP(J))
  12 CONTINUE
DO 13 J=1,N                    Ditto RB.
    WKSP(J)=RB(J)
  13 CONTINUE
DO 14 J=1,N
    RB(J)=WKSP(IWKSP(J))
  14 CONTINUE
DO 15 J=1,N                    Ditto RC.
    WKSP(J)=RC(J)
  15 CONTINUE
DO 16 J=1,N
    RC(J)=WKSP(IWKSP(J))
  16 CONTINUE
RETURN
END
```

The generalization to any other number of arrays is obviously straightforward.

A *rank table* is different from an index table. A rank table's j^{th} entry gives the rank of the jth element of the original array of keys, ranging from 1 (if that element was the smallest) to N (if that element was the largest). One can easily construct a rank table from an index table, however:

```
SUBROUTINE RANK(N,INDX,IRANK)
    Given INDX of length N as output from the routine INDEXX, this routine returns an array
    IRANK, the corresponding table of ranks.
DIMENSION INDX(N),IRANK(N)
DO 11 J=1,N
    IRANK(INDX(J))=J
  11 CONTINUE
RETURN
END
```

Figure 8.3.1 summarizes the concepts discussed in this section.

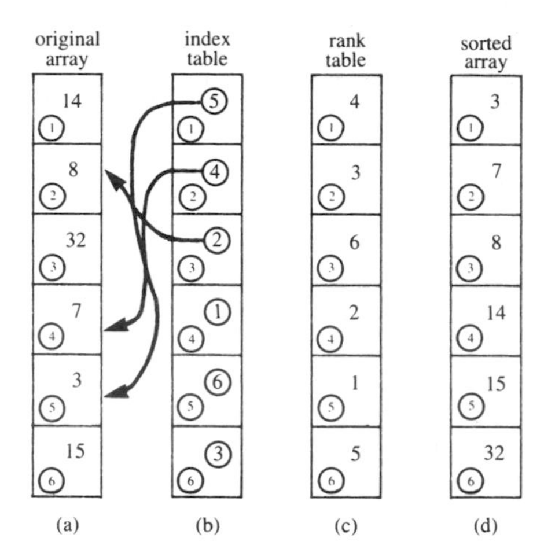

Figure 8.3.1. (a) An unsorted array of six numbers. (b) Index table, whose entries are pointers to the elements of (a) in ascending order. (c) Rank table, whose entries are the ranks of the corresponding elements of (a). (d) Sorted array of the elements in (a).

8.4 Quicksort

For reasons already mentioned in §8.0, we prefer Heapsort to Quicksort. Nevertheless, Quicksort is, on most machines, on average, for large N the fastest known sorting algorithm.

Quicksort is a "partition-exchange" sorting method: by pairwise exchanges of elements, the original array is partitioned into two subarrays that can be sorted independently. In turn these are partitioned by the same technique, and so on. The reader is referred to Knuth for a full description of the method.

Quicksort requires an auxiliary array of storage, of length $2\log_2 N$, which it uses as a push-down stack for keeping track of the pending subarrays. When a subarray has gotten down to some size M, it becomes faster to sort it by straight insertion (§8.1), so we will do this, following Knuth. The optimal setting of M is machine dependent, but $M = 7$ is not too far wrong.

As already mentioned, Quicksort's *average* running time is fast, but its *worst case* running time can be very slow: for the worst case it is, in fact, an N^2 method! And for the most straightforward implementation of Quicksort it turns out that the worst case is achieved for an input array that is already in order! This ordering of the input array might easily occur in practice. Therefore it is necessary to build into the implementation a little random number generator, just to *spoil* any regularity that might be in the input array. We do just that in the following Quicksort implementation, which closely follows Knuth:

```
      SUBROUTINE QCKSRT(N,ARR)
C        Sorts an array ARR of length N into ascending numerical order using the Quicksort algo-
C        rithm. N is input; ARR is replaced on output by its sorted rearrangement.
      PARAMETER (M=7,NSTACK=50,FM=7875.,FA=211.,FC=1663.,FMI=1./FM)
C        Here M is the size of subarrays sorted by straight insertion, NSTACK is the required aux-
C        iliary storage, and the remaining constants are used by the random number generating
C        statements.
      DIMENSION ARR(N),ISTACK(NSTACK)
      JSTACK=0
      L=1
      IR=N
      FX=0.
10    IF(IR-L.LT.M)THEN                Sort by straight insertion:
        DO 13 J=L+1,IR
          A=ARR(J)
          DO 11 I=J-1,1,-1
            IF(ARR(I).LE.A)GO TO 12
            ARR(I+1)=ARR(I)
11        CONTINUE
          I=0
12        ARR(I+1)=A
13      CONTINUE
        IF(JSTACK.EQ.0)RETURN
        IR=ISTACK(JSTACK)
        L=ISTACK(JSTACK-1)
        JSTACK=JSTACK-2
      ELSE
        I=L
        J=IR
        FX=MOD(FX*FA+FC,FM)            Generate a random integer IQ between L and IR, inclusive.
        IQ=L+(IR-L+1)*(FX*FMI)
        A=ARR(IQ)
        ARR(IQ)=ARR(L)
20      CONTINUE
21        IF(J.GT.0)THEN
            IF(A.LT.ARR(J))THEN
              J=J-1
              GO TO 21
            ENDIF
          ENDIF
          IF(J.LE.I)THEN
            ARR(I)=A
            GO TO 30
          ENDIF
          ARR(I)=ARR(J)
          I=I+1
22        IF(I.LE.N)THEN
            IF(A.GT.ARR(I))THEN
              I=I+1
              GO TO 22
            ENDIF
          ENDIF
          IF(J.LE.I)THEN
            ARR(J)=A
            I=J
            GO TO 30
          ENDIF
          ARR(J)=ARR(I)
          J=J-1
        GO TO 20
30      JSTACK=JSTACK+2
        IF(JSTACK.GT.NSTACK)PAUSE 'NSTACK must be made larger.'
        IF(IR-I.GE.I-L)THEN
          ISTACK(JSTACK)=IR
```

```
        ISTACK(JSTACK-1)=I+1
        IR=I-1
    ELSE
        ISTACK(JSTACK)=I-1
        ISTACK(JSTACK-1)=L
        L=I+1
    ENDIF
ENDIF
GO TO 10
END
```

8.5 Determination of Equivalence Classes

A number of techniques for sorting and searching relate to data structures whose details are beyond the scope of this book, for example, trees, linked lists, etc. These structures and their manipulations are the bread and butter of computer science, as distinct from numerical analysis, and there is no shortage of books on the subject.

In working with experimental data, we have found that one particular such manipulation, namely the determination of equivalence classes, arises sufficiently often to justify inclusion here.

The problem is this: There are N "elements" (or "data points" or whatever), numbered $1, \ldots, N$. You are given pairwise information about whether elements are in the same *equivalence class* of "sameness," by whatever criterion happens to be of interest. For example, you may have a list of facts like: "Element 3 and element 7 are in the same class; element 19 and element 4 are in the same class; element 7 and element 12 are in the same class, ..." Alternatively, you may have a procedure, given the numbers of two elements j and k, for deciding whether they are in the same class or different classes. (Recall that an equivalence relation can be anything satisfying the *RST properties*: reflexive, symmetric, transitive. This is compatible with any intuitive definition of "sameness.")

The desired output is an assignment to each of the N elements of an equivalence class number, such that two elements are in the same class if and only if they are assigned the same class number.

Efficient algorithms work like this: Let $F(j)$ be the class or "family" number of element j. Start off with each element in its own family, so that $F(j) = j$. The array $F(j)$ can be interpreted as a tree structure, where $F(j)$ denotes the parent of j. If we arrange for each family to be its own tree, disjoint from all the other "family trees," then we can label each family (equivalence class) by its most senior great-great-...grandparent. The detailed topology of the tree doesn't matter at all, as long as we graft each related element onto it *somewhere*.

Therefore, we process each elemental datum "j is equivalent to k" by tracking j up to its highest ancestor, (ii) tracking k up to its highest

ancestor, (iii) giving j to k as a new parent, or vice versa (it makes no difference). After processing all the relations, we go through all the elements j and reset their $F(j)$'s to their highest possible ancestors, which then label the equivalence classes.

The following routine, based on Knuth, assumes that there are M elemental pieces of information, stored in two arrays of length M, LISTA, LISTB, the interpretation being that LISTA(J) and LISTB(J), J=1...M, are the numbers of two elements which (we are thus told) are related.

```
      SUBROUTINE ECLASS(NF,N,LISTA,LISTB,M)
C     Given M equivalences between pairs of N individual elements in the form of the input arrays
C     LISTA and LISTB, this routine returns in NF the number of the equivalence class of each
C     of the N elements, integers between 1 and N (not all such integers used).
      DIMENSION NF(N),LISTA(M),LISTB(M)
      DO 11 K=1,N                         Initialize each element its own class.
        NF(K)=K
11    CONTINUE
      DO 12 L=1,M                         For each piece of input information...
        J=LISTA(L)
1       IF(NF(J).NE.J)THEN                Track first element up to its ancestor.
          J=NF(J)
        GOTO 1
        ENDIF
        K=LISTB(L)
2       IF(NF(K).NE.K)THEN                Track second element up to its ancestor.
          K=NF(K)
        GOTO 2
        ENDIF
        IF(J.NE.K)NF(J)=K                 If they are not already related, make them so.
12    CONTINUE
      DO 13 J=1,N                         Final sweep up to highest ancestors.
3       IF(NF(J).NE.NF(NF(J)))THEN
          NF(J)=NF(NF(J))
        GOTO 3
        ENDIF
13    CONTINUE
      RETURN
      END
```

Alternatively, we may be able to construct a procedure EQUIV(J,K) that returns a value .TRUE. if elements J and K are related, or .FALSE. if they are not. Then we want to loop over all pairs of elements to get the complete picture. D. Eardley has devised a clever way of doing this while simultaneously sweeping the tree up to high ancestors in a manner that keeps it current and obviates most of the final sweep phase:

```
      SUBROUTINE ECLAZZ(NF,N,EQUIV)
C     Given a user-supplied logical function EQUIV which tells whether a pair of elements, each
C     in the range 1...N, are related, return in NF equivalence class numbers for each element.
      LOGICAL EQUIV
      DIMENSION NF(N)
      NF(1)=1
      DO 12 JJ=2,N                        Loop over first element of all pairs.
        NF(JJ)=JJ
        DO 11 KK=1,JJ-1                   Loop over second element of all pairs.
          NF(KK)=NF(NF(KK))               Sweep it up this much.
          IF (EQUIV(JJ,KK)) NF(NF(NF(KK)))=JJ    Good exercise for the reader to figure out
                                          why this much ancestry is necessary!
```

```
      11 CONTINUE
    12 CONTINUE
DO 13 JJ=1,N                  Only this much sweeping is needed finally.
      NF(JJ)=NF(NF(JJ))
    13 CONTINUE
RETURN
END
```

REFERENCES AND FURTHER READING:

Knuth, Donald E. 1968, *Fundamental Algorithms*, vol. 1 of *The Art of Computer Programming* (Reading, Mass.: Addison-Wesley), §2.3.3.

Chapter 9. Root Finding and Nonlinear Sets of Equations

9.0 Introduction

We now consider that most basic of tasks, solving equations numerically. While most equations are born with both a right-hand side and a left-hand side, one traditionally subtracts the two sides, leaving

$$f(x) = 0 \tag{9.0.1}$$

whose solution or solutions are desired. When there is only one independent variable, the problem is *one-dimensional*, namely to find the root or roots of a function.

With more than one independent variable, more than one equation can be satisfied simultaneously. You likely once learned the *implicit function theorem* which (in this context) gives us the hope of satisfying N equations in N unknowns simultaneously. Note that we have only hope, not certainty. A nonlinear set of equations may have no (real) solutions at all. Contrariwise, it may have more than one solution. The implicit function theorem tells us that "generically" the solutions will be distinct, pointlike, and separated from each other. If, however, life is so unkind as to present you with a non-generic, i.e. degenerate, case, then you can get a continuous family of solutions. In vector notation, we want to find one or more N-dimensional solution vectors $\mathbf{x}$ such that

$$\mathbf{f}(\mathbf{x}) = \mathbf{0} \tag{9.0.2}$$

where $\mathbf{f}$ is the N-dimensional vector-valued function whose components are the individual equations to be satisfied simultaneously.

Don't be fooled by the apparent notational similarity of equations (9.0.2) and (9.0.1). Simultaneous solution of equations in N dimensions is *much* more difficult than finding roots in the one-dimensional case. The principal

difference between one and many dimensions is that, in one dimension, it is possible to bracket or "trap" a root between bracketing values, and then hunt it down like a rabbit. In multidimensions, you can never be sure that the root is there at all until you have found it.

Except in linear problems, root finding invariably proceeds by iteration, and this is equally true in one or in many dimensions. Starting from some approximate trial solution, a useful algorithm will improve the solution until some predetermined convergence criterion is satisfied. For smoothly varying functions, good algorithms will always converge, *provided* that the initial guess is good enough. Indeed one can even determine in advance the rate of convergence of most algorithms.

It cannot be overemphasized, however, how crucially success depends on having a good first-guess for the solution, especially for multidimensional problems. This crucial beginning usually depends on analysis rather than numerics. Carefully crafted initial estimates reward you not only with reduced computational effort, but also with understanding and increased self-esteem. Hamming's motto, "the purpose of computing is insight, not numbers," is particularly apt in the area of finding roots. You should say this motto aloud whenever your program converges, with ten-digit accuracy, to the wrong root of a problem, or whenever it fails to converge because there is actually *no* root, or because there is a root but your initial estimate was not sufficiently close to it.

"This talk of insight is all very well, but what do I actually do?" For one-dimensional root finding, it is possible to give some straightforward answers: You should try to get some idea of what your function looks like before trying to find its roots. If you need to mass-produce roots for many different functions, then you should at least know what some typical members of the ensemble look like. Next, you should always bracket a root, that is, know that the function changes sign in an identified interval, before trying to converge to the root's value.

Finally (this is advice with which some daring souls might disagree, but we give it nonetheless) never let your iteration method get outside of the best bracketing bounds obtained at any stage. We will see below that some pedagogically important algorithms, such as *secant method* or *Newton-Raphson*, can violate this last constraint, and are thus not recommended unless certain fixups are implemented.

Multiple roots, or very close roots, are a real problem, especially if the multiplicity is an even number. In that case, there may be no readily apparent sign change in the function, so the notion of bracketing a root – and maintaining the bracket – becomes difficult. We are hard-liners: we nevertheless insist on bracketing a root, even if it takes the minimum-searching techniques of Chapter 11 to determine whether a tantalizing dip in the function really does cross zero or not. (You can easily modify the simple golden section routine of §10.1 to return early if it detects a sign change in the function. And, if the minimum of the function is exactly zero, then you have found a *double* root.)

As usual, we want to discourage you from using routines as black boxes without understanding them. However, as a guide to beginners, here are some reasonable starting points:

- Brent's algorithm in §9.3 is the method of choice to find a bracketed root of a general one-dimensional function, when you cannot easily compute the function's derivative.
- When you can compute the function's derivative, the routine **RTSAFE** in §9.4, which combines the Newton-Raphson method with some bookkeeping on bounds, is recommended. Again, you must first bracket your root.
- Roots of polynomials are a special case. Laguerre's method, in §9.5, is recommended as a starting point. Beware: some polynomials are ill-conditioned!
- Finally, for multidimensional problems, there is really nothing within the scope of this book except Newton-Raphson, which works *very* well if you can supply a reasonable first guess of the solution. Try it. Then read §9.6 more carefully to find out where next to turn.

Avoiding implementations for specific computers, this book must generally steer clear of interactive or graphics-related routines. We make an exception right now. The following routine, which produces a crude function plot with interactively scaled axes, can save you a lot of grief as you enter the world of root finding.

```
      SUBROUTINE SCRSHO(FX)
         For interactive CRT terminal use. Produce a crude graph of the function FX over the
         interval X1,X2. Query for another plot interval until the user signals satisfaction.
      PARAMETER (ISCR=60,JSCR=21)  Number of horizontal and vertical positions in display.
      CHARACTER*1 SCR(ISCR,JSCR),BLANK,ZERO,YY,XX,FF
      DIMENSION Y(ISCR)
      DATA BLANK,ZERO,YY,XX,FF/' ','-','l','-','x'/
1     CONTINUE
      WRITE (*,*) ' Enter X1,X2 (= to stop)'  Query for another plot, quit if X1=X2
      READ (*,*) X1,X2
      IF(X1.EQ.X2) RETURN
      DO 11 J=1,JSCR                    Fill vertical sides with character Y.
         SCR(1,J)=YY
         SCR(ISCR,J)=YY
11    CONTINUE
      DO 13 I=2,ISCR-1
         SCR(I,1)=XX                    Fill top, bottom with character X.
         SCR(I,JSCR)=XX
         DO 12 J=2,JSCR-1               Fill interior with blanks.
            SCR(I,J)=BLANK
12       CONTINUE
13    CONTINUE
      DX=(X2-X1)/(ISCR-1)
      X=X1
      YBIG=0.                           Limits will include 0.
      YSML=YBIG
      DO 14 I=1,ISCR                    Evaluate the function at equal intervals. Find the largest and smallest
         Y(I)=FX(X)                         values.
         IF(Y(I).LT.YSML) YSML=Y(I)
         IF(Y(I).GT.YBIG) YBIG=Y(I)
         X=X+DX
14    CONTINUE
      IF(YBIG.EQ.YSML) YBIG=YSML+1.     Be sure to separate top and bottom.
```

```
DYJ=(JSCR-1)/(YBIG-YSML)
JZ=1-YSML*DYJ                     Note which row corresponds to 0.
DO 15 I=1,ISCR                    Place an indicator at function height and 0.
    SCR(I,JZ)=ZERO
    J=1+(Y(I)-YSML)*DYJ
    SCR(I,J)=FF
 15 CONTINUE
WRITE (*,'(1X,1PE10.3,1X,80A1)') YBIG,(SCR(I,JSCR),I=1,ISCR)
DO 16 J=JSCR-1,2,-1               Display.
    WRITE (*,'(12X,80A1)') (SCR(I,J),I=1,ISCR)
 16 CONTINUE
WRITE (*,'(1X,1PE10.3,1X,80A1)') YSML,(SCR(I,1),I=1,ISCR)
WRITE (*,'(12X,1PE10.3,40X,E10.3)') X1,X2
GOTO 1
END
```

REFERENCES AND FURTHER READING:

Stoer, J., and Bulirsch, R. 1980, *Introduction to Numerical Analysis* (New York: Springer-Verlag), Chapter 5.

Acton, Forman S. 1970, *Numerical Methods That Work* (New York: Harper and Row), Chapters 2, 7, and 14.

Ralston, Anthony, and Rabinowitz, Philip. 1978, *A First Course in Numerical Analysis*, 2nd ed. (New York: McGraw-Hill), Chapter 8.

Householder, A.S. 1970, *The Numerical Treatment of a Single Nonlinear Equation* (New York: McGraw-Hill).

9.1 Bracketing and Bisection

We will say that a root is *bracketed* in the interval (a, b) if $f(a)$ and $f(b)$ have opposite signs. If the function is continuous, then at least one root must lie in that interval (the *intermediate value theorem*). If the function is discontinuous, but bounded, then instead of a root there might be a step discontinuity which crosses zero (see Figure 9.1.1). For numerical purposes, that might as well be a root, since the behavior is indistinguishable from the case of a continuous function whose zero crossing occurs in between two "adjacent" floating point numbers in a machine's finite-precision representation. Only for functions with singularities is there the possibility that a bracketed root is not really there, as for example

$$f(x) = \frac{1}{x - c} \tag{9.1.1}$$

Some root-finding algorithms (e.g. bisection in this section) will readily converge to c in (9.1.1). Luckily there is not much possibility of your mistaking c, or any number x close to it, for a root, since mere evaluation of $|f(x)|$ will give a very large, rather than a very small, result.

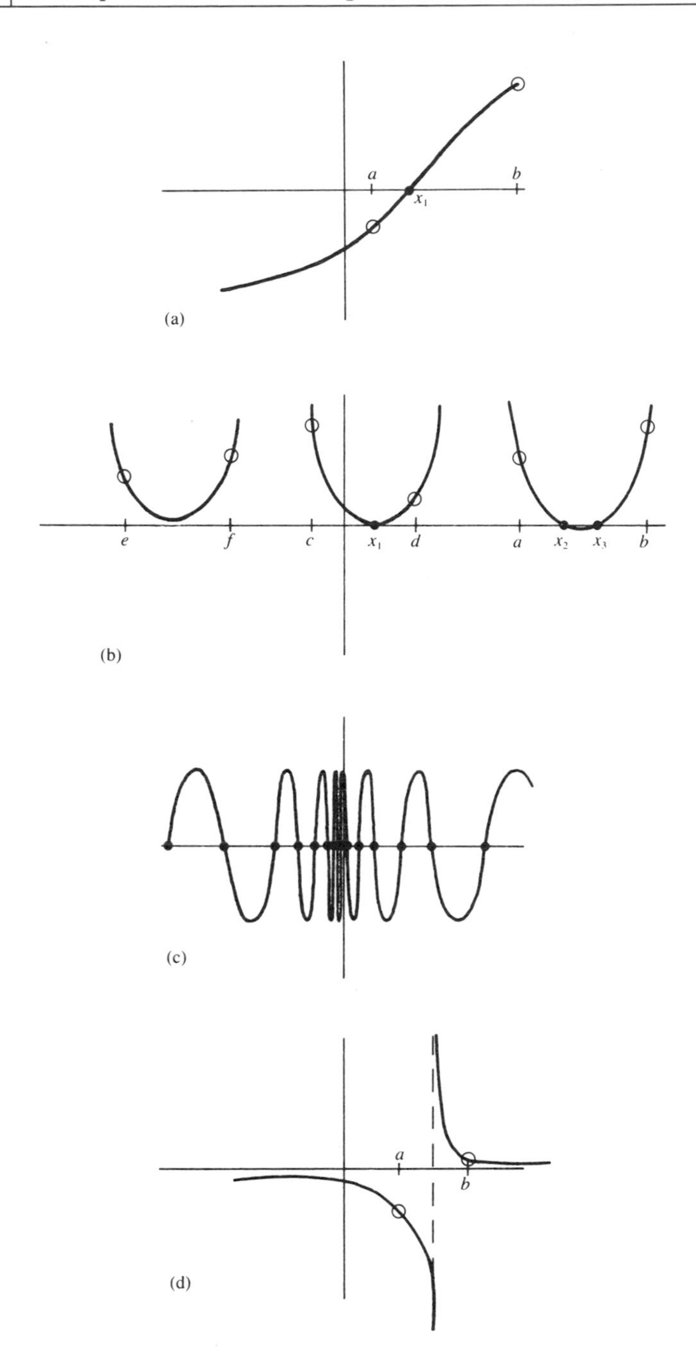

Figure 9.1.1. Some situations encountered while root finding: (a) shows an isolated root x_1 bracketed by two points a and b at which the function has opposite signs; (b) illustrates that there is not necessarily a sign change in the function near a double root (in fact, there is not necessarily a root!); (c) is a pathological function with many roots; in (d) the function has opposite signs at points a and b, but the points bracket a singularity, not a root.

If you are given a function in a black box, there is no sure way of bracketing its roots, or of even determining that it has roots. If you like pathological examples, think about the problem of locating the two real roots of equation (3.0.1), which dips below zero only in the ridiculously small interval of about $x = \pi \pm 10^{-667}$.

In the next chapter we will deal with the related problem of bracketing a function's minimum. There it is possible to give a procedure that always succeeds; in essence, "Go downhill, taking steps of increasing size, until your function starts back uphill." There is no analogous procedure for roots. The procedure "go downhill until your function changes sign," can be foiled by a function that has a simple extremum. Nevertheless, if you are prepared to deal with a "failure" outcome, this procedure is often a good first start; success is usual if your function has opposite signs in the limit $x \to \pm\infty$.

```
SUBROUTINE ZBRAC(FUNC,X1,X2,SUCCES)
    Given a function FUNC and an initial guessed range X1 to X2, the routine expands the range
    geometrically until a root is bracketed by the returned values X1 and X2 (in which case
    SUCCES returns as .TRUE.) or until the range becomes unacceptably large (in which case
    SUCCES returns as .FALSE.).
PARAMETER (FACTOR=1.6,NTRY=50)
LOGICAL SUCCES
IF(X1.EQ.X2)PAUSE 'You have to guess an initial range'
F1=FUNC(X1)
F2=FUNC(X2)
SUCCES=.TRUE.
DO 11 J=1,NTRY
    IF(F1*F2.LT.0.)RETURN
    IF(ABS(F1).LT.ABS(F2))THEN
        X1=X1+FACTOR*(X1-X2)
        F1=FUNC(X1)
    ELSE
        X2=X2+FACTOR*(X2-X1)
        F2=FUNC(X2)
    ENDIF
 11 CONTINUE
SUCCES=.FALSE.
RETURN
END
```

Alternatively, you might want to "look inward" on an initial interval, rather than "look outward" from it, asking if there are any roots of the function $f(x)$ in the interval from x_1 to x_2 when a search is carried out by subdivision into N equal intervals. The following subroutine returns brackets for up to NB distinct intervals which each contain one or more roots.

```
SUBROUTINE ZBRAK(FX,X1,X2,N,XB1,XB2,NB)
    Given a function FX defined on the interval from X1-X2 subdivide the interval into N equally
    spaced segments, and search for zero crossings of the function. NB is input as the
    maximum number of roots sought, and is reset to the number of bracketing pairs XB1,
    XB2 that are found.
DIMENSION XB1(1),XB2(1)
NBB=NB
NB=0
X=X1
DX=(X2-X1)/N                        Determine the spacing appropriate to the mesh.
```

```
FP=FX(X)
DO 11 I=1,N                          Loop over all intervals
   X=X+DX
   FC=FX(X)
   IF(FC*FP.LT.0.) THEN              If a sign change occurs then record values for the bounds.
      NB=NB+1
      XB1(NB)=X-DX
      XB2(NB)=X
   ENDIF
   FP=FC
   IF(NBB.EQ.NB)RETURN
11 CONTINUE
RETURN
END
```

Bisection Method

Once we know that an interval contains a root, several classical procedures are available to refine it. These proceed with varying degrees of speed and sureness toward the answer. Unfortunately, the methods that are guaranteed to converge plod along most slowly, while those that rush to the solution in the best cases can also dash rapidly to infinity without warning if measures are not taken to avoid such behavior.

The *bisection method* is one which cannot fail. It is thus not to be sneered at as a method for otherwise badly behaved problems. The idea is simple. Over some interval the function is known to pass through zero because it changes sign. Evaluate the function at the interval's midpoint and examine its sign. Use the midpoint to replace whichever limit has the same sign. After each iteration the bounds containing the root decrease by a factor of two. If after n iterations the root is known to be within an interval of size ϵ_n, then after the next iteration it will be bracketed within an interval of size

$$\epsilon_{n+1} = \epsilon_n/2 \tag{9.1.2}$$

neither more nor less. Thus, we know in advance the number of iterations required to achieve a given tolerance in the solution,

$$n = \log_2 \frac{\epsilon_0}{\epsilon} \tag{9.1.3}$$

where ϵ_0 is the size of the initially bracketing interval, ϵ is the desired ending tolerance.

Bisection *must* succeed. If the interval happens to contain two or more roots, bisection will find one of them. If the interval contains no roots and merely straddles a singularity, it will converge on the singularity.

When a method converges as a factor (less than 1) times the previous uncertainty to the first power (as is the case for bisection), it is said to converge *linearly*. Methods which converge as a higher power,

$$\epsilon_{n+1} = \text{constant} \times (\epsilon_n)^m \qquad m > 1 \tag{9.1.4}$$

are said to converge superlinearly. In other contexts "linear" convergence would be termed "exponential," or "geometrical." That is not too bad at all: Linear convergence means that successive significant figures are won linearly with computational effort.

It remains to discuss practical criteria for convergence. It is crucial to keep in mind that computers use a fixed number of binary digits to represent floating point numbers. While your function might analytically pass through zero, it is possible that its computed value is never zero, for any floating point argument. One must decide what accuracy on the root is attainable: convergence to within 10^{-6} in absolute value is reasonable when the root lies near 1, but certainly unachievable if the root lies near 10^{26}. One might thus think to specify convergence by a relative (fractional) criterion, but this becomes unworkable for roots near zero. To be most general, the routines below will require you to specify an absolute tolerance, such that iterations continue until the interval becomes smaller than this tolerance in absolute units. Usually you may wish to take the tolerance to be $\epsilon(x_1 + x_2)/2$ where ϵ is the machine precision and x_1 and x_2 are the initial brackets. When the root lies near zero you ought to consider carefully what reasonable tolerance means for your function. The following routine quits after 40 bisections in any event, with $2^{-40} \approx 10^{-12}$.

```
FUNCTION RTBIS(FUNC,X1,X2,XACC)
    Using bisection, find the root of a function FUNC known to lie between X1 and X2. The
    root, returned as RTBIS, will be refined until its accuracy is ±XACC.
PARAMETER (JMAX=40)              Maximum allowed number of bisections.
FMID=FUNC(X2)
F=FUNC(X1)
IF(F*FMID.GE.0.)  PAUSE 'Root must be bracketed for bisection.'
IF(F.LT.0.)THEN                  Orient the search so that F>0 lies at X+DX.
    RTBIS=X1
    DX=X2-X1
ELSE
    RTBIS=X2
    DX=X1-X2
ENDIF
DO 11 J=1,JMAX                   Bisection loop
    DX=DX*.5
    XMID=RTBIS+DX
    FMID=FUNC(XMID)
    IF(FMID.LE.0.)RTBIS=XMID
    IF(ABS(DX).LT.XACC .OR. FMID.EQ.0.)  RETURN
 11 CONTINUE
PAUSE 'too many bisections'
END
```

9.2 Secant Method and False Position Method

For functions that are smooth near a root, the methods known respectively as *false position* (or *regula falsa*) and *secant method* generally converge faster than bisection. In both of these methods the function is assumed to be approximately linear in the local region of interest, and the next improvement in the root is taken as the point where the approximating line crosses the axis. After each iteration one of the previous boundary points is discarded in favor of the latest estimate of the root.

The *only* difference between the methods is that false position retains that prior estimate for which the function value has opposite sign from the function value at the current best estimate of the root, so that the two points continue to bracket the root (Figure 9.2.2), while secant retains the most recent of the prior estimates (Figure 9.2.1; this requires an arbitrary choice on the first iteration). Mathematically, the secant method converges more rapidly near a root of a sufficiently continuous function. Its order of convergence can be shown to be the "golden ratio" 1.618..., so that

$$\lim_{k\to\infty} |\epsilon_{k+1}| \approx \text{const} \times |\epsilon_k|^{1.618} \tag{9.2.1}$$

The secant method has, however, the disadvantage that the root does not necessarily remain bracketed. For functions that are *not* sufficiently continuous, the algorithm can therefore not be guaranteed to converge: Local behavior might send it off towards infinity.

False position, since it sometimes keeps an older rather than newer function evaluation, has a lower order of convergence. Since the newer function value will *sometimes* be kept, the method is often superlinear, but estimation of its exact order is not so easy.

Here are sample implementations of these two related methods. While these methods are concise, *Brent's method*, in the next section, is more often the method of choice. Figure 9.2.3 shows the behavior of secant and false-position methods in a difficult situation.

```
FUNCTION RTFLSP(FUNC,X1,X2,XACC)
   Using the false position method, find the root of a function FUNC known to lie between X1
   and X2. The root, returned as RTFLSP, is refined until its accuracy is ±XACC.
PARAMETER (MAXIT=30)          Set MAXIT to the maximum allowed number of iterations.
FL=FUNC(X1)
FH=FUNC(X2)                   Be sure the interval brackets a root.
IF(FL*FH.GT.0.)  PAUSE 'Root must be bracketed for false position.'
IF(FL.LT.0.)THEN              Identify the limits so that XL corresponds to the low side.
    XL=X1
    XH=X2
ELSE
    XL=X2
    XH=X1
    SWAP=FL
    FL=FH
    FH=SWAP
ENDIF
DX=XH-XL
```

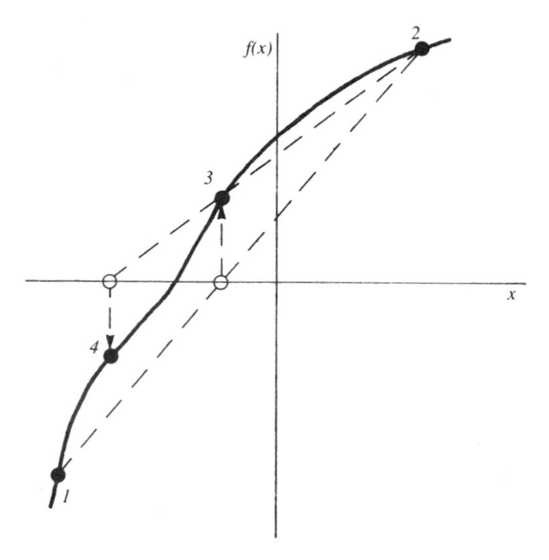

Figure 9.2.1. Secant method. Extrapolation or interpolation lines (dashed) are drawn through the two most recently evaluated points, whether or not they bracket the function. The points are numbered in the order that they are used.

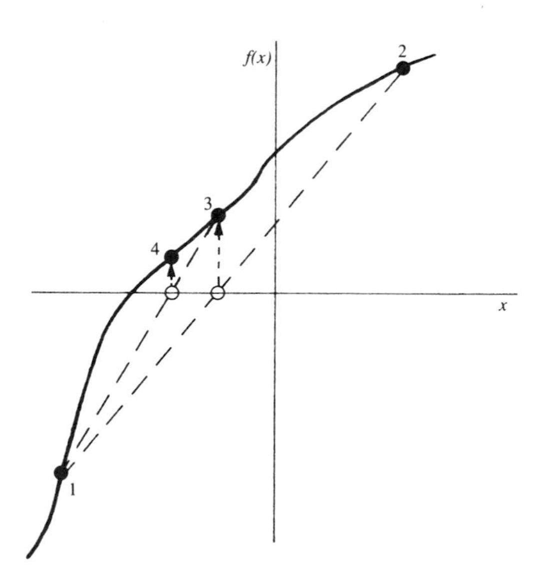

Figure 9.2.2. False position method. Interpolation lines (dashed) are drawn through the most recent points *that bracket the root.* In this example, point 1 thus remains "active" for many steps. False position converges less rapidly than the secant method, but it is more certain.

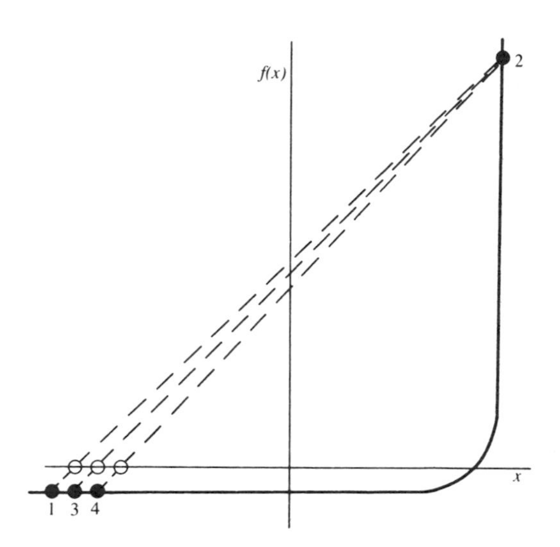

Figure 9.2.3. Example where both the secant and false position methods will take many iterations to arrive at the true root. This function would be difficult for many other root-finding methods.

```
DO 11 J=1,MAXIT                    False position loop.
    RTFLSP=XL+DX*FL/(FL-FH)        Increment with respect to latest value.
    F=FUNC(RTFLSP)
    IF(F.LT.0.)  THEN              Replace appropriate limit.
        DEL=XL-RTFLSP
        XL=RTFLSP
        FL=F
    ELSE
        DEL=XH-RTFLSP
        XH=RTFLSP
        FH=F
    ENDIF
    DX=XH-XL                       Convergence.
    IF(ABS(DEL).LT.XACC.OR.F.EQ.0.)RETURN
 11 CONTINUE
PAUSE 'RTFLSP exceed maximum iterations'
END
```

```
FUNCTION RTSEC(FUNC,X1,X2,XACC)
```

Using the secant method, find the root of a function **FUNC** thought to lie between **X1** and **X2**. The root, returned as **RTSEC**, is refined until its accuracy is ±**XACC**.

```
PARAMETER (MAXIT=30)               Maximum allowed number of iterations.
FL=FUNC(X1)
F=FUNC(X2)
IF(ABS(FL).LT.ABS(F))THEN          Pick the bound with the smaller function value as the most recent
    RTSEC=X1                           guess.
    XL=X2
    SWAP=FL
    FL=F
    F=SWAP
ELSE
    XL=X1
```

```
          RTSEC=X2
        ENDIF
        DO 11 J=1,MAXIT                       Secant loop.
          DX=(XL-RTSEC)*F/(F-FL)              Increment with respect to latest value.
          XL=RTSEC
          FL=F
          RTSEC=RTSEC+DX
          F=FUNC(RTSEC)                       Convergence.
          IF(ABS(DX).LT.XACC.OR.F.EQ.0.)RETURN
11      CONTINUE
        PAUSE 'RTSEC exceed maximum iterations'
        END
```

REFERENCES AND FURTHER READING:

Ralston, Anthony, and Rabinowitz, Philip. 1978, *A First Course in Numerical Analysis*, 2nd ed. (New York: McGraw-Hill), §8.3.

Ostrowski, A.M. 1966, *Solutions of Equations and Systems of Equations*, 2nd ed. (New York: Academic Press), Chapter 12.

9.3 Van Wijngaarden–Dekker–Brent Method

While secant and false position formally converge faster than bisection, one finds in practice pathological functions for which bisection converges more rapidly. These can be choppy, discontinuous functions, or even smooth functions if the second derivative changes sharply near the root. Bisection always halves the interval, while secant and false position can sometimes spend many cycles slowly pulling distant bounds closer to a root. Is there anything we can do to get the best of both worlds?

Yes. We can keep track of whether a supposedly superlinear method is actually converging the way it is supposed to, and, if it is not, we can intersperse bisection steps so as to guarantee *at least* linear convergence. This kind of super-strategy requires attention to bookkeeping detail, and also careful consideration of how roundoff errors can affect the guiding strategy. Also, we must be able to determine reliably when convergence has been achieved.

An excellent algorithm that pays close attention to these matters was developed in the 1960s by van Wijngaarden, Dekker, and others at the Mathematical Center in Amsterdam, and later improved by Brent (reference below). For brevity, we refer to the final form of the algorithm as *Brent's method.* The method is *guaranteed* (by Brent) to converge, so long as the function can be evaluated within the initial interval known to contain a root.

Brent's method combines root bracketing, bisection, and *inverse quadratic interpolation* to converge from the neighborhood of a zero crossing. While the false position and secant methods assume approximately linear behavior between two prior root estimates, inverse quadratic interpolation uses three prior points to fit an inverse quadratic function (x as a quadratic function of y) whose value at $y = 0$ is taken as the next estimate of the root x. Of course

one must have contingency plans for what to do if the root falls outside of the brackets. Brent's method takes care of all that. If the three point pairs are $[a, f(a)], [b, f(b)], [c, f(c)]$ then the interpolation formula (cf. equation 3.1.1) is

$$
\begin{aligned}
x = &\frac{[y - f(a)][y - f(b)]c}{[f(c) - f(a)][f(c) - f(b)]} + \frac{[y - f(b)][y - f(c)]a}{[f(a) - f(b)][f(a) - f(c)]} \\
&+ \frac{[y - f(c)][y - f(a)]b}{[f(b) - f(c)][f(b) - f(a)]}
\end{aligned}
\tag{9.3.1}
$$

Setting y to zero gives a result for the next root estimate, which can be written as

$$x = b + P/Q \tag{9.3.2}$$

where, in terms of

$$R \equiv f(b)/f(c), \qquad S \equiv f(b)/f(a), \qquad T \equiv f(a)/f(c) \tag{9.3.3}$$

we have

$$P = S\,[T(R - T)(c - b) - (1 - R)(b - a)] \tag{9.3.4}$$

$$Q = (T - 1)(R - 1)(S - 1) \tag{9.3.5}$$

In practice b is the current best estimate of the root and P/Q ought to be a "small" correction. Quadratic methods work well only when the function behaves smoothly; they run the serious risk of giving very bad estimates of the next root or causing machine failure by an inappropriate division by a very small number ($Q \approx 0$). Brent's method guards against this problem by maintaining brackets on the root and checking where the interpolation would land before carrying out the division. When the correction P/Q would not land within the bounds, or when the bounds are not collapsing rapidly enough, the algorithm takes a bisection step. Thus, Brent's method combines the sureness of bisection with the speed of a higher-order method when appropriate. We recommend it as the method of choice for general one-dimensional root finding where a function's values only (and not its derivative or functional form) are available.

```
FUNCTION ZBRENT(FUNC,X1,X2,TOL)
    Using Brent's method, find the root of a function FUNC known to lie between X1 and X2.
    The root, returned as ZBRENT, will be refined until its accuracy is TOL.
PARAMETER (ITMAX=100,EPS=3.E-8)  Maximum allowed number of iterations, and machine floating
A=X1                                 point precision.
B=X2
FA=FUNC(A)
FB=FUNC(B)
IF(FB*FA.GT.0.)  PAUSE 'Root must be bracketed for ZBRENT.'
FC=FB
DO 11 ITER=1,ITMAX
    IF(FB*FC.GT.0.)  THEN
        C=A                         Rename A,B,C and adjust bounding interval D.
        FC=FA
        D=B-A
        E=D
    ENDIF
    IF(ABS(FC).LT.ABS(FB)) THEN
        A=B
        B=C
        C=A
        FA=FB
        FB=FC
        FC=FA
    ENDIF
    TOL1=2.*EPS*ABS(B)+0.5*TOL        Convergence check.
    XM=.5*(C-B)
    IF(ABS(XM).LE.TOL1 .OR. FB.EQ.0.)THEN
        ZBRENT=B
        RETURN
    ENDIF
    IF(ABS(E).GE.TOL1 .AND. ABS(FA).GT.ABS(FB)) THEN
        S=FB/FA                     Attempt inverse quadratic interpolation.
        IF(A.EQ.C) THEN
            P=2.*XM*S
            Q=1.-S
        ELSE
            Q=FA/FC
            R=FB/FC
            P=S*(2.*XM*Q*(Q-R)-(B-A)*(R-1.))
            Q=(Q-1.)*(R-1.)*(S-1.)
        ENDIF
        IF(P.GT.0.)  Q=-Q           Check whether in bounds.
        P=ABS(P)
        IF(2.*P .LT. MIN(3.*XM*Q-ABS(TOL1*Q),ABS(E*Q))) THEN
            E=D                     Accept interpolation.
            D=P/Q
        ELSE
            D=XM                    Interpolation failed, use bisection.
            E=D
        ENDIF
    ELSE                            Bounds decreasing too slowly, use bisection.
        D=XM
        E=D
    ENDIF
    A=B                             Move last best guess to A.
    FA=FB
    IF(ABS(D) .GT. TOL1) THEN          Evaluate new trial root.
        B=B+D
    ELSE
        B=B+SIGN(TOL1,XM)
    ENDIF
    FB=FUNC(B)
11  CONTINUE
```

```
PAUSE 'ZBRENT exceeding maximum iterations.'
ZBRENT=B
RETURN
END
```

REFERENCES AND FURTHER READING:

Brent, Richard P. 1973, *Algorithms for Minimization without Derivatives* (Englewood Cliffs, N.J.: Prentice-Hall), Chapters 3, 4.

Forsythe, George E., Malcolm, Michael A., and Moler, Cleve B. 1977, *Computer Methods for Mathematical Computations* (Englewood Cliffs, N.J.: Prentice-Hall), §7.2.

9.4 Newton-Raphson Method Using Derivative

Perhaps the most celebrated of all one-dimensional root-finding routines is *Newton's method*, also called the *Newton-Raphson method.* This method is distinguished from the methods of previous sections by the fact that it requires the evaluation of both the function $f(x)$, *and* the derivative $f'(x)$, at arbitrary points x. The Newton-Raphson formula consists geometrically of extending the tangent line at a current point x_i until it crosses zero, then setting the next guess x_{i+1} to the abscissa of that zero-crossing (see Figure 9.4.1). Algebraically, the method derives from the familiar Taylor series expansion of a function in the neighborhood of a point,

$$f(x+\delta) \approx f(x) + f'(x)\delta + \frac{f''(x)}{2}\delta^2 + \ldots. \tag{9.4.1}$$

For small enough values of δ, and for well-behaved functions, the terms beyond linear are unimportant, hence $f(x+\delta) = 0$ implies

$$\delta = -\frac{f(x)}{f'(x)}. \tag{9.4.2}$$

Newton-Raphson is not restricted to one dimension. The method readily generalizes to multiple dimensions, as we shall see in §9.6 below.

Far from a root, where the higher order terms in the series *are* important, the Newton-Raphson formula can give grossly inaccurate, meaningless corrections. For instance, the initial guess for the root might be so far from the true root as to let the search interval include a local maximum or minimum of the function. This can be death to the method (see Figure 9.4.2). If an iteration places a trial guess near such a local extremum, so that the first derivative nearly vanishes, then Newton-Raphson sends its solution off to limbo,

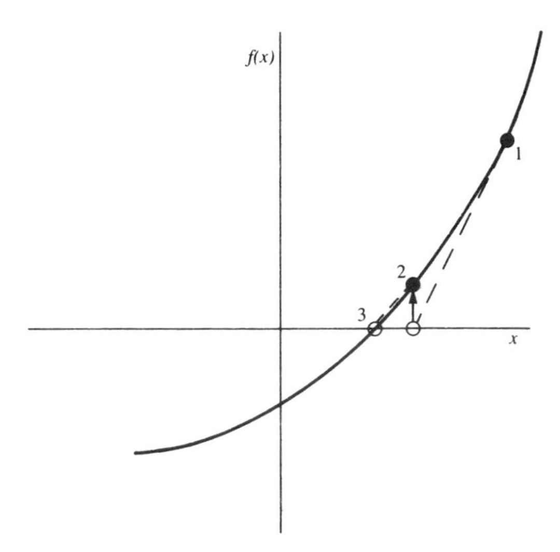

Figure 9.4.1. Newton's method extrapolates the local derivative to find the next estimate of the root. In this example it works well and converges quadratically.

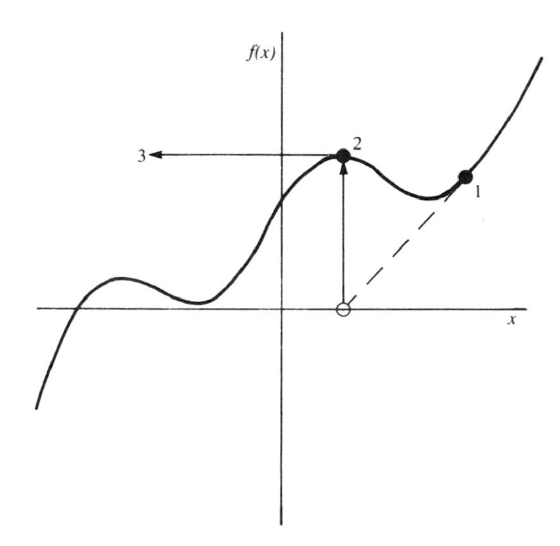

Figure 9.4.2. Unfortunate case where Newton's method encounters a local extremum and shoots off to outer space. Here bracketing bounds, as in **RTSAFE**, would save the day.

with vanishingly small hope of recovery. Like most powerful tools, Newton-Raphson can be destructive used in inappropriate circumstances. Figure 9.4.3 demonstrates another possible pathology.

Why do we call Newton-Raphson powerful? The answer lies in its rate of convergence: Within a small distance ϵ of x the function and its derivative

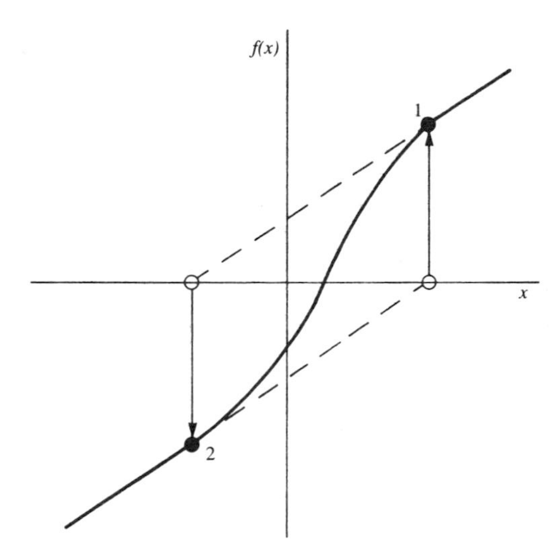

Figure 9.4.3. Unfortunate case where Newton's method enters a nonconvergent cycle. This behavior is often encountered when the function f is obtained, in whole or in part, by table interpolation. With a better initial guess, the method would have succeeded.

are approximately:

$$\begin{aligned} f(x+\epsilon) &= f(x) + \epsilon f'(x) + \epsilon^2 \frac{f''(x)}{2} + \cdots, \\ f'(x+\epsilon) &= f'(x) + \epsilon f''(x) + \cdots \end{aligned} \tag{9.4.3}$$

By the Newton-Raphson formula,

$$x_{i+1} = x_i - \frac{f(x_i)}{f'(x_i)}, \tag{9.4.4}$$

so that

$$\epsilon_{i+1} = \epsilon_i - \frac{f(x_i)}{f'(x_i)}. \tag{9.4.5}$$

When a trial solution x_i differs from the true root by ϵ_i, we can use (9.4.3) to express $f(x_i), f'(x_i)$ in (9.4.4) in terms of ϵ_i and derivatives at the root itself. The result is a recurrence relation for the deviations of the trial solutions

$$\epsilon_{i+1} = -\epsilon_i^2 \frac{f''(x)}{2f'(x)}. \tag{9.4.6}$$

Equation (9.4.6) says that Newton-Raphson converges *quadratically* (cf. equation 9.2.3). Near a root, the number of significant digits approximately *doubles* with each step. This very strong convergence property makes Newton-Raphson the method of choice for any function whose derivative can be evaluated efficiently, and whose derivative is continuous in the neighborhood of a root.

Even where Newton-Raphson is rejected for the early stages of convergence (because of its poor global convergence properties), it is very common to "polish up" a root with one or two steps of Newton-Raphson, which can multiply by two or four its number of significant figures!

For an efficient realization of Newton-Raphson the user provides a function subroutine which evaluates both $f(x)$ and its first derivative $f'(x)$ at the point x. The Newton-Raphson formula can also be applied using a numerical difference to approximate the true local derivative,

$$f'(x) \approx \frac{f(x+dx) - f(x)}{dx}. \tag{9.4.7}$$

This is not, however, a recommended procedure for the following reasons: (i) You are doing two function evaluations per step, so *at best* the superlinear order of convergence will be only $\sqrt{2}$. (ii) If you take dx too small you will be wiped out by roundoff, while if you take it too large your order of convergence will be only linear, no better than using the *initial* evaluation $f'(x_0)$ for all subsequent steps. Therefore, Newton-Raphson with numerical derivatives is (in one dimension) always dominated by the secant method of §9.2. (In multidimensions, where there is a paucity of available methods, Newton-Raphson with numerical derivatives must be taken more seriously. See §9.6.)

The following subroutine calls a user supplied subroutine FUNCD(X,FN,DF) which returns the function value as FN and the derivative as DF. We have included input bounds on the root simply to be consistent with previous root-finding routines: Newton does not adjust bounds, and works only on local information at the point X. The bounds are only used to pick the midpoint as the first guess, and to reject the solution if it wanders outside of the bounds.

```
FUNCTION RTNEWT(FUNCD,X1,X2,XACC)
    Using the Newton-Raphson method, find the root of a function known to lie in the interval
    X1-X2. The root RTNEWT will be refined until its accuracy is known within ±XACC. FUNCD
    is a user supplied subroutine that returns both the function value and the first derivative
    of the function at the point X.
PARAMETER (JMAX=20)              Set to maximum number of iterations.
RTNEWT=.5*(X1+X2)                Initial guess.
DO 11 J=1,JMAX
    CALL FUNCD(RTNEWT,F,DF)
    DX=F/DF
    RTNEWT=RTNEWT-DX
    IF((X1-RTNEWT)*(RTNEWT-X2).LT.0.)PAUSE 'jumped out of brackets'
    IF(ABS(DX).LT.XACC) RETURN          Convergence.
11  CONTINUE
PAUSE 'RTNEWT exceeding maximum iterations'
END
```

While Newton-Raphson's global convergence properties are poor, it is fairly easy to design a fail-safe routine that utilizes a combination of bisection and Newton-Raphson. The hybrid algorithm takes a bisection step whenever Newton-Raphson would take the solution out of bounds, or whenever Newton-Raphson is not reducing the size of the brackets rapidly enough.

```
FUNCTION RTSAFE(FUNCD,X1,X2,XACC)
    Using a combination of Newton-Raphson and bisection, find the root of a function brack-
    eted between X1 and X2. The root, returned as the function value RTSAFE, will be refined
    until its accuracy is known within ±XACC. FUNCD is a user supplied subroutine which
    returns both the function value and the first derivative of the function.
PARAMETER (MAXIT=100)          Maximum allowed number of iterations.
CALL FUNCD(X1,FL,DF)
CALL FUNCD(X2,FH,DF)
IF(FL*FH.GE.0.)  PAUSE 'root must be bracketed'
IF(FL.LT.0.)THEN               Orient the search so that f(XL) < 0.
    XL=X1
    XH=X2
ELSE
    XH=X1
    XL=X2
ENDIF
RTSAFE=.5*(X1+X2)              Initialize the guess for root,
DXOLD=ABS(X2-X1)               the "step-size before last,"
DX=DXOLD                       and the last step.
CALL FUNCD(RTSAFE,F,DF)
DO 11 J=1,MAXIT                Loop over allowed iterations.
    IF(((RTSAFE-XH)*DF-F)*((RTSAFE-XL)*DF-F).GE.0.   Bisect if Newton out of range,
*       .OR. ABS(2.*F).GT.ABS(DXOLD*DF) ) THEN       or not decreasing fast enough.
        DXOLD=DX
        DX=0.5*(XH-XL)
        RTSAFE=XL+DX
        IF(XL.EQ.RTSAFE)RETURN        Change in root is negligible.
    ELSE                       Newton step acceptable. Take it.
        DXOLD=DX
        DX=F/DF
        TEMP=RTSAFE
        RTSAFE=RTSAFE-DX
        IF(TEMP.EQ.RTSAFE)RETURN
    ENDIF
    IF(ABS(DX).LT.XACC) RETURN        Convergence criterion.
    CALL FUNCD(RTSAFE,F,DF)    The one new function evaluation per iteration.
    IF(F.LT.0.)  THEN          Maintain the bracket on the root.
        XL=RTSAFE
    ELSE
        XH=RTSAFE
    ENDIF
11  CONTINUE
PAUSE 'RTSAFE exceeding maximum iterations'
RETURN
END
```

For many functions the derivative $f'(x)$ often converges to machine accuracy before the function $f(x)$ itself does. When that is the case one need not subsequently update $f'(x)$. This shortcut is recommended only when you confidently understand the generic behavior of your function, but it speeds computations when the derivative calculation is laborious. (Formally this makes the convergence only linear, but if the derivative isn't changing anyway, you can do no better.)

REFERENCES AND FURTHER READING:

Acton, Forman S. 1970, *Numerical Methods That Work* (New York: Harper and Row), Chapter 2.

Ralston, Anthony, and Rabinowitz, Philip. 1978, *A First Course in Numerical Analysis*, 2nd ed. (New York: McGraw-Hill), §8.4.

Ortega, J., and Rheinboldt, W. 1970, *Iterative Solution of Nonlinear Equations in Several Variables* (New York: Academic Press).

9.5 Roots of Polynomials

Here we present a few methods for finding roots of polynomials. These will serve for most practical problems involving polynomals of low-to-moderate degree or for well-conditioned polynomials of higher degree. Not as well appreciated as it ought to be is the fact that some polynomials are exceedingly ill-conditioned. The tiniest changes in a polynomial's coefficients can, in the worst case, send its roots sprawling all over the complex plane. (An infamous example due to Wilkinson is detailed in Acton's book.)

Recall that a polynomial of degree n will have n roots. The roots can be real or complex, and they might not be distinct. If the coefficients of the polynomial are real, then complex roots will occur in pairs that are conjugate, i.e. if $x_1 = a + bi$ is a root then $x_2 = a - bi$ will also be a root. When the coefficients are complex, the complex roots need not be related.

Multiple roots, or closely spaced roots, produce the most difficulty for numerical algorithms (see Figure 9.5.1). For example, $P(x) = (x - a)^2$, has a double real root at $x = a$. However, we cannot bracket the root by the usual technique of identifying neighborhoods where the function changes sign, nor will slope-following methods such as Newton-Raphson work well, because both the function and its derivative vanish at a multiple root. Newton-Raphson *may* work, but slowly, since large roundoff errors can occur. When a root is known in advance to be multiple, then special methods of attack are readily devised. Problems arise when (as is generally the case) we do not know in advance what pathology a root will display.

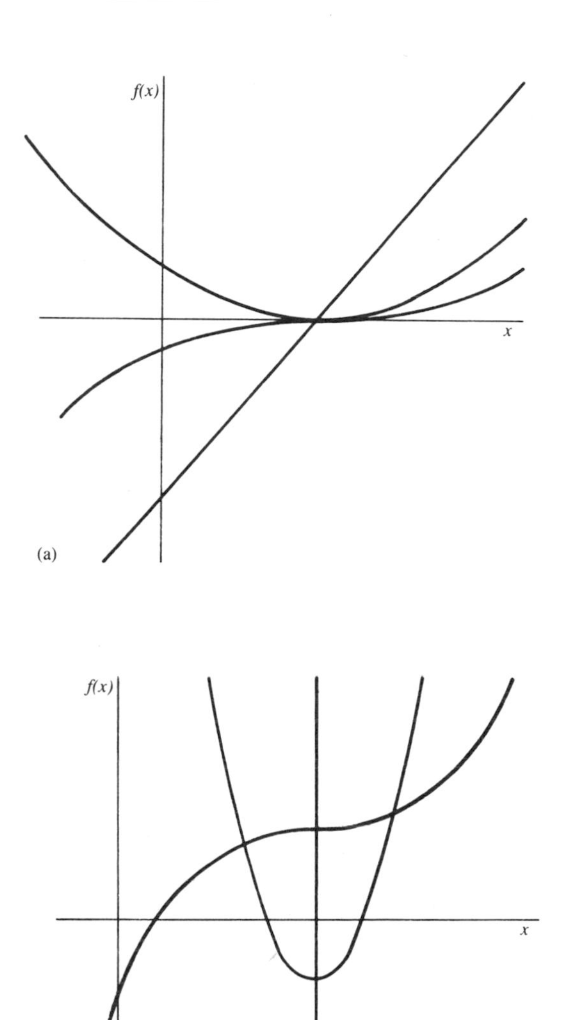

Figure 9.5.1. (a) Linear, quadratic, and cubic behavior at the roots of polynomials. Only under high magnification (b) does it become apparent that the cubic has one, not three, roots, and that the quadratic has two roots rather than none.

Deflation of Polynomials

When seeking several or all roots of a polynomial, the total effort can be significantly reduced by the use of *deflation*. As each root r is found, the polynomial is factored into a product involving the root and a reduced

polynomial of degree one less than the original, i.e. $P(x) = (x - r)Q(x)$. Since the roots of Q are exactly the remaining roots of P, the effort of finding additional roots decreases, because we work with polynomials of lower and lower degree as we find successive roots. Even more important, with deflation we can avoid the blunder of having our iterative method converge twice to the same (nonmultiple) root instead of separately to two different roots.

Deflation, which amounts to synthetic division, is a simple operation that acts on the array of polynomial coefficients. The concise code for synthetic division by a monomial factor was given in §5.3 above. You can deflate complex roots either by converting that code to complex data type, or else – in the case of a polynomial with real coefficients but possibly complex roots – by deflating by a quadratic factor,

$$[x - (a + ib)]\,[x - (a - ib)] = x^2 - 2ax + (a^2 + b^2) \tag{9.5.1}$$

The routine POLDIV in §5.3 can be used to divide the polynomial by this factor.

Deflation must, however, be utilized with care. Because each new root is only known with finite accuracy, errors creep into the determination of the coefficients of the successively deflated polynomial. Consequently, the roots can become more and more inaccurate. It matters a lot whether the inaccuracy creeps in stably (plus or minus a few multiples of the machine precision at each stage) or unstably (erosion of successive significant figures until the results become meaningless). Which behavior occurs depends on just how the root is divided out. *Forward deflation*, where the new polynomial coefficients are computed in the order from the highest power of x down to the constant term, was illustrated in §5.3. This turns out to be stable if the root of smallest absolute value is divided out at each stage. Alternatively, one can do *backward deflation*, where new coefficients are computed in order from the constant term up to the coefficient of the highest power of x. This is stable if the remaining root of *largest* absolute value is divided out at each stage.

A polynomial whose coefficients are interchanged "end-to-end," so that the constant becomes the highest coefficient, etc., has its roots mapped into their reciprocals. (Proof: Divide the whole polynomial by its highest power x^n and rewrite it as a polynomial in $1/x$.) The algorithm for backward deflation is therefore virtually identical to that of forward deflation, except that the original coefficients are taken in reverse order and the reciprocal of the deflating root is used. Since we will use forward deflation below, we leave to you the exercise of writing a concise coding for backward deflation (as in §5.3). For more on the stability of deflation, consult the article by Peters and Wilkinson.

To minimize the impact of increasing errors (even stable ones) when using deflation, it is advisable to treat roots of the successively deflated polynomials as only *tentative* roots of the original polynomial. One then *polishes* these tentative roots by taking them as initial guesses that are to be re-solved for, using the *nondeflated* original polynomial P. Again you must beware lest two deflated roots are inaccurate enough that, under polishing, they both converge to the same undeflated root; in that case you gain a spurious root-multiplicity and lose a distinct root. This is detectable, since you can compare

each polished root for equality to previous ones from distinct tentative roots. When it happens, you are advised to deflate the polynomial just once (and for this root only), then again polish the tentative root, or to use Maehly's procedure (see equation 9.5.21 below).

Below we say more about techniques for polishing real and complex-conjugate tentative roots. First, let's get back to overall strategy.

There are two schools of thought about how to proceed when faced with a polynomial of real coefficients. One school says to go after the easiest quarry, the real, distinct roots, by the same kinds of methods that we have discussed in previous sections for general functions, i.e. trial-and-error bracketing followed by a safe Newton-Raphson as in `RTSAFE`. Sometimes you are *only* interested in real roots, in which case the strategy is complete. Otherwise, you then go after quadratic factors of the form (9.5.1) by any of a variety of methods. One such is Bairstow's method, which we will discuss below in the context of root polishing. Another is Muller's method, which we here briefly discuss.

Muller's Method

Muller's method generalizes the secant method, but uses quadratic interpolation among three points instead of linear interpolation between two. Solving for the zeros of the quadratic allows the method to find complex pairs of roots. Given *three* previous guesses for the root x_{i-2}, x_{i-1}, x_i, and the values of the polynomial $P(x)$ at those points, the next approximation x_{i+1} is produced by the following formulas,

$$\begin{aligned}
q &\equiv \frac{x_i - x_{i-1}}{x_{i-1} - x_{i-2}} \\
A &\equiv qP(x_i) - q(1+q)P(x_{i-1}) + q^2 P(x_{i-2}) \\
B &\equiv (2q+1)P(x_i) - (1+q)^2 P(x_{i-1}) + q^2 P(x_{i-2}) \\
C &\equiv (1+q)P(x_i)
\end{aligned} \tag{9.5.2}$$

followed by

$$x_{i+1} = x_i - (x_i - x_{i-1}) \left[\frac{2C}{B \pm \sqrt{B^2 - 4AC}} \right] \tag{9.5.3}$$

where the sign in the denominator is chosen to make its absolute value or modulus as large as possible. You can start the iterations with any three values of x that you like, e.g. three equally spaced values on the real axis. Note that you must allow for the possibility of a complex denominator, and subsequent complex arithmetic, in implementing the method.

Muller's method is sometimes also used for finding complex zeros of analytic functions (not just polynomials) in the complex plane, for example in the IMSL routine `ZANLYT`.

Laguerre's Method

The second school regarding overall strategy happens to be the one to which we belong. That school advises you to use one of a very small number of methods which will converge (though with greater or lesser efficiency) to all types of roots: real, complex, single or multiple. Use such a method to get tentative values for all n roots of your n^{th} degree polynomial. Then go back and polish them as you desire.

Laguerre's method is by far the most straightforward of these sure-fire methods. It does require that you perform complex arithmetic (even while converging to real roots), but it is guaranteed to converge to a root from any starting point. In some instances the complex arithmetic is no disadvantage, since the polynomial itself may have complex coefficients.

To motivate (although not rigorously derive) the Laguerre formulas we can note the following relations between the polynomial and its roots and derivatives

$$P_n(x) = (x - x_1)(x - x_2)\dots(x - x_n) \tag{9.5.4}$$

$$\ln|P_n(x)| = \ln|x - x_1| + \ln|x - x_2| + \dots + \ln|x - x_n| \tag{9.5.5}$$

$$\frac{d\ln|P_n(x)|}{dx} = +\frac{1}{x - x_1} + \frac{1}{x - x_2} + \dots + \frac{1}{x - x_n} = \frac{P'_n}{P_n} \equiv G \tag{9.5.6}$$

$$\begin{aligned} -\frac{d^2\ln|P_n(x)|}{dx^2} &= +\frac{1}{(x - x_1)^2} + \frac{1}{(x - x_2)^2} + \dots + \frac{1}{(x - x_n)^2} \\ &= \left[\frac{P'_n}{P_n}\right]^2 - \frac{P''_n}{P_n} \equiv H \end{aligned} \tag{9.5.7}$$

Starting from these relations, the Laguerre formulas make what Acton nicely calls "a rather drastic set of assumptions": The root x_1 that we seek is assumed to be located some distance a from our current guess x, while *all other roots* are assumed to be located at a distance b

$$a = x - x_1 \qquad b = x - x_i \quad i = 2, 3, \dots, n \tag{9.5.8}$$

Then we can express (9.5.6), (9.5.7) as

$$\frac{1}{a} + \frac{n-1}{b} = G \tag{9.5.9}$$

$$\frac{1}{a^2} + \frac{n-1}{b^2} = H \tag{9.5.10}$$

which yields as the solution for a

$$a = \frac{n}{G \pm \sqrt{(n-1)(nH - G^2)}} \tag{9.5.11}$$

where the sign should be taken to yield the largest magnitude for the denominator. Since the factor inside the square root can be negative, a can be complex. (A more rigorous justification of equation 9.5.11 is in Ralston and Rabinowitz.)

The method operates iteratively: For a trial value x, a is calculated by equation (9.5.11). Then $x - a$ becomes the next trial value. This continues until a is sufficiently small.

The following routine implements the Laguerre method to find one root of a given polynomial of degree M, whose coefficients can be complex. As usual, the first coefficient A(1) is the constant term, while A(M+1) is the coefficient of the highest power of x. The routine implements a simplified version of an elegant stopping criterion due to Adams (*Communications of the A.C.M.*, **10**, 655, [1967]). This criterion neatly balances the desire to achieve full machine accuracy, on the one hand, with the danger of iterating forever in the presence of roundoff error, on the other.

```
SUBROUTINE LAGUER(A,M,X,EPS,POLISH)
   Given the degree M and the M+1 complex coefficients A of the polynomial ∑_{i=1}^{M+1} A(i)x^{i-1},
   and given EPS the desired fractional accuracy, and given a complex value X, this routine
   improves X by Laguerre's method until it converges to a root of the given polynomial.
   For normal use POLISH should be input as .FALSE.. When POLISH is input as .TRUE.,
   the routine ignores EPS and instead attempts to improve X (assumed to be a good initial
   guess) to the achievable roundoff limit.
COMPLEX A(M+1),X,DX,X1,B,D,F,G,H,SQ,GP,GM,G2,ZERO
LOGICAL POLISH
PARAMETER (ZERO=(0.,0.),EPSS=6.E-8,MAXIT=100)
DXOLD=CABS(X)
DO 12 ITER=1,MAXIT                    Loop over iterations up to allowed maximum.
   B=A(M+1)
   ERR=CABS(B)
   D=ZERO
   F=ZERO
   ABX=CABS(X)
   DO 11 J=M,1,-1                     Efficient computation of the polynomial and its first two derivatives.
      F=X*F+D
      D=X*D+B
      B=X*B+A(J)
      ERR=CABS(B)+ABX*ERR
11 CONTINUE
   ERR=EPSS*ERR                       Estimate of roundoff error in evaluating polynomial.
   IF(CABS(B).LE.ERR) THEN            We are on the root.
      DX=ZERO
      RETURN
   ELSE                               The generic case: use Laguerre's formula.
      G=D/B
      G2=G*G
      H=G2-2.*F/B
```

```
        SQ=CSQRT((M-1)*(M*H-G2))
        GP=G+SQ
        GM=G-SQ
        IF(CABS(GP).LT.CABS(GM)) GP=GM
        DX=M/GP
    ENDIF
    X1=X-DX
    IF(X.EQ.X1)RETURN                              Converged.
    X=X1
    CDX=CABS(DX)
    IF(ITER.GT.6.AND.CDX.GE.DXOLD)RETURN           Reached roundoff limit.
    DXOLD=CDX
    IF(.NOT.POLISH)THEN
        IF(CDX.LE.EPS*CABS(X))RETURN               Converged.
    ENDIF
12  CONTINUE
PAUSE 'too many iterations'
RETURN
END
```

Here is a driver routine which calls LAGUER in succession for each root, performs the deflation, optionally polishes the roots by the same Laguerre method — if you are not going to polish in some other way — and finally sorts the roots by their real parts. (We will use this routine as a tool in Chapter 12.)

```
SUBROUTINE ZROOTS(A,M,ROOTS,POLISH)
```

Given the degree M and the M+1 complex coefficients A of the polynomial $\sum_{i=1}^{M+1} A(i)x^{i-1}$, this routine successively calls **LAGUER** and finds all M complex **ROOTS**. The logical variable **POLISH** should be input as **.TRUE.** if polishing (also by Laguerre's method) is desired, **.FALSE.** if the roots will be subsequently polished by other means.

```
PARAMETER (EPS=1.E-6,MAXM=101)
```

Desired accuracy and maximum anticipated value of M+1.

```
COMPLEX A(M+1),ROOTS(M),AD(MAXM),X,B,C
LOGICAL POLISH
DO 11 J=1,M+1                     Copy of coefficients for successive deflation.
    AD(J)=A(J)
11  CONTINUE
DO 13 J=M,1,-1                    Loop over each root to be found.
    X=CMPLX(0.,0.)                Start at zero to favor convergence to smallest remaining root.
    CALL LAGUER(AD,J,X,EPS,.FALSE.)    Find the root.
    IF(ABS(AIMAG(X)).LE.2.*EPS**2*ABS(REAL(X))) X=CMPLX(REAL(X),0.)
    ROOTS(J)=X
    B=AD(J+1)                     Forward deflation.
    DO 12 JJ=J,1,-1
        C=AD(JJ)
        AD(JJ)=B
        B=X*B+C
12      CONTINUE
13  CONTINUE
IF (POLISH) THEN
    DO 14 J=1,M                   Polish the roots using the undeflated coefficients.
        CALL LAGUER(A,M,ROOTS(J),EPS,.TRUE.)
14      CONTINUE
ENDIF
DO 16 J=2,M                       Sort roots by their real parts by straight insertion.
    X=ROOTS(J)
    DO 15 I=J-1,1,-1
        IF(REAL(ROOTS(I)).LE.REAL(X))GO TO 10
        ROOTS(I+1)=ROOTS(I)
15      CONTINUE
    I=0
10  ROOTS(I+1)=X
```

```
16 CONTINUE
   RETURN
   END
```

Other Sure-Fire Techniques

The *Jenkins-Traub method* has become practically a standard in black-box polynomial root-finders, e.g. in the IMSL library. The method is too complicated to discuss here, but is detailed, with references to the primary literature, in Ralston and Rabinowitz.

The *Lehmer-Schur algorithm* is one of a class of methods which isolate roots in the complex plane by generalizing the notion of one-dimensional bracketing. It is possible to determine efficiently whether there are any polynomial roots within a circle of given center and radius. From then on it is a matter of bookkeeping to hunt down all the roots by a series of decisions regarding where to place new trial circles. Consult Acton for an introduction.

Techniques for Root-Polishing

Newton-Raphson works very well for real roots once the neighborhood of a root has been identified. The polynomial and its derivative can be efficiently simultaneously evaluated as in §5.3. For a polynomial of degree `N-1` with coefficients `C(1)...C(N)`, the following segment of code embodies one cycle of Newton-Raphson:

```
P=C(N)*X+C(N-1)
P1=C(N)
DO I=N-2,1,-1
    P1=P+P1*X
    P=C(I)+P*X
ENDDO
IF (P1.EQ.0.)  PAUSE 'derivative should not vanish'
X=X-P/P1
```

Once all real roots of a polynomial have been polished, one must polish the complex roots, either directly, or by looking for quadratic factors.

Direct polishing by Newton-Raphson is straightforward for complex roots if the above code is converted to complex data types. With real polynomial coefficients, note that your starting guess (tentative root) *must* be off the real axis, otherwise you will never get off that axis – and may get shot off to infinity by a minimum or maximum of the polynomial.

For real polynomials, the alternative means of polishing complex roots (or, for that matter, double real roots) is *Bairstow's method*, which seeks quadratic factors. The advantage of going after quadratic factors is that it avoids all complex arithmetic. Bairstow's method seeks a quadratic factor that embodies the two roots $x = a \pm ib$, namely

$$x^2 - 2ax + (a^2 + b^2) \equiv x^2 + Bx + C \tag{9.5.12}$$

In general if we divide a polynomial by a quadratic factor, there will be a linear remainder

$$P(x) = (x^2 + Bx + C)Q(x) + Rx + S. \tag{9.5.13}$$

Given B and C, R and S can be readily found, by polynomial division (§5.3). We can consider R, S to be adjustable functions of B, C which will be zero if the quadratic factor is a root.

In the neighborhood of a root a first order Taylor series expansion approximates the variation of R, S with respect to small changes in B, C

$$R(B + \delta B, C + \delta C) \approx R(B, C) + \frac{\partial R}{\partial B}\delta B + \frac{\partial R}{\partial C}\delta C \tag{9.5.14}$$

$$S(B + \delta B, C + \delta C) \approx S(B, C) + \frac{\partial S}{\partial B}\delta B + \frac{\partial S}{\partial C}\delta C \tag{9.5.15}$$

To evaluate the partial derivatives, consider the derivative of (9.5.13) with respect to C. Since $P(x)$ is a fixed polynomial, it is independent of C, hence

$$0 = (x^2 + Bx + C)\frac{\partial Q}{\partial C} + Q(x) + \frac{\partial R}{\partial C}x + \frac{\partial S}{\partial C} \tag{9.5.16}$$

which can be rewritten as

$$-Q(x) = (x^2 + Bx + C)\frac{\partial Q}{\partial C} + \frac{\partial R}{\partial C}x + \frac{\partial S}{\partial C} \tag{9.5.17}$$

Similarly, $P(x)$ is independent of B, so differentiating (9.5.13) with respect to B gives

$$-xQ(x) = (x^2 + Bx + C)\frac{\partial Q}{\partial B} + \frac{\partial R}{\partial B}x + \frac{\partial S}{\partial B} \tag{9.5.18}$$

Note that (9.5.17) and (9.5.18) match (9.5.13) in form, so that the partial derivatives of R, S with respect to B, C can be found by performing synthetic division of $-Q(x), -xQ(x)$ by the trial quadratic factor.

Bairstow's method now consists of using Newton-Raphson in two dimensions (which is actually the subject of the *next* section) to find a simultaneous zero of R and S. Synthetic division is used three times per cycle to evaluate R, S and their partial derivatives with respect to B, C. Like one-dimensional Newton-Raphson, the method works well in the vicinity of a root pair (real or complex), but it can fail miserably when started at a random point. We therefore recommend it in the context of polishing tentative complex roots.

```
      SUBROUTINE QROOT(P,N,B,C,EPS)
C     Given N coefficients P of a polynomial of degree N-1, and trial values for the coefficients
C     of a quadratic factor X*X+B*X+C, improve the solution until the coefficients B,C change by
C     less than EPS. The routine POLDIV §5.3 is used.
      PARAMETER (NMAX=20,ITMAX=20,TINY=1.0E-6)     At most NMAX coefficients, ITMAX iterations.
      DIMENSION P(N),Q(NMAX),D(3),REM(NMAX),QQ(NMAX)
      D(3)=1.
      DO 12 ITER=1,ITMAX
        D(2)=B
        D(1)=C
        CALL POLDIV(P,N,D,3,Q,REM)
        S=REM(1)                         First division R,S.
        R=REM(2)
        CALL POLDIV(Q,N-1,D,3,QQ,REM)
        SC=-REM(1)                       Second division partial R,S with respect to C.
        RC=-REM(2)
        DO 11 I=N-1,1,-1
          Q(I+1)=Q(I)
11      CONTINUE
        Q(1)=0.
        CALL POLDIV(Q,N,D,3,QQ,REM)
        SB=-REM(1)                       Third division partial R,S with respect to B.
        RB=-REM(2)
        DIV=1./(SB*RC-SC*RB)             Solve 2x2 equation.
        DELB=(R*SC-S*RC)*DIV
        DELC=(-R*SB+S*RB)*DIV
        B=B+DELB
        C=C+DELC
        IF((ABS(DELB).LE.EPS*ABS(B).OR.ABS(B).LT.TINY)
     *      .AND.(ABS(DELC).LE.EPS*ABS(C)
     *      .OR.ABS(C).LT.TINY)) RETURN              Coefficients converged.
12    CONTINUE
      PAUSE 'too many iterations in QROOT'
      END
```

We have already remarked on the annoyance of having two tentative roots collapse to one value under polishing. You are left not knowing whether your polishing procedure has lost a root, or whether there *is* actually a double root, which was split only by roundoff errors in your previous deflation. One solution is deflate-and-repolish; but deflation is what we are trying to avoid at the polishing stage. An alternative is *Maehly's procedure.* Maehly pointed out that the derivative of the reduced polynomial

$$P_j(x) \equiv \frac{P(x)}{(x-x_1)\cdots(x-x_j)} \tag{9.5.19}$$

can be written as

$$P'_j(x) = \frac{P'(x)}{(x-x_1)\cdots(x-x_j)} - \frac{P(x)}{(x-x_1)\cdots(x-x_j)}\sum_{i=1}^{j}(x-x_i)^{-1} \tag{9.5.20}$$

Hence one step of Newton-Raphson, taking a guess x_k into a new guess x_{k+1}, can be written as

$$x_{k+1} = x_k - \frac{P(x_k)}{P'(x_k) - P(x_k)\sum_{i=1}^{j}(x_k - x_i)^{-1}} \tag{9.5.21}$$

This equation, if used with i ranging over the roots already polished, will prevent a tentative root from spuriously hopping to another one's true root. It is an example of so-called *zero suppression* as an alternative to true deflation.

Muller's method, which was described above, can also be useful at the polishing stage.

REFERENCES AND FURTHER READING:

Stoer, J., and Bulirsch, R. 1980, *Introduction to Numerical Analysis* (New York: Springer-Verlag) §5.5–5.9.

Acton, Forman S. 1970, *Numerical Methods That Work* (New York: Harper and Row), Chapter 7.

Ralston, Anthony, and Rabinowitz, Philip. 1978, *A First Course in Numerical Analysis*, 2nd ed. (New York: McGraw-Hill) §8.9–8.13.

Henrici, P. 1974, *Applied and Computational Complex Analysis*, vol. 1 (New York: Wiley).

Peters G., and Wilkinson, J.H. 1971, *J. Inst. Math. Appl.*, vol. 8, pp. 16-35.

IMSL Library Reference Manual, 1980, ed. 8 (IMSL Inc., 7500 Bellaire Boulevard, Houston TX 77036).

9.6 Newton-Raphson Method for Nonlinear Systems of Equations

We make an extreme, but wholly defensible, statement: There are *no* good, general methods for solving systems of more than one nonlinear equation. Furthermore, it is not hard to see why (very likely) there *never will be* any good, general methods: Consider the case of two dimensions, where we want to solve simultaneously

$$\begin{aligned} f(x,y) &= 0 \\ g(x,y) &= 0 \end{aligned} \tag{9.6.1}$$

The functions f and g are two arbitrary functions, each of which has zero contour lines that divide the (x,y) plane into regions where their respective function is positive or negative. These zero contour boundaries are of interest to us. The solutions that we seek are those points (if any) which are common to the zero contours of f and g (see Figure 9.6.1). Unfortunately, the functions

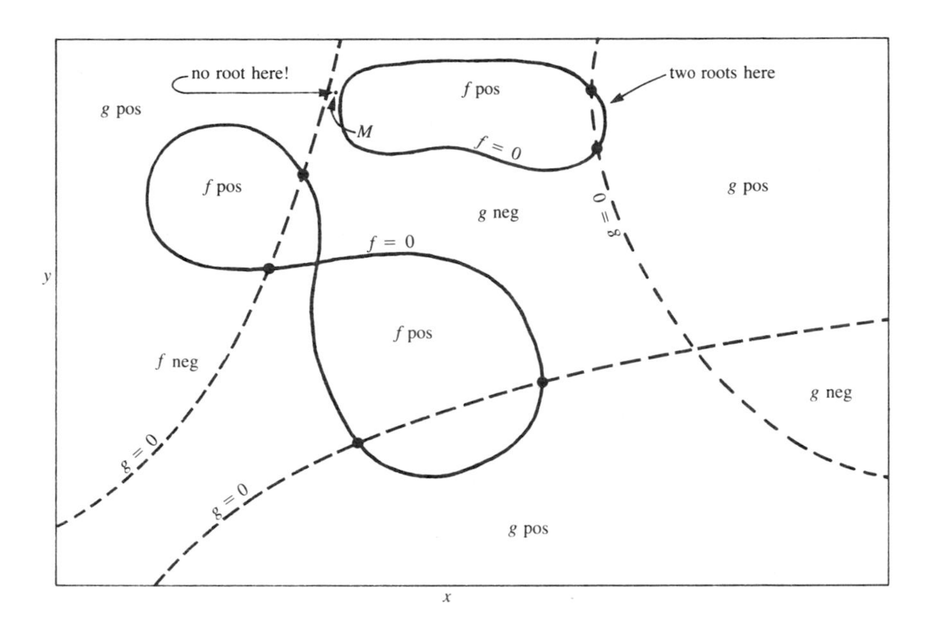

Figure 9.6.1. Solution of two nonlinear equations in two unknowns. Solid curves refer to $f(x,y)$, dashed curves to $g(x,y)$. Each equation divides the (x,y) plane into positive and negative regions, bounded by zero curves. The desired solutions are the intersections of these unrelated zero curves. The number of solutions is *a priori* unknown.

f and g have, in general, no relation to each other at all! There is nothing special about a common point from either f's point of view, or from g's. In order to find all common points, which are the solutions of our nonlinear equations, we will (in general) have to do neither more nor less than map out the full zero contours of both functions. Note further that the zero contours will (in general) consist of an unknown number of disjoint closed curves. How can we ever hope to know when we have found all such disjoint pieces?

For problems in more than two dimensions, we need to find points mutually common to N unrelated zero-contour hyperplanes, each of dimension $N-1$. You see that root finding becomes virtually impossible without insight! You will almost always have to use additional information, specific to your particular problem, to answer such basic questions as, "Do I expect a unique solution?" and "Approximately where?" Acton has a good discussion of some of the particular strategies that can be tried.

Once, however, you identify the neighborhood of a root, or of a place where there *might* be a root, then the problem firms up considerably: It is time to turn to Newton-Raphson, which readily generalizes to multiple dimensions. This method gives you a very efficient means of converging to the root, if it exists, or of spectacularly failing to converge, indicating (though not proving) that your putative root does not exist nearby.

A typical problem gives N functional relations to be zeroed, involving variables $x_i, i = 1, 2, \ldots, N$:

$$f_i(x_1, x_2, \ldots, x_N) = 0 \qquad i = 1, 2, \ldots, N. \tag{9.6.2}$$

If we let $\mathbf{X}$ denote the entire vector of values x_i then, in the neighborhood of $\mathbf{X}$, each of the functions f_i can be expanded in Taylor series

$$f_i(\mathbf{X} + \delta\mathbf{X}) = f_i(\mathbf{X}) + \sum_{j=1}^{N} \frac{\partial f_i}{\partial x_j} \delta x_j + O(\delta\mathbf{X}^2). \tag{9.6.3}$$

By neglecting terms of order $\delta\mathbf{X}^2$ and higher, we obtain a set of linear equations for the corrections $\delta\mathbf{X}$ that move each function closer to zero simultaneously, namely

$$\sum_{j=1}^{N} \alpha_{ij} \delta x_j = \beta_i, \tag{9.6.4}$$

where

$$\alpha_{ij} \equiv \frac{\partial f_i}{\partial x_j} \qquad \beta_i \equiv -f_i \tag{9.6.5}$$

Matrix equation (9.6.4) can be solved by LU decomposition as described in §2.3. The corrections are then added to the solution vector,

$$x_i^{new} = x_i^{old} + \delta x_i \quad i = 1, \ldots, N \tag{9.6.6}$$

and the process is iterated to convergence. In general it is a good idea to check the degree to which both functions and variables have converged. Once either reaches machine accuracy, the other won't change.

The following routine MNEWT performs NTRIAL iterations starting from an initial guess at the solution vector X of length N variables. Iteration stops if either the sum of the magnitudes of the functions f_i is less than some tolerance TOLF, or the sum of the absolute values of the corrections to δx_i is less than some tolerance TOLX. MNEWT calls a user supplied subroutine USRFUN which must return the matrix of partial derivatives α, and the negative of the function values β, as defined in (9.6.5). You should not make NTRIAL too big; rather inspect to see what is happening before continuing for some further iterations.

```
SUBROUTINE MNEWT(NTRIAL,X,N,TOLX,TOLF)
    Given an initial guess X for a root in N dimension, take NTRIAL Newton-Raphson steps to
    improve the root. Stop if the root converges in either summed variable increments TOLX
    or summed function values TOLF.
PARAMETER (NP=15)                   Up to NP variables
DIMENSION X(NP),ALPHA(NP,NP),BETA(NP),INDX(NP)
DO 13 K=1,NTRIAL
    CALL USRFUN(X,ALPHA,BETA)       User subroutine supplies matrix coefficients.
    ERRF=0.                         Check function convergence.
    DO 11 I=1,N
        ERRF=ERRF+ABS(BETA(I))
11  CONTINUE
    IF(ERRF.LE.TOLF)RETURN
    CALL LUDCMP(ALPHA,N,NP,INDX,D)  Solve linear equations using LU decomposition.
    CALL LUBKSB(ALPHA,N,NP,INDX,BETA)
    ERRX=0.                         Check root convergence.
    DO 12 I=1,N                     Update solution.
        ERRX=ERRX+ABS(BETA(I))
        X(I)=X(I)+BETA(I)
12  CONTINUE
    IF(ERRX.LE.TOLX)RETURN
13 CONTINUE
RETURN
END
```

Multidimensional Root Finding Versus Multidimensional Minimization

In the next chapter, we will find that there *are* efficient general techniques for finding a minimum of a function of many variables. Why is that task (relatively) easy, while multidimensional root finding is often quite hard? Isn't minimization equivalent to finding a zero of an N-dimensional gradient vector, not so different from zeroing an N-dimensional function? No! The components of a gradient vector are not independent, arbitrary functions. Rather, they obey so-called integrability conditions that are highly restrictive. Put crudely, you can always find a minimum by sliding downhill on a single surface. The test of "downhillness" is thus one-dimensional. There is no analogous conceptual procedure for finding a multidimensional root, where "downhill" must mean simultaneously downhill in N separate function spaces, thus allowing a multitude of trade-offs, as to how much progress in one dimension is worth compared with progress in another.

A popular idea for multidimensional root finding is to collapse all these dimensions into one, by adding up the sums of squares of the individual functions f_i to get a master function F which (i) is positive definite, and (ii) has a global minimum of zero exactly at all solutions of the original set of nonlinear equations. Unfortunately, as you will see in the next chapter, the efficient algorithms for finding minima come to rest on global and local minima indiscriminately. You will almost surely find, to your great dissatisfaction, that your function F has a great number of local minima. In Figure 9.6.1, for example, there is likely to be a local minimum wherever the zero contours of f and g make a close approach to each other. The point labeled M is such a point, and one sees that there are no nearby roots.

REFERENCES AND FURTHER READING:

Acton, Forman S. 1970, *Numerical Methods That Work* (New York: Harper and Row), Chapter 14.

Ostrowski, A.M. 1966, *Solutions of Equations and Systems of Equations*, 2nd ed. (New York: Academic Press).

Ortega, J., and Rheinboldt, W. 1970, *Iterative Solution of Nonlinear Equations in Several Variables* (New York: Academic Press).

Chapter 10. Minimization or Maximization of Functions

10.0 Introduction

In a nutshell: You are given a single function f which depends on one or more independent variables. You want to find the value of those variables where f takes on a maximum or a minimum value. You can then calculate what value of f is achieved at the maximum or minimum. The tasks of maximization and minimization are trivially related to each other, since one person's function f could just as well be another person's $-f$. The computational desiderata are the usual ones: do it quickly, cheaply, and in small memory. Often the computational effort is dominated by the cost of evaluating f (and also perhaps its partial derivatives with respect to all variables, if the chosen algorithm requires them). In such cases the desiderata are sometimes replaced by the simple surrogate: evaluate f as few times as possible.

An extremum (maximum or minimum point) can be either *global* (truly the highest or lowest function value) or *local* (the highest or lowest in a finite neighborhood and not on the boundary of that neighborhood). (See Figure 10.0.1.) Virtually nothing is known about finding global extrema in general. There are two standard heuristics that everyone uses: (i) find local extrema starting from widely varying starting values of the independent variables, and then pick the most extreme of these (if they are not all the same), or (ii) perturb a local extremum by taking a finite amplitude step away from it, and then see if your routine returns you to a better point, or "always" to the same one. There are tantalizing hints that so-called "annealing methods" may lead to important progress on the global extremization problem (see below).

Our chapter title could just as well be *optimization*, which is the usual name for this very large field of numerical research. The importance ascribed to the various tasks in this field depends strongly on the particular interests of whom you talk to. Economists, and some engineers, are particularly concerned with *constrained optimization*, where there are *a priori* limitations on the allowed values of independent variables. For example, the production of wheat in the U.S. must be a non-negative number. One particularly well-developed area of constrained optimization is *linear programming*, where both the function to be optimized and the constraints happen to be linear functions of the independent variables. Section 10.8, which is otherwise somewhat

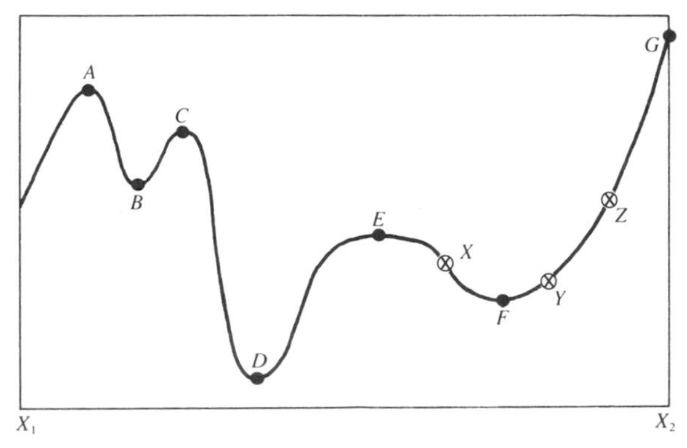

Figure 10.0.1. Extrema of a function in an interval. Points A, C, and E are local, but not global maxima. Points B and F are local, but not global minima. The global maximum occurs at G, which is on the boundary of the interval so that the derivative of the function need not vanish there. The global minimum is at D. At point E, derivatives higher than the first vanish, a situation which can cause difficulty for some algorithms. The points X, Y, and Z are said to "bracket" the minimum F, since Y is less than both X and Z.

disconnected from the rest of the material that we have chosen to include in this chapter, implements the so-called "simplex algorithm" for linear programming problems.

One other section, §10.9, also lies outside of our main thrust, but for exactly the opposite reason: so-called "annealing methods" are so new that we do not yet know where they will ultimately fit into the scheme of things. We do know that they have solved some problems previously thought to be practically insoluble, and that they have a direct bearing on the problem of finding global extrema in the presence of large numbers of undesired local extrema. We have therefore chosen to include some discussion of this new area.

The other sections in this chapter constitute a selection of the best established algorithms in unconstrained minimization. (For definiteness, we will henceforth regard the optimization problem as that of minimization.) These sections are connected, with later ones depending on earlier ones. If you are just looking for the one "perfect" algorithm to solve your particular application, you may feel that we are telling you more than you want to know. Unfortunately, there is *no* perfect optimization algorithm. This is a case where we strongly urge you to try more than one method in comparative fashion. Your initial choice of method can be based on the following considerations:

- You must choose between methods that need only evaluations of the function to be minimized and methods that also require evaluations of the derivative of that function. In the multidimensional case, this derivative is the gradient, a vector quantity. Algorithms using the derivative are somewhat more powerful than those using only the function, but not always enough so as to compensate for the additional calculations of derivatives. We can easily construct examples favoring one approach or favoring the other. However, if you *can* compute derivatives, be prepared to try using them.

- For one-dimensional minimization (minimize a function of one variable) *without* calculation of the derivative, bracket the minimum as described in §10.1, and then use *Brent's method* as described in §10.2. If your function has a discontinuous second (or lower) derivative, then the parabolic interpolations of Brent's method are of no advantage, and you might wish to use the simplest form of *golden section search*, as described in §10.1.
- For one-dimensional minimization *with* calculation of the derivative, §10.3 supplies a variant of Brent's method which makes limited use of the first derivative information. We shy away from the alternative of using derivative information to construct high-order interpolating polynomials. In our experience the improvement in convergence very near a smooth, analytic minimum does not make up for the tendency of polynomials sometimes to give wildly wrong interpolations at early stages, especially for functions which may have sharp, "exponential" features.

We now turn to the multidimensional case, both with and without computation of first derivatives.

- You must choose between methods that require storage of order N^2 and those that require only of order N, where N is the number of dimensions. For moderate values of N and reasonable memory sizes this is not a serious constraint. There will be, however, the occasional application where storage may be critical.
- We give in §10.4 a somewhat neglected *downhill simplex method* due to Nelder and Mead. (This use of the word "simplex" is not to be confused with the simplex method of linear programming.) This method just crawls downhill in a straightforward fashion that makes almost no special assumptions about your function. This can be extremely slow, but it can also, in some cases, be extremely robust. Not to be overlooked is the fact that the code is concise and completely self-contained: a general N-dimensional minimization program in under 100 program lines! This method is most useful when the minimization calculation is only an incidental part of your overall problem. The storage requirement is of order N^2, and derivative calculations are not required.
- Section 10.5 deals with *direction-set methods*, of which *Powell's method* is the prototype. These are the methods of choice when you cannot easily calculate derivatives, and are not necessarily to be sneered at even if you can. Although derivatives are not needed, the method does require a one-dimensional minimization sub-algorithm such as Brent's method (see above). Storage is of order N^2.

There are two major families of algorithms for multidimensional minimization *with* calculation of first derivatives. Both families require a one-dimensional minimization sub-algorithm, which can itself either use, or not

use, the derivative information, as you see fit (depending on the relative effort of computing the function and of its gradient vector). We do not think that either family dominates the other in all applications; you should think of them as available alternatives:

- The first family goes under the name *conjugate gradient methods*, as typified by the *Fletcher-Reeves algorithm* and the closely related and probably superior *Polak-Ribiere algorithm*. Conjugate gradient methods require only of order a few times N storage, require derivative calculations and one-dimensional sub-minimization. Turn to §10.6 for detailed discussion and implementation.
- The second family goes under the names *quasi-Newton* or *variable metric* methods, as typified by the *Davidon-Fletcher-Powell (DFP)* algorithm (sometimes referred to just as *Fletcher-Powell*) or the closely related *Broyden-Fletcher-Goldfarb-Shanno (BFGS)* algorithm. These methods require of order N^2 storage, require derivative calculations and one-dimensional sub-minimization. Details are in §10.7.

You are now ready to proceed with scaling the peaks (and/or plumbing the depths) of practical optimization.

REFERENCES AND FURTHER READING:

Acton, Forman S. 1970, *Numerical Methods That Work* (New York: Harper and Row), Chapter 17.

Jacobs, David A.H., ed. 1977, *The State of the Art in Numerical Analysis* (London: Academic Press), Chapter III.1.

Brent, Richard P. 1973, *Algorithms for Minimization without Derivatives* (Englewood Cliffs, N.J.: Prentice-Hall).

Dahlquist, Germund, and Bjorck, Ake. 1974, *Numerical Methods* (Englewood Cliffs, N.J.: Prentice-Hall), Chapter 10.

Polak, E. 1971, *Computational Methods in Optimization* (New York: Academic Press).

10.1 Golden Section Search in One Dimension

Recall how the bisection method finds roots of functions in one dimension (§9.2): The root is supposed to have been bracketed in an interval (a, b). One then evaluates the function at an intermediate point x and obtains a new, smaller bracketing interval, either (a, x) or (x, b). The process continues until the bracketing interval is acceptably small. It is optimal to choose x to be the midpoint of (a, b) so that the decrease in the interval length is maximized when the function is as uncooperative as it can be, i.e., when the luck of the draw forces you to take the bigger bisected segment.

There is a precise, though slightly subtle, translation of these considerations to the minimization problem: What does it mean to *bracket* a minimum?

A root of a function is known to be bracketed by a pair of points, $a < b$, when the function has opposite sign at those two points. A minimum, by contrast, is known to be bracketed only when there is a *triplet* of points, $a < b < c$, such that $f(b)$ is less than *both* $f(a)$ and $f(c)$. In this case we know that the function (if it is nonsingular) has a minimum in the interval (a, c).

The analog of bisection is to choose a new point x, either between a and b or between b and c. Suppose, to be specific, that we make the latter choice. Then we evaluate $f(x)$. If $f(b) < f(x)$, then the new bracketing triplet of points is $a < b < x$; contrariwise, if $f(b) > f(x)$, then the new bracketing triplet is $b < x < c$. In all cases the middle point of the new triplet is the abscissa whose ordinate is the best minimum achieved so far; see Figure 10.1.1. We continue the process of bracketing until the distance between the two outer points of the triplet is tolerably small.

How small is "tolerably" small? For a minimum located at a value b, you might naively think that you will be able to bracket it in as small a range as $(1 - \epsilon)b < b < (1 + \epsilon)b$, where ϵ is your computer's floating point precision, a number like 3×10^{-8} (single precision) or 10^{-15} (double precision). Not so! In general, the shape of your function $f(x)$ near b will be given by Taylor's theorem

$$f(x) \approx f(b) + \frac{1}{2} f''(b)(x - b)^2 \tag{10.1.1}$$

The second term will be negligible compared to the first (that is, will be a factor ϵ smaller and will act just like zero when added to it) whenever

$$|x - b| < \sqrt{\epsilon}\, b \sqrt{\frac{2\,|f(b)|}{b^2 f''(b)}} \tag{10.1.2}$$

The reason for writing the right-hand side in this way is that, for most functions, the final square root is a number of order unity. Therefore, as a rule of thumb, it is hopeless to ask for a bracketing interval of width less than $\sqrt{\epsilon}$ times its central value, a fractional width of only about 10^{-4} (single precision) or 3×10^{-8} (double precision). Knowing this inescapable fact will save you a lot of useless bisections!

The minimum-finding routines of this chapter will often call for a user-supplied argument `TOL`, and return with an abscissa whose fractional precision is about $\pm$`TOL` (bracketing interval of fractional size about 2×`TOL`). Unless you have a better estimate for the right-hand side of equation (10.1.2), you should set `TOL` equal to (not much less than) the square root of your machine's floating precision, since smaller values will gain you nothing.

It remains to decide on a strategy for choosing the new point x, given a, b, c. Suppose that b is a fraction W of the way between a and c, i.e.

$$\frac{b - a}{c - a} = W \qquad \frac{c - b}{c - a} = 1 - W \tag{10.1.3}$$

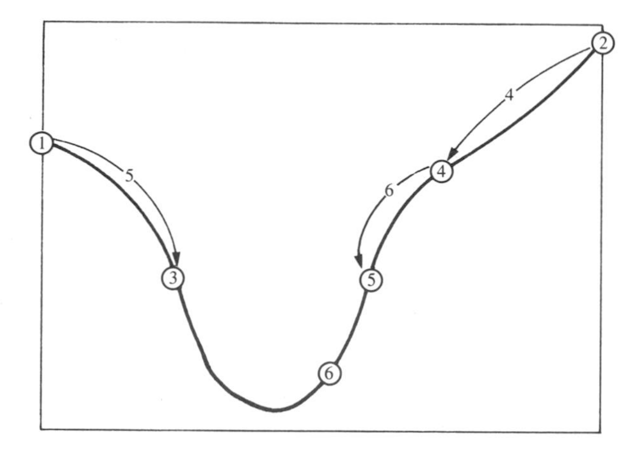

Figure 10.1.1. Successive bracketing of a minimum. The minimum is originally bracketed by points 1,3,2. The function is evaluated at 4, which replaces 2; then at 5, which replaces 1; then at 6, which replaces 4. The rule at each stage is to keep a center point that is lower than the two outside points. After the steps shown, the minimum is bracketed by points 3,6,5.

Also suppose that our next trial point x is an additional fraction Z beyond b,

$$\frac{x-b}{c-a} = Z \tag{10.1.4}$$

Then the next bracketing segment will either be of length $W + Z$ relative to the current one, or else of length $1 - W$. If we want to minimize the worst case possibility, then we will choose Z to make these equal, namely

$$Z = 1 - 2W \tag{10.1.5}$$

We see at once that the new point is the symmetric point to b in the interval, namely with $|b - a|$ equal to $|x - c|$. This implies that the point x lies in the larger of the two segments (Z is positive only if $W < 1/2$).

But where in the larger segment? Where did the value of W itself come from? Presumably from the previous stage of applying our same strategy. Therefore, if Z is chosen to be optimal, then so was W before it. This *scale similarity* implies that x should be the same fraction of the way from b to c (if that is the bigger segment) as was b from a to c, in other words,

$$\frac{Z}{1-W} = W \tag{10.1.6}$$

Equations (10.1.5) and (10.1.6) yield the quadratic equation

$$W^2 - 3W + 1 = 0 \qquad \text{yielding} \qquad W = \frac{3-\sqrt{5}}{2} \approx 0.38197 \tag{10.1.7}$$

In other words, the optimal bracketing interval $a < b < c$ has its middle point b a fractional distance 0.38197 from one end (say, a), and 0.61803 from the other end (say, b). These fractions are those of the so-called *golden mean* or *golden section*, whose supposedly aesthetic properties hark back to the ancient Pythagoreans. This optimal method of function minimization, the analog of the bisection method for finding zeros, is thus called the *golden section search*, summarized as follows:

Given, at each stage, a bracketing triplet of points, the next point to be tried is that which is a fraction 0.38197 into the larger of the two intervals (measuring from the central point of the triplet). If you start out with a bracketing triplet whose segments are not in the golden ratios, the procedure of choosing successive points at the golden mean point of the larger segment will quickly converge you to the proper, self-replicating ratios.

The golden section search guarantees that each new function evaluation will (after self-replicating ratios have been achieved) bracket the minimum to an interval just 0.61803 times the size of the preceding interval. This is comparable to, but not quite as good as, the 0.50000 which holds when finding roots by bisection. Note that the convergence is *linear* (in the language of Chapter 9), meaning that successive significant figures are won linearly with additional function evaluations. In the next section we will give a superlinear method, where the rate at which successive significant figures are liberated increases with each successive function evaluation.

Routine for Initially Bracketing a Minimum

The preceding discussion has assumed that you are able to bracket the minimum in the first place. We consider this initial bracketing to be an essential part of any one-dimensional minimization. There are some one-dimensional algorithms that do not require a rigorous initial bracketing. However, we would *never* trade the secure feeling of *knowing* that a minimum is "in there somewhere" for the dubious reduction of function evaluations that these non-bracketing routines may promise. Please bracket your minima (or, for that matter, your zeros) before isolating them!

There is not much theory as to how to do this bracketing. Obviously you want to step downhill. But how far? We like to take larger and larger steps, starting with some (wild?) initial guess and then increasing the step size at each step either by a constant factor, or else by the result of a parabolic extrapolation of the preceding points that is designed to take us to the extrapolated turning point. It doesn't much matter if the steps get big. After all, we are stepping downhill, so we already have the left and middle points of the bracketing triplet. We just need to take a big enough step to stop the downhill trend and get a high third point.

Our standard routine is this:

```
      SUBROUTINE MNBRAK(AX,BX,CX,FA,FB,FC,FUNC)
```

Given a function FUNC, and given distinct initial points AX and BX, this routine searches in the downhill direction (defined by the function as evaluated at the initial points) and returns new points AX, BX, CX which bracket a minimum of the function. Also returned are the function values at the three points, FA, FB, and FC.

```
      PARAMETER (GOLD=1.618034, GLIMIT=100., TINY=1.E-20)
```

The first parameter is the default ratio by which successive intervals are magnified; the second is the maximum magnification allowed for a parabolic-fit step.

```
      FA=FUNC(AX)
      FB=FUNC(BX)
      IF(FB.GT.FA)THEN              Switch roles of A and B so that we can go downhill in the direction
        DUM=AX                          from A to B.
        AX=BX
        BX=DUM
        DUM=FB
        FB=FA
        FA=DUM
      ENDIF
      CX=BX+GOLD*(BX-AX)            First guess for C.
      FC=FUNC(CX)
1     IF(FB.GE.FC)THEN              "DO WHILE": keep returning here until we bracket.
        R=(BX-AX)*(FB-FC)           Compute U by parabolic extrapolation from A,B,C. TINY is used to
        Q=(BX-CX)*(FB-FA)               prevent any possible division by zero.
        U=BX-((BX-CX)*Q-(BX-AX)*R)/(2.*SIGN(MAX(ABS(Q-R),TINY),Q-R))
        ULIM=BX+GLIMIT*(CX-BX)      We won't go farther than this. Now to test various possibilities:
        IF((BX-U)*(U-CX).GT.0.)THEN     Parabolic U is between B and C: try it.
          FU=FUNC(U)
          IF(FU.LT.FC)THEN          Got a minimum between B and C.
            AX=BX
            FA=FB
            BX=U
            FB=FU
            GO TO 1                 (which will exit).
          ELSE IF(FU.GT.FB)THEN         Got a minimum between between A and U.
            CX=U
            FC=FU
            GO TO 1                 (which will exit).
          ENDIF
          U=CX+GOLD*(CX-BX)         Parabolic fit was no use. Use default magnification.
          FU=FUNC(U)
        ELSE IF((CX-U)*(U-ULIM).GT.0.)THEN      Parabolic fit is between C and its allowed limit.
          FU=FUNC(U)
          IF(FU.LT.FC)THEN
            BX=CX
            CX=U
            U=CX+GOLD*(CX-BX)
            FB=FC
            FC=FU
            FU=FUNC(U)
          ENDIF
        ELSE IF((U-ULIM)*(ULIM-CX).GE.0.)THEN       Limit parabolic U to maximum allowed value.
          U=ULIM
          FU=FUNC(U)
        ELSE                        Reject parabolic U, use default magnification.
          U=CX+GOLD*(CX-BX)
          FU=FUNC(U)
        ENDIF
        AX=BX                       Eliminate oldest point and continue.
        BX=CX
        CX=U
        FA=FB
        FB=FC
        FC=FU
        GO TO 1
```

```
ENDIF
RETURN
END
```

(Because of the housekeeping involved in moving around three or four points and their function values, the above program ends up looking deceptively formidable. That is true of several other programs in this chapter, be advised. The underlying ideas are quite simple.)

Routine for Golden Section Search

```
      FUNCTION GOLDEN(AX,BX,CX,F,TOL,XMIN)
         Given a function F, and given a bracketing triplet of abscissas AX, BX, CX (such that BX is
         between AX and CX, and F(BX) is less than both F(AX) and F(CX)), this routine performs
         a golden section search for the minimum, isolating it to a fractional precision of about
         TOL. The abscissa of the minimum is returned as XMIN, and the minimum function value
         is returned as GOLDEN, the returned function value.
      PARAMETER (R=.61803399,C=1.-R)   Golden ratios.
      X0=AX                            At any given time we will keep track of four points, X0,X1,X2,X3.
      X3=CX
      IF(ABS(CX-BX).GT.ABS(BX-AX))THEN    Make X0 to X1 the smaller segment,
          X1=BX
          X2=BX+C*(CX-BX)                 and fill in the new point to be tried.
      ELSE
          X2=BX
          X1=BX-C*(BX-AX)
      ENDIF
      F1=F(X1)                         The initial function evaluations. Note that we never need to evaluate
      F2=F(X2)                            the function at the original endpoints.
1     IF(ABS(X3-X0).GT.TOL*(ABS(X1)+ABS(X2)))THEN   Do-while loop: we keep returning here.
          IF(F2.LT.F1)THEN                One possible outcome,
              X0=X1                       its housekeeping,
              X1=X2
              X2=R*X1+C*X3
              F0=F1
              F1=F2
              F2=F(X2)                    and a new function evaluation.
          ELSE                            The other outcome,
              X3=X2
              X2=X1
              X1=R*X2+C*X0
              F3=F2
              F2=F1
              F1=F(X1)                    and its new function evaluation.
          ENDIF
      GOTO 1                           Back to see if we are done.
      ENDIF
      IF(F1.LT.F2)THEN                 We are done. Output the best of the two current values.
          GOLDEN=F1
          XMIN=X1
      ELSE
          GOLDEN=F2
          XMIN=X2
      ENDIF
      RETURN
      END
```

10.2 Parabolic Interpolation and Brent's Method in One-Dimension

We already tipped our hand about the desirability of parabolic interpolation in the previous section's MNBRAK routine, but it is now time to be more explicit. A golden section search is designed to handle, in effect, the worst possible case of function minimization, with the uncooperative minimum hunted down and cornered like a scared rabbit. But why assume the worst? If the function is nicely parabolic near to the minimum — surely the generic case for sufficiently smooth functions — then the parabola fitted through any three points ought to take us in a single leap to the minimum, or at least very near to it (see Figure 10.2.1). Since we want to find an abscissa rather than an ordinate, the procedure is technically called *inverse parabolic interpolation.*

The formula for the abscissa x which is the minimum of a parabola through three points $f(a)$, $f(b)$, and $f(c)$ is

$$x = b + \frac{1}{2}\frac{(b-a)^2[f(b)-f(c)]-(b-c)^2[f(b)-f(a)]}{(b-a)[f(b)-f(c)]-(b-c)[f(b)-f(a)]} \tag{10.2.1}$$

as you can easily derive. This formula fails only if the three points are collinear, in which case the denominator is zero (minimum of the parabola is infinitely far away). Note, however, that (10.2.1) is as happy jumping to a parabolic maximum as to a minimum. No minimization scheme that depends solely on (10.2.1) is likely to succeed in practice.

The exacting task is to invent a scheme which relies on a sure-but-slow technique, like golden section search, when the function is not cooperative, but which switches over to (10.2.1) when the function allows. The task is nontrivial for several reasons, including these: (i) The housekeeping needed to avoid unnecessary function evaluations in switching between the two methods can be complicated. (ii) Careful attention must be given to the "endgame," where the function is being evaluated very near to the roundoff limit of equation (10.1.2). (iii) The scheme for detecting a cooperative versus noncooperative function must be very robust.

Brent's method (Brent, 1973) is up to the task in all particulars. At any particular stage, it is keeping track of six function points (not necessarily all distinct), a, b, u, v, w and x, defined as follows: the minimum is bracketed between a and b; x is the point with the very least function value found so far (or the most recent one in case of a tie); w is the point with the second least function value; v is the previous value of w; u is the point at which the function was evaluated most recently. Also appearing in the algorithm is the point x_m, the midpoint between a and b; however the function is not evaluated there.

You can read the code below to understand the method's logical organization. Mention of a few general principles here may, however, be helpful: Parabolic interpolation is attempted, fitting through the points x, v, and w. To be acceptable, the parabolic step must (i) fall within the bounding interval (a, b), and (ii) imply a movement from the best current value x that is *less*

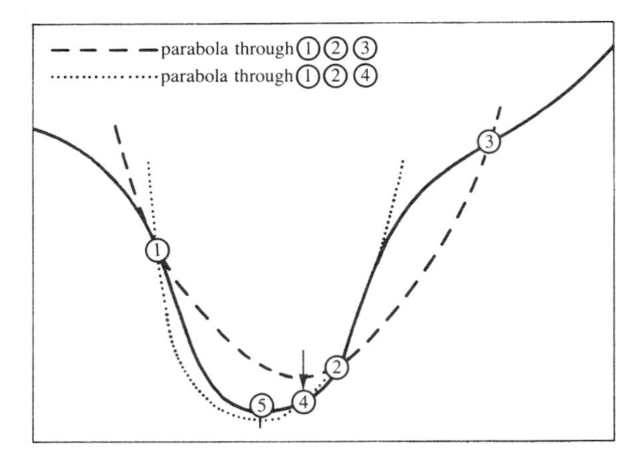

Figure 10.2.1. Convergence to a minimum by inverse parabolic interpolation. A parabola (dashed line) is drawn through the three original points 1,2,3 on the given function (solid line). The function is evaluated at the parabola's minimum, 4, which replaces point 3. A new parabola (dotted line) is drawn through points 1,4,2. The minimum of this parabola is at 5, which is close to the minimum of the function.

than half the movement of the *step before last.* This second criterion insures that the parabolic steps are actually converging to something, rather than, say, bouncing around in some nonconvergent limit cycle. In the worst possible case, where the parabolic steps are acceptable but useless, the method will approximately alternate between parabolic steps and golden sections, converging in due course by virtue of the latter. The reason for comparing to the step *before* last seems essentially heuristic: experience shows that it is better not to "punish" the algorithm for a single bad step if it can make it up on the next one.

Another principle exemplified in the code is never to evaluate the function less than a distance TOL from a point already evaluated (or from a known bracketing point). The reason is that, as we saw in equation (10.1.2), there is simply no information content in doing so: the function will differ from the value already evaluated only by an amount of order the roundoff error. Therefore in the code below you will find several tests and modifications of a potential new point, imposing this restriction. This restriction also interacts subtly with the test for "doneness," which the method takes into account.

A typical ending configuration for Brent's method is that a and b are $2 \times x \times$ TOL apart, with x (the best abscissa) at the midpoint of a and b, and therefore fractionally accurate to $\pm$TOL.

Indulge us a final reminder that TOL should generally be no smaller than the square root of your machine's floating point precision.

```
FUNCTION BRENT(AX,BX,CX,F,TOL,XMIN)
   Given a function F, and given a bracketing triplet of abscissas AX, BX, CX (such that BX is
   between AX and CX, and F(BX) is less than both F(AX) and F(CX)), this routine isolates
   the minimum to a fractional precision of about TOL using Brent's method. The abscissa of
   the minimum is returned as XMIN, and the minimum function value is returned as BRENT,
   the returned function value.
PARAMETER (ITMAX=100,CGOLD=.3819660,ZEPS=1.0E-10)
```

```
   Maximum allowed number of iterations; golden ratio; and a small number which protects
   against trying to achieve fractional accuracy for a minimum that happens to be exactly
   zero.
      A=MIN(AX,CX)                  A and B must be in ascending order, though the input abscissas need
      B=MAX(AX,CX)                      not be.
      V=BX                          Initializations...
      W=V
      X=V
      E=0.                          This will be the distance moved on the step before last.
      FX=F(X)
      FV=FX
      FW=FX
      DO 11 ITER=1,ITMAX            Main program loop.
        XM=0.5*(A+B)
        TOL1=TOL*ABS(X)+ZEPS
        TOL2=2.*TOL1
        IF(ABS(X-XM).LE.(TOL2-.5*(B-A))) GOTO 3 Test for done here.
        IF(ABS(E).GT.TOL1) THEN            Construct a trial parabolic fit.
          R=(X-W)*(FX-FV)
          Q=(X-V)*(FX-FW)
          P=(X-V)*Q-(X-W)*R
          Q=2.*(Q-R)
          IF(Q.GT.0.) P=-P
          Q=ABS(Q)
          ETEMP=E
          E=D
          IF(ABS(P).GE.ABS(.5*Q*ETEMP).OR.P.LE.Q*(A-X).OR.
     *        P.GE.Q*(B-X)) GOTO 1
   The above conditions determine the acceptability of the parabolic fit. Here it is o.k.:
          D=P/Q                     Take the parabolic step.
          U=X+D
          IF(U-A.LT.TOL2 .OR. B-U.LT.TOL2) D=SIGN(TOL1,XM-X)
          GOTO 2                    Skip over the golden section step.
        ENDIF
1       IF(X.GE.XM) THEN            We arrive here for a golden section step, which we take into the larger
          E=A-X                         of the two segments.
        ELSE
          E=B-X
        ENDIF
        D=CGOLD*E                   Take the golden section step.
2       IF(ABS(D).GE.TOL1) THEN     Arrive here with D computed either from parabolic fit, or else from
          U=X+D                         golden section.
        ELSE
          U=X+SIGN(TOL1,D)
        ENDIF
        FU=F(U)                     This is the one function evaluation per iteration,
        IF(FU.LE.FX) THEN           and now we have to decide what to do with our function evaluation.
          IF(U.GE.X) THEN               Housekeeping follows:
            A=X
          ELSE
            B=X
          ENDIF
          V=W
          FV=FW
          W=X
          FW=FX
          X=U
          FX=FU
        ELSE
          IF(U.LT.X) THEN
            A=U
          ELSE
            B=U
          ENDIF
```

```
            IF(FU.LE.FW .OR. W.EQ.X) THEN
                V=W
                FV=FW
                W=U
                FW=FU
            ELSE IF(FU.LE.FV .OR. V.EQ.X .OR. V.EQ.W) THEN
                V=U
                FV=FU
            ENDIF
          ENDIF                        Done with housekeeping. Back for another iteration.
11      CONTINUE
        PAUSE 'Brent exceed maximum iterations.'
3       XMIN=X                         Arrive here ready to exit with best values.
        BRENT=FX
        RETURN
        END
```

REFERENCES AND FURTHER READING:

Brent, Richard P. 1973, *Algorithms for Minimization without Derivatives* (Englewood Cliffs, N.J.: Prentice-Hall), Chapter 5.

Forsythe, George E., Malcolm, Michael A., and Moler, Cleve B. 1977, *Computer Methods for Mathematical Computations* (Englewood Cliffs, N.J.: Prentice-Hall), §8.2.

10.3 One-Dimensional Search with First Derivatives

Here we want to accomplish precisely the same goal as in the previous section, namely to isolate a functional minimum that is bracketed by the triplet of abscissas $a < b < c$, but utilizing an additional capability to compute the function's first derivative as well as its value.

In principle, we might simply search for a zero of the derivative, ignoring the function value information, using a root finder like RTFLSP or ZBRENT (§9.2). It doesn't take long to reject *that* idea: How do we distinguish maxima from minima? Where do we go from initial conditions where the derivatives on one or both of the outer bracketing points indicate that "downhill" is in the direction *out* of the bracketed interval?

We don't want to give up our strategy of maintaining a rigorous bracket on the minimum at all times. The only way to keep such a bracket is to update it using function (not derivative) information, with the central point in the bracketing triplet always that with the lowest function value. Therefore the role of the derivatives can only be to help us choose new trial points within the bracket.

One school of thought is to "use everything you've got": Compute a polynomial of relatively high order (cubic or above) which agrees with some number of previous function and derivative evaluations. For example, there is a unique cubic that agrees with function and derivative at two points,

and one can jump to the interpolated minimum of that cubic (if there is a minimum within the bracket). Suggested by Davidon and others, formulas for this tactic are given in Acton.

We like to be more conservative than this. Once superlinear convergence sets in, it hardly matters whether its order is moderately lower or higher. In practical problems that we have met, most function evaluations are spent in getting globally close enough to the minimum for superlinear convergence to commence. So we are more worried about all the funny "stiff" things that high order polynomials can do (cf. Figure 3.0.1b), and about their sensitivities to roundoff error.

This leads us to use derivative information only as follows: The sign of the derivative at the central point of the bracketing triplet $a < b < c$ indicates uniquely whether the next test point should be taken in the interval (a, b) or in the interval (b, c). The value of this derivative and of the derivative at the second-best-so-far point are extrapolated to zero by the secant method (inverse linear interpolation), which by itself is superlinear of order 1.618. (The golden mean again: see Acton, p. 57.) We impose the same sort of restrictions on this new trial point as in Brent's method. If the trial point must be rejected, we *bisect* the interval under scrutiny.

Yes, we are fuddy-duddies when it comes to making flamboyant use of derivative information in one-dimensional minimization. But we have had a bellyful of functions whose computed "derivatives" *don't* integrate up to the function value and *don't* accurately point the way to the minimum, usually because of roundoff errors, sometimes because of truncation error in the method of derivative evaluation.

You will see that the following routine is closely modeled on BRENT in the previous section.

```
FUNCTION DBRENT(AX,BX,CX,F,DF,TOL,XMIN)
```

Given a function **F** and its derivative function **DF**, and given a bracketing triplet of abscissas **AX**, **BX**, **CX** [such that **BX** is between **AX** and **CX**, and **F(BX)** is less than both **F(AX)** and **F(CX)**], this routine isolates the minimum to a fractional precision of about **TOL** using a modification of Brent's method that uses derivatives. The abscissa of the minimum is returned as **XMIN**, and the minimum function value is returned as **DBRENT**, the returned function value.

```
PARAMETER (ITMAX=100,ZEPS=1.0E-10)
```

Comments following will point out only differences from the routine **BRENT**. Read that routine first.

```
LOGICAL OK1,OK2                 Will be used as flags for whether proposed steps are acceptable or
A=MIN(AX,CX)                        not.
B=MAX(AX,CX)
V=BX
W=V
X=V
E=0.
FX=F(X)
FV=FX
FW=FX
DX=DF(X)                        All our housekeeping chores are doubled by the necessity of moving
DV=DX                               derivative values around as well as function values.
DW=DX
DO 11 ITER=1,ITMAX
    XM=0.5*(A+B)
```

```
          TOL1=TOL*ABS(X)+ZEPS
          TOL2=2.*TOL1
          IF(ABS(X-XM).LE.(TOL2-.5*(B-A))) GOTO 3
          IF(ABS(E).GT.TOL1) THEN
              D1=2.*(B-A)                Initialize these D's to an out-of-bracket value.
              D2=D1
              IF(DW.NE.DX) D1=(W-X)*DX/(DX-DW)     Secant method.
              IF(DV.NE.DX) D2=(V-X)*DX/(DX-DV)     Secant method with the other stored point.
```

Which of these two estimates of D shall we take? We will insist that they be within the bracket, and on the side pointed to by the derivative at X:

```
              U1=X+D1
              U2=X+D2
              OK1=((A-U1)*(U1-B).GT.O.).AND.(DX*D1.LE.O.)
              OK2=((A-U2)*(U2-B).GT.O.).AND.(DX*D2.LE.O.)
              OLDE=E                     Movement on the step before last.
              E=D
              IF(.NOT.(OK1.OR.OK2))THEN     Take only an acceptable D, and if both are acceptable, then
                  GO TO 1                   take the smallest one.
              ELSE IF (OK1.AND.OK2)THEN
                  IF(ABS(D1).LT.ABS(D2))THEN
                      D=D1
                  ELSE
                      D=D2
                  ENDIF
              ELSE IF (OK1)THEN
                  D=D1
              ELSE
                  D=D2
              ENDIF
              IF(ABS(D).GT.ABS(0.5*OLDE))GO TO 1
              U=X+D
              IF(U-A.LT.TOL2 .OR. B-U.LT.TOL2) D=SIGN(TOL1,XM-X)
              GOTO 2
          ENDIF
1         IF(DX.GE.O.)  THEN             Decide which segment by the sign of the derivative.
              E=A-X
          ELSE
              E=B-X
          ENDIF
          D=0.5*E                        Bisect, not golden section.
2         IF(ABS(D).GE.TOL1) THEN
              U=X+D
              FU=F(U)
          ELSE
              U=X+SIGN(TOL1,D)
              FU=F(U)
              IF(FU.GT.FX)GO TO 3   If the minimum step in the downhill direction takes us uphill, then we
          ENDIF                           are done.
          DU=DF(U)                       Now all the housekeeping, sigh.
          IF(FU.LE.FX) THEN
              IF(U.GE.X) THEN
                  A=X
              ELSE
                  B=X
              ENDIF
              V=W
              FV=FW
              DV=DW
              W=X
              FW=FX
              DW=DX
              X=U
              FX=FU
              DX=DU
```

```
        ELSE
            IF(U.LT.X) THEN
                A=U
            ELSE
                B=U
            ENDIF
            IF(FU.LE.FW .OR. W.EQ.X) THEN
                V=W
                FV=FW
                DV=DW
                W=U
                FW=FU
                DW=DU
            ELSE IF(FU.LE.FV .OR. V.EQ.X .OR. V.EQ.W) THEN
                V=U
                FV=FU
                DV=DU
            ENDIF
        ENDIF
11    CONTINUE
      PAUSE 'DBRENT exceeded maximum iterations.'
3     XMIN=X
      DBRENT=FX
      RETURN
      END
```

REFERENCES AND FURTHER READING:

Acton, Forman S. 1970, *Numerical Methods That Work* (New York: Harper and Row), p. 55, pp. 454–458.

Brent, Richard P. 1973, *Algorithms for Minimization without Derivatives* (Englewood Cliffs, N.J.: Prentice-Hall), p. 78.

10.4 Downhill Simplex Method in Multidimensions

With this section we begin consideration of multidimensional minimization, that is, finding the minimum of a function of more than one independent variable. This section stands apart from those which follow, however: All of the algorithms after this section will make explicit use of the one-dimensional minimization algorithms of §10.1, §10.2, or §10.3 as a part of their computational strategy. This section implements an entirely self-contained strategy, in which one-dimensional minimization does not figure.

The *downhill simplex method* is due to Nelder and Mead (1965). The method requires only function evaluations, not derivatives. It is not very efficient in terms of the number of function evaluations that it requires. Powell's method (§10.5) is almost surely faster in all likely applications. However the downhill simplex method may frequently be the *best* method to use if the figure of merit is "get something working quickly" for a problem whose computational burden is small.

The method has a geometrical naturalness about it which makes it delightful to describe or work through:

A *simplex* is the geometrical figure consisting, in N dimensions, of $N+1$ points (or vertices) and all their interconnecting line segments, polygonal faces, etc. In two dimensions, a simplex is a triangle. In three dimensions it is a tetrahedron, not necessarily the regular tetrahedron. (The *simplex method* of linear programming also makes use of the geometrical concept of a simplex. Otherwise it is completely unrelated to the algorithm that we are describing in this section.) In general we are only interested in simplexes that are nondegenerate, i.e. which enclose a finite inner N-dimensional volume. If any point of a nondegenerate simplex is taken as the origin, then the N other points define vector directions that span the N-dimensional vector space.

In one-dimensional minimization, it was possible to bracket a minimum, so that the success of a subsequent isolation was guaranteed. Alas! There is no analogous procedure in multidimensional space. For multidimensional minimization, the best we can do is give our algorithm a starting guess, that is, an N-vector of independent variables as the first point to try. The algorithm is then supposed to make its own way downhill through the unimaginable complexity of an N-dimensional topography, until it encounters an (at least local) minimum.

The downhill simplex method must be started not just with a single point, but with $N+1$ points, defining an initial simplex. If you think of one of these points (it matters not which) as being your initial starting point $\mathbf{P}_0$, then you can take the other N points to be

$$\mathbf{P}_i = \mathbf{P}_0 + \lambda \mathbf{e}_i \tag{10.4.1}$$

where the $\mathbf{e}_i$'s are N unit vectors, and where λ is a constant which is your guess of the problem's characteristic length scale. (Or, you could have different λ_i's for each vector direction.)

The downhill simplex method now takes a series of steps, most steps just moving the point of the simplex where the function is largest ("highest point") through the opposite face of the simplex to a lower point. These steps are called reflections, and they are constructed to conserve the volume of the simplex (hence maintain its nondegeneracy). When it can do so, the method expands the simplex in one or another direction to take larger steps. When it reaches a "valley floor," the method contracts itself in the transverse direction and tries to ooze down the valley. If there is a situation where the simplex is trying to "pass through the eye of a needle," it contracts itself in all directions, pulling itself in around its lowest (best) point. The routine name `AMOEBA` is intended to be descriptive of this kind of behavior; the basic moves are summarized in Figure 10.4.1.

Termination criteria can be delicate in any multidimensional minimization routine. Without bracketing, and with more than one independent variable, we no longer have the option of requiring a certain tolerance for a single independent variable. We typically can identify one "cycle" or "step" of our multidimensional algorithm. It is then possible to terminate when the vector

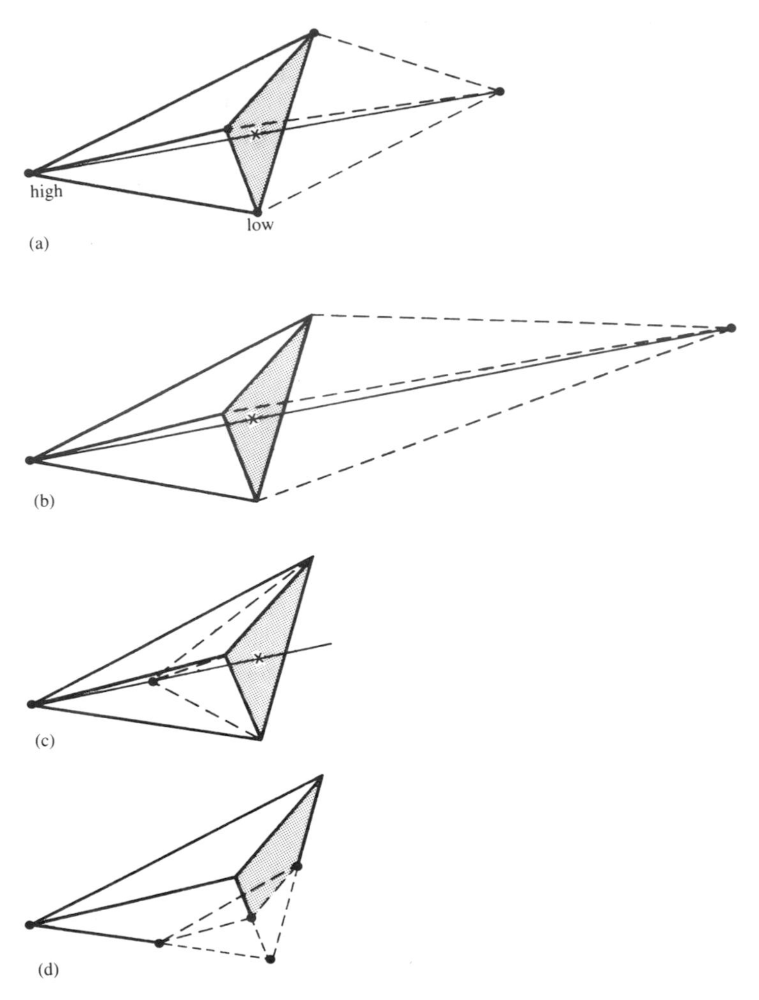

Figure 10.4.1. Possible outcomes for a step in the downhill simplex method. The simplex at the beginning of the step, here a tetrahedron, is drawn with solid lines. The simplex at the end of the step (drawn dashed) can be either (a) a reflection away from the high point, (b) a reflection and expansion away from the high point, (c) a contraction along one dimension from the high point, or (d) a contraction along all dimensions toward the low point. An appropriate sequence of such steps will always converge to a minimum of the function.

distance moved in that step is fractionally smaller in magnitude than some tolerance TOL. Alternatively, we could require that the decrease in the function value in the terminating step be fractionally smaller than some tolerance FTOL. Note that while TOL should not usually be smaller than the square root of the machine precision, it is perfectly appropriate to let FTOL be of order the machine precision (or perhaps slightly larger so as not to be diddled by roundoff).

Note well that either of the above criteria might be fooled by a single anomalous step that, for one reason or another, failed to get anywhere. Therefore, it is frequently a good idea to *restart* a multidimensional minimiza-

tion routine at a point where it claims to have found a minimum. For this restart, you should reinitialize any ancillary input quantities. In the downhill simplex method, for example, you should reinitialize N of the $N+1$ vertices of the simplex again by equation (10.4.1), with $\mathbf{P}_0$ being one of the vertices of the claimed minimum.

Restarts should never be very expensive; your algorithm did, after all, converge to the restart point once, and now you are starting the algorithm already there.

Consider, then, our N-dimensional amoeba:

```
      SUBROUTINE AMOEBA(P,Y,MP,NP,NDIM,FTOL,FUNK,ITER)
```

Multidimensional minimization of the function FUNK(X) where X is an NDIM-dimensional vector, by the downhill simplex method of Nelder and Mead. Input is a matrix P whose NDIM+1 rows are NDIM-dimensional vectors which are the vertices of the starting simplex. [Logical dimensions of P are P(NDIM+1,NDIM); physical dimensions are input as P(MP,NP)]. Also input is the vector Y of length NDIM+1, whose components must be pre-initialized to the values of FUNK evaluated at the NDIM+1 vertices (rows) of P; and FTOL the fractional convergence tolerance to be achieved in the function value (n.b.!). On output, P and Y will have been reset to NDIM+1 new points all within FTOL of a minimum function value, and ITER gives the number of iterations taken.

```
      PARAMETER (NMAX=20,ALPHA=1.0,BETA=0.5,GAMMA=2.0,ITMAX=500)
```

Expected maximum number of dimensions, three parameters which define the expansions and contractions, and maximum allowed number of iterations.

```
      DIMENSION P(MP,NP),Y(MP),PR(NMAX),PRR(NMAX),PBAR(NMAX)
      MPTS=NDIM+1              Note that MP is the physical dimension corresponding to the logical
      ITER=0                       dimension MPTS, NP to NDIM.
1     ILO=1                    First we must determine which point is the highest (worst), next-
      IF(Y(1).GT.Y(2))THEN         highest, and lowest (best),
        IHI=1
        INHI=2
      ELSE
        IHI=2
        INHI=1
      ENDIF
      DO 11 I=1,MPTS           by looping over the points in the simplex.
        IF(Y(I).LT.Y(ILO)) ILO=I
        IF(Y(I).GT.Y(IHI))THEN
          INHI=IHI
          IHI=I
        ELSE IF(Y(I).GT.Y(INHI))THEN
          IF(I.NE.IHI) INHI=I
        ENDIF
11    CONTINUE
```

Compute the fractional range from highest to lowest and return if satisfactory.

```
      RTOL=2.*ABS(Y(IHI)-Y(ILO))/(ABS(Y(IHI))+ABS(Y(ILO)))
      IF(RTOL.LT.FTOL)RETURN
      IF(ITER.EQ.ITMAX) PAUSE 'Amoeba exceeding maximum iterations.'
      ITER=ITER+1
      DO 12 J=1,NDIM
        PBAR(J)=0.
12    CONTINUE                 Begin a new iteration. Compute the vector average of all points except
      DO 14 I=1,MPTS               the highest, i.e. the center of the "face" of the simplex across
        IF(I.NE.IHI)THEN           from the high point. We will subsequently explore along the
          DO 13 J=1,NDIM           ray from the high point through that center.
            PBAR(J)=PBAR(J)+P(I,J)
13        CONTINUE
        ENDIF
14    CONTINUE
      DO 15 J=1,NDIM           Extrapolate by a factor ALPHA through the face, i.e. reflect the simplex
                                   from the high point.
```

```
    PBAR(J)=PBAR(J)/NDIM
    PR(J)=(1.+ALPHA)*PBAR(J)-ALPHA*P(IHI,J)
  15 CONTINUE
YPR=FUNK(PR)                     Evaluate the function at the reflected point.
IF(YPR.LE.Y(ILO))THEN            Gives a result better than the best point, so
    DO 16 J=1,NDIM               try an additional extrapolation by a factor GAMMA,
      PRR(J)=GAMMA*PR(J)+(1.-GAMMA)*PBAR(J)
    16 CONTINUE
    YPRR=FUNK(PRR)               and check out the function there.
    IF(YPRR.LT.Y(ILO))THEN       The additional extrapolation succeeded,
      DO 17 J=1,NDIM             and replaces the high point.
        P(IHI,J)=PRR(J)
      17 CONTINUE
      Y(IHI)=YPRR
    ELSE                         The additional extrapolation failed,
      DO 18 J=1,NDIM             but we can still use the reflected point.
        P(IHI,J)=PR(J)
      18 CONTINUE
      Y(IHI)=YPR
    ENDIF
ELSE IF(YPR.GE.Y(INHI))THEN      The reflected point is worse than the second-highest.
    IF(YPR.LT.Y(IHI))THEN        If it's better than the highest, then replace the highest,
      DO 19 J=1,NDIM
        P(IHI,J)=PR(J)
      19 CONTINUE
      Y(IHI)=YPR
    ENDIF
    DO 21 J=1,NDIM               but look for an intermediate lower point,
      PRR(J)=BETA*P(IHI,J)+(1.-BETA)*PBAR(J)
    21 CONTINUE                  in other words, perform a contraction of the simplex along one di-
    YPRR=FUNK(PRR)                   mension. Then evaluate the function.
    IF(YPRR.LT.Y(IHI))THEN       Contraction gives an improvement,
      DO 22 J=1,NDIM             so accept it.
        P(IHI,J)=PRR(J)
      22 CONTINUE
      Y(IHI)=YPRR
    ELSE                         Can't seem to get rid of that high point. Better contract around the
      DO 24 I=1,MPTS                 lowest (best) point.
        IF(I.NE.ILO)THEN
          DO 23 J=1,NDIM
            PR(J)=0.5*(P(I,J)+P(ILO,J))
            P(I,J)=PR(J)
          23 CONTINUE
          Y(I)=FUNK(PR)
        ENDIF
      24 CONTINUE
    ENDIF
ELSE                             We arrive here if the original reflection gives a middling point. Replace
    DO 25 J=1,NDIM                   the old high point and continue
      P(IHI,J)=PR(J)
    25 CONTINUE
    Y(IHI)=YPR
ENDIF
GO TO 1                          for the test of doneness and the next iteration.
END
```

REFERENCES AND FURTHER READING:

Nelder, J.A., and Mead, R. 1965, *Computer Journal*, vol. 7, p. 308.

Yarbro, L.A., and Deming, S.N. 1974, *Analytica Chim. Acta*, vol. 73, p. 391.

Jacoby, S.L.S, Kowalik, J.S., and Pizzo, J.T. 1972, *Iterative Methods for Nonlinear Optimization Problems* (Englewood Cliffs, N.J.: Prentice-Hall).

10.5 Direction Set (Powell's) Methods in Multidimensions

We know (§10.1–§10.3) how to minimize a function of one variable. If we start at a point $\mathbf{P}$ in N-dimensional space, and proceed from there in some vector direction $\mathbf{n}$, then any function of N variables $f(\mathbf{P})$ can be minimized along the line $\mathbf{n}$ by our one-dimensional methods. One can dream up various multidimensional minimization methods which consist of sequences of such line minimizations. Different methods will differ only by how, at each stage, they choose the next direction $\mathbf{n}$ to try. All such methods presume the existence of a "black-box" subalgorithm, which we might call `LINMIN` (given as an explicit routine at the end of this section), whose definition can be taken for now as

> `LINMIN`: Given as input the vectors $\mathbf{P}$ and $\mathbf{n}$, and the function f, find the scalar λ that minimizes $f(\mathbf{P} + \lambda\mathbf{n})$. Replace $\mathbf{P}$ by $\mathbf{P} + \lambda\mathbf{n}$. Replace $\mathbf{n}$ by $\lambda\mathbf{n}$. Done.

All the minimization methods in this section and in the two sections following fall under this general schema of successive line minimizations. In this section we consider a class of methods whose choice of successive directions does not involve explicit computation of the function's gradient; the next two sections do require such gradient calculations. You will note that we need not specify whether `LINMIN` uses gradient information or not. That choice is up to you, and its optimization depends on your particular function. You would be crazy, however, to use gradients in `LINMIN` and *not* use them in the choice of directions, since in this latter role they can drastically reduce the total computational burden.

But what if, in your application, calculation of the gradient is out of the question. You might first think of this simple method: Take the unit vectors $\mathbf{e}_1, \mathbf{e}_2, \dots \mathbf{e}_N$ as a *set of directions*. Using `LINMIN`, move along the first direction to its minimum, then *from there* along the second direction to *its* minimum, and so on, cycling through the whole set of directions as many times as necessary, until the function stops decreasing.

This dumb method is actually not too bad for many functions. Even more interesting is why it *is* bad, i.e. very inefficient, for some other functions. Consider a function of two dimensions whose contour map (level lines) happens to define a long, narrow valley at some angle to the coordinate basis vectors (see Figure 10.5.1). Then the only way "down the length of the valley" going along the basis vectors at each stage is by a series of many tiny steps. More generally, in N dimensions, if the function's second derivatives are much larger in magnitude in some directions than in others, then many cycles through all N basis vectors will be required in order to get anywhere. This condition is not all that unusual; by Murphy's Law, you should count on it.

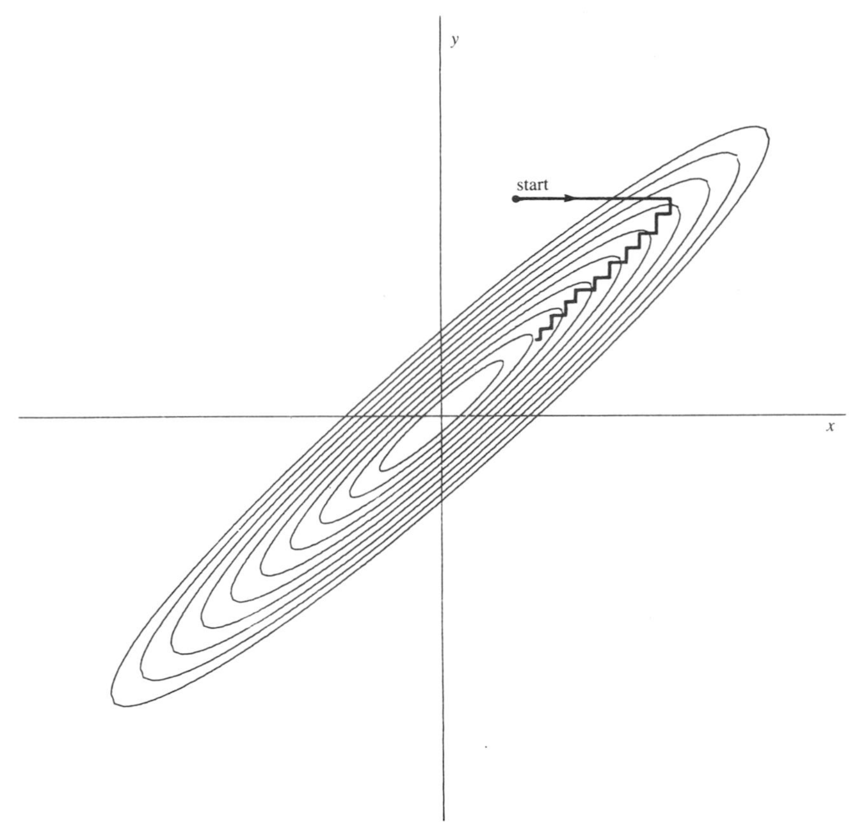

Figure 10.5.1. Successive minimizations along coordinate directions in a long, narrow "valley" (shown as contour lines). Unless the valley is optimally oriented, this method is extremely inefficient, taking many tiny steps to get to the minimum, crossing and re-crossing the principal axis.

Obviously what we need is a better set of directions than the $\mathbf{e}_i$'s. All *direction set methods* consist of prescriptions for updating the set of directions as the method proceeds, attempting to come up with a set which either (i) includes some very good directions that will take us far along narrow valleys, or else (more subtly) (ii) includes some number of "non-interfering" directions with the special property that minimization along one is not "spoiled" by subsequent minimization along another, so that interminable cycling through the set of directions can be avoided.

Conjugate Directions

This concept of "non-interfering" directions, more conventionally called *conjugate directions*, is worth making mathematically explicit.

First, note that if we minimize a function along some direction $\mathbf{u}$, then the gradient of the function must be perpendicular to $\mathbf{u}$ at the line minimum; if not, then there would still be a nonzero directional derivative along $\mathbf{u}$.

Next take some particular point **P** as the origin of the coordinate system with coordinates **x**. Then any function f can be approximated by its Taylor series

$$\begin{aligned} f(\mathbf{x}) &= f(\mathbf{P}) + \sum_i \frac{\partial f}{\partial x_i} x_i + \frac{1}{2} \sum_{i,j} \frac{\partial^2 f}{\partial x_i \partial x_j} x_i x_j + \cdots \\ &\approx c - \mathbf{b}\cdot\mathbf{x} + \frac{1}{2}\mathbf{x}\cdot\mathbf{A}\cdot\mathbf{x} \end{aligned} \tag{10.5.1}$$

where

$$c \equiv f(\mathbf{P}) \qquad \mathbf{b} \equiv -\nabla f|_{\mathbf{P}} \qquad [\mathbf{A}]_{ij} \equiv \left.\frac{\partial^2 f}{\partial x_i \partial x_j}\right|_{\mathbf{P}} \tag{10.5.2}$$

The matrix **A** whose components are the second partial derivative matrix of the function is called the *Hessian matrix* of the function at **P**.

In the approximation of (10.5.1), the gradient of f is easily calculated as

$$\nabla f = \mathbf{A}\cdot\mathbf{x} - \mathbf{b} \tag{10.5.3}$$

(This implies that the gradient will vanish — the function will be at an extremum — at a value of **x** obtained by solving $\mathbf{A}\cdot\mathbf{x} = \mathbf{b}$. This idea we will return to in §10.7!)

How does the gradient ∇f *change* as we move along some direction? Evidently

$$\delta(\nabla f) = \mathbf{A}\cdot(\delta\mathbf{x}) \tag{10.5.4}$$

Suppose that we have moved along some direction **u** to a minimum and now propose to move along some new direction **v**. The condition that motion along **v** not *spoil* our minimization along **u** is just that the gradient stay perpendicular to **u**, i.e. that the change in the gradient be perpendicular to **u**. By equation (10.5.4) this is just

$$0 = \mathbf{u}\cdot\delta(\nabla f) = \mathbf{u}\cdot\mathbf{A}\cdot\mathbf{v} \tag{10.5.5}$$

When (10.5.5) holds for two vectors **u** and **v**, they are said to be *conjugate*. When the relation holds pairwise for all members of a set of vectors, they are said to be a conjugate set. If you do successive line minimization of a function along a conjugate set of directions, then you don't need to redo any of those directions (unless, of course, you spoil things by minimizing along a direction that they are *not* conjugate to).

A triumph for a direction set method is to come up with a set of N linearly independent, mutually conjugate directions. Then, one pass of N line minimizations will put it exactly at the minimum of a quadratic form like (10.5.1). For functions f which are not exactly quadratic forms, it won't be exactly at the minimum; but repeated cycles of N line minimizations will in due course converge *quadratically* to the minimum.

Powell's Quadratically Convergent Method

Powell first discovered a direction set method which does produce N mutually conjugate directions. Here is how it goes: Initialize the set of directions $\mathbf{u}_i$ to the basis vectors,

$$\mathbf{u}_i = \mathbf{e}_i \qquad i = 1, \ldots, N \tag{10.5.6}$$

Now repeat the following sequence of steps ("basic procedure") until your function stops decreasing:

- Save your starting position as $\mathbf{P}_0$.
- For $i = 1, \ldots, N$, move $\mathbf{P}_{i-1}$ to the minimum along direction $\mathbf{u}_i$ and call this point $\mathbf{P}_i$.
- For $i = 1, \ldots, N-1$, set $\mathbf{u}_i \leftarrow \mathbf{u}_{i+1}$.
- Set $\mathbf{u}_N \leftarrow \mathbf{P}_N - \mathbf{P}_0$.
- Move $\mathbf{P}_N$ to the minimum along direction $\mathbf{u}_N$ and call this point $\mathbf{P}_0$.

Powell, in 1964, showed that, for a quadratic form like (10.5.1), k iterations of the above basic procedure produce a set of directions $\mathbf{u}_i$ whose last k members are mutually conjugate. Therefore, N iterations of the basic procedure, amounting to $N(N+1)$ line minimizations in all, will exactly minimize a quadratic form. Brent (1973) gives proofs of these statements in accessible form.

Unfortunately, there is a problem with Powell's quadratically convergent algorithm. The procedure of throwing away, at each stage, $\mathbf{u}_1$ in favor of $\mathbf{P}_N - \mathbf{P}_0$ tends to produce sets of directions that "fold up on each other" and become linearly dependent. Once this happens, then the procedure finds the minimum of the function f only over a subspace of the full N-dimensional case; in other words, it gives the wrong answer. Therefore, the algorithm must not be used in the form given above.

There are a number of ways to fix up the problem of linear dependence in Powell's algorithm, among them:

1. You can reinitialize the set of directions $\mathbf{u}_i$ to the basis vectors $\mathbf{e}_i$ after every N or $N+1$ iterations of the basic procedure. This produces a serviceable method, which we commend to you if quadratic convergence is important for your application (i.e. if your functions are close to quadratic forms and if you desire high accuracy).

2. Brent points out that the set of directions can equally well be reset to the columns of any orthogonal matrix. Rather than throw away the information on conjugate directions already built up, he resets the direction set to calculated principal directions of the matrix **A** (which he gives a procedure for determining). The calculation is essentially a singular value decomposition algorithm (see §2.9). Brent has a number of other cute tricks up his sleeve, and his modification of Powell's method is probably the best presently known. Consult his book for a detailed description and listing of the program. Unfortunately it is rather too elaborate for us to include here.

3. You can give up the property of quadratic convergence in favor of a more heuristic scheme (due to Powell) which tries to find a few good directions along narrow valleys instead of N necessarily conjugate directions. This is the method which we now implement. (It is also the version of Powell's method given in Acton, from which parts of the following discussion are drawn.)

Powell's Method Discarding the Direction of Largest Decrease

The fox and the grapes: Now that we are going to give up the property of quadratic convergence, was it so important after all? That depends on the function that you are minimizing. Some applications produce functions with long, twisty valleys. Quadratic convergence is of no particular advantage to a program which must slalom down the length of a valley floor that twists one way and another (and another, and another, ... – there are N dimensions!). Along the long direction, a quadratically convergent method is trying to extrapolate to the minimum of a parabola which just isn't (yet) there; while the conjugacy of the $N-1$ transverse directions keeps getting spoiled by the twists.

Sooner or later, however, we do arrive at an approximately ellipsoidal minimum (cf. equation 10.5.1 when $\mathbf{b}$, the gradient, is zero). Then, depending on how much accuracy we require, a method with quadratic convergence can save us several times N^2 extra line minimizations, since quadratic convergence *doubles* the number of significant figures at each iteration.

The basic idea of our now-modified Powell's method is still to take $\mathbf{P}_N - \mathbf{P}_0$ as a new direction; it is, after all, the average direction moved after trying all N possible directions. For a valley whose long direction is twisting slowly, this direction is likely to give us a good run along the new long direction. The change is to discard the old direction along which the function f made its *largest decrease*. This seems paradoxical, since that direction was the *best* of the previous iteration. However, it is also likely to be a major component of the new direction that we are adding, so dropping it gives us the best chance of avoiding a buildup of linear dependence.

There are a couple of exceptions to this basic idea. Sometimes it is better *not* to add a new direction at all. Define

$$f_0 \equiv f(\mathbf{P}_0) \qquad f_N \equiv f(\mathbf{P}_N) \qquad f_E \equiv f(2\mathbf{P}_N - \mathbf{P}_0) \tag{10.5.7}$$

Here f_E is the function value at an "extrapolated" point somewhat further along the proposed new direction. Also define Δf to be the magnitude of the largest decrease along one particular direction of the present basic procedure iteration. (Δf is a positive number.) Then:

1. If $f_E \geq f_0$, then keep the old set of directions for the next basic procedure, because the average direction $\mathbf{P}_N - \mathbf{P}_0$ is all played out.

2. If $2(f_0 - 2f_N + f_E)[(f_0 - f_N) - \Delta f]^2 \geq (f_0 - f_E)^2 \Delta f$, then keep the old set of directions for the next basic procedure, because either (i) the decrease along the average direction was not primarily due to any single direction's decrease, or (ii) there is a substantial second derivative along the average direction and we seem to be near to the bottom of its minimum.

The following routine implements Powell's method in the version just described. In the routine, XI is the matrix whose columns are the set of directions $\mathbf{n}_i$; otherwise the correspondence of notation should be self-evident.

```
      SUBROUTINE POWELL(P,XI,N,NP,FTOL,ITER,FRET)
         Minimization of a function FUNC of N variables. (FUNC is not an argument, it is a fixed
         function name.) Input consists of an initial starting point P that is a vector of length N;
         an initial matrix XI whose logical dimensions are N by N, physical dimensions NP by NP,
         and whose columns contain the initial set of directions (usually the N unit vectors); and
         FTOL, the fractional tolerance in the function value such that failure to decrease by more
         than this amount on one iteration signals doneness. On output, P is set to the best point
         found, XI is the then-current direction set, FRET is the returned function value at P, and
         ITER is the number of iterations taken. The routine LINMIN is used.
      PARAMETER (NMAX=20,ITMAX=200)     Maximum expected value of N, and maximum allowed iterations.
      DIMENSION P(NP),XI(NP,NP),PT(NMAX),PTT(NMAX),XIT(NMAX)
      FRET=FUNC(P)
      DO 11 J=1,N                       Save the initial point.
        PT(J)=P(J)
11    CONTINUE
      ITER=0
1     ITER=ITER+1
      FP=FRET
      IBIG=0
      DEL=0.                            Will be the biggest function decrease.
      DO 13 I=1,N                       In each iteration, loop over all directions in the set.
        DO 12 J=1,N                     Copy the direction,
          XIT(J)=XI(J,I)
12      CONTINUE
        FPTT=FRET
        CALL LINMIN(P,XIT,N,FRET)       minimize along it,
        IF(ABS(FPTT-FRET).GT.DEL)THEN   and record it if it is the largest decrease so far.
          DEL=ABS(FPTT-FRET)
          IBIG=I
        ENDIF
13    CONTINUE
      IF(2.*ABS(FP-FRET).LE.FTOL*(ABS(FP)+ABS(FRET)))RETURN   Termination criterion.
      IF(ITER.EQ.ITMAX) PAUSE 'Powell exceeding maximum iterations.'
      DO 14 J=1,N                       Construct the extrapolated point and the average direction moved.
        PTT(J)=2.*P(J)-PT(J)            Save the old starting point.
        XIT(J)=P(J)-PT(J)
        PT(J)=P(J)
14    CONTINUE
      FPTT=FUNC(PTT)                    Function value at extrapolated point.
      IF(FPTT.GE.FP)GO TO 1             One reason not to use new direction.
      T=2.*(FP-2.*FRET+FPTT)*(FP-FRET-DEL)**2-DEL*(FP-FPTT)**2
      IF(T.GE.0.)GO TO 1                Other reason not to use new direction.
      CALL LINMIN(P,XIT,N,FRET)         Move to the minimum of the new direction,
      DO 15 J=1,N                       and save the new direction.
```

```
    XI(J,IBIG)=XIT(J)
 15 CONTINUE
GO TO 1                          Back for another iteration.
END
```

Implementation of Line Minimization

In the above routine, you might have wondered why we didn't make the function name FUNC an argument of the routine. The reason is buried in a slightly dirty FORTRAN practicality in our implementation of LINMIN.

Make no mistake, there is a *right* way to implement LINMIN: It is to use the *methods* of one-dimensional minimization described in §10.1–§10.3, but to rewrite the programs of those sections so that their bookkeeping is done on vector-valued points $\mathbf{P}$ (all lying along a given direction $\mathbf{n}$) rather than scalar-valued abscissas x. That straightforward task produces long routines densely populated with "DO K=1,N" loops.

We do not have space to include such routines in this book. Our LINMIN, which works just fine, is instead a kind of bookkeeping swindle. It constructs an "artificial" function of one variable called F1DIM, which is the value of your function FUNC along the line going through the point P in the direction XI. LINMIN communicates with F1DIM through a common block. (Woe betide the Pascal programmer!) It then calls our familiar one-dimensional routines MNBRAK (§10.1) and BRENT (§10.2) and instructs them to minimize F1DIM.

Still following? Then try this: BRENT is passed the function name F1DIM, which it dutifully calls. But there is no way to signal to F1DIM that it is supposed to use your function name, which could have been passed to LINMIN as an argument. Therefore, we have to make F1DIM use a *fixed* function name, namely FUNC. The situation is reminiscent of Henry Ford's black automobile: POWELL will minimize any function, as long as it is named FUNC. Needed to remedy this situation is a way to pass a function name through a common block; this is lacking in FORTRAN. And Pascal is even worse; the Pascal programmer probably has no idea what we are talking about!

The only thing inefficient about LINMIN is this: Its use as an interface between a multidimensional minimization strategy and a one-dimensional minimization routine results in some unnecessary copying of vectors from hither to yon and back again. That should not normally be a significant addition to the overall computational burden, but we cannot disguise its inelegance.

```
SUBROUTINE LINMIN(P,XI,N,FRET)
   Given an N dimensional point P and an N dimensional direction XI, moves and resets P
   to where the function FUNC(P) takes on a minimum along the direction XI from P, and
   replaces XI by the actual vector displacement that P was moved. Also returns as FRET
   the value of FUNC at the returned location P. This is actually all accomplished by calling
   the routines MNBRAK and BRENT.
PARAMETER (NMAX=50,TOL=1.E-4)
   Maximum anticipated N, and TOL passed to BRENT.
EXTERNAL F1DIM
DIMENSION P(N),XI(N)
COMMON /F1COM/ NCOM,PCOM(NMAX),XICOM(NMAX)
NCOM=N                           Set up the common block.
DO 11 J=1,N
```

```
    PCOM(J)=P(J)
    XICOM(J)=XI(J)
11  CONTINUE
AX=0.                           Initial guess for brackets.
XX=1.
CALL MNBRAK(AX,XX,BX,FA,FX,FB,F1DIM)
FRET=BRENT(AX,XX,BX,F1DIM,TOL,XMIN)
DO 12 J=1,N                     Construct the vector results to return.
    XI(J)=XMIN*XI(J)
    P(J)=P(J)+XI(J)
12  CONTINUE
RETURN
END
```

```
FUNCTION F1DIM(X)
   Must accompany LINMIN.
PARAMETER (NMAX=50)
COMMON /F1COM/ NCOM,PCOM(NMAX),XICOM(NMAX)
DIMENSION XT(NMAX)
DO 11 J=1,NCOM
    XT(J)=PCOM(J)+X*XICOM(J)
11  CONTINUE
F1DIM=FUNC(XT)
RETURN
END
```

REFERENCES AND FURTHER READING:

Brent, Richard P. 1973, *Algorithms for Minimization without Derivatives* (Englewood Cliffs, N.J.: Prentice-Hall), Chapter 7.

Acton, Forman S. 1970, *Numerical Methods That Work* (New York: Harper and Row), pp. 464–467.

Jacobs, David A.H., ed. 1977, *The State of the Art in Numerical Analysis* (London: Academic Press), pp. 259–262.

10.6 Conjugate Gradient Methods in Multidimensions

We consider now the case where you are able to calculate, at a given N-dimensional point $\mathbf{P}$, not just the value of a function $f(\mathbf{P})$ but also the gradient (vector of first partial derivatives) $\nabla f(\mathbf{P})$.

A rough counting argument will show how advantageous it is to use the gradient information: Suppose that the function f is roughly approximated as a quadratic form, as above in equation (10.5.1),

$$f(\mathbf{x}) \approx c - \mathbf{b}\cdot\mathbf{x} + \frac{1}{2}\mathbf{x}\cdot\mathbf{A}\cdot\mathbf{x} \tag{10.6.1}$$

Then the number of unknown parameters in f is equal to the number of free parameters in $\mathbf{A}$ and $\mathbf{b}$, which is $\frac{1}{2}N(N+1)$, which we see to be of order N^2. Changing any one of these parameters can move the location of the minimum. Therefore, we should not expect to be able to *find* the minimum until we have collected an equivalent information content, of order N^2 numbers.

In the direction set methods of §10.5, we collected the necessary information by making on the order of N^2 separate line minimizations, each requiring "a few" (but sometimes a *big* few!) function evaluations. Now, each evaluation of the gradient will bring us N new components of information. If we use them wisely, we should need to make only of order N separate line minimizations. That is in fact the case for the algorithms in this section and the next.

A factor of N improvement in computational speed is not necessarily implied. As a rough estimate, we might imagine that the calculation of *each component* of the gradient takes about as long as evaluating the function itself. In that case there will be of order N^2 equivalent function evaluations both with and without gradient information. Even if the advantage is not of order N, however, it is nevertheless quite substantial: (i) Each calculated component of the gradient will typically save not just one function evaluation, but a number of them, equivalent to, say, a whole line minimization. (ii) There is often a high degree of redundancy in the formulas for the various components of a function's gradient; when this is so, especially when there is also redundancy with the calculation of the function, then the calculation of the gradient may cost significantly less than N function evaluations.

A common beginner's error is to assume that any reasonable way of incorporating gradient information should be about as good as any other. This line of thought leads to the following *not very good* algorithm, the *steepest descent method*:

Steepest Descent: Start at a point $\mathbf{P}_0$. As many times as needed, move from point $\mathbf{P}_i$ to the point $\mathbf{P}_{i+1}$ by minimizing along the line from $\mathbf{P}_i$ in the direction of the local downhill gradient $-\nabla f(\mathbf{P}_i)$.

The problem with the steepest descent method (which, incidentally, goes all the way back to Cauchy), is similar to the problem which was shown in Figure 10.5.1. The method will perform many small steps in going down a long, narrow valley, even if the valley is a perfect quadratic form. You might have hoped that, say in two dimensions, your first step would take you to the valley floor, the second step directly down the long axis; but remember that the new gradient at the minimum point of any line minimization is perpendicular to the direction just traversed. Therefore, with the steepest descent method, you *must* make a right angle turn, which does *not*, in general, take you to the minimum. (See Figure 10.6.1.)

Just as in the discussion that led up to equation (10.5.5), we really want a way of proceeding not down the new gradient, but rather in a direction that is somehow constructed to be *conjugate* to the old gradient, and, insofar as

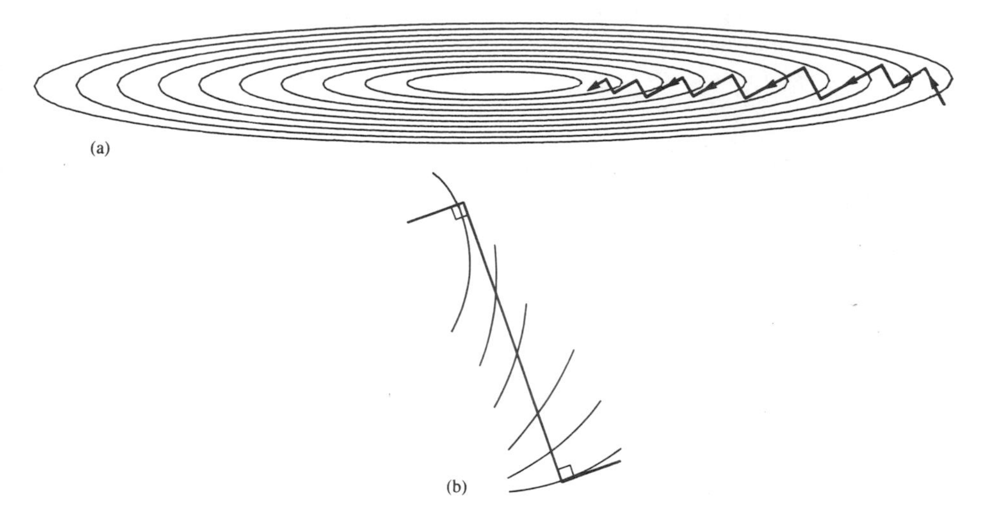

Figure 10.6.1. (a) Steepest descent method in a long, narrow "valley". While more efficient than the strategy of Figure 10.5.1, steepest descent is nonetheless an inefficient strategy, taking many steps to reach the valley floor. (b) Magnified view of one step: a step starts off in the local gradient direction, perpendicular to the contour lines, and traverses a straight line until a local minimum is reached, where the traverse is parallel to the local contour lines.

possible, to all previous directions traversed. Methods which accomplish this construction are called *conjugate gradient* methods.

The two most important conjugate gradient methods are the *Fletcher-Reeves method* and the *Polak-Ribiere method.* These two methods are closely related. Both are based on the following, rather remarkable, theorem: Let $\mathbf{A}$ be a symmetric, positive-definite, $n \times n$ matrix. Let $\mathbf{g}_0$ be an arbitrary vector; let $\mathbf{h}_0 = \mathbf{g}_0$. For $i = 0, 1, 2, \ldots$ define the two sequences of vectors

$$\mathbf{g}_{i+1} = \mathbf{g}_i - \lambda_i \mathbf{A} \cdot \mathbf{h}_i \qquad \mathbf{h}_{i+1} = \mathbf{g}_{i+1} + \gamma_i \mathbf{h}_i \tag{10.6.2}$$

where λ_i, γ_i are chosen to make $\mathbf{g}_{i+1} \cdot \mathbf{g}_i = 0$ and $\mathbf{h}_{i+1} \cdot \mathbf{A} \cdot \mathbf{h}_i = 0$, i.e.

$$\lambda_i = \frac{\mathbf{g}_i \cdot \mathbf{g}_i}{\mathbf{g}_i \cdot \mathbf{A} \cdot \mathbf{h}_i} \qquad \gamma_i = \frac{\mathbf{g}_{i+1} \cdot \mathbf{A} \cdot \mathbf{h}_i}{\mathbf{h}_i \cdot \mathbf{A} \cdot \mathbf{h}_i} \tag{10.6.3}$$

(unless the denominators equal zero, in which case take $\lambda_i = 0$, $\gamma_i = 0$). Then, for *all* $i \neq j$,

$$\mathbf{g}_i \cdot \mathbf{g}_j = 0 \qquad \mathbf{h}_i \cdot \mathbf{A} \cdot \mathbf{h}_j = 0 \tag{10.6.4}$$

In other words, the procedure (10.6.2), which is a kind of *Gram-Schmidt bi-orthogonalization* making each $\mathbf{g}_i$ *orthogonal* to its immediate predecessor,

and each $\mathbf{h}_i$ *conjugate* to its immediate predecessor, actually produces a sequence of $\mathbf{g}$'s that are all *mutally* orthogonal, and a sequence of $\mathbf{h}$'s that are all *mutually* conjugate!

The proof of this theorem proceeds by straightforward induction. For details, consult the book by Polak.

Knowing that (10.6.4) holds, you can fiddle around with (10.6.2) and verify that the following expressions for γ_i and λ_i are equivalent to (10.6.3),

$$\gamma_i = \frac{\mathbf{g}_{i+1} \cdot \mathbf{g}_{i+1}}{\mathbf{g}_i \cdot \mathbf{g}_i} = \frac{(\mathbf{g}_{i+1} - \mathbf{g}_i) \cdot \mathbf{g}_{i+1}}{\mathbf{g}_i \cdot \mathbf{g}_i} \tag{10.6.5}$$

$$\lambda_i = \frac{\mathbf{g}_i \cdot \mathbf{h}_i}{\mathbf{h}_i \cdot \mathbf{A} \cdot \mathbf{h}_i} \tag{10.6.6}$$

We are now ready to apply this formalism to the problem of minimizing a function approximated by a quadratic form (10.6.1). Suppose that we knew the Hessian matrix $\mathbf{A}$. Then we could use the construction (10.6.2) to find successively conjugate directions $\mathbf{h}_i$ along which to line-minimize. After N such, we would efficiently have arrived at the minimum of the quadratic form. But we don't know $\mathbf{A}$.

Here is another remarkable theorem to save the day: Let $\mathbf{g}_i$ and $\mathbf{h}_i$ be the same sequence of vectors as before. Suppose we happen to have $\mathbf{g}_i = -\nabla f(\mathbf{P}_i)$, for some point $\mathbf{P}_i$, where f is of the form (10.6.1). Suppose that we proceed from $\mathbf{P}_i$ along the direction $\mathbf{h}_i$ to the local minimum of f located at some point $\mathbf{P}_{i+1}$ and then set $\mathbf{g}_{i+1} = -\nabla f(\mathbf{P}_{i+1})$. Then, this $\mathbf{g}_{i+1}$ is the same vector as would have been constructed by equation (10.6.2). (And we have constructed it without knowledge of $\mathbf{A}$!).

Proof: By equation (10.5.3), $\mathbf{g}_i = -\mathbf{A} \cdot \mathbf{P}_i + \mathbf{b}$, and

$$\mathbf{g}_{i+1} = -\mathbf{A} \cdot (\mathbf{P}_i + \lambda \mathbf{h}_i) + \mathbf{b} = \mathbf{g}_i - \lambda \mathbf{A} \cdot \mathbf{h}_i \tag{10.6.7}$$

with λ chosen to take us to the line minimum. But at the line minimum $\mathbf{h}_i \cdot \nabla f = -\mathbf{h}_i \cdot \mathbf{g}_{i+1} = 0$. This latter condition is easily combined with (10.6.7) to solve for λ. The result is exactly the expression (10.6.6). But with this value of λ, (10.6.7) is the same as (10.6.2), q.e.d.

We have, then, the basis of an algorithm which requires neither knowledge of the Hessian matrix $\mathbf{A}$, nor even the storage necessary to store such a matrix. A sequence of directions $\mathbf{h}_i$ is constructed, using only line minimizations, evaluations of the gradient vector, and an auxiliary vector to store the latest in the sequence of $\mathbf{g}$'s.

Thus far, everything said is applicable both to the Fletcher-Reeves and to the Polak-Ribiere methods. There is only one tiny, but sometimes significant, difference between the two methods. Fletcher and Reeves originally used the first expression for γ_i of equation (10.6.5) above. Polak and Ribiere, later,

proposed using the second expression in the same equation. "Wait," you say, "aren't they equal?" They are equal for exact quadratic forms. In the real world, however, your function is not exactly a quadratic form. Arriving at the supposed minimum of the quadratic form, you may still need to proceed for another set of iterations. There is some evidence (see Jacobs) that the Polak-Ribiere formula accomplishes the transition to further iterations more gracefully: When it runs out of steam, it tends to reset **h** to be down the local gradient, which is equivalent to beginning the conjugate-gradient procedure anew.

The following routine implements the Polak-Ribiere variant, which we recommend; but changing one program line, as shown, will give you Fletcher-Reeves. The routine presumes the existence of a function `FUNC(P)`, where `P` is a vector of length `N`, and also presumes the existence of a subroutine `DFUNC(P,DF)` that returns the vector gradient `DF` evaluated at the input point `P`.

The routine calls `LINMIN` to do the line minimizations. As already discussed, you may wish to use a modified version of `LINMIN` which uses `DBRENT` instead of `BRENT`, i.e., which uses the gradient in doing the line minimizations. See note below.

```
      SUBROUTINE FRPRMN(P,N,FTOL,ITER,FRET)
```
Given a starting point **P** that is a vector of length **N**, Fletcher-Reeves-Polak-Ribiere minimization is performed on a function **FUNC**, using its gradient as calculated by a routine **DFUNC**. The convergence tolerance on the function value is input as **FTOL**. Returned quantities are **P** (the location of the minimum), **ITER** (the number of iterations that were performed), and **FRET** (the minimum value of the function). The routine **LINMIN** is called to perform line minimizations.
```
      PARAMETER (NMAX=50,ITMAX=200,EPS=1.E-10)
```
Maximum anticipated value of **N**; maximum allowed number of iterations; small number to rectify special case of converging to exactly zero function value.
```
      DIMENSION P(N),G(NMAX),H(NMAX),XI(NMAX)
      FP=FUNC(P)                          Initializations.
      CALL DFUNC(P,XI)
      DO 11 J=1,N
        G(J)=-XI(J)
        H(J)=G(J)
        XI(J)=H(J)
11    CONTINUE
      DO 14 ITS=1,ITMAX                   Loop over iterations.
        ITER=ITS
        CALL LINMIN(P,XI,N,FRET)          Next statement is the normal return:
        IF(2.*ABS(FRET-FP).LE.FTOL*(ABS(FRET)+ABS(FP)+EPS))RETURN
        FP=FUNC(P)
        CALL DFUNC(P,XI)
        GG=0.
        DGG=0.
        DO 12 J=1,N
          GG=GG+G(J)**2
C         DGG=DGG+XI(J)**2                This statement for Fletcher-Reeves.
          DGG=DGG+(XI(J)+G(J))*XI(J)      This statement for Polak-Ribiere.
12      CONTINUE
        IF(GG.EQ.0.)RETURN                Unlikely. If gradient is exactly zero then we are already done.
        GAM=DGG/GG
        DO 13 J=1,N
          G(J)=-XI(J)
          H(J)=G(J)+GAM*H(J)
          XI(J)=H(J)
13      CONTINUE
```

```
 14 CONTINUE
PAUSE 'FRPR maximum iterations exceeded'
RETURN
END
```

Note on Line Minimization Using Derivatives

Kindly reread the last part of §10.5. We here want to do the same thing, but using derivative information in performing the line minimization.

Rather than reprint the whole routine LINMIN just to show one modified statement, let us just tell you what the change is: The statement

```
FRET=BRENT(AX,XX,BX,F1DIM,TOL,XMIN)
```

should be replaced by

```
FRET=DBRENT(AX,XX,BX,F1DIM,DF1DIM,TOL,XMIN)
```

You must also include the following function, which is analogous in function to F1DIM as discussed in §10.5. And remember, your function must be named FUNC, and its gradient calculation must be named DFUNC.

```
FUNCTION DF1DIM(X)
PARAMETER (NMAX=50)
COMMON /F1COM/ NCOM,PCOM(NMAX),XICOM(NMAX)
DIMENSION XT(NMAX),DF(NMAX)
DO 11 J=1,NCOM
    XT(J)=PCOM(J)+X*XICOM(J)
 11 CONTINUE
CALL DFUNC(XT,DF)
DF1DIM=0.
DO 12 J=1,NCOM
    DF1DIM=DF1DIM+DF(J)*XICOM(J)
 12 CONTINUE
RETURN
END
```

Note on Sparse Linear Systems

In §2.10 we gave a routine SPARSE for solving a sparse linear systems of equations

$$\mathbf{A} \cdot \mathbf{x} = \mathbf{b} \tag{10.6.8}$$

by using the conjugate gradient method of minimization. You should now be in a position to understand the inner workings of that program.

It was remarked in §2.10 that high accuracy was difficult to obtain because the condition number of the quadratic form that was minimized was the *square* of the condition number of $\mathbf{A}$. This was the case because $\mathbf{A}$ was not known to be symmetric and positive definite, so it had to be "squared" before use.

From this section, you can see that if your matrix $\mathbf{A}$ *is* known to be symmetric and positive definite (or *negative* definite, in which case just use $-\mathbf{A}$), then you can use $\mathbf{A}$ directly:

To solve (10.6.8), you write subroutines which calculate, for a given $\mathbf{x}$,

$$f(\mathbf{x}) \equiv \frac{1}{2}\mathbf{x}\cdot\mathbf{A}\cdot\mathbf{x} - \mathbf{b}\cdot\mathbf{x} \qquad \nabla f \equiv \mathbf{A}\cdot\mathbf{x} - \mathbf{b} \tag{10.6.9}$$

(compare equations 10.5.1 and 10.5.3). Take care to write these routines as cleverly as you can, taking advantage of the sparseness of your matrix.

Now call `FRPRMN` with any starting value of $\mathbf{x}$, e.g. $\mathbf{x} = 0$. The value of $\mathbf{x}$ returned should be the solution of (10.6.8).

REFERENCES AND FURTHER READING:

Polak, E. 1971, *Computational Methods in Optimization* (New York: Academic Press), §2.3.

Jacobs, David A.H., ed. 1977, *The State of the Art in Numerical Analysis* (London: Academic Press), Chapter III.1.7 (by K.W. Brodlie).

Stoer, J., and Bulirsch, R. 1980, *Introduction to Numerical Analysis* (New York: Springer-Verlag), §8.7.

10.7 Variable Metric Methods in Multidimensions

The goal of *variable metric* methods, which are sometimes called *quasi-Newton* methods, is not different from the goal of conjugate gradient methods: to accumulate information from successive line minimizations so that N such line minimizations lead to the exact minimum of a quadratic form in N dimensions. In that case, the method will also be quadratically convergent for more general smooth functions.

Both variable metric and conjugate gradient methods require that you are able to compute your function's gradient, or first partial derivatives, at arbitrary points. The variable metric approach differs from the conjugate gradient in the way that it stores and updates the information that is accumulated. Instead of requiring intermediate storage on the order of N, the number of dimensions, it requires a matrix of size $N \times N$. Generally, for any moderate N, this is an entirely trivial disadvantage.

On the other hand, there is not, as far as we know, any overwhelming advantage that the variable metric methods hold over the conjugate gradient

techniques, except perhaps an historical one. Developed somewhat earlier, and more widely propagated, the variable metric methods have by now developed a wider constituency of satisfied users. Likewise, some fancier implementations of variable metric methods (going beyond the scope of this book, see below) have been developed to a greater level of sophistication on issues like the minimization of roundoff error, handling of special conditions, and so on. *We* tend to use variable metric rather than conjugate gradient, but we have no reason to urge this habit on you.

The most frequently implemented variable metric method is the *Davidon-Fletcher-Powell (DFP)* algorithm (sometimes referred to as simply *Fletcher-Powell*). A related method which goes by the name *Broyden-Fletcher-Goldfarb-Shanno (BFGS)* algorithm also has proved important. Dixon has shown that the BFGS and DFP schemes differ only in details of their roundoff error, convergence tolerances, and similar "dirty" issues which are outside of our scope (see Jacobs). However, it has become generally recognized that, empirically, the BFGS scheme is superior in these details. We will implement BFGS in this section.

As before, we imagine that our arbitrary function $f(\mathbf{x})$ can be locally approximated by the quadratic form of equation (10.6.1). We don't, however, have any information about the values of the quadratic form's parameters $\mathbf{A}$ and $\mathbf{b}$, except insofar as we can glean such information from our function evaluations and line minimizations.

The basic idea of the variable metric method is to build up, iteratively, a good approximation to the inverse Hessian matrix $\mathbf{A}^{-1}$, that is, to construct a sequence of matrices $\mathbf{H}_i$ with the property,

$$\lim_{i\to\infty} \mathbf{H}_i = \mathbf{A}^{-1} \tag{10.7.1}$$

Even better if the limit is achieved after N iterations instead of ∞.

If we are successful in achieving (10.7.1), then we can use $\mathbf{H}$ as follows: The minimum point $\mathbf{x}_m$ satisfies

$$\mathbf{A}\cdot\mathbf{x}_m = \mathbf{b} \tag{10.7.2}$$

(compare equation 10.5.3). At the current point $\mathbf{x}_i$, whatever it is, we have

$$\mathbf{A}\cdot\mathbf{x}_i = \nabla f(\mathbf{x}_i) + \mathbf{b} \tag{10.7.3}$$

(also by 10.5.3). Subtracting these two equations and multiplying by the inverse matrix $\mathbf{A}^{-1}$, we have

$$\mathbf{x}_m - \mathbf{x}_i = \mathbf{A}^{-1}\cdot[-\nabla f(\mathbf{x}_i)] \tag{10.7.4}$$

The left-hand side is the finite step we need take to get to the exact minimum; the right-hand side is known once we have accumulated an accurate $\mathbf{H} \approx \mathbf{A}^{-1}$.

We won't rigorously derive the DFP algorithm for taking $\mathbf{H}_i$ into $\mathbf{H}_{i+1}$; you can consult Polak for clear derivations. Following Brodlie (in Jacobs), we will give the following heuristic motivation of the procedure.

Subtracting equation (10.7.4) at $\mathbf{x}_{i+1}$ from that same equation at $\mathbf{x}_i$ gives

$$\mathbf{x}_{i+1} - \mathbf{x}_i = \mathbf{A}^{-1} \cdot (\nabla f_{i+1} - \nabla f_i) \tag{10.7.5}$$

where $\nabla f_j \equiv \nabla f(\mathbf{x}_j)$. Having made the step from $\mathbf{x}_i$ to $\mathbf{x}_{i+1}$, we might reasonably want to require that the new approximation $\mathbf{H}_{i+1}$ satisfy (10.7.5) as if it were actually $\mathbf{A}^{-1}$, that is,

$$\mathbf{x}_{i+1} - \mathbf{x}_i = \mathbf{H}_{i+1} \cdot (\nabla f_{i+1} - \nabla f_i) \tag{10.7.6}$$

We might also imagine that the updating formula should be of the form $\mathbf{H}_{i+1} = \mathbf{H}_i + \text{correction}$.

What "objects" are around out of which to construct a correction term? Most notable are the two vectors $\mathbf{x}_{i+1} - \mathbf{x}_i$ and $\nabla f_{i+1} - \nabla f_i$; and there is also $\mathbf{H}_i$. There are not infinitely many natural ways of making a matrix out of these objects, especially if (10.7.6) must hold! One such way, the *DFP updating formula* is

$$\begin{aligned}\mathbf{H}_{i+1} = \mathbf{H}_i &+ \frac{(\mathbf{x}_{i+1} - \mathbf{x}_i) \otimes (\mathbf{x}_{i+1} - \mathbf{x}_i)}{(\mathbf{x}_{i+1} - \mathbf{x}_i) \cdot (\nabla f_{i+1} - \nabla f_i)} \\ &- \frac{[\mathbf{H}_i \cdot (\nabla f_{i+1} - \nabla f_i)] \otimes [\mathbf{H}_i \cdot (\nabla f_{i+1} - \nabla f_i)]}{(\nabla f_{i+1} - \nabla f_i) \cdot \mathbf{H}_i \cdot (\nabla f_{i+1} - \nabla f_i)}\end{aligned} \tag{10.7.7}$$

where $\otimes$ denotes the "outer" or "direct" product of two vectors, a matrix: the ij component of $\mathbf{u} \otimes \mathbf{v}$ is $u_i v_j$. (You might want to verify that 10.7.7 does satisfy 10.7.6.)

The *BFGS updating formula* is exactly the same, but with one additional term,

$$\cdots + [(\nabla f_{i+1} - \nabla f_i) \cdot \mathbf{H}_i \cdot (\nabla f_{i+1} - \nabla f_i)]\; \mathbf{u} \otimes \mathbf{u} \tag{10.7.8}$$

where $\mathbf{u}$ is defined as the vector

$$\begin{aligned}\mathbf{u} \equiv &\frac{(\mathbf{x}_{i+1} - \mathbf{x}_i)}{(\mathbf{x}_{i+1} - \mathbf{x}_i) \cdot (\nabla f_{i+1} - \nabla f_i)} \\ &- \frac{\mathbf{H}_i \cdot (\nabla f_{i+1} - \nabla f_i)}{(\nabla f_{i+1} - \nabla f_i) \cdot \mathbf{H}_i \cdot (\nabla f_{i+1} - \nabla f_i)}\end{aligned} \tag{10.7.9}$$

(You might also verify that this satisfies 10.7.6.)

You will have to take on faith — or else consult Polak for details of — the "deep" result that equation (10.7.7), with or without (10.7.8), does in fact converge to $\mathbf{A}^{-1}$ in N steps, if f is a quadratic form.

Without further ado, we have the following implementation, which uses the line minimization routine LINMIN as in the previous section.

```
SUBROUTINE DFPMIN(P,N,FTOL,ITER,FRET)
   Given a starting point P that is a vector of length N, The Broyden-Fletcher-Goldfarb-
   Shanno variant of Davidon-Fletcher-Powell minimization is performed on a function FUNC,
   using its gradient as calculated by a routine DFUNC. The convergence requirement on the
   function value is input as FTOL. Returned quantities are P (the location of the minimum),
   ITER (the number of iterations that were performed), and FRET (the minimum value of the
   function). The routine LINMIN is called to perform line minimizations.
PARAMETER (NMAX=50,ITMAX=200,EPS=1.E-10)
   Maximum anticipated value of N; maximum allowed number of iterations; small number to
   rectify special case of converging to exactly zero function value.
DIMENSION P(N),HESSIN(NMAX,NMAX),XI(NMAX),G(NMAX),DG(NMAX),HDG(NMAX)
FP=FUNC(P)                          Calculate starting function value and gradient,
CALL DFUNC(P,G)
DO 12 I=1,N                         and initialize the inverse Hessian to the unit matrix.
   DO 11 J=1,N
      HESSIN(I,J)=0.
11    CONTINUE
   HESSIN(I,I)=1.
   XI(I)=-G(I)                      Initial line direction.
12 CONTINUE
DO 24 ITS=1,ITMAX                   Main loop over the iterations.
   ITER=ITS
   CALL LINMIN(P,XI,N,FRET)            Next statement is the normal return:
   IF(2.*ABS(FRET-FP).LE.FTOL*(ABS(FRET)+ABS(FP)+EPS))RETURN
   FP=FRET                          Save the old function value
   DO 13 I=1,N                      and the old gradient.
      DG(I)=G(I)
13    CONTINUE
   FRET=FUNC(P)                     Get new function value and gradient.
   CALL DFUNC(P,G)
   DO 14 I=1,N                      Compute difference of gradients,
      DG(I)=G(I)-DG(I)
14    CONTINUE
   DO 16 I=1,N                      and difference times current matrix.
      HDG(I)=0.
      DO 15 J=1,N
         HDG(I)=HDG(I)+HESSIN(I,J)*DG(J)
15       CONTINUE
16    CONTINUE
   FAC=0.                           Calculate dot products for the denominators,
   FAE=0.
   DO 17 I=1,N
      FAC=FAC+DG(I)*XI(I)
      FAE=FAE+DG(I)*HDG(I)
17    CONTINUE
   FAC=1./FAC                       and make the denominators multiplicative.
   FAD=1./FAE
   DO 18 I=1,N                      The vector which makes BFGS different from DFP:
      DG(I)=FAC*XI(I)-FAD*HDG(I)
18    CONTINUE
   DO 21 I=1,N                      The BFGS updating formula:
      DO 19 J=1,N
         HESSIN(I,J)=HESSIN(I,J)+FAC*XI(I)*XI(J)
*           -FAD*HDG(I)*HDG(J)+FAE*DG(I)*DG(J)
19       CONTINUE
21    CONTINUE
   DO 23 I=1,N                      Now calculate the next direction to go,
```

```
          XI(I)=0.
          DO 22 J=1,N
            XI(I)=XI(I)-HESSIN(I,J)*G(J)
22        CONTINUE
23      CONTINUE
24    CONTINUE                         and go back for another iteration.
      PAUSE 'too many iterations in DFPMIN'
      RETURN
      END
```

Advanced Implementations of Variable Metric Methods

Although rare, it can conceivably happen that roundoff errors cause the matrix $\mathbf{H}_i$ to become nearly singular or non-positive-definite. This can be serious, because the supposed search directions might then not lead downhill, and because nearly singular $\mathbf{H}_i$'s tend to give subsequent $\mathbf{H}_i$'s which are also nearly singular.

There is a simple fix for this rare problem, the same as was mentioned in §10.4: In case of any doubt, you should *restart* the algorithm at the claimed minimum point, and see if it goes anywhere. Simple, but not very elegant. Modern implementations of variable metric methods deal with the problem in a more sophisticated way.

Instead of building up an approximation to $\mathbf{A}^{-1}$, it is possible to build up an approximation of $\mathbf{A}$ itself. Then, instead of calculating the left-hand side of (10.7.4) directly, one solves the set of linear equations

$$\mathbf{A} \cdot (\mathbf{x}_m - \mathbf{x}_i) = -\nabla f(\mathbf{x}_i) \tag{10.7.10}$$

At first glance this seems like a bad idea, since solving (10.7.10) is a process of order N^3 – and anyway, how does this help the roundoff problem? The trick is not to store $\mathbf{A}$ but rather a triangular decomposition of $\mathbf{A}$, its *Cholesky decomposition*, somewhat similar to the *LU* decomposition discussed in Chapter 2 but specialized to the case of symmetric, positive-definite matrices. The updating formula for the Cholesky decomposition of $\mathbf{A}$ is such as to guarantee that the matrix remains positive definite and nonsingular, even in the presence of finite roundoff. This method is due to Gill and Murray (see Jacobs).

Another issue is the question of how exact the line searches need to be in order to retain good convergence for the overall method. It seems wasteful to converge so accurately on each line minimization to a way-point that has no particular significance to the N-dimensional function. There are some known results in this area, but beyond our present scope. One useful fact is that the BFGS updating formula is more tolerant of inexactitude in the line minimization than is the DFP updating formula.

Oddly, there seems to be no known minimization method which makes efficient use of *all* the function and/or gradient evaluations which are performed during each line minimization. One might hope for future progress in this area.

REFERENCES AND FURTHER READING:

Polak, E. 1971, *Computational Methods in Optimization* (New York: Academic Press), pp. 56ff.

Jacobs, David A.H., ed. 1977, *The State of the Art in Numerical Analysis* (London: Academic Press), Chapter III.1, §§3–6 (by K. W. Brodlie).

Acton, Forman S. 1970, *Numerical Methods That Work* (New York: Harper and Row), pp. 467–468 (note, however, missing minus sign in the DFP algorithm).

10.8 Linear Programming and the Simplex Method

The subject of *linear programming*, sometimes called *linear optimization*, concerns itself with the following problem: For N independent variables $x_1, \ldots, x_N$, *maximize* the function

$$z = a_{01}x_1 + a_{02}x_2 + \cdots + a_{0N}x_N \tag{10.8.1}$$

subject to the primary constraints

$$x_1 \geq 0, \quad x_2 \geq 0, \quad \ldots \quad x_N \geq 0 \tag{10.8.2}$$

and simultaneously subject to $M = m_1 + m_2 + m_3$ additional constraints, m_1 of them of the form

$$a_{i1}x_1 + a_{i2}x_2 + \cdots + a_{iN}x_N \leq b_i \qquad (b_i \geq 0) \qquad i = 1, \ldots, m_1 \tag{10.8.3}$$

m_2 of them of the form

$$a_{j1}x_1 + a_{j2}x_2 + \cdots + a_{jN}x_N \geq b_j \geq 0 \qquad j = m_1 + 1, \ldots, m_1 + m_2 \tag{10.8.4}$$

and m_3 of them of the form

$$a_{k1}x_1 + a_{k2}x_2 + \cdots + a_{kN}x_N = b_k \geq 0 \qquad k = m_1 + m_2 + 1, \ldots, m_1 + m_2 + m_3 \tag{10.8.5}$$

The various a_{ij}'s can have either sign, or be zero. The fact that the b's must all be nonnegative (as indicated by the final inequality in the above three equations) is a matter of convention only, since you can multiply any contrary inequality by -1. There is no particular significance in the number

of constraints M being less than, equal to, or greater than the number of unknowns N.

A set of values $x_1 \dots x_N$ that satisfies the constraints (10.8.2)–(10.8.5) is called a *feasible vector.* The function that we are trying to maximize is called the *objective function.* The feasible vector that maximizes the objective function is called the *optimal feasible vector.* An optimal feasible vector can fail to exist for two distinct reasons: (i) there are *no* feasible vectors, i.e. the given constraints are incompatible, or (ii) there is no maximum, i.e. there is a direction in N space where one or more of the variables can be taken to infinity while still satisfying the constraints, giving an unbounded value for the objective function.

As you see, the subject of linear programming is surrounded by notational and terminological thickets. Both of these thorny defenses are lovingly cultivated by a coterie of stern acolytes who have devoted themselves to the field. Actually, the basic ideas of linear programming are quite simple. Avoiding the shrubbery, we want to teach you the basics by means of a couple of specific examples; it should then be quite obvious how to generalize.

Why is linear programming so important? (i) Because "nonnegativity" is the usual constraint on any variable x_i that represents the tangible amount of some physical commodity, like guns, butter, dollars, units of vitamin E, food calories, kilowatt hours, mass, etc. Hence equation (10.8.2). (ii) Because one is often interested in additive (linear) limitations or bounds imposed by man or nature: minimum nutritional requirement, maximum affordable cost, maximum on available labor or capital, minimum tolerable level of voter approval, etc. Hence equations (10.8.3)–(10.8.5). (iii) Because the function that one wants to optimize may be linear, or else may at least be approximated by a linear function — since that is the problem that linear programming *can* solve. Hence equation (10.8.1). For a short, semipopular survey of linear programming applications, see Bland (1981).

Here is a specific example of a problem in linear programming, which has $N = 4$, $m_1 = 2$, $m_2 = m_3 = 1$, hence $M = 4$:

$$\text{Maximize} \quad z = x_1 + x_2 + 3x_3 - \tfrac{1}{2}x_4 \tag{10.8.6}$$

with all the x's non-negative and also with

$$\begin{aligned} x_1 + 2x_3 &\le 740 \\ 2x_2 - 7x_4 &\le 0 \\ x_2 - x_3 + 2x_4 &\ge \tfrac{1}{2} \\ x_1 + x_2 + x_3 + x_4 &= 9 \end{aligned} \tag{10.8.7}$$

The answer turns out to be (to 2 significant figures) $x_1 = 0$, $x_2 = 3.33$, $x_3 = 4.73$, $x_4 = 0.95$. In the rest of this section we will learn how this answer is obtained. Figure 10.8.1 summarizes some of the terminology thus far.

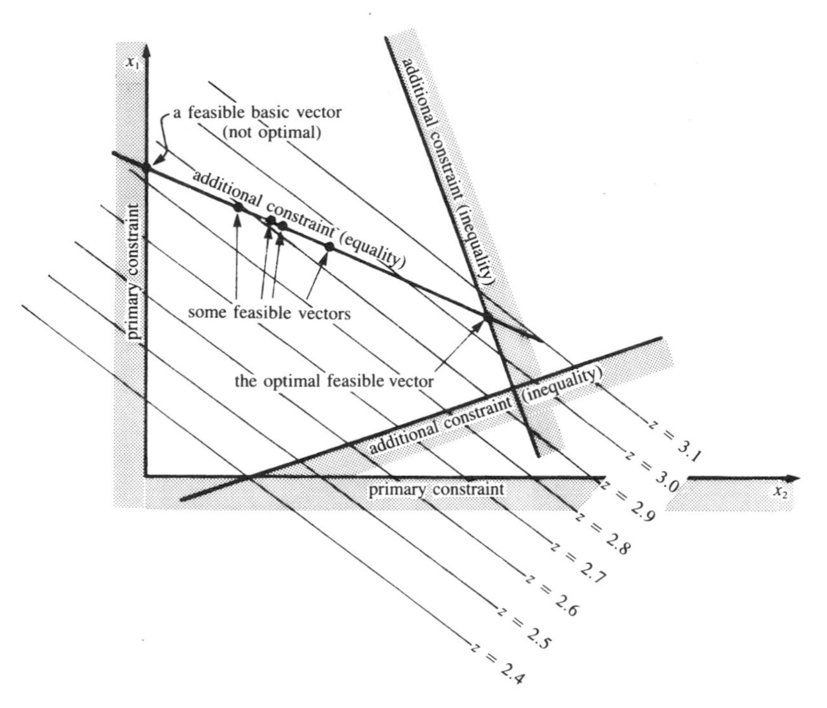

Figure 10.8.1. Basic concepts of linear programming. The case of only two independent variables, x_1, x_2, is shown. The linear function z, to be maximized, is represented by its contour lines. Primary constraints require x_1 and x_2 to be positive. Additional constraints may restrict the solution to regions (inequality constraints) or to surfaces of lower dimensionality (equality constraints). Feasible vectors satisfy all constraints. Feasible basic vectors also lie on the boundary of the allowed region. The simplex method steps among feasible basic vectors until the optimal feasible vector is found.

Fundamental Theorem of Linear Optimization

Imagine that we start with a full N-dimensional space of candidate vectors. Then (in mind's eye, at least) we carve away the regions that are eliminated in turn by each imposed constraint. Since the constraints are linear, every boundary introduced by this process is a plane, or rather hyperplane. Equality constraints of the form (10.8.5) force the feasible region onto hyperplanes of smaller dimension, while inequalities simply divide the then-feasible region into allowed and non-allowed pieces.

When all the constraints are imposed, either we are left with some feasible region or else there are no feasible vectors. Since the feasible region is bounded by hyperplanes, it is geometrically a kind of convex polyhedron or simplex (cf. §10.4). If there is a feasible region, can the optimal feasible vector be somewhere in its interior, away from the boundaries? No, because the objective function is linear. This means that it always has a nonzero vector gradient. This, in turn, means that we could always increase the objective function by running up the gradient until we hit a boundary wall.

The boundary of any geometrical region has one less dimension than its interior. Therefore, we can now run up the gradient projected into the

boundary wall until we reach an edge of that wall. We can then run up that edge, and so on, down through whatever number of dimensions, until we finally arrive at a point, a *vertex* of the original simplex. Since this point has all N of its coordinates defined, it must be the solution of N simultaneous *equalities* drawn from the original set of equalities and inequalities (10.8.2)–(10.8.5).

Points which are feasible vectors and which satisfy this property, that they satisfy N of the original constraints as equalities, are termed *feasible basic vectors.* If $N > M$, then a feasible basic vector has *at least* $N - M$ of its components equal to zero, since at least that many of the constraints (10.8.2) will be needed to make up the total of N. Put the other way, *at most* M components of a feasible basic vector are nonzero. In the example (10.8.6)–(10.8.7), you can check that the solution as given satisfies as equalities the last three constraints of (10.8.7) and the constraint $x_1 \geq 0$, for the required total of 4.

Put together the two preceding paragraphs and you have the *Fundamental Theorem of Linear Optimization*: If an optimal feasible vector exists, then there is a feasible basic vector which is optimal. (Didn't we warn you about the terminological thicket?)

The importance of the fundamental theorem is that it reduces the optimization problem to a "combinatorial" problem, that of determining which N constraints (out of the $M + N$ constraints in 10.8.2–10.8.5) should be satisfied by the optimal feasible vector. We have only to keep trying different combinations, and computing the objective function for each trial, until we find the best.

Doing this blindly would take halfway to forever. The *simplex method*, first published by Dantzig in 1948, is a way of organizing the procedure so that (i) a series of combinations is tried for which the objective function increases at each step, and (ii) the optimal feasible vector is reached after a number of iterations that is almost always no larger than of order M or N, whichever is larger. An interesting mathematical sidelight is that this second property, although known empirically ever since the simplex method was devised, was not proved to be true until the 1982 work of Stephen Smale. (For a contemporary account, see the Kolata reference.)

Simplex Method for a Restricted Normal Form

A linear programming problem is said to be in *normal form* if it has no constraints in the form (10.8.3) or (10.8.4), but rather only equality constraints of the form (10.8.5) and nonnegativity constraints of the form (10.8.2).

For our purposes it will be useful to consider an even more restricted set of cases, with this additional property: Each equality constraint of the form (10.8.5) must have at least one variable that has a positive coefficient and *that appears uniquely in that one constraint only.* We can then choose one such variable in each constraint equation, and solve that constraint equation for it. The variables thus chosen are called *left-hand variables* or *basic variables*, and there are exactly M $(= m_3)$ of them. The remaining $N - M$ variables are called *right-hand variables* or *nonbasic variables.* Obviously this *restricted*

normal form can only be achieved in the case $M \leq N$, so that is the case that we will consider.

You may be thinking that our restricted normal form is so specialized that it is unlikely to include the linear programming problem that you wish to solve. Not at all! We will presently show how *any* linear programming problem can be transformed into restricted normal form. Therefore bear with us and learn how to apply the simplex method to a restricted normal form.

Here is an example of a problem in restricted normal form:

$$\text{Maximize} \quad z = 2x_2 - 4x_3 \tag{10.8.8}$$

with x_1, x_2, x_3, and x_4 all non-negative and also with

$$\begin{aligned} x_1 &= 2 - 6x_2 + x_3 \\ x_4 &= 8 + 3x_2 - 4x_3 \end{aligned} \tag{10.8.9}$$

This example has $N = 4$, $M = 2$; the left-hand variables are x_1 and x_4; the right-hand variables are x_2 and x_3. The objective function (10.8.8) is written so as to depend only on right-hand variables; note, however, that this is not an actual restriction on objective functions in restricted normal form, since any left-hand variables appearing in the objective function could be eliminated algebraically by use of (10.8.9) or its analogs.

For any problem in restricted normal form, we can instantly read off a feasible basic vector (although not necessarily the *optimal* feasible basic vector). Simply set all right-hand variables equal to zero, and equation (10.8.9) then gives the values of the left-hand variables for which the constraints are satisfied. The idea of the simplex method is to proceed by a series of exchanges. In each exchange, a right-hand variable and a left-hand variable change places. At each stage we maintain a problem in restricted normal form that is equivalent to the original problem.

It is notationally convenient to record the information content of equations (10.8.8) and (10.8.9) in a so-called *tableau*, as follows:

		x_2	x_3
z	0	2	−4
x_1	2	−6	1
x_4	8	3	−4

(10.8.10)

You should study (10.8.10) to be sure that you understand where each entry comes from, and how to translate back and forth between the tableau and equation formats of a problem in restricted normal form.

The first step in the simplex method is to examine the top row of the tableau, which we will call the "z-row." Look at the entries in columns labeled

by right-hand variables (we will call these "right-columns"). We want to imagine in turn the effect of increasing each right-hand variable from its present value of zero, while leaving all the other right-hand variables at zero. Will the objective function increase or decrease? The answer is given by the sign of the entry in the z-row. Since we want to increase the objective function, only right columns having positive z-row entries are of interest. In (10.8.10) there is only one such column, whose z-row entry is 2.

The second step is to examine the column entries below each z-row entry which were selected by step one. We want to ask how much we can increase the right-hand variable before one of the left-hand variables is driven negative, which is not allowed. If the tableau element at the intersection of the right-hand column and the left-hand variable's row is positive, then it poses no restriction: the corresponding left-hand variable will just be driven more and more positive. If *all* the entries in any right-hand column are positive, then there is no bound on the objective function and (having said so) we are done with the problem.

If one or more entries below a positive z-row entry are negative, then we have to figure out which such entry first limits the increase of that column's right-hand variable. Evidently the limiting increase is given by dividing the element in the right-hand column (which is called the *pivot element*) into the element in the "constant column" (leftmost column) of the pivot element's row. A value that is small in magnitude is most restrictive. The increase in the objective function for this choice of pivot element is then that value multiplied by the z-row entry of that column. We repeat this procedure on all possible right-hand columns to find the pivot element with the largest such increase. That completes our "choice of a pivot element."

In the above example, the only positive z-row entry is 2. There is only one negative entry below it, namely -6, so this is the pivot element. Its constant-column entry is 2. This pivot will therefore allow x_2 to be increased by $2 \div |6|$, which results in an increase of the objective function by an amount $(2 \times 2) \div |6|$.

The third step is to *do* the increase of the selected right-hand variable, thus making it a left-hand variable; and simultaneously to modify the left-hand variables, reducing the pivot-row element to zero and thus making it a right-hand variable. For our above example let's do this first by hand: We first solve the pivot-row equation for the new left-hand variable x_2 in favor of the old one x_1, namely

$$x_1 = 2 - 6x_2 + x_3 \qquad \rightarrow \qquad x_2 = \tfrac{1}{3} - \tfrac{1}{6}x_1 + \tfrac{1}{6}x_3 \tag{10.8.11}$$

We then substitute this into the old z-row,

$$z = 2x_2 - 4x_3 = 2\left[\tfrac{1}{3} - \tfrac{1}{6}x_1 + \tfrac{1}{6}x_3\right] - 4x_3 = \tfrac{2}{3} - \tfrac{1}{3}x_1 - \tfrac{11}{3}x_3 \tag{10.8.12}$$

and into all other left-variable rows, in this case only x_4,

$$x_4 = 8 + 3\left[\tfrac{1}{3} - \tfrac{1}{6}x_1 + \tfrac{1}{6}x_3\right] - 4x_3 = 9 - \tfrac{1}{2}x_1 - \tfrac{7}{2}x_3 \tag{10.8.13}$$

Equations (10.8.11)–(10.8.13) form the new tableau

		x_1	x_3
z	$\frac{2}{3}$	$-\frac{1}{3}$	$-\frac{11}{3}$
x_2	$\frac{1}{3}$	$-\frac{1}{6}$	$\frac{1}{6}$
x_4	9	$-\frac{1}{2}$	$-\frac{7}{2}$

(10.8.14)

The fourth step is to go back and repeat the first step, looking for another possible increase of the objective function. We do this as many times as possible, that is, until all the right-hand entries in the z-row are negative, signaling that no further increase is possible. In the present example, this already occurs in (10.8.14), so we are done.

The answer can now be read from the constant column of the final tableau. In (10.8.14) we see that the objective function is maximized to a value of 2/3 for the solution vector $x_2 = 1/3$, $x_4 = 9$, $x_1 = x_3 = 0$.

Now look back over the procedure which led from (10.8.10) to (10.8.14). You will find that it could be summarized entirely in tableau format as a series of prescribed elementary matrix operations:

- Locate the pivot element and save it.
- Save the whole pivot column.
- Replace each row, except the pivot row, by that linear combination of itself and the pivot row which makes its pivot-column entry zero.
- Divide the pivot row by the negative of the pivot.
- Replace the pivot element by the reciprocal of its saved value.
- Replace the rest of pivot column by its saved values divided by the saved pivot element.

This is the sequence of operations actually performed by a linear programming routine, such as the one that we will presently give.

You should now be able to solve almost any linear programming problem that starts in restricted normal form. The only special case which might stump you is if an entry in the constant column turns out to be zero at some stage, so that a left-hand variable is zero at the same time as all the right-hand variables are zero. This is called a *degenerate feasible vector*. To proceed, you may need to exchange the degenerate left-hand variable for one of the right-hand variables, perhaps even making several such exchanges.

Writing the General Problem in Restricted Normal Form

Here is a pleasant surprise. There exist a couple of clever tricks which render trivial the task of translating a general linear programming problem into restricted normal form!

First, we need to get rid of the inequalities of the form (10.8.3) or (10.8.4), for example, the first three constraints in (10.8.7). We do this by adding to the problem so-called *slack variables* which, when their nonnegativity is required, convert the inequalities to equalities. We will denote slack variables as y_i. There will be $m_1 + m_2$ of them. Once they are introduced, you treat them on an equal footing with the original variables x_i; then, at the very end, you simply ignore them.

For example, introducing slack variables leaves (10.8.6) unchanged but turns (10.8.7) into

$$\begin{aligned} x_1 + 2x_3 + y_1 &= 740 \\ 2x_2 - 7x_4 + y_2 &= 0 \\ x_2 - x_3 + 2x_4 - y_3 &= \tfrac{1}{2} \\ x_1 + x_2 + x_3 + x_4 &= 9 \end{aligned} \tag{10.8.15}$$

(Notice how the sign of the coefficient of the slack variable is determined by which sense of inequality it is replacing.)

Second, we need to insure that there is a set of M left-hand vectors, so that we can set up a starting tableau in restricted normal form. (In other words, we need to find a "feasible basic starting vector".) The trick is again to invent new variables! There are M of these, and they are called *artificial variables*; we denote them by z_i. You put exactly one artificial variable into each constraint equation on the following model for the example (10.8.15):

$$\begin{aligned} z_1 &= 740 - x_1 - 2x_3 - y_1 \\ z_2 &= -2x_2 + 7x_4 - y_2 \\ z_3 &= \tfrac{1}{2} - x_2 + x_3 - 2x_4 + y_3 \\ z_4 &= 9 - x_1 - x_2 - x_3 - x_4 \end{aligned} \tag{10.8.16}$$

Our example is now in restricted normal form.

Now you may object that (10.8.16) is not the same problem as (10.8.15) or (10.8.7) *unless all the z_i's are zero.* Right you are! There is some subtlety here! We must proceed to solve our problem in two phases. First phase: We replace our objective function (10.8.6) by a so-called *auxiliary objective function*

$$z' \equiv -z_1 - z_2 - z_3 - z_4 = -(749\tfrac{1}{2} - 2x_1 - 4x_2 - 2x_3 + 4x_4 - y_1 - y_2 + y_3) \tag{10.8.17}$$

(where the last equality follows from using 10.8.16). We now perform the simplex method on the auxiliary objective function (10.8.17) with the constraints (10.8.16). Obviously the auxiliary objective function will be maximized for nonnegative z_i's if all the z_i's are zero. We therefore expect the simplex method in this first phase to produce a set of left-hand variables drawn from the x_i's and y_i's only, with all the z_i's being right-hand variables. Aha! We then cross out the z_i's, leaving a problem involving only x_i's and y_i's in restricted normal form. In other words, the first phase produces an initial feasible basic vector. Second phase: Solve the problem produced by the first phase, using the original objective function, not the auxiliary.

And what if the first phase *doesn't* produce zero values for all the z_i's? That signals that there is *no* initial feasible basic vector, i.e. that the constraints given to us are inconsistent among themselves. Report that fact, and you are done.

Here is how to translate into tableau format the information needed for both the first and second phases of the overall method. As before, the underlying problem to be solved is as posed in equations (10.8.6)–(10.8.7).

		x_1	x_2	x_3	x_4	y_1	y_2	y_3
z	0	1	1	3	$-\frac{1}{2}$	0	0	0
z_1	740	-1	0	-2	0	-1	0	0
z_2	0	0	-2	0	7	0	-1	0
z_3	$\frac{1}{2}$	0	-1	1	-2	0	0	1
z_4	9	-1	-1	-1	-1	0	0	0
z'	$-749\frac{1}{2}$	2	4	2	-4	1	1	-1

(10.8.18)

This is not as daunting as it may, at first sight, appear. The table entries inside the box of double lines are no more than the coefficients of the original problem (10.8.6)–(10.8.7) organized into a tabular form. In fact, these entries, along with the values of N, M, m_1, m_2, and m_3, are the only input that is needed by the simplex method routine below. The columns under the slack variables y_i simply record whether each of the M constraints is of the form $\leq$, $\geq$, or $=$; this is redundant information with the values m_1, m_2, m_3, as long as we are sure to enter the rows of the tableau in the correct respective order. The coefficients of the auxiliary objective function (bottom row) are just the negatives of the column sums of the rows above, so these are easily calculated automatically.

The output from a simplex routine will be (i) a flag telling whether a finite solution, no solution, or an unbounded solution was found, and (ii) an updated tableau. The output tableau that derives from (10.8.18), given to

two significant figures, is

		x_1	y_2	y_3	$\cdots$
z	17.03	−.95	−.05	−1.05	$\cdots$
x_2	3.33	−.35	−.15	.35	$\cdots$
x_3	4.73	−.55	.05	−.45	$\cdots$
x_4	.95	−.10	.10	.10	$\cdots$
y_1	730.55	.10	−.10	.90	$\cdots$

(10.8.19)

A little counting of the x_i's and y_i's will convince you that there are $M+1$ rows (including the z-row) in both the input and the output tableaux, but that only $N+1-m_3$ columns of the output tableau (including the constant column) contain any useful information, the other columns belonging to now-discarded artificial variables. In the output, the first numerical column contains the solution vector, along with the maximum value of the objective function. Where a slack variable (y_i) appears on the left, the corresponding value is the amount by which its inequality is safely satisfied. Variables which are not left-hand variables in the output tableau have zero values. Slack variables with zero values represent constraints which are satisfied as equalities.

Routine Implementing the Simplex Method

The following routine is based algorithmically on the implementation of Kuenzi, Tzschach, and Zehnder. Aside from input values of M, N, m_1, m_2, m_3, the principal input to the routine is a two-dimensional array `A` containing the portion of the tableau (10.8.18) that is contained between the double lines. This input occupies the first $M+1$ rows and $N+1$ columns of `A`. Note, however, that reference is made internally to row $M+2$ of `A` (used for the auxiliary objective function, just as in 10.8.18). Therefore the physical dimensions of `A`,

```
DIMENSION A(MP,NP)
```
(10.8.20)

must have `NP`$\geq N+1$ and `MP`$\geq M+2$. You will suffer endless agonies if you fail to understand this simple point. Also do not neglect to order the rows of `A` in the same order as equations (10.8.1), (10.8.3), (10.8.4), and (10.8.5), that is, objective function, $\leq$-constraints, $\geq$-constraints, $=$-constraints.

On output, the tableau `A` is indexed by two returned arrays of integers. `IPOSV(J)` contains, for `J`$=1\ldots M$, the number i whose original variable x_i is now represented by row `J+1` of `A`. These are thus the left-hand variables in the solution. (The first row of `A` is of course the z-row.) A value $i>N$ indicates that the variable is a y_i rather than an x_i, $x_{N+j}\equiv y_j$. Likewise, `IZROV(J)` contains, for `J`$=1\ldots N$, the number i whose original variable x_i is

now a right-hand variable, represented by column J+1 of A. These variables are all zero in the solution. The meaning of $i > N$ is the same as above, except that $i > N + m_1 + m_2$ denotes an artificial or slack variable which was used only internally and should now be entirely ignored.

The flag ICASE is returned as zero if a finite solution is found, +1 if the objective function is unbounded, −1 if no solution satisfies the given constraints.

The routine treats the case of degenerate feasible vectors, so don't worry about them. You may also wish to admire the fact that the routine does not require storage for the columns of the tableau (10.8.18) which are to the right of the double line; it keeps track of slack variables by more efficient bookkeeping.

Please note that, as given, the routine is only "semi-sophisticated" in its tests for convergence. While the routine properly implements tests for inequality with zero as tests against some small parameter EPS, it does not adjust this parameter to reflect the scale of the input data. This is adequate for many problems, where the input data do not differ from unity by too many orders of magnitude. If, however, you encounter endless cycling, then you should modify EPS in the routines SIMPLX and SIMP2. Permuting your variables can also help. Finally, consult Wilkinson and Reinsch.

```
      SUBROUTINE SIMPLX(A,M,N,MP,NP,M1,M2,M3,ICASE,IZROV,IPOSV)
C        Simplex method for linear programming. Input parameters A, M, N, MP, NP, M1, M2, and M3,
C        and output parameters A, ICASE, IZROV, and IPOSV are described above.
      PARAMETER(MMAX=100,EPS=1.E-6)         MMAX is the maximum number of constraints expected.
      DIMENSION A(MP,NP),IZROV(N),IPOSV(M),L1(MMAX),L2(MMAX),L3(MMAX)
      IF(M.NE.M1+M2+M3)PAUSE 'Bad input constraint counts.'
      NL1=N                                 Initially make all variables right-hand.
      DO 11 K=1,N                           Initialize index lists.
          L1(K)=K
          IZROV(K)=K
11    CONTINUE
      NL2=M                                 Make all artificial variables left-hand,
      DO 12 I=1,M                           and initialize those lists.
          IF(A(I+1,1).LT.0.)PAUSE 'Bad input tableau.'   Constants b_i must be nonnegative.
          L2(I)=I
          IPOSV(I)=N+I
12    CONTINUE
      DO 13 I=1,M2                          Used later, but initialized here.
          L3(I)=1
13    CONTINUE
      IR=0                                  This flag setting means we are in phase two, i.e have a feasible
      IF(M2+M3.EQ.0)GO TO 30                    starting solution. GOTO 30 if origin is a feasible solution.
      IR=1                                  Flag meaning that we must start out in phase one.
      DO 15 K=1,N+1                         Compute the auxiliary objective function.
          Q1=0.
          DO 14 I=M1+1,M
              Q1=Q1+A(I+1,K)
14        CONTINUE
          A(M+2,K)=-Q1
15    CONTINUE
10    CALL SIMP1(A,MP,NP,M+1,L1,NL1,0,KP,BMAX)    Find max. coeff. of auxiliary objective fn.
      IF(BMAX.LE.EPS.AND.A(M+2,1).LT.-EPS)THEN
          ICASE=-1                          Auxiliary objective function is still negative and can't be improved,
          RETURN                                hence no feasible solution exists.
      ELSE IF(BMAX.LE.EPS.AND.A(M+2,1).LE.EPS)THEN
          M12=M1+M2+1                       Auxiliary objective function is zero and can't be improved. This sig-
          IF(M12.LE.M)THEN                      nals that we have a feasible starting vector. Clean out the
                                                artificial variables by GOTO 1's and then move on to phase two
                                                by GOTO 30.
```

```
        DO 16 IP=M12,M
            IF(IPOSV(IP).EQ.IP+N)THEN
             CALL SIMP1(A,MP,NP,IP,L1,NL1,1,KP,BMAX)
             IF(BMAX.GT.0.)GO TO 1
            ENDIF
          16 CONTINUE
    ENDIF
    IR=0                               Set flag indicating we have reached phase two.
    M12=M12-1
    IF(M1+1.GT.M12)GO TO 30
    DO 18 I=M1+1,M12
        IF(L3(I-M1).EQ.1)THEN
            DO 17 K=1,N+1
                A(I+1,K)=-A(I+1,K)
              17 CONTINUE
        ENDIF
      18 CONTINUE
    GO TO 30                           Go to phase two.
ENDIF
CALL SIMP2(A,M,N,MP,NP,L2,NL2,IP,KP,Q1)      Locate a pivot element (phase one).
IF(IP.EQ.0)THEN                        Maximum of auxiliary objective function is unbounded, so no feasible
    ICASE=-1                                 solution exists.
    RETURN
ENDIF
1   CALL SIMP3(A,MP,NP,M+1,N,IP,KP)         Exchange a left- and a right-hand variable (phase one), then
IF(IPOSV(IP).GE.N+M1+M2+1)THEN        update lists.
    DO 19 K=1,NL1
        IF(L1(K).EQ.KP)GO TO 2
      19 CONTINUE
2       NL1=NL1-1
    DO 21 IS=K,NL1
        L1(IS)=L1(IS+1)
      21 CONTINUE
ELSE
    IF(IPOSV(IP).LT.N+M1+1)GO TO 20
    KH=IPOSV(IP)-M1-N
    IF(L3(KH).EQ.0)GO TO 20
    L3(KH)=0
ENDIF
A(M+2,KP+1)=A(M+2,KP+1)+1.
DO 22 I=1,M+2
    A(I,KP+1)=-A(I,KP+1)
  22 CONTINUE
20  IS=IZROV(KP)
IZROV(KP)=IPOSV(IP)
IPOSV(IP)=IS
IF(IR.NE.0)GO TO 10                    If still in phase one, go back to 10.
    End of phase one code for finding an initial feasible solution. Now, in phase two, optimize
    it.
30  CALL SIMP1(A,MP,NP,0,L1,NL1,0,KP,BMAX)   Test the z-row for doneness.
IF(BMAX.LE.0.)THEN                     Done. Solution found. Return with the good news.
    ICASE=0
    RETURN
ENDIF
CALL SIMP2(A,M,N,MP,NP,L2,NL2,IP,KP,Q1)      Locate a pivot element (phase two).
IF(IP.EQ.0)THEN                        Objective function is unbounded. Report and return.
    ICASE=1
    RETURN
ENDIF
CALL SIMP3(A,MP,NP,M,N,IP,KP)          Exchange a left- and a right-hand variable (phase two),
GO TO 20                               and return for another iteration.
END
```

The preceding routine makes use of the following utility subroutines.

```
      SUBROUTINE SIMP1(A,MP,NP,MM,LL,NLL,IABF,KP,BMAX)
          Determines the maximum of those elements whose index is contained in the supplied list
          LL, either with or without taking the absolute value, as flagged by IABF.
      DIMENSION A(MP,NP),LL(NP)
      KP=LL(1)
      BMAX=A(MM+1,KP+1)
      IF(NLL.LT.2)RETURN
      DO 11 K=2,NLL
          IF(IABF.EQ.0)THEN
              TEST=A(MM+1,LL(K)+1)-BMAX
          ELSE
              TEST=ABS(A(MM+1,LL(K)+1))-ABS(BMAX)
          ENDIF
          IF(TEST.GT.0.)THEN
              BMAX=A(MM+1,LL(K)+1)
              KP=LL(K)
          ENDIF
11    CONTINUE
      RETURN
      END
```

```
      SUBROUTINE SIMP2(A,M,N,MP,NP,L2,NL2,IP,KP,Q1)
          Locate a pivot element, taking degeneracy into account.
      PARAMETER (EPS=1.E-6)
      DIMENSION A(MP,NP),L2(MP)
      IP=0
      IF(NL2.LT.1)RETURN
      DO 11 I=1,NL2
          IF(A(L2(I)+1,KP+1).LT.-EPS)GO TO 2
11    CONTINUE
      RETURN                                    No possible pivots. Return with message.
2     Q1=-A(L2(I)+1,1)/A(L2(I)+1,KP+1)
      IP=L2(I)
      IF(I+1.GT.NL2)RETURN
      DO 13 I=I+1,NL2
          II=L2(I)
          IF(A(II+1,KP+1).LT.-EPS)THEN
              Q=-A(II+1,1)/A(II+1,KP+1)
              IF(Q.LT.Q1)THEN
                  IP=II
                  Q1=Q
              ELSE IF (Q.EQ.Q1) THEN            We have a degeneracy.
                  DO 12 K=1,N
                      QP=-A(IP+1,K+1)/A(IP+1,KP+1)
                      Q0=-A(II+1,K+1)/A(II+1,KP+1)
                      IF(Q0.NE.QP)GO TO 6
12                CONTINUE
6                 IF(Q0.LT.QP)IP=II
              ENDIF
          ENDIF
13    CONTINUE
      RETURN
      END
```

```
SUBROUTINE SIMP3(A,MP,NP,I1,K1,IP,KP)
    Matrix operations to exchange a left-hand and right-hand variable (see text).
DIMENSION A(MP,NP)
PIV=1./A(IP+1,KP+1)
IF(I1.GE.0)THEN
    DO 12 II=1,I1+1
        IF(II-1.NE.IP)THEN
            A(II,KP+1)=A(II,KP+1)*PIV
            DO 11 KK=1,K1+1
                IF(KK-1.NE.KP)THEN
                    A(II,KK)=A(II,KK)-A(IP+1,KK)*A(II,KP+1)
                ENDIF
11          CONTINUE
        ENDIF
12  CONTINUE
ENDIF
DO 13 KK=1,K1+1
    IF(KK-1.NE.KP)A(IP+1,KK)=-A(IP+1,KK)*PIV
13 CONTINUE
A(IP+1,KP+1)=PIV
RETURN
END
```

Other Topics Briefly Mentioned

Every linear programming problem in normal form with N variables and M constraints has a corresponding *dual* problem with M variables and N constraints. The tableau of the dual problem is, in essence, the transpose of the tableau of the original (sometimes called *primal*) problem. It is possible to go from a solution of the dual to a solution of the primal. This can occasionally be computationally useful, but generally it is no big deal.

The *revised simplex method* is exactly equivalent to the simplex method in its choice of which left-hand and right-hand variables are exchanged. Its computational effort is not significantly less than that of the simplex method. It does differ in the organization of its storage, requiring only a matrix of size $M \times M$, rather than $M \times N$, in its intermediate stages. If you have a lot of constraints and memory size is one of them, then you should look into it.

The *primal-dual algorithm* and the *composite simplex algorithm* are two different methods for avoiding the two phases of the usual simplex method: Progress is made simultaneously towards finding a feasible solution and finding an optimal solution. There seems to be no clearcut evidence that these methods are superior to the usual method by any factor substantially larger than the "tender-loving-care factor" (which reflects the programming effort of the proponents).

Problems where the objective function and/or one or more of the constraints are replaced by expressions nonlinear in the variables are called *nonlinear programming problems.* The literature on such problems is vast, but outside our scope. The special case of quadratic expressions is called *quadratic programming.* Optimization problems where the variables take on only integer values are called *integer programming* problems, a special case of *discrete*

optimization generally. The next section looks at a particular kind of discrete optimization problem.

REFERENCES AND FURTHER READING:

Bland, R.G. 1981, *Scientific American*, vol. 244, (June) p. 126.

Kolata, G. 1982, *Science*, vol. 217, p. 39.

Cooper, L., and Steinberg, D. 1970, *Introduction to Methods of Optimization* (Philadelphia: Saunders).

Dantzig, G.B. 1963, *Linear Programming and Extensions* (Princeton, N.J.: Princeton University Press).

Gass, S.T. 1969, *Linear Programming*, 3rd ed. (New York: McGraw-Hill).

Murty, K.G. 1976, *Linear and Combinatorial Programming* (New York: Wiley).

Land, A.H., and Powell, S. 1973, *Fortran Codes for Mathematical Programming* (London: Wiley-Interscience).

Kuenzi, H.P., Tzschach, H.G., and Zehnder, C.A. 1971 *Numerical Methods of Mathematical Optimization* (New York: Academic Press).

Stoer, J., and Bulirsch, R. 1980, *Introduction to Numerical Analysis* (New York: Springer-Verlag), §4.10.

Wilkinson, J.H., and Reinsch, C. 1971, *Linear Algebra*, vol. II of *Handbook for Automatic Computation* (New York: Springer-Verlag).

10.9 Combinatorial Minimization: Method of Simulated Annealing

The *method of simulated annealing* is a technique that has recently attracted significant attention as suitable for optimization problems of very large scale. For practical purposes, it has effectively "solved" the famous *traveling salesman problem* of finding the shortest cyclical itinerary for a traveling salesman who must visit each of N cities in turn. The method has also been used successfully for designing complex integrated circuits: The arrangement of several hundred thousand circuit elements on a tiny silicon substrate is optimized so as to minimize interference among their connecting wires. Amazingly, the implementation of the algorithm is quite simple.

Notice that the two applications cited are both examples of *combinatorial minimization*. There is an objective function to be minimized, as usual; but the space over which that function is defined is not simply the N-dimensional space of N continuously variable parameters. Rather, it is a discrete, but very large, configuration space, like the set of possible orders of cities, or the set of possible allocations of silicon "real estate" to circuit elements. The number of elements in the configuration space is factorially large, so that they cannot be explored exhaustively. Furthermore, since the set is discrete, we are deprived of any notion of "continuing downhill in a favorable direction." The concept of "direction" may not have any meaning in the configuration space.

At the heart of the method of simulated annealing is an analogy with thermodynamics, specifically with the way that liquids freeze and crystallize,

or metals cool and anneal. At high temperatures, the molecules of a liquid move freely with respect to one another. If the liquid is cooled slowly, thermal mobility is lost. The atoms are often able to line themselves up and form a pure crystal that is completely ordered over a distance up to billions of times the size of an individual atom in all directions. This crystal is the state of minimum energy for this system. The amazing fact is that, for slowly cooled systems, nature is able to find this minimum energy state. In fact, if a liquid metal is cooled quickly or "quenched," it does not reach this state but rather ends up in a polycrystalline or amorphous state having somewhat higher energy.

So the essence of the process is *slow* cooling, allowing ample time for redistribution of the atoms as they lose mobility. This is the technical definition of *annealing*, and it is essential for ensuring that a low energy state will be achieved.

Although the analogy is not perfect, there is a sense in which all of the minimization algorithms thus far in this chapter correspond to rapid cooling or quenching. In all cases, we have gone greedily for the quick, nearby solution: from the starting point, go immediately downhill as far as you can go. This, as often remarked above, leads to a local, but not necessarily a global, minimum. Nature's own minimization algorithm is based on quite a different procedure. The so-called Boltzmann probability distribution,

$$\text{Prob}\,(E) \sim \exp(-E/kT) \tag{10.9.1}$$

expresses the idea that a system in thermal equilibrium at temperature T has its energy probabilistically distributed among all different energy states E. Even at low temperature, there is a chance, albeit very small, of a system being in a high energy state. Therefore, there is a corresponding chance for the system to get out of a local energy minimum in favor of finding a better, more global, one. The quantity k (Boltzmann's constant) is a constant of nature which relates temperature to energy. In other words, the system sometimes goes *uphill* as well as downhill; but the lower the temperature, the less likely is any significant uphill excursion.

In 1953, Metropolis and coworkers first incorporated these kinds of principles into numerical calculations. Offered a succession of options, a simulated thermodynamic system was assumed to change its configuration from energy E_1 to energy E_2 with probability $p = \exp[-(E_2 - E_1)/kT]$. Notice that if $E_2 < E_1$, this probability is greater than unity; in such cases the change is arbitrarily assigned a probability $p = 1$, i.e. the system *always* took such an option. This general scheme, of always taking a downhill step while *sometimes* taking an uphill step, has come to be known as the Metropolis algorithm.

To make use of the Metropolis algorithm for other than thermodynamic systems, one must provide the following elements:

1. A description of possible system configurations.

2. A generator of random changes in the configuration; these changes are the "options" presented to the system.

3. An objective function E (analog of energy) whose minimization is the goal of the procedure.

4. A control parameter T (analog of temperature) and an *annealing schedule* which tells how it is lowered from high to low values, e.g., after how many random changes in configuration is each downward step in T taken, and how large is that step. The meaning of "high" and "low" in this context, and the assignment of a schedule, may require physical insight and/or trial-and-error experiments.

The Traveling Salesman Problem

A concrete illustration is provided by the traveling salesman problem. The salesperson visits N cities with given positions (x_i, y_i), returning finally to his or her city of origin. Each city is to be visited only once, and the route is to be made as short as possible. This problem belongs to a class known as *NP-complete* problems, whose computation time for an *exact* solution increases with N as exp(const. $\times N$), becoming rapidly prohibitive in cost as N increases. The traveling salesman problem also belongs to a class of minimization problems for which the objective function E has many local minima. In practical cases, it is often enough to be able to choose from these a minimum which, even if not absolute, cannot be significantly improved upon. The annealing method manages to achieve this, while limiting its calculations to scale as a small power of N.

As a problem in simulated annealing, the traveling salesman problem is handled as follows:

1. *Configuration.* The cities are numbered $i = 1 \ldots N$ and each has coordinates (x_i, y_i). A configuration is a permutation of the number $1 \ldots N$, interpreted as the order in which the cities are visited.

2. *Rearrangements.* An efficient set of moves has been suggested by Lin. The moves consist of two types: (a) A section of path is removed and then replaced with the same cities running in the opposite order; or (b) a section of path is removed and then replaced in between two cities on another, randomly chosen, part of the path.

3. *Objective Function.* In the simplest form of the problem, E is taken just as the total length of journey,

$$E = L \equiv \sum_{i=1}^{N} \sqrt{(x_i - x_{i+1})^2 + (y_i - y_{i+1})^2} \qquad (10.9.2)$$

with the convention that point $N+1$ is identified with point 1. To illustrate the flexibility of the method, however, we can add the following additional wrinkle: suppose that the salesman has an irrational fear of flying over the Mississippi River. In that case, we would assign each city a parameter μ_i,

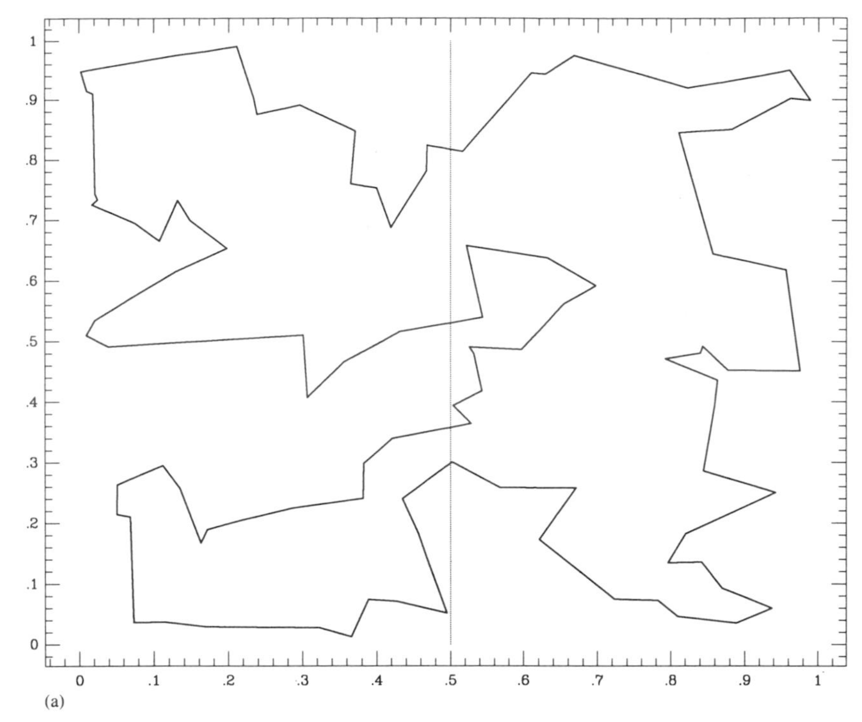

Figure 10.9.1. Traveling salesman problem solved by simulated annealing. The (nearly) shortest path among 100 randomly positioned cities is shown in (a). The dotted line is a river, but there is no penalty in crossing. In (b) the river-crossing penalty is made large, and the solution restricts itself to the minimum number of crossings, two. In (c) the penalty has been made negative: the salesman is actually a smuggler who crosses the river on the flimsiest excuse!

equal to $+1$ if it is east of the Mississippi, -1 if it is west, and take the objective function to be

$$E = \sum_{i=1}^{N} \sqrt{(x_i - x_{i+1})^2 + (y_i - y_{i+1})^2} + \lambda(\mu_i - \mu_{i+1})^2 \qquad (10.9.3)$$

A penalty 4λ is thereby assigned to any river crossing. The algorithm now finds the shortest path that avoids crossings. The relative importance that it assigns to length of path versus river crossings is determined by our choice of λ. Figure 10.9.1 shows the results obtained. Clearly, this technique can be generalized to include many conflicting goals in the minimization.

4. *Annealing schedule.* This requires experimentation. We first generate some random rearrangements, and use them to determine the range of values of ΔE that will be encountered from move to move. Choosing a starting value for the parameter T which is considerably larger than the largest ΔE normally enountered, we proceed downward in multiplicative steps each amounting to a 10 percent decrease in T. We hold each new value of T constant for, say, $100N$

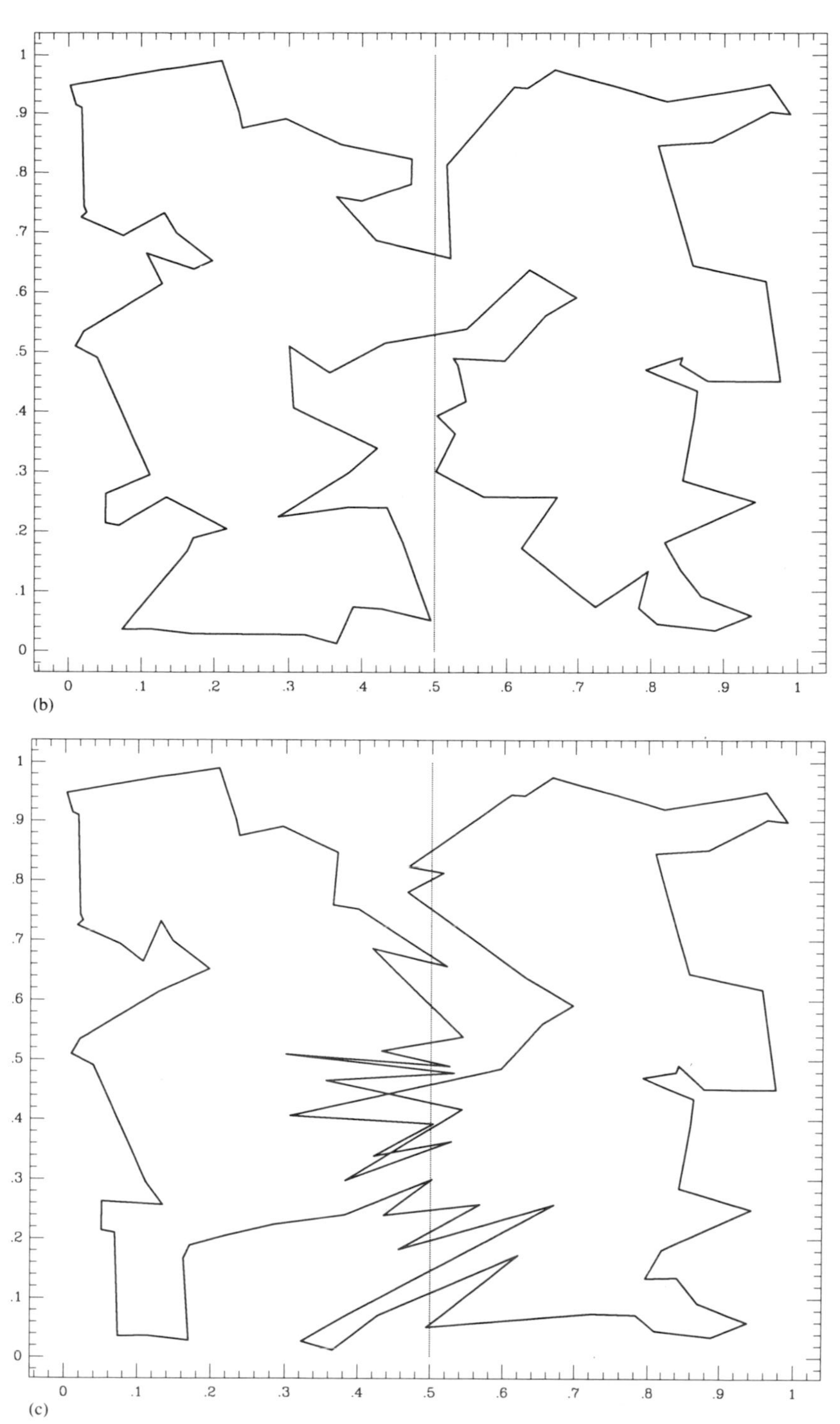

Figure 10.9.1. Traveling salesman problem solved by simulated annealing (see caption on previous page).

reconfigurations, or for $10N$ successful reconfigurations, whichever comes first. When efforts to reduce E further become sufficiently discouraging, we stop.

The following traveling salesman program, using the Metropolis algorithm, should illustrate for you the important aspects of the simulated annealing technique.

```
      SUBROUTINE ANNEAL(X,Y,IORDER,NCITY)
         This algorithm finds the shortest round-trip path to NCITY cities whose coordinates are
         in the arrays X(I),Y(I). The array IORDER(I) specifies the order in which the cities are
         visited. On input, the elements of IORDER may be set to any permutation of the numbers
         1 to NCITY. This routine will return the best alternative path it can find.
      DIMENSION X(NCITY),Y(NCITY),IORDER(NCITY),N(6)
      LOGICAL ANS
      ALEN(X1,X2,Y1,Y2)=SQRT((X2-X1)**2+(Y2-Y1)**2)
      NOVER=100*NCITY             Maximum number of paths tried at any temperature.
      NLIMIT=10*NCITY             Maximum number of successful path changes before continuing.
      TFACTR=0.9                  Annealing schedule – T is reduced by this factor on each step.
      PATH=0.0
      T=0.5
      DO 11 I=1,NCITY-1           Calculate initial path length.
        I1=IORDER(I)
        I2=IORDER(I+1)
        PATH=PATH+ALEN(X(I1),X(I2),Y(I1),Y(I2))
11    CONTINUE
      I1=IORDER(NCITY)            Close the loop by tying path ends together.
      I2=IORDER(1)
      PATH=PATH+ALEN(X(I1),X(I2),Y(I1),Y(I2))
      IDUM=-1
      ISEED=111
      DO 13 J=1,100               Try up to 100 temperature steps.
        NSUCC=0
        DO 12 K=1,NOVER
1         N(1)=1+INT(NCITY*RAN3(IDUM))       Choose beginning of segment ..
          N(2)=1+INT((NCITY-1)*RAN3(IDUM)) ..and end of segment.
          IF (N(2).GE.N(1)) N(2)=N(2)+1
          NN=1+MOD((N(1)-N(2)+NCITY-1),NCITY) NN is the number of cities not on the segment.
          IF (NN.LT.3) GOTO 1
          IDEC=IRBIT1(ISEED)        Decide whether to do a segment reversal or transport.
          IF (IDEC.EQ.0) THEN       Do a transport.
            N(3)=N(2)+INT(ABS(NN-2)*RAN3(IDUM))+1
            N(3)=1+MOD(N(3)-1,NCITY)         Transport to a location not on the path.
            CALL TRNCST(X,Y,IORDER,NCITY,N,DE)  Calculate cost.
            CALL METROP(DE,T,ANS)       Consult the oracle.
            IF (ANS) THEN
              NSUCC=NSUCC+1
              PATH=PATH+DE
              CALL TRNSPT(IORDER,NCITY,N)  Carry out the transport.
            ENDIF
          ELSE                      Do a path reversal
            CALL REVCST(X,Y,IORDER,NCITY,N,DE)  Calculate cost.
            CALL METROP(DE,T,ANS)       Consult the oracle.
            IF (ANS) THEN
              NSUCC=NSUCC+1
              PATH=PATH+DE
              CALL REVERS(IORDER,NCITY,N)  Carry out the reversal.
            ENDIF
          ENDIF
          IF (NSUCC.GE.NLIMIT) GOTO 2  Finish early if we have enough successful changes.
12      CONTINUE
2       WRITE(*,*)
        WRITE(*,*) 'T =',T,' Path Length =',PATH
        WRITE(*,*) 'Successful Moves:  ',NSUCC
```

```
    T=T*TFACTR                   Annealing schedule.
    IF (NSUCC.EQ.0) RETURN       If no success, we are done.
13  CONTINUE
RETURN
END
```

```
SUBROUTINE REVCST(X,Y,IORDER,NCITY,N,DE)
```

This subroutine returns the value of the cost function for a proposed path reversal. NCITY is the number of cities, and arrays X(I),Y(I) give the coordinates of these cities. IORDER(I) holds the present itinerary. The first two values N(1) and N(2) of array N give the starting and ending cities along the path segment which is to be reversed. On output, DE is the cost of making the reversal. The actual reversal is not performed by this routine.

```
DIMENSION X(NCITY),Y(NCITY),IORDER(NCITY),N(6),XX(4),YY(4)
ALEN(X1,X2,Y1,Y2)=SQRT((X2-X1)**2+(Y2-Y1)**2)
N(3)=1+MOD((N(1)+NCITY-2),NCITY)     Find the city before N(1) ..
N(4)=1+MOD(N(2),NCITY)               .. and the city after N(2)
DO 11 J=1,4
    II=IORDER(N(J))              Find coordinates for the four cities involved.
    XX(J)=X(II)
    YY(J)=Y(II)
11  CONTINUE
DE=-ALEN(XX(1),XX(3),YY(1),YY(3))    Calculate cost of disconnecting the segment at both ends
*     -ALEN(XX(2),XX(4),YY(2),YY(4))   and reconnecting in the opposite order.
*     +ALEN(XX(1),XX(4),YY(1),YY(4))
*     +ALEN(XX(2),XX(3),YY(2),YY(3))
RETURN
END
```

```
SUBROUTINE REVERS(IORDER,NCITY,N)
```

This routine performs a path segment reversal. IORDER(I) is an input array giving the present itinerary. The vector N has as its first four elements the first and last cities N(1),N(2) of the path segment to be reversed, and the two cities N(3) and N(4) which immediately precede and follow this segment. N(3) and N(4) are found by subroutine REVCST. On output, IORDER(I) contains the segment from N(1) to N(2) in reversed order.

```
DIMENSION IORDER(NCITY),N(6)
NN=(1+MOD(N(2)-N(1)+NCITY,NCITY))/2 This many cities must be swapped to effect the reversal.
DO 11 J=1,NN
    K=1+MOD((N(1)+J-2),NCITY)      Start at the ends of the segment and swap pairs of cities, moving
    L=1+MOD((N(2)-J+NCITY),NCITY)  toward the center.
    ITMP=IORDER(K)
    IORDER(K)=IORDER(L)
    IORDER(L)=ITMP
11  CONTINUE
RETURN
END
```

```
SUBROUTINE TRNCST(X,Y,IORDER,NCITY,N,DE)
```

This subroutine returns the value of the cost function for a proposed path segment transport. NCITY is the number of cities, and arrays X(I) and Y(I) give the city coordinates. IORDER is an array giving the present itinerary. The first three elements of array N give the starting and ending cities of the path to be transported, and the point among the remaining cities after which it is to be inserted. On output, DE is the cost of the change. The actual transport is not performed by this routine.

```
DIMENSION X(NCITY),Y(NCITY),IORDER(NCITY),N(6),XX(6),YY(6)
ALEN(X1,X2,Y1,Y2)=SQRT((X2-X1)**2+(Y2-Y1)**2)
N(4)=1+MOD(N(3),NCITY)               Find the city following N(3)..
N(5)=1+MOD((N(1)+NCITY-2),NCITY)     ..and the one preceding N(1)..
N(6)=1+MOD(N(2),NCITY)               ..and the one following N(2).
DO 11 J=1,6
    II=IORDER(N(J))              Determine coordinates for the six cities involved.
    XX(J)=X(II)
```

```
      YY(J)=Y(II)
11    CONTINUE
      DE=-ALEN(XX(2),XX(6),YY(2),YY(6))     Calculate the cost of disconnecting the path segment from
     *    -ALEN(XX(1),XX(5),YY(1),YY(5))      N(1) to N(2), opening a space between N(3) and N(4),
     *    -ALEN(XX(3),XX(4),YY(3),YY(4))      connecting the segment in the space, and connecting
     *    +ALEN(XX(1),XX(3),YY(1),YY(3))      N(5) to N(6).
     *    +ALEN(XX(2),XX(4),YY(2),YY(4))
     *    +ALEN(XX(5),XX(6),YY(5),YY(6))
      RETURN
      END

      SUBROUTINE TRNSPT(IORDER,NCITY,N)
```

This routine does the actual path transport, once METROP has approved. IORDER is an input array of length NCITY giving the present itinerary. The array N has as its six elements the beginning N(1) and end N(2) of the path to be transported, the adjacent cities N(3) and N(4) between which the path is to be placed, and the cities N(5) and N(6) which precede and follow the path. N(4), N(5) and N(6) are calculated by subroutine TRNCST. On output, IORDER is modified to reflect the movement of the path segment.

```
      PARAMETER(MXCITY=1000)            Maximum number of cities anticipated.
      DIMENSION IORDER(NCITY),JORDER(MXCITY),N(6)
      M1=1+MOD((N(2)-N(1)+NCITY),NCITY)   Find the number of cities from N(1) to N(2) ..
      M2=1+MOD((N(5)-N(4)+NCITY),NCITY)   ..and the number from N(4) to N(5) ..
      M3=1+MOD((N(3)-N(6)+NCITY),NCITY)   ..and the number from N(6) to N(3).
      NN=1
      DO 11 J=1,M1
        JJ=1+MOD((J+N(1)-2),NCITY)      Copy the chosen segment.
        JORDER(NN)=IORDER(JJ)
        NN=NN+1
11    CONTINUE
      IF (M2.GT.0) THEN
        DO 12 J=1,M2                    Then copy the segment from N(4) to N(5).
          JJ=1+MOD((J+N(4)-2),NCITY)
          JORDER(NN)=IORDER(JJ)
          NN=NN+1
12      CONTINUE
      ENDIF
      IF (M3.GT.0) THEN
        DO 13 J=1,M3                    Finally, the segment from N(6) to N(3).
          JJ=1+MOD((J+N(6)-2),NCITY)
          JORDER(NN)=IORDER(JJ)
          NN=NN+1
13      CONTINUE
      ENDIF
      DO 14 J=1,NCITY
        IORDER(J)=JORDER(J)             Copy JORDER back into IORDER.
14    CONTINUE
      RETURN
      END

      SUBROUTINE METROP(DE,T,ANS)
```

Metropolis algorithm. ANS is a logical variable which issues a verdict on whether to accept a reconfiguration which leads to a change DE in the objective function E. If DE<0, ANS=.TRUE., while if DE>0, ANS is only .TRUE. with probability exp(-DE/T), where T is a temperature determined by the annealing schedule.

```
      PARAMETER(JDUM=1)
      LOGICAL ANS
      ANS=(DE.LT.0.0).OR.(RAN3(JDUM).LT.EXP(-DE/T))
      RETURN
      END
```

Assessing the Promise of Simulated Annealing

There is not yet enough practical experience with the method of simulated annealing to say definitively that it will realize its current promise. The method has several extremely attractive features, rather unique when compared with other optimization techniques.

First, it is not "greedy", in the sense that it is not easily fooled by the quick payoff achieved by falling into unfavorable local minima. Provided that sufficiently general reconfigurations are given, it wanders freely among local minima of depth less than about T. As T is lowered, the number of such minima qualifying for frequent visits is gradually reduced.

Second, configuration decisions tend to proceed in a logical order. Changes which cause the greatest energy differences are sifted over when the control parameter T is large. These decisions become more permanent as T is lowered, and attention then shifts more to smaller refinements in the solution. For example, in the traveling salesman problem with the Mississippi River twist, if λ is large, a decision to cross the Mississippi only twice is made at high T, while the specific routes on each side of the river are determined only at later stages.

The analogies to thermodynamics may be pursued to a greater extent than we have done here. Quantities analogous to specific heat and entropy may be defined, and these can be useful in monitoring the progress of the algorithm toward an acceptable solution. Information on this subject is found in the references by Kirkpatrick et al.

REFERENCES AND FURTHER READING:

Kirkpatrick, S., Gelatt, C.D., and Vecchi, M.P. 1983, *Science*, vol. 220, pp. 671–680.

Kirkpatrick, S. 1984, *Journal of Statistical Physics*, vol. 34, p. 975.

Vecchi, M.P. and Kirkpatrick, S. 1983, *IEEE Transactions on Computer Aided Design*, vol. CAD-2, p. 215.

Metropolis, N., Rosenbluth, A., Rosenbluth, M., Teller A., and Teller, E. 1953 *J. Chem. Phys.*, vol. 21, p. 1087.

Lin, S. 1965, *Bell Syst. Tech. Journ.*, vol. 44, p. 2245.

Christofides, N., Mingozzi, A., Toth, P., and Sandi, C., eds. 1979, *Combinatorial Optimization* (London and New York: Wiley-Interscience) [not simulated annealing, but other topics and algorithms].

Chapter 11. Eigensystems

11.0 Introduction

An $N \times N$ matrix $\mathbf{A}$ is said to have an *eigenvector* $\mathbf{x}$ and corresponding *eigenvalue* λ if

$$\mathbf{A} \cdot \mathbf{x} = \lambda \mathbf{x} \tag{11.0.1}$$

Obviously any multiple of an eigenvector $\mathbf{x}$ will also be an eigenvector, but we won't consider such multiples as being distinct eigenvectors. (The zero vector is not considered to be an eigenvector at all.) Evidently (11.0.1) can hold only if

$$\det |\mathbf{A} - \lambda \mathbf{1}| = 0 \tag{11.0.2}$$

which, if expanded out, is an Nth degree polynomial in λ whose roots are the eigenvalues. This proves that there are always N (not necessarily distinct) eigenvalues. Equal eigenvalues coming from multiple roots are called *degenerate*. Root-searching in the characteristic equation (11.0.2) is usually a very poor computational method for finding eigenvalues. We will learn much better ways in this chapter, as well as efficient ways for finding corresponding eigenvectors.

The above two equations also prove that every one of the N eigenvalues has a (not necessarily distinct) corresponding eigenvector: If λ is set to an eigenvalue, then the matrix $\mathbf{A} - \lambda \mathbf{1}$ is singular, and we know that every singular matrix has at least one nonzero vector in its nullspace (see §2.9 on singular value decomposition).

If you add $\tau \mathbf{x}$ to both sides of (11.0.1), you will easily see that the eigenvalues of any matrix can be changed or *shifted* by an additive constant τ by adding to the matrix that constant times the identity matrix. The eigenvectors are unchanged by this shift. Shifting, as we will see, is an important part of many algorithms for computing eigenvalues. We see also that there is no special significance to a zero eigenvalue. Any eigenvalue can be shifted to zero, or any zero eigenvalue can be shifted away from zero.

Definitions and Basic Facts

A matrix is called *symmetric* if it is equal to its transpose,

$$\mathbf{A} = \mathbf{A}^T \qquad \text{or} \qquad a_{ij} = a_{ji} \tag{11.0.3}$$

It is called *Hermitian* or *self-adjoint* if it equals the complex-conjugate of its transpose (its *Hermitian conjugate*, denoted by "†")

$$\mathbf{A} = \mathbf{A}^\dagger \qquad \text{or} \qquad a_{ij} = {a_{ji}}^* \tag{11.0.4}$$

It is termed *orthogonal* if its transpose equals its inverse,

$$\mathbf{A}^T \cdot \mathbf{A} = \mathbf{A} \cdot \mathbf{A}^T = \mathbf{1} \tag{11.0.5}$$

and *unitary* if its Hermitian conjugate equals its inverse. Finally, a matrix is called *normal* if it *commutes* with its Hermitian conjugate,

$$\mathbf{A} \cdot \mathbf{A}^\dagger = \mathbf{A}^\dagger \cdot \mathbf{A} \tag{11.0.6}$$

For real matrices, Hermitian means the same as symmetric, unitary means the same as orthogonal, and *both* of these distinct classes are normal.

The reason that "Hermitian" is an important concept has to do with eigenvalues. The eigenvalues of a Hermitian matrix are all real. In particular, the eigenvalues of a real symmetric matrix are all real. Contrariwise, the eigenvalues of a real nonsymmetric matrix may include real values, but may also include pairs of complex conjugate values; and the eigenvalues of a complex matrix that is not Hermitian will in general be complex.

The reason that "normal" is an important concept has to do with the eigenvectors. The eigenvectors of a normal matrix with nondegenerate (i.e. distinct) eigenvalues are complete and orthogonal, spanning the N-dimensional vector space. For a normal matrix with degenerate eigenvalues, we have the additional freedom of replacing the eigenvectors corresponding to a degenerate eigenvalue by linear combinations of themselves. Using this freedom, we can always perform Gram-Schmidt orthogonalization (consult any linear algebra text) and *find* a set of eigenvectors that are complete and orthogonal, just as in the nondegenerate case. The matrix whose columns are an orthonormal set of eigenvectors is evidently unitary. A special case is that the matrix of eigenvectors of a real, symmetric matrix is orthogonal, since the eigenvectors of that matrix are all real.

When a matrix is not normal, as typified by any random, nonsymmetric, real matrix, then in general we cannot find *any* orthonormal set of eigenvectors, nor even any pairs of eigenvectors that are orthogonal (except perhaps

by rare chance). While the N non-orthonormal eigenvectors will "usually" span the N-dimensional vector space, they do not always do so; that is, the eigenvectors are not always complete. Such a matrix is said to be *defective.*

Left and Right Eigenvectors

While the eigenvectors of a non-normal matrix are not particularly orthogonal among themselves, they *do* have an orthogonality relation with a different set of vectors, which we must now define. Up to now our eigenvectors have been column vectors which are multiplied to the right of a matrix $\mathbf{A}$, as in (11.0.1). These, more explicitly, are termed *right eigenvectors.* We could also, however, try to find row vectors, which multiply $\mathbf{A}$ to the left and satisfy

$$\mathbf{x} \cdot \mathbf{A} = \lambda \mathbf{x} \tag{11.0.7}$$

These are called *left eigenvectors.* Taking the transpose of equation (11.0.7), it should be obvious that every left eigenvector is the transpose of a right eigenvector *of the transpose of* $\mathbf{A}$. Now comparing to (11.0.2), and using the fact that the determinant of a matrix equals the determinant of its transpose, it becomes obvious that the left and right eigen*values* are identical.

If the matrix $\mathbf{A}$ is symmetrical, then the left and right eigenvectors are just transposes of each other, that is, have the same numerical values as components. Likewise, if the matrix is self-adjoint, the left and right eigenvectors are Hermitian conjugates of each other. For the general nonnormal case, however, we have the following calculation: Let $\mathbf{X}_R$ be the matrix formed by columns from the right eigenvectors, $\mathbf{X}_L$ be the matrix formed by rows from the left eigenvectors. Then (11.0.1) and (11.0.7) can be rewritten as

$$\mathbf{A} \cdot \mathbf{X}_R = \mathbf{X}_R \cdot \operatorname{diag}(\lambda_1 \ldots \lambda_N) \qquad \mathbf{X}_L \cdot \mathbf{A} = \operatorname{diag}(\lambda_1 \ldots \lambda_N) \cdot \mathbf{X}_L \tag{11.0.8}$$

Multiplying the first of these equations on the left by $\mathbf{X}_L$, the second on the right by $\mathbf{X}_R$, and subtracting the two, gives

$$(\mathbf{X}_L \cdot \mathbf{X}_R) \cdot \operatorname{diag}(\lambda_1 \ldots \lambda_N) = \operatorname{diag}(\lambda_1 \ldots \lambda_N) \cdot (\mathbf{X}_L \cdot \mathbf{X}_R) \tag{11.0.9}$$

This says that the matrix of dot products of the left and right eigenvectors commutes with the diagonal matrix of eigenvalues. But the only matrices that commute with a diagonal matrix *of distinct elements* are themselves diagonal. Thus, if the eigenvalues are nondegenerate, each left eigenvector is orthogonal to all right eigenvectors except its corresponding one, and vice versa. By choice of normalization, the dot products of corresponding left and right eigenvectors can always be made unity for any matrix with nondegenerate eigenvalues.

If some eigenvalues are degenerate, then either the left or the right eigenvectors corresponding to a degenerate eigenvalue must be linearly combined among themselves to achieve orthogonality with the right or left ones respectively. This can always be done by a procedure akin to Gram-Schmidt orthogonalization. The normalization can then be adjusted to give unity for the nonzero dot products between corresponding left and right eigenvectors. If the dot product of corresponding left and right eigenvectors is zero at this stage, then you have a case where the eigenvectors are incomplete! Note that incomplete eigenvectors can occur only where there are degenerate eigenvalues, but do not always occur in such cases (in fact, never occur for the class of "normal" matrices). See Stoer and Bulirsch for a clear discussion.

In both the degenerate and nondegenerate cases, the final normalization to unity of all nonzero dot products produces the result: The matrix whose rows are left eigenvectors is the inverse matrix of the matrix whose columns are right eigenvectors, *if the inverse exists.*

Diagonalization of a Matrix

Multiplying the first equation in (11.0.8) by $\mathbf{X}_L$, and using the fact that $\mathbf{X}_L$ and $\mathbf{X}_R$ are matrix inverses, we get

$$\mathbf{X}_R^{-1} \cdot \mathbf{A} \cdot \mathbf{X}_R = \mathrm{diag}(\lambda_1 \ldots \lambda_N) \tag{11.0.10}$$

This is a particular case of a *similarity transform* of the matrix $\mathbf{A}$,

$$\mathbf{A} \quad \rightarrow \quad \mathbf{Z}^{-1} \cdot \mathbf{A} \cdot \mathbf{Z} \tag{11.0.11}$$

for some transformation matrix $\mathbf{Z}$. Similarity transformations play a crucial role in the computation of eigenvalues, because they leave the eigenvalues of a matrix unchanged. This is easily seen from

$$\begin{aligned} \det\left|\mathbf{Z}^{-1} \cdot \mathbf{A} \cdot \mathbf{Z} - \lambda \mathbf{1}\right| &= \det\left|\mathbf{Z}^{-1} \cdot (\mathbf{A} - \lambda \mathbf{1}) \cdot \mathbf{Z}\right| \\ &= \det|\mathbf{Z}| \, \det|\mathbf{A} - \lambda\mathbf{1}| \, \det\left|\mathbf{Z}^{-1}\right| \\ &= \det|\mathbf{A} - \lambda\mathbf{1}| \end{aligned} \tag{11.0.12}$$

Equation (11.0.10) shows that any matrix with complete eigenvectors (which includes all normal matrices and "most" random nonnormal ones) can be diagonalized by a similarity transformation, that the columns of the transformation matrix that effects the diagonalization are the right eigenvectors, and that the rows of its inverse are the left eigenvectors.

For real, symmetric matrices, the eigenvectors are real and orthonormal, so the transformation matrix is orthogonal. The similarity transformation is then also an *orthogonal transformation* of the form

$$\mathbf{A} \quad \rightarrow \quad \mathbf{Z}^T \cdot \mathbf{A} \cdot \mathbf{Z} \tag{11.0.13}$$

While real nonsymmetric matrices can be diagonalized in their usual case of complete eigenvectors, the transformation matrix is not necessarily real. It turns out, however, that a real similarity transformation can "almost" do the job. It can reduce the matrix down to a form with little two-by-two blocks along the diagonal, all other elements zero. Each two-by-two block corresponds to a complex-conjugate pair of complex eigenvalues. We will see this idea exploited in some routines given later in the chapter.

The "grand strategy" of virtually all modern eigensystem routines is to nudge the matrix $\mathbf{A}$ towards diagonal form by a sequence of similarity transformations,

$$\begin{aligned} \mathbf{A} \quad &\rightarrow \quad \mathbf{P}_1^{-1} \cdot \mathbf{A} \cdot \mathbf{P}_1 \quad \rightarrow \quad \mathbf{P}_2^{-1} \cdot \mathbf{P}_1^{-1} \cdot \mathbf{A} \cdot \mathbf{P}_1 \cdot \mathbf{P}_2 \\ &\rightarrow \quad \mathbf{P}_3^{-1} \cdot \mathbf{P}_2^{-1} \cdot \mathbf{P}_1^{-1} \cdot \mathbf{A} \cdot \mathbf{P}_1 \cdot \mathbf{P}_2 \mathbf{P}_3 \quad \rightarrow \quad \text{etc.} \end{aligned} \tag{11.0.14}$$

If we get all the way to diagonal form, then the eigenvectors are the columns of the accumulated transformation

$$\mathbf{X}_R = \mathbf{P}_1 \cdot \mathbf{P}_2 \cdot \mathbf{P}_3 \cdot \ldots \tag{11.0.15}$$

Sometimes we do not want to go all the way to diagonal form. For example, if we are interested only in eigenvalues, not eigenvectors, it is enough to transform the matrix $\mathbf{A}$ to be triangular, with all elements below (or above) the diagonal zero. In this case the diagonal elements are already the eigenvalues, as you can see by mentally evaluating (11.0.2) using expansion by minors.

There are two rather different sets of techniques for implementing the grand strategy (11.0.14). It turns out that they work rather well in combination, so most modern eigensystem routines use both. The first set of techniques constructs individual $\mathbf{P}_i$'s as explicit "atomic" transformations designed to perform specific tasks, for example zeroing a particular off-diagonal element (Jacobi transformation, §11.1), or a whole particular row or column (Householder transformation, §11.2; elimination method, §11.5). In general, a finite sequence of these simple transformations cannot completely diagonalize a matrix. There are then two choices: either use the finite sequence of transformations to go most of the way (e.g., to some special form like *tridiagonal* or *Hessenberg*, see §11.2 and §11.5 below) and follow up with the second set of techniques about to be mentioned; or else iterate the finite sequence of simple transformations over and over until the deviation of the matrix from diagonal is negligibly small. This latter approach is conceptually simplest, so

we will discuss it in the next section; however, for N greater than ~ 10, it is computationally inefficient by a roughly constant factor ~ 5.

The second set of techniques, called *factorization methods*, is more subtle. Suppose that the matrix $\mathbf{A}$ can be factored into a left factor $\mathbf{F}_L$ and a right factor $\mathbf{F}_R$. Then

$$\mathbf{A} = \mathbf{F}_L \cdot \mathbf{F}_R \qquad \text{or equivalently} \qquad \mathbf{F}_L^{-1} \cdot \mathbf{A} = \mathbf{F}_R \tag{11.0.16}$$

If we now multiply back together the factors in the reverse order, and use the second equation in (11.0.16) we get

$$\mathbf{F}_R \cdot \mathbf{F}_L = \mathbf{F}_L^{-1} \cdot \mathbf{A} \cdot \mathbf{F}_L \tag{11.0.17}$$

which we recognize as having effected a similarity transformation on $\mathbf{A}$ with the transformation matrix being $\mathbf{F}_L$! In §11.3 and §11.6 we will discuss the *QR method* which exploits this idea.

Factorization methods also do not converge exactly in a finite number of transformations. But the better ones do converge rapidly and reliably, and, when following an appropriate initial reduction by simple similarity transformations, they are the methods of choice.

"Eigenpackages of Canned Eigenroutines"

You have probably gathered by now that the solution of eigensystems is a fairly complicated business. It is. It is one of the few subjects covered in this book for which we do *not* recommend that you avoid canned routines. On the contrary, the purpose of this chapter is precisely to give you some appreciation of what is going on inside such canned routines, so that you can make intelligent choices about using them, and intelligent diagnoses when something goes wrong.

You will find that almost all canned routines in use nowadays trace their ancestry back to routines published in Wilkinson and Reinsch's *Handbook for Automatic Computation, Vol. II, Linear Algebra.* This excellent reference, containing papers by a number of authors, is the Bible of the field. A public-domain implementation of the *Handbook* routines in FORTRAN is the EISPACK set of programs (see references). The routines in this chapter are translations of either the *Handbook* or EISPACK routines, so understanding these will take you a lot of the way towards understanding those canonical packages.

IMSL and NAG each provide proprietary implementations of what are essentially the Handbook routines.

A good "eigenpackage" will provide separate routines, or separate paths through sequences of routines, for the following desired calculations:

- all eigenvalues and no eigenvectors
- all eigenvalues and some corresponding eigenvectors

- all eigenvalues and all corresponding eigenvectors

The purpose of these distinctions is to save compute time and storage; it is wasteful to calculate eigenvectors that you don't need. Often one is interested only in the eigenvectors corresponding to the largest few eigenvalues, or largest few in magnitude, or few that are negative. The method usually used to calculate "some" eigenvectors is typically more efficient than calculating all eigenvectors if you desire fewer than about a quarter of the eigenvectors.

A good eigenpackage also provides separate paths for each of the above calculations for each of the following special forms of the matrix:

- real, symmetric, tridiagonal
- real, symmetric, banded (only a small number of sub- and superdiagonals are nonzero)
- real, symmetric,
- real, nonsymmetric
- complex, Hermitian
- complex, non-Hermitian

Again, the purpose of these distinctions is to save time and storage by using the *least* general routine that will serve in any particular application.

Many eigenpackages also deal with the so-called *generalized eigenproblem,*

$$\mathbf{A} \cdot \mathbf{x} = \lambda \mathbf{B} \cdot \mathbf{x} \tag{11.0.18}$$

where **A** and **B** are both matrices. This problem can in fact be handled by ordinary eigensystem techniques, taking the "square root" of whichever of the matrices is known to be positive definite, in its diagonal frame (see a good linear algebra text or Stoer and Bulirsch §6.8). Nevertheless it is convenient and efficient to have a package routine which attacks the problem directly.

In this chapter, as a bare introduction, we give good routines for the following paths:

- all eigenvalues and eigenvectors of a real, symmetric, tridiagonal matrix (§11.3)
- all eigenvalues and eigenvectors of a real, symmetric, matrix (§11.1–§11.3)
- all eigenvalues and eigenvectors of a complex, Hermitian matrix (§11.4)
- all eigenvalues and no eigenvectors of a real, nonsymmetric matrix (§11.5–§11.6)

We also discuss, in §11.7, how to obtain some eigenvectors of nonsymmetric matrices by the method of inverse iteration.

REFERENCES AND FURTHER READING:

Wilkinson, J.H., and Reinsch, C. 1971, *Linear Algebra*, vol. II of *Handbook for Automatic Computation* (New York: Springer-Verlag).

Wilkinson, J.H. 1965, *The Algebraic Eigenvalue Problem* (New York: Oxford University Press).

Smith, B.T., et al. 1976, *Matrix Eigensystem Routines — EISPACK Guide*, 2nd ed., vol. 6 of Lecture Notes in Computer Science (New York: Springer-Verlag).

Golub, Gene H., and Van Loan, Charles F. 1983, *Matrix Computations* (Baltimore: Johns Hopkins University Press), §7.7 [extensive references on the generalized eigenvalue problem].

Acton, Forman S. 1970, *Numerical Methods That Work* (New York: Harper and Row), Chapter 13.

Stoer, J., and Bulirsch, R. 1980, *Introduction to Numerical Analysis* (New York: Springer-Verlag), Chapter 6.

NAG Fortran Library Manual Mark 8, 1980 (NAG Central Office, 7 Banbury Road, Oxford OX26NN, U.K.), Chapter F02.

IMSL Library Reference Manual, 1980, ed. 8 (IMSL Inc., 7500 Bellaire Boulevard, Houston TX 77036).

11.1 Jacobi Transformations of a Symmetric Matrix

The Jacobi method consists of a sequence of orthogonal similarity transformations of the form of equation (11.0.14). Each transformation (a *Jacobi rotation*) is just a plane rotation designed to annihilate one of the off-diagonal matrix elements. Successive transformations undo previously set zeros, but the off-diagonal elements nevertheless get smaller and smaller, until the matrix is diagonal to machine precision. Accumulating the product of the transformations as you go gives the matrix of eigenvectors, equation (11.0.15), while the elements of the final diagonal matrix are the eigenvalues.

The Jacobi method is absolutely foolproof for all real symmetric matrices. For matrices of order greater than about 10, say, the algorithm is slower, by a significant constant factor, than the QR method we shall give in §11.3. However, the Jacobi algorithm is much simpler than the more efficient methods. We thus recommend it for matrices of moderate order, where expense is not a major consideration.

The basic Jacobi rotation $\mathbf{P}_{pq}$ is a matrix of the form

$$\mathbf{P}_{pq} = \begin{bmatrix} 1 & & & & & & \\ & \cdots & & & & & \\ & & c & \cdots & s & & \\ & & \vdots & 1 & \vdots & & \\ & & -s & \cdots & c & & \\ & & & & & \cdots & \\ & & & & & & 1 \end{bmatrix} \tag{11.1.1}$$

Here all the diagonal elements are unity except for the two elements c in rows (and columns) p and q. All off-diagonal elements are zero except the two elements s and $-s$. The numbers c and s are the cosine and sine of a rotation angle ϕ, so $c^2 + s^2 = 1$.

A plane rotation such as (11.1.1) is used to transform the matrix **A** according to

$$\mathbf{A}' = \mathbf{P}_{pq}^T \cdot \mathbf{A} \cdot \mathbf{P}_{pq} \tag{11.1.2}$$

Now, $\mathbf{P}_{pq}^T \cdot \mathbf{A}$ changes only rows p and q of **A**, while $\mathbf{A} \cdot \mathbf{P}_{pq}$ changes only columns p and q. Notice that the subscripts p and q do not denote components of $\mathbf{P}_{pq}$, but rather label which kind of rotation the matrix is, i.e. which rows and columns it affects. Thus the changed elements of **A** in (11.1.2) are only in the p and q rows and columns indicated below:

$$\mathbf{A}' = \begin{bmatrix} & \cdots & a'_{1p} & \cdots & a'_{1q} & \cdots & \\ \vdots & & \vdots & & \vdots & & \vdots \\ a'_{p1} & \cdots & a'_{pp} & \cdots & a'_{pq} & \cdots & a'_{pn} \\ \vdots & & \vdots & & \vdots & & \vdots \\ a'_{q1} & \cdots & a'_{qp} & \cdots & a'_{qq} & \cdots & a'_{qn} \\ \vdots & & \vdots & & \vdots & & \vdots \\ & \cdots & a'_{np} & \cdots & a'_{nq} & \cdots & \end{bmatrix} \tag{11.1.3}$$

Multiplying out equation (11.1.2) and using the symmetry of **A**, we get the explicit formulas

$$\begin{aligned} a'_{rp} &= ca_{rp} - sa_{rq} \\ a'_{rq} &= ca_{rq} + sa_{rp} \qquad\qquad r \neq p,\ r \neq q && (11.1.4) \\ a'_{pp} &= c^2 a_{pp} + s^2 a_{qq} - 2sca_{pq} && (11.1.5) \\ a'_{qq} &= s^2 a_{pp} + c^2 a_{qq} + 2sca_{pq} && (11.1.6) \\ a'_{pq} &= (c^2 - s^2)a_{pq} + sc(a_{pp} - a_{qq}). && (11.1.7) \end{aligned}$$

The idea of the Jacobi method is to try to zero the off-diagonal elements by a series of plane rotations. Accordingly, to set $a'_{pq} = 0$, equation (11.1.7) gives the following expression for the rotation angle

$$\theta \equiv \cot 2\phi \equiv \frac{c^2 - s^2}{2sc} = \frac{a_{qq} - a_{pp}}{2a_{pq}}. \tag{11.1.8}$$

If we let $t \equiv s/c$, the definition of θ can be rewritten

$$t^2 + 2t\theta - 1 = 0. \tag{11.1.9}$$

The smaller root of this equation corresponds to a rotation angle less than $\pi/4$ in magnitude; this choice at each stage gives the most stable reduction. Using the form of the quadratic formula with the discriminant in the denominator, we can write this smaller root as

$$t = \frac{\operatorname{sgn}(\theta)}{|\theta| + \sqrt{\theta^2 + 1}}. \tag{11.1.10}$$

If θ is so large that θ^2 would overflow on the computer, we set $t = 1/(2\theta)$. It now follows that

$$c = \frac{1}{\sqrt{\theta^2 + 1}}, \tag{11.1.11}$$

$$s = tc \tag{11.1.12}$$

When we actually use equations (11.1.4) - (11.1.7) numerically, we rewrite them to minimize roundoff error. Equation (11.1.7) is replaced by

$$a'_{pq} = 0. \tag{11.1.13}$$

The idea in the remaining equations is to set the new quantity equal to the old quantity plus a small correction. Thus we can use (11.1.7) and (11.1.13) to eliminate a_{qq} from (11.1.5), giving

$$a'_{pp} = a_{pp} - ta_{pq}. \tag{11.1.14}$$

Similarly,

$$a'_{qq} = a_{qq} + ta_{pq}, \tag{11.1.15}$$

$$a'_{rp} = a_{rp} - s(a_{rq} + \tau a_{rp}), \tag{11.1.16}$$

$$a'_{rq} = a_{rq} + s(a_{rp} - \tau a_{rq}), \tag{11.1.17}$$

where τ $(= \tan\phi/2)$ is defined by

$$\tau \equiv \frac{s}{1+c}. \tag{11.1.18}$$

One can see the convergence of the Jacobi method by considering the sum of the squares of the off-diagonal elements

$$S = \sum_{r \neq s} |a_{rs}|^2. \tag{11.1.19}$$

Equations (11.1.4) - (11.1.7) imply that

$$S' = S - 2|a_{pq}|^2. \tag{11.1.20}$$

(Since the transformation is orthogonal, the sum of the squares of the diagonal elements increases correspondingly by $2|a_{pq}|^2$.) The sequence of S's thus decreases monotonically. Since the sequence is bounded below by zero, and since we can choose a_{pq} to be whatever element we want, the sequence can be made to converge to zero.

Eventually one obtains a matrix $\mathbf{D}$ that is diagonal to machine precision. The diagonal elements give the eigenvalues of the original matrix $\mathbf{A}$, since

$$\mathbf{D} = \mathbf{V}^T \cdot \mathbf{A} \cdot \mathbf{V}, \tag{11.1.21}$$

where

$$\mathbf{V} = \mathbf{P}_1 \cdot \mathbf{P}_2 \cdot \mathbf{P}_3 \cdots \tag{11.1.22}$$

the $\mathbf{P}_i$'s being the successive Jacobi rotation matrices. The columns of $\mathbf{V}$ are the eigenvectors (since $\mathbf{A} \cdot \mathbf{V} = \mathbf{V} \cdot \mathbf{D}$). They can be computed by applying

$$\mathbf{V}' = \mathbf{V} \cdot \mathbf{P}_i \tag{11.1.23}$$

at each stage of calculation, where initially $\mathbf{V}$ is the identity matrix. In detail, equation (11.1.23) is

$$\begin{aligned}
v'_{rs} &= v_{rs} \qquad (s \neq p,\ s \neq q) \\
v'_{rp} &= c v_{rp} - s v_{rq}, \\
v'_{rq} &= s v_{rp} + c v_{rq}.
\end{aligned} \tag{11.1.24}$$

We rewrite these equations in terms of τ as in equations (11.1.16) and (11.1.17) to minimize roundoff.

The only remaining question is the strategy one should adopt for the order in which the elements are to be annihilated. Jacobi's original algorithm of 1846 searched the whole upper triangle at each stage and set the largest off-diagonal element to zero. This is a reasonable strategy for hand calculation, but it is prohibitive on a computer since the search alone makes each Jacobi rotation a process of order N^2 instead of N.

A better strategy for our purposes is the *cyclic Jacobi method*, where one annihilates elements in strict order. For example, one can simply proceed down the rows: $\mathbf{P}_{12}, \mathbf{P}_{13}, ..., \mathbf{P}_{1n}$; then $\mathbf{P}_{23}, \mathbf{P}_{24}$, etc. One can show that

convergence is generally quadratic for both the original or the cyclic Jacobi methods, for nondegenerate eigenvalues. One such set of $n(n-1)/2$ Jacobi rotations is called a *sweep*.

The program below, based on the *Handbook* and EISPACK implementations, uses two further refinements:

- In the first three sweeps, we carry out the pq rotation only if $|a_{pq}| > \epsilon$ for some threshold value

$$\epsilon = \frac{1}{5}\frac{S_o}{n^2}, \tag{11.1.25}$$

where S_o is the sum of the off-diagonal moduli,

$$S_o = \sum_{r<s} |a_{rs}| \tag{11.1.26}$$

- After four sweeps, if $|a_{pq}| \ll |a_{pp}|$ and $|a_{pq}| \ll |a_{qq}|$, we set $|a_{pq}| = 0$ and skip the rotation. The criterion used in the comparison is $|a_{pq}| < 10^{-(D+2)}|a_{pp}|$, where D is the number of significant decimal digits on the machine, and similarly for $|a_{qq}|$.

In the following routine the N×N symmetric matrix A is stored in an NP×NP array. On output, the superdiagonal elements of A are destroyed, but the diagonal and subdiagonal are unchanged and give full information on the original symmetric matrix A. The parameter D is a vector of length NP. On output, it returns the eigenvalues of A in its first N elements. During the computation, it contains the current diagonal of A. The matrix V outputs the normalized eigenvector belonging to D(K) in its Kth column. The parameter NROT is the number of Jacobi rotations that were needed to achieve convergence.

Typical matrices require 6 to 10 sweeps to achieve convergence, or $3n^2$ to $5n^2$ Jacobi rotations. Each rotation requires of order $4n$ operations, each consisting of a multiply and an add, so the total labor is of order $12n^3$ to $20n^3$ operations. Calculation of the eigenvectors as well as the eigenvalues changes the operation count from $4n$ to $6n$ per rotation, which is only a 50 percent overhead.

```
SUBROUTINE JACOBI(A,N,NP,D,V,NROT)
   Computes all eigenvalues and eigenvectors of a real symmetric matrix A, which is of size N
   by N, stored in a physical NP by NP array. On output, elements of A above the diagonal are
   destroyed. D returns the eigenvalues of A in its first N elements. V is a matrix with the same
   logical and physical dimensions as A whose columns contain, on output, the normalized
   eigenvectors of A. NROT returns the number of Jacobi rotations which were required.
PARAMETER (NMAX=100)
DIMENSION A(NP,NP),D(NP),V(NP,NP),B(NMAX),Z(NMAX)
DO 12 IP=1,N                          Initialize to the identity matrix.
   DO 11 IQ=1,N
      V(IP,IQ)=0.
11 CONTINUE
   V(IP,IP)=1.
```

```
   12 CONTINUE
DO 13 IP=1,N
     B(IP)=A(IP,IP)              Initialize B and D to the diagonal of A.
     D(IP)=B(IP)
     Z(IP)=0.                    This vector will accumulate terms of the form ta_pq as in equation
   13 CONTINUE                          (11.1.14).
NROT=0
DO 24 I=1,50
     SM=0.
     DO 15 IP=1,N-1              Sum off-diagonal elements.
         DO 14 IQ=IP+1,N
             SM=SM+ABS(A(IP,IQ))
           14 CONTINUE
       15 CONTINUE
     IF(SM.EQ.O.)RETURN          The normal return, which relies on quadratic convergence to machine
     IF(I.LT.4)THEN                     underflow.
         TRESH=0.2*SM/N**2                    ...on the first three sweeps.
     ELSE
         TRESH=0.                             ...thereafter.
     ENDIF
     DO 22 IP=1,N-1
         DO 21 IQ=IP+1,N
             G=100.*ABS(A(IP,IQ))
     After four sweeps, skip the rotation if the off-diagonal element is small.
             IF((I.GT.4).AND.(ABS(D(IP))+G.EQ.ABS(D(IP)))
*               .AND.(ABS(D(IQ))+G.EQ.ABS(D(IQ))))THEN
                 A(IP,IQ)=0.
             ELSE IF(ABS(A(IP,IQ)).GT.TRESH)THEN
                 H=D(IQ)-D(IP)
                 IF(ABS(H)+G.EQ.ABS(H))THEN
                     T=A(IP,IQ)/H                     t = 1/(2θ)
                 ELSE
                     THETA=0.5*H/A(IP,IQ)             Equation (11.1.10).
                     T=1./(ABS(THETA)+SQRT(1.+THETA**2))
                     IF(THETA.LT.0.)T=-T
                 ENDIF
                 C=1./SQRT(1+T**2)
                 S=T*C
                 TAU=S/(1.+C)
                 H=T*A(IP,IQ)
                 Z(IP)=Z(IP)-H
                 Z(IQ)=Z(IQ)+H
                 D(IP)=D(IP)-H
                 D(IQ)=D(IQ)+H
                 A(IP,IQ)=0.
                 DO 16 J=1,IP-1               Case of rotations 1 ≤ j < p.
                     G=A(J,IP)
                     H=A(J,IQ)
                     A(J,IP)=G-S*(H+G*TAU)
                     A(J,IQ)=H+S*(G-H*TAU)
                   16 CONTINUE
                 DO 17 J=IP+1,IQ-1            Case of rotations p < j < q.
                     G=A(IP,J)
                     H=A(J,IQ)
                     A(IP,J)=G-S*(H+G*TAU)
                     A(J,IQ)=H+S*(G-H*TAU)
                   17 CONTINUE
                 DO 18 J=IQ+1,N               Case of rotations q < j ≤ n.
                     G=A(IP,J)
                     H=A(IQ,J)
                     A(IP,J)=G-S*(H+G*TAU)
                     A(IQ,J)=H+S*(G-H*TAU)
                   18 CONTINUE
                 DO 19 J=1,N
```

```
                  G=V(J,IP)
                  H=V(J,IQ)
                  V(J,IP)=G-S*(H+G*TAU)
                  V(J,IQ)=H+S*(G-H*TAU)
19              CONTINUE
                NROT=NROT+1
              ENDIF
21          CONTINUE
22        CONTINUE
          DO 23 IP=1,N
            B(IP)=B(IP)+Z(IP)
            D(IP)=B(IP)             Update D with the sum of ta_pq,
            Z(IP)=0.                and reinitialize Z.
23        CONTINUE
24      CONTINUE
        PAUSE '50 iterations should never happen'
        RETURN
        END
```

Note that the above routine assumes that underflows are set to zero. On machines where this is not true, the program must be modified.

The eigenvalues are not ordered on output. If sorting is desired, the following routine can be invoked to reorder the output of JACOBI or of later routines in this chapter. (The method, straight insertion, is N^2 rather than $N \log N$; but since you have just done an N^3 procedure to get the eigenvalues, you can afford yourself this little indulgence.)

```
      SUBROUTINE EIGSRT(D,V,N,NP)
        Given the eigenvalues D and eigenvectors V as output from JACOBI (§11.1) or TQLI (§11.3),
        this routine sorts the eigenvalues into descending order, and rearranges the columns of V
        correspondingly. The method is straight insertion.
      DIMENSION D(NP),V(NP,NP)
      DO 13 I=1,N-1
        K=I
        P=D(I)
        DO 11 J=I+1,N
          IF(D(J).GE.P)THEN
            K=J
            P=D(J)
          ENDIF
11      CONTINUE
        IF(K.NE.I)THEN
          D(K)=D(I)
          D(I)=P
          DO 12 J=1,N
            P=V(J,I)
            V(J,I)=V(J,K)
            V(J,K)=P
12        CONTINUE
        ENDIF
13    CONTINUE
      RETURN
      END
```

REFERENCES AND FURTHER READING:

Golub, Gene H., and Van Loan, Charles F. 1983, *Matrix Computations* (Baltimore: Johns Hopkins University Press), §8.4.

Smith, B.T., et al. 1976, *Matrix Eigensystem Routines — EISPACK Guide*, 2nd ed., vol. 6 of Lecture Notes in Computer Science (New York: Springer-Verlag).

Wilkinson, J.H., and Reinsch, C. 1971, *Linear Algebra*, vol. II of *Handbook for Automatic Computation* (New York: Springer-Verlag).

11.2 Reduction of a Symmetric Matrix to Tridiagonal Form: Givens and Householder Reductions

As already mentioned, the optimum strategy for finding eigenvalues and eigenvectors is, first, to reduce the matrix to a simple form, only then beginning an iterative procedure. For symmetric matrices, the preferred simple form is tridiagonal. The *Givens reduction* is a modification of the Jacobi method. Instead of trying to reduce the matrix all the way to diagonal form, we are content to stop when the matrix is tridiagonal. This allows the procedure to be carried out *in a finite number of steps*, unlike the Jacobi method which requires iteration to convergence.

Givens Method

For the Givens method, we choose the rotation angle in equation (11.1.1) so as to zero an element that is *not* at one of the four "corners," i.e., not a_{pp}, a_{pq}, or a_{qq} in equation (11.1.3). Specifically, we first choose $\mathbf{P}_{23}$ to annihilate a_{31} (and, by symmetry, a_{13}). Then we choose $\mathbf{P}_{24}$ to annihilate a_{41}. In general, we choose the sequence

$$\mathbf{P}_{23}, \mathbf{P}_{24}, \ldots, \mathbf{P}_{2n}; \mathbf{P}_{34}, \ldots, \mathbf{P}_{3n}; \ldots; \mathbf{P}_{n-1,n}$$

where $\mathbf{P}_{jk}$ annihilates $a_{k,j-1}$. The method works because elements such as a'_{rp} and a'_{rq}, with $r \neq p$ $r \neq q$, are linear combinations of the old quantities a_{rp} and a_{rq}, by equation (11.1.4). Thus, if a_{rp} and a_{rq} have already been set to zero, they remain zero as the reduction proceeds. Evidently, of order $n^2/2$ rotations are required, and the number of multiplications in a straightforward implementation is of order $4n^3/3$, not counting those for keeping track of the product of the transformation matrices, required for the eigenvectors.

The Householder method, to be discussed next, is just as stable as the Givens reduction and it is a factor of 2 more efficient, so the Givens method is not generally used. Recent work (see Jacobs reference) has shown that the Givens reduction can be reformulated to reduce the number of operations by

a factor of 2, and also avoid the necessity of taking square roots. This makes the algorithm competitive with the Householder reduction. However, there is not yet enough experience with the method to cause us to recommend it over the Householder method.

Householder Method

The Householder algorithm reduces an $n \times n$ symmetric matrix $\mathbf{A}$ to tridiagonal form by $n-2$ orthogonal transformations. Each transformation annihilates the required part of a whole column and whole corresponding row. The basic ingredient is a Householder matrix $\mathbf{P}$, which has the form

$$\mathbf{P} = \mathbf{1} - 2\mathbf{w} \cdot \mathbf{w}^T \tag{11.2.1}$$

where $\mathbf{w}$ is a real vector with $|\mathbf{w}|^2 = 1$. (In the present notation, the *outer* or matrix product of two vectors, $\mathbf{a}$ and $\mathbf{b}$ is written $\mathbf{a} \cdot \mathbf{b}^T$, while the *inner* or scalar product of the vectors is written as $\mathbf{a}^T \cdot \mathbf{b}$.) The matrix $\mathbf{P}$ is orthogonal, because

$$\begin{aligned} \mathbf{P}^2 &= (\mathbf{1} - 2\mathbf{w} \cdot \mathbf{w}^T) \cdot (\mathbf{1} - 2\mathbf{w} \cdot \mathbf{w}^T) \\ &= \mathbf{1} - 4\mathbf{w} \cdot \mathbf{w}^T + 4\mathbf{w} \cdot (\mathbf{w}^T \cdot \mathbf{w}) \cdot \mathbf{w}^T \\ &= \mathbf{1} \end{aligned} \tag{11.2.2}$$

Therefore $\mathbf{P} = \mathbf{P}^{-1}$. But $\mathbf{P}^T = \mathbf{P}$, and so $\mathbf{P}^T = \mathbf{P}^{-1}$, proving orthogonality.

Rewrite $\mathbf{P}$ as

$$\mathbf{P} = \mathbf{1} - \frac{\mathbf{u} \cdot \mathbf{u}^T}{H}, \tag{11.2.3}$$

where the scalar H is

$$H \equiv \frac{1}{2} |\mathbf{u}|^2 \tag{11.2.4}$$

and $\mathbf{u}$ can now be any vector. Suppose $\mathbf{x}$ is the vector composed of the first column of $\mathbf{A}$. Choose

$$\mathbf{u} = \mathbf{x} \mp |\mathbf{x}| \mathbf{e}_1 \tag{11.2.5}$$

where $\mathbf{e}_1$ is the unit vector $[1, 0, \ldots, 0]^T$, and the choice of signs will be made later. Then

$$\begin{aligned}
\mathbf{P} \cdot \mathbf{x} &= \mathbf{x} - \frac{\mathbf{u}}{H} \cdot (\mathbf{x} \mp |\mathbf{x}|\mathbf{e}_1)^T \cdot \mathbf{x} \\
&= \mathbf{x} - \frac{2\mathbf{u} \cdot (|\mathbf{x}|^2 \mp |\mathbf{x}| x_1)}{2|\mathbf{x}|^2 \mp 2|\mathbf{x}| x_1} \\
&= \mathbf{x} - \mathbf{u} \\
&= \pm |\mathbf{x}| \mathbf{e}_1
\end{aligned} \tag{11.2.6}$$

This shows that the Householder matrix $\mathbf{P}$ acts on a given vector $\mathbf{x}$ to zero all its elements except the first one.

To reduce a symmetric matrix $\mathbf{A}$ to tridiagonal form, we choose the vector $\mathbf{x}$ for the first Householder matrix to be the lower $n-1$ elements of the first column. Then the lower $n-2$ elements will be zeroed:

$$\begin{aligned}
\mathbf{P}_1 \cdot \mathbf{A} &= \left[\begin{array}{c|cccc} 1 & 0 & 0 & \cdots & 0 \\ \hline 0 & & & & \\ 0 & & & & \\ \vdots & & {}^{(n-1)}\mathbf{P}_1 & & \\ 0 & & & & \end{array}\right] \cdot \left[\begin{array}{c|cccc} a_{11} & a_{12} & a_{13} & \cdots & a_{1n} \\ \hline a_{21} & & & & \\ a_{31} & & & & \\ \vdots & & & \text{irrelevant} & \\ a_{n1} & & & & \end{array}\right] \\
&= \left[\begin{array}{c|cccc} a_{11} & a_{12} & a_{13} & \cdots & a_{1n} \\ \hline k & & & & \\ 0 & & & & \\ \vdots & & & \text{irrelevant} & \\ 0 & & & & \end{array}\right]
\end{aligned} \tag{11.2.7}$$

Here we have written the matrices in partitioned form, with ${}^{(n-1)}\mathbf{P}$ denoting a Householder matrix with dimensions $(n-1) \times (n-1)$. The quantity k is simply plus or minus the magnitude of the vector $[a_{21}, \ldots, a_{n1}]^T$.

The complete orthogonal transformation is now

$$\mathbf{A}' = \mathbf{P} \cdot \mathbf{A} \cdot \mathbf{P} = \left[\begin{array}{c|cccc} a_{11} & k & 0 & \cdots & 0 \\ \hline k & & & & \\ 0 & & & & \\ \vdots & & & \text{irrelevant} & \\ 0 & & & & \end{array}\right] \tag{11.2.8}$$

We have used the fact that $\mathbf{P}^T = \mathbf{P}$.

Now choose the vector $\mathbf{x}$ for the second Householder matrix to be the bottom $n-2$ elements of the second column, and from it construct

$$\mathbf{P}_2 \equiv \left[\begin{array}{cc|ccc} 1 & 0 & 0 & \cdots & 0 \\ 0 & 1 & 0 & \cdots & 0 \\ \hline 0 & 0 & & & \\ \vdots & \vdots & & ^{(n-2)}\mathbf{P}_2 & \\ 0 & 0 & & & \end{array}\right] \tag{11.2.9}$$

The identity block in the upper left corner insures that the tridiagonalization achieved in the first step will not be spoiled by this one, while the $(n-2)$-dimensional Householder matrix $^{(n-2)}\mathbf{P}_2$ creates one additional row and column of the tridiagonal output. Clearly, a sequence of $n-2$ such transformations will reduce the matrix $\mathbf{A}$ to tridiagonal form.

Instead of actually carrying out the matrix multiplications in $\mathbf{P}\cdot\mathbf{A}\cdot\mathbf{P}$, we compute a vector

$$\mathbf{p} \equiv \frac{\mathbf{A}\cdot\mathbf{u}}{H} \tag{11.2.10}$$

Then

$$\begin{aligned} \mathbf{A}\cdot\mathbf{P} &= \mathbf{A}\cdot(1-\frac{\mathbf{u}\cdot\mathbf{u}^T}{H}) = \mathbf{A}-\mathbf{p}\cdot\mathbf{u}^T \\ \mathbf{A}' &= \mathbf{P}\cdot\mathbf{A}\cdot\mathbf{P} = \mathbf{A}-\mathbf{p}\cdot\mathbf{u}^T-\mathbf{u}\cdot\mathbf{p}^T+2K\mathbf{u}\cdot\mathbf{u}^T \end{aligned}$$

where the scalar K is defined by

$$K = \frac{\mathbf{u}^T\cdot\mathbf{p}}{2H} \tag{11.2.11}$$

If we write

$$\mathbf{q} \equiv \mathbf{p}-K\mathbf{u} \tag{11.2.12}$$

then we have

$$\mathbf{A}' = \mathbf{A}-\mathbf{q}\cdot\mathbf{u}^T-\mathbf{u}\cdot\mathbf{q}^T \tag{11.2.13}$$

This is the computationally useful formula.

Following Wilkinson and Reinsch, the routine for Householder reduction given below actually starts in the nth column of $\mathbf{A}$, not the first as in the explanation above. In detail, the equations are as follows: At stage m ($m =$

$1, 2, \ldots, n-2$) the vector $\mathbf{u}$ has the form

$$\mathbf{u}^T = [a_{i1}, a_{i2}, \ldots, a_{i,i-2},\, a_{i,i-1} \pm \sqrt{\sigma},\, 0, \ldots, 0] \tag{11.2.14}$$

Here

$$i \equiv n - m + 1 = n, n-1, \ldots, 3 \tag{11.2.15}$$

and the quantity σ ($|x|^2$ in our earlier notation) is

$$\sigma = (a_{i1})^2 + \cdots + (a_{i,i-1})^2 \tag{11.2.16}$$

We choose the sign of σ in (11.2.14) to be the same as the sign of $a_{i,i-1}$ to lessen roundoff error.

Variables are thus computed in the following order: $\sigma, \mathbf{u}, H, \mathbf{p}, K, \mathbf{q}, \mathbf{A}'$. At any stage m, $\mathbf{A}$ is tridiagonal in its last $m-1$ rows and columns.

If the eigenvectors of the final tridiagonal matrix are found (for example, by the routine in the next section), then the eigenvectors of $\mathbf{A}$ can be obtained by applying the accumulated transformation

$$\mathbf{Q} = \mathbf{P}_1 \cdot \mathbf{P}_2 \cdots \mathbf{P}_{n-2} \tag{11.2.17}$$

to those eigenvectors. We therefore form $\mathbf{Q}$ by recursion after all the $\mathbf{P}$'s have been determined:

$$\begin{aligned} \mathbf{Q}_{n-2} &= \mathbf{P}_{n-2}, \\ \mathbf{Q}_j &= \mathbf{P}_j \cdot \mathbf{Q}_{j+1}, \qquad j = n-3, \ldots, 1, \\ \mathbf{Q} &= \mathbf{Q}_1 \end{aligned} \tag{11.2.18}$$

The input parameters for the routine below are the `N`×`N` real, symmetric matrix `A`, stored in an `NP`×`NP` array. On output, `A` contains the elements of the orthogonal matrix `Q`. The vector `D` returns the diagonal elements of the tridiagonal matrix $\mathbf{A}'$, while the vector `E` returns the off-diagonal elements in its components 2 through `N`, with `E(1)=0`. Note that since `A` is overwritten, you should copy it before calling the routine, if it is required for subsequent computations.

No extra storage arrays are needed for the intermediate results. At stage m, the vectors $\mathbf{p}$ and $\mathbf{q}$ are nonzero only in elements $1, \ldots, i$ (recall that $i = n - m + 1$), while $\mathbf{u}$ is nonzero only in elements $1, \ldots, i-1$. The elements of the vector `E` are being determined in the order $n, n-1, \ldots$, so we can store $\mathbf{p}$ in the elements of `E` not already determined. The vector $\mathbf{q}$ can overwrite $\mathbf{p}$ once $\mathbf{p}$ is no longer needed. We store $\mathbf{u}$ in the ith row of `A` and $\mathbf{u}/H$ in the

i^{th} column of A. Once the reduction is complete, we compute the matrices $\mathbf{Q}_j$ using the quantities $\mathbf{u}$ and $\mathbf{u}/H$ that have been stored in A. Since $\mathbf{Q}_j$ is an identity matrix in the last $n-j+1$ rows and columns, we only need compute its elements up to row and column $n-j$. These can overwrite the $\mathbf{u}$'s and $\mathbf{u}/H$'s in the corresponding rows and columns of A, which are no longer required for subsequent $\mathbf{Q}$'s.

The routine TRED2, given below, includes one further refinement. If the quantity σ is zero or "small" at any stage, one can skip the corresponding transformation. A simple criterion, such as

$$\sigma < \frac{\text{smallest positive number representable on machine}}{\text{machine precision}}$$

would be fine most of the time. A more careful criterion is actually used. Define the quantity

$$\epsilon = \sum_{k=1}^{i-1} |a_{ik}| \tag{11.2.19}$$

If $\epsilon = 0$ to machine precision, we skip the transformation. Otherwise we redefine

$$a_{ik} \quad \text{becomes} \quad a_{ik}/\epsilon \tag{11.2.20}$$

and use the scaled variables for the transformation. (A Householder transformation depends only on the ratios of the elements.)

Note that when dealing with a matrix whose elements vary over many orders of magnitude, it is important that the matrix be permuted, insofar as possible, so that the smaller elements are in the top left-hand corner. This is because the reduction is performed starting from the bottom right-hand corner, and a mixture of small and large elements there can lead to considerable rounding errors.

The routine TRED2 is designed for use with the routine TQLI of the next section. TQLI finds the eigenvalues and eigenvectors of a symmetric, tridiagonal matrix. The combination of TRED2 and TQLI is the most efficient known technique for finding all the eigenvalues and eigenvectors (or just all the eigenvalues) of a real, symmetric matrix.

In the listing below, the commented (C) statements are required only for subsequent computation of eigenvectors. If only eigenvalues are required, omission of the commented statements speeds up the execution time of TRED2 by a factor of 2 for large n. In the limit of large n, the operation count of the Householder reduction is $2n^3/3$ for eigenvalues only, and $4n^3/3$ for both eigenvalues and eigenvectors.

```
      SUBROUTINE TRED2(A,N,NP,D,E)
```

Householder reduction of a real, symmetric, N by N matrix **A**, stored in an NP by NP physical array. On output, **A** is replaced by the orthogonal matrix **Q** effecting the transformation. D returns the diagonal elements of the tridiagonal matrix, and E the off-diagonal elements, with E(1)=0. Several statements, as noted in comments, can be omitted if only eigenvalues are to be found, in which case **A** contains no useful information on output. Otherwise they are to be included.

```
      DIMENSION A(NP,NP),D(NP),E(NP)
      IF(N.GT.1)THEN
        DO 18 I=N,2,-1
          L=I-1
          H=0.
          SCALE=0.
          IF(L.GT.1)THEN
            DO 11 K=1,L
              SCALE=SCALE+ABS(A(I,K))
11          CONTINUE
            IF(SCALE.EQ.0.)THEN           Skip transformation.
              E(I)=A(I,L)
            ELSE
              DO 12 K=1,L
                A(I,K)=A(I,K)/SCALE       Use scaled a's for transformation.
                H=H+A(I,K)**2             Form σ in H.
12            CONTINUE
              F=A(I,L)
              G=-SIGN(SQRT(H),F)
              E(I)=SCALE*G
              H=H-F*G                     Now H is equation (11.2.4).
              A(I,L)=F-G                  Store u in the Ith row of A.
              F=0.
              DO 15 J=1,L
C     Omit following line if finding only eigenvalues
                A(J,I)=A(I,J)/H           Store u/H in Ith column of A.
                G=0.                      Form an element of A · u in G.
                DO 13 K=1,J
                  G=G+A(J,K)*A(I,K)
13              CONTINUE
                IF(L.GT.J)THEN
                  DO 14 K=J+1,L
                    G=G+A(K,J)*A(I,K)
14                CONTINUE
                ENDIF
                E(J)=G/H                  Form element of p in temporarily unused element
                F=F+E(J)*A(I,J)           of E.
15            CONTINUE
              HH=F/(H+H)                  Form K, equation (11.2.11).
              DO 17 J=1,L                 Form q and store in E overwriting p.
                F=A(I,J)                  Note that E(L)=E(I-1) survives.
                G=E(J)-HH*F
                E(J)=G
                DO 16 K=1,J               Reduce A, equation (11.2.13).
                  A(J,K)=A(J,K)-F*E(K)-G*A(I,K)
16              CONTINUE
17            CONTINUE
            ENDIF
          ELSE
            E(I)=A(I,L)
          ENDIF
          D(I)=H
18      CONTINUE
      ENDIF
C     Omit following line if finding only eigenvalues.
      D(1)=0.
      E(1)=0.
```

```
      DO 23 I=1,N                  Begin accumulation of transformation matrices.
C       Delete lines from here ...
          L=I-1
          IF(D(I).NE.0.)THEN       This block skipped when I=1.
            DO 21 J=1,L
              G=0.
              DO 19 K=1,L          Use u and u/H stored in A to form P · Q.
                G=G+A(I,K)*A(K,J)
19            CONTINUE
              DO 20 K=1,L
                A(K,J)=A(K,J)-G*A(K,I)
20            CONTINUE
21          CONTINUE
          ENDIF
C       ... to here when finding only eigenvalues.
          D(I)=A(I,I)              This statement remains.
C       Also delete lines from here ...
          A(I,I)=1.                Reset row and column of A to identity matrix for next iteration.
          IF(L.GE.1)THEN
            DO 22 J=1,L
              A(I,J)=0.
              A(J,I)=0.
22          CONTINUE
          ENDIF
C       ... to here when finding only eigenvalues.
23    CONTINUE
      RETURN
      END
```

REFERENCES AND FURTHER READING:

Jacobs, David A.H., ed. 1977, *The State of the Art in Numerical Analysis* (London: Academic Press), Chapter I.1 by J.H. Wilkinson.

Golub, Gene H., and Van Loan, Charles F. 1983, *Matrix Computations* (Baltimore: Johns Hopkins University Press), §8.2.

Smith, B.T., et al. 1976, *Matrix Eigensystem Routines — EISPACK Guide*, 2nd ed., vol. 6 of Lecture Notes in Computer Science (New York: Springer-Verlag).

Wilkinson, J.H., and Reinsch, C. 1971, *Linear Algebra*, vol. II of *Handbook for Automatic Computation* (New York: Springer-Verlag).

11.3 Eigenvalues and Eigenvectors of a Tridiagonal Matrix

Evaluation of the Characteristic Polynomial

Once our original, real, symmetric matrix has been reduced to tridiagonal form, one possible way to determine its eigenvalues is to find the roots of the characteristic polynomial $p_n(\lambda)$ directly. The characteristic polynomial of a tridiagonal matrix can be evaluated for any trial value of λ by an efficient recursion relation (see, for example, Acton). The polynomials of lower degree produced during the recurrence form a Sturmian sequence that can be

used to localize the eigenvalues to intervals on the real axis. A root-finding method such as bisection or Newton's method can then be employed to refine the intervals. The corresponding eigenvectors can then be found by inverse iteration (see §11.7).

Procedures based on these ideas can be found in the *Handbook* and in EISPACK. If, however, more than a small fraction of all the eigenvalues and eigenvectors are required, then the factorization method next considered is much more efficient.

The QR and QL Algorithms

The basic idea behind the QR algorithm is that any real matrix can be decomposed in the form

$$\mathbf{A} = \mathbf{Q} \cdot \mathbf{R}, \tag{11.3.1}$$

where $\mathbf{Q}$ is orthogonal and $\mathbf{R}$ is upper triangular. For a general matrix, the decomposition is constructed by applying Householder transformations to annihilate successive columns of $\mathbf{A}$ below the diagonal.

Now consider the matrix formed by writing the factors in (11.3.1) in the opposite order:

$$\mathbf{A}' = \mathbf{R} \cdot \mathbf{Q}. \tag{11.3.2}$$

Since $\mathbf{Q}$ is orthogonal, equation (11.3.1) gives $\mathbf{R} = \mathbf{Q}^T \cdot \mathbf{A}$. Thus equation (11.3.2) becomes

$$\mathbf{A}' = \mathbf{Q}^T \cdot \mathbf{A} \cdot \mathbf{Q} \tag{11.3.3}$$

We see that $\mathbf{A}'$ is an orthogonal transformation of $\mathbf{A}$.

You can verify that a QR transformation preserves the following properties of a matrix: symmetry, tridiagonal form, and Hessenberg form (to be defined in §11.5).

There is nothing special about choosing one of the factors of $\mathbf{A}$ to be upper triangular; one could equally well make it lower triangular. This is called the QL algorithm, since

$$\mathbf{A} = \mathbf{Q} \cdot \mathbf{L} \tag{11.3.4}$$

where $\mathbf{L}$ is lower triangular. (The standard, but confusing, nomenclature R and L stands for whether the *right* or *left* of the matrix is nonzero.)

Recall that in the Householder reduction to tridiagonal form in §11.2, we started in the nth (last) column of the original matrix. To minimize roundoff, we then exhorted you to put the biggest elements of the matrix in the lower

right-hand corner, if you can. If we now wish to diagonalize the resulting tridiagonal matrix, the QL algorithm will have smaller roundoff than the QR algorithm, so we shall use QL henceforth.

The QL algorithm consists of a *sequence* of orthogonal transformations:

$$\begin{aligned} \mathbf{A}_s &= \mathbf{Q}_s \cdot \mathbf{L}_s \\ \mathbf{A}_{s+1} &= \mathbf{L}_s \cdot \mathbf{Q}_s \qquad (= \mathbf{Q}_s^T \cdot \mathbf{A}_s \cdot \mathbf{Q}_s) \end{aligned} \tag{11.3.5}$$

The following (nonobvious!) theorem is the basis of the algorithm for a general matrix $\mathbf{A}$: (i) If $\mathbf{A}$ has eigenvalues of different absolute value $|\lambda_i|$, then $\mathbf{A}_s \to$ [lower triangular form] as $s \to \infty$. The eigenvalues appear on the diagonal in increasing order of absolute magnitude. (ii) If $\mathbf{A}$ has an eigenvalue $|\lambda_i|$ of multiplicity p, $\mathbf{A}_s \to$ [lower triangular form] as $s \to \infty$, except for a diagonal block matrix of order p, whose eigenvalues $\to \lambda_i$. The proof of this theorem is fairly lengthy; see, for example, Stoer and Bulirsch.

The workload in the QL algorithm is $O(n^3)$ per iteration for a general matrix, which is prohibitive for a general matrix. However, the workload is only $O(n)$ per iteration for a tridiagonal matrix and $O(n^2)$ for a Hessenberg matrix, which makes it highly efficient on these forms.

In this section we are concerned only with the case where $\mathbf{A}$ is a real, symmetric, tridiagonal matrix. All the eigenvalues λ_i are thus real. According to the theorem, if any λ_i has a multiplicity p, then there must be at least $p-1$ zeros on the sub- and superdiagonal. Thus the matrix can be split into submatrices that can be diagonalized separately, and the complication of diagonal blocks that can arise in the general case is irrelevant.

In the proof of the theorem quoted above, one finds that in general a superdiagonal element converges to zero like

$$a_{ij}^{(s)} \sim \left(\frac{\lambda_i}{\lambda_j}\right)^s \tag{11.3.6}$$

Although $\lambda_i < \lambda_j$, convergence can be slow if λ_i is close to λ_j. Convergence can be accelerated by the technique of *shifting*: If k is any constant, then $\mathbf{A} - k\mathbf{1}$ has eigenvalues $\lambda_i - k$. If we decompose

$$\mathbf{A}_s - k_s\mathbf{1} = \mathbf{Q}_s \cdot \mathbf{L}_s \tag{11.3.7}$$

so that

$$\begin{aligned} \mathbf{A}_{s+1} &= \mathbf{L}_s \cdot \mathbf{Q}_s + k_s\mathbf{1} \\ &= \mathbf{Q}_s^T \cdot \mathbf{A}_s \cdot \mathbf{Q}_s \end{aligned} \tag{11.3.8}$$

then the convergence is determined by the ratio

$$\frac{\lambda_i - k_s}{\lambda_j - k_s}. \tag{11.3.9}$$

The idea is to choose the shift k_s at each stage to maximize the rate of convergence. A good choice for the shift initially would be k_s close to λ_1, the smallest eigenvalue. Then the first row of off-diagonal elements would tend rapidly to zero. However, λ_1 is not usually known *a priori.* A very effective strategy in practice (although there is no proof that it is optimal) is to compute the eigenvalues of the leading 2×2 diagonal submatrix of $\mathbf{A}$. Then set k_s equal to the eigenvalue closer to a_{11}.

More generally, suppose you have already found $r-1$ eigenvalues of $\mathbf{A}$. Then you can *deflate* the matrix by crossing out the first $r-1$ rows and columns, leaving

$$\mathbf{A} = \begin{bmatrix} 0 & & & \cdots & \cdots & & & 0 \\ & \cdots & & & & & & \\ & & 0 & & & & & \\ \vdots & & & d_r & e_r & & & \vdots \\ \vdots & & & e_r & d_{r+1} & & & \\ & & & & & \cdots & & 0 \\ & & & & & & d_{n-1} & e_{n-1} \\ 0 & & & \cdots & & 0 & e_{n-1} & d_n \end{bmatrix} \tag{11.3.10}$$

Choose k_s equal to the eigenvalue of the leading 2×2 submatrix that is closer to d_r. One can show that the convergence of the algorithm with this strategy is generally cubic (and at worst quadratic for degenerate eigenvalues). This rapid convergence is what makes the algorithm so attractive.

Note that with shifting, the eigenvalues no longer necessarily appear on the diagonal in order of increasing absolute magnitude. The routine `EIGSRT` (§11.1) can be used if required.

As we mentioned earlier, the QL decomposition of a general matrix is effected by a sequence of Householder transformations. For a tridiagonal matrix, however, it is more efficient to use plane rotations $\mathbf{P}_{pq}$. One uses the sequence $\mathbf{P}_{12}, \mathbf{P}_{23}, \ldots, \mathbf{P}_{n-1,n}$ to annihilate the elements $a_{12}, a_{23}, \ldots, a_{n-1,n}$. By symmetry, the subdiagonal elements $a_{21}, a_{32}, \ldots, a_{n,n-1}$ will be annihilated too. Thus each $\mathbf{Q}_s$ is a product of plane rotations:

$$\mathbf{Q}_s^T = \mathbf{P}_1^{(s)} \cdot \mathbf{P}_2^{(s)} \cdots \mathbf{P}_{n-1}^{(s)} \tag{11.3.11}$$

where $\mathbf{P}_i$ annihilates $a_{i,i+1}$. Note that it is $\mathbf{Q}^T$ in equation (11.3.11), not $\mathbf{Q}$, because we defined $\mathbf{L} = \mathbf{Q}^T \cdot \mathbf{A}$.

QL Algorithm with Implicit Shifts

The algorithm as described so far can be very successful. However when the elements of $\mathbf{A}$ differ widely in order of magnitude, subtracting a large k_s from the diagonal elements can lead to loss of accuracy for the small eigenvalues. This difficulty is avoided by the QL algorithm with *implicit shifts.* The implicit QL algorithm is mathematically equivalent to the original QL algorithm, but the computation does not require $k_s\mathbf{1}$ to be actually subtracted from $\mathbf{A}$.

The algorithm is based on the following lemma: If $\mathbf{A}$ is a symmetric nonsingular matrix and $\mathbf{B} = \mathbf{Q}^T \cdot \mathbf{A} \cdot \mathbf{Q}$, where $\mathbf{Q}$ is orthogonal and $\mathbf{B}$ is tridiagonal with positive off-diagonal elements, then $\mathbf{Q}$ and $\mathbf{B}$ are fully determined when the last row of $\mathbf{Q}^T$ is specified. Proof: Let $\mathbf{q}_i^T$ denote the ith row vector of the matrix $\mathbf{Q}^T$. Then $\mathbf{q}_i$ is the ith column vector of the matrix $\mathbf{Q}$. The relation $\mathbf{B} \cdot \mathbf{Q}^T = \mathbf{Q}^T \cdot \mathbf{A}$ can be written

$$\begin{bmatrix} \beta_1 & \gamma_1 & & & & \\ \alpha_2 & \beta_2 & \gamma_2 & & & \\ & & \vdots & & & \\ & & & \alpha_{n-1} & \beta_{n-1} & \gamma_{n-1} \\ & & & & \alpha_n & \beta_n \end{bmatrix} \cdot \begin{bmatrix} \mathbf{q}_1^T \\ \mathbf{q}_2^T \\ \vdots \\ \mathbf{q}_{n-1}^T \\ \mathbf{q}_n^T \end{bmatrix} = \begin{bmatrix} \mathbf{q}_1^T \\ \mathbf{q}_2^T \\ \vdots \\ \mathbf{q}_{n-1}^T \\ \mathbf{q}_n^T \end{bmatrix} \cdot \mathbf{A} \tag{11.3.12}$$

The nth row of this matrix equation is

$$\alpha_n \mathbf{q}_{n-1}^T + \beta_n \mathbf{q}_n^T = \mathbf{q}_n^T \cdot \mathbf{A} \tag{11.3.13}$$

Since $\mathbf{Q}$ is orthogonal,

$$\mathbf{q}_n^T \cdot \mathbf{q}_m = \delta_{nm} \tag{11.3.14}$$

Thus if we postmultiply equation (11.3.13) by $\mathbf{q}_n$, we find

$$\beta_n = \mathbf{q}_n^T \cdot \mathbf{A} \cdot \mathbf{q}_n \tag{11.3.15}$$

which is known since $\mathbf{q}_n$ is known. Then equation (11.3.13) gives

$$\alpha_n \mathbf{q}_{n-1}^T = \mathbf{z}_{n-1}^T (11.3.16)$$

where

$$\mathbf{z}_{n-1}^T \equiv \mathbf{q}_n^T \cdot \mathbf{A} - \beta_n \mathbf{q}_n^T \tag{11.3.17}$$

is known. Therefore

$$\alpha_n^2 = \mathbf{z}_{n-1}^T \mathbf{z}_{n-1}, \tag{11.3.18}$$

or

$$\alpha_n = |\mathbf{z}_{n-1}| \tag{11.3.19}$$

and

$$\mathbf{q}_{n-1}^T = \mathbf{z}_{n-1}^T / \alpha_n \tag{11.3.20}$$

(where α_n is nonzero by hypothesis). Similarly, one can show by induction that if we know $\mathbf{q}_n, \mathbf{q}_{n-1}, \ldots, \mathbf{q}_{n-j}$ and the α's, β's and γ's up to level $n-j$, one can determine the quantities at level $n-(j+1)$.

To apply the lemma in practice, suppose one can somehow find a tridiagonal matrix $\overline{\mathbf{A}}_{s+1}$ such that

$$\overline{\mathbf{A}}_{s+1} = \overline{\mathbf{Q}}_s^T \cdot \overline{\mathbf{A}}_s \cdot \overline{\mathbf{Q}}_s \tag{11.3.21}$$

where $\overline{\mathbf{Q}}_s^T$ is orthogonal and has the same last row as $\mathbf{Q}_s^T$ in the original QL algorithm. Then $\overline{\mathbf{Q}}_s = \mathbf{Q}_s$ and $\overline{\mathbf{A}}_{s+1} = \mathbf{A}_{s+1}$.

Now, in the original algorithm, from equation (11.3.11) we see that the last row of $\mathbf{Q}_s^T$ is the same as the last row of $\mathbf{P}_{n-1}^{(s)}$. But recall that $\mathbf{P}_{n-1}^{(s)}$ is a plane rotation designed to annihilate the $(n-1,n)$ element of $\mathbf{A}_s - k_s\mathbf{1}$. A simple calculation using the expression (11.1.1) shows that it has parameters

$$c = \frac{d_n - k_s}{\sqrt{e_n^2 + (d_n - k_s)^2}}\,, \qquad s = \frac{-e_{n-1}}{\sqrt{e_n^2 + (d_n - k_s)^2}} \tag{11.3.22}$$

The matrix $\mathbf{P}_{n-1}^{(s)} \cdot \mathbf{A}_s \cdot \mathbf{P}_{n-1}^{(s)T}$ is tridiagonal with 2 extra elements:

$$\begin{bmatrix} \cdots & & & & & \\ & \times & \times & \times & & \\ & & \times & \times & \times & \mathbf{x} \\ & & & \times & \times & \times \\ & & & \mathbf{x} & \times & \times \end{bmatrix} \tag{11.3.23}$$

We must now reduce this to tridiagonal form with an orthogonal matrix whose last row is $[0, 0, \ldots, 0, 1]$ so that the last row of $\overline{\mathbf{Q}}_s^T$ will stay equal to $\mathbf{P}_{n-1}^{(s)}$. This can be done by a sequence of Householder or Givens transformations. For the special form of the matrix (11.3.23), Givens is better. We rotate in the plane $(n-2, n-1)$ to annihilate the $(n-2, n)$ element. [By symmetry, the $(n, n-2)$ element will also be zeroed.] This leaves us with tridiagonal form except for extra elements $(n-3, n-1)$ and $(n-1, n-3)$. We annihilate

these with a rotation in the $(n-3, n-2)$ plane, and so on. Thus a sequence of $n-2$ Givens rotations are required. The result is that

$$\mathbf{Q}_s^T = \overline{\mathbf{Q}}_s^T = \overline{\mathbf{P}}_1^{(s)} \cdot \overline{\mathbf{P}}_2^{(s)} \cdots \overline{\mathbf{P}}_{n-2}^{(s)} \cdot \mathbf{P}_{n-1}^{(s)} \tag{11.3.24}$$

where the $\overline{\mathbf{P}}$'s are the Givens rotations and $\mathbf{P}_{n-1}$ is the same plane rotation as in the original algorithm. Then equation (11.3.21) gives the next iterate of $\mathbf{A}$. Note that the shift k_s enters implicitly through the parameters (11.3.22).

The following routine TQLI ("Tridiagonal QL Implicit"), based algorithmically on the *Handbook* and EISPACK implementations, works extremely well in practice. The number of iterations for the first few eigenvalues might be 4 or 5 say, but meanwhile the off-diagonal elements in the lower right-hand corner have been reduced too. The later eigenvalues are liberated with very little work. The average number of iterations per eigenvalue is typically $1.3-1.6$. The operation count per iteration is $O(n)$, with a fairly large effective coefficient, say $\sim 20n$. The total operation count for the diagonalization is then $\sim 20n \times (1.3-1.6)n \sim 30n^2$. If the eigenvectors are required, the commented statements are included and there is an additional, much larger, workload of about $3n^3$ operations.

```
      SUBROUTINE TQLI(D,E,N,NP,Z)
```

QL algorithm with implicit shifts, to determine the eigenvalues and eigenvectors of a real, symmetric, tridiagonal matrix, or of a real, symmetric matrix previously reduced by **TRED2** §11.2. D is a vector of length NP. On input, its first N elements are the diagonal elements of the tridiagonal matrix. On output, it returns the eigenvalues. The vector E inputs the subdiagonal elements of the tridiagonal matrix, with E(1) arbitrary. On output E is destroyed. When finding only the eigenvalues, several lines may be omitted, as noted in the comments. If the eigenvectors of a tridiagonal matrix are desired, the matrix Z (N by N matrix stored in NP by NP array) is input as the identity matrix. If the eigenvectors of a matrix that has been reduced by **TRED2** are required, then Z is input as the matrix output by **TRED2**. In either case, the Kth column of Z returns the normalized eigenvector corresponding to D(K).

```
      DIMENSION D(NP),E(NP),Z(NP,NP)
      IF (N.GT.1) THEN
        DO 11 I=2,N                   Convenient to renumber the elements of E.
          E(I-1)=E(I)
11      CONTINUE
        E(N)=0.
        DO 15 L=1,N
          ITER=0
1         DO 12 M=L,N-1               Look for a single small subdiagonal element to split the
            DD=ABS(D(M))+ABS(D(M+1))      matrix.
            IF (ABS(E(M))+DD.EQ.DD) GO TO 2
12        CONTINUE
          M=N
2         IF(M.NE.L)THEN
            IF(ITER.EQ.30)PAUSE 'too many iterations'
            ITER=ITER+1
            G=(D(L+1)-D(L))/(2.*E(L))     Form shift.
            R=SQRT(G**2+1.)
            G=D(M)-D(L)+E(L)/(G+SIGN(R,G))    This is d_m - k_s.
            S=1.
            C=1.
            P=0.
```

```
              DO 14 I=M-1,L,-1              A plane rotation as in the original QL, followed by
                F=S*E(I)                    Givens rotations to restore tridiagonal form.
                B=C*E(I)
                IF(ABS(F).GE.ABS(G))THEN
                  C=G/F
                  R=SQRT(C**2+1.)
                  E(I+1)=F*R
                  S=1./R
                  C=C*S
                ELSE
                  S=F/G
                  R=SQRT(S**2+1.)
                  E(I+1)=G*R
                  C=1./R
                  S=S*C
                ENDIF
                G=D(I+1)-P
                R=(D(I)-G)*S+2.*C*B
                P=S*R
                D(I+1)=G+P
                G=C*R-B
C     Omit lines from here ...
                DO 13 K=1,N                 Form eigenvectors.
                  F=Z(K,I+1)
                  Z(K,I+1)=S*Z(K,I)+C*F
                  Z(K,I)=C*Z(K,I)-S*F
13              CONTINUE
C     ... to here when finding only eigenvalues.
14            CONTINUE
              D(L)=D(L)-P
              E(L)=G
              E(M)=0.
              GO TO 1
            ENDIF
15        CONTINUE
      ENDIF
      RETURN
      END
```

REFERENCES AND FURTHER READING:

Wilkinson, J.H., and Reinsch, C. 1971, *Linear Algebra*, vol. II of *Handbook for Automatic Computation* (New York: Springer-Verlag).

Smith, B.T., et al. 1976, *Matrix Eigensystem Routines — EISPACK Guide*, 2nd ed., vol. 6 of Lecture Notes in Computer Science (New York: Springer-Verlag).

Stoer, J., and Bulirsch, R. 1980, *Introduction to Numerical Analysis* (New York: Springer-Verlag), §6.6.6.

Acton, Forman S. 1970, *Numerical Methods That Work* (New York: Harper and Row), pp. 331–335.

11.4 Hermitian Matrices

The complex analog of a real, symmetric matrix is a Hermitian matrix, satisfying equation (11.0.4). Jacobi transformations can be used to find eigenvalues and eigenvectors, as also can Householder reduction to tridiagonal form followed by QL iteration. Complex versions of the previous routines `JACOBI`, `TRED2` and `TQLI` are quite analogous to their real counterparts. For working routines, consult the *Handbook* or EISPACK.

An alternative, using the routines in this book, is to convert the Hermitian problem to a real, symmetric one: If $\mathbf{C} = \mathbf{A} + i\mathbf{B}$ is a Hermitian matrix, then the $n \times n$ complex eigenvalue problem

$$(\mathbf{A} + i\mathbf{B}) \cdot (\mathbf{u} + i\mathbf{v}) = \lambda(\mathbf{u} + i\mathbf{v}) \tag{11.4.1}$$

is equivalent to the $2n \times 2n$ real problem

$$\begin{bmatrix} \mathbf{A} & -\mathbf{B} \\ \mathbf{B} & \mathbf{A} \end{bmatrix} \cdot \begin{bmatrix} \mathbf{u} \\ \mathbf{v} \end{bmatrix} = \lambda \begin{bmatrix} \mathbf{u} \\ \mathbf{v} \end{bmatrix} \tag{11.4.2}$$

Note that the $2n \times 2n$ matrix in (11.4.2) is symmetric: $\mathbf{A}^T = \mathbf{A}$ and $\mathbf{B}^T = -\mathbf{B}$ if $\mathbf{C}$ is Hermitian.

Corresponding to a given eigenvalue λ, the vector

$$\begin{bmatrix} -\mathbf{v} \\ \mathbf{u} \end{bmatrix} \tag{11.4.3}$$

is also an eigenvector, as you can verify by writing out the two matrix equations implied by (11.4.2). Thus if $\lambda_1, \lambda_2, \ldots, \lambda_n$ are the eigenvalues of $\mathbf{C}$, then the $2n$ eigenvalues of the augmented problem (11.4.2) are $\lambda_1, \lambda_1, \lambda_2, \lambda_2, \ldots,$ λ_n, λ_n; each, in other words, is repeated twice. The eigenvectors are pairs of the form $\mathbf{u} + i\mathbf{v}$ and $i(\mathbf{u} + i\mathbf{v})$; that is, they are the same up to an inessential phase. Thus we solve the augmented problem (11.4.2), and choose one eigenvalue and eigenvector from each pair. These give the eigenvalues and eigenvectors of the original matrix $\mathbf{C}$.

Working with the augmented matrix requires a factor of 2 more storage than the original complex matrix. In principle, a complex algorithm is also a factor of 2 more efficient in computer time than is the solution of the augmented problem. In practice, most complex implementations do not achieve this factor unless they are written entirely in real arithmetic. (Good library routines always do this.)

For simplicity, we prefer the formulation (11.4.2).

REFERENCES AND FURTHER READING:

Wilkinson, J.H., and Reinsch, C. 1971, *Linear Algebra*, vol. II of *Handbook for Automatic Computation* (New York: Springer-Verlag).

Smith, B.T., et al. 1976, *Matrix Eigensystem Routines — EISPACK Guide*, 2nd ed., vol. 6 of Lecture Notes in Computer Science (New York: Springer-Verlag).

11.5 Reduction of a General Matrix to Hessenberg Form

The algorithms for symmetric matrices, given in the preceding sections, are highly satisfactory in practice. By contrast, it is impossible to design equally satisfactory algorithms for the nonsymmetric case. There are two reasons for this. First, the eigenvalues of a nonsymmetric matrix can be very sensitive to small changes in the matrix elements. Second, the matrix itself can be defective, so that there is no complete set of eigenvectors. We emphasize that these difficulties are intrinsic properties of certain nonsymmetric matrices, and no numerical procedure can "cure" them. The best we can hope for are procedures which don't exacerbate such problems.

The presence of rounding error can only make the situation worse. With finite-precision arithmetic, one cannot even design a foolproof algorithm to determine whether a given matrix is defective or not. Thus current algorithms generally *try* to find a *complete* set of eigenvectors, and rely on the user to inspect the results. If any eigenvectors are almost parallel, the matrix is probably defective.

Apart from referring you to the literature, and to the collected *Handbook* or EISPACK routines, we are going to sidestep the problem of eigenvectors, giving algorithms for eigenvalues only. If you require just a few eigenvectors, you can read §11.7 and consider finding them by inverse iteration. We consider the problem of finding *all* eigenvectors of a nonsymmetric matrix as lying beyond the scope of this book.

Balancing

The sensitivity of eigenvalues to rounding errors during the execution of some algorithms can be reduced by the procedure of *balancing*. The errors in the eigensystem found by a numerical procedure are generally proportional to the Euclidean norm of the matrix, that is, to the square root of the sum of the squares of the elements. The idea of balancing is to use similarity transformations to make corresponding rows and columns of the matrix have comparable norms, thus reducing the overall norm of the matrix while leaving the eigenvalues unchanged. A symmetric matrix is already balanced.

Balancing is a procedure with of order N^2 operations. Thus, the time taken by the procedure BALANC, given below, should be never more than a few percent of the total time required to find the eigenvalues. It is therefore recommended that you *always* balance nonsymmetric matrices. It never hurts, and it can substantially improve the accuracy of the eigenvalues computed for a badly balanced matrix.

The actual algorithm used is due to Osborne, as discussed in the *Handbook.* It consists of a sequence of similarity transformations by diagonal matrices **D**. To avoid introducing rounding errors during the balancing process, the elements of **D** are restricted to be exact powers of the radix base employed for floating point arithmetic (i.e. 2 for most machines, but 16 for IBM mainframe architectures). The output is a matrix that is balanced in the norm given by summing the absolute magnitudes of the matrix elements. This is more efficient than using the Euclidean norm, and equally effective: A large reduction in one norm implies a large reduction in the other.

Note that if the off-diagonal elements of any row or column of a matrix are all zero, then the diagonal element is an eigenvalue. If the eigenvalue happens to be ill-conditioned (sensitive to small changes in the matrix elements), it will have relatively large errors when determined by the routine HQR (§11.6). Had we merely inspected the matrix beforehand, we could have determined the isolated eigenvalue exactly and then deleted the corresponding row and column from the matrix. You should consider whether such a pre-inspection might be useful in your application. (For symmetric matrices, the routines we gave will determine isolated eigenvalues accurately in all cases.)

The routine BALANC does not keep track of the accumulated similarity transformation of the original matrix since we will only be concerned with finding eigenvalues of nonsymmetric matrices, not eigenvectors. Consult the references if you want to keep track of the transformation.

```
      SUBROUTINE BALANC(A,N,NP)
         Given an N by N matrix A stored in an array of physical dimensions NP by NP, this rou-
         tine replaces it by a balanced matrix with identical eigenvalues. A symmetric matrix is
         already balanced and is unaffected by this procedure. The parameter RADIX should be the
         machine's floating point radix.
      PARAMETER (RADIX=2.,SQRDX=RADIX**2)
      DIMENSION A(NP,NP)
1     CONTINUE
        LAST=1
        DO 14 I=1,N
          C=0.
          R=0.
          DO 11 J=1,N
            IF(J.NE.I)THEN
              C=C+ABS(A(J,I))
              R=R+ABS(A(I,J))
            ENDIF
11        CONTINUE
          IF(C.NE.0..AND.R.NE.0.)THEN
            G=R/RADIX
            F=1.
            S=C+R
2           IF(C.LT.G)THEN
              F=F*RADIX
              C=C*SQRDX
            GO TO 2
            ENDIF
            G=R*RADIX
3           IF(C.GT.G)THEN
              F=F/RADIX
              C=C/SQRDX
            GO TO 3
            ENDIF
```

```
            IF((C+R)/F.LT.0.95*S)THEN
                LAST=0
                G=1./F
                DO 12 J=1,N
                    A(I,J)=A(I,J)*G
12              CONTINUE
                DO 13 J=1,N
                    A(J,I)=A(J,I)*F
13              CONTINUE
            ENDIF
        ENDIF
14  CONTINUE
    IF(LAST.EQ.0)GO TO 1
    RETURN
    END
```

Reduction to Hessenberg Form

The strategy for finding the eigensystem of a general matrix parallels that of the symmetric case. First we reduce the matrix to a simpler form, and then we perform an iterative procedure on the simplified matrix. The simpler structure we use here is called *Hessenberg* form. An *upper Hessenberg* matrix has zeros everywhere below the diagonal except for the first subdiagonal row. For example, in the 6×6 case, the nonzero elements are:

$$\begin{bmatrix} \times & \times & \times & \times & \times & \times \\ \times & \times & \times & \times & \times & \times \\ & \times & \times & \times & \times & \times \\ & & \times & \times & \times & \times \\ & & & \times & \times & \times \\ & & & & \times & \times \end{bmatrix}$$

By now you should be able to tell at a glance that such a structure can be achieved by a sequence of Householder transformations, each one zeroing the required elements in a column of the matrix. Householder reduction to Hessenberg form is in fact an accepted technique. An alternative, however, is a procedure analogous to Gaussian elimination with pivoting. We will use this elimination procedure since it is about a factor of 2 more efficient than the Householder method, and also since we want to teach you the method. It is possible to construct matrices for which the Householder reduction, being orthogonal, is stable and elimination is not, but such matrices are extremely rare in practice.

Straight Gaussian elimination is not a similarity transformation of the matrix. Accordingly, the actual elimination procedure used is slightly different. Before the rth stage, the original matrix $\mathbf{A} \equiv \mathbf{A}_1$ has become $\mathbf{A}_r$, which is upper Hessenberg in its first $r-1$ rows and columns. The rth stage then consists of the following sequence of operations:

- Find the element of maximum magnitude in the rth column below the diagonal. If it is zero, skip the next two "bullets" and the stage is done. Otherwise, suppose the maximum element was in row r'.
- Interchange rows r' and $r+1$. This is the pivoting procedure. To make the permutation a similarity transformation, also interchange columns r' and $r+1$.
- For $i = r+2, r+3, \ldots, N$, compute the multiplier

$$n_{i,r+1} \equiv \frac{a_{ir}}{a_{r+1,r}}$$

Subtract $n_{i,r+1}$ times row $r+1$ from row i. To make the elimination a similarity transformation, also *add* $n_{i,r+1}$ times column i to column $r+1$.

A total of $N-2$ such stages are required.

When the magnitudes of the matrix elements vary over many orders, you should try to rearrange the matrix so that the largest elements are in the top left-hand corner. This reduces the roundoff error, since the reduction proceeds from left to right.

Since we are only concerned with eigenvalues, the routine **ELMHES** does not keep track of the accumulated similarity transformation. The operation count is about $5N^3/6$ for large N.

```
SUBROUTINE ELMHES(A,N,NP)
    Reduction to Hessenberg form by the elimination method. The real, nonsymmetric, N by
    N matrix A, stored in an array of physical dimensions NP by NP, is replaced by an upper
    Hessenberg matrix with identical eigenvalues. Recommended, but not required, is that
    this routine be preceded by BALANC. On output, the Hessenberg matrix is in elements
    A(I,J) with I ≤J+1. Elements with I > J+1 are to be thought of as zero, but are
    returned with random values.
DIMENSION A(NP,NP)
IF(N.GT.2)THEN
    DO 17 M=2,N-1                   M is called r + 1 in the text.
        X=0.
        I=M
        DO 11 J=M,N                 Find the pivot.
            IF(ABS(A(J,M-1)).GT.ABS(X))THEN
                X=A(J,M-1)
                I=J
            ENDIF
11      CONTINUE
        IF(I.NE.M)THEN              Interchange rows and columns.
            DO 12 J=M-1,N
                Y=A(I,J)
                A(I,J)=A(M,J)
                A(M,J)=Y
12          CONTINUE
            DO 13 J=1,N
                Y=A(J,I)
                A(J,I)=A(J,M)
                A(J,M)=Y
13          CONTINUE
        ENDIF
        IF(X.NE.0.)THEN             Carry out the elimination.
```

```
            DO 16 I=M+1,N
              Y=A(I,M-1)
              IF(Y.NE.0.)THEN
                Y=Y/X
                A(I,M-1)=Y
                DO 14 J=M,N
                  A(I,J)=A(I,J)-Y*A(M,J)
14              CONTINUE
                DO 15 J=1,N
                  A(J,M)=A(J,M)+Y*A(J,I)
15              CONTINUE
              ENDIF
16          CONTINUE
          ENDIF
17      CONTINUE
      ENDIF
      RETURN
      END
```

REFERENCES AND FURTHER READING:

Wilkinson, J.H., and Reinsch, C. 1971, *Linear Algebra*, vol. II of *Handbook for Automatic Computation* (New York: Springer-Verlag).

Smith, B.T., et al. 1976, *Matrix Eigensystem Routines — EISPACK Guide*, 2nd ed., vol. 6 of Lecture Notes in Computer Science (New York: Springer-Verlag).

Stoer, J., and Bulirsch, R. 1980, *Introduction to Numerical Analysis* (New York: Springer-Verlag), §6.5.4.

11.6 The QR Algorithm for Real Hessenberg Matrices

Recall the following relations for the QR algorithm with shifts:

$$\mathbf{Q}_s \cdot (\mathbf{A}_s - k_s \mathbf{1}) = \mathbf{R}_s \tag{11.6.1}$$

where $\mathbf{Q}$ is orthogonal and $\mathbf{R}$ is upper triangular, and

$$\begin{aligned} \mathbf{A}_{s+1} &= \mathbf{R}_s \cdot \mathbf{Q}_s^T + k_s \mathbf{1} \\ &= \mathbf{Q}_s \cdot \mathbf{A}_s \cdot \mathbf{Q}_s^T \end{aligned} \tag{11.6.2}$$

The QR transformation preserves the upper Hessenberg form of the original matrix $\mathbf{A} \equiv \mathbf{A}_1$, and the workload on such a matrix is $O(n^2)$ per iteration as opposed to $O(n^3)$ on a general matrix. As $s \to \infty$, $\mathbf{A}_s$ converges to a form where the eigenvalues are either isolated on the diagonal or are eigenvalues of a 2×2 submatrix on the diagonal.

As we pointed out in §11.3, shifting is essential for rapid convergence. A key difference here is that a nonsymmetric real matrix can have complex eigenvalues. This means that good choices for the shifts k_s may be complex, apparently necessitating complex arithmetic.

Complex arithmetic can be avoided, however, by a clever trick. The trick depends on a result analogous to the lemma we used for implicit shifts in §11.3. The lemma we need here states that if $\mathbf{B}$ is a nonsingular matrix such that

$$\mathbf{B} \cdot \mathbf{Q} = \mathbf{Q} \cdot \mathbf{H} \tag{11.6.3}$$

where $\mathbf{Q}$ is orthogonal and $\mathbf{H}$ is upper Hessenberg, then $\mathbf{Q}$ and $\mathbf{H}$ are fully determined by the first column of $\mathbf{Q}$. (The determination is unique if $\mathbf{H}$ has positive subdiagonal elements.) The lemma can be proved by induction analogously to the proof given for tridiagonal matrices in §11.3.

The lemma is used in practice by taking two steps of the QR algorithm, either with two real shifts k_s and k_{s+1}, or with complex conjugate values k_s and $k_{s+1} = k_s{}^*$. This gives a real matrix $\mathbf{A}_{s+2}$, where

$$\mathbf{A}_{s+2} = \mathbf{Q}_{s+1} \cdot \mathbf{Q}_s \cdot \mathbf{A}_s \cdot \mathbf{Q}_s^T \cdot \mathbf{Q}_{s+1}^T \cdot \tag{11.6.4}$$

The $\mathbf{Q}$'s are determined by

$$\mathbf{A}_s - k_s \mathbf{1} = \mathbf{Q}_s^T \cdot \mathbf{R}_s \tag{11.6.5}$$

$$\mathbf{A}_{s+1} = \mathbf{Q}_s \cdot \mathbf{A}_s \cdot \mathbf{Q}_s^T \tag{11.6.6}$$

$$\mathbf{A}_{s+1} - k_{s+1} \mathbf{1} = \mathbf{Q}_{s+1}^T \cdot \mathbf{R}_{s+1} \tag{11.6.7}$$

Using (11.6.6), equation (11.6.7) can be rewritten

$$\mathbf{A}_s - k_{s+1} \mathbf{1} = \mathbf{Q}_s^T \cdot \mathbf{Q}_{s+1}^T \cdot \mathbf{R}_{s+1} \cdot \mathbf{Q}_s \tag{11.6.8}$$

Hence, if we define

$$\mathbf{M} = (\mathbf{A}_s - k_{s+1} \mathbf{1}) \cdot (\mathbf{A}_s - k_s \mathbf{1}) \tag{11.6.9}$$

equations (11.6.5) and (11.6.8) give

$$\mathbf{R} = \mathbf{Q} \cdot \mathbf{M} \tag{11.6.10}$$

where

$$\mathbf{Q} = \mathbf{Q}_{s+1} \cdot \mathbf{Q}_s \tag{11.6.11}$$

$$\mathbf{R} = \mathbf{R}_{s+1} \cdot \mathbf{R}_s \tag{11.6.12}$$

Equation (11.6.4) can be rewritten

$$\mathbf{A}_s \cdot \mathbf{Q}^T = \mathbf{Q}^T \cdot \mathbf{A}_{s+2} \tag{11.6.13}$$

Thus suppose we can somehow find an upper Hessenberg matrix $\mathbf{H}$ such that

$$\mathbf{A}_s \cdot \overline{\mathbf{Q}}^T = \overline{\mathbf{Q}}^T \cdot \mathbf{H} \tag{11.6.14}$$

where $\overline{\mathbf{Q}}$ is orthogonal. If $\overline{\mathbf{Q}}^T$ has the same first column as $\mathbf{Q}^T$ (i.e., $\overline{\mathbf{Q}}$ has the same first row as $\mathbf{Q}$), then $\overline{\mathbf{Q}} = \mathbf{Q}$ and $\mathbf{A}_{s+2} = \mathbf{H}$.

The first row of $\mathbf{Q}$ is found as follows. Equation (11.6.10) shows that $\mathbf{Q}$ is the orthogonal matrix that triangularizes the real matrix $\mathbf{M}$. Any real matrix can be triangularized by premultiplying it by a sequence of Householder matrices $\mathbf{P}_1$ (acting on the first column), $\mathbf{P}_2$ (acting on the second column), ..., $\mathbf{P}_{n-1}$. Thus $\mathbf{Q} = \mathbf{P}_{n-1} \cdots \mathbf{P}_2 \cdot \mathbf{P}_1$, and the first row of $\mathbf{Q}$ is the first row of $\mathbf{P}_1$ since $\mathbf{P}_i$ is an $(i-1) \times (i-1)$ identity matrix in the top left-hand corner. We now must find $\overline{\mathbf{Q}}$ satisfying (11.6.14) whose first row is that of $\mathbf{P}_1$.

The Householder matrix $\mathbf{P}_1$ is determined by the first column of $\mathbf{M}$. Since $\mathbf{A}_s$ is upper Hessenberg, equation (11.6.9) shows that the first column of $\mathbf{M}$ has the form $[p_1, q_1, r_1, 0, ..., 0]^T$, where

$$\begin{aligned} p_1 &= a_{11}^2 - a_{11}(k_s + k_{s+1}) + k_s k_{s+1} + a_{12}a_{21} \\ q_1 &= a_{21}(a_{11} + a_{22} - k_s - k_{s+1}) \\ r_1 &= a_{21}a_{32} \end{aligned} \tag{11.6.15}$$

Hence

$$\mathbf{P}_1 = 1 - 2\mathbf{w}_1 \cdot \mathbf{w}_1^T \tag{11.6.16}$$

where $\mathbf{w}_1$ has only its first 3 elements nonzero (cf. equation 11.2.5). The matrix $\mathbf{P}_1 \cdot \mathbf{A}_s \cdot \mathbf{P}_1^T$ is therefore upper Hessenberg with 3 extra elements:

$$\mathbf{P}_1 \cdot \mathbf{A}_1 \cdot \mathbf{P}_1^T = \begin{bmatrix} \times & \times & \times & \times & \times & \times & \times \\ \times & \times & \times & \times & \times & \times & \times \\ \mathbf{x} & \times & \times & \times & \times & \times & \times \\ \mathbf{x} & \mathbf{x} & \times & \times & \times & \times & \times \\ & & & \times & \times & \times & \times \\ & & & & \times & \times & \times \\ & & & & & \times & \times \end{bmatrix} \tag{11.6.17}$$

This matrix can be restored to upper Hessenberg form without affecting the first row by a sequence of Householder similarity transformations. The first such Householder matrix, $\mathbf{P}_2$, acts on elements 2, 3 and 4 in the first column, annihilating elements 3 and 4. This produces a matrix of the same form as (11.6.17), with the 3 extra elements appearing one column over:

$$\begin{bmatrix} \times & \times & \times & \times & \times & \times & \times \\ \times & \times & \times & \times & \times & \times & \times \\ & \times & \times & \times & \times & \times & \times \\ & \mathbf{x} & \times & \times & \times & \times & \times \\ & \mathbf{x} & \mathbf{x} & \times & \times & \times & \times \\ & & & & \times & \times & \times \\ & & & & & \times & \times \end{bmatrix} \tag{11.6.18}$$

Proceeding in this way up to $\mathbf{P}_{n-1}$, we see that at each stage the Householder matrix $\mathbf{P}_r$ has a vector $\mathbf{w}_r$ that is nonzero only in elements r, $r+1$ and $r+2$. These elements are determined by the elements r, $r+1$ and $r+2$ in the $(r-1)^{st}$ column of the current matrix. Note that the preliminary matrix $\mathbf{P}_1$ has the same structure as $\mathbf{P}_2, \ldots, \mathbf{P}_{n-1}$.

The result is that

$$\mathbf{P}_{n-1} \cdots \mathbf{P}_2 \cdot \mathbf{P}_1 \cdot \mathbf{A}_s \cdot \mathbf{P}_1^T \cdot \mathbf{P}_2^T \cdots \mathbf{P}_{n-1} = \mathbf{H} \tag{11.6.19}$$

where $\mathbf{H}$ is upper Hessenberg. Thus

$$\overline{\mathbf{Q}} = \mathbf{Q} = \mathbf{P}_{n-1} \cdots \mathbf{P}_2 \cdot \mathbf{P}_1 \tag{11.6.20}$$

and

$$\mathbf{A}_{s+2} = \mathbf{H} \tag{11.6.21}$$

The shifts of origin at each stage are taken to be the eigenvalues of the 2×2 matrix in the bottom right-hand corner of the current $\mathbf{A}_s$. This gives

$$\begin{aligned} k_s + k_{s+2} &= a_{n-1,n-1} + a_{nn} \\ k_s k_{s+1} &= a_{n-1,n-1} a_{nn} - a_{n-1,n} a_{n,n-1} \end{aligned} \tag{11.6.22}$$

Substituting (11.6.22) in (11.6.15), we get

$$\begin{aligned} p_1 &= a_{21} \left\{ [(a_{nn} - a_{11})(a_{n-1,n-1} - a_{11}) - a_{n-1,n} a_{n,n-1}]/a_{21} + a_{12} \right\} \\ q_1 &= a_{21} [a_{22} - a_{11} - (a_{nn} - a_n) - (a_{n-1,n-1} - a_{11})] \\ r_1 &= a_{21} a_{32} \end{aligned} \tag{11.6.23}$$

We have judiciously grouped terms to reduce possible roundoff when there are small off-diagonal elements. Since only the ratios of elements are relevant for a Householder transformation, we can omit the factor a_{21} from (11.6.23).

In summary, to carry out a double QR step we construct the Householder matrices $\mathbf{P}_r$, $r = 1, \ldots, n-1$. For $\mathbf{P}_1$ we use p_1, q_1, and r_1 given by (11.6.23). For the remaining matrices, p_r, q_r and r_r are determined by the $(r, r-1)$, $(r+1, r-1)$ and $(r+2, r-1)$ elements of the current matrix. The number of arithmetic operations can be reduced by writing the nonzero elements of the $2\mathbf{w} \cdot \mathbf{w}^T$ part of the Householder matrix in the form

$$2\mathbf{w} \cdot \mathbf{w}^T = \begin{bmatrix} (p \pm s)/(\pm s) \\ q/(\pm s) \\ r/(\pm s) \end{bmatrix} \cdot [1 \quad q/(p \pm s) \quad r/(p \pm s)] \tag{11.6.24}$$

where

$$s^2 = p^2 + q^2 + r^2 \tag{11.6.25}$$

(We have simply divided each element by a piece of the normalizing factor; cf. the equations in §11.2.)

Proceeding in this way, convergence is usually very fast. There are two possible ways of terminating the iteration for an eigenvalue. First, if $a_{n,n-1}$ becomes "negligible," then a_{nn} is an eigenvalue. We can then delete the nth row and column of the matrix and look for the next eigenvalue. Alternatively, $a_{n-1,n-2}$ may become negligible. In this case the eigenvalues of the 2×2 matrix in the lower right-hand corner may be taken to be eigenvalues. We delete the nth and $(n-1)^{st}$ rows and columns of the matrix and continue.

The test for convergence to an eigenvalue is combined with a test for negligible subdiagonal elements that allows splitting of the matrix into submatrices. We find the largest i such that $a_{i,i-1}$ is negligible. If $i = n$, we have found a single eigenvalue. If $i = n-1$, we have found two eigenvalues. Otherwise we continue the iteration on the submatrix in rows i to n (i being set to unity if there is no small subdiagonal element).

After determining i, the submatrix in rows i to n is examined to see if the *product* of any two consecutive subdiagonal elements is small enough that we can work with an even smaller submatrix, starting say in row m. We start with $m = n-2$ and decrease it down to $i+1$, computing p, q and r according to equations (11.6.23) with 1 replaced by m and 2 by $m+1$. If these were indeed the elements of the special "first" Householder matrix in a double QR step, then applying the Householder matrix would lead to nonzero elements in positions $(m+1, m-1)$, $(m+2, m-1)$ and $(m+2, m)$. We require that the first two of these elements be small compared with the local diagonal elements $a_{m-1,m-1}$, a_{mm} and $a_{m+1,m+1}$. A satisfactory approximate criterion is

$$|a_{m,m-1}|(|q| + |r|) \ll |p|(|a_{m+1,m+1}| + |a_{mm}| + |a_{m-1,m-1}|) \tag{11.6.26}$$

Very rarely, the procedure described so far will fail to converge. On such matrices, experience shows that if one double step is performed with any shifts that are of order the norm of the matrix, convergence is subsequently very rapid. Accordingly, if ten iterations occur without determining an eigenvalue, the usual shifts are replaced for the next iteration by shifts defined by

$$\begin{aligned} k_s + k_{s+1} &= 1.5 \times \left(|a_{n,n-1}| + |a_{n-1,n-2}|\right), \\ k_s k_{s+1} &= \left(|a_{n,n-1}| + |a_{n-1,n-2}|\right)^2 \end{aligned} \tag{11.6.27}$$

The factor 1.5 was arbitrarily chosen to lessen the likelihood of an "unfortunate" choice of shifts. This strategy is repeated after 20 unsuccessful iterations. After 30 unsuccessful iterations, the routine reports failure.

The operation count for the QR algorithm described here is $\sim 5k^2$ per iteration, where k is the current size of the matrix. The typical average number of iterations per eigenvalue is ~ 1.8, so the total operation count for all the eigenvalues is $\sim 3n^3$. This estimate neglects any possible efficiency due to splitting or sparseness of the matrix.

The following routine HQR is based algorithmically on the above description, in turn following the *Handbook* and EISPACK implementations.

```
      SUBROUTINE HQR(A,N,NP,WR,WI)
         Finds all eigenvalues of an N by N upper Hessenberg matrix A that is stored in an NP by NP
         array. On input A can be exactly as output from ELMHES §11.5; on output it is destroyed.
         The real and imaginary parts of the eigenvalues are returned in WR and WI respectively.
      DIMENSION A(NP,NP),WR(NP),WI(NP)
      ANORM=ABS(A(1,1))               Compute matrix norm for possible use in locating single small subdi-
      DO 12 I=2,N                         agonal element.
         DO 11 J=I-1,N
            ANORM=ANORM+ABS(A(I,J))
11       CONTINUE
12    CONTINUE
      NN=N
      T=0.                            ...gets changed only by an exceptional shift.
1     IF(NN.GE.1)THEN                 Begin search for next eigenvalue.
         ITS=0
2        DO 13 L=NN,2,-1              Begin iteration: look for single small subdiagonal element.
            S=ABS(A(L-1,L-1))+ABS(A(L,L))
            IF(S.EQ.0.)S=ANORM
            IF(ABS(A(L,L-1))+S.EQ.S)GO TO 3
13       CONTINUE
         L=1
3        X=A(NN,NN)
         IF(L.EQ.NN)THEN              One root found.
            WR(NN)=X+T
            WI(NN)=0.
            NN=NN-1
         ELSE
            Y=A(NN-1,NN-1)
            W=A(NN,NN-1)*A(NN-1,NN)
            IF(L.EQ.NN-1)THEN         Two roots found...
               P=0.5*(Y-X)
               Q=P**2+W
               Z=SQRT(ABS(Q))
               X=X+T
               IF(Q.GE.0.)THEN        ...a real pair.
```

```
            Z=P+SIGN(Z,P)
            WR(NN)=X+Z
            WR(NN-1)=WR(NN)
            IF(Z.NE.0.)WR(NN)=X-W/Z
            WI(NN)=0.
            WI(NN-1)=0.
        ELSE                    ...a complex pair.
            WR(NN)=X+P
            WR(NN-1)=WR(NN)
            WI(NN)=Z
            WI(NN-1)=-Z
        ENDIF
        NN=NN-2
    ELSE                        No roots found.  Continue iteration.
        IF(ITS.EQ.30)PAUSE 'too many iterations'
        IF(ITS.EQ.10.OR.ITS.EQ.20)THEN        Form exceptional shift.
            T=T+X
            DO 14 I=1,NN
                A(I,I)=A(I,I)-X
14          CONTINUE
            S=ABS(A(NN,NN-1))+ABS(A(NN-1,NN-2))
            X=0.75*S
            Y=X
            W=-0.4375*S**2
        ENDIF
        ITS=ITS+1
        DO 15 M=NN-2,L,-1      Form shift and then look for 2 consecutive small subdiagonal
            Z=A(M,M)              elements.
            R=X-Z
            S=Y-Z
            P=(R*S-W)/A(M+1,M)+A(M,M+1)        Equation 11.6.23.
            Q=A(M+1,M+1)-Z-R-S
            R=A(M+2,M+1)
            S=ABS(P)+ABS(Q)+ABS(R)     Scale to prevent overflow or underflow.
            P=P/S
            Q=Q/S
            R=R/S
            IF(M.EQ.L)GO TO 4
            U=ABS(A(M,M-1))*(ABS(Q)+ABS(R))
            V=ABS(P)*(ABS(A(M-1,M-1))+ABS(Z)+ABS(A(M+1,M+1)))
            IF(U+V.EQ.V)GO TO 4        Equation 11.6.26.
15      CONTINUE
4       DO 16 I=M+2,NN
            A(I,I-2)=0.
            IF (I.NE.M+2) A(I,I-3)=0.
16      CONTINUE
        DO 19 K=M,NN-1      Double QR step on rows L to NN and columns M to NN.
            IF(K.NE.M)THEN
                P=A(K,K-1)     Begin setup of Householder vector.
                Q=A(K+1,K-1)
                R=0.
                IF(K.NE.NN-1)R=A(K+2,K-1)
                X=ABS(P)+ABS(Q)+ABS(R)
                IF(X.NE.0.)THEN
                    P=P/X      Scale to prevent overflow or underflow.
                    Q=Q/X
                    R=R/X
                ENDIF
            ENDIF
            S=SIGN(SQRT(P**2+Q**2+R**2),P)
            IF(S.NE.0.)THEN
                IF(K.EQ.M)THEN
                    IF(L.NE.M)A(K,K-1)=-A(K,K-1)
                ELSE
```

```
                        A(K,K-1)=-S*X
                    ENDIF
                    P=P+S       Equations 11.6.24.
                    X=P/S
                    Y=Q/S
                    Z=R/S
                    Q=Q/P
                    R=R/P
                    DO 17 J=K,NN           Row modification.
                        P=A(K,J)+Q*A(K+1,J)
                        IF(K.NE.NN-1)THEN
                            P=P+R*A(K+2,J)
                            A(K+2,J)=A(K+2,J)-P*Z
                        ENDIF
                        A(K+1,J)=A(K+1,J)-P*Y
                        A(K,J)=A(K,J)-P*X
17                  CONTINUE
                    DO 18 I=L,MIN(NN,K+3)     Column modification.
                        P=X*A(I,K)+Y*A(I,K+1)
                        IF(K.NE.NN-1)THEN
                            P=P+Z*A(I,K+2)
                            A(I,K+2)=A(I,K+2)-P*R
                        ENDIF
                        A(I,K+1)=A(I,K+1)-P*Q
                        A(I,K)=A(I,K)-P
18                  CONTINUE
                ENDIF
19          CONTINUE
            GO TO 2                    ...for next iteration on current eigenvalue.
          ENDIF
        ENDIF
      GO TO 1                          ...for next eigenvalue.
      ENDIF
      RETURN
      END
```

REFERENCES AND FURTHER READING:

Wilkinson, J.H., and Reinsch, C. 1971, *Linear Algebra*, vol. II of *Handbook for Automatic Computation* (New York: Springer-Verlag).

Golub, Gene H., and Van Loan, Charles F. 1983, *Matrix Computations* (Baltimore: Johns Hopkins University Press), §7.5.

Smith, B.T., et al. 1976, *Matrix Eigensystem Routines — EISPACK Guide*, 2nd ed., vol. 6 of Lecture Notes in Computer Science (New York: Springer-Verlag).

11.7 Improving Eigenvalues and/or Finding Eigenvectors by Inverse Iteration

The basic idea behind inverse iteration is quite simple. Let $\mathbf{y}$ be the solution of the linear system

$$(\mathbf{A} - \tau\mathbf{1}) \cdot \mathbf{y} = \mathbf{b} \tag{11.7.1}$$

where $\mathbf{b}$ is a random vector and τ is close to some eigenvalue λ of $\mathbf{A}$. Then the solution $\mathbf{y}$ will be close to the eigenvector corresponding to λ. The procedure can be iterated: replace $\mathbf{b}$ by $\mathbf{y}$ and solve for a new $\mathbf{y}$, which will be even closer to the true eigenvector.

We can see why this works by expanding both $\mathbf{y}$ and $\mathbf{b}$ as linear combinations of the eigenvectors $\mathbf{x}_j$ of $\mathbf{A}$:

$$\mathbf{y} = \sum_j \alpha_j \mathbf{x}_j \qquad \mathbf{b} = \sum_j \beta_j \mathbf{x}_j \tag{11.7.2}$$

Then (11.7.1) gives

$$\sum_j \alpha_j(\lambda_j - \tau)\mathbf{x}_j = \sum_j \beta_j \mathbf{x}_j \tag{11.7.3}$$

so that

$$\alpha_j = \frac{\beta_j}{\lambda_j - \tau} \tag{11.7.4}$$

and

$$\mathbf{y} = \sum_j \frac{\beta_j \mathbf{x}_j}{\lambda_j - \tau} \tag{11.7.5}$$

If τ is close to λ_n, say, then provided β_n is not accidentally too small, $\mathbf{y}$ will be approximately $\mathbf{x}_n$, up to a normalization. Moreover, each iteration of this procedure gives another power of $\lambda_j - \tau$ in the denominator of (11.7.5). Thus the convergence is rapid for well-separated eigenvalues.

Suppose at the ith stage of iteration we are solving the equation

$$(\mathbf{A} - \lambda_i \mathbf{1}) \cdot \mathbf{y} = \mathbf{x}_i \tag{11.7.6}$$

where $\mathbf{x}_i$ and λ_i are our current guesses for some eigenvector and eigenvalue of interest (we shall see below how to update λ_i). The exact eigenvector and eigenvalue satisfy

$$\mathbf{A} \cdot \mathbf{x} = \lambda \mathbf{x} \tag{11.7.7}$$

so

$$(\mathbf{A} - \lambda_i \mathbf{1}) \cdot \mathbf{x} = (\lambda - \lambda_i)\mathbf{x} \tag{11.7.8}$$

Since $\mathbf{y}$ of (11.7.6) is an improved approximation to $\mathbf{x}$, we normalize it and set

$$\mathbf{x}_{i+1} = \frac{\mathbf{y}}{|\mathbf{y}|} \tag{11.7.9}$$

We get an improved estimate of the eigenvalue by substituting our improved guess $\mathbf{y}$ in (11.7.8). By (11.7.6), the left-hand side is $\mathbf{x}_i$, so calling λ our new value λ_{i+1}, we find

$$\lambda_{i+1} = \lambda_i + \frac{|\mathbf{x}_i|^2}{|\mathbf{x}_i \cdot \mathbf{y}|} \tag{11.7.10}$$

While the above formulas look simple enough, in practice the implementation can be quite tricky. The first question to be resolved is *when* to use inverse iteration. Most of the computational load occurs in solving the linear system (11.7.6). Thus a possible strategy is first to reduce the matrix $\mathbf{A}$ to a special form that allows easy solution of (11.7.6). Tridiagonal form for symmetric matrices or Hessenberg for nonsymmetric are the obvious choices. Then apply inverse iteration to generate all the eigenvectors. While this is an $O(N^3)$ method for symmetric matrices, it is many times less efficient than the QL method given earlier. In fact, even the best inverse iteration packages are less efficient than the QL method as soon as more than about 25 percent of the eigenvectors are required. Accordingly, inverse iteration is generally used when one already has good eigenvalues and only wants a few selected eigenvectors.

You can write a simple inverse iteration routine yourself using LU decomposition to solve (11.7.6). You can decide whether to use the general LU algorithm we gave in Chapter 2 or whether to take advantage of tridiagonal or Hessenberg form. Note that, since the linear system (11.7.6) is nearly singular, you must be careful to use a version of LU decomposition like that in §2.3 which replaces a zero pivot with a very small number.

We have chosen not to give a general inverse iteration routine in this book, because it is quite cumbersome to take account of all the cases that can arise. Routines are given, for example, in the *Handbook* and in EISPACK. If you use these, or write your own routine, you may appreciate the following pointers.

One starts by supplying a value for the eigenvalue λ. Choose a random normalized vector as the initial guess for the eigenvector, $\mathbf{x}_0$, and solve (11.7.6). The new vector $\mathbf{y}$ is bigger than $\mathbf{x}_0$ by a "growth factor" $|\mathbf{y}|$, which ideally should be large. Equivalently, the change in the eigenvalue, which by (11.7.10) is essentially $1/|\mathbf{y}|$, should be small. The following cases can arise:

- If the growth factor is too small initially, then we assume we have made a "bad" choice of random vector. This can happen not just because of a small β_n in (11.7.5), but also in the case of a defective matrix, when (11.7.5) does not even apply (see e.g. Wilkinson and Reinsch or Stoer and Bulirsch for details). We go back to the beginning and choose a new initial vector.
- The change $|\mathbf{x}_1 - \mathbf{x}_0|$ might be less than some tolerance ϵ. We can use this as a criterion for stopping, iterating until it is satisfied, with a maximum of 5 – 10 iterations, say.
- After a few iterations, if $|\mathbf{x}_{i+1} - \mathbf{x}_i|$ is not decreasing rapidly enough, we can try updating the eigenvalue according to (11.7.10). If $\lambda_{i+1} = \lambda_i$ to machine accuracy, we are not going to improve the eigenvector much more and can quit. Otherwise start another cycle of iterations with the new eigenvalue.

The reason we do not update the eigenvalue at every step is that when we solve the linear system (11.7.6) by LU decomposition, we can save the decomposition if λ_i is fixed. We only need do the backsubstitution step each time we update $\mathbf{x}_i$. The number of iterations with a fixed λ_i is a trade off between the quadratic convergence but $O(N^3)$ workload for updating λ_i at each step and the linear convergence but $O(N^2)$ load for keeping λ_i fixed. If you have determined the eigenvalue by one of the routines given earlier in the chapter, it is probably correct to machine accuracy anyway, and you can omit updating it.

There are two different pathologies that can arise during inverse iteration. The first is multiple or closely-spaced roots. This is more often a problem with symmetric matrices. Inverse iteration will find only one eigenvector for a given initial guess λ. A good strategy is to perturb the last few significant digits in λ and then repeat the iteration. Usually this provides an independent eigenvector. Special steps generally have to be taken to ensure orthogonality of the linearly independent eigenvectors, whereas the Jacobi and QL algorithms automatically yield orthogonal eigenvectors even in the case of multiple eigenvalues.

The second problem, peculiar to nonsymmetric matrices, is the defective case. Unless one makes a "good" initial guess, the growth factor is small. Moreover, iteration does not improve matters. In this case, the remedy is to choose random initial vectors, solve (11.7.6) once, and quit as soon as *any* vector gives an acceptably large growth factor. Typically only a few trials are necessary.

One further complication in the nonsymmetric case is that a real matrix can have complex-conjugate pairs of eigenvalues. You will then have to use complex arithmetic to solve (11.7.6) for the complex eigenvectors. For any moderate sized (or larger) nonsymmetric matrix, our recommendation is to avoid inverse iteration in favor of a QR method that includes the eigenvector computation in complex arithmetic. You will find routines for this in the *Handbook*, EISPACK, and other places.

REFERENCES AND FURTHER READING:

Acton, Forman S. 1970, *Numerical Methods That Work* (New York: Harper and Row).

Wilkinson, J.H., and Reinsch, C. 1971, *Linear Algebra*, vol. II of *Handbook for Automatic Computation* (New York: Springer-Verlag), p. 418.

Stoer, J., and Bulirsch, R. 1980, *Introduction to Numerical Analysis* (New York: Springer-Verlag), p. 356.

Smith, B.T., et al. 1976, *Matrix Eigensystem Routines — EISPACK Guide*, 2nd ed., vol. 6 of Lecture Notes in Computer Science (New York: Springer-Verlag).

Chapter 12. Fourier Transform Spectral Methods

12.0 Introduction

A very large class of important computational problems falls under the general rubric of "Fourier transform methods" or "spectral methods." For some of these problems, the Fourier transform is simply an efficient computational tool for accomplishing certain common manipulations of data. In other cases, we have problems for which the Fourier transform (or the related "power spectrum") is itself of intrinsic interest. These two kinds of problems share a common methodology.

Largely for historical reasons the literature on Fourier and spectral methods has been disjoint from the literature on "classical" numerical analysis. In this day and age there is no justification for such a split. Fourier methods are commonplace in research and we shall not treat them as specialized or arcane. At the same time, we realize that many computer users have had relatively less experience with this field than with, say, differential equations or numerical integration. Therefore our summary of analytical results will be more complete. Numerical algorithms, per se, begin in §12.2.

A physical process can be described either in the *time domain*, by the values of some quantity h as a function of time t, e.g. $h(t)$, or else in the *frequency domain*, where the process is specified by giving its amplitude H (generally a complex number indicating phase also) as a function of frequency f, that is $H(f)$, with $-\infty < f < \infty$. For many purposes it is useful to think of $h(t)$ and $H(f)$ as being two different *representations* of the *same* function. One goes back and forth between these two representations by means of the *Fourier transform* equations,

$$
\begin{aligned}
H(f) &= \int_{-\infty}^{\infty} h(t)e^{2\pi ift}dt \\
h(t) &= \int_{-\infty}^{\infty} H(f)e^{-2\pi ift}df
\end{aligned}
\tag{12.0.1}
$$

If t is measured in seconds, then f in equation (12.0.1) is in cycles per second, or Hertz (the unit of frequency). However, the equations work with

other units. If h is a function of position x (in meters), H will be a function of inverse wavelength (cycles per meter), and so on. If you are trained as a physicist or mathematician, you are probably more used to using *angular frequency* ω, which is given in *radians* per sec. The relation between ω and f, $H(\omega)$ and $H(f)$ is

$$\omega \equiv 2\pi f \qquad H(\omega) \equiv [H(f)]_{f=\omega/2\pi} \tag{12.0.2}$$

and equation (12.0.1) looks like this

$$\begin{aligned} H(\omega) &= \int_{-\infty}^{\infty} h(t) e^{i\omega t} dt \\ h(t) &= \frac{1}{2\pi} \int_{-\infty}^{\infty} H(\omega) e^{-i\omega t} d\omega \end{aligned} \tag{12.0.3}$$

We were raised on the ω-convention, but we changed! There are fewer factors of 2π to remember if you use the f-convention, especially when we get to discretely sampled data in §12.1.

From equation (12.0.1) it is evident at once that Fourier transformation is a *linear* operation. The transform of the sum of two functions is equal to the sum of the transforms. The transform of a constant times a function is that same constant times the transform of the function.

In the time domain, function $h(t)$ may happen to have one or more special symmetries It might be *purely real* or *purely imaginary* or it might be *even*, $h(t) = h(-t)$, or *odd*, $h(t) = -h(-t)$. In the frequency domain, these symmetries lead to relationships between $H(f)$ and $H(-f)$. The following table gives the correspondence between symmetries in the two domains:

If...	then...
$h(t)$ is real	$H(-f) = [H(f)]^*$
$h(t)$ is imaginary	$H(-f) = -[H(f)]^*$
$h(t)$ is even	$H(-f) = H(f)$ [i.e. $H(f)$ is even]
$h(t)$ is odd	$H(-f) = -H(f)$ [i.e. $H(f)$ is odd]
$h(t)$ is real and even	$H(f)$ is real and even
$h(t)$ is real and odd	$H(f)$ is imaginary and odd
$h(t)$ is imaginary and even	$H(f)$ is imaginary and even
$h(t)$ is imaginary and odd	$H(f)$ is real and odd

In subsequent sections we shall see how to use these symmetries to increase computational efficiency.

Here are some other elementary properties of the Fourier transform. (We'll use the "$\Longleftrightarrow$" symbol to indicate transform pairs.) If

$$h(t) \Longleftrightarrow H(f)$$

is such a pair, then other transform pairs are

$$h(at) \Longleftrightarrow \frac{1}{|a|}H(\frac{f}{a}) \qquad \text{“time scaling”} \tag{12.0.4}$$

$$\frac{1}{|b|}h(\frac{t}{b}) \Longleftrightarrow H(bf) \qquad \text{“frequency scaling”} \tag{12.0.5}$$

$$h(t-t_0) \Longleftrightarrow H(f)\, e^{2\pi i f t_0} \qquad \text{“time shifting”} \tag{12.0.6}$$

$$h(t)\, e^{-2\pi i f_0 t} \Longleftrightarrow H(f-f_0) \qquad \text{“frequency shifting”} \tag{12.0.7}$$

With two functions $h(t)$ and $g(t)$, and their corresponding Fourier transforms $H(f)$ and $G(f)$, we can form two combinations of special interest. The *convolution* of the two functions, denoted $g * h$, is defined by

$$g * h \equiv \int_{-\infty}^{\infty} g(\tau)h(t-\tau)\, d\tau \tag{12.0.8}$$

Note that $g * h$ is a function in the time domain and that $g * h = h * g$. It turns out that the function $g * h$ is one member of a simple transform pair

$$g * h \Longleftrightarrow G(f)H(f) \qquad \text{“Convolution Theorem”} \tag{12.0.9}$$

In other words, the Fourier transform of the convolution is just the product of the individual Fourier transforms.

The *correlation* of two functions, denoted $\mathrm{Corr}(g,h)$, is defined by

$$\mathrm{Corr}(g,h) \equiv \int_{-\infty}^{\infty} g(\tau+t)h(\tau)\, d\tau \tag{12.0.10}$$

The correlation is a function of t, which is called the *lag*. It therefore lies in the time domain, and it turns out to be one member of the transform pair:

$$\mathrm{Corr}(g,h) \Longleftrightarrow G(f)H^*(f) \qquad \text{“Correlation Theorem”} \tag{12.0.11}$$

[More generally, the second member of the pair is $G(f)H(-f)$, but we are restricting ourselves to the usual case in which g and h are real functions, so we take the liberty of setting $H(-f) = H^*(f)$.] This result shows that multiplying the Fourier transform of one function by the complex conjugate of the Fourier Transform of the other gives the Fourier transform of their correlation. The correlation of a function with itself is called its *autocorrelation*. In this case (12.0.11) becomes the transform pair

$$\mathrm{Corr}(g,g) \Longleftrightarrow |G(f)|^2 \qquad \text{“Wiener-Khinchin Theorem”} \tag{12.0.12}$$

The *total power* in a signal is the same whether we compute it in the time domain or in the frequency domain. This result is known as *Parseval's theorem*:

$$\text{Total Power} \equiv \int_{-\infty}^{\infty} |h(t)|^2 \, dt = \int_{-\infty}^{\infty} |H(f)|^2 \, df \qquad (12.0.13)$$

Frequently one wants to know "how much power" is contained in the frequency interval between f and $f + df$. In such circumstances one does not usually distinguish between positive and negative f, but rather regards f as varying from 0 ("zero frequency" or D.C.) to $+\infty$. In such cases, one defines the *one-sided power spectral density (PSD)* of the function h as

$$P_h(f) \equiv |H(f)|^2 + |H(-f)|^2 \qquad 0 \le f < \infty \qquad (12.0.14)$$

so that the total power is just the integral of $P_h(f)$ from $f = 0$ to $f = \infty$. When the function $h(t)$ is real, then the two terms in (12.0.14) are equal, so $P_h(f) = 2\,|H(f)|^2$. Be warned that one occasionally sees PSDs defined without this factor two. These, strictly speaking, are called *two-sided power spectral densities*, but some books are not careful about stating whether one- or two-sided is to be assumed. We will always use the one-sided density given by equation (12.0.14). Figure 12.0.1 contrasts the two conventions.

If the function $h(t)$ goes endlessly from $-\infty < t < \infty$, then its total power and power spectral density will, in general, be infinite. Of interest then is the *(one- or two-sided) power spectral density per unit time.* This is computed by taking a long, but finite, stretch of the function $h(t)$, computing its PSD [that is, the PSD of a function which equals $h(t)$ in the finite stretch but is zero everywhere else], and then dividing the resulting PSD by the length of the stretch used. Parseval's theorem in this case states that the integral of the one-sided PSD-per-unit-time over positive frequency is equal to the mean-square amplitude of the signal $h(t)$.

You might well worry about how the PSD-per-unit-time, which is a function of frequency f, converges as one evaluates it using longer and longer stretches of data. This interesting question is the content of the subject of "power spectrum estimation," and will be considered below in §12.8–§12.9. A crude answer for now is: the PSD-per-unit-time converges to finite values at all frequencies *except* those where $h(t)$ has a discrete sine-wave (or cosine-wave) component of finite amplitude. At those frequencies, it becomes a delta-function, i.e. a sharp spike, whose width gets narrower and narrower, but whose area converges to be the mean-square amplitude of the discrete sine or cosine component at that frequency.

We have by now stated all of the analytical formalism that we will need in this chapter with one exception: In computational work, especially with experimental data, we are almost never given a continuous function $h(t)$ to work with, but are given, rather, a list of measurements of $h(t_i)$ for a discrete

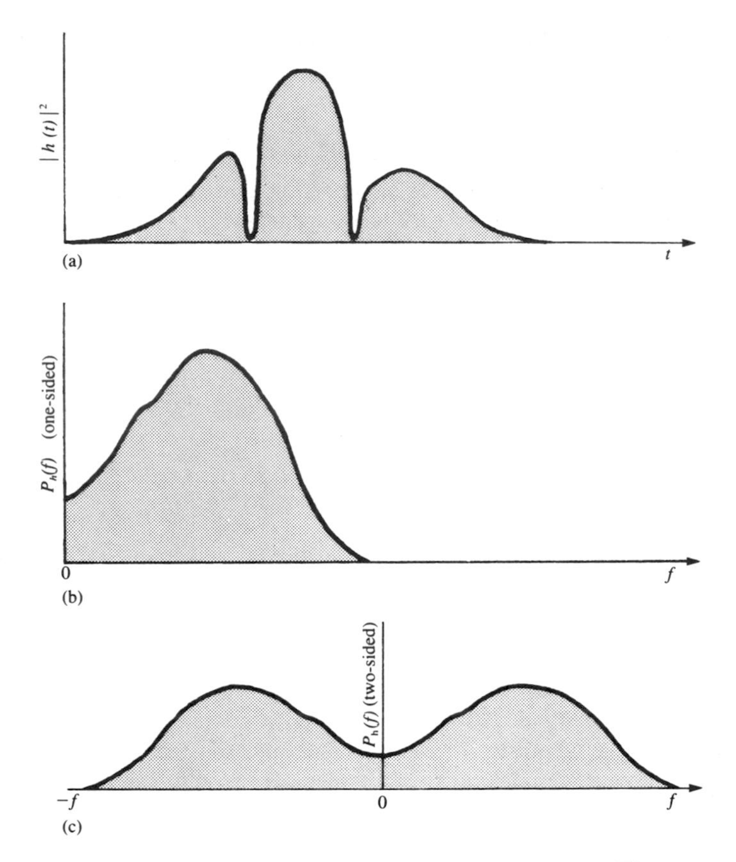

Figure 12.0.1 Normalizations of one- and two-sided power spectra. The area under the square of the function, (a), equals the area under its one-sided power spectrum at positive frequencies, (b), and also equals the area under its two-sided power spectrum at positive and negative frequencies, (c).

set of t_i's. The profound implications of this seemingly unimportant fact are the subject of the next section.

REFERENCES AND FURTHER READING:

Champeney, D.C. 1973, *Fourier Transforms and Their Physical Applications* (New York: Academic Press).

Elliott, D.F., and Rao, K.R. 1982, *Fast Transforms: Algorithms, Analyses, Applications* (New York: Academic Press).

12.1 Fourier Transform of Discretely Sampled Data

In the most common situations, function $h(t)$ is sampled (i.e., its value is recorded) at evenly spaced intervals in time. Let Δ denote the time interval between consecutive samples, so that the sequence of sampled values is

$$h_n = h(n\Delta) \qquad n = \dots, -3, -2, -1, 0, 1, 2, 3, \dots \tag{12.1.1}$$

The reciprocal of the time interval Δ is called the *sampling rate*; if Δ is measured in seconds, for example, then the sampling rate is the number of samples recorded per second.

Sampling Theorem and Aliasing

For any sampling interval Δ, there is also a special frequency f_c, called the *Nyquist critical frequency*, given by

$$f_c \equiv \frac{1}{2\Delta} \tag{12.1.2}$$

If a sine wave of the Nyquist critical frequency is sampled at its positive peak value, then the next sample will be at its negative trough value, the sample after that at the positive peak again, and so on. Expressed otherwise: *Critical sampling of a sine wave is two sample points per cycle.* One frequently chooses to measure time in units of the sampling interval Δ. In this case the Nyquist critical frequency is just the constant $1/2$.

The Nyquist critical frequency is important for two related, but distinct, reasons. One is good news, and the other bad news. First the good news. It is the remarkable fact known as the *sampling theorem*: If a continuous function $h(t)$, sampled at an interval Δ, happens to be *band-width limited* to frequencies smaller in magnitude than f_c, i.e., if $H(f) = 0$ for all $|f| > f_c$, then the function $h(t)$ is *completely determined* by its samples h_n. In fact, $h(t)$ is given explicitly by the formula

$$h(t) = \Delta \sum_{n=-\infty}^{+\infty} h_n \frac{\sin[2\pi f_c(t - n\Delta)]}{\pi(t - n\Delta)} \tag{12.1.3}$$

This is a remarkable theorem for many reasons, among them that it shows that the "information content" of a band-width limited function is, in some sense, infinitely smaller than that of a general continuous function. Fairly often, one is dealing with a signal which is known on physical grounds to be band-width limited (or at least approximately band-width limited). For example, the signal may have passed through an amplifier with a known, finite

frequency response. In this case, the sampling theorem tells us that the entire information content of the signal can be recorded by sampling it at a rate Δ^{-1} equal to twice the maximum frequency passed by the amplifier (cf. 12.1.2).

Now the bad news. The bad news concerns the effect of sampling a continuous function that is *not* band-width limited to less than the Nyquist critical frequency. In that case, it turns out that all of the power spectral density which lies outside of the frequency range $-f_c < f < f_c$ is spuriously moved into that range. This phenomenon is called *aliasing*. Any frequency component outside of the frequency range $(-f_c, f_c)$ is *aliased* (falsely translated) into that range by the very act of discrete sampling. You can readily convince yourself that two waves $\exp(2\pi i f_1 t)$ and $\exp(2\pi i f_2 t)$ give the same samples at an interval Δ if and only if f_1 and f_2 differ by a multiple of $1/\Delta$, which is just the width in frequency of the range $(-f_c, f_c)$. There is little that you can do to remove aliased power once you have discretely sampled a signal. The way to overcome aliasing is to (i) know the natural band-width limit of the signal—or else enforce a known limit by analog filtering of the continuous signal, and then (ii) sample at a rate sufficiently rapid to give two points per cycle of the highest frequency present. Figure 12.1.1 illustrates these considerations.

To put the best face on this, we can take the alternative point of view: If a continuous function has been competently sampled, then, when we come to estimate its Fourier transform from the discrete samples, we can *assume* (or rather we *might as well* assume) that its Fourier transform is equal to zero outside of the frequency range in between $-f_c$ and f_c. Then we look to the Fourier transform to tell whether the continuous function *has* been competently sampled (aliasing effects minimized). We do this by looking to see whether the Fourier transform is already approaching zero as the frequency approaches f_c from below, or $-f_c$ from above. If, on the contrary, the transform is going towards some finite value, then chances are that components outside of the range have been folded back over onto the critical range.

Discrete Fourier Transform

We now estimate the Fourier transform of a function from a finite number of its sampled points. Suppose that we have N consecutive sampled values

$$h_k \equiv h(t_k), \qquad t_k \equiv k\Delta, \qquad k = 0, 1, 2, \ldots N-1 \tag{12.1.4}$$

so that the sampling interval is Δ. To make things simpler, let us also suppose that N is even. If the function $h(t)$ is nonzero only in a finite interval of time, then that whole interval of time is supposed to be contained in the range of the N points given. Alternatively, if the function $h(t)$ goes on forever, then the sampled points are supposed to be at least "typical" of what $h(t)$ looks like at all other times.

With N numbers of input, we will evidently be able to produce no more than N independent numbers of output. So, instead of trying to estimate the

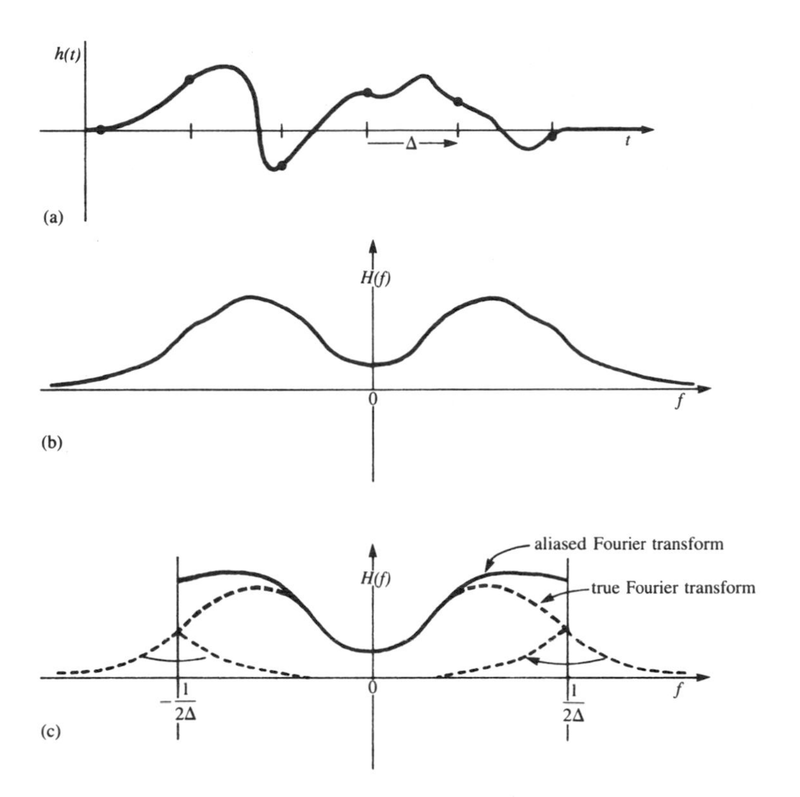

Figure 12.1.1. The continuous function shown in (a) is nonzero only for a finite interval of time T. It follows that its Fourier transform, shown schematically in (b), is not bandwidth limited but has finite amplitude for all frequencies. If the original function is sampled with a sampling interval Δ, as in (a), then the Fourier transform (c) is defined only between plus and minus the Nyquist critical frequency. Power outside that range is folded over or "aliased" into the range. The effect can be eliminated only by low-pass filtering the original function *before sampling*.

Fourier transform $H(f)$ at all values of f in the range $-f_c$ to f_c, let us seek estimates only at the discrete values

$$f_n \equiv \frac{n}{N\Delta}, \qquad n = -\frac{N}{2}, \ldots, \frac{N}{2} \tag{12.1.5}$$

The extreme values of n in (12.1.5) correspond exactly to the lower and upper limits of the Nyquist critical frequency range. If you are really on the ball, you will have noticed that there are $N + 1$, not N, values of n in (12.1.5); it will turn out that the two extreme values of n are not independent (in fact they are equal), but all the others are. This reduces the count to N.

The remaining step is to approximate the integral in (12.0.1) by a discrete

sum:

$$H(f_n) = \int_{-\infty}^{\infty} h(t)e^{2\pi i f_n t}dt \approx \sum_{k=0}^{N-1} h_k\, e^{2\pi i f_n t_k}\Delta = \Delta \sum_{k=0}^{N-1} h_k\, e^{2\pi i k n/N} \tag{12.1.6}$$

Here equations (12.1.4) and (12.1.5) have been used in the final equality. The final summation in equation (12.1.6) is called the *discrete Fourier transform* of the N points h_k. Let us denote it by H_n,

$$H_n \equiv \sum_{k=0}^{N-1} h_k\, e^{2\pi i k n/N} \tag{12.1.7}$$

The discrete Fourier transform maps N complex numbers (the h_k's) into N complex numbers (the H_n's). It does not depend on any dimensional parameter, such as the time scale Δ. The relation (12.1.6) between the discrete Fourier transform of a set of numbers and their continuous Fourier transform when they are viewed as samples of a continuous function sampled at an interval Δ can be rewritten as

$$H(f_n) \approx \Delta H_n \tag{12.1.8}$$

where f_n is given by (12.1.5).

Up to now we have taken the view that the index n in (12.1.7) varies from $-N/2$ to $N/2$ (cf. 12.1.5). You can easily see, however, that (12.1.7) is periodic in n, with period N. Therefore, $H_{-n} = H_{N-n} \quad n = 1, 2, \ldots$. With this conversion in mind, one generally lets the n in H_n vary from 0 to $N-1$ (one complete period). Then n and k (in h_k) vary exactly over the same range, so the mapping of N numbers into N numbers is manifest. When this convention is followed, you must remember that zero frequency corresponds to $n = 0$, positive frequencies $0 < f < f_c$ correspond to values $1 \leq n \leq N/2 - 1$, while negative frequencies $-f_c < f < 0$ correspond to $N/2 + 1 \leq n \leq N - 1$. The value $n = N/2$ corresponds to *both* $f = f_c$ and $f = -f_c$.

The discrete Fourier transform has symmetry properties almost exactly the same as the continuous Fourier transform. For example, all the symmetries in the table following equation (12.0.3) hold if we read h_k for $h(t)$, H_n for $H(f)$, and H_{N-n} for $H(-f)$. (Likewise, "even" and "odd" in time refer to whether the values h_k at k and $N-k$ are identical or the negative of each other.)

The formula for the discrete *inverse* Fourier transform, which recovers the set of h_k's exactly from the H_n's is:

$$h_k = \frac{1}{N} \sum_{n=0}^{N-1} H_n \, e^{-2\pi ikn/N} \tag{12.1.9}$$

Notice that the only differences between (12.1.9) and (12.1.7) are (i) changing the sign in the exponential, and (ii) dividing the answer by N. This means that a routine for calculating discrete Fourier transforms can also, with slight modification, calculate the inverse transforms.

The discrete form of Parseval's theorem is

$$\sum_{k=0}^{N-1} |h_k|^2 = \frac{1}{N} \sum_{n=0}^{N-1} |H_n|^2 \tag{12.1.10}$$

There are also discrete analogs to the convolution and correlation theorems (equations 12.0.9 and 12.0.11), but we shall defer them to §12.4 and §12.5, respectively.

REFERENCES AND FURTHER READING:

Brigham, E. Oran. 1974, *The Fast Fourier Transform* (Englewood Cliffs, N.J.: Prentice-Hall).

Elliott, D.F., and Rao, K.R. 1982, *Fast Transforms: Algorithms, Analyses, Applications* (New York: Academic Press).

12.2 Fast Fourier Transform (FFT)

How much computation is involved in computing the discrete Fourier transform (12.1.7) of N points? For many years, until the mid-1960s, the standard answer was this: Define W as the complex number

$$W \equiv e^{2\pi i/N} \tag{12.2.1}$$

Then (12.1.7) can be written as

$$H_n = \sum_{k=0}^{N-1} W^{nk} h_k \tag{12.2.2}$$

In other words, the vector of h_k's is multiplied by a matrix whose $(n,k)^{th}$ element is the constant W to the power $n \times k$. The matrix multiplication produces a vector result whose components are the H_n's. This matrix multiplication evidently requires N^2 complex multiplications, plus a smaller number of operations to generate the required powers of W. So, the discrete Fourier transform appears to be an $O(N^2)$ process. These appearances are deceiving! The discrete Fourier transform can, in fact, be computed in $O(N \log_2 N)$ operations with an algorithm called the *Fast Fourier Transform*, or *FFT*. The difference between $N \log_2 N$ and N^2 is immense. With $N = 10^6$, for example, it is the difference between, roughly, 30 seconds of CPU time and 2 weeks of CPU time on a microsecond cycle time computer. The existence of an FFT algorithm became generally known only in the mid-1960s, from the work of J.W. Cooley and J.W. Tukey, who in turn had been prodded by R.L. Garwin of IBM Yorktown Heights Research Center. Retrospectively, we now know that a few clever individuals had independently discovered, and in some cases implemented, fast Fourier transforms as many as 20 years previously (see Brigham for references).

One of the earliest "discoveries" of the FFT, that of Danielson and Lanczos in 1942, still provides one of the clearest derivations of the algorithm. Danielson and Lanczos showed that a discrete Fourier transform of length N can be rewritten as the sum of two discrete Fourier transforms, each of length $N/2$. One of the two is formed from the even-numbered points of the original N, the other from the odd-numbered points. The proof is simply this:

$$\begin{aligned}
F_k &= \sum_{j=0}^{N-1} e^{2\pi ijk/N} f_j \\
&= \sum_{j=0}^{N/2-1} e^{2\pi ik(2j)/N} f_{2j} + \sum_{j=0}^{N/2-1} e^{2\pi ik(2j+1)/N} f_{2j+1} \\
&= \sum_{j=0}^{N/2-1} e^{2\pi ikj/(N/2)} f_{2j} + W^k \sum_{j=0}^{N/2-1} e^{2\pi ikj/(N/2)} f_{2j+1} \\
&= F_k^e + W^k F_k^o
\end{aligned} \tag{12.2.3}$$

In the last line, W is the same complex constant as in (12.2.1), F_k^e denotes the kth component of the Fourier transform of length $N/2$ formed from the even components of the original f_j's, while F_k^o is the corresponding transform of length $N/2$ formed from the odd components. Notice also that k in the last line of (12.2.3) varies from 0 to N, not just to $N/2$. Nevertheless, the transforms F_k^e and F_k^o are periodic in k with length $N/2$. So each is repeated through two cycles to obtain F_k.

The wonderful thing about the *Danielson-Lanczos Lemma* is that it can be used recursively. Having reduced the problem of computing F_k to that of computing F_k^e and F_k^o, we can do the same reduction of F_k^e to the problem of computing the transform of *its* $N/4$ even-numbered input data and $N/4$

odd-numbered data. In other words, we can define F_k^{ee} and F_k^{eo} to be the discrete Fourier transforms of the points which are respectively even-even and even-odd on the successive subdivisions of the data.

Although there are ways of treating other cases, by far the easiest case is the one in which the original N is an integer power of 2. In fact, we categorically recommend that you *only* use FFTs with N a power of two. If the length of your data set is not a power of two, pad it with zeros up to the next power of two. (We will give more sophisticated suggestions in subsequent sections below.) With this restriction on N, it is evident that we can continue applying the Danielson-Lanczos Lemma until we have subdivided the data all the way down to transforms of length 1. What is the Fourier transform of length one? It is just the identity operation that copies its one input number into its one output slot! In other words, for every pattern of e's and o's (numbering $\log_2 N$ in all), there is a one-point transform that is just one of the input numbers f_n

$$F_k^{eoeeoeo\cdots oee} = f_n \qquad \text{for some } n \tag{12.2.4}$$

(Of course this one-point transform actually does not depend on k, since it is periodic in k with period 1.)

The next trick is to figure out which value of n corresponds to which pattern of e's and o's in equation (12.2.4). The answer is: reverse the pattern of e's and o's, then let $e = 0$ and $o = 1$, and you will have, *in binary* the value of n. Do you see why it works? It is because the successive subdivisions of the data into even and odd are tests of successive low-order (least significant) bits of n. This idea of *bit reversal* can be exploited in a very clever way which, along with the Danielson-Lanczos Lemma, makes FFTs practical: Suppose we take the original vector of data f_j and rearrange it into bit-reversed order (see Figure 12.2.1), so that the individual numbers are in the order not of j, but of the number obtained by bit-reversing j. Then the bookkeeping on the recursive application of the Danielson-Lanczos Lemma becomes extraordinarily simple. The points as given are the one-point transforms. We combine adjacent pairs to get two-point transforms, then combine adjacent pairs of pairs to get 4-point transforms, and so on, until the first and second halves of the whole data set are combined into the final transform. Each combination takes of order N operations, and there are evidently $\log_2 N$ combinations, so the whole algorithm is of order $N \log_2 N$ (assuming, as is the case, that the process of sorting into bit-reversed order is no greater in order than $N \log_2 N$).

This, then, is the structure of an FFT algorithm: It has two sections. The first section sorts the data into bit-reversed order. Luckily this takes no additional storage, since it involves only swapping pairs of elements. (If k_1 is the bit reverse of k_2, then k_2 is the bit reverse of k_1.) The second section has an outer loop which is executed $\log_2 N$ times and calculates, in turn, transforms of length $2, 4, 8, \ldots, N$. For each stage of this process, two nested inner loops range over the subtransforms already computed and the elements of each transform, implementing the Danielson-Lanczos Lemma. The operation is made more efficient by restricting external calls for trigonometric sines and

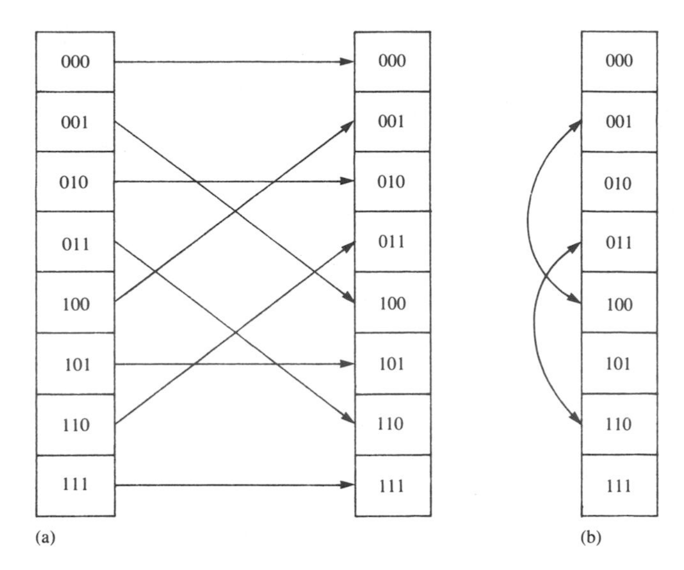

Figure 12.2.1. Reordering an array (here of length 8) by bit reversal, (a) between two arrays, versus (b) in place. Bit reversal reordering is a necessary part of the Fast Fourier Transform (FFT) algorithm.

cosines to the outer loop, where they are made only $\log_2 N$ times. Computation of the sines and cosines of multiple angles is through simple recurrence relations in the inner loops.

The FFT routine given below is based on one originally written by N. Brenner of Lincoln Laboratories. The input quantities are the number of complex data points (NN), the data array (DATA), and ISIGN, which should be set to either ± 1 and is the sign of i in the exponential of equation (12.1.7). When ISIGN is set to -1, the routine thus calculates the inverse transform (12.1.9) — except that it does not multiply by the normalizing factor $1/N$ that appears in that equation. You can do that yourself.

Notice that the argument NN is the number of complex data points, although we avoid the use of complex arithmetic due to the inefficient implementations found on many computers. The actual length of the real array DATA is 2 times NN, with each complex value occupying two consecutive locations. In other words, DATA(1) is the real part of f_0, DATA(2) is the imaginary part of f_0, and so on up to DATA(2*NN-1), which is the real part of f_{N-1}, and DATA(2*NN) which is the imaginary part of f_{N-1}. The FFT routine returns the F_n's packed in exactly the same fashion, as NN complex numbers. The real and imaginary parts of the zero frequency component F_0 are in DATA(1) and DATA(2); the smallest nonzero positive frequency has real and imaginary parts in DATA(3) and DATA(4); the smallest (in magnitude) nonzero negative frequency has real and imaginary parts in DATA(2*NN-1) and DATA(2*NN). Positive frequencies increasing in magnitude are stored in the real-imaginary pairs DATA(5), DATA(6) up to DATA(NN-1), DATA(NN). Negative frequencies of increasing magnitude are stored in DATA(2*NN-3), DATA(2*NN-2) down to DATA(NN+3), DATA(NN+4). Finally, the pair DATA(NN+1), DATA(NN+2) con-

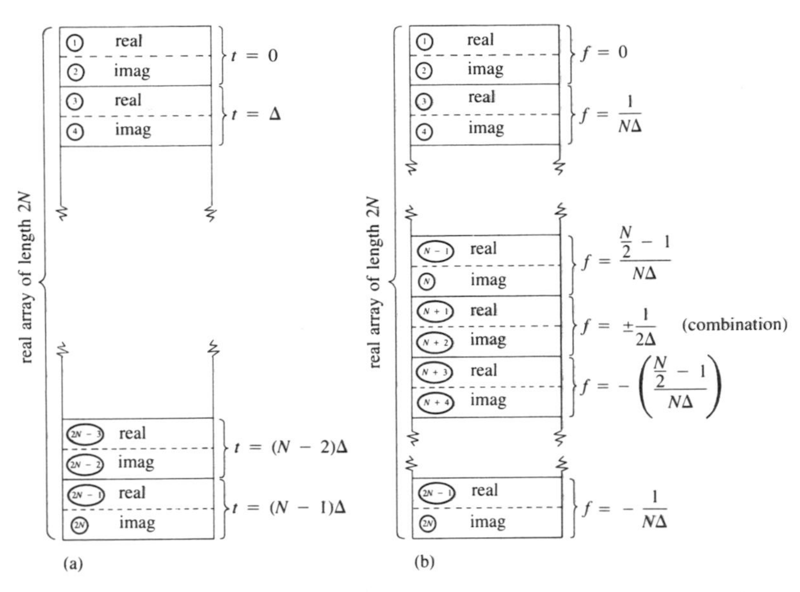

Figure 12.2.2. Input and output arrays for FFT. (a) The input array contains N (a power of 2) complex time samples in a real array of length $2N$, with real and imaginary parts alternating. (b) The output array contains the complex Fourier spectrum at N values of frequency. Real and imaginary parts again alternate. The array starts with zero frequency, works up to the most positive frequency (which is ambiguous with the most negative frequency). Negative frequencies follow, from the second-most negative up to the frequency just below zero.

tain the real and imaginary parts of the one aliased point which contains the most positive and the most negative frequency. You should try to develop a familiarity with this storage arrangement of complex spectra, also shown in the Figure 12.2.2, since it is the practical standard.

```
      SUBROUTINE FOUR1(DATA,NN,ISIGN)
```

Replaces **DATA** by its discrete Fourier transform, if **ISIGN** is input as 1; or replaces **DATA** by **NN** times its inverse discrete Fourier transform, if **ISIGN** is input as −1. **DATA** is a complex array of length **NN** or, equivalently, a real array of length **2*NN**. **NN** MUST be an integer power of 2 (this is not checked for!).

```
      REAL*8 WR,WI,WPR,WPI,WTEMP,THETA      Double precision for the trigonometric recurrences.
      DIMENSION DATA(2*NN)
      N=2*NN
      J=1
      DO 11 I=1,N,2                         This is the bit-reversal section of the routine.
        IF(J.GT.I)THEN
          TEMPR=DATA(J)                     Exchange the two complex numbers.
          TEMPI=DATA(J+1)
          DATA(J)=DATA(I)
          DATA(J+1)=DATA(I+1)
          DATA(I)=TEMPR
          DATA(I+1)=TEMPI
        ENDIF
        M=N/2
1       IF ((M.GE.2).AND.(J.GT.M)) THEN
          J=J-M
          M=M/2
        GO TO 1
```

```
        ENDIF
        J=J+M
      11 CONTINUE
      MMAX=2                             Here begins the Danielson-Lanczos section of the routine.
2     IF (N.GT.MMAX) THEN                Outer loop executed log2 NN times.
        ISTEP=2*MMAX
        THETA=6.28318530717959D0/(ISIGN*MMAX)   Initialize for the trigonometric recurrence.
        WPR=-2.D0*DSIN(0.5D0*THETA)**2
        WPI=DSIN(THETA)
        WR=1.D0
        WI=0.D0
        DO 13 M=1,MMAX,2                 Here are the two nested inner loops.
          DO 12 I=M,N,ISTEP
            J=I+MMAX                     This is the Danielson-Lanczos formula:
            TEMPR=SNGL(WR)*DATA(J)-SNGL(WI)*DATA(J+1)
            TEMPI=SNGL(WR)*DATA(J+1)+SNGL(WI)*DATA(J)
            DATA(J)=DATA(I)-TEMPR
            DATA(J+1)=DATA(I+1)-TEMPI
            DATA(I)=DATA(I)+TEMPR
            DATA(I+1)=DATA(I+1)+TEMPI
12        CONTINUE
          WTEMP=WR                       Trigonometric recurrence.
          WR=WR*WPR-WI*WPI+WR
          WI=WI*WPR+WTEMP*WPI+WI
13      CONTINUE
        MMAX=ISTEP
      GO TO 2                            Not yet done.
      ENDIF                              All done.
      RETURN
      END
```

Other FFT Algorithms

We should mention that there are a number of variants on the basic FFT algorithm given above. As we have seen, that algorithm first rearranges the input elements into bit-reverse order, then builds up the output transform in $\log_2 N$ iterations. In the literature, this sequence is called a *decimation-in-time* or *Cooley-Tukey* FFT algorithm. It is also possible to derive FFT algorithms which first go through a set of $\log_2 N$ iterations on the input data, and rearrange the *output* values into bit-reverse order. These are called *decimation-in-frequency* or *Sande-Tukey* FFT algorithms. For some applications, such as convolution (§12.4), one takes a data set into the Fourier domain and then, after some manipulation, back out again. In these cases it is possible to avoid all bit reversing. You use a decimation-in-frequency algorithm (without its bit reversing) to get into the "scrambled" Fourier domain, do your operations there, and then use an inverse algorithm (without *its* bit reversing) to get back to the time domain. While elegant in principle, this procedure does not in practice save much computation time, since the bit reversals represent only a small fraction of an FFT's operations count, and since most useful operations in the frequency domain require a knowledge of which points correspond to which frequencies.

Another class of FFTs subdivides the initial data set of length N not all the way down to the trivial transform of length 1, but rather only down to

some other small power of 2, for example $N = 4$, *base-4 FFTs*, or $N = 8$, *base-8 FFTs*. These small transforms are then done by small sections of highly optimized coding which take advantage of special symmetries of that particular small N. For example, for $N = 4$, the trigonometric sines and cosines that enter are all ± 1 or 0, so many multiplications are eliminated, leaving largely additions and subtractions. These can be faster than simpler FFTs by some significant, but not overwhelming, factor, e.g. 20 or 30 percent.

There are also FFT algorithms for data sets of length N not a power of two. They work by using relations analogous to the Danielson-Lanczos Lemma to subdivide the initial problem into successively smaller problems, not by factors of 2, but by whatever small prime factors happen to divide N. The larger that the largest prime factor of N is, the worse this method works. If N is prime, then no subdivision is possible, and the user (whether he knows it or not) is taking a *slow* Fourier transform, of order N^2 instead of order $N \log_2 N$. Our advice is to stay clear of such FFT implementations, with perhaps one class of exceptions, the *Winograd Fourier transform algorithms.* Winograd algorithms are in some ways analogous to the base-4 and base-8 FFTs. Winograd has derived highly optimized codings for taking small-N discrete Fourier transforms, e.g., for $N = 2, 3, 4, 5, 7, 8, 11, 13, 16$. The algorithms also use a new and clever way of combining the subfactors. The method involves a reordering of the data both before the hierarchical processing and after it, but it allows a significant reduction in the number of multiplications in the algorithm. For some especially favorable values of N, the Winograd algorithms can be significantly (e.g., up to a factor of 2) faster than the simpler FFT algorithms of the nearest integer power of 2. This advantage in speed, however, must be weighed against the considerably more complicated data indexing involved in these transforms, and the fact that the Winograd transform cannot be done "in place."

Finally, an interesting class of transforms for doing convolutions quickly are number theoretic transforms. These schemes replace floating point arithmetic with integer arithmetic modulo some large prime $N+1$, and the Nth root of 1 by the modulo arithmetic equivalent. Strictly speaking, these are not *Fourier* transforms at all, but the properties are quite similar and computational speed can be far superior. On the other hand, their use is somewhat restricted to quantities like correlations and convolutions since the transform itself is not easily interpretable as a "frequency" spectrum.

REFERENCES AND FURTHER READING:

Nussbaumer, H.J. 1982, *Fast Fourier Transform and Convolution Algorithms* (New York: Springer-Verlag).

Elliott, D.F., and Rao, K.R. 1982, *Fast Transforms: Algorithms, Analyses, Applications* (New York: Academic Press).

Brigham, E. Oran. 1974, *The Fast Fourier Transform* (Englewood Cliffs, N.J.: Prentice-Hall).

Bloomfield, P. 1976, *Fourier Analysis of Time Series – An Introduction* (New York: Wiley).

Beauchamp, K.G. 1975, *Walsh Functions and Their Applications* (New York: Academic Press) [a non-Fourier transform of recent interest].

12.3 FFT of Real Functions, Sine and Cosine Transforms

It happens frequently that the data whose FFT is desired consist of real-valued samples f_j, $j = 0 \ldots N-1$. To use FOUR1, we put these into a complex array with all imaginary parts set to zero. The resulting transform F_n, $n = 0 \ldots N-1$ satisfies $F_{N-n}{}^* = F_n$. Since this complex-valued array has real values for F_0 and $F_{N/2}$, and $(N/2)-1$ other independent values $F_1 \ldots F_{N/2-1}$, it has the same $2(N/2-1)+2 = N$ "degrees of freedom" as the original, real data set. However, the use of the full complex FFT algorithm for real data is inefficient, both in execution time and in storage required. You would think that there is a better way.

There are *two* better ways. The first is "mass production": Pack two separate real functions into the input array in such a way that their individual transforms can be separated from the result. This is implemented in the program TWOFFT below. This may remind you of a one-cent sale, at which you are coerced to purchase two of an item when you only need one. However, remember that for correlations and convolutions the Fourier transforms of two functions are involved, and this is a handy way to do them both at once. The second method is to pack the real input array cleverly, without extra zeros, into a complex array of half its length. One then performs a complex FFT on this shorter length; the trick is then to get the required answer out of the result. This is done in the program REALFT below.

Transform of Two Real Functions Simultaneously

First we show how to exploit the symmetry of the transform F_n to handle two real functions at once: Since the input data f_j are real, the components of the discrete Fourier transform satisfy

$$F_{N-n} = (F_n)^* \tag{12.3.1}$$

where the asterisk denotes complex conjugation. By the same token, the discrete Fourier transform of a purely imaginary set of g_j's has the opposite symmetry.

$$G_{N-n} = -(G_n)^* \tag{12.3.2}$$

Therefore we can take the discrete Fourier transform of two real functions each of length N simultaneously by packing the two data arrays as the real and imaginary parts respectively of the complex input array of FOUR1. Then the resulting transform array can be unpacked into two complex arrays with the aid of the two symmetries. Routine TWOFFT works out these ideas.

```
SUBROUTINE TWOFFT(DATA1,DATA2,FFT1,FFT2,N)
   Given two real input arrays DATA1 and DATA2, each of length N, this routine calls FOUR1
   and returns two complex output arrays, FFT1 and FFT2, each of complex length N (i.e.
   real length 2*N), which contain the discrete Fourier transforms of the respective DATAs.
   N MUST be an integer power of 2.
DIMENSION DATA1(N),DATA2(N)
COMPLEX FFT1(N),FFT2(N),H1,H2,C1,C2
C1=CMPLX(0.5,0.0)
C2=CMPLX(0.0,-0.5)
DO 11 J=1,N
    FFT1(J)=CMPLX(DATA1(J),DATA2(J))      Pack the two real arrays into one complex array.
11  CONTINUE
CALL FOUR1(FFT1,N,1)              Transform the complex array.
FFT2(1)=CMPLX(AIMAG(FFT1(1)),0.0)
FFT1(1)=CMPLX(REAL(FFT1(1)),0.0)
N2=N+2
DO 12 J=2,N/2+1
    H1=C1*(FFT1(J)+CONJG(FFT1(N2-J)))    Use symmetries to separate the two transforms.
    H2=C2*(FFT1(J)-CONJG(FFT1(N2-J)))
    FFT1(J)=H1                    Ship them out in two complex arrays.
    FFT1(N2-J)=CONJG(H1)
    FFT2(J)=H2
    FFT2(N2-J)=CONJG(H2)
12  CONTINUE
RETURN
END
```

What about the reverse process? Suppose you have two complex transform arrays, each of which has the symmetry (12.3.1), so that you know that the inverses of both transforms are real functions. Can you invert both in a single FFT? This is even easier than the other direction. Use the fact that the FFT is linear and form the sum of the first transform plus i times the second. Invert using FOUR1 with ISIGN=-1. The real and imaginary parts of the resulting complex array are the two desired real functions.

FFT of Single Real Function

To implement the second method, which allows us to perform the FFT of a *single* real function without redundancy, we split the data set in half, thereby forming two real arrays of half the size. We can apply the program above to these two, but of course the result will not be the transform of the original data. It will be a schizophrenic set of two transforms each of which has half of the information we need. Fortunately, this is schizophrenia of a treatable form. It works like this:

The right way to split the original data is to take the even-numbered f_j as one data set, and the odd-numbered f_j as the other. The beauty of this is that we can take the original real array and treat it as a complex array h_j of half the length. The first data set is the real part of this array, and the second is the imaginary part, as prescribed for TWOFFT. No repacking is required. In other words $h_j = f_{2j} + if_{2j+1}, \quad j = 0 \ldots N/2-1$. We submit this to FOUR1

and it will return a complex array $H_n = F_n^e + iF_n^o, \quad n = 0 \ldots N/2-1$ with

$$F_n^e = \sum_{k=0}^{N/2-1} f_{2k}\, e^{2\pi ikn/(N/2)}$$
$$F_n^o = \sum_{k=0}^{N/2-1} f_{2k+1}\, e^{2\pi ikn/(N/2)} \tag{12.3.3}$$

The discussion of program TWOFFT tells you how to separate the two transforms F_n^e and F_n^o out of H_n. How do you work them into the transform F_n of the original data set f_j? We recommend a quick glance back at equation (12.2.3):

$$F_n = F_n^e + e^{2\pi in/N} F_n^o \qquad n = 0 \ldots N-1 \tag{12.3.4}$$

Expressed directly in terms of the transform H_n of our real (masquerading as complex) data set, the result is

$$F_n = \frac{1}{2}(H_n + H_{N/2-n}{}^*) - \frac{i}{2}(H_n - H_{N/2-n}{}^*)e^{2\pi in/N} \qquad n = 0, \ldots, N-1 \tag{12.3.5}$$

A few remarks:

- Since $F_{N-n}{}^* = F_n$ there is no point in saving the entire spectrum. The positive frequency half is sufficient and can be stored in the same array as the original data. The operation can, in fact, be done in place.
- Even so, we need values H_n, $n = 0 \ldots N/2$ whereas FOUR1 returns only the values $n = 0 \ldots N/2-1$. Symmetry to the rescue, $H_{N/2} = H_0$.
- The values F_0 and $F_{N/2}$ are real and independent. In order to actually get the entire F_n in the original array space, it is convenient to return $F_{N/2}$ as the imaginary part of F_0.
- Despite its complicated form, the process above is invertible. First peel $F_{N/2}$ out of F_0. Then construct

$$F_n^e = \frac{1}{2}(F_n + F_{N/2-n}^*)$$
$$F_n^o = \frac{1}{2}e^{-2\pi in/N}(F_n - F_{N/2-n}^*) \qquad n = 0 \ldots N/2-1 \tag{12.3.6}$$

and use FOUR1 to find the inverse transform of $H_n = F_n^{(1)} + iF_n^{(2)}$. Surprisingly, the actual algebraic steps are virtually identical to those of the forward transform. Here is a representation of what we have said:

```
SUBROUTINE REALFT(DATA,N,ISIGN)
   Calculates the Fourier Transform of a set of 2N real-valued data points. Replaces this
   data (which is stored in array DATA) by the positive frequency half of its complex Fourier
   Transform. The real-valued first and last components of the complex transform are
   returned as elements DATA(1) and DATA(2) respectively. N must be a power of 2. This
   routine also calculates the inverse transform of a complex data array if it is the transform
   of real data. (Result in this case must be multiplied by 1/N.)
REAL*8 WR,WI,WPR,WPI,WTEMP,THETA        Double precision for the trigonometric recurrences.
DIMENSION DATA(2*N)
THETA=3.141592653589793D0/DBLE(N)       Initialize the recurrence.
C1=0.5
IF (ISIGN.EQ.1) THEN
    C2=-0.5
    CALL FOUR1(DATA,N,+1)          The forward transform is here.
ELSE
    C2=0.5                         Otherwise set up for an inverse transform.
    THETA=-THETA
ENDIF
WPR=-2.0D0*DSIN(0.5D0*THETA)**2
WPI=DSIN(THETA)
WR=1.0D0+WPR
WI=WPI
N2P3=2*N+3
DO 11 I=2,N/2                      Case I=1 done separately below.
    I1=2*I-1
    I2=I1+1
    I3=N2P3-I2
    I4=I3+1
    WRS=SNGL(WR)
    WIS=SNGL(WI)
    H1R=C1*(DATA(I1)+DATA(I3))     The two separate transforms are separated out of DATA.
    H1I=C1*(DATA(I2)-DATA(I4))
    H2R=-C2*(DATA(I2)+DATA(I4))
    H2I=C2*(DATA(I1)-DATA(I3))
    DATA(I1)=H1R+WRS*H2R-WIS*H2I   Here they are recombined to form the true transform of the
    DATA(I2)=H1I+WRS*H2I+WIS*H2R   original real data.
    DATA(I3)=H1R-WRS*H2R+WIS*H2I
    DATA(I4)=-H1I+WRS*H2I+WIS*H2R
    WTEMP=WR                       The recurrence.
    WR=WR*WPR-WI*WPI+WR
    WI=WI*WPR+WTEMP*WPI+WI
11 CONTINUE
IF (ISIGN.EQ.1) THEN
    H1R=DATA(1)
    DATA(1)=H1R+DATA(2)
    DATA(2)=H1R-DATA(2)            Squeeze the first and last data together to get
ELSE                               them all within the original array.
    H1R=DATA(1)
    DATA(1)=C1*(H1R+DATA(2))
    DATA(2)=C1*(H1R-DATA(2))
    CALL FOUR1(DATA,N,-1)          This is the inverse transform for the case ISIGN=-1.
ENDIF
RETURN
END
```

Fast Sine and Cosine Transforms

Among their other uses, the Fourier transforms of functions can be used to solve differential equations (see Chapter 17). The most common boundary conditions for the solutions are 1) they have the value zero at the boundaries, or 2) their derivatives are zero at the boundaries. In these instances, two more transforms arise naturally, the *sine* transform and the *cosine* transform, given by

$$F_k = \sum_{j=1}^{N-1} f_j \sin(\pi jk/N) \qquad \text{sine transform}$$
$$F_k = \sum_{j=0}^{N-1} f_j \cos(\pi jk/N) \qquad \text{cosine transform} \tag{12.3.7}$$

where f_j, $j = 0 \ldots N-1$ is the data array.

At first blush these appear to be simply the imaginary and real parts respectively of the discrete Fourier transform. However, the argument of the sine and cosine differ by a factor of two from the value that would make that so. The sine transform uses *sines only* as a complete set of functions in the interval from 0 to 2π, and the cosine transform uses *cosines only*. By contrast, the normal FFT uses both sines and cosines. (See Figure 12.3.1.)

However, the sine and cosine transforms can be "force-fit" into a form which allows their calculation via the FFT. The idea is to extend the given function rightward past its last tabulated value. In the case of the sine transform, we extend the data to twice their length in such a way as to make them an *odd* function about $j = N$, with $f_N = 0$,

$$f_{2N-j} \equiv -f_j \qquad j = 0, \ldots, N-1 \tag{12.3.8}$$

When a FFT is performed on this extended function, it reduces to the sine transform by symmetry:

$$F_k = \sum_{j=0}^{2N-1} f_j e^{2\pi ijk/(2N)} \tag{12.3.9}$$

The half of this sum from $j = N$ to $j = 2N-1$ can be rewritten with the substitution $j' = 2N - j$

$$\begin{aligned} \sum_{j=N}^{2N-1} f_j e^{2\pi ijk/(2N)} &= \sum_{j'=1}^{N} f_{2N-j'} e^{2\pi i(2N-j')k/(2N)} \\ &= -\sum_{j'=0}^{N-1} f'_j e^{-2\pi ij'k/(2N)} \end{aligned} \tag{12.3.10}$$

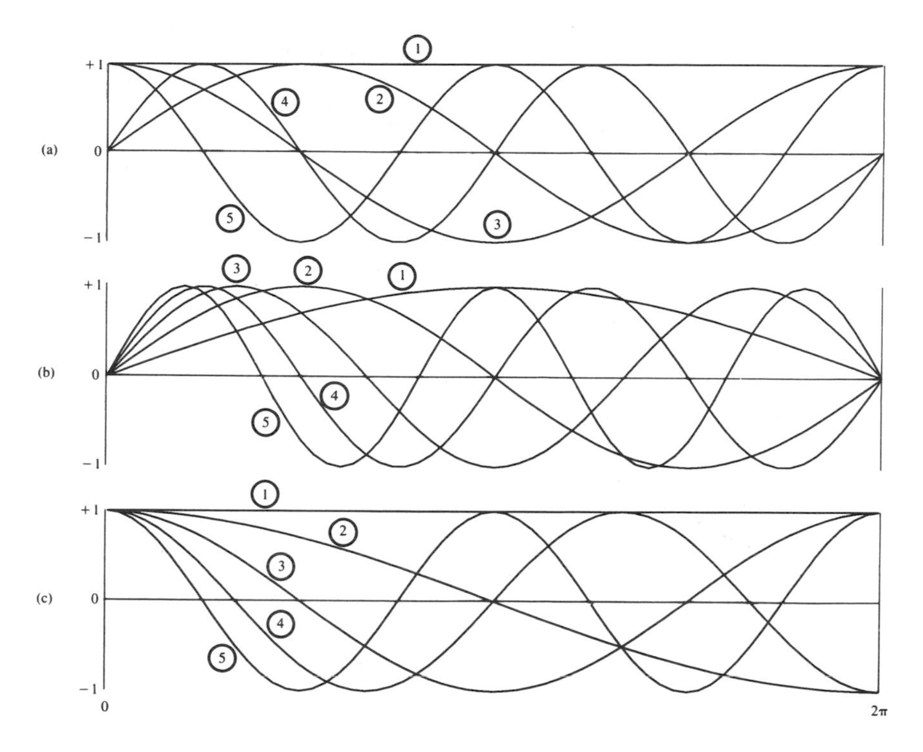

Figure 12.3.1. Basis functions used by the Fourier transform (a), sine transform (b), and cosine transform (c), are plotted. The first five basis functions are shown in each case. (For the Fourier transform, the real and imaginary parts of the basis functions are both shown.) While some basis functions occur in more than one transform, the basis sets are distinct. For example, the sine transform functions labeled (1), (3), (5) are not present in the Fourier basis. Any of the three sets can expand any function in the interval shown; however the sine or cosine transform best expands functions matching the boundary conditions of the respective basis functions, namely zero function values for sine, zero derivatives for cosine.

so that

$$\begin{aligned} f_k &= \sum_{j=0}^{N-1} f_j \left[e^{2\pi ijk/(2N)} - e^{-2\pi ijk/(2N)} \right] \\ &= 2i \sum_{j=0}^{N-1} f_j \sin(\pi jk/N) \end{aligned} \tag{12.3.11}$$

Thus, up to a factor $2i$ we get the sine transform from the FFT of the extended function. The same procedure applies to the cosine transform with the exception that the data are extended as an *even* function.

In both cases, however, this method introduces a factor of two inefficiency into the computation by extending the data. This inefficiency shows up in the FFT output, which has zeros for the real part of every element of the transform (for the sine transform). For a one-dimensional problem, the factor of two may be bearable, especially in view of the simplicity of the

method. When we work with partial differential equations in two or three dimensions, though, the factor becomes four or eight, so we are inspired to eliminate the inefficiency.

From the original real data array f_j we will construct an auxiliary array y_j and apply to it the routine **REALFT**. The output will then be used to construct the desired transform. For the sine transform of data f_j, $j = 1, \ldots, N$ the auxiliary array is

$$\begin{aligned} y_0 &= 0 \\ y_j &= \sin(j\pi/N)(f_j + f_{N-j}) + \frac{1}{2}(f_j - f_{N-j}) \\ j &= 1, \ldots, N-1 \end{aligned} \tag{12.3.12}$$

This array is of the same dimension as the original. Notice that the first term is symmetric about $j = N/2$ and the second is antisymmetric. Consequently, when **REALFT** is applied to y_j, the result has real parts R_k and imaginary parts I_k given by

$$\begin{aligned} R_k &= \sum_{j=0}^{N-1} y_j \cos(2\pi jk/N) \\ &= \sum_{j=1}^{N-1} (f_j + f_{N-j}) \sin(j\pi/N) \cos(2\pi jk/N) \\ &= \sum_{j=0}^{N-1} 2f_j \sin(j\pi/N) \cos(2\pi jk/N) \\ &= \sum_{j=0}^{N-1} f_j \left\{ \sin(\frac{2k+1}{N} j\pi) - \sin(\frac{2k-1}{N} j\pi) \right\} \\ &= F_{2k+1} - F_{2k-1} \end{aligned} \tag{12.3.13}$$

$$\begin{aligned} I_k &= \sum_{j=0}^{N-1} y_j \sin(2\pi jk/N) \\ &= \sum_{j=1}^{N-1} (f_j - f_{N-j}) \frac{1}{2} \sin(2\pi jk/N) \\ &= \sum_{j=0}^{N-1} f_j \sin(2\pi jk/N) \\ &= F_{2k} \end{aligned} \tag{12.3.14}$$

Therefore F_k can be determined as follows,

$$F_{2k} = I_k \qquad F_{2k+1} = F_{2k-1} + R_k \qquad k = 0, \ldots, (N/2 - 1) \tag{12.3.15}$$

The even terms of F_k are thus determined very directly. The odd terms require a recursion, the starting point of which is

$$F_1 = \sum_{j=0}^{N-1} f_j \sin(j\pi/N) \tag{12.3.16}$$

The implementing program is

```
SUBROUTINE SINFT(Y,N)
    Calculates the sine transform of a set of N real-valued data points stored in array Y. The
    number N must be a power of 2. On exit Y is replaced by its transform. This program,
    without changes, also calculates the inverse sine transform, but in this case the output
    array should be multiplied by 2/N.
REAL*8 WR,WI,WPR,WPI,WTEMP,THETA       Double precision in the trigonometric recurrences.
DIMENSION Y(N)
THETA=3.14159265358979D0/DBLE(N)        Initialize the recurrence.
WR=1.0D0
WI=0.0D0
WPR=-2.0D0*DSIN(0.5D0*THETA)**2
WPI=DSIN(THETA)
Y(1)=0.0
M=N/2
DO 11 J=1,M
    WTEMP=WR
    WR=WR*WPR-WI*WPI+WR            Calculate the sine for the auxiliary array.
    WI=WI*WPR+WTEMP*WPI+WI         The cosine is needed to continue the recurrence.
    Y1=WI*(Y(J+1)+Y(N-J+1))        Construct the auxiliary array.
    Y2=0.5*(Y(J+1)-Y(N-J+1))
    Y(J+1)=Y1+Y2                   Terms j and N - j are related
    Y(N-J+1)=Y1-Y2
11  CONTINUE
CALL REALFT(Y,M,+1)                Transform the auxiliary array.
SUM=0.0
Y(1)=0.5*Y(1)                      Initialize the sum used for odd terms below.
Y(2)=0.0
DO 12 J=1,N-1,2
    SUM=SUM+Y(J)
    Y(J)=Y(J+1)                    Even terms in the transform are determined directly.
    Y(J+1)=SUM                     Odd terms are determined by this running sum.
12  CONTINUE
RETURN
END
```

The sine transform, curiously, is its own inverse. If you apply it twice, you get the original data, but multiplied by a factor of $N/2$.

The cosine transform is slightly more difficult, but the idea is the same. Now, the auxiliary function is

$$y_0 = f_0 \qquad y_j = \frac{1}{2}(f_j + f_{N-j}) - \sin(j\pi/N)(f_j - f_{N-j}) \tag{12.3.17}$$

and the same analysis leads to

$$F_{2k} = R_k \qquad F_{2k+1} = F_{2k-1} + I_k \qquad k = 0, \ldots, (N/2 - 1) \tag{12.3.18}$$

The starting value for the recursion in this case is

$$F_1 = \sum_{j=0}^{N-1} f_j \cos(j\pi/N) \tag{12.3.19}$$

This sum does not appear naturally among the R_k and I_k, and so we accumulate it during the generation of the array y_i.

An additional complication is that the cosine transform is not its own inverse. To derive the inverse of an array F_k, we first compute an array $\widetilde{f}_l$ as if the transform *were* its own inverse,

$$\widetilde{f}_l = \sum_{k=0}^{N-1} F_k \cos(\pi k l/N) \tag{12.3.20}$$

One easily verifies the relations between $\widetilde{f}_l$ and the desired inverse f_l,

$$\begin{aligned} \widetilde{f}_0 &= N f_0 + \sum_{j \text{ odd}} f_j \\ \widetilde{f}_l &= \frac{N}{2} f_l + \sum_{j \text{ even}} f_j \qquad \text{for } l \text{ odd} \\ \widetilde{f}_l &= \frac{N}{2} f_l + \sum_{j \text{ odd}} f_j \qquad \text{for } l \text{ even}, l \neq 0 \end{aligned} \tag{12.3.21}$$

The unknown sums on the right of these equations are determined as follows

$$C_1 \equiv \sum_{l \text{ even}} \widetilde{f}_l = \frac{N}{2}\left(f_0 + \sum_{j=0}^{N-1} f_j\right)$$
$$C_2 \equiv \sum_{l \text{ odd}} \widetilde{f}_l = \frac{N}{2}\sum_{j=0}^{N-1} f_j \tag{12.3.22}$$

It follows that

$$\frac{N}{2} f_0 = C_1 - C_2 \tag{12.3.23}$$

so,

$$\sum_{j \text{ odd}} f_j = \widetilde{f}_0 - 2(C_1 - C_2) \qquad \sum_{j \text{ even}} f_j = \frac{2}{N}C_2 - \sum_{j \text{ odd}} f_j \tag{12.3.24}$$

Knowing these sums, the desired f_j are now easily recovered from equation (12.3.21). This is implemented in the following routine.

```
SUBROUTINE COSFT(Y,N,ISIGN)
    Calculates the cosine transform of a set Y of N real-valued data points. The transformed
    data replace the original data in array Y. N must be a power of 2. Set ISIGN to +1 for
    a transform, and to -1 for an inverse transform. For an inverse transform, the output
    array should be multiplied by 2/N.
REAL*8 WR,WI,WPR,WPI,WTEMP,THETA        Double precision for the trigonometric recurrences.
DIMENSION Y(N)
THETA=3.14159265358979D0/DBLE(N)          Initialize the recurrence.
WR=1.0D0
WI=0.0D0
WPR=-2.0D0*DSIN(0.5D0*THETA)**2
WPI=DSIN(THETA)
SUM=Y(1)
M=N/2
DO 11 J=1,M-1                       J=M unnecessary since Y(N/2+1) unchanged
    WTEMP=WR
    WR=WR*WPR-WI*WPI+WR             Carry out the recurrence.
    WI=WI*WPR+WTEMP*WPI+WI
    Y1=0.5*(Y(J+1)+Y(N-J+1))            Calculates the auxiliary function.
    Y2=(Y(J+1)-Y(N-J+1))
    Y(J+1)=Y1-WI*Y2                 The values for j and N - j are related.
    Y(N-J+1)=Y1+WI*Y2
    SUM=SUM+WR*Y2                   Carry along this sum for later use in unfolding the transform.
11  CONTINUE
CALL REALFT(Y,M,+1)                 Calculate the transform of the auxiliary function.
Y(2)=SUM                            SUM is the value in equation (12.3.19).
DO 12 J=4,N,2
    SUM=SUM+Y(J)                    Equation (12.3.18).
    Y(J)=SUM
12  CONTINUE
IF (ISIGN.EQ.-1) THEN               This code applies only to the inverse transform.
    EVEN=Y(1)
```

```
          ODD=Y(2)
          DO 13 I=3,N-1,2
              EVEN=EVEN+Y(I)              Sum up the even and odd transform values as in equation (12.3.22).
              ODD=ODD+Y(I+1)
13        CONTINUE
          ENFO=2.0*(EVEN-ODD)
          SUMO=Y(1)-ENFO                  Next, implement equation (12.3.24).
          SUME=(2.0*ODD/FLOAT(N))-SUMO
          Y(1)=0.5*ENFO
          Y(2)=Y(2)-SUME
          DO 14 I=3,N-1,2
              Y(I)=Y(I)-SUMO              Finally, equation (12.3.21) gives us the true inverse cosine trans-
              Y(I+1)=Y(I+1)-SUME            form (excepting the factor 2/N).
14        CONTINUE
      ENDIF
      RETURN
      END
```

REFERENCES AND FURTHER READING:

Brigham, E. Oran. 1974, *The Fast Fourier Transform* (Englewood Cliffs, N.J.: Prentice-Hall), §10-10.

Hockney, R.W. 1971, in *Methods in Computational Physics*, vol. 9 (New York: Academic Press).

Temperton, C. 1980, *Journal of Computational Physics*, vol. 34, pp. 314-329.

12.4 Convolution and Deconvolution Using the FFT

We have defined the *convolution* of two functions for the continuous case in equation (12.0.8), and have given the *convolution theorem* as equation (12.0.9). The theorem says that the Fourier transform of the convolution of two functions is equal to the product of their individual Fourier transforms. Now, we want to deal with the discrete case. We will mention first the context in which convolution is a useful procedure, and then discuss how to compute it efficiently using the FFT.

The convolution of two functions $r(t)$ and $s(t)$, denoted $r * s$, is mathematically equal to their convolution in the opposite order, $s * r$. Nevertheless, in most applications the two functions have quite different meanings and characters. One of the functions, say s, is typically a signal or data stream, which goes on indefinitely in time (or in whatever the appropriate independent variable may be). The other function r is a "response function," typically a peaked function that falls to zero in both directions from its maximum. The effect of convolution is to smear the signal $s(t)$ in time according to the recipe provided by the response function $r(t)$, as shown in Figure 12.4.1. In particular, a spike or delta-function of unit area in s which occurs at some time t_0 is supposed

to be smeared into the shape of the response function itself, but translated from time 0 to time t_0 as $r(t - t_0)$.

In the discrete case, the signal $s(t)$ is represented by its sampled values at equal time intervals s_j. The response function is also a discrete set of numbers r_k, with the following interpretation: r_0 tells what multiple of the input signal in one channel (one particular value of j) is copied into the identical output channel (same value of j); r_1 tells what multiple of input signal in channel j is additionally copied into output channel $j + 1$; r_{-1} tells the multiple that is copied into channel $j - 1$; and so on for both positive and negative values of k in r_k. Figure 12.4.2 illustrates the situation.

Example: a response function with $r_0 = 1$ and all other r_k's equal to zero is just the identity filter: convolution of a signal with this response function gives identically the signal. Another example is the response function with $r_{14} = 1.5$ and all other r_k's equal to zero. This produces convolved output which is the input signal multiplied by 1.5 and delayed by 14 sample intervals.

Evidently, we have just described in words the following definition of discrete convolution with a response function of finite duration M:

$$(r * s)_j \equiv \sum_{k=-M/2+1}^{M/2} s_{j-k}\, r_k \tag{12.4.1}$$

If a discrete response function is nonzero only in some range $-M/2 < k \leq M/2$, where M is a sufficiently large even integer, then the response function is called a *finite impulse response (FIR)*, and its *duration* is M. (Notice that we are defining M as the number of nonzero *values* of r_k; these values span a time interval of $M - 1$ sampling times.) In most practical circumstances the case of finite M is the case of interest, either because the response really has a finite duration, or because we choose to truncate it at some point and approximate it by a finite-duration response function.

The *discrete convolution theorem* is this: If a signal s_j is *periodic* with period N, so that it is completely determined by the N values $s_0, \ldots, s_{N-1}$, then its discrete convolution with a response function *of finite duration* N is a member of the discrete Fourier transform pair,

$$\sum_{k=-N/2+1}^{N/2} s_{j-k}\, r_k \iff S_n R_n \tag{12.4.2}$$

Here S_n, $(n = 0, \ldots, N - 1)$ is the discrete Fourier transform of the values s_j, $(j = 0, \ldots, N - 1)$, while R_n, $(n = 0, \ldots, N - 1)$ is the discrete Fourier transform of the values r_k, $(k = 0, \ldots, N - 1)$. These values of r_k are the same ones as for the range $k = -N/2+1, \ldots, N/2$, but in wrap-around order, exactly as was described at the end of §12.2.

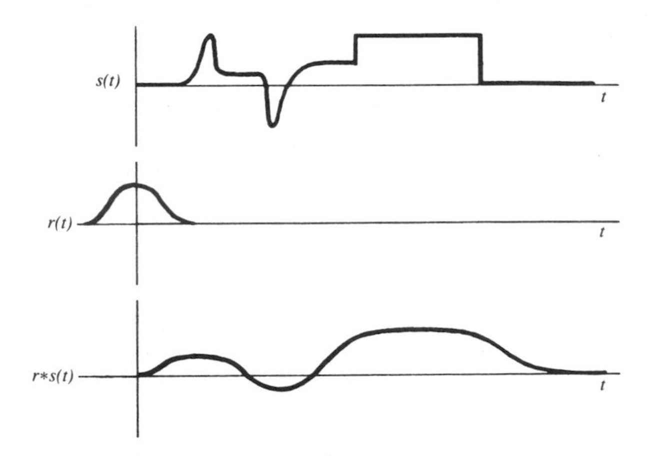

Figure 12.4.1. Example of the convolution of two functions. A signal $s(t)$ is convolved with a response function $r(t)$. Since the response function is broader than some features in the original signal, these are "washed out" in the convolution. In the absence of any additional noise, the process can be reversed by deconvolution.

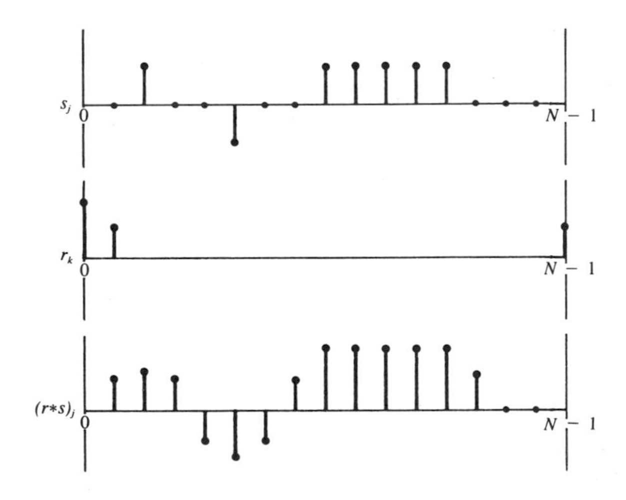

Figure 12.4.2. Convolution of discretely sampled functions. Note how the response function for negative times is wrapped around and stored at the extreme right end of the array r_k.

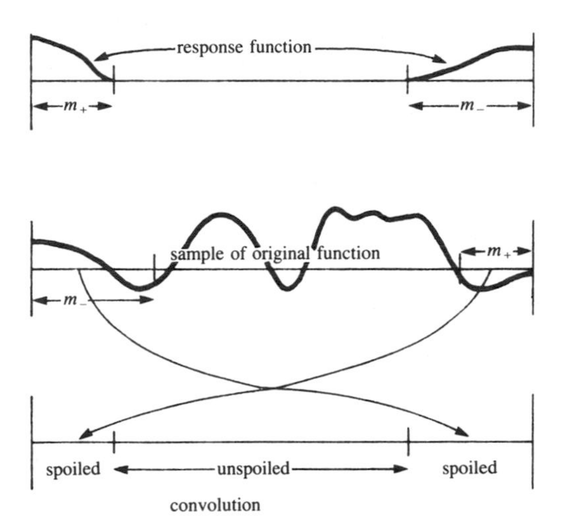

Figure 12.4.3. The wraparound problem in convolving finite segments of a function. Not only must the response function wrap be viewed as cyclic, but so must the sampled original function. Therefore a portion at each end of the original function is erroneously wrapped around by convolution with the response function.

Treatment of End Effects by Zero Padding

The discrete convolution theorem presumes a set of two circumstances which are not universal. First, it assumes that the input signal is periodic, whereas real data often either go forever without repetition or else consist of one non-periodic stretch of finite length. Second, the convolution theorem takes the duration of the response to be the same as the period of the data; they are both N. We need to work around these two constraints.

The second is very straightforward. Almost always, one is interested in a response function whose duration M is much shorter than the length of the data set N. In this case, one simply extends the response function to length N by padding it with zeros, i.e. defining $r_k = 0$ for $M/2 \leq k \leq N/2$ and also for $-N/2+1 \leq k \leq -M/2+1$. Dealing with the first constraint is more challenging. Since the convolution theorem rashly assumes that the data are periodic, it will falsely "pollute" the first output channel $(r*s)_0$ with some wrapped-around data from the far end of the data stream s_{N-1}, s_{N-2}, etc. (See Figure 12.4.3.) So, we need to set up a buffer zone of zero-padded values at the end of the s_j vector, in order to make this pollution zero. How many zero values do we need in this buffer? Exactly as many as the most negative index for which the response function is nonzero. For example, if r_{-3} is nonzero, while $r_{-4}, r_{-5}, \ldots$ are all zero, then we need three zero pads at the end of the data: $s_{N-3} = s_{N-2} = s_{N-1} = 0$. These zeros will protect the first output channel $(r*s)_0$ from wrap-around pollution. It should be obvious that the second output channel $(r*s)_1$ and subsequent ones will also be protected by these same zeros. Let K denote the number of padding zeros, so that the last actual input data point is s_{N-K-1}.

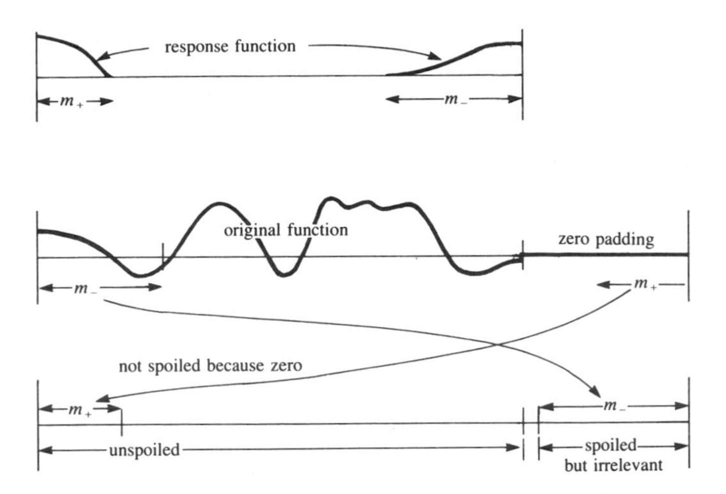

Figure 12.4.4. Zero padding as solution to the wraparound problem. The original function is extended by zeros, serving a dual purpose: when the zeros wrap around, they do not disturb the true convolution; and while the original function wraps around onto the zero region, that region can be discarded.

What now about pollution of the very *last* output channel? Since the data now end with s_{N-K-1}, the last output channel of interest is $(r * s)_{N-K-1}$. This channel can be polluted by wrap around from input channel s_0 unless the number K is also large enough to take care of the most positive index k for which the response function r_k is nonzero. For example, if r_0 through r_6 are nonzero, while $r_7, r_8 \dots$ are all zero, then we need at least $K = 6$ padding zeros at the end of the data: $s_{N-6} = \dots = s_{N-1} = 0$.

To summarize – we need to pad the data with a number of zeros *on one end* equal to the maximum positive duration *or* maximum negative duration of the response function, *whichever is larger.* (For a symmetric response function of duration M, you will need only $M/2$ zero pads.) Combining this operation with the padding of the response r_k described above, we effectively insulate the data from artifacts of undesired periodicity. Figure 12.4.4 illustrates matters.

Use of FFT for Convolution

The data, complete with zero padding, are now a set of real numbers s_j, $j = 0, \dots, N-1$, and the response function is zero padded out to duration N and arranged in wrap-around order. (Generally this means that a large contiguous section of the r_k's, in the middle of that array, is zero, with nonzero values clustered at the two extreme ends of the array.) You now compute the discrete convolution as follows: Use the FFT algorithm to compute the discrete Fourier transform of s and of r. Multiply the two transforms together component by component, remembering that the transforms consist of complex numbers. Then use the FFT algorithm to take the inverse discrete Fourier transform of the products. The answer is the convolution $r * s$.

What about *deconvolution*? Deconvolution is the process of *undoing* the smearing in a data set which has occurred under the influence of a known response function, for example, due to the known effect of a less-than-perfect measuring apparatus. The defining equation of deconvolution is the same as that for convolution, namely (12.4.1), except now the left-hand side is taken to be known, and (12.4.1) is to be considered as a set of N linear equations for the unknown quantities s_j. Solving these simultaneous linear equations in the time domain of (12.4.1) is unrealistic in most cases, but the FFT renders the problem almost trivial. Instead of multiplying the transform of the signal and response to get the transform of the convolution, we just divide the transform of the (known) convolution by the transform of the response to get the transform of the deconvolved signal.

This procedure can go wrong *mathematically* if the transform of the response function is exactly zero for some value R_n, so that we can't divide by it. This indicates that the original convolution has truly lost all information at that one frequency, so that a reconstruction of that frequency component is not possible. You should be aware, however, that apart from mathematical problems, the process of deconvolution has other practical shortcomings. The process is generally quite sensitive to noise in the input data, and to the accuracy to which the response function r_k is known. Perfectly reasonable attempts at deconvolution can sometimes produce nonsense for these reasons. In such cases you may want to make use of the additional process of *optimal filtering*, which is discussed in §12.6.

Here is our routine for convolution and deconvolution, using the FFT as implemented in **FOUR1** of §12.2. Since the data and response functions are real, not complex, both of their transforms can be taken simultaneously by the technique described in §12.3. The routine thus makes just one call to compute an FFT and one call to compute an inverse FFT. The data are assumed to be stored in a real array **DATA** of length N, which must be an integer power of two. The response function is assumed to be stored in wrap around order in a real array **RESPNS** of length M. The value of M can be any *odd* integer less than or equal to N, since the first thing the program does is to recopy the response function into the appropriate wrap around order in an array of length N. The answer is returned in **ANS**, which is also used as working space.

```
SUBROUTINE CONVLV(DATA,N,RESPNS,M,ISIGN,ANS)
   Convolves or deconvolves a real data set DATA of length N (including any user-supplied
   zero padding) with a response function RESPNS, stored in wrap around order in a real
   array of length M ≤ N. (M should be an odd integer.) Wrap around order means that the
   first half of the array RESPNS contains the impulse response function at positive times,
   while the second half of the array contains the impulse response function at negative
   times, counting down from the highest element RESPNS(M). On input ISIGN is +1 for
   convolution, −1 for deconvolution. The answer is returned in the first N components of
   ANS. However, ANS must be supplied in the calling program with length at least 2*N, for
   consistency with TWOFFT. N MUST be an integer power of two.
PARAMETER(NMAX=8192)          Maximum anticipated size of FFT.
DIMENSION DATA(N),RESPNS(N)
COMPLEX FFT(NMAX),ANS(N)
DO 11 I=1,(M-1)/2             Put RESPNS in array of length N.
   RESPNS(N+1-I)=RESPNS(M+1-I)
11 CONTINUE
DO 12 I=(M+3)/2,N-(M-1)/2     Pad with zeros.
```

```
    RESPNS(I)=0.0
  12 CONTINUE
CALL TWOFFT(DATA,RESPNS,FFT,ANS,N)        FFT both at once.
NO2=N/2
DO 13 I=1,NO2+1
    IF (ISIGN.EQ.1) THEN
        ANS(I)=FFT(I)*ANS(I)/NO2          Multiply FFTs to convolve.
    ELSE IF (ISIGN.EQ.-1) THEN
        IF (CABS(ANS(I)).EQ.0.0) PAUSE 'deconvolving at a response zero'
        ANS(I)=FFT(I)/ANS(I)/NO2          Divide FFTs to deconvolve.
    ELSE
        PAUSE 'no meaning for ISIGN'
    ENDIF
  13 CONTINUE
ANS(1)=CMPLX(REAL(ANS(1)),REAL(ANS(NO2+1))) Pack last element with first for REALFT.
CALL REALFT(ANS,NO2,-1)       Inverse transform back to time domain.
RETURN
END
```

Convolving or Deconvolving Very Large Data Sets

If your data set is so long that you do not want to fit it into memory all at once, then you must break it up into sections and convolve each section separately. Now, however, the treatment of end effects is a bit different. You have to worry not only about spurious wrap-around effects, but also about the fact that the ends of each section of data *should* have been influenced by data at the nearby ends of the immediately preceding and following sections of data, but were not so influenced since only one section of data is in the machine at a time.

There are two, related, standard solutions to this problem. Both are fairly obvious, so with a few words of description here, you ought to be able to implement them for yourself. The first solution is called the *overlap-save method.* In this technique you pad only the very beginning of the data with enough zeros to avoid wrap around pollution. After this initial padding, you forget about zero padding altogether. Bring in a section of data and convolve or deconvolve it. Then throw out the points at each end that are polluted by wrap around end effects. Output only the remaining good points in the middle. Now bring in the next section of data, but not all new data. The first points the next section are to overlap points from the preceding section of data. The sections are to be overlapped sufficiently so that the polluted output points at the end of one section are recomputed as the first of the unpolluted output points from the subsequent section. With a bit of thought you can easily determine how many points to overlap and save.

The second solution, called the *overlap-add method*, is illustrated in Figure 12.4.5. Here you *don't* overlap the input data. Each section of data is disjoint from the others and is used exactly once. However, you carefully zero-pad it at both ends so that there is no wrap-around ambiguity in the output convolution or deconvolution. Now you overlap *and add* these sections of output. Thus, an output point near the end of one section will have the response due to the input points at the beginning of the next section of data

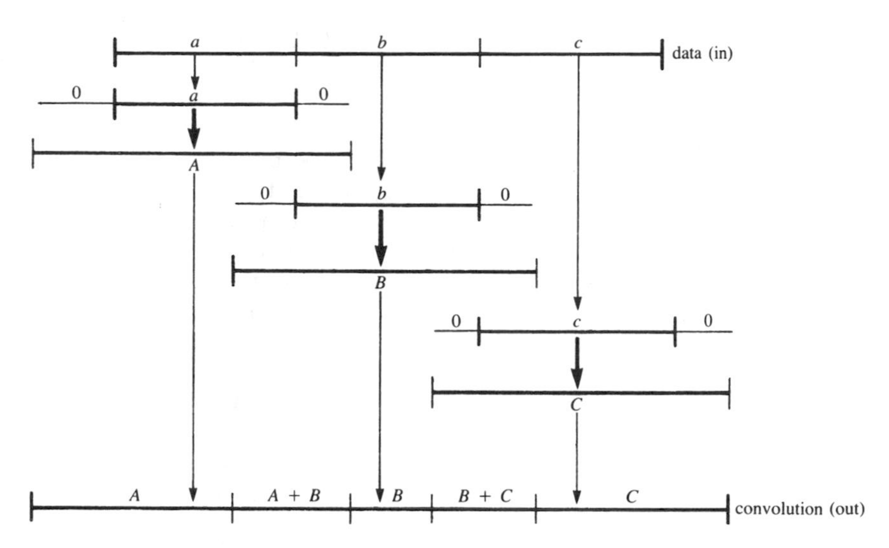

Figure 12.4.5. The overlap-add method for convolving a response with a very long signal. The signal data is broken up into smaller pieces. Each is zero padded at both ends and convolved (denoted by bold arrows in the figure). Finally the pieces are added back together, including the overlapping regions formed by the zero pads.

properly added in to it, and likewise for an output point near the beginning of a section, *mutatis mutandis*.

Even when computer memory is available, there is some slight gain in computing speed in segmenting a long data set, since the FFTs' $N \log_2 N$ is slightly slower than linear in N. However, the log term is so slowly varying that you will often be much happier to avoid the bookkeeping complexities of the overlap-add or overlap-save methods: if it is practical to do so, just cram the whole data set into memory and FFT away. Then you will have more time for the finer things in life, some of which are described in succeeding sections of this chapter.

REFERENCES AND FURTHER READING:

Nussbaumer, H.J. 1982, *Fast Fourier Transform and Convolution Algorithms* (New York: Springer-Verlag).

Elliott, D.F., and Rao, K.R. 1982, *Fast Transforms: Algorithms, Analyses, Applications* (New York: Academic Press).

Brigham, E. Oran. 1974, *The Fast Fourier Transform* (Englewood Cliffs, N.J.: Prentice-Hall), Chapter 13.

12.5 Correlation and Autocorrelation Using the FFT

Correlation is the close mathematical cousin of convolution. It is in some ways simpler, however, because the two functions that go into a correlation are not as conceptually distinct as were the data and response functions which entered into convolution. Rather, in correlation, the functions are represented by different, but generally similar, data sets. We investigate their "correlation," by comparing them both directly superposed, and with one of them shifted left or right.

We have already defined in equation (12.0.10) the correlation between two continuous functions $g(t)$ and $h(t)$, which is denoted $\text{Corr}(g, h)$, and is a function of *lag* t. We will occasionally show this time dependence explicitly, with the rather awkward notation $\text{Corr}(g, h)(t)$. The correlation will be large at some value of t if the first function (g) is a close copy of the second (h) but lags it in time by t, i.e., if the first function is shifted to the right of the second. Likewise, the correlation will be large for some negative value of t if the first function *leads* the second, i.e., is shifted to the left of the second. The relation that holds when the two functions are interchanged is

$$\text{Corr}(g, h)(t) = \text{Corr}(h, g)(-t) \tag{12.5.1}$$

The discrete correlation of two sampled functions g_k and h_k, each periodic with period N, is defined by

$$\text{Corr}(g, h)_j \equiv \sum_{k=0}^{N-1} g_{j+k} h_k \tag{12.5.2}$$

The *discrete correlation theorem* says that this discrete correlation of two real functions g and h is one member of the discrete Fourier transform pair

$$\text{Corr}(g, h)_j \Longleftrightarrow G_k H_k{}^* \tag{12.5.3}$$

where G_k and H_k are the discrete Fourier transforms of g_j and h_j, and asterisk denotes complex conjugation. This theorem makes the same presumptions about the functions as those encountered for the discrete convolution theorem.

We can compute correlations using the FFT as follows: FFT the two data sets, multiply one resulting transform by the complex conjugate of the other, and inverse transform the product. The result (call it r_k) will formally be a complex vector of length N. However, it will turn out to have all its imaginary parts zero since the original data sets were both real. The components of r_k are the values of the correlation at different lags, with positive and negative lags stored in the by now familiar wrap-around order: The correlation at zero

lag is in r_0, the first component; the correlation at lag 1 is in r_1, the second component; the correlation at lag -1 is in r_{N-1}, the last component; etc.

Just as in the case of convolution we have to consider end effects, since our data will not, in general, be periodic as intended by the correlation theorem. Here again, we can use zero padding. If you are interested in the correlation for lags as large as $\pm K$, then you must append a buffer zone of K zeros at the end of both input data sets. If you want all possible lags from N data points (not a usual thing), then you will need to pad the data with an equal number of zeros; this is the extreme case. So here is the program:

```
SUBROUTINE CORREL(DATA1,DATA2,N,ANS)
   Computes the correlation of two real data sets DATA1 and DATA2, each of length N (including
   any user-supplied zero padding). N MUST be an integer power of two. The answer is
   returned as the first N points in ANS stored in wraparound order, i.e. correlations at
   increasingly negative lags are in ANS(N) on down to ANS(N/2+1), while correlations at
   increasingly positive lags are in ANS(1) (zero lag) on up to ANS(N/2). Note that ANS must
   be supplied in the calling program with length at least 2*N, since it is also used as working
   space. Sign convention of this routine: if DATA1 lags DATA2, i.e. is shifted to the right of
   it, then ANS will show a peak at positive lags.
PARAMETER(NMAX=8192)          Maximum anticipated FFT size
DIMENSION DATA1(N),DATA2(N)
COMPLEX FFT(NMAX),ANS(N)
CALL TWOFFT(DATA1,DATA2,FFT,ANS,N)   Transform both DATA vectors at once.
NO2=N/2                       Normalization for inverse FFT.
DO 11 I=1,NO2+1
    ANS(I)=FFT(I)*CONJG(ANS(I))/FLOAT(NO2)   Multiply to find FFT of their correlation.
 11 CONTINUE
ANS(1)=CMPLX(REAL(ANS(1)),REAL(ANS(NO2+1)))  Pack first and last into one element.
CALL REALFT(ANS,NO2,-1)       Inverse transform gives correlation.
RETURN
END
```

The *discrete autocorrelation* of a sampled function g_j is just the discrete correlation of the function with itself. Obviously this is always symmetric with respect to positive and negative lags. Feel free to use the above routine CORREL to obtain autocorrelations, simply calling it with the same DATA vector in both arguments. If the inefficiency bothers you, routine REALFT can, of course, be used to transform the DATA vector instead.

REFERENCES AND FURTHER READING:

Brigham, E. Oran. 1974, *The Fast Fourier Transform* (Englewood Cliffs, N.J.: Prentice-Hall), §13-2.

12.6 Optimal (Wiener) Filtering with the FFT

There are a number of other tasks in numerical processing which are routinely handled with Fourier techniques. One of these is filtering for the removal of noise from a "corrupted" signal. The particular situation we consider is this: There is some underlying, uncorrupted signal $u(t)$ that we want to measure. The measurement process is imperfect, however, and what comes out of our measurement device is a corrupted signal $c(t)$. The signal $c(t)$ may be less than perfect in either or both of two respects. First, the apparatus may not have a perfect "delta-function" response, so that the true signal $u(t)$ is convolved with (smeared out by) some known response function $r(t)$ to give a smeared signal $s(t)$,

$$s(t) = \int_{-\infty}^{\infty} r(\tau)u(t-\tau)\,d\tau \quad \text{or} \quad S(f) = R(f)U(f) \tag{12.6.1}$$

where S, R, U are the Fourier transforms of s, r, u respectively. Second, the measured signal $c(t)$ may contain an additional component of noise $n(t)$,

$$c(t) = s(t) + n(t) \tag{12.6.2}$$

We already know how to deconvolve the effects of the response function r in the absence of any noise (§12.4); we just divide $C(f)$ by $R(f)$ to get a deconvolved signal. We now want to treat the analogous problem when noise is present. Our task is to find the *optimal filter*, $\phi(t)$ or $\Phi(f)$ which, when applied to the measured signal $c(t)$ or $C(f)$, and then deconvolved by $r(t)$ or $R(f)$, produces a signal $\widetilde{u}(t)$ or $\widetilde{U}(f)$ that is as close as possible to the uncorrupted signal $u(t)$ or $U(f)$. In other words we will estimate the true signal U by

$$\widetilde{U}(f) = \frac{C(f)\Phi(f)}{R(f)} \tag{12.6.3}$$

In what sense is $\widetilde{U}$ to be close to U? We ask that they be *close in the least-square sense*

$$\int_{-\infty}^{\infty} |\widetilde{u}(t) - u(t)|^2\,dt = \int_{-\infty}^{\infty} \left|\widetilde{U}(f) - U(f)\right|^2\,df \quad \text{is minimized.} \tag{12.6.4}$$

Substituting equations (12.6.3) and (12.6.2), the right-hand side of (12.6.4) becomes

$$\begin{aligned}&\int_{-\infty}^{\infty}\left|\frac{[S(f)+N(f)]\Phi(f)}{R(f)}-\frac{S(f)}{R(f)}\right|^2 df\\ =&\int_{-\infty}^{\infty}|R(f)|^{-2}\left\{|S(f)|^2\,|1-\Phi(f)|^2+|N(f)|^2\,|\Phi(f)|^2\right\} df\end{aligned} \tag{12.6.5}$$

The signal S and the noise N are *uncorrelated*, so their cross product, when integrated over frequency f gave zero. (This is practically the *definition* of what we mean by noise!). Obviously (12.6.5) will be a minimum if and only if the integrand is minimized with respect to $\Phi(f)$ at every value of f. Let us search for such a solution where $\Phi(f)$ is a real function. Differentiating with respect to Φ, and setting the result equal to zero gives

$$\Phi(f)=\frac{|S(f)|^2}{|S(f)|^2+|N(f)|^2} \tag{12.6.6}$$

This is the formula for the optimal filter $\Phi(f)$.

Notice that equation (12.6.6) involves S, the smeared signal, and N, the noise. The two of these add up to be C, the measured signal. Equation (12.6.6) does not contain U the "true" signal. This makes for an important simplification: The optimal filter can be determined independently of the determination of the deconvolution function that relates S and U.

To determine the optimal filter from equation (12.6.6) we need some way of separately estimating $|S|^2$ and $|N|^2$. There is no way to do this from the measured signal C alone without some other information, or some assumption or guess. Luckily, the extra information is often easy to obtain. For example, we can sample a long stretch of data $c(t)$ and plot its power spectral density using equations (12.0.14), (12.1.8) and (12.1.5). This quantity is proportional to the sum $|S|^2+|N|^2$, so we have

$$|S(f)|^2+|N(f)|^2 \approx P_c(f)=|C(f)|^2 \qquad 0\le f<f_c \tag{12.6.7}$$

(More sophisticated methods of estimating the power spectral density will be discussed in §§12.7 and 12.8, but the estimation above is almost always good enough for the optimal filter problem.) The resulting plot (see Figure 12.6.1) will often immediately show the spectral signature of a signal sticking up above a continuous noise spectrum. The noise spectrum may be flat, or tilted, or smoothly varying; it doesn't matter, as long as we can guess a reasonable hypothesis as to what it is. Draw a smooth curve through the noise spectrum, extrapolating it into the region dominated by the signal as well. Now draw a smooth curve through the signal plus noise power. The difference between these two curves is your smooth "model" of the signal power. The

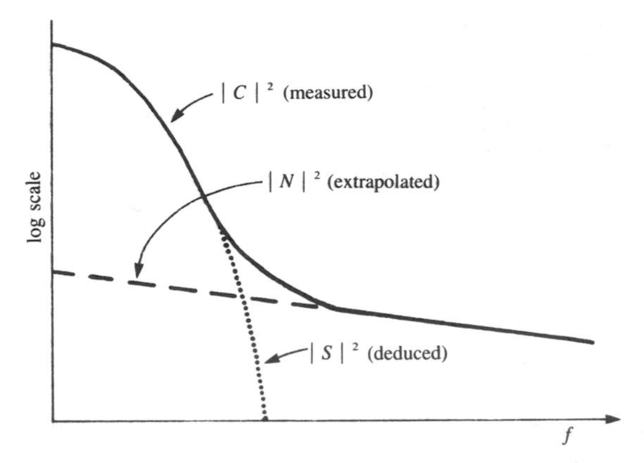

Figure 12.6.1. Optimal (Wiener) filtering. The power spectrum of signal plus noise shows a signal peak added to a noise tail. The tail is extrapolated back into the signal region as a "noise model." Subtracting gives the "signal model." The models need not be accurate for the method to be useful. A simple algebraic combination of the models gives the optimal filter (see text).

quotient of your model of signal power to your model of signal plus noise power is the optimal filter $\Phi(f)$. [Extend it to negative values of f by the formula $\Phi(-f) = \Phi(f)$.] Notice that $\Phi(f)$ will be close to unity where the noise is negligible, and close to zero where the noise is dominant. That is how it does its job! The intermediate dependence given by equation (12.6.6) just turns out to be the optimal way of going in between these two extremes.

Because the optimal filter results from a minimization problem, the quality of the results obtained by optimal filtering differs from the true optimum by an amount that is *second order* in the precision to which the optimal filter is determined. In other words, even a fairly crudely determined optimal filter (sloppy, say, at the 10 percent level) can give excellent results when it is applied to data. That is why the separation of the measured signal C into signal and noise components S and N can usefully be done "by eye" from a crude plot of power spectral density. All of this may give you thoughts about iterating the procedure we have just described. For example, after designing a filter with response $\Phi(f)$ and using it to make a respectable guess at the signal $\widetilde{U}(f) = \Phi(f)C(f)/R(f)$, you might turn about and regard $\widetilde{U}(f)$ as a fresh new signal which you could improve even further with the same filtering technique. Don't waste your time on this line of thought. The scheme converges to a signal of $S(f) = 0$. Converging iterative methods do exist; this just isn't one of them.

You can use the routine `FOUR1` (§12.2) or `REALFT` (§12.3) to FFT your data when you are constructing an optimal filter. To apply the filter to your data, you can use the methods described in §12.4. The specific routine `CONVLV` is not needed for optimal filtering, since your filter is constructed in the frequency domain to begin with. If you are also deconvolving your data with a known response function, however, you can modify `CONVLV` to multiply by your optimal filter just before it takes the inverse Fourier transform.

REFERENCES AND FURTHER READING:

Rabiner, L.R., and Gold B. 1975, *Theory and Application of Digital Signal Processing* (Englewood Cliffs, N.J.: Prentice-Hall).

Nussbaumer, H.J. 1982, *Fast Fourier Transform and Convolution Algorithms* (New York: Springer-Verlag).

Elliott, D.F., and Rao, K.R. 1982, *Fast Transforms: Algorithms, Analyses, Applications* (New York: Academic Press).

12.7 Power Spectrum Estimation Using the FFT

In the previous section we "informally" estimated the power spectral density of a function $c(t)$ by taking the modulus-squared of the discrete Fourier transform of some finite, sampled stretch of it. In this section we'll do roughly the same thing, but with considerably greater attention to details. Our attention will uncover some surprises.

The first detail is power spectrum (also called a power spectral density or PSD) normalization. In general there is *some* relation of proportionality between a measure of the squared amplitude of the function and a measure of the amplitude of the PSD. Unfortunately there are several different conventions for describing the normalization in each domain, and many opportunities for getting wrong the relationship between the two domains. Suppose that our function $c(t)$ is sampled at N points to produce values $c_0 \dots c_{N-1}$, and that these points span a range of time T, that is $T = (N-1)\Delta$, where Δ is the sampling interval. Then here are several different descriptions of the total power:

$$\sum_{j=0}^{N-1} |c_j|^2 \equiv \text{“sum squared amplitude”} \tag{12.7.1}$$

$$\frac{1}{T}\int_0^T |c(t)|^2\, dt \approx \frac{1}{N}\sum_{j=0}^{N-1} |c_j|^2 \equiv \text{“mean squared amplitude”} \tag{12.7.2}$$

$$\int_0^T |c(t)|^2\, dt \approx \Delta \sum_{j=0}^{N-1} |c_j|^2 \equiv \text{“time-integral squared amplitude”} \tag{12.7.3}$$

PSD estimators, as we shall see, have an even greater variety. In this section, we consider a class of them that give estimates at discrete values of frequency f_i, where i will range over integer values. In the next section, we will learn about a different class of estimators that produce estimates that are

continuous functions of frequency f. Even if it is agreed always to relate the PSD normalization to a particular description of the function normalization (e.g. 12.7.2), there are at least the following possibilities: The PSD is

- defined for discrete positive, zero, and negative frequencies, and its sum over these is the function mean squared amplitude
- defined for zero and discrete positive frequencies only, and its sum over these is the function mean squared amplitude
- defined in the Nyquist interval from $-f_c$ to f_c, and its integral over this range is the function mean squared amplitude
- defined from 0 to f_c, and its integral over this range is the function mean squared amplitude

It *never* makes sense to integrate the PSD of a sampled function outside of the Nyquist interval $-f_c$ and f_c since, according to the sampling theorem, power there will have been aliased into the Nyquist interval.

It is hopeless to define enough notation to distinguish all possible combinations of normalizations. In what follows, we use the notation $P(f)$ to mean *any* of the above PSDs, stating in each instance how the particular $P(f)$ is normalized. Beware the inconsistent notation in the literature.

The method of power spectrum estimation used in the previous section is a simple version of an estimator called, historically, the *periodogram.* If we take an N-point sample of the function $c(t)$ at equal intervals and use the FFT to compute its discrete Fourier transform

$$C_k = \sum_{j=0}^{N-1} c_j \, e^{2\pi i jk/N} \qquad k = 0, \dots, N-1 \tag{12.7.4}$$

then the periodogram estimate of the power spectrum is defined at $N/2+1$ frequencies as

$$\begin{aligned}
P(0) = P(f_0) &= \frac{1}{N^2} |C_0|^2 \\
P(f_k) &= \frac{1}{N^2} \left[|C_k|^2 + |C_{N-k}|^2 \right] \qquad k = 1, 2, \dots, \left(\frac{N}{2} - 1 \right) \\
P(f_c) = P(f_{N/2}) &= \frac{1}{N^2} \left| C_{N/2} \right|^2
\end{aligned} \tag{12.7.5}$$

where f_k is defined only for the zero and positive frequencies

$$f_k \equiv \frac{k}{N\Delta} = 2 f_c \frac{k}{N} \qquad k = 0, 1, \dots, \frac{N}{2} \tag{12.7.6}$$

By Parseval's theorem, equation (12.1.10), we see immediately that equation (12.7.5) is normalized so that the sum of the $N/2+1$ values of P is equal to the mean squared amplitude of the function c_j.

We must now ask this question. In what sense is the periodogram estimate (12.7.5) a "true" estimator of the power spectrum of the underlying function $c(t)$? You can find the answer treated in considerable detail in the literature cited (see, e.g., Oppenheim and Schafer for an introduction). Here is a summary.

First, is the *expectation value* of the periodogram estimate equal to the power spectrum, i.e., is the estimator correct on average? Well, yes and no. We wouldn't really expect one of the $P(f_k)$'s to equal the continuous $P(f)$ at *exactly* f_k, since f_k is supposed to be representative of a whole frequency "bin" extending from halfway from the preceding discrete frequency to halfway to the next one. We *should* be expecting the $P(f_k)$ to be some kind of average of $P(f)$ over a narrow window function centered on its f_k. For the periodogram estimate (12.7.6) that window function, as a function of s the frequency offset *in bins*, is

$$W(s) = \frac{1}{N^2}\left[\frac{\sin(\pi s)}{\sin(\pi s/N)}\right]^2 \tag{12.7.7}$$

Notice that $W(s)$ has oscillatory lobes but, apart from these, falls off only about as $W(s) \approx (\pi s)^{-2}$. This is not a very rapid fall-off, and it results in significant *leakage* (that is the technical term) from one frequency to another in the periodogram estimate. Notice also that $W(s)$ happens to be zero for s equal to a nonzero integer. This means that if the function $c(t)$ is a pure sine wave of frequency exactly equal to one of the f_k's, then there will be *no* leakage to adjacent f_k's. But this is not the characteristic case! If the frequency is, say, one-third of the way between two adjacent f_k's, then the leakage will extend *well* beyond those two adjacent bins. The solution to the problem of leakage is called *data windowing*, and we will discuss it below.

Turn now to another question about the periodogram estimate. What is the variance of that estimate as N goes to infinity? In other words, as we take more sampled points from the original function (either sampling a longer stretch of data at the same sampling rate, or else by resampling the same stretch of data with a faster sampling rate), then how much more accurate do the estimates P_k become? The unpleasant answer is that the periodogram estimates *do not become more accurate at all!* In fact, the variance of the periodogram estimate at a frequency f_k is always equal to the square of its expectation value at that frequency. In other words, the standard deviation is always 100 percent of the value, independent of N! How can this be? Where did all the information go as we added points? It all went into producing estimates at a greater number of discrete frequencies f_k. If we sample a longer run of data using the same sampling rate, then the Nyquist critical frequency f_c is unchanged, but we now have finer frequency resolution (more f_k's) within the Nyquist frequency interval; alternatively, if we sample the same length of data with a finer sampling interval, then our frequency resolution is unchanged, but the Nyquist range now extends up to a higher frequency. In neither case do the additional samples reduce the variance of any one particular frequency's estimated PSD.

You don't have to live with PSD estimates with 100 percent standard deviations, however. You simply have to know some techniques for reducing the variance of the estimates. Here are two techniques that are very nearly identical mathematically, though different in implementation. The first is to compute a periodogram estimate with finer discrete frequency spacing than you really need, and then to sum the periodogram estimates at K consecutive discrete frequencies to get one "smoother" estimate at the mid frequency of those K. The variance of that summed estimate will be smaller than the estimate itself by a factor of exactly $1/K$, i.e. the standard deviation will be smaller than 100 percent by a factor $1/\sqrt{K}$. Thus, to estimate the power spectrum at $M+1$ discrete frequencies between 0 and f_c inclusive, you begin by taking the FFT of $2MK$ points (which number had better be an integer power of two!). You then take the modulus square of the resulting coefficients, add positive and negative frequency pairs and divide by $(2MK)^2$, all according to equation (12.7.5) with $N = 2MK$. Finally, you "bin" the results into summed (not averaged) groups of K. This procedure is very easy to program, so we will not bother to give a routine for it. The reason that you sum, rather than average, K consecutive points is so that your final PSD estimate will preserve the normalization property that the sum of its $M+1$ values equals the mean square value of the function.

A second technique for estimating the PSD at $M+1$ discrete frequencies in the range 0 to f_c is to partition the original sampled data into K segments each of $2M$ consecutive sampled points. Each segment is separately FFT'd to produce a periodogram estimate (equation 12.7.5 with $N \equiv 2M$). Finally, the K periodogram estimates are averaged at each frequency. It is this final averaging that reduces the variance of the estimate by a factor K (standard deviation by $\sqrt{K}$). This second technique is computationally more efficient than the first technique above by a modest factor, since it is logarithmically more efficient to take many shorter FFTs than one longer one. The principal advantage of the second technique, however, is that only $2M$ data points are manipulated at a single time, not $2KM$ as in the first technique. This means that the second technique is the natural choice for processing long runs of data, as from a magnetic tape or other data record. We will give a routine later for implementing this second technique, but we need first to return to the matters of leakage and data windowing which were brought up after equation (12.7.7) above.

Data Windowing

The purpose of data windowing is to modify equation (12.7.7), which expresses the relation between the spectral estimate P_k at a discrete frequency and the actual underlying continuous spectrum $P(f)$ at nearby frequencies. In general, the spectral power in one "bin" k contains leakage from frequency components that are actually s bins away, where s is the independent variable in equation (12.7.7). There is, as we pointed out, quite substantial leakage even from moderately large values of s.

When we select a run of N sampled points for periodogram spectral estimation, we are in effect multiplying an infinite run of sampled data c_j

by a window function in time, one which is zero except during the total sampling time $N\Delta$, and is unity during that time. In other words, the data are windowed by a square window function. By the convolution theorem (12.0.9; but interchanging the roles of f and t), the Fourier transform of the product of the data with this square window function is equal to the convolution of the data's Fourier transform with the window's Fourier transform. In fact, we determined equation (12.7.7) as nothing more than the square of the discrete Fourier transform of the unity window function.

$$W(s) = \frac{1}{N^2}\left[\frac{\sin(\pi s)}{\sin(\pi s/N)}\right]^2 = \frac{1}{N^2}\left|\sum_{k=0}^{N-1} e^{2\pi i s k/N}\right|^2 \qquad (12.7.8)$$

The reason for the leakage at large values of s, is that the square window function turns on and off so rapidly. Its Fourier transform has substantial components at high frequencies. To remedy this situation, we can multiply the input data c_j, $j = 0, \ldots, N-1$ by a window function w_j that changes more gradually from zero to a maximum and then back to zero as j ranges from 0 to $N-1$. In this case, the equations for the periodogram estimator (12.7.4–12.7.5) become

$$D_k \equiv \sum_{j=0}^{N-1} c_j w_j \, e^{2\pi i j k/N} \qquad k = 0, \ldots, N-1 \qquad (12.7.9)$$

$$\begin{aligned} P(0) = P(f_0) &= \frac{1}{W_{ss}} \left|D_0\right|^2 \\ P(f_k) &= \frac{1}{W_{ss}} \left[\left|D_k\right|^2 + \left|D_{N-k}\right|^2\right] \qquad k = 1, 2, \ldots, \left(\frac{N}{2} - 1\right) \\ P(f_c) = P(f_{N/2}) &= \frac{1}{W_{ss}} \left|D_{N/2}\right|^2 \end{aligned} \qquad (12.7.10)$$

where W_{ss} stands for "window squared and summed,"

$$W_{ss} \equiv N \sum_{j=0}^{N} w_j^2 \qquad (12.7.11)$$

and f_k is given by (12.7.6). The more general form of (12.7.7) can now be written in terms of the window function w_j as

$$\begin{aligned} W(s) &= \frac{1}{W_{ss}} \left|\sum_{k=0}^{N-1} e^{2\pi i s k/N} w_k\right|^2 \\ &\approx \frac{1}{W_{ss}} \left|\int_{-N/2}^{N/2} \cos(2\pi s k/N) w(k - N/2)\, dk\right|^2 \end{aligned} \qquad (12.7.12)$$

Here the approximate equality is useful for practical estimates, and holds for any window that is left-right symmetric (the usual case), and for $s \ll N$ (the case of interest for estimating leakage into nearby bins). The continuous function $w(k - N/2)$ in the integral is meant to be some smooth function that passes through the points w_k.

There is a lot of perhaps unnecessary lore about choice of a window function, and practically every function which rises from zero to a peak and then falls again has been named after someone. A few of the more common (also shown in Figure 12.7.1) are:

$$w_j = 1 - \left| \frac{j - \frac{1}{2}(N-1)}{\frac{1}{2}(N+1)} \right| \equiv \text{“Parzen window”} \tag{12.7.13}$$

(The “Bartlett window” is very similar to this.)

$$w_j = \frac{1}{2}\left[1 - \cos\left(\frac{2\pi j}{N-1}\right)\right] \equiv \text{“Hanning window”} \tag{12.7.14}$$

(The “Hamming window” is similar but does not go exactly to zero at the ends.)

$$w_j = 1 - \left(\frac{j - \frac{1}{2}(N-1)}{\frac{1}{2}(N+1)}\right)^2 \equiv \text{“Welch window”} \tag{12.7.15}$$

We are inclined to follow Welch in recommending that you use either (12.7.13) or (12.7.15) in practical work. However, at the level of this book, there is effectively *no difference* between any of these (or similar) window functions. Their difference lies in subtle tradeoffs among the various figures of merit that can be used to describe the narrowness or peakedness of the spectral leakage functions computed by (12.7.12). These figures of merit have such names as: *highest sidelobe level (db), sidelobe fall-off (db per octave), equivalent noise bandwidth (bins), 3-db bandwidth (bins), scallop loss (db), worst case process loss (db).* Roughly speaking, the principal tradeoff is between making the central peak as narrow as possible versus making the tails of the distribution fall off as rapidly as possible. For details, see (e.g.) Harris. Figure 12.7.2 plots the leakage amplitudes for several windows already discussed.

There is particularly a lore about window functions which rise smoothly from zero to unity in the first small fraction (say 10 percent) of the data, then stay at unity until the last small fraction (again say 10 percent) of the data, during which the window function falls smoothly back to zero. These windows will squeeze a little bit of extra narrowness out of the main lobe of the leakage function (never as much as a factor of two, however), but trade this off by widening the leakage tail by a significant factor (e.g., the reciprocal of 10 percent, a factor of ten). If we distinguish between the *width* of a window (number of samples for which it is at its maximum value) and its

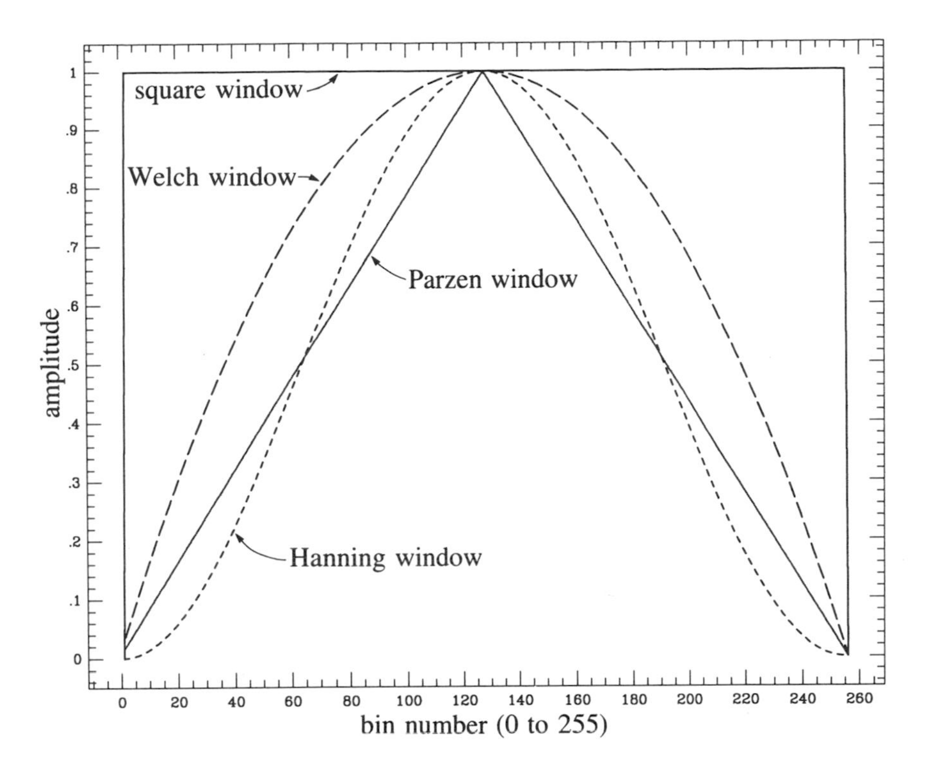

Figure 12.7.1. Window functions commonly used in FFT power spectral estimation. The data segment, here of length 256, is multiplied (bin by bin) by the window function before the FFT is computed. The square window, which is equivalent to no windowing, is least recommended. The Welch and Parzen windows are good choices.

rise/fall time (number of samples during which it rises and falls); and if we distinguish between the *FWHM* (full width to half maximum value) of the leakage function's main lobe and the *leakage width* (full width that contains half of the spectral power that is not contained in the main lobe); then these quantities are related roughly by

$$(\text{FWHM in bins}) \approx \frac{N}{(\text{window width})} \tag{12.7.16}$$

$$(\text{leakage width in bins}) \approx \frac{N}{(\text{window rise/fall time})} \tag{12.7.17}$$

For the windows given above in (12.7.13)–(12.7.15), the effective window widths and the effective window rise/fall times are both of order $\frac{1}{2}N$. Generally speaking, we feel that the advantages of windows whose rise and fall times are only small fractions of the data length are minor or nonexistent, and we avoid using them. One sometimes hears it said that flat-topped windows "throw away less of the data," but we will now show you a better way of dealing with that problem by use of overlapping data segments.

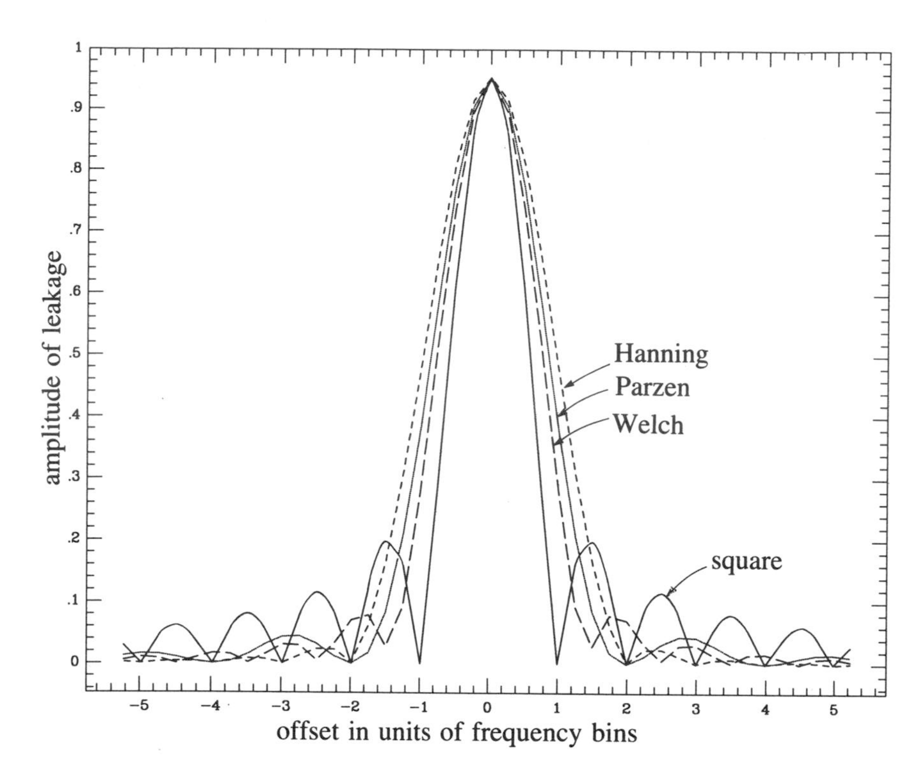

Figure 12.7.2. Leakage functions for the window functions of Figure 12.7.1. A signal whose frequency is actually located at zero offset "leaks" into neighboring bins with the amplitude shown. The purpose of windowing is to reduce the leakage at large offsets, where square (no) windowing has large sidelobes. Offset can have a fractional value, since the actual signal frequency can be located between two frequency bins of the FFT.

Let us now suppose that we have chosen a window function, and that we are ready to segment the data into K segments of $N = 2M$ points. Each segment will be FFT'd, and the resulting K periodograms will be averaged together to obtain a PSD estimate at M frequency values between 0 and f_c. We must now distinguish between two possible situations. We might want to obtain the smallest variance from a fixed amount of computation, without regard to the number of data points used. This will generally be the goal when the data are being gathered in real time, with the data-reduction being computer-limited. Alternatively, we might want to obtain the smallest variance from a fixed number of available sampled data points. This will generally be the goal in cases where the data are already recorded and we are analyzing it after the fact.

In the first situation (smallest spectral variance per computer operation), it is best to segment the data without any overlapping. The first $2M$ data points constitute segment number 1; the next $2M$ data points constitute segment number 2; and so on, up to segment number K, for a total of $2KM$ sampled points. The variance in this case, relative to a single segment, is reduced by a factor K.

In the second situation (smallest spectral variance per data point), it turns out to be optimal, or very nearly optimal, to overlap the segments by one half of their length. The first and second sets of M points are segment number 1; the second and third sets of M points are segment number 2; and so on, up to segment number K, which is made of the Kth and $K+1^{st}$ sets of M points. The total number of sampled points is therefore $(K+1)M$, just over half as many as with nonoverlapping segments. The reduction in the variance is not a full factor of K, since the segments are not statistically independent. It can be shown that the variance is instead reduced by a factor of about $9K/11$ (see the paper by Welch in Childers). This is, however, significantly better than the reduction of about $K/2$ which would have resulted if the same *number* of data points were segmented without overlapping.

We can now codify these ideas into a routine for spectral estimation. While we generally avoid input/output coding, we make an exception here to show how data are read sequentially in one pass through a data file (here FORTRAN Unit 9). Only a small fraction of the data is in memory at any one time.

```
      SUBROUTINE SPCTRM(P,M,K,OVRLAP,W1,W2)
C       Reads data from input unit 9 and returns as P(J) the data's power (mean square
C       amplitude) at frequency (J-1)/(2*M) cycles per gridpoint, for J=1,2,...,M, based on
C       (2*K+1)*M data points (if OVRLAP is set .TRUE.) or 4*K*M data points (if OVRLAP is set
C       .FALSE.). The number of segments of the data is 2*K in both cases: the routine calls
C       FOUR1 K-times, each call with 2 partitions each of 2*M real data points. W1 and W2 are
C       user-supplied workspaces of length 4*M and M respectively.
      LOGICAL OVRLAP                  True for overlapping segments, false otherwise.
      DIMENSION P(M),W1(4*M),W2(M)
      WINDOW(J)=(1.-ABS(((J-1)-FACM)*FACP))      Statement function defines Parzen window.
C     WINDOW(J)=1.                                  Alternative for square window.
C     WINDOW(J)=(1.-(((J-1)-FACM)*FACP)**2)         Alternative for Welch window.
      MM=M+M                          Useful factors.
      M4=MM+MM
      M44=M4+4
      M43=M4+3
      DEN=0.
      FACM=M-0.5                      Factors used by the window statement function.
      FACP=1./(M+0.5)
      SUMW=0.                         Accumulate the squared sum of the weights.
      DO 11 J=1,MM
        SUMW=SUMW+WINDOW(J)**2
11    CONTINUE
      DO 12 J=1,M                     Initialize the spectrum to zero.
        P(J)=0.
12    CONTINUE
      IF(OVRLAP)THEN                  Initialize the "save" half-buffer.
        READ (9,*) (W2(J),J=1,M)
      ENDIF
      DO 18 KK=1,K                    Loop over data set segments in groups of two.
        DO 15 JOFF=-1,0,1             Get two complete segments into workspace.
          IF (OVRLAP) THEN
            DO 13 J=1,M
              W1(JOFF+J+J)=W2(J)
13          CONTINUE
            READ (9,*) (W2(J),J=1,M)
            JOFFN=JOFF+MM
            DO 14 J=1,M
              W1(JOFFN+J+J)=W2(J)
14          CONTINUE
```

```
          ELSE
             READ (9,*) (W1(J),J=JOFF+2,M4,2)
          ENDIF
15      CONTINUE
        DO 16 J=1,MM                  Apply the window to the data.
          J2=J+J
          W=WINDOW(J)
          W1(J2)=W1(J2)*W
          W1(J2-1)=W1(J2-1)*W
16      CONTINUE
        CALL FOUR1(W1,MM,1)           Fourier transform the windowed data.
        P(1)=P(1)+W1(1)**2+W1(2)**2   Sum results into previous segments.
        DO 17 J=2,M
          J2=J+J
          P(J)=P(J)+W1(J2)**2+W1(J2-1)**2
*             +W1(M44-J2)**2+W1(M43-J2)**2
17      CONTINUE
        DEN=DEN+SUMW
18    CONTINUE
      DEN=M4*DEN                      Correct normalization.
      DO 19 J=1,M
        P(J)=P(J)/DEN                 Normalize the output.
19    CONTINUE
      RETURN
      END
```

REFERENCES AND FURTHER READING:

Harris, F.J. 1978, *Proceedings of the IEEE*, vol. 66, p. 51.

Childers, Donald G. (ed.). 1978, *Modern Spectrum Analysis* (New York: IEEE Press), paper by P.D. Welch.

Oppenheim, A.V., and Schafer, R.W. 1975, *Digital Signal Processing* (Englewood Cliffs, N.J.: Prentice Hall).

Champeney, D.C. 1973, *Fourier Transforms and Their Physical Applications* (New York: Academic Press).

Elliott, D.F., and Rao, K.R. 1982, *Fast Transforms: Algorithms, Analyses, Applications* (New York: Academic Press).

Bloomfield, P. 1976, *Fourier Analysis of Time Series – An Introduction* (New York: Wiley).

Rabiner, L.R., and Gold B. 1975 *Theory and Application of Digital Signal Processing* (Englewood Cliffs, N.J.: Prentice-Hall).

12.8 Power Spectrum Estimation by the Maximum Entropy (All Poles) Method

The FFT is not the only way to estimate the power spectrum of a process, nor is it necessarily the best way for all purposes. To see how one might devise another method, let us enlarge our view for a moment, so that it includes not only real frequencies in the Nyquist interval $-f_c < f < f_c$, but also the entire complex frequency plane. From that vantage point, let us transform the complex f-plane to a new plane, called the *z-transform plane* or *z-plane*, by the relation

$$z \equiv e^{2\pi i f \Delta} \tag{12.8.1}$$

where Δ is, as usual, the sampling interval in the time domain. Notice that the Nyquist interval on the real axis of the f-plane maps one-to-one onto the unit circle in the complex z-plane.

If we now compare (12.8.1) to equations (12.7.4) and (12.7.6), we see that the FFT power spectrum estimate (12.7.5) for any real sampled function $c_k \equiv c(t_k)$ can be written, except for normalization convention, as

$$P(f) = \left| \sum_{k=-N/2}^{N/2-1} c_k z^k \right|^2 \tag{12.8.2}$$

Of course, (12.8.2) is not the *true* power spectrum of the underlying function $c(t)$, but only an estimate. We can see in two related ways why the estimate is not likely to be exact. First, in the time domain, the estimate is based on only a finite range of the function $c(t)$ which may, for all we know, have continued from $t = -\infty$ to ∞. Second, in the z-plane of equation (12.8.2), the finite Laurent series offers, in general, only an approximation to a general analytic function of z. In fact, a formal expression for representing "true" power spectra (up to normalization) is

$$P(f) = \left| \sum_{k=-\infty}^{\infty} c_k z^k \right|^2 \tag{12.8.3}$$

This is an infinite Laurent series which depends on an infinite number of values c_k. Equation (12.8.2) is just one kind of analytic approximation to the analytic function of z represented by (12.8.3); the kind, in fact, that is implicit in the use of FFTs to estimate power spectra by periodogram methods. It goes under several names, including *direct method, all-zero model,* and *moving average (MA) model.* The term "all-zero" in particular refers to the fact that the model spectrum can have zeros in the z-plane, but not poles.

If we look at the problem of approximating (12.8.3) more generally it seems clear that we could do a better job with a rational function, one with a series of type (12.8.2) in both the numerator and the denominator. Less obviously, it turns out that there are some advantages in an approximation whose free parameters all lie in the *denominator*, namely,

$$P(f) \approx \frac{1}{\left|\sum_{k=-M/2}^{M/2} b_k z^k\right|^2} = \frac{a_0}{\left|1 + \sum_{k=1}^{M} a_k z^k\right|^2} \tag{12.8.4}$$

Here the second equality brings in a new set of coefficients a_k's, which can be determined from the b_k's using the fact that z lies on the unit circle. The b_k's can be thought of as being determined by the condition that power series expansion of (12.8.4) agree with the first $M+1$ terms of (12.8.3). In practice, as we shall see, one determines the b_k's or a_k's by another method.

The differences between the approximations (12.8.2) and (12.8.4) are not just cosmetic. They are approximations with very different character. Most notable is the fact that (12.8.4) can have *poles*, corresponding to infinite power spectral density, on the unit z-circle, i.e. at real frequencies in the Nyquist interval. Such poles can provide an accurate representation for underlying power spectra which have sharp, discrete "lines" or delta-functions. By contrast, (12.8.2) can have only zeros, not poles, at real frequencies in the Nyquist interval, and must thus attempt to fit sharp spectral features with, essentially, a polynomial. The approximation (12.8.4) goes under several names: *all-poles model, maximum entropy method (MEM), autoregressive model (AR)*. Let us now learn how to compute the coefficients a_0 and the a_k's from a data set, so that we can actually use (12.8.4) to obtain spectral estimates.

Consider the *autocorrelation at lag j* of the sampled function c_k, namely

$$\phi_j \equiv \langle c_i c_{i+j} \rangle \qquad j = \ldots -3, -2, -1, 0, 1, 2, 3 \ldots \tag{12.8.5}$$

where the angle brackets denote averaging over i. Given a finite run of data c_0 to c_N, the most natural estimate of (12.8.5) is

$$\phi_j = \phi_{-j} \approx \frac{1}{N+1-j} \sum_{i=0}^{N-j} c_i c_{i+j} \qquad j = 0, 1, 2, \ldots, N \tag{12.8.6}$$

In other words, from $N+1$ data points, we can estimate the autocorrelation at $N+1$ different lags j.

The Wiener-Khinchin theorem (12.0.12) says that the Fourier transform of the autocorrelation is equal to the power spectrum. In z-transform language, this Fourier transform is just a Laurent series in z. The equation that

is to be satisfied by the coefficients in equation (12.8.4) is thus

$$\frac{a_0}{\left|1 + \sum_{k=1}^{M} a_k z^k\right|^2} \approx \sum_{j=-M}^{M} \phi_j z^j \tag{12.8.7}$$

The approximately equal sign in (12.8.7) has a somewhat special interpretation. It means that the series expansion of the left-hand side is supposed to agree with the right-hand side term by term from z^{-M} to z^M. Outside this range of terms, the right-hand side is obviously zero, while the left-hand side will still have nonzero terms. Notice that M, the number of coefficients in the approximation on the left-hand side, can be any integer up to N, the total number of autocorrelations available. (In practice, one often chooses M much smaller than N.) M is called the *order* or *number of poles* of the approximation.

Whatever the chosen value of M, the series expansion of the left-hand side of (12.8.7) defines a certain sort of *extrapolation* of the autocorrelation function to lags larger than M, in fact even to lags larger than N, i.e., *larger than the run of data can actually measure.* It turns out that this particular extrapolation can be shown to have, among all possible extrapolations, the maximum *entropy* in a definable information-theoretic sense. Hence the name *maximum entropy method* or MEM. The maximum entropy property has caused MEM to acquire a certain "cult" popularity; one sometimes hears that it gives an intrinsically "better" estimate than is given by other methods. Don't believe it. MEM has the very cute property of being able to fit sharp spectral features, but there is nothing else magical about its power spectrum estimates.

There is, however, some magic in how (12.8.7) is actually solved for the coefficients on the left-hand side, in terms of the known autocorrelations on the right: Although it is by no means obvious by inspection, (12.8.7) implies a *linear* set of relations between the autocorrelations and the coefficients a_0 and a_k. In fact, these coefficients satisfy the matrix equation

$$\begin{bmatrix} \phi_0 & \phi_1 & \phi_2 & \cdots & \phi_M \\ \phi_1 & \phi_0 & \phi_1 & \cdots & \phi_{M-1} \\ \phi_2 & \phi_1 & \phi_0 & \cdots & \phi_{M-2} \\ \cdots & & & & \cdots \\ \phi_M & \phi_{M-1} & \phi_{M-2} & \cdots & \phi_0 \end{bmatrix} \cdot \begin{bmatrix} 1 \\ a_1 \\ a_2 \\ \cdots \\ a_M \end{bmatrix} = \begin{bmatrix} a_0 \\ 0 \\ 0 \\ \cdots \\ 0 \end{bmatrix} \tag{12.8.8}$$

The matrix in (12.8.8) is a symmetric *Toeplitz matrix*, i.e., one whose elements are constant along diagonals. Information on these matrices may be found in §2.8. Here we present an efficient, recursive algorithm that exploits the symmetry of the set. This algorithm was originated by Burg, and also implemented by Andersen.

```
SUBROUTINE MEMCOF(DATA,N,M,PM,COF,WK1,WK2,WKM)
    Given a real vector of DATA of length N, and given M, this routine returns a vector COF of
    length M with COF(J) = a_j, and a scalar PM = b_0, which are the coefficients for Maximum
    Entropy Method spectral estimation. The user must provide workspace vectors WK1, WK2,
    and WKM of lengths N, N, and M, respectively.
DIMENSION DATA(N),COF(M),WK1(N),WK2(N),WKM(M)
P=0.
DO 11 J=1,N
    P=P+DATA(J)**2
11  CONTINUE
PM=P/N
WK1(1)=DATA(1)
WK2(N-1)=DATA(N)
DO 12 J=2,N-1
    WK1(J)=DATA(J)
    WK2(J-1)=DATA(J)
12  CONTINUE
DO 17 K=1,M
    PNEUM=0.
    DENOM=0.
    DO 13 J=1,N-K
        PNEUM=PNEUM+WK1(J)*WK2(J)
        DENOM=DENOM+WK1(J)**2+WK2(J)**2
13      CONTINUE
    COF(K)=2.*PNEUM/DENOM
    PM=PM*(1.-COF(K)**2)
    IF(K.NE.1)THEN
        DO 14 I=1,K-1
            COF(I)=WKM(I)-COF(K)*WKM(K-I)
14          CONTINUE
    ENDIF
    The algorithm is recursive, building up the answer for larger and larger values of M until
    the desired value is reached. At this point in the algorithm, one could return the vector
    COF and scalar PM for an MEM spectral estimate of K (rather than M) terms.
    IF(K.EQ.M)RETURN
    DO 15 I=1,K
        WKM(I)=COF(I)
15      CONTINUE
    DO 16 J=1,N-K-1
        WK1(J)=WK1(J)-WKM(K)*WK2(J)
        WK2(J)=WK2(J+1)-WKM(K)*WK1(J+1)
16      CONTINUE
17  CONTINUE
PAUSE 'never get here'
END
```

The operations count in MEMCOF scales as the product of N (the number of data points) and M (the desired order of the MEM approximation). If M were chosen to be as large as N, then the method would be much slower than the $N \log N$ FFT methods of the previous section. In practice, however, one usually wants to limit the order (or number of poles) of the MEM approximation to a few times the number of sharp spectral features that one desires it to fit. With this restricted number of poles, the method will smooth the spectrum somewhat, but this is often a desirable property. While exact values depend on the application, one might take $M = 10$ or 20 or 50 for $N = 1000$ or 10000. In that case MEM estimation is not much slower than FFT estimation.

We feel obliged to warn you that MEMCOF can be a bit quirky at times. If the number of poles or number of data points is too large, roundoff error can

be a problem, even in double precision. With "peaky" data (i.e. data with extremely sharp spectral features), the algorithm may suggest split peaks even at modest orders, and the peaks may shift with the phase of the sine wave. Also, with noisy input functions, if you choose too high an order, you will find spurious peaks galore! Some experts recommend the use of this algorithm in conjunction with more conservative methods, like periodograms, to help choose the correct model order, and to avoid getting too fooled by spurious spectral features. MEM can be finicky, but it can also do remarkable things. We recommend that you try it out, cautiously, on your own problems. We now turn to the evaluation of the MEM spectral estimate from its coefficients.

The MEM estimation (12.8.4) is a function of continuously varying frequency f. There is no special significance to specific equally spaced frequencies as there was in the FFT case. In fact, since the MEM estimate may have very sharp spectral features, one wants to be able to evaluate it on a very fine mesh near to those features, but perhaps only more coarsely farther away from them. Here is a function subroutine which, given the coefficients already computed, evaluates (12.8.4) and returns the estimated power spectrum as a function of $f\Delta$ (the frequency times the sampling interval). Of course, $f\Delta$ should lie in the Nyquist range between $-1/2$ and $1/2$.

```
FUNCTION EVLMEM(FDT,COF,M,PM)
    Given COF, M, PM as returned by MEMCOF, this function returns the power spectrum esti-
    mate P(f) as a function of FDT = fΔ.
DIMENSION COF(M)
REAL*8 WR,WI,WPR,WPI,WTEMP,THETA      Do trigonometric recurrences in double precision.
THETA=6.28318530717959D0*FDT
WPR=DCOS(THETA)                       Set up for recurrence relations.
WPI=DSIN(THETA)
WR=1.D0
WI=0.D0
SUMR=1.                               These will accumulate the denominator of (12.8.4).
SUMI=0.
DO 11 I=1,M                           Loop over the terms in the sum.
    WTEMP=WR
    WR=WR*WPR-WI*WPI
    WI=WI*WPR+WTEMP*WPI
    SUMR=SUMR-COF(I)*SNGL(WR)
    SUMI=SUMI-COF(I)*SNGL(WI)
11  CONTINUE
EVLMEM=PM/(SUMR**2+SUMI**2)   Equation (12.8.4).
RETURN
END
```

Be sure to evaluate $P(f)$ on a fine enough grid to *find* any narrow features that may be there! Such narrow features, if present, can contain virtually all of the power in the data. You might also wish to know how the $P(f)$ produced by the routines MEMCOF and EVLMEM is normalized with respect to the mean

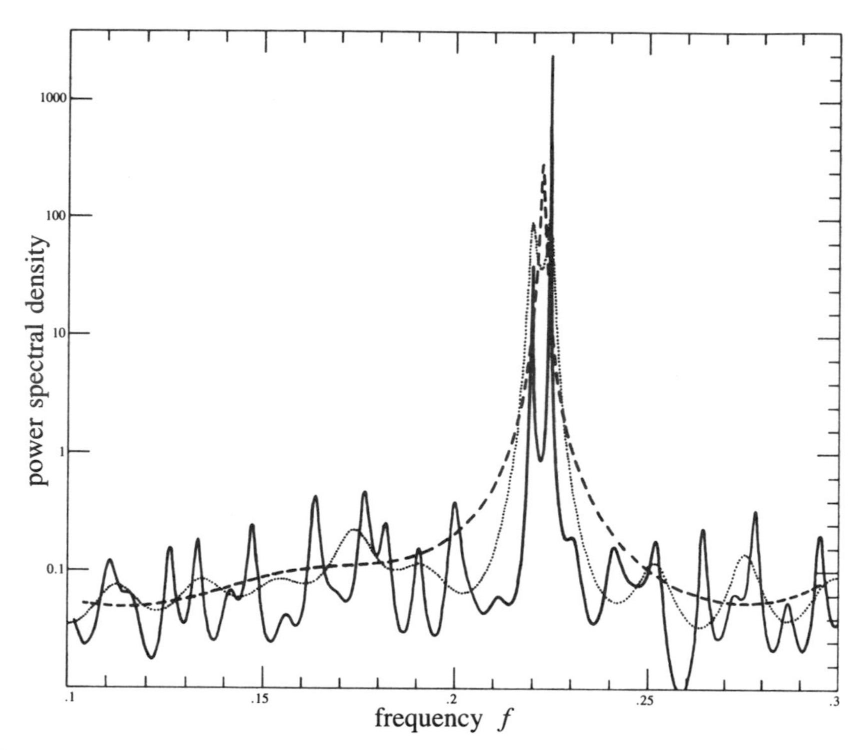

Figure 12.8.1. Sample output of maximum entropy spectral estimation. The input signal consists of 512 samples of the sum of two sinusoids of very nearly the same frequency, plus white noise with about equal power. Shown is an expanded portion of the full Nyquist frequency interval (which would extend from zero to 0.5). The dashed spectral estimate uses 20 poles; the dotted, 40; the solid, 150. With the larger number of poles, the method can resolve the distinct sinusoids; but the flat noise background is beginning to show spurious peaks. (Note logarithmic scale.)

square value of the input data vector. The answer is

$$\int_{-1/2}^{1/2} P(f\Delta)d(f\Delta) = 2\int_{0}^{1/2} P(f\Delta)d(f\Delta) = \text{mean square value of data} \tag{12.8.9}$$

Sample spectra produced by the routines MEMCOF and EVLMEM are shown in Figure 12.8.1.

REFERENCES AND FURTHER READING:

Childers, Donald G. (ed.). 1978, *Modern Spectrum Analysis* (New York: IEEE Press), Chapter II.

Kay, S.M., and Marple, S.L. 1981, *Proceedings of the IEEE*, vol. 69, p. 1380.

12.9 Digital Filtering in the Time Domain

We'll next suppose that you have a signal that you want to filter digitally. For example, perhaps you want to apply ***high-pass*** or ***low-pass*** filtering, to eliminate noise at low or high frequencies respectively; or perhaps the interesting part of your signal lies only in a certain frequency band, so that you need a ***band-pass*** filter. Or, if your measurements are contaminated by 60 Hz power-line interference, you may need a ***notch filter*** to remove only a narrow band around that frequency. This section speaks particularly about the case in which you have chosen to do such filtering in the time domain.

Before continuing, we hope you will reconsider this choice. Remember how convenient it is to filter in the Fourier domain. You just take your whole data record, FFT it, multiply the FFT output by a filter function $\mathcal{H}(f)$, and then do an inverse FFT to get back a filtered data set in time domain. Here is some additional background on the Fourier technique that you will want to take into account.

- Remember that you must define your filter function $\mathcal{H}(f)$ for both positive and negative frequencies, and that the magnitude of the frequency extremes is always the Nyquist frequency $1/(2\Delta)$, where Δ is the sampling interval. The magnitude of the smallest nonzero frequencies in the FFT is $\pm 1/(N\Delta)$, where N is the number of (complex) points in the FFT. The positive and negative frequencies to which this filter are applied are arranged in wrap-around order.
- If the measured data are real, and you want the filtered output also to be real, then your arbitrary filter function should obey $\mathcal{H}(-f) = \mathcal{H}(f)^*$. You can arrange this most easily by picking an $\mathcal{H}$ that is real and even in f.
- If your chosen $\mathcal{H}(f)$ has sharp vertical edges in it, then the ***impulse response*** of your filter (the output arising from a short impulse as input) will have damped "ringing" at frequencies corresponding to these edges. There is nothing wrong with this, but if you don't like it, then pick a smoother $\mathcal{H}(f)$. To get a first-hand look at the impulse response of your filter, just take the inverse FFT of your $\mathcal{H}(f)$. If you smooth all edges of the filter function over some number k of points, then the impulse response function of your filter will have a span on the order of a fraction $1/k$ of the whole data record.
- If your data set is too long to FFT all at once, then break it up into segments of any convenient size, as long as they are much longer than the impulse response function of the filter. Use zero-padding, if necessary.
- You should probably remove any trend from the data, by subtracting from it a straight line through the first and last points (i.e. make the first and last points equal to zero). If you are segmenting the data, then you can pick overlapping segments and use only the middle section of each, comfortably distant from edge effects.

- A digital filter is said to be *causal* or *physically realizable* if its output for a particular time-step depends only on inputs at that particular time-step or earlier. It is said to be *acausal* if its output can depend on both earlier and later inputs. Filtering in the Fourier domain is, in general, acausal, since the data are processed "in a batch," without regard to time ordering. Don't let this bother you! Acausal filters can generally give superior performance (e.g. less dispersion of phases, sharper edges, less asymmetric impulse response functions). People use causal filters not because they are better, but because some situations just don't allow access to out-of-time-order data. Time domain filters can, in principle, be either causal or acausal, but they are most often used in applications where physical realizability is a constraint. For this reason we will restrict ourselves to the causal case in what follows.

If you are still favoring time-domain filtering after all we have said, it is probably because you have a real-time application, for which you must process a continuous data stream and wish to output filtered values at the same rate as you receive raw data. Otherwise, it may be that the quantity of data to be processed is so large that you can only afford a very small number of floating operations on each data point and cannot afford even a modest-sized FFT (with a number of floating operations per data point several times the logarithm of the number of points in the data set or segment).

Linear Filters

The most general linear filter takes a sequence x_k of input points and produces a sequence y_n of output points by the formula

$$y_n = \sum_{k=0}^{M} c_k \, x_{n-k} + \sum_{j=1}^{N} d_j \, y_{n-j} \qquad (12.9.1)$$

Here the $M+1$ coefficients c_k and the N coefficients d_j are fixed and define the filter response. The filter (12.9.1) produces each new output value from the current and M previous input values, and from its own N previous output values. If $N = 0$, so that there is no second sum in (12.9.1), then the filter is called *nonrecursive* or *finite impulse response (FIR)*. If $N \neq 0$, then it is called *recursive* or *infinite impulse response (IIR)*. (The term "IIR" connotes only that such filters are *capable* of having infinitely long impulse responses, not that their impulse response is necessarily long in a particular application. Typically the response of an IIR filter will drop off exponentially at late times, rapidly becoming negligible.)

The relation between the c_k's and d_j's and the filter response function $\mathcal{H}(f)$ is

$$\mathcal{H}(f) = \frac{\sum_{k=0}^{M} c_k e^{-2\pi i k(f\Delta)}}{1 - \sum_{j=1}^{N} d_k e^{-2\pi i k(f\Delta)}} \tag{12.9.2}$$

where Δ is, as usual, the sampling interval. The Nyquist interval corresponds to $f\Delta$ between $-1/2$ and $1/2$. For FIR filters the denominator of (12.9.2) is just unity.

Equation (12.9.2) tells how to determine $\mathcal{H}(f)$ from the c's and d's. To design a filter, though, we need a way of doing the inverse, getting a suitable set of c's and d's – as small a set as possible, to minimize the computational burden – from a desired $\mathcal{H}(f)$. Entire books are devoted to this issue. Like many other "inverse problems," it has no all-purpose solution. One clearly has to make compromises, since $\mathcal{H}(f)$ is a full continuous function, while the short list of c's and d's represents only a few adjustable parameters. The subject of digital filter design concerns itself with the various ways of making these compromises. We cannot hope to give any sort of complete treatment of the subject. We can, however, sketch a couple of basic techniques to get you started. For further details, you will have to consult some specialized books (see references).

FIR (Nonrecursive) Filters

When the denominator in (12.9.2) is unity, the right-hand side is just a discrete Fourier transform. The transform is easily invertible, giving the desired small number of c_k coefficients in terms of the same small number of values of $\mathcal{H}(f_i)$ at some discrete frequencies f_i. This fact, however, is not very useful. The reason is that, for values of c_k computed in this way, $\mathcal{H}(f)$ will tend to oscillate wildly in between the discrete frequencies where it is pinned down to specific values.

A better strategy, and one which is the basis of several formal methods in the literature, is this: Start by pretending that you are willing to have a relatively large number of filter coefficients, that is, a relatively large value of M. Then $\mathcal{H}(f)$ can be fixed to desired values on a relatively fine mesh, and the M coefficients c_k, $k = 0, \ldots, M-1$ can be found by a FFT. Next, truncate (set to zero) most of the c_k's, leaving nonzero only the first, say, K, $(c_0, c_1, \ldots, c_{K-1})$ and last $K-1$, $(c_{M-K+1}, \ldots, c_{M-1})$. The last few c_k's are filter coefficients at *negative lag*, because of the wrap-around property of the FFT. But we don't want coefficients at negative lag. Therefore we cyclically shift the array of c_k's, to bring everything to positive lag. (This corresponds to introducing a time-delay into the filter.) Do this by copying the c_k's into a new array of length M in the following order:

$$(c_{M-K+1}, \ldots, c_{M-1},\ c_0,\ c_1, \ldots, c_{K-1},\ 0,\ 0, \ldots, 0) \tag{12.9.3}$$

To see if your truncation is acceptable, take the FFT of the array (12.9.3), giving an approximation to your original $\mathcal{H}(f)$. You will generally want to compare the *modulus* $|\mathcal{H}(f)|$ to your original function, since the time-delay will have introduced complex phases into the filter response.

If the new filter function is acceptable, then you are done and have a set of $2K-1$ filter coefficients. If it is not acceptable, then you can either (i) increase K and try again, or (ii) do something fancier to improve the acceptability for the same K. An example of something fancier is to modify the magnitudes (but not the phases) of the unacceptable $\mathcal{H}(f)$ to bring it more in line with your ideal, and then to FFT to get new c_k's. Once again set to zero all but the first $2K-1$ values of these (no need to cyclically shift since you have preserved the time-delaying phases), then inverse transform to get a new $\mathcal{H}(f)$, which will often be more acceptable. You can iterate this procedure. Note, however, that the procedure will not converge if your requirements for acceptability are more stringent than your $2K-1$ coefficients can handle.

The key idea, in other words, is to iterate between the space of coefficients and the space of functions $\mathcal{H}(f)$, until a Fourier conjugate pair that satisfies the imposed constraints *in both spaces* is found. A more formal technique for this kind of iteration is the *Remez Exchange Algorithm* which produces the best Chebyshev approximation to a given desired frequency response with a fixed number of filter coefficients.

IIR (Recursive) Filters

Recursive filters, whose output at a given time depends both on the current and previous inputs and on previous outputs, can generally have performance that is superior to nonrecursive filters with the same total number of coefficients (or same number of floating operations per input point). The reason is fairly clear by inspection of (12.9.2): A nonrecursive filter has a frequency response that is a polynomial in the variable $1/z$, where

$$z \equiv e^{2\pi i(f\Delta)} \tag{12.9.4}$$

By contrast, a recursive filter's frequency response is a *rational function* in $1/z$. The class of rational functions is especially good at fitting functions with sharp edges or narrow features, and most desired filter functions are in this category.

Nonrecursive filters are always stable. If you turn off the sequence of incoming x_i's, then after no more than M steps the sequence of y_j's produced by (12.9.1) will also turn off. Recursive filters, feeding as they do on their own output, are not necessarily stable. If the coefficients d_j are badly chosen, a recursive filter can have exponentially growing, so-called *homogeneous*, modes, which become huge even after the input sequence has been turned off. This is not good. The problem of designing recursive filters, therefore, is not just an inverse problem; it is an inverse problem with an additional stability constraint.

How do you tell if the filter (12.9.1) is stable for a given set of c_k and d_j coefficients? Stability depends only on the d_j's. The filter is stable if and only if all N complex roots of the *characteristic polynomial* equation

$$z^N - \sum_{j=1}^{N} d_j z^{N-j} = 0 \tag{12.9.5}$$

are inside the unit circle, i.e., satisfy

$$|z| \leq 1 \tag{12.9.6}$$

The various methods for constructing stable recursive filters again form a subject area for which you will need more specialized books. One very useful technique, however, is the *bilinear transformation method.* For this topic we define a new variable w which reparametrizes the frequency f,

$$w \equiv \tan[\pi(f\Delta)] = i\left(\frac{1 - e^{2\pi i(f\Delta)}}{1 + e^{2\pi i(f\Delta)}}\right) = i\left(\frac{1-z}{1+z}\right) \tag{12.9.7}$$

Don't be fooled by the i's in (12.9.7). This equation maps real frequencies f into real values of w. In fact, it maps the Nyquist interval $-\frac{1}{2} < f\Delta < \frac{1}{2}$ onto the real w axis $-\infty < w < +\infty$. The inverse equation to (12.9.7) is

$$z = e^{2\pi i(f\Delta)} = \frac{1 + iw}{1 - iw} \tag{12.9.8}$$

In reparametrizing f, w also reparametrizes z, of course. Therefore, the condition for stability (12.9.5)–(12.9.6) can be rephrased in terms of w: If the filter response $\mathcal{H}(f)$ is written as a function of w, then the filter is stable if and only if the poles of the filter function (zeros of its denominator) are all in the upper half complex plane,

$$\text{Im}(w) \geq 0 \tag{12.9.9}$$

The idea of the bilinear transformation method is that instead of specifying your desired $\mathcal{H}(f)$, you specify only its desired modulus square, $|\mathcal{H}(f)|^2 = \mathcal{H}(f)\mathcal{H}(f)^* = \mathcal{H}(f)\mathcal{H}(-f)$. Pick this to be approximated by some rational function in w^2. Then find all the poles of this function in the w complex plane. Every pole in the lower half-plane will have a corresponding pole in the upper half-plane, by symmetry. The idea is to form a product only of the factors with good poles, ones in the upper half-plane. This product is your *stably realizable* $\mathcal{H}(f)$. Now substitute equation (12.9.7) to write the function

as a rational function in z, and compare with equation (12.9.2) to read off the c's and d's.

The procedure becomes more clear when we go through an example. Suppose we want to design a simple bandpass filter, whose lower cutoff frequency corresponds to a value $w = a$, and whose upper cutoff frequency corresponds to a value $w = b$, with a and b both positive numbers. A simple rational function which accomplishes this is

$$|\mathcal{H}(f)|^2 = \left(\frac{w^2}{w^2+a^2}\right)\left(\frac{b^2}{w^2+b^2}\right) \tag{12.9.10}$$

This function does not have a very sharp cutoff, but it is illustrative of the more general case. To obtain sharper edges, one could take the function (12.9.10) to some positive integer power, or, equivalently, run the data sequentially through some number of copies of the filter which we will obtain from (12.9.10).

The poles of (12.9.10) are evidently at $w = \pm ia$ and $w = \pm ib$. Therefore the stably realizable $\mathcal{H}(f)$ is

$$\mathcal{H}(f) = \left(\frac{w}{w-ia}\right)\left(\frac{ib}{w-ib}\right) = \frac{\left(\frac{1-z}{1+z}\right)b}{\left[\left(\frac{1-z}{1+z}\right)-a\right]\left[\left(\frac{1-z}{1+z}\right)-b\right]} \tag{12.9.11}$$

We put the i in the numerator of the second factor in order to end up with real-valued coefficients. If we multiply out all the denominators, (12.9.11) can be rewritten in the form

$$\mathcal{H}(f) = \frac{-\frac{b}{(1+a)(1+b)} + \frac{b}{(1+a)(1+b)}z^{-2}}{1 - \frac{(1+a)(1-b)+(1-a)(1+b)}{(1+a)(1+b)}z^{-1} + \frac{(1-a)(1-b)}{(1+a)(1+b)}z^{-2}} \tag{12.9.12}$$

from which one reads off the filter coefficients for equation (12.9.1),

$$c_0 = -\frac{b}{(1+a)(1+b)}$$
$$c_1 = 0$$
$$c_2 = \frac{b}{(1+a)(1+b)}$$

$$d_1 = \frac{(1+a)(1-b)+(1-a)(1+b)}{(1+a)(1+b)}$$
$$d_2 = -\frac{(1-a)(1-b)}{(1+a)(1+b)} \qquad (12.9.13)$$

This completes the design of the bandpass filter.

Sometimes you can figure out how to construct directly a rational function in w for $\mathcal{H}(f)$, rather than having to start with its modulus square. The function that you construct has to have its poles only in the upper half-plane, for stability. It should also have the property of going into its own complex conjugate if you substitute $-w$ for w, so that the filter coefficients will be real.

For example, here is a function for a notch filter, designed to remove only a narrow frequency band around some fiducial frequency $w = w_0$, where w_0 is a positive number,

$$\begin{aligned}\mathcal{H}(f) &= \left(\frac{w-w_0}{w-w_0-i\epsilon w_0}\right)\left(\frac{w+w_0}{w+w_0-i\epsilon w_0}\right)\\ &= \frac{w^2-w_0^2}{(w-i\epsilon w_0)^2-w_0^2}\end{aligned} \qquad (12.9.14)$$

In (12.9.14) the parameter ϵ is a small positive number which is the desired width of the notch, as a fraction of w_0. Going through the arithmetic of substituting z for w gives the filter coefficients

$$\begin{aligned}c_0 &= \frac{1+w_0^2}{(1+\epsilon w_0)^2+w_0^2}\\ c_1 &= -2\frac{1-w_0^2}{(1+\epsilon w_0)^2+w_0^2}\\ c_2 &= \frac{1+w_0^2}{(1+\epsilon w_0)^2+w_0^2}\\ d_1 &= 2\frac{1-\epsilon^2 w_0^2-w_0^2}{(1+\epsilon w_0)^2+w_0^2}\\ d_2 &= -\frac{(1-\epsilon w_0)^2+w_0^2}{(1+\epsilon w_0)^2+w_0^2}\end{aligned} \qquad (12.9.15)$$

Figure 12.9.1 shows the results of using a filter of the form (12.9.15) on a "chirp" input signal, one which glides upwards in frequency, crossing the notch frequency along the way.

While the bilinear transformation may seem very general, its applications are limited by some features of the resulting filters. The method is good at getting the general shape of the desired filter, and good where "flatness"

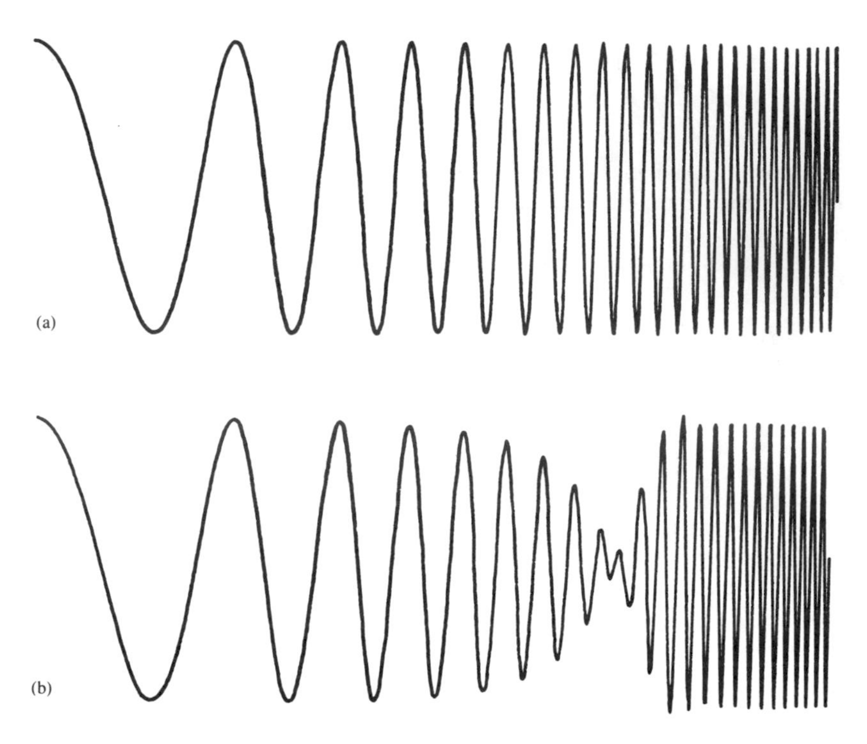

Figure 12.9.1. (a) A "chirp", or signal whose frequency increases continuously with time. (b) Same signal after it has passed through the notch filter (12.9.15). The parameter ϵ is here 0.2.

a desired goal. However, the nonlinear mapping between w and f makes it difficult to design to a desired shape for a cutoff, and may move cutoff frequencies (defined by a certain number of dB) from their desired places. Consequently, practitioners of the art of digital filter design reserve the bilinear transformation for specific situations, and arm themselves with a variety of other tricks. We suggest that you do likewise, as your projects demand.

REFERENCES AND FURTHER READING:

Hamming, R.W. 1977, *Digital Filters* (Englewood Cliffs, N.J.: Prentice Hall).

Oppenheim, A.V., and Schafer, R.W. 1975, *Digital Signal Processing* (Englewood Cliffs, N.J.: Prentice Hall).

Rice, J.R. 1964, *The Approximation of Functions* (Reading, Mass.: Addison-Wesley); also 1969, *op. cit.*, Vol. 2.

Rabiner, L.R., and Gold, B., 1975, *Theory and Application of Digital Signal Processing* (Englewood Cliffs, N.J.: Prentice Hall).

12.10 Linear Prediction and Linear Predictive Coding

The material in this section is closely related to that of §12.8 (maximum entropy method) and §12.9 (IIR filters). Review those sections before continuing.

Consider the special case of equation (12.9.1) with $M = 0$,

$$y_n = \sum_{j=1}^{N} d_j \, y_{n-j} + x_n \tag{12.10.1}$$

We can give this equation a somewhat different interpretation from that of the previous section: Equation (12.10.1) is an equation for *predicting* the next value y_n of a time series from the previous N values y_{n-j}, $j = 1 \dots N$. In this interpretation, x_n is the *discrepancy* of the prediction at timestep n, i.e. the amount which must be added to the predicted value (the sum) to give the true value y_n. If the predicted values are themselves pretty good, then the corrections will, on the average, be small. That is

$$\sum_n |x_n|^2 \ll \sum_n |y_n|^2 \tag{12.10.2}$$

In fact, our ideal is to have $|x_n| \ll |y_n|$ for all n.

To make use of (12.10.1) we need to have some way of finding good *linear prediction (LP) coefficients* $d_1, \dots, d_N$. We will return to this below. Suppose, for now, that we have the LP coefficients. What can we do with them?

Linear Prediction as an Extrapolation Technique

The most straightforward use of the coefficients is to predict the future of a time series from a record of its past. We first apply (12.10.1) to the known record to get an idea of how large are the discrepancies x_i. If (12.10.2) is in fact well satisfied, then we can continue applying (12.10.1) right on into the future, imagining the unknown "future" discrepancies x_i to be zero. In this application, (12.10.1) is a kind of extrapolation formula. In many situations, this extrapolation turns out to be vastly more powerful than any kind of simple polynomial extrapolation. (By the way, you should not confuse the terms "linear prediction" and "linear extrapolation"; the general functional form used by linear prediction is *much* more complex than a straight line, or even a low-order polynomial!)

Linear prediction is especially successful at extrapolating signals which are smooth and oscillatory, though not necessarily periodic. In such cases, linear prediction often extrapolates accurately through *many cycles* of the signal. By contrast, polynomial extrapolation in general becomes seriously

inaccurate after at most a cycle or two. A prototypical example of a signal that can successfully be linearly predicted is the height of ocean tides, for which the fundamental 12-hour period is modulated in phase and amplitude over the course of the month and year, and for which local hydrodynamic effects may make even one cycle of the curve look rather different in shape from a sine wave.

If you want to know, mathematically, what characterizes the class of signals for which linear prediction will be successful [i.e. what class of functions can be produced by equation (12.10.1) with the x_i's set to zero and with N arbitrary starting values y_i], then you have only to go back and compare equations (12.8.4) and (12.9.2). Identifying $P(f)$, the all-poles or maximum entropy estimate of a power spectrum, with $|\mathcal{H}(f)|^2$, the modulus square of the filter function of an IIR filter with $M = 0$ (an "all-poles filter"), we can begin to see what linear prediction is doing for us. It is generating some particular realization of a function whose power spectrum is defined by N conjugate pairs of poles in the complex z-plane. Question: *Which* realization? That is, what amplitudes and phases are associated with each pole? Answer: The unique one which agrees with the N initial values of y_i that (12.10.1) starts with.

Linear prediction (LP) and maximum entropy method spectral estimation (MEM) are thus two sides of the same coin: MEM *characterizes* a known signal in terms of a finite number of poles that best represent its spectrum in the complex z-plane. LP extrapolates the signal *using* its characterization in terms of these same poles. You should now not be surprised to learn, therefore, that we already have a routine for obtaining LP coefficients from a data set. The routine is `MEMCOF` in §12.8. Before you use `MEMCOF` in this application there is one more issue to face, and that is *stability.* The condition that (12.10.1) be stable as a linear predictor is precisely that given in equations (12.9.5) and (12.9.6), namely that the characteristic polynomial

$$z^N - \sum_{j=1}^{N} d_j z^{N-j} = 0 \tag{12.10.3}$$

have all N of its roots inside the unit circle,

$$|z| \le 1 \tag{12.10.4}$$

There is no guarantee that the coefficients produced by `MEMCOF` will have this property. If the signal data supplied to `MEMCOF` contain many oscillations without any particular trend towards increasing or decreasing amplitude, then the complex roots of (12.10.3) will generally all be rather close to the unit circle. The finite length of the data set will cause some of these roots to be inside the unit circle, others outside. In some applications, where the resulting instabilities are slowly growing and the linear prediction is not pushed too far, it is best to use the "unmassaged" LP coefficients that come directly out of `MEMCOF`. For example, one might be extrapolating to fill a short gap in a data

set; then one might extrapolate both forwards across the gap and backwards from the data beyond the gap. If the two extrapolations agree tolerably well, then instability is not a problem.

When instability *is* a problem, you have to "massage" the LP coefficients. You do this by (i) solving (numerically) equation (12.10.3) for its N complex roots; (ii) moving the roots to where you think they ought to be inside or on the unit circle; (iii) reconstituting the now-modified LP coefficients. You may think that step (ii) sounds a little vague. It is. There is no "best" procedure. If you think that your signal is truly a sum of undamped sine and cosine waves (perhaps with incommensurate periods), then you will want simply to move each root z_i onto the unit circle,

$$z_i \rightarrow z_i / |z_i| \tag{12.10.5}$$

In other circumstances it may seem appropriate to reflect a bad root across the unit circle

$$z_i \rightarrow 1/{z_i}^* \tag{12.10.6}$$

This alternative has the property that it preserves the amplitude of the output of (12.10.1) when it is driven by a sinusoidal set of x_i's. It assumes that MEMCOF has correctly identified the spectral width of a resonance, but only slipped up on identifying its time sense so that signals which should be damped as time proceeds end up growing in amplitude. The choice between (12.10.5) and (12.10.6) sometimes might as well be based on voodoo. We prefer (12.10.6).

Also magical is the choice of N, the number of poles. You should choose N to be as small as works for you, that is, you should choose it by experimenting with your data. Try $N = 5, 10, 20, 40$. If you need larger N's than this, be aware that the procedure of "massaging" all those complex roots is quite sensitive to roundoff error. Use double precision. We must also again note MEMCOF's susceptibility to roundoff error if the number of data points or poles is too large. No amount of massaging will fix roots that are simply wrong.

Here, then, are procedures for estimating the LP coefficients of a data set; for rendering them stable (if you choose to do so); and for extrapolating the data set by linear prediction, using the original or massaged LP coefficients. Two previously given routines are used: MEMCOF (§12.8) and ZROOTS (§9.5, finds all complex roots of a polynomial).

```
DIMENSION DATA(NDATA),D(NPOLES),WK1(NDATA),WK2(NDATA),WKM(NPOLES)
...
CALL MEMCOF(DATA,NDATA,NPOLES,DUM,D,WK1,WK2,WKM)
```

This fragment shows how to call MEMCOF to obtain linear prediction coefficients D(I), I=1...NPOLES from the data set DATA(J), J=1...NDATA. Input to the call is DATA, NDATA, and NPOLES. Output is D. Note the lengths of the required work spaces, and that a value DUM is returned that has no use in this application.

```
      SUBROUTINE FIXRTS(D,NPOLES)
```
Given the LP coefficients D(J), J=1...NPOLES, this routine finds all roots of the characteristic polynomial (12.10.3), reflects any roots that are outside the unit circle back inside, and then returns a modified set of D(J)'s. The routine ZROOTS of §9.5 is referenced.
```
      PARAMETER (NPMAX=100)              Largest expected value of NPOLES.
      DIMENSION D(NPOLES)
      LOGICAL POLISH
      COMPLEX A(NPMAX),ROOTS(NPMAX)
      A(NPOLES+1)=CMPLX(1.,0.)
      DO 11 J=NPOLES,1,-1                Set up complex coefficients for polynomial root finder.
        A(J)=CMPLX(-D(NPOLES+1-J),0.)
11    CONTINUE
      POLISH=.TRUE.
      CALL ZROOTS(A,NPOLES,ROOTS,POLISH)       Find all the roots.
      DO 12 J=1,NPOLES                   Look for a...
        IF(CABS(ROOTS(J)).GT.1.)THEN           root outside the unit circle,
          ROOTS(J)=1./CONJG(ROOTS(J))          and reflect it back inside.
        ENDIF
12    CONTINUE
      A(1)=-ROOTS(1)                     Now reconstruct the polynomial coefficients,
      A(2)=CMPLX(1.,0.)
      DO 14 J=2,NPOLES                   by looping over the roots
        A(J+1)=CMPLX(1.,0.)
        DO 13 I=J,2,-1                   and synthetically multiplying.
          A(I)=A(I-1)-ROOTS(J)*A(I)
13      CONTINUE
        A(1)=-ROOTS(J)*A(1)
14    CONTINUE
      DO 15 J=1,NPOLES                   The polynomial coefficients are guaranteed to be real,
        D(NPOLES+1-J)=-REAL(A(J))          so we need only return the real part as new LP coefficients.
15    CONTINUE
      RETURN
      END
```

```
      SUBROUTINE PREDIC(DATA,NDATA,D,NPOLES,FUTURE,NFUT)
```
Given DATA(J), J=1...NDATA, and given the data's LP coefficients D(I), I=1...NPOLES, this routine applies equation (12.10.1) to predict the next NFUT data points, which it returns in the array FUTURE. Note that the routine references only the last NPOLES values of DATA, as initial values for the prediction.
```
      PARAMETER (NPMAX=100)              Largest expected value of NPOLES.
      DIMENSION DATA(NDATA),D(NPOLES),FUTURE(NFUT),REG(NPMAX)
      DO 11 J=1,NPOLES
        REG(J)=DATA(NDATA+1-J)
11    CONTINUE
      DO 14 J=1,NFUT
        DISCRP=0.                        This is where you would put in a known discrepancy if you were
        SUM=DISCRP                           reconstructing a function by linear predictive coding rather
        DO 12 K=1,NPOLES                     than extrapolating a function by linear prediction. See text
          SUM=SUM+D(K)*REG(K)                below.
12      CONTINUE
        DO 13 K=NPOLES,2,-1              [If you know how to implement circular arrays, you can avoid this
          REG(K)=REG(K-1)                    shifting of coefficients!]
13      CONTINUE
        REG(1)=SUM
        FUTURE(J)=SUM
14    CONTINUE
      RETURN
      END
```

Linear Predictive Coding (LPC)

A different, though related, method to which the formalism above can be applied is the "compression" of a sampled signal so that it can be stored more compactly. The original form should be *exactly* recoverable from the compressed version. Obviously, compression can only be accomplished if there is redundancy in the signal. Equation (12.10.2) describes one kind of redundancy: it says that the signal, except for a small discrepancy, is predictable from its previous values and from a small number of LP coefficients. Compression of a signal by the use of (12.10.1) is thus called *linear predictive coding* or *LPC*.

The basic idea of LPC (in its simplest form) is to record as a compressed file (i) the number of LP coefficients N, (ii) their N values, e.g. as obtained by MEMCOF, (iii) the first N data points, and then (iv) for each subsequent data point only its residual discrepancy x_i (equation 12.10.1). When you are creating the compressed file, you find the residual by applying (12.10.1) to the previous N points, subtracting the sum from the actual value of the current point. When you are reconstructing the original file, you add the residual back in, at the point indicated in the routine PREDIC.

It may not be obvious why there is any compression at all in this scheme. After all, we are storing one value of residual per data point! Why not just store the original data point? The answer depends on the relative sizes of the numbers involved. The residual is obtained by subtracting two very nearly equal numbers (the data and the linear prediction). Therefore, the discrepancy typically has only a very small number of nonzero bits. These can be stored in a compressed file. How do you do it in a high level language? Here is one way: Scale your data to have integer values, say between $+1000000$ and -1000000 (supposing that you need six significant figures). Modify equation (12.10.1) by enclosing the sum term in an "integer part of" operator. The discrepancy will now, by definition, be an integer. Experiment with different values of N, to find LP coefficients that make the range of the discrepancy as small as you can. If you can get to within a range of ± 127 (and in our experience this is not at all difficult) then you can write it to a file as a single byte. This is a compression factor of 4, compared to 4-byte integer or floating formats.

Notice that the LP coefficients are computed using the *quantized* data, and that the discrepency is also quantized, i.e. quantization is done both outside and inside the LPC loop. If you are careful in following this prescription, then, apart from the initial quantization of the data, you will not introduce even a single bit of roundoff error into the compression-reconstruction process: While the evaluation of the sum in (12.10.1) may have roundoff errors, the residual that you store is the value which, when added back to the sum, gives *exactly* the original (quantized) data value. Notice also that you do not need to massage the LP coefficients for stability; by adding the residual back in to each point, you never depart from the original data, so instabilities cannot grow. There is therefore no need for FIXRTS, above.

This is not the place to try to tell you about *Huffman coding*, which will further compress the residuals by taking advantage of the fact that smaller

values of discrepancy will occur more often than larger values. Consult your local computer scientist, or visit your library if you find the idea intriguing. A very primitive version of Huffman coding would be this: If most of the discrepancies are in the range ± 127, but an occasional one is outside, then reserve the value 127 to mean "out of range," and then record on the file (immediately following the 127) a full-word value of the out-of-range discrepancy.

There are many variant procedures which all fall under the rubric of LPC.

- If the spectral character of the data is time-variable, then it is best not to use a single set of LP coefficients for the whole data set, but rather to partition the data into segments, computing and storing different LP coefficients for each segment.
- If the data are really well characterized by its LP coefficients, and you can tolerate some small amount of error, then don't bother storing all of the residuals. Just do linear prediction until you are outside of tolerances, then reinitialize (using N sequential stored residuals) and continue predicting.
- In some applications, most notably speech synthesis, one cares only about the spectral content of the reconstructed signal, not the relative phases. In this case, one need not store any starting values at all, but only the LP coefficients for each segment of the data. The output is reconstructed by driving these coefficients with initial conditions consisting of all zeros except for one non-zero spike. A speech synthesizer chip may have of order 10 LP coefficients, which change perhaps 20 to 50 times per second.
- Some people believe that it is interesting to analyze a signal by LPC, even when the residuals x_i are *not* small. The x_i's are then interpreted as the underlying "input signal" which, when filtered through the all-poles filter defined by the LP coefficients, produces the observed "output signal." LPC reveals simultaneously, it is said, the nature of the filter *and* the particular input that is driving it. We are skeptical of these applications; the literature, however, is full of extravagant claims.

REFERENCES AND FURTHER READING:

Childers, Donald G. (ed.). 1978, *Modern Spectrum Analysis* (New York: IEEE Press), especially the paper by J. Makhoul (reprinted from *Proceedings of the IEEE*, vol. 63, p. 561, 1975).

12.11 FFT in Two or More Dimensions

Given a complex function $h(k_1, k_2)$ defined over the two-dimensional grid $0 \le k_1 \le N_1 - 1$, $0 \le k_2 \le N_2 - 1$, we can define its two-dimensional discrete

Fourier transform as a complex function $H(n_1, n_2)$, defined over the same grid,

$$H(n_1, n_2) \equiv \sum_{k_2=0}^{N_2-1} \sum_{k_1=0}^{N_1-1} \exp(2\pi i k_2 n_2/N_2)\ \exp(2\pi i k_1 n_1/N_1)\ h(k_1, k_2) \tag{12.11.1}$$

By pulling the "subscripts 2" exponential outside of the sum over k_1, or by reversing the order of summation and pulling the "subscripts 1" outside of the sum over k_2, we can see instantly that the two-dimensional FFT can be computed by taking one-dimensional FFTs sequentially on each index of the original function. Symbolically,

$$\begin{aligned} H(n_1, n_2) &= \text{FFT-on-index-1}\,(\text{FFT-on-index-2}\,[h(k_1, k_2)]) \\ &= \text{FFT-on-index-2}\,(\text{FFT-on-index-1}\,[h(k_1, k_2)]) \end{aligned} \tag{12.11.2}$$

For this to be practical, of course, both N_1 and N_2 should be some efficient length for an FFT, usually a power of 2. Programming a two-dimensional FFT, using (12.11.2) with a one-dimensional FFT routine, is a bit clumsier than it seems at first. Because the one-dimensional routine requires that its input be in consecutive order as a one-dimensional complex array, you find that you are endlessly copying things out of the multidimensional input array and then copying things back into it. This is not recommended technique. Rather, you should use a multidimensional FFT routine, such as the one we give below.

The generalization of (12.11.1) to more than two dimensions, say to L-dimensions, is evidently

$$\begin{aligned} H(n_1, \ldots, n_L) \equiv \sum_{k_L=0}^{N_L-1} \cdots \sum_{k_1=0}^{N_1-1} & \exp(2\pi i k_L n_L/N_L) \times \cdots \\ & \times \exp(2\pi i k_1 n_1/N_1)\ h(k_1, \ldots, k_L) \end{aligned} \tag{12.11.3}$$

where n_1 and k_1 range from 0 to $N_1 - 1$, ... , n_L and k_L range from 0 to $N_L - 1$. How many calls to a one-dimensional FFT are in (12.11.3)? Quite a few! For each value of $k_1, k_2, \ldots, k_{L-1}$ you FFT to transform the L index. Then for each value of $k_1, k_2, \ldots, k_{L-2}$ and n_L you FFT to transform the $L - 1$ index. And so on. It is best to rely on someone else having done the bookkeeping for once and for all.

The inverse transforms of (12.11.1) or (12.11.3) are just what you would expect them to be: Change the i's in the exponentials to $-i$'s, and put an overall factor of $1/(N_1 \times \cdots \times N_L)$ in front of the whole thing. Most other features of multidimensional FFTs are also analogous to features already discussed in the one-dimensional case:

- Frequencies are arranged in wrap-around order in the transform, but now for each separate dimension.
- The input data are also treated as if they were wrapped around. If they are discontinuous across this periodic identification (in any dimension) then the spectrum will have some excess power at high frequencies due to the discontinuity. The fix, if you care, is to remove multidimensional linear trends.
- If you are doing spatial filtering and are worried about wrap-around effects, then you need to zero-pad all around the border of the multidimensional array. However, be sure to notice how costly zero-padding is in multidimensional transforms. If you use too thick a zero-pad, you are going to waste a *lot* of storage, especially in 3 or more dimensions!
- Aliasing occurs as always if sufficient bandwidth limiting does not exist along one or more of the dimensions of the transform.

The routine FOURN that we furnish herewith is a descendant of one written by N.M. Brenner of Lincoln Laboratories. It requires as input (i) a scalar, telling the number of dimensions, e.g. 2; (ii) a vector, telling the length of the array in each dimension, e.g. (32,64). Note that these lengths *must all* be powers of 2, and are the numbers of *complex* values in each direction; (iii) the usual scalar equal to ± 1 indicating whether you want the transform or its inverse; and, finally (iv) the array of data.

A few words about the data array: FOURN accesses it as a one-dimensional array of real numbers, of length equal to twice the product of the lengths of the L dimensions. It assumes that the array represents an L-dimensional complex array, in normal FORTRAN order. Normal FORTRAN order means: (i) each complex value occupies two sequential locations, real part followed by imaginary; (ii) the first subscript changes most rapidly as one goes through the array; the last subscript changes least rapidly; (iii) subscripts range from 1 to their maximum values ($N_1, N_2, \ldots, N_L$, respectively), rather than from 0 to $N_1 - 1$, $N_2 - 1, \ldots$. Almost all failures to get FOURN to work result from improper understanding of the normal FORTRAN ordering of the data array, so take care! (Figure 12.11.1 illustrates the format of the output array.)

```
SUBROUTINE FOURN(DATA,NN,NDIM,ISIGN)
    Replaces DATA by its NDIM-dimensional discrete Fourier transform, if ISIGN is input as 1.
    NN is an integer array of length NDIM, containing the lengths of each dimension (number of
    complex values), which MUST all be powers of 2. DATA is a real array of length twice the
    product of these lengths, in which the data are stored as in a multidimensional complex
    FORTRAN array. If ISIGN is input as -1, DATA is replaced by its inverse transform times
    the product of the lengths of all dimensions.
REAL*8 WR,WI,WPR,WPI,WTEMP,THETA      Double precision for trigonometric recurrences.
DIMENSION NN(NDIM),DATA(*)
NTOT=1
DO 11 IDIM=1,NDIM                     Compute total number of complex values.
    NTOT=NTOT*NN(IDIM)
 11 CONTINUE
NPREV=1
DO 18 IDIM=1,NDIM                     Main loop over the dimensions.
    N=NN(IDIM)
    NREM=NTOT/(N*NPREV)
```

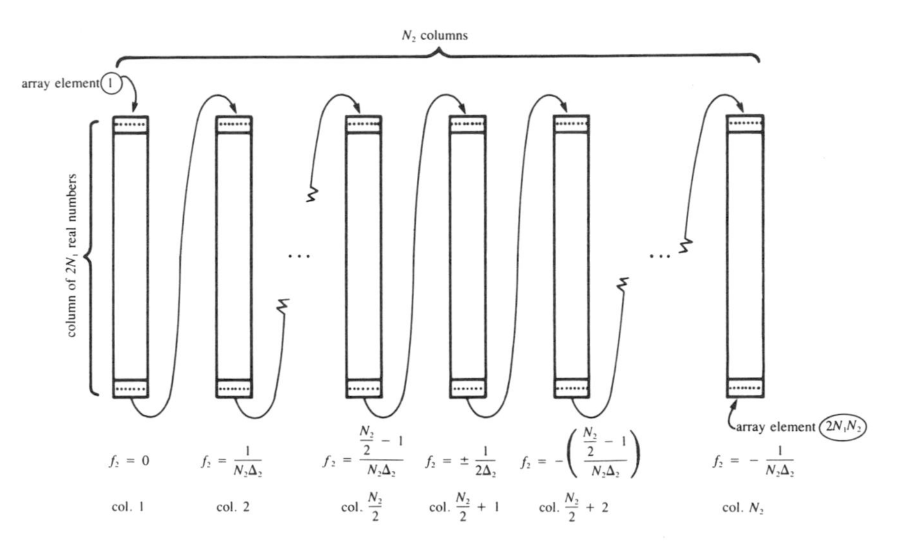

Figure 12.11.1. Storage arrangement of frequencies in the output $H(f_1, f_2)$ of a two-dimensional FFT. The input data is a two-dimensional $N_1 \times N_2$ array $h(t_1, t_2)$ (stored by columns of complex numbers). The output is also stored by complex columns. Each column corresponds to a particular value of f_2, as shown in the figure. Within each column, the arrangement of frequencies f_1 is exactly as shown in Figure 12.2.2. Δ_1 and Δ_2 are the sampling intervals in the 1 and 2 directions, respectively. The total number of (real) array elements is $2N_1N_2$. The program **FOURN** can also do more than two dimensions, and the storage arrangement generalizes in the obvious way.

```
          IP1=2*NPREV
          IP2=IP1*N
          IP3=IP2*NREM
          I2REV=1
          DO 14 I2=1,IP2,IP1           This is the bit reversal section of the routine.
            IF(I2.LT.I2REV)THEN
              DO 13 I1=I2,I2+IP1-2,2
                DO 12 I3=I1,IP3,IP2
                  I3REV=I2REV+I3-I2
                  TEMPR=DATA(I3)
                  TEMPI=DATA(I3+1)
                  DATA(I3)=DATA(I3REV)
                  DATA(I3+1)=DATA(I3REV+1)
                  DATA(I3REV)=TEMPR
                  DATA(I3REV+1)=TEMPI
12              CONTINUE
13            CONTINUE
            ENDIF
            IBIT=IP2/2
1           IF ((IBIT.GE.IP1).AND.(I2REV.GT.IBIT)) THEN
              I2REV=I2REV-IBIT
              IBIT=IBIT/2
            GO TO 1
            ENDIF
            I2REV=I2REV+IBIT
14        CONTINUE
          IFP1=IP1                     Here begins the Danielson-Lanczos section of the routine.
2         IF(IFP1.LT.IP2)THEN
            IFP2=2*IFP1
```

```
        THETA=ISIGN*6.28318530717959D0/(IFP2/IP1)   Initialize for the trig. recurrence.
        WPR=-2.D0*DSIN(0.5D0*THETA)**2
        WPI=DSIN(THETA)
        WR=1.D0
        WI=0.D0
        DO 17 I3=1,IFP1,IP1
          DO 16 I1=I3,I3+IP1-2,2
            DO 15 I2=I1,IP3,IFP2
              K1=I2                              Danielson-Lanczos formula:
              K2=K1+IFP1
              TEMPR=SNGL(WR)*DATA(K2)-SNGL(WI)*DATA(K2+1)
              TEMPI=SNGL(WR)*DATA(K2+1)+SNGL(WI)*DATA(K2)
              DATA(K2)=DATA(K1)-TEMPR
              DATA(K2+1)=DATA(K1+1)-TEMPI
              DATA(K1)=DATA(K1)+TEMPR
              DATA(K1+1)=DATA(K1+1)+TEMPI
15          CONTINUE
16        CONTINUE
          WTEMP=WR                               Trigonometric recurrence.
          WR=WR*WPR-WI*WPI+WR
          WI=WI*WPR+WTEMP*WPI+WI
17      CONTINUE
        IFP1=IFP2
      GO TO 2
      ENDIF
      NPREV=N*NPREV
18  CONTINUE
RETURN
END
```

REFERENCES AND FURTHER READING:

Nussbaumer, H.J. 1982, *Fast Fourier Transform and Convolution Algorithms* (New York: Springer-Verlag).

Chapter 13. Statistical Description of Data

13.0 Introduction

In this chapter and the next, the concept of *data* enters the discussion more prominently than before.

Data consist of numbers, of course. But these numbers are fed into the computer, not produced by it. These are numbers to be treated with considerable respect, never to be tampered with, nor subjected to a numerical process whose character you do not completely understand. You are well advised to acquire a reverence for data that is rather different from the "sporty" attitude which is sometimes allowable, or even commendable, in other numerical tasks.

The analysis of data inevitably involves some trafficking with the field of *statistics*, that gray area which is as surely not a branch of mathematics as it is neither a branch of science. In the following sections, you will repeatedly encounter the following paradigm:

- apply some formula to the data to compute "a statistic"
- compute where the value of that statistic falls in a probability distribution that is computed on the basis of some "null hypothesis"
- if it falls in a very unlikely spot, way out on a tail of the distribution, conclude that the null hypothesis is *false* for your data set

If a statistic falls in a *reasonable* part of the distribution, you must not make the mistake of concluding that the null hypothesis is "verified" or "proved." That is the curse of statistics, that it can never prove things, only disprove them! At best, you can substantiate a hypothesis by ruling out, statistically, a whole long list of competing hypotheses, every one that has ever been proposed. After a while your adversaries and competitors will give up trying to think of alternative hypotheses, or else they will grow old and die, and *then your hypothesis will become accepted.* Sounds crazy, we know, but that's how science works!

In this book we make a somewhat arbitrary distinction between data analysis procedures that are *model-independent* and those that are *model-dependent.* In the former category, we include so-called *descriptive statistics* that characterize a data set in general terms: its mean, variance, and so on. We also include statistical tests which seek to establish the "sameness" or "differentness" of two or more data sets, or which seek to establish and

measure a degree of *correlation* between two data sets. These subjects are discussed in this chapter.

In the other category, model-dependent statistics, we lump the whole subject of fitting data to a theory, parameter estimation, least-squares fits, and so on. Those subjects are introduced in Chapter 14.

Sections 13.1–13.3 of this chapter deal with so-called *measures of central tendency*, the moments of a distribution, the median and mode. In §13.4 we learn to test whether different data sets are drawn from distributions with different values of these measures of central tendency. This leads naturally, in §13.5 to the more general question of whether two distributions can be shown to be (significantly) different.

In §13.6–§13.8, we deal with *measures of association* for two distributions. We want to determine whether two variables are "correlated" or "dependent" on one another. If they are, we want to characterize the degree of correlation in some simple ways. The distinction between parametric and nonparametric (rank) methods is emphasized.

Section 13.9 touches briefly on the murky area of data smoothing.

This chapter draws mathematically on the material on special functions that was presented in Chapter 6, especially §6.1–§6.3. You may wish, at this point, to review those sections.

REFERENCES AND FURTHER READING:

Bevington, Philip R. 1969, *Data Reduction and Error Analysis for the Physical Sciences* (New York: McGraw-Hill).

Kendall, Maurice, and Stuart, Alan. 1977, *The Advanced Theory of Statistics*, 4th ed. (London: Griffin and Co.).

SPSS: Statistical Package for the Social Sciences, 2nd ed., by Norman H. Nie, et al. (New York: McGraw-Hill).

13.1 Moments of a Distribution: Mean, Variance, Skewness, and so forth

When a set of values has a sufficiently strong central tendency, that is, a tendency to cluster around some particular value, then it may be useful to characterize the set by a few numbers that are related to its *moments*, the sums of integer powers of the values.

Best known is the *mean* of the values $x_1, \ldots, x_N$,

$$\overline{x} = \frac{1}{N} \sum_{j=1}^{N} x_j \tag{13.1.1}$$

which estimates the value around which central clustering occurs. Note the use of an overbar to denote the mean; angle brackets are an equally common

notation, e.g. $\langle x \rangle$. You should be aware of the fact that the mean is not the only available estimator of this quantity, nor is it necessarily the best one. For values drawn from a probability distribution with very broad "tails," the mean may converge poorly, or not at all, as the number of sampled points is increased. Alternative estimators, the *median* and the *mode*, are discussed in §13.2 and §13.3.

Having characterized a distribution's central value, one conventionally next characterizes its "width" or "variability" around that value. Here again, more than one measure is available. Most common is the *variance*,

$$\mathrm{Var}(x_1 \ldots x_N) = \frac{1}{N-1} \sum_{j=1}^{N} (x_j - \overline{x})^2 \tag{13.1.2}$$

or its square root, the *standard deviation*,

$$\sigma(x_1 \ldots x_N) = \sqrt{\mathrm{Var}(x_1 \ldots x_N)} \tag{13.1.3}$$

Equation (13.1.2) estimates the mean squared-deviation of x from its mean value. There is a long story about why the denominator of (13.1.2) is $N-1$ instead of N. If you have never heard that story, you may consult any good statistics text. Here we will be content to note that the $N-1$ *should* be changed to N if you are ever in the situation of measuring the variance of a distribution whose mean $\overline{x}$ is known *a priori* rather than being estimated from the data. (We might also comment that if the difference between N and $N-1$ ever matters to you, then you are probably up to no good anyway – e.g., trying to substantiate a questionable hypothesis with marginal data.)

As the mean depends on the first moment of the data, so do the variance and standard deviation depend on the second moment. It is not uncommon, in real life, to be dealing with a distribution whose second moment does not exist (i.e. is infinite). In this case, the variance or standard deviation is useless as a measure of the data's width around its central value: the values obtained from equations (13.1.2) or (13.1.3) will not converge with increased numbers of points, nor show any consistency from data set to data set drawn from the same distribution. This can occur even when the width of the peak looks, by eye, perfectly finite. A more robust estimator of the width is the *average deviation* or *mean absolute deviation*, defined by

$$\mathrm{ADev}(x_1 \ldots x_N) = \frac{1}{N} \sum_{j=1}^{N} |x_j - \overline{x}| \tag{13.1.4}$$

Statisticians have historically sniffed at the use of (13.1.4) instead of (13.1.2), since the absolute value brackets in (13.1.4) are "nonanalytic" and make theorem-proving difficult. In recent years, however, the fashion has changed, and the subject of *robust estimation* (meaning, estimation for broad

distributions with significant numbers of "outlier" points) has become a popular and important one. Higher moments, or statistics involving higher powers of the input data, are almost always less robust than lower moments or statistics that involve only linear sums or (the lowest moment of all) counting.

That being the case, the *skewness* or *third moment*, and the *kurtosis* or *fourth moment* should be used with caution or, better yet, not at all.

The skewness characterizes the degree of asymmetry of a distribution around its mean. While the mean, standard deviation, and average deviation are *dimensional* quantities, that is, have the same units as the measured quantities x_j, the skewness is conventionally defined in such a way as to make it *nondimensional*. It is a pure number that characterizes only the shape of the distribution. The usual definition is

$$\text{Skew}(x_1 \ldots x_N) = \frac{1}{N} \sum_{j=1}^{N} \left[\frac{x_j - \overline{x}}{\sigma} \right]^3 \tag{13.1.5}$$

where $\sigma = \sigma(x_1 \ldots x_N)$ is the distribution's standard deviation (13.1.3). A positive value of skewness signifies a distribution with an asymmetric tail extending out towards more positive x; a negative value signifies a distribution whose tail extends out towards more negative x (see Figure 13.1.1).

Of course, any set of N measured values is likely to give a nonzero value for (13.1.5), even if the underlying distribution is in fact symmetrical (has zero skewness). For (13.1.5) to be meaningful, we need to have some idea of *its* standard deviation as an estimator of the skewness of the underlying distribution. Unfortunately, that depends on the shape of the underlying distribution, and rather critically on its tails! For the idealized case of a normal (Gaussian) distribution, the standard deviation of (13.1.5) is approximately $\sqrt{6/N}$. In real life it is good practice to believe in skewnesses only when they are several or many times as large as this.

The kurtosis is also a nondimensional quantity. It measures the relative peakedness or flatness of a distribution. Relative to what? A normal distribution, what else! A distribution with positive kurtosis is termed *leptokurtic*; the outline of the Matterhorn is an example. A distribution with negative kurtosis is termed *platykurtic*; the outline of a loaf of bread is an example. (See Figure 13.1.1.) And, you will no doubt be pleased to hear, an in-between distribution is termed *mesokurtic*.

The conventional definition of the kurtosis is

$$\text{Kurt}(x_1 \ldots x_N) = \left\{ \frac{1}{N} \sum_{j=1}^{N} \left[\frac{x_j - \overline{x}}{\sigma} \right]^4 \right\} - 3 \tag{13.1.6}$$

where the -3 term makes the value zero for a normal distribution.

The standard deviation of (13.1.6) as an estimator of the kurtosis of an underlying normal distribution is $\sqrt{24/N}$. However, the kurtosis depends on

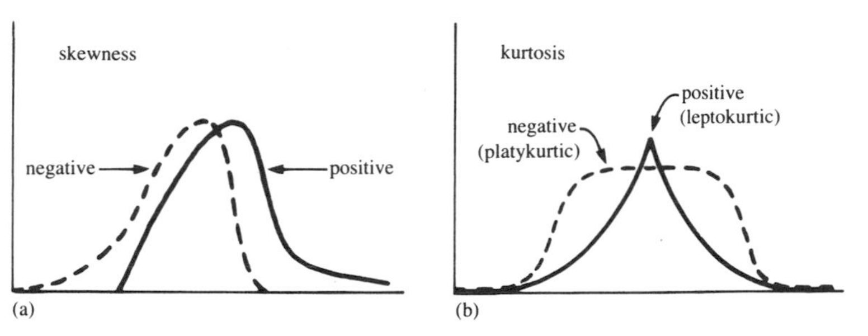

Figure 13.1.1. Distributions whose third and fourth moments are significantly different from a normal (Gaussian) distribution. (a) Skewness or third moment. (b) Kurtosis or fourth moment.

such a high moment that there are many real-life distributions for which the standard deviation of (13.1.6) as an estimator is effectively infinite.

Calculation of the quantities defined in this section is perfectly straightforward. Many textbooks use the binomial theorem to expand out the definitions into sums of various powers of the data, e.g. the familiar

$$\mathrm{Var}(x_1 \ldots x_N) = \frac{1}{N-1}\left[\left(\sum_{j=1}^{N} x_j^2\right) - N\overline{x}^2\right] \approx \overline{x^2} - \overline{x}^2 \qquad (13.1.7)$$

but this is generally unjustifiable in terms of computing speed and/or roundoff error.

```
SUBROUTINE MOMENT(DATA,N,AVE,ADEV,SDEV,VAR,SKEW,CURT)
    Given an array of DATA of length N, this routine returns its mean AVE, average deviation
    ADEV, standard deviation SDEV, variance VAR, skewness SKEW, and kurtosis CURT.
DIMENSION DATA(N)
IF(N.LE.1)PAUSE 'N must be at least 2'
S=0.                                First pass to get the mean.
DO 11 J=1,N
    S=S+DATA(J)
11 CONTINUE
AVE=S/N
ADEV=0.                             Second pass to get the first (absolute), second, third, and fourth
VAR=0.                                  moments of the deviation from the mean.
SKEW=0.
CURT=0.
DO 12 J=1,N
    S=DATA(J)-AVE
    ADEV=ADEV+ABS(S)
    P=S*S
    VAR=VAR+P
    P=P*S
    SKEW=SKEW+P
    P=P*S
    CURT=CURT+P
12 CONTINUE
ADEV=ADEV/N                         Put the pieces together according to the conventional definitions.
VAR=VAR/(N-1)
SDEV=SQRT(VAR)
IF(VAR.NE.0.)THEN
```

```
    SKEW=SKEW/(N*SDEV**3)
    CURT=CURT/(N*VAR**2)-3.
ELSE
    PAUSE 'no skew or kurtosis when zero variance'
ENDIF
RETURN
END
```

REFERENCES AND FURTHER READING:

Downie, N.M., and Heath, R.W. 1965, *Basic Statistical Methods*, 2nd ed. (New York: Harper and Row), Chapters 4 and 5.

Bevington, Philip R. 1969, *Data Reduction and Error Analysis for the Physical Sciences* (New York: McGraw-Hill), Chapter 2.

Kendall, Maurice, and Stuart, Alan. 1977, *The Advanced Theory of Statistics*, 4th ed. (London: Griffin and Co.), vol. 1, §10.15

SPSS: Statistical Package for the Social Sciences, 2nd ed., by Norman H. Nie, et al. (New York: McGraw-Hill), §14.1

13.2 Efficient Search for the Median

The median of a probability distribution function $p(x)$ is the value x_{med} for which larger and smaller values of x are equally probable:

$$\int_{-\infty}^{x_{med}} p(x)\,dx = \frac{1}{2} = \int_{x_{med}}^{\infty} p(x)\,dx \tag{13.2.1}$$

The median of a distribution is estimated from a sample of values $x_1, \ldots,$ x_N by finding that value x_i which has equal numbers of values above it and below it. Of course, this is not possible when N is even. In that case it is conventional to estimate the median as the mean of the unique *two* central values. If the values $x_j\ j = 1, \ldots, N$ are sorted into ascending (or, for that matter, descending) order, then the formula for the median is

$$\begin{aligned} x_{med} &= x_{(N+1)/2} && N \text{ odd} \\ &= \frac{1}{2}(x_{N/2} + x_{(N/2)+1}) && N \text{ even} \end{aligned} \tag{13.2.2}$$

If a distribution has a strong central tendency, so that most of its area is under a single peak, then the median is an estimator of the central value. It is a more robust estimator than the mean is: the median fails as an estimator only if the area in the tails is large, while the mean fails if the first moment of the tails is large; it is easy to construct examples where the first moment of the tails is large even though their area is negligible.

To find the median of a set of values, one can proceed by sorting the set and then applying (13.2.2). This is a process of order $N \log N$. You might think that this is wasteful, since it provides much more information than just the median (e.g. the upper and lower quartile points, the deciles, etc.). In fact, there are faster methods known for finding the median only, in (e.g.) order $N \log\log N$ or $N^{2/3} \log N$ operations. For details, see Knuth. As a practical matter, these differences in speed are not usually substantial enough to be important. All of the "combinatorial" methods share the same disadvantage: they rearrange all the data, either in place or else in additional scratch memory.

```
SUBROUTINE MDIAN1(X,N,XMED)
    Given an array X of N numbers, returns their median value XMED. The array X is modified
    and returned sorted into ascending order.
DIMENSION X(N)
CALL SORT(N,X)                This routine is in §8.2.
N2=N/2
IF(2*N2.EQ.N)THEN
    XMED=0.5*(X(N2)+X(N2+1))
ELSE
    XMED=X(N2+1)
ENDIF
RETURN
END
```

There are times, however, when it is quite inconvenient to rearrange all the data just to find the median value. For example, you might have a very long tape of values. While you may be willing to make a number of consecutive passes through the tape, sorting the whole thing is a task of considerably greater magnitude, requiring an additional external storage device, etc. The nice thing about the *mean* is that it can be computed in one pass through the data. Can we do anything like that for the median?

Yes, actually, if you are willing to make on the order of $\log N$ passes through the data. The method for doing so does not seem to be generally known, so we will explain it in some detail: The median (in the case of even N, say) satisfies the equation

$$\sum_{j=1}^{N} \frac{x_j - x_{med}}{|x_j - x_{med}|} = 0 \tag{13.2.3}$$

since each term in the sum is ± 1, depending on whether x_j is larger or smaller than the median value. Equation (13.2.3) can be rewritten as

$$x_{med} = \frac{\sum_{j=1}^{N} \frac{x_j}{|x_j - x_{med}|}}{\sum_{j=1}^{N} \frac{1}{|x_j - x_{med}|}} \tag{13.2.4}$$

which shows that the median is a kind of weighted average of the points, weighted by the reciprocal of their distance from the median.

Equation (13.2.4) is an implicit equation for x_{med}. It can be solved iteratively, using the previous guess on the right-hand side to compute the next guess. With this procedure (supplemented by a couple of tricks explained in comments below), convergence is obtained in of order $\log N$ passes through the data. The following routine pretends that the data array is in memory, but can easily be modified to make sequential passes through an external tape.

```
      SUBROUTINE MDIAN2(X,N,XMED)
         Given an array X of N numbers, returns their median value XMED. The array X is not
         modified, and is accessed sequentially in each consecutive pass.
      DIMENSION X(N)
      PARAMETER (BIG=1.E30,AFAC=1.5,AMP=1.5)
         Here, AMP is an overconvergence factor: on each iteration, we move the guess by this
         factor more than (13.2.4) would naively indicate. AFAC is a factor used to optimize the
         size of the "smoothing constant" EPS at each iteration.
      A=0.5*(X(1)+X(N))              This can be any first guess for the median.
      EPS=ABS(X(N)-X(1))             This can be any first guess for the characteristic spacing of the data
      AP=BIG                                points near the median.
      AM=-BIG                        AP and AM are upper and lower bounds on the median.
1     SUM=0.                         Here we start one pass through the data.
      SUMX=0.
      NP=0                           Number of points above the current guess,
      NM=0                               and below it.
      XP=BIG                         Value of the point above and closest to the guess,
      XM=-BIG                            and below and closest.
      DO 11 J=1,N                    Go throught the points,
        XX=X(J)
        IF(XX.NE.A)THEN              omit a zero denominator in the sums,
          IF(XX.GT.A)THEN            update the diagnostics,
            NP=NP+1
            IF(XX.LT.XP)XP=XX
          ELSE IF(XX.LT.A)THEN
            NM=NM+1
            IF(XX.GT.XM)XM=XX
          ENDIF
          DUM=1./(EPS+ABS(XX-A))          The smoothing constant is used here.
          SUM=SUM+DUM                accumulate the sums.
          SUMX=SUMX+XX*DUM
        ENDIF
11    CONTINUE
      IF(NP-NM.GE.2)THEN             Guess is too low; make another pass,
        AM=A                                 with a new lower bound,
        AA=XP+MAX(0.,SUMX/SUM-A)*AMP          a new best guess
        IF(AA.GT.AP)AA=0.5*(A+AP)             (but no larger than the upper bound)
        EPS=AFAC*ABS(AA-A)                    and a new smoothing factor.
        A=AA
        GO TO 1
      ELSE IF(NM-NP.GE.2)THEN        Guess is too high; make another pass,
        AP=A                                 with a new upper bound,
        AA=XM+MIN(0.,SUMX/SUM-A)*AMP          a new best guess
        IF(AA.LT.AM)AA=0.5*(A+AM)             (but no smaller than the lower bound)
        EPS=AFAC*ABS(AA-A)                    and a new smoothing factor.
        A=AA
        GO TO 1
      ELSE                           Got it!
        IF(MOD(N,2).EQ.0)THEN                For even N median is always an average.
          IF(NP.EQ.NM)THEN
            XMED=0.5*(XP+XM)
          ELSE IF(NP.GT.NM)THEN
```

```
            XMED=0.5*(A+XP)
        ELSE
            XMED=0.5*(XM+A)
        ENDIF
    ELSE                                    For odd N median is always one point.
        IF(NP.EQ.NM)THEN
            XMED=A
        ELSE IF(NP.GT.NM)THEN
            XMED=XP
        ELSE
            XMED=XM
        ENDIF
    ENDIF
ENDIF
RETURN
END
```

It is typical for MDIAN2 to take about 12 ± 2 passes for 10,000 points. and (presumably) about double this number for 100,000,000 points. One nice thing is that you can quit after a smaller number of passes if NP and NM are close enough for your purposes. For example, it will generally take only about 6 passes (independent of the number of points) for them to be within a percent or so of each other, putting the guess A in the 50 ± 0.5^{th} percentile, close enough for Government work. Note also the possibility of doing the "endgame" by (on one pass) copying to a relatively small array only those points between the bounds AP and AM, along with the counts of how many points are above and below these bounds.

REFERENCES AND FURTHER READING:

Knuth, Donald E. 1973, *Sorting and Searching*, vol. 3 of *The Art of Computer Programming* (Reading, Mass.: Addison-Wesley), pp. 216ff.

13.3 Estimation of the Mode for Continuous Data

The *mode* of a probability distribution function $p(x)$ is the value of x where it takes on a maximum value. The mode is useful primarily when there is a single, sharp maximum, in which case it estimates the central value. Occasionally, a distribution will be *bimodal*, with two relative maxima; then one may wish to know the two modes individually. Note that, in such cases, the mean and median are not very useful, since they may give only a "compromise" value between the two peaks.

Elementary statistics texts seem to have the notion that the only way to estimate the mode is to bin the data into a *histogram* or bar-graph, selecting as the estimated value the midpoint of the bin with the largest number of points. If the data are intrinsically integer-valued, then the binning is automatic, and in some sense "natural," so this procedure is not unreasonable.

Real-life data are more often not integer-valued, however, but real-valued. (True, data stored with finite precision can be viewed as integer valued; but the number of possible integers is a number so large, e.g. 2^{32}, that you will rarely have more than one data point per "integer" bin.) It is almost always a bad idea to bin continuous data, since the binning throws away information.

The right way to estimate the mode of a distribution from a set of sampled points goes by the name "estimating the rate of an inhomogeneous Poisson process by J^{th} waiting times":

- First, sort all the points x_j $j = 1, \ldots, N$ into ascending order.
- Second, select an integer J as your "window size," the resolution (measured in number of consecutive points) over which the distribution function will be smeared. Smaller J gives better resolution in x and the chance of finding a high, but very narrow peak; but smaller J gives poorer accuracy in finding the true maximum as opposed to a chance fluctuation in the data. We will discuss this below. J should never be smaller than 3 and should be as large as you can tolerate.
- Third, for every $i = 1, \ldots, N - J$, estimate $p(x)$ as follows:

$$p\left(\tfrac{1}{2}[x_i + x_{i+J}]\right) \approx \frac{J}{N(x_{i+J} - x_i)} \tag{13.3.1}$$

- Fourth, take the value $\frac{1}{2}(x_i + x_{i+J})$ of the largest of these estimates to be the estimated mode.

The standard deviation of (13.3.1) as an estimator of $p(x)$ is

$$\sigma\left[p\left(\tfrac{1}{2}[x_i + x_{i+J}]\right)\right] \approx \frac{\sqrt{J}}{N(x_{i+J} - x_i)} \tag{13.3.2}$$

In other words, the fractional error is on the order of $1/\sqrt{J}$. So, the selection between two possible candidates for the mode is not significant unless their p's differ by several times (13.3.2).

A more subtle problem is how to compare mode candidate values obtained with different values of J. This problem occurs in practice quite often: the data may have a narrow peak containing only a few points, which, for larger values of J, "washes out" in favor of a broader peak at some different value of x. We know of no rigorous solution to this problem, but we do know at least a high-brow heuristic technique:

Suppose we have some set of mode candidates with different J's. Let x_j be the value of the mode for the case $J = j$, and let p_j be the estimate of $p(x)$ at that point. Let Δx_j denote the range of the window around candidate x_j; for example, it might be $\Delta x_7 = x_{19} - x_{12}$, if the mode candidate with window length 7 occurred between points 12 and 19.

Let H_j denote the hypothesis "the true mode is at x_j, while all the other mode candidates are chance fluctuations of a distribution function no larger than the true mode." If H_j is true, then we need to know how unlikely each

of the other Δx_n's are, in other words, how likely is it that the range around x_n should be as short or shorter than observed if the true probability density is p_j rather than p_n? This is given by the gamma probability distribution previously introduced in §7.3,

$$\begin{aligned}\int_0^{\Delta x_n} (Np_j)\frac{(Np_j x)^{n-1}}{(n-1)!}e^{-Np_j x}dx \\ &= P(n, Np_j\Delta x_n) \\ &= P(n, \frac{np_j}{p_n}) \\ &= P(n, \frac{j\Delta x_n}{\Delta x_j})\end{aligned} \tag{13.3.3}$$

In this equation, $P(a, x)$ is the incomplete gamma function (see equation 6.2.1), and the last two equalities follow from equation (13.3.1).

The *likelihood* of hypothesis H_j is now the product of these probabilities over all the *other* n's, not including j:

$$\text{Likelihood}(H_j) = \prod_{n\neq j} P(n, \frac{j\Delta x_n}{\Delta x_j}) \tag{13.3.4}$$

You select the hypothesis H_j which has the the (so-called) *maximum likelihood.* This method is straightforward to program, using the incomplete gamma function routine `GAMMP` in §6.2.

REFERENCES AND FURTHER READING:

Parzen, Emanual, 1962, *Stochastic Processes* (San Francisco: Holden Day).

13.4 Do Two Distributions Have the Same Means or Variances?

Not uncommonly we want to know whether two distributions have the same mean. For example, a first set of measured values may have been gathered before some event, a second set after it. We want to know whether the event, a "treatment" or a "change in a control parameter," made a difference.

Our first thought is to ask "how many standard deviations" one sample mean is from the other. That number may in fact be a useful thing to know. It does relate to the strength or "importance" of a difference of means *if that difference is genuine.* However, by itself, it says nothing about whether the difference *is* genuine, that is, statistically significant. A difference of means

can be very small compared to the standard deviation, and yet very significant, if the number of data points is large. Conversely, a difference may be moderately large but not significant, if the data are sparse. We will be meeting these distinct concepts of *strength* and *significance* several times in the next few sections.

A quantity that measures the significance of a difference of means is not the number of standard deviations that they are apart, but the number of so-called *standard errors* that they are apart. The standard error of a set of values is their standard deviation divided by the square root of their number; the standard error estimates the standard deviation of the sample mean as an estimator of the population (or "true") mean.

Student's t-test for Significantly Different Means

Applying the concept of standard error, the conventional statistic for measuring the significance of a difference of means is termed *Student's t.* When the two distributions are thought to have the same variance, but possibly different means, then Student's t is computed as follows: First, estimate the standard error of the difference of the means, s_D, from the "pooled variance" by the formula

$$s_D = \sqrt{\frac{\sum_{one}(x_i - \overline{x_{one}})^2 + \sum_{two}(x_i - \overline{x_{two}})^2}{N_1 + N_2 - 2}\left(\frac{1}{N_1} + \frac{1}{N_2}\right)} \qquad (13.4.1)$$

where each sum is over the points in one sample, the first or second, each mean likewise refers to one sample or the other, and N_1 and N_2 are the numbers of points in the first and second samples respectively. Second, compute t by

$$t = \frac{\overline{x_{one}} - \overline{x_{two}}}{s_D} \qquad (13.4.2)$$

Third, evaluate the significance of this value of t for Student's distribution with $N_1 + N_2 - 2$ degrees of freedom, by equations (6.3.7) and (6.3.9), and by the routine `BETAI` (incomplete beta function) of §6.3.

The significance is a number between zero and one, and is the probability that $|t|$ could be this large or larger just by chance, for distributions with equal means. Therefore, a small numerical value of the significance (0.05 or 0.01) means that the observed difference is "very significant." The function $A(t|\nu)$ in equation (6.3.7) is one minus the significance.

As a routine, we have

```
      SUBROUTINE TTEST(DATA1,N1,DATA2,N2,T,PROB)
C     Given the arrays DATA1 of length N1 and DATA2 of length N2, this routine returns Student's
C     t as T, and its significance as PROB, small values of PROB indicating that the arrays have
C     significantly different means. The data arrays are assumed to be drawn from populations
C     with the same true variance.
      DIMENSION DATA1(N1),DATA2(N2)
      CALL AVEVAR(DATA1,N1,AVE1,VAR1)
      CALL AVEVAR(DATA2,N2,AVE2,VAR2)
      DF=N1+N2-2                            Degrees of freedom.
      VAR=((N1-1)*VAR1+(N2-1)*VAR2)/DF      Pooled variance.
      T=(AVE1-AVE2)/SQRT(VAR*(1./N1+1./N2))
      PROB=BETAI(0.5*DF,0.5,DF/(DF+T**2))   See equation (6.3.9).
      RETURN
      END
```

which makes use of the following routine for computing the mean and variance of a set of numbers,

```
      SUBROUTINE AVEVAR(DATA,N,AVE,VAR)
C     Given array DATA of length N, returns its mean as AVE and its variance as VAR.
      DIMENSION DATA(N)
      AVE=0.0
      VAR=0.0
      DO 11 J=1,N
        AVE=AVE+DATA(J)
11    CONTINUE
      AVE=AVE/N
      DO 12 J=1,N
        S=DATA(J)-AVE
        VAR=VAR+S*S
12    CONTINUE
      VAR=VAR/(N-1)
      RETURN
      END
```

The next case to consider is where the two distributions have significantly different variances, but we nevertheless want to know if their means are the same or different. (A treatment for baldness has caused some patients to *lose* all their hair and turned others into werewolves, but we want to know if it helps cure baldness *on the average*!) Be suspicious of the unequal-variance t-test: if two distributions have very different variances, then they may also be substantially different in shape; in that case, the difference of the means may not be a particularly useful thing to know.

To find out whether the two data sets have variances that are significantly different, you use the *F-test*, described later on in this section.

The relevant statistic for the unequal variance t-test is

$$t = \frac{\overline{x_{one}} - \overline{x_{two}}}{[\text{Var}(x_{one})/N_1 + \text{Var}(x_{two})/N_2]^{1/2}} \qquad (13.4.3)$$

This statistic is distributed *approximately* as Student's t with a number of degrees of freedom equal to

$$\frac{\left[\frac{\mathrm{Var}(x_{one})}{N_1}+\frac{\mathrm{Var}(x_{two})}{N_2}\right]^2}{\frac{[\mathrm{Var}(x_{one})/N_1]^2}{N_1-1}+\frac{[\mathrm{Var}(x_{two})/N_2]^2}{N_2-1}} \tag{13.4.4}$$

Expression (13.4.4) is in general not an integer, but equation (6.3.7) doesn't care.

The routine is

```
SUBROUTINE TUTEST(DATA1,N1,DATA2,N2,T,PROB)
    Given the arrays DATA1 of length N1 and DATA2 of length N2, this routine returns Student's
    t as T, and its significance as PROB, small values of PROB indicating that the arrays have
    significantly different means. The data arrays are allowed to be drawn from populations
    with unequal variances.
DIMENSION DATA1(N1),DATA2(N2)
CALL AVEVAR(DATA1,N1,AVE1,VAR1)
CALL AVEVAR(DATA2,N2,AVE2,VAR2)
T=(AVE1-AVE2)/SQRT(VAR1/N1+VAR2/N2)
DF=(VAR1/N1+VAR2/N2)**2/((VAR1/N1)**2/(N1-1)+(VAR2/N2)**2/(N2-1))
PROB=BETAI(0.5*DF,0.5,DF/(DF+T**2))
RETURN
END
```

Our final example of a Student's t test is the case of *paired samples*. Here we imagine that much of the variance in *both* samples is due to effects which are point-by-point identical in the two samples. For example, we might have two job candidates who have each been rated by the same ten members of a hiring committee. We want to know if the means of the ten scores differ significantly. We first try TTEST above, and obtain a value of PROB which is not especially significant (e.g., > 0.05). But perhaps the significance is being washed out by the tendency of some committee members always to give high scores, others always to give low scores, which increases the apparent variance and thus decreases the significance of any difference in the means. We thus try the paired-sample formulas,

$$\mathrm{Cov}(x_{one}, x_{two}) \equiv \frac{1}{N-1}\sum_{i=1}^{N}(x_{one\,i} - \overline{x_{one}})(x_{two\,i} - \overline{x_{two}}) \tag{13.4.5}$$

$$s_D = \sqrt{\frac{\mathrm{Var}(x_{one}) + \mathrm{Var}(x_{two}) - 2\mathrm{Cov}(x_{one}, x_{two})}{N}} \tag{13.4.6}$$

$$t = \frac{\overline{x_{one}} - \overline{x_{two}}}{s_D} \tag{13.4.7}$$

where N is the number in each sample (number of pairs). Notice that it is important that a particular value of i label the corresponding points in each sample, that is, the ones that are paired. The significance of the t statistic in (13.4.5) is evaluated for $N - 1$ degrees of freedom.

The routine is

```
SUBROUTINE TPTEST(DATA1,DATA2,N,T,PROB)
    Given the paired arrays DATA1 and DATA2, both of length N, this routine returns Student's
    t for paired data as T, and its significance as PROB, small values of PROB indicating a
    significant difference of means.
DIMENSION DATA1(N),DATA2(N)
CALL AVEVAR(DATA1,N,AVE1,VAR1)
CALL AVEVAR(DATA2,N,AVE2,VAR2)
COV=0.
DO 11 J=1,N
    COV=COV+(DATA1(J)-AVE1)*(DATA2(J)-AVE2)
11  CONTINUE
DF=N-1
COV=COV/DF
SD=SQRT((VAR1+VAR2-2.*COV)/N)
T=(AVE1-AVE2)/SD
PROB=BETAI(0.5*DF,0.5,DF/(DF+T**2))
RETURN
END
```

F-Test for Significantly Different Variances

The *F-test* tests the hypothesis that two samples have different variances by trying to reject the null hypothesis that their variances are actually consistent. The statistic F is the ratio of one variance to the other, so values either $\gg 1$ or $\ll 1$ will indicate very significant differences. The distribution of F in the null case is given in equation (6.3.11), which is evaluated using the routine BETAI. In the most common case, we are willing to disprove the null hypothesis (of equal variances) by either very large or very small values of F, so the correct significance is *two-tailed*, the sum of two incomplete beta functions. These considerations and equation(6.3.3) give the routine

```
SUBROUTINE FTEST(DATA1,N1,DATA2,N2,F,PROB)
    Given the arrays DATA1 of length N1 and DATA2 of length N2, this routine returns the value
    of F, and its significance as PROB. Small values of PROB indicate that the two arrays have
    significantly different variances.
DIMENSION DATA1(N1),DATA2(N2)
CALL AVEVAR(DATA1,N1,AVE1,VAR1)
CALL AVEVAR(DATA2,N2,AVE2,VAR2)
IF(VAR1.GT.VAR2)THEN          Make F the ratio of the larger variance to the smaller one.
    F=VAR1/VAR2
    DF1=N1-1
    DF2=N2-1
ELSE
    F=VAR2/VAR1
    DF1=N2-1
    DF2=N1-1
```

```
ENDIF
PROB = 2.*BETAI(0.5*DF2,0.5*DF1,DF2/(DF2+DF1*F))
IF(PROB.GT.1.)PROB=2.-PROB
RETURN
END
```

REFERENCES AND FURTHER READING:

von Mises, Richard. 1964, *Mathematical Theory of Probability and Statistics* (New York: Academic Press), Chapter IX(B).

SPSS: Statistical Package for the Social Sciences, 2nd ed., by Norman H. Nie, et al. (New York: McGraw-Hill), §17.2.

13.5 Are Two Distributions Different?

Given two sets of data, we can generalize the questions asked in the previous section and ask the single question: Are the two sets drawn from the same distribution function, or from different distribution functions? Equivalently, in proper statistical language, "Can we disprove, to a certain required level of significance, the null hypothesis that two data sets are drawn from the same population distribution function?" Disproving the null hypothesis in effect proves that the data sets are from different distributions. Failing to disprove the null hypothesis, on the other hand, only shows that the data sets can be *consistent* with a single distribution function. One can never *prove* that two data sets come from a single distribution, since (e.g.) no practical amount of data can distinguish between two distributions which differ only by one part in 10^{10}.

Proving that two distributions are different, or showing that they are consistent, is a task that comes up all the time in many areas of research: Are the visible stars distributed uniformly in the sky? (That is, is the distribution of stars as a function of declination — position in the sky — the same as the distribution of sky area as a function of declination?) Are educational patterns the same in Brooklyn as in the Bronx? (That is, are the distributions of people as a function of last-grade-attended the same?) Do two brands of fluorescent lights have the same distribution of burn-out times? Is the incidence of chicken pox the same for first-born, second-born, third-born children, etc.?

These four examples illustrate the four combinations arising from two different dichotomies: (1) Either the data is continuous or binned. (2) Either we wish to compare one data set to a known distribution, or we wish to compare two equally unknown data sets. The data sets on fluorescent lights and on stars are continuous, since we can be given lists of individual burnout times or of stellar positions. The data sets on chicken pox and educational level are binned, since we are given tables of numbers of events in discrete categories: first-born, second-born, etc.; or 6th Grade, 7th Grade, etc. Stars and chicken pox, on the other hand, share the property that the null hypothesis is

a known distribution (distribution of area in the sky, or incidence of chicken pox in the general population). Fluorescent lights and educational level involve the comparison of two equally unknown data sets (the two brands, or Brooklyn and the Bronx).

One can always turn continuous data into binned data, by grouping the events into specified ranges of the continuous variable(s): Declinations between 0 and 10 degrees, 10 and 20, 20 and 30, etc. Binning involves a loss of information, however. Also, there is often considerable arbitrariness as to how the bins should be chosen. Along with many other investigators, we prefer to avoid unnecessary binning of data.

The accepted test for differences between binned distributions is the *chi-square test.* For continuous data as a function of a single variable, the most generally accepted test is the *Kolmogorov-Smirnov test.* We consider each in turn.

Chi-Square Test

Suppose that N_i is the number of events observed in the ith bin, and that n_i is the number expected according to some known distribution. Note that the N_i's are integers, while the n_i's may not be. Then the chi-square statistic is

$$\chi^2 = \sum_i \frac{(N_i - n_i)^2}{n_i} \qquad (13.5.1)$$

where the sum is over all bins. A large value of χ^2 indicates that the null hypothesis (that the N_i's are drawn from the population represented by the n_i's) is rather unlikely.

Any term j in (13.5.1) with $0 = n_j = N_j$ should be omitted from the sum. A term with $n_j = 0,\ N_j \neq 0$ gives an infinite χ^2, as it should, since in this case the N_i's cannot possibly be drawn from the n_i's!

The *chi-square probability function* $Q(\chi^2|\nu)$ is an incomplete gamma function, and was already discussed in §6.2 (see equation 6.2.18). Strictly speaking $Q(\chi^2|\nu)$ is the probability that the sum of the squares of ν random *normal* variables of unit variance will be greater than χ^2. The terms in the sum (13.5.1) are not individually normal. However, if either the number of bins is large ($\gg 1$), or the number of events in each bin is large ($\gg 1$), then the chi-square probability function is a good approximation to the distribution of (13.5.1) in the case of the null hypothesis. Its use to estimate the significance of the chi-square test is standard.

The appropriate value of ν, the number of degrees of freedom, bears some additional discussion. If the data were collected in such a way that the n_i's were all determined in advance, and there were no *a priori* constraints on any of the N_i's, then ν equals the number of bins N_B (note that this is *not* the total number of *events*!). Much more commonly, the n_i's are normalized after the fact so that their sum equals the sum of the N_i's, the total number of events measured. In this case the correct value for ν is $N_B - 1$. If the

model that gives the n_i's had additional free parameters that were adjusted after the fact to agree with the data, then each of these additional "fitted" parameters reduces ν by one additional unit. The number of these additional fitted parameters (*not* including the normalization of the n_i's) is commonly called the "number of constraints," so the number of degrees of freedom is $\nu = N_B - 1$ when there are "zero constraints."

We have, then, the following program:

```
SUBROUTINE CHSONE(BINS,EBINS,NBINS,KNSTRN,DF,CHSQ,PROB)
    Given the array BINS of length NBINS, containing the observed numbers of events, and an
    array EBINS of length NBINS containing the expected numbers of events, and given the
    number of constraints KNSTRN (normally zero), this routine returns (trivially) the number
    of degrees of freedom DF, and (nontrivially) the chi-square CHSQ and the significance PROB.
    A small value of PROB indicates a significant difference between the distributions BINS
    and EBINS. Note that BINS and EBINS are both real arrays, although BINS will normally
    contain integer values.
DIMENSION BINS(NBINS),EBINS(NBINS)
DF=NBINS-1-KNSTRN
CHSQ=0.
DO 11 J=1,NBINS
    IF(EBINS(J).LE.0.)PAUSE 'bad expected number'
    CHSQ=CHSQ+(BINS(J)-EBINS(J))**2/EBINS(J)
11  CONTINUE
PROB=GAMMQ(0.5*DF,0.5*CHSQ)  Chi-square probability function. See §6.2.
RETURN
END
```

Next we consider the case of comparing *two* binned data sets. Let R_i be the number of events in bin i for the first data set, S_i the number of events in the same bin i for the second data set. Then the chi-square statistic is

$$\chi^2 = \sum_i \frac{(R_i - S_i)^2}{R_i + S_i} \tag{13.5.2}$$

Comparing (13.5.2) to (13.5.1), you should note that the denominator of (13.5.2) is *not* just the average of R_i and S_i (which would be an estimator of n_i in 13.5.1). Rather, it is twice the average, the sum. The reason is that each term in a chi-square sum is supposed to approximate the square of a normally distributed quantity with unit variance. The variance of the difference of two normal quantities is the sum of their individual variances, not the average.

If the data were collected in such a way that the sum of the R_i's is necessarily equal to the sum of S_i's, then the number of degrees of freedom is equal to one less than the number of bins, $N_B - 1$ (that is, KNSTRN=0), the usual case. If this requirement were absent, then the number of degrees of freedom would be N_B. Example: A birdwatcher wants to know whether the distribution of sighted birds as a function of species is the same this year as last. Each bin corresponds to one species. If the birdwatcher takes his data to be the first 1000 birds that he saw in each year, then the number of degrees of freedom is $N_B - 1$. If he takes his data to be all the birds he saw on a random sample of days, the same days in each year, then the number of degrees of

freedom is N_B (KNSTRN= -1). In this latter case, note that he is also testing whether the birds were more numerous overall in one year or the other: that is the extra degree of freedom. Of course, any additional constraints on the data set lower the number of degrees of freedom (i.e., increase KNSTRN to *positive* values) in accordance with their number.

The program is

```
SUBROUTINE CHSTWO(BINS1,BINS2,NBINS,KNSTRN,DF,CHSQ,PROB)
    Given the arrays BINS1 and BINS2, both of length NBINS, containing two sets of binned
    data, and given the number of additional constraints KNSTRN (normally 0 or -1), this rou-
    tine returns the number of degrees of freedom DF, the chi-square CHSQ and the significance
    PROB. A small value of PROB indicates a significant difference between the distributions
    BINS1 and BINS2. Note that BINS1 and BINS2 are both real arrays, although they will
    normally contain integer values.
DIMENSION BINS1(NBINS),BINS2(NBINS)
DF=NBINS-1-KNSTRN
CHSQ=0.
DO 11 J=1,NBINS
    IF(BINS1(J).EQ.0..AND.BINS2(J).EQ.0.)THEN
        DF=DF-1.                    No data means one less degree of freedom.
    ELSE
        CHSQ=CHSQ+(BINS1(J)-BINS2(J))**2/(BINS1(J)+BINS2(J))
    ENDIF
11  CONTINUE
PROB=GAMMQ(0.5*DF,0.5*CHSQ)  Chi-square probability function. See §6.2.
RETURN
END
```

Kolmogorov-Smirnov Test

The Kolmogorov-Smirnov (or *K–S*) test is applicable to unbinned distributions that are functions of a single independent variable, that is, to data sets where each data point can be associated with a single number (lifetime of each lightbulb when it burns out, or declination of each star). In such cases, the list of data points can be easily converted to an unbiased estimator $S_N(x)$ of the *cumulative* distribution function of the probability distribution from which it was drawn: If the N events are located at values x_i, $i = 1, \ldots, N$, then $S_N(x)$ is the function giving the fraction of data points to the left of a given value x. This function is obviously constant between consecutive (i.e. sorted into ascending order) x_i's, and jumps by the same constant $1/N$ at each x_i. (See Figure 13.5.1).

Different distribution functions, or sets of data, give different cumulative distribution function estimates by the above procedure. However, all cumulative distribution functions agree at the smallest allowable value of x (where they are zero), and at the largest allowable value of x (where they are unity). (The smallest and largest values might of course be $\pm\infty$.) So it is the behavior between the largest and smallest values that distinguishes distributions.

One can think of any number of statistics to measure the overall difference between two cumulative distribution functions: The absolute value of the area between them, for example. Or their integrated mean square difference. The Kolmogorov-Smirnov D is a particularly simple measure: It is defined as the

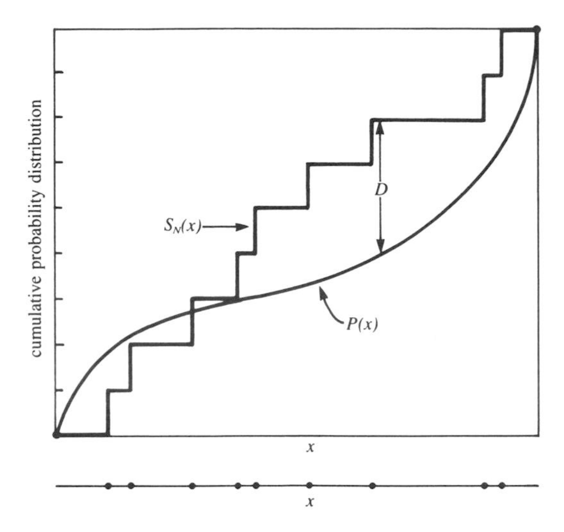

Figure 13.5.1. Kolmogorov-Smirnov statistic D. A measured distribution of values in x (shown as N dots on the lower abscissa) is to be compared with a theoretical distribution whose cumulative probability distribution is plotted as $P(x)$. A step-function cumulative probability distribution $S_N(x)$ is constructed, one which rises an equal amount at each measured point. D is the greatest distance between the two cumulative distributions.

maximum value of the absolute difference between two cumulative distribution functions. Thus, for comparing one data set's $S_N(x)$ to a known cumulative distribution function $P(x)$, the K–S statistic is

$$D = \max_{-\infty<x<\infty} |S_N(x) - P(x)| \tag{13.5.3}$$

while for comparing two different cumulative distribution functions $S_{N_1}(x)$ and $S_{N_2}(x)$, the K–S statistic is

$$D = \max_{-\infty<x<\infty} |S_{N_1}(x) - S_{N_2}(x)| \tag{13.5.4}$$

What makes the K–S statistic useful is that *its* distribution in the case of the null hypothesis (data sets drawn from the same distribution) can be calculated, at least to useful approximation, thus giving the significance of any observed nonzero value of D.

The function which enters into the calculation of the significance can be written as the following sum,

$$Q_{KS}(\lambda) = 2\sum_{j=1}^{\infty}(-)^{j-1}\, e^{-2j^2\lambda^2} \tag{13.5.5}$$

which is a monotonic function with the limiting values

$$Q_{KS}(0) = 1 \qquad Q_{KS}(\infty) = 0 \tag{13.5.6}$$

In terms of this function, the significance level of an observed value of D (as a disproof of the null hypothesis that the distributions are the same) is given approximately by the formulas

$$\text{Probability } (D > \text{observed }) = Q_{KS}(\sqrt{N}\, D) \tag{13.5.7}$$

for the case (13.5.3) of one distribution, where N is the number of data points, and

$$\text{Probability } (D > \text{observed }) = Q_{KS}\left(\sqrt{\frac{N_1 N_2}{N_1 + N_2}}\, D\right) \tag{13.5.8}$$

for the case (13.5.4) of two distributions, where N_1 is the number of data points in the first distribution, N_2 the number in the second.

The nature of the approximation involved in (13.5.7) and (13.5.8) is that it becomes asymptotically accurate as the N's become large. In practice, $N = 20$ is large enough, especially if you are being conservative and requiring a strong significance level (0.01 or smaller).

So, we have the following routines for the cases of one and two distributions:

```
SUBROUTINE KSONE(DATA,N,FUNC,D,PROB)
    Given an array of N values, DATA, and given a user-supplied function of a single variable
    FUNC which is a cumulative distribution function ranging from 0 (for smallest values of its
    argument) to 1 (for largest values of its argument), this routine returns the K–S statistic D,
    and the significance level PROB. Small values of PROB show that the cumulative distribution
    function of DATA is significantly different from FUNC. The array DATA is modified by being
    sorted into ascending order.
DIMENSION DATA(N)
CALL SORT(N,DATA)              If the data are already sorted into ascending order, then this call can
EN=N                                 be omitted.
D=0.
FO=0.                          Data's c.d.f. before the next step.
DO 11 J=1,N                    Loop over the sorted data points.
    FN=J/EN                    Data's c.d.f. after this step.
    FF=FUNC(DATA(J))           Compare to the user-supplied function.
    DT=AMAX1(ABS(FO-FF),ABS(FN-FF))      Maximum distance.
    IF(DT.GT.D)D=DT
    FO=FN
11  CONTINUE
PROB=PROBKS(SQRT(EN)*D)        Compute significance.
RETURN
END
```

```
      SUBROUTINE KSTWO(DATA1,N1,DATA2,N2,D,PROB)
C     Given an array DATA1 of N1 values, and an array DATA2 of N2 values, this routine returns
C     the K–S statistic D, and the significance level PROB for the null hypothesis that the data
C     sets are drawn from the same distribution. Small values of PROB show that the cumulative
C     distribution function of DATA1 is significantly different from that of DATA2. The arrays
C     DATA1 and DATA2 are modified by being sorted into ascending order.
      DIMENSION DATA1(N1),DATA2(N2)
      CALL SORT(N1,DATA1)
      CALL SORT(N2,DATA2)
      EN1=N1
      EN2=N2
      J1=1                             Next value of DATA1 to be processed.
      J2=1                             Ditto, DATA2.
      FN1=0.
      FN2=0.
      D=0.
1     IF(J1.LE.N1.AND.J2.LE.N2)THEN    If we are not done...
          D1=DATA1(J1)
          D2=DATA2(J2)
          IF(D1.LE.D2)THEN                     Next step is in DATA1.
              FN1=J1/EN1
              J1=J1+1
          ENDIF
          IF(D2.LE.D1)THEN                     Next step is in DATA2.
              FN2=J2/EN2
              J2=J2+1
          ENDIF
          DT=ABS(FN2-FN1)
          IF(DT.GT.D)D=DT
      GO TO 1
      ENDIF
      PROB=PROBKS(SQRT(EN1*EN2/(EN1+EN2))*D)   Compute significance.
      RETURN
      END
```

Both of the above routines use the following routine for calculating the function Q_{KS}:

```
      FUNCTION PROBKS(ALAM)
      PARAMETER (EPS1=0.001, EPS2=1.E-8)
      A2=-2.*ALAM**2
      FAC=2.
      PROBKS=0.
      TERMBF=0.                        Previous term in sum.
      DO 11 J=1,100
          TERM=FAC*EXP(A2*J**2)
          PROBKS=PROBKS+TERM
          IF(ABS(TERM).LE.EPS1*TERMBF.OR.ABS(TERM).LE.EPS2*PROBKS)RETURN
          FAC=-FAC                     Alternating signs in sum.
          TERMBF=ABS(TERM)
11    CONTINUE
      PROBKS=1.                        Get here only by failing to converge.
      RETURN
      END
```

REFERENCES AND FURTHER READING:

von Mises, Richard. 1964, *Mathematical Theory of Probability and Statistics* (New York: Academic Press), Chapters IX(C) and IX(E).

13.6 Contingency Table Analysis of Two Distributions

In this section, and the next two sections, we deal with *measures of association* for two distributions. The situation is this: Each data point has two or more different quantities associated with it, and we want to know whether knowledge of one quantity gives us any demonstrable advantage in predicting the value of another quantity. In many cases, one variable will be an "independent" or "control" variable, and another will be a "dependent" or "measured" variable. Then, we want to know if the latter variable *is* in fact dependent on or *associated* with the former variable. If it is, we want to have some quantitative measure of the strength of the association. One often hears this loosely stated as the question of whether two variables are *correlated* or *uncorrelated*, but we will reserve those terms for a particular kind of association (linear, or at least monotonic), as discussed in §13.7 and §13.8.

Notice that, as in previous sections, the different concepts of significance and strength appear: The association between two distributions may be very significant even if that association is weak – if the quantity of data is large enough.

It is useful to distinguish among some different kinds of variables, with different categories forming a loose hierarchy.

- A variable is called *nominal* if its values are the members of some unordered set. For example, "state of residence" is a nominal variable that (in the U.S.) takes on one of 50 values; in astrophysics, "type of galaxy" is a nominal variable with the three values "spiral," "elliptical," and "irregular."
- A variable is termed *ordinal* if its values are the members of a discrete, but ordered, set. Examples are: grade in school, planetary order from the Sun (Mercury = 1, Venus = 2, ...), number of offspring. There need not be any concept of "equal metric distance" between the values of an ordinal variable, only that they be intrinsically ordered.
- We will call a variable *continuous* if its values are real numbers, as are times, distances, temperatures, etc. (Social scientists sometimes distinguish between *interval* and *ratio* continuous variables, but we do not find that distinction very compelling.)

A continuous variable can always be made into an ordinal one by binning it into ranges. If we choose to ignore the ordering of the bins, then we can turn it into a nominal variable. Nominal variables constitute the lowest type of the hierarchy, and therefore the most general. For example, a set of *several* continuous or ordinal variables can be turned, if crudely, into a single nominal variable, by coarsely binning each variable and then taking each distinct combination of bin assignments as a single nominal value. When multidimensional data are sparse, this is often the only sensible way to proceed.

The remainder of this section will deal with measures of association between *nominal* variables. For any pair of nominal variables, the data can be

	1. red	2. green	. . .	
1. male	# of red males N_{11}	# of green males N_{12}	. . .	# of males $N_{1\cdot}$
2. female	# of red females N_{21}	# of green females N_{22}	. . .	# of females $N_{2\cdot}$
⋮	⋮	⋮	. . .	⋮
	# of red $N_{\cdot 1}$	# of green $N_{\cdot 2}$	. . .	total # N

Figure 13.6.1. Example of a contingency table for two nominal variables, here sex and color. The row and column marginals (totals) are shown. The variables are "nominal," i.e. the order in which their values are listed is arbitrary and does not affect the result of the contingency table analysis. If the ordering of values has some intrinsic meaning, then the variables are "ordinal" or "continuous," and correlation techniques (§13.7 – §13.8) can be utilized.

displayed as a *contingency table*, a table whose rows are labeled by the values of one nominal variable, whose columns are labeled by the values of the other nominal variable, and whose entries are nonnegative integers giving the number of observed events for each combination of row and column (see Figure 13.6.1). The analysis of association between nominal variables is thus called *contingency table analysis* or *crosstabulation analysis.*

We will introduce two different approaches. The first approach, based on the chi-square statistic, does a good job of characterizing the significance of association, but is only so-so as a measure of the strength (principally because its numerical values have no very direct interpretations). The second approach, based on the information-theoretic concept of *entropy*, says nothing at all about the significance of association (use chi-square for that!), but is capable of very elegantly characterizing the strength of an association already known to be significant.

Measures of Association Based on Chi-Square

Some notation first: Let N_{ij} denote the number of events which occur with the first variable x taking on its ith value, and the second variable y taking on its jth value. Let N denote the total number of events, the sum of all the N_{ij}'s. Let $N_{i\cdot}$ denote the number of events for which the first variable x takes on its ith value regardless of the value of y; $N_{\cdot j}$ is the number of events with the jth value of y regardless of x. So we have

$$\begin{aligned} N_{i\cdot} &= \sum_j N_{ij} \qquad N_{\cdot j} = \sum_i N_{ij} \\ N &= \sum_i N_{i\cdot} = \sum_j N_{\cdot j} \end{aligned} \tag{13.6.1}$$

$N_{\cdot j}$ and $N_{i\cdot}$ are sometimes called the *row and column totals* or *marginals*, but we will use these terms cautiously since we can never keep straight which are the rows and which are the columns!

The null hypothesis is that the two variables x and y have no association. In this case, the probability of a particular value of x given a particular value of y should be the same as the probability of that value of x regardless of y. Therefore, in the null hypothesis, the expected number for any N_{ij}, which we will denote n_{ij}, can be calculated from only the row and column totals,

$$\frac{n_{ij}}{N_{\cdot j}} = \frac{N_{i\cdot}}{N} \qquad \text{which implies} \qquad n_{ij} = \frac{N_{i\cdot}N_{\cdot j}}{N} \tag{13.6.2}$$

Notice that if a column or row total is zero, then the expected number for all the entries in that column or row is also zero; in that case, the never-occurring bin of x or y should simply be removed from the analysis.

The chi-square statistic is now given by equation (13.5.1) which, in the present case, is summed over all entries in the table,

$$\chi^2 = \sum_{i,j} \frac{(N_{ij} - n_{ij})^2}{n_{ij}} \tag{13.6.3}$$

The number of degrees of freedom is equal to the number of entries in the table (product of its row size and column size) minus the number of constraints that have arisen from our use of the data itself to determine the n_{ij}. Each row total and column total is a constraint, except that this overcounts by one, since the total of the column totals and the total of the row totals both equal N, the total number of data points. Therefore, if the table is of size I by J, the number of degrees of freedom is $IJ - I - J + 1$. Equation (13.6.3), along with the chi-square probability function (§6.2) now give the significance of an association between the variables x and y.

Suppose there is a significant association. How do we quantify its strength, so that (e.g.) we can compare the strength of one association to another? The idea here is to find some reparametrization of χ^2 which maps it into some convenient interval, like 0 to 1, where the result is not dependent on the quantity of data that we happen to sample, but rather depends only on the underlying population from which the data were drawn. There are several different ways of doing this. Two of the more common are called *Cramer's V* and the *contingency coefficient C.*

The formula for Cramer's V is

$$V = \sqrt{\frac{\chi^2}{N \min(I-1, J-1)}} \tag{13.6.4}$$

where I and J are again the numbers of rows and columns, N is the total number of events. Cramer's V has the pleasant property that it lies between

zero and one inclusive, equals zero when there is no association, and equals one only when the association is perfect: All the events in any row lie in one unique column, and vice versa. (In chess parlance, no two rooks, placed on a nonzero table entry, can capture each other.)

In the case of $I = J = 2$, Cramer's V is also referred to as the *phi* statistic.

The contingency coefficient C is defined as

$$C = \sqrt{\frac{\chi^2}{\chi^2 + N}} \tag{13.6.5}$$

It also lies between zero and one, but (as is apparent from the formula) it can never achieve the upper limit. While it can be used to compare the strength of association of two tables with the same I and J, its upper limit depends on I and J. Therefore it can never be used to compare tables of different sizes.

The trouble with both Cramer's V and with the contingency coefficient C is that, when they take on values in between their extremes, there is no very direct interpretation of what that value means. For example, you are in Las Vegas, and a friend tells you that there is a small, but significant, association between the color of a croupier's eyes and the occurrence of red and black on his roulette wheel. Cramer's V is about 0.028, your friend tells you. You know what the usual odds against you are (due to the green zero and double zero on the wheel). Is this association sufficient for you to make money? Don't ask us!

```
SUBROUTINE CNTAB1(NN,NI,NJ,CHISQ,DF,PROB,CRAMRV,CCC)
    Given a two-dimensional contingency table in the form of an integer array NN(I,J), where
    I ranges from 1 to NI, J ranges from 1 to NJ, this routine returns the chi-square CHISQ,
    the number of degrees of freedom DF, the significance level PROB (small values indicating a
    significant association), and two measures of association, Cramer's V (CRAMRV) and the
    contingency coefficient C (CCC).
PARAMETER (MAXI=100,MAXJ=100,TINY=1.E-30)    Maximum table size, and a small number.
DIMENSION NN(NI,NJ),SUMI(MAXI),SUMJ(MAXJ)
SUM=0                              Will be total number of events.
NNI=NI                             Number of rows
NNJ=NJ                             and columns.
DO 12 I=1,NI                       Get the row totals.
    SUMI(I)=0.
    DO 11 J=1,NJ
        SUMI(I)=SUMI(I)+NN(I,J)
        SUM=SUM+NN(I,J)
   11 CONTINUE
    IF(SUMI(I).EQ.0.)NNI=NNI-1           Eliminate any zero rows by reducing the number.
 12 CONTINUE
DO 14 J=1,NJ                       Get the column totals.
    SUMJ(J)=0.
    DO 13 I=1,NI
        SUMJ(J)=SUMJ(J)+NN(I,J)
   13 CONTINUE
    IF(SUMJ(J).EQ.0.)NNJ=NNJ-1           Eliminate any zero columns.
 14 CONTINUE
DF=NNI*NNJ-NNI-NNJ+1                   Corrected number of degrees of freedom.
CHISQ=0.
DO 16 I=1,NI                       Do the chi-square sum.
    DO 15 J=1,NJ
```

```
          EXPCTD=SUMJ(J)*SUMI(I)/SUM
          CHISQ=CHISQ+(NN(I,J)-EXPCTD)**2/(EXPCTD+TINY)   Here TINY guarantees that any elim-
15      CONTINUE                                          inated row or column will not contribute to the sum.
16    CONTINUE
      PROB=GAMMQ(0.5*DF,0.5*CHISQ)     Chi-square probability function.
      CRAMRV=SQRT(CHISQ/(SUM*MIN(NNI-1,NNJ-1)))
      CCC=SQRT(CHISQ/(CHISQ+SUM))
      RETURN
      END
```

Measures of Association Based on Entropy

Consider the game of "twenty questions," where by repeated yes/no questions you try to eliminate all except one correct possibility for an unknown object. Better yet, consider a generalization of the game, where you are allowed to ask multiple choice questions as well as binary (yes/no) ones. The categories in your multiple choice questions are supposed to be mutually exclusive and exhaustive (as are "yes" and "no").

The value to you of an answer increases with the number of possibilities that it eliminates. More specifically, an answer that eliminates all except a fraction p of the remaining possibilities can be assigned a value $-\ln p$ (a positive number, since $p < 1$). The purpose of the logarithm is to make the value additive, since (e.g.) one question which eliminates all but 1/6 of the possibilities is considered as good as two questions that, in sequence, reduce the number by factors 1/2 and 1/3.

So that is the value of an answer; but what is the value of a question? If there are I possible answers to the question ($i = 1, \ldots, I$) and the fraction of possibilities consistent with the ith answer is p_i (with the sum of the p_i's equal to one), then the value of the question is the expectation value of the value of the answer, denoted H,

$$H = -\sum_{i=1}^{I} p_i \ln(p_i) \tag{13.6.6}$$

In evaluating (13.6.6), note that

$$\lim_{p \to 0} p \ln(p) = 0 \tag{13.6.7}$$

The value H lies between 0 and $\ln I$. It is zero only when one of the p_i's is one, all the others zero: in this case, the question is valueless, since its answer is preordained. H takes on its maximum value when all the p_i's are equal, in which case the question is sure to eliminate all but a fraction $1/I$ of the remaining possibilities.

The value H is conventionally termed the *entropy* of the distribution given by the p_i's, a terminology borrowed from statistical physics.

So far we have said nothing about the association of two variables; but suppose we are deciding what question to ask next in the game and have to choose between two candidates, or possibly want to ask both in one order or another. Suppose that one question, x, has I possible answers, labeled by i, and that the other question, y, as J possible answers, labeled by j. Then the possible outcomes of asking both questions form a contingency table whose entries N_{ij}, when normalized by dividing by the total number of remaining possibilities N, give all the information about the p's. In particular, we can make contact with the notation (13.6.1) by identifying

$$\begin{aligned} p_{ij} &= \frac{N_{ij}}{N} \\ p_{i\cdot} &= \frac{N_{i\cdot}}{N} \qquad \text{(outcomes of question } x \text{ alone)} \\ p_{\cdot j} &= \frac{N_{\cdot j}}{N} \qquad \text{(outcomes of question } y \text{ alone)} \end{aligned} \tag{13.6.8}$$

The entropies of the questions x and y are respectively

$$H(x) = -\sum_i p_{i\cdot} \ln p_{i\cdot} \qquad H(y) = -\sum_j p_{\cdot j} \ln p_{\cdot j} \tag{13.6.9}$$

The entropy of the two questions together is

$$H(x,y) = -\sum_{i,j} p_{ij} \ln p_{ij} \tag{13.6.10}$$

Now what is the entropy of the question y *given* x (that is, if x is asked first)? It is the expectation value over the answers to x of the entropy of the restricted y distribution that lies in a single column of the contingency table (corresponding to the x answer):

$$H(y|x) = \sum_i p_{i\cdot} \sum_j \frac{p_{ij}}{p_{i\cdot}} \ln \frac{p_{ij}}{p_{i\cdot}} = \sum_{i,j} p_{ij} \ln \frac{p_{ij}}{p_{i\cdot}} \tag{13.6.11}$$

Correspondingly, the entropy of x given y is

$$H(x|y) = \sum_j p_{\cdot j} \sum_i \frac{p_{ij}}{p_{\cdot j}} \ln \frac{p_{ij}}{p_{\cdot j}} = \sum_{i,j} p_{ij} \ln \frac{p_{ij}}{p_{\cdot j}} \tag{13.6.12}$$

We can readily prove that the entropy of y given x is never more than the entropy of y alone, i.e. that asking x first can only reduce the usefulness

of asking y (in which case the two variables are *associated*!):

$$\begin{aligned}
H(y|x) - H(y) &= -\sum_{i,j} p_{ij} \ln \frac{p_{ij}/p_{i\cdot}}{p_{\cdot j}} \\
&= \sum_{i,j} p_{ij} \ln \frac{p_{\cdot j} p_{i\cdot}}{p_{ij}} \\
&\le \sum_{i,j} p_{ij} \left(\frac{p_{\cdot j} p_{i\cdot}}{p_{ij}} - 1 \right) \\
&= \sum_{i,j} p_{i\cdot} p_{\cdot j} - \sum_{i,j} p_{ij} \\
&= 1 - 1 = 0
\end{aligned} \tag{13.6.13}$$

where the inequality follows from the fact

$$\ln w \le w - 1 \tag{13.6.14}$$

We now have everything we need to define a measure of the "dependency" of y on x, that is to say a measure of association. This measure is sometimes called the *uncertainty coefficient* of y. We will denote it as $U(y|x)$,

$$U(y|x) \equiv \frac{H(y) - H(y|x)}{H(y)} \tag{13.6.15}$$

This measure lies between zero and one, with the value 0 indicating that x and y have no association, the value 1 indicating that knowledge of x completely predicts y. For in-between values, $U(y|x)$ gives the fraction of y's entropy $H(y)$ that is lost if x is already known (i.e., that is redundant with the information in x). In our game of "twenty questions," $U(y|x)$ is the fractional loss in the utility of question y if question x is to be asked first.

If we wish to view x as the dependent variable, y as the independent one, then interchanging x and y we can of course define the dependency of x on y,

$$U(x|y) \equiv \frac{H(x) - H(x|y)}{H(x)} \tag{13.6.16}$$

If we want to treat x and y symmetrically, then the useful combination turns out to be

$$U(x, y) \equiv 2 \left[\frac{H(y) + H(x) - H(x, y)}{H(x) + H(y)} \right] \tag{13.6.17}$$

If the two variables are completely independent, then $H(x,y) = H(x)+H(y)$, so (13.6.17) vanishes. If the two variables are completely dependent, then $H(x) = H(y) = H(x,y)$, so (13.6.16) equals unity. In fact, you can use the identities (easily proved from equations 13.6.9–13.6.12)

$$H(x,y) = H(x) + H(y|x) = H(y) + H(x|y) \tag{13.6.18}$$

to show that

$$U(x,y) = \frac{H(x)U(x|y) + H(y)U(y|x)}{H(x) + H(y)} \tag{13.6.19}$$

i.e. that the symmetrical measure is just a weighted average of the two asymmetrical measures (13.6.15) and (13.6.16), weighted by the entropy of each variable separately.

Here is a program for computing all the quantities discussed, $H(x)$, $H(y)$, $H(x|y)$, $H(y|x)$, $H(x,y)$, $U(x|y)$, $U(y|x)$, and $U(x,y)$:

```
SUBROUTINE CNTAB2(NN,NI,NJ,H,HX,HY,HYGX,HXGY,UYGX,UXGY,UXY)
    Given a two-dimensional contingency table in the form of an integer array NN(I,J), where
    I labels the x variable and ranges from 1 to NI, J labels the y variable and ranges from
    1 to NJ, this routine returns the entropy H of the whole table, the entropy HX of the x
    distribution, the entropy HY of the y distribution, the entropy HYGX of y given x, the
    entropy HXGY of x given y, the dependency UYGX of y on x (eq. 13.6.15), the dependency
    UXGY of x on y (eq. 13.6.16), and the symmetrical dependency UXY (eq. 13.6.17).
PARAMETER (MAXI=100,MAXJ=100,TINY=1.E-30)    Maximum size of table, and a small number.
DIMENSION NN(NI,NJ),SUMI(MAXI),SUMJ(MAXJ)
SUM=0
DO 12 I=1,NI                             Get the row totals.
    SUMI(I)=0.0
    DO 11 J=1,NJ
        SUMI(I)=SUMI(I)+NN(I,J)
        SUM=SUM+NN(I,J)
    11 CONTINUE
  12 CONTINUE
DO 14 J=1,NJ                             Get the column totals.
    SUMJ(J)=0.
    DO 13 I=1,NI
        SUMJ(J)=SUMJ(J)+NN(I,J)
    13 CONTINUE
  14 CONTINUE
HX=0.
DO 15 I=1,NI                             Entropy of the x distribution,
    IF(SUMI(I).NE.0.)THEN
        P=SUMI(I)/SUM
        HX=HX-P*ALOG(P)
    ENDIF
  15 CONTINUE
HY=0.                                        and of the y distribution.
DO 16 J=1,NJ
    IF(SUMJ(J).NE.0.)THEN
        P=SUMJ(J)/SUM
        HY=HY-P*ALOG(P)
    ENDIF
  16 CONTINUE
H=0.
```

```
DO 18 I=1,NI                            Total entropy: loop over both x
    DO 17 J=1,NJ                            and y.
        IF(NN(I,J).NE.0)THEN
            P=NN(I,J)/SUM
            H=H-P*ALOG(P)
        ENDIF
17  CONTINUE
18  CONTINUE
HYGX=H-HX                               Uses equation (13.6.18),
HXGY=H-HY                                   as does this.
UYGX=(HY-HYGX)/(HY+TINY)                    Equation (13.6.15).
UXGY=(HX-HXGY)/(HX+TINY)                    Equation (13.6.16).
UXY=2.*(HX+HY-H)/(HX+HY+TINY)               Equation (13.6.17).
RETURN
END
```

REFERENCES AND FURTHER READING:

SPSS: Statistical Package for the Social Sciences, 2nd ed., by Norman H. Nie, et al. (New York: McGraw-Hill), §16.1.

Downie, N.M., and Heath, R.W. 1965, *Basic Statistical Methods*, 2nd ed. (New York: Harper and Row), pp. 196, 210.

Fano, Robert M. 1961, *Transmission of Information* (New York: Wiley and MIT Press), Chapter 2.

13.7 Linear Correlation

We next turn to measures of association between variables that are ordinal or continuous, rather than nominal. Most widely used is the *linear correlation coefficient.* For pairs of quantities (x_i, y_i), $i = 1, \ldots, N$, the linear correlation coefficient r (also called *Pearson's* r) is given by the formula

$$r = \frac{\sum\limits_i (x_i - \overline{x})(y_i - \overline{y})}{\sqrt{\sum\limits_i (x_i - \overline{x})^2}\sqrt{\sum\limits_i (y_i - \overline{y})^2}} \tag{13.7.1}$$

where, as usual, $\overline{x}$ is the mean of the x_i's, $\overline{y}$ is the mean of the y_i's.

The value of r lies between -1 and 1, inclusive. It takes on a value of 1, termed "complete positive correlation," when the data points lie on a perfect straight line with positive slope, with x and y increasing together. The value 1 holds independent of the magnitude of the slope. If the data points lie on a perfect straight line with negative slope, y decreasing as x increases, then r has the value -1; this is called "complete negative correlation." A value of r near zero indicates that the variables x and y are *uncorrelated.*

When a correlation is known to be significant, r is one conventional way of summarizing its strength. In fact, the value of r can be translated into

a statement about what residuals (root means square deviations) are to be expected if the data are fitted to a straight line by the least-squares method (see §14.2, especially equations 14.2.11 and 14.2.12). Unfortunately, r is a rather poor statistic for deciding *whether* an observed correlation is statistically significant, and/or whether one observed correlation is significantly stronger than another. The reason is that r is ignorant of the individual distributions of x and y, so there is no universal way to compute its distribution in the case of the null hypothesis.

About the only general statement that can be made is this: If the null hypothesis is that x and y are uncorrelated, and if the distributions for x and y each have enough convergent moments ("tails" die off sufficiently rapidly), and if N is large (typically > 20), then r is distributed approximately normally, with a mean of zero and a standard deviation of $1/\sqrt{N}$. In that case, the (double-sided) significance of the correlation, that is, the probability that $|r|$ should be larger than its observed value in the null hypothesis, is

$$\text{erfc}\left(\frac{|r|\sqrt{N}}{\sqrt{2}}\right) \tag{13.7.2}$$

where $\text{erfc}(x)$ is the complementary error function, equation (6.2.8), computed by the routines ERFC or ERFCC of §6.2. A small value of (13.7.2) indicates that the two distributions are significantly correlated.

Most statistics books try to go beyond (13.7.2) and give additional statistical tests which can be made using r. In almost all cases, however, these tests are valid only for a very special class of hypotheses, namely that the distributions of x and y jointly form a *binormal* or *two-dimensional Gaussian* distribution around their mean values, with joint probability density

$$p(x,y)\,dxdy = \text{const.} \times \exp\left[-\frac{1}{2}(a_{11}x^2 - 2a_{12}xy + a_{22}y^2)\right]\,dxdy \tag{13.7.3}$$

where a_{11}, a_{12}, and a_{22} are arbitrary constants. For this distribution r has the value

$$r = -\frac{a_{12}}{\sqrt{a_{11}a_{22}}} \tag{13.7.4}$$

There are occasions when (13.7.3) may be known to be a good model of the data. There may be other occasions when we are willing to take (13.7.3) as at least a rough and ready guess, since many two-dimensional distributions do resemble a binormal distribution, at least not too far out on their tails. In either situation, we can use (13.7.3) to go beyond (13.7.2) in any of several directions:

First, we can allow for the possibility that the number N of data points is not large. Here, it turns out that the statistic

$$t = r\sqrt{\frac{N-2}{1-r^2}} \tag{13.7.5}$$

is distributed in the null case (of no correlation) like Student's t-distribution with $\nu = N-2$ degrees of freedom, whose two-sided significance level is given by $1 - A(t|\nu)$ (equation 6.3.7). As N becomes large, this significance and (13.7.2) become asymptotically the same, so that one never does worse by using (13.7.5), even if the binormal assumption is not well substantiated.

Second, when N is moderately large (≥ 10), we can compare whether the difference of two significantly nonzero r's, e.g., from different experiments, is itself significant. In other words, we can quantify whether a change in some control variable significantly alters an existing correlation between two other variables. This is done by using *Fisher's z-transformation* to associate each measured r with a corresponding z,

$$z = \frac{1}{2}\ln\left(\frac{1+r}{1-r}\right) \tag{13.7.6}$$

Then, each z is approximately normally distributed with a mean value

$$\overline{z} = \frac{1}{2}\left[\ln\left(\frac{1+r_{true}}{1-r_{true}}\right) + \frac{r_{true}}{N-1}\right] \tag{13.7.7}$$

where r_{true} is the actual or population value of the correlation coefficient, and with a standard deviation

$$\sigma(z) \approx \frac{1}{\sqrt{N-3}} \tag{13.7.8}$$

Equations (13.7.7) and (13.7.8), when they are valid, give several useful statistical tests. For example, the significance level at which a measured value of r differs from some hypothesized value r_{true} is given by

$$\text{erfc}\left(\frac{|z-\overline{z}|\sqrt{N-3}}{\sqrt{2}}\right) \tag{13.7.9}$$

where z and $\overline{z}$ are given by (13.7.6) and (13.7.7), with small values of (13.7.9) indicating a significant difference. Similarly, the significance of a difference

between two measured correlation coefficients r_1 and r_2 is

$$\operatorname{erfc}\left(\frac{|z_1 - z_2|}{\sqrt{2}\sqrt{\frac{1}{N_1-3} + \frac{1}{N_2-3}}}\right) \tag{13.7.10}$$

where z_1 and z_2 are obtained from r_1 and r_2 using (13.7.6), and where N_1 and N_2 are respectively the number of data points in the measurement of r_1 and r_2.

All of the significances above are two-sided. If you wish to disprove the null hypothesis in favor of a one-sided hypothesis, such as that $r_1 > r_2$ (where the sense of the inequality was decided *a priori*), then (i) if your measured r_1 and r_2 have the *wrong* sense, you have failed to demonstrate your one-sided hypothesis, but (ii) if they have the right ordering, you can multiply the significances given above by 0.5, which makes them more significant.

But keep in mind: These interpretations of the r statistic can be completely meaningless if the joint probability distribution of your variables x and y is too different from a binormal distribution.

```
      SUBROUTINE PEARSN(X,Y,N,R,PROB,Z)
```

Given two arrays **X** and **Y** of length N, this routine computes their correlation coefficient r (returned as **R**), the significance level at which the null hypothesis of zero correlation is disproved (**PROB** whose small value indicates a significant correlation), and Fisher's z (returned as **Z**), whose value can be used in further statistical tests as described above.

```
      PARAMETER (TINY=1.E-20)         Will regularize the unusual case of complete correlation.
      DIMENSION X(N),Y(N)
      AX=0.
      AY=0.
      DO 11 J=1,N                     Find the means.
          AX=AX+X(J)
          AY=AY+Y(J)
11    CONTINUE
      AX=AX/N
      AY=AY/N
      SXX=0.
      SYY=0.
      SXY=0.
      DO 12 J=1,N                     Compute the correlation coefficient.
          XT=X(J)-AX
          YT=Y(J)-AY
          SXX=SXX+XT**2
          SYY=SYY+YT**2
          SXY=SXY+XT*YT
12    CONTINUE
      R=SXY/SQRT(SXX*SYY)
      Z=0.5*ALOG(((1.+R)+TINY)/((1.-R)+TINY))  Fisher's z transformation.
      DF=N-2
      T=R*SQRT(DF/(((1.-R)+TINY)*((1.+R)+TINY)))  Equation (13.7.5).
      PROB=BETAI(0.5*DF,0.5,DF/(DF+T**2))      Student's t probability.
C     PROB=ERFCC(ABS(Z*SQRT(N-1.))/1.4142136)  For large N, this easier computation of PROB, using
      RETURN                                   the short routine ERFCC, would give approximately the same
      END                                      value.
```

REFERENCES AND FURTHER READING:

Downie, N.M., and Heath, R.W. 1965, *Basic Statistical Methods*, 2nd ed. (New York: Harper and Row), Chapters 7 and 13.

Hoel, P.G. 1971, *Introduction to Mathematical Statistics*, 4th ed. (New York: Wiley), Chapter 7.

von Mises, Richard. 1964, *Mathematical Theory of Probability and Statistics* (New York: Academic Press), Chapters IX(A) and IX(B).

Korn, G.A., and Korn, T.M. 1968, *Mathematical Handbook for Scientists and Engineers*, 2nd ed. (New York: McGraw-Hill), §19.7.

SPSS: Statistical Package for the Social Sciences, 2nd ed., by Norman H. Nie, et al. (New York: McGraw-Hill), §18.2.

13.8 Nonparametric or Rank Correlation

It is precisely the uncertainty in interpreting the significance of the linear correlation coefficient r that leads us to the important concepts of *nonparametric* or *rank correlation*. As before, we are given N pairs of measurements (x_i, y_i). Before, difficulties arose from the fact that we did not necessarily know the probability distribution function from which the x_i's or y_i's were drawn.

The key concept of nonparametric correlation is this: If we replace the value of each x_i by the value of its *rank* among all the other x_i's in the sample, that is, $1, 2, 3, \ldots, N$, then the resulting list of numbers will be drawn from a perfectly known distribution function, namely uniformly from the integers between 1 and N, inclusive. Better than uniformly, in fact, since if the x_i's are all distinct, then each integer will occur precisely once. If some of the x_i's have identical values, it is conventional to assign to all these "ties" the mean of the ranks that they would have had if their values had been slightly different. This *midrank* will sometimes be an integer, sometimes a half-integer. In all cases the sum of all assigned ranks will be the same as the sum of the integers from 1 to N, namely $\frac{1}{2}N(N+1)$.

Of course we do exactly the same procedure for the y_i's, replacing each value by its rank among the other y_i's in the sample.

Now we are free to invent statistics for detecting correlation between uniform sets of integers between 1 and N, keeping in mind the possibility of ties in the ranks. There is, of course, some loss of information in replacing the original numbers by ranks. We could construct some rather artificial examples where a correlation could be detected parametrically (e.g. in the linear correlation coefficient r), but cannot be detected nonparametrically. Such examples are very rare in real life, however, and the slight loss of information in ranking is a small price to pay for a very major advantage: When a correlation is demonstrated to be present nonparametrically, then it is really there! (That is, to a certainty level which depends on the significance chosen.) Nonparametric correlation is more robust than linear correlation, more resistant to unplanned defects in the data, in the same sort of sense that the median is more robust than the mean. For more on the concept of robustness, see §14.6.

As always in statistics, some particular choices of a statistic have already been invented for us and consecrated, if not beatified, by popular use. We will discuss two, the *Spearman rank-order correlation coefficient* (r_s), and *Kendall's tau* (τ).

Spearman Rank-Order Correlation Coefficient

Let R_i be the rank of x_i among the other x's, S_i be the rank of y_i among the other y's, ties being assigned the appropriate midrank as described above. Then the rank-order correlation coefficient is defined to be the linear correlation coefficient of the ranks, namely,

$$r_s = \frac{\sum_i (R_i - \overline{R})(S_i - \overline{S})}{\sqrt{\sum_i (R_i - \overline{R})^2}\sqrt{\sum_i (S_i - \overline{S})^2}} \tag{13.8.1}$$

The significance of a nonzero value of r_s is tested by computing

$$t = r_s \sqrt{\frac{N-2}{1-r_s^2}} \tag{13.8.2}$$

which is distributed approximately as Student's distribution with $N-2$ degrees of freedom. A key point is that this approximation does not depend on the original distribution of the x's and y's; it is always the same approximation, and always pretty good.

It turns out that r_s is closely related to another conventional measure of nonparametric correlation, the so-called *sum squared difference of ranks*, defined as

$$D = \sum_{i=1}^{N} (R_i - S_i)^2 \tag{13.8.3}$$

(This D is sometimes denoted D^{**}, where the asterisks are used to indicate that ties are treated by midranking.)

When there are no ties in the data, then the exact relation between D and r_s is

$$r_s = 1 - \frac{6D}{N^3 - N} \tag{13.8.4}$$

When there are ties, then the exact relation is slightly more complicated: Let f_k be the number of ties in the kth group of ties among the R_i's, and let

g_m be the number of ties in the mth group of ties among the S_i's. Then it turns out that

$$r_s = \frac{1 - \frac{6}{N^3 - N}\left[D + \frac{1}{2}\sum_k (f_k^3 - f_k) + \frac{1}{2}\sum_m (g_m^3 - g_m)\right]}{\left[1 - \frac{\sum_k (f_k^3 - f_k)}{N^3 - N}\right]\left[1 - \frac{\sum_m (g_m^3 - g_m)}{N^3 - N}\right]} \tag{13.8.5}$$

holds exactly. Notice that if all the f_k's and all the g_m's are equal to one, meaning that there are no ties, then equation (13.8.5) reduces to equation (13.8.4).

In (13.8.2) we gave a t-statistic which tests the significance of a nonzero r_s. It is also possible to test the significance of D directly. The expectation value of D in the null hypothesis of uncorrelated data sets is

$$\overline{D} = \frac{1}{6}(N^3 - N) - \frac{1}{12}\sum_k (f_k^3 - f_k) - \frac{1}{12}\sum_m (g_m^3 - g_m) \tag{13.8.6}$$

its variance is

$$\begin{aligned}\mathrm{Var}(D) = &\frac{(N-1)N^2(N+1)^2}{36} \\ &\times \left[1 - \frac{\sum_k (f_k^3 - f_k)}{N^3 - N}\right]\left[1 - \frac{\sum_m (g_m^3 - g_m)}{N^3 - N}\right]\end{aligned} \tag{13.8.7}$$

and it is approximately normally distributed, so that the significance level is a complementary error function (cf. equation 13.7.2). In the program that follows, we return both the significance level obtained by using (13.8.2) and the significance level obtained by using (13.8.7); their discrepancy will give you an idea of how good the approximations are. You will also notice that we break off the task of assigning ranks (including tied midranks) into a separate routine, **CRANK**.

```
SUBROUTINE SPEAR(DATA1,DATA2,N,WKSP1,WKSP2,D,ZD,PROBD,RS,PROBRS)
```

Given two data arrays, **DATA1** and **DATA2**, each of length N, and given two workspaces of equal length, this routine returns their sum-squared difference of ranks as D, the number of standard deviations by which D deviates from its null-hypothesis expected value as **ZD**, the two-sided significance level of this deviation as **PROBD**, Spearman's rank correlation r_s as **RS**, and the two-sided significance level of its deviation from zero as **PROBRS**. The work spaces can be identical to the data arrays, but in that case the data arrays are destroyed. The external routines **CRANK** (below) and **SORT2** (§8.2) are used. A small value of either **PROBD** or **PROBRS** indicates a significant correlation (**RS** positive) or anticorrelation (**RS** negative).

```
DIMENSION DATA1(N),DATA2(N),WKSP1(N),WKSP2(N)
DO 11 J=1,N
    WKSP1(J)=DATA1(J)
    WKSP2(J)=DATA2(J)
11  CONTINUE
CALL SORT2(N,WKSP1,WKSP2)
```

Sort each of the data arrays, and convert the entries to ranks. The values SF and SG return the sums $\sum(f_k^3 - f_k)$ and $\sum(g_m^3 - g_m)$ respectively.

```
      CALL CRANK(N,WKSP1,SF)
      CALL SORT2(N,WKSP2,WKSP1)
      CALL CRANK(N,WKSP2,SG)
      D=0.
      DO 12 J=1,N                      Sum the squared difference of ranks.
        D=D+(WKSP1(J)-WKSP2(J))**2
12    CONTINUE
      EN=N
      EN3N=EN**3-EN
      AVED=EN3N/6.-(SF+SG)/12.              Expectation value of D,
      FAC=(1.-SF/EN3N)*(1.-SG/EN3N)
      VARD=((EN-1.)*EN**2*(EN+1.)**2/36.)*FAC  and variance of D give
      ZD=(D-AVED)/SQRT(VARD)                    number of standard deviations,
      PROBD=ERFCC(ABS(ZD)/1.4142136)           and significance.
      RS=(1.-(6./EN3N)*(D+0.5*(SF+SG)))/FAC    Rank correlation coefficient,
      T=RS*SQRT((EN-2.)/((1.+RS)*(1.-RS)))     and its t value,
      DF=EN-2.
      PROBRS=BETAI(0.5*DF,0.5,DF/(DF+T**2))    give its significance.
      RETURN
      END
```

```
      SUBROUTINE CRANK(N,W,S)
         Given a sorted array W of N elements, replaces the elements by their rank, including
         midranking of ties, and returns as S the sum of f^3 - f, where f is the number of elements
         in each tie.
      DIMENSION W(N)
      S=0.
      J=1                              The next rank to be assigned.
1     IF(J.LT.N)THEN                   "DO WHILE" structure.
        IF(W(J+1).NE.W(J))THEN               Not a tie.
          W(J)=J
          J=J+1
        ELSE                                 A tie:
          DO 11 JT=J+1,N                          How far does it go?
            IF(W(JT).NE.W(J))GO TO 2
11        CONTINUE
          JT=N+1                                 If here, it goes all the way to the last element.
2         RANK=0.5*(J+JT-1)                       This is the mean rank of the tie,
          DO 12 JI=J,JT-1                         so enter it into all the tied entries,
            W(JI)=RANK
12        CONTINUE
          T=JT-J
          S=S+T**3-T                              and update S.
          J=JT
        ENDIF
      GO TO 1
      ENDIF
      IF(J.EQ.N)W(N)=N                 If the last element was not tied, this is its rank.
      RETURN
      END
```

Kendall's Tau

Kendall's τ is even more nonparametric than Spearman's r_s or D. Instead of using the numerical difference of ranks, it only uses the relative ordering of ranks: higher in rank, lower in rank, or the same in rank. But in that

case we don't even have to rank the data! Ranks will be higher, lower, or the same if and only if the values are larger, smaller, or equal, respectively. Since it uses a "weaker" property of the data there can be applications in which τ is more robust than r_s; however, since it throws away information that is available to r_s, there can also be applications where τ is less powerful than r_s (i.e., τ fails to find an actual correlation that r_s does find). On balance, we prefer r_s as being the more straightforward nonparametric test, but both statistics are in general use.

To define τ, we start with the N data points (x_i, y_i). Now consider all $\frac{1}{2}N(N-1)$ *pairs* of data points, where a data point cannot be paired with itself, and where the points in either order count as one pair. We call a pair *concordant* if the relative ordering of the ranks of the two x's (or for that matter the two x's themselves) is the same as the relative ordering of the ranks of the two y's (or for that matter the two y's themselves). We call a pair *discordant* if the relative ordering of the ranks of the two x's is opposite from the relative ordering of the ranks of the two y's. If there is a tie in either the ranks of the two x's or the ranks of the two y's, then we don't call the pair either concordant or discordant. If the tie is in the x's, we will call the pair an "extra y" pair. If the tie is in the y's, we will call the pair an "extra x pair." If the tie is in both the x's and the y's, we don't call the pair anything at all. Are you still with us?

Kendall's τ is now the following simple combination of these various counts:

$$\tau = \frac{\text{concordant} - \text{discordant}}{\sqrt{\text{concordant} + \text{discordant} + \text{extra-}y}\ \sqrt{\text{concordant} + \text{discordant} + \text{extra-}x}} \tag{13.8.8}$$

You can easily convince yourself that this must lie between 1 and -1, and that it takes on the extreme values only for complete rank agreement or complete rank reversal, respectively.

More important, Kendall has worked out, from the combinatorics, the approximate distribution of τ in the null hypothesis of no association between x and y. In this case τ is approximately normally distributed, with zero expectation value and a variance of

$$\text{Var}(\tau) = \frac{4N + 10}{9N(N-1)} \tag{13.8.9}$$

The following program proceeds according to the above description, and therefore loops over all pairs of data points. Beware: This is an $O(N^2)$ algorithm, unlike the algorithm for r_s, whose dominant sort operations are of order $N \log N$. If you are routinely computing Kendall's τ for data sets of more than a few thousand points, you may be in for some serious computing. If, however, you are willing to bin your data into a moderate number of bins, then keep reading.

```
SUBROUTINE KENDL1(DATA1,DATA2,N,TAU,Z,PROB)
    Given data arrays DATA1 and DATA2, each of length N, this program returns Kendall's τ
    as TAU, its number of standard deviations from zero as Z, and its two-sided significance
    level as PROB. Small values of PROB indicate a significant correlation (TAU positive) or
    anticorrelation (TAU negative).
DIMENSION DATA1(N),DATA2(N)
N1=0                              This will be the argument of one square root in (13.8.8),
N2=0                                  and this the other.
IS=0                              This will be the numerator in (13.8.8).
DO 12 J=1,N-1                     Loop over first member of pair,
    DO 11 K=J+1,N                         and second member.
        A1=DATA1(J)-DATA1(K)
        A2=DATA2(J)-DATA2(K)
        AA=A1*A2
        IF(AA.NE.0.)THEN                      Neither array has a tie.
            N1=N1+1
            N2=N2+1
            IF(AA.GT.0.)THEN
                IS=IS+1
            ELSE
                IS=IS-1
            ENDIF
        ELSE                                  One or both arrays have ties.
            IF(A1.NE.0.)N1=N1+1                       An "extra x" event.
            IF(A2.NE.0.)N2=N2+1                       An "extra y" event.
        ENDIF
11  CONTINUE
12 CONTINUE
TAU=FLOAT(IS)/SQRT(FLOAT(N1)*FLOAT(N2))     Equation (13.8.8).
VAR=(4.*N+10.)/(9.*N*(N-1.))                Equation (13.8.9).
Z=TAU/SQRT(VAR)
PROB=ERFCC(ABS(Z)/1.4142136)                Significance.
RETURN
END
```

Sometimes it happens that there are only a few possible values each for x and y. In that case, the data can be recorded as a contingency table (see §13.6) which gives the number of data points for each contingency of x and y.

Spearman's rank-order correlation coefficient is not a very natural statistic under these circumstances, since it assigns to each x and y bin a not-very-meaningful midrank value and then totals up vast numbers of identical rank differences. Kendall's tau, on the other hand, with its simple counting, remains quite natural. Furthermore, its $O(N^2)$ algorithm is no longer a problem, since we can arrange for it to loop over pairs of contingency table entries (each containing many data points) instead of over pairs of data points. This is implemented in the program that follows.

Note that Kendall's tau can only be applied to contingency tables where both variables are *ordinal*, i.e. well-ordered, and that it looks specifically for monotonic correlations, not for arbitrary associations. These two properties make it less general than the methods of §13.6, which applied to *nominal*, i.e. unordered, variables and arbitrary associations.

Comparing KENDL1 above with KENDL2 below, you will see that we have "floated" a number of variables. This is because the number of events in a contingency table might be sufficiently large as to cause overflows in some of

the integer arithmetic, while the number of individual data points in a list could not possibly be that large [for an $O(N^2)$ routine!].

```
SUBROUTINE KENDL2(TAB,I,J,IP,JP,TAU,Z,PROB)
    Given a two-dimensional table TAB of physical dimension (IP,JP) and logical dimension
    (I,J), such that TAB(K,L) contains the number of events falling in bin K of one variable
    and bin L of another, this program returns Kendall's τ as TAU, its number of standard
    deviations from zero as Z, and its two-sided significance level as PROB. Small values of PROB
    indicate a significant correlation (TAU positive) or anticorrelation (TAU negative) between
    the two variables. Although TAB is a real array, it will normally contain integral values.
DIMENSION TAB(IP,JP)
EN1=0.                              See KENDL1 above.
EN2=0.
S=0.
NN=I*J                              Total number of entries in contingency table.
POINTS=TAB(I,J)
DO 12 K=0,NN-2                      Loop over entries in table,
    KI=K/J                              decoding a row
    KJ=K-J*KI                           and a column.
    POINTS=POINTS+TAB(KI+1,KJ+1)            Increment the total count of events.
    DO 11 L=K+1,NN-1                    Loop over other member of the pair,
        LI=L/J                              decoding its row
        LJ=L-J*LI                           and column.
        M1=LI-KI
        M2=LJ-KJ
        MM=M1*M2
        PAIRS=TAB(KI+1,KJ+1)*TAB(LI+1,LJ+1)
        IF(MM.NE.0)THEN                     Not a tie.
            EN1=EN1+PAIRS
            EN2=EN2+PAIRS
            IF(MM.GT.0)THEN                 Concordant, or
                S=S+PAIRS
            ELSE                            discordant.
                S=S-PAIRS
            ENDIF
        ELSE
            IF(M1.NE.0)EN1=EN1+PAIRS
            IF(M2.NE.0)EN2=EN2+PAIRS
        ENDIF
11      CONTINUE
12  CONTINUE
TAU=S/SQRT(EN1*EN2)
VAR=(4.*POINTS+10.)/(9.*POINTS*(POINTS-1.))
Z=TAU/SQRT(VAR)
PROB=ERFCC(ABS(Z)/1.4142136)
RETURN
END
```

REFERENCES AND FURTHER READING:

Lehmann, E.L. 1975, *Nonparametrics: Statistical Methods Based on Ranks* (San Francisco: Holden-Day).

Downie, N.M., and Heath, R.W. 1965, *Basic Statistical Methods*, 2nd ed. (New York: Harper and Row), pp. 206–209.

SPSS: Statistical Package for the Social Sciences, 2nd ed., by Norman H. Nie, et al. (New York: McGraw-Hill), §18.3.

13.9 Smoothing of Data

The concept of "smoothing" data lies in a murky area, just beyond the fringe of these better posed and more highly recommended techniques:

- Least squares fitting to a parametric model, which is covered in some detail in Chapter 14.
- Optimal filtering of a noisy signal, see §12.6.
- Letting it all hang out. (Have you considered showing the data as they actually are?)

On the other hand, it is useful to have some techniques available that are more objective than

- Draftsman's license. "The smooth curve was drawn by eye through the original data." (Through each individual data point? Or through the forest of scattered points? By a draftsman? Or by someone who knows what hypothesis the data are supposed to substantiate?)

Data smoothing is probably most justified when it is used simply as a graphical technique, to guide the eye through a forest of data points all with large error bars. In this case, the individual points and their error bars should be plotted on the same graph, and no quantitative claims should be made on the basis of the smoothed curve. Data smoothing is least justified when it is used subjectively to massage the data this way and that, until some feature in the smoothed curve emerges and is pounced on in support of an hypothesis.

Data smoothing is what we would call "semi-parametric." It clearly involves some notion of "averaging" the measured dependent (y) variable, which is parametric. Smoothing a set of values will not, in general, be the same as smoothing their logarithms. You have to think about which is closer to your needs. On the other hand, smoothing is not supposed to be tied to any particular functional form $y(x)$, or to any particular parametrization of the x axis. Smoothing is art, not science.

Here is a program for smoothing an array of ordinates (y's) that are in order of increasing abscissas (x's), but without using the abscissas themselves. The program pretends that the abscissas are equally spaced, as they are if reparametrized to the variable "point number." It removes any linear trend, and then uses a Fast Fourier Transform to low-pass filter the data. The linear trend is reinserted at the end. One user-specified constant enters: the "amount of smoothing," specified as the number of points over which the data should be smoothed (not necessarily an integer). Zero gives no smoothing at all, while any value larger than about half the number of data points will render the data virtually featureless. The program gives results that are generally in accord with the notion "draw a smooth curve through these scattered points," and is at least arguably objective in doing so. A sample of its output is shown in Figure 13.9.1.

```
SUBROUTINE SMOOFT(Y,N,PTS)
    Smooths an array Y of length N, with a window whose full width is of order PTS neighboring
    points, a user supplied value.  Y is modified.
PARAMETER(MMAX=1024)          Maximum size of padded array.
DIMENSION Y(MMAX)
```

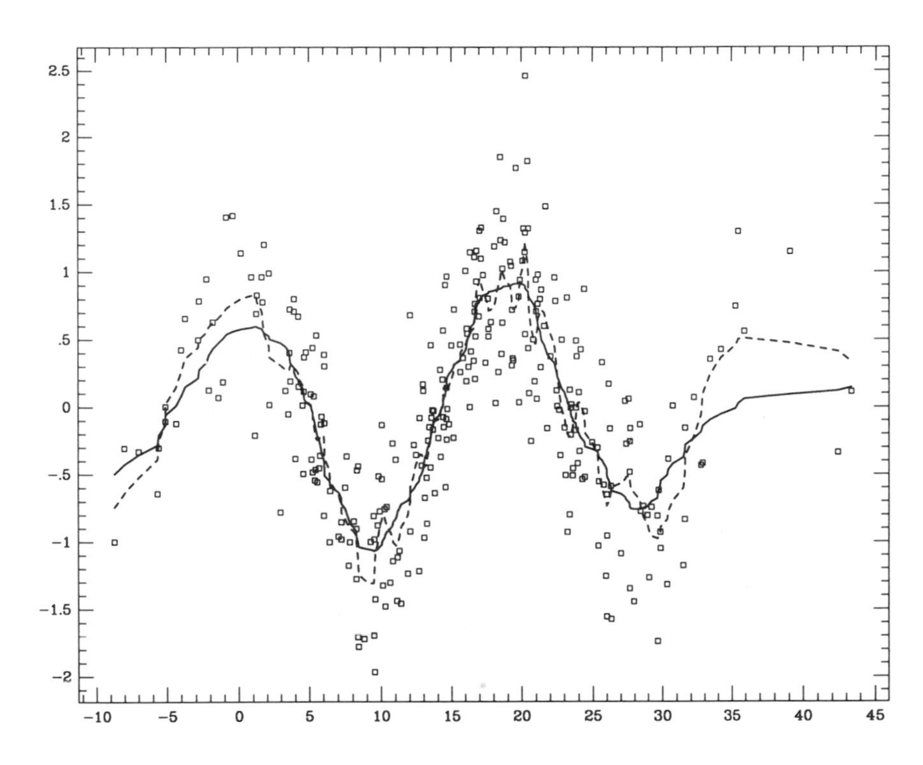

Figure 13.9.1. Smoothing of data with the routine **SMOOFT**. The open squares are noisy data points, nonuniformly sampled with the greatest sampling density towards the middle of the abscissa. The dotted curve is obtained with smoothing parameter **PTS**= 10.0 (averaging approximately 10 points); the solid curve has **PTS**= 30.0.

```
      M=2
      NMIN=N+2.*PTS                    Minimum size including buffer against wrap around.
1     IF(M.LT.NMIN)THEN                Find the next larger power of 2.
        M=2*M
      GO TO 1
      ENDIF
      IF(M.GT.MMAX) PAUSE 'MMAX to small'
      CONST=(PTS/M)**2                 Useful constants below.
      Y1=Y(1)
      YN=Y(N)
      RN1=1./(N-1.)
      DO 11 J=1,N                      Remove the linear trend and transfer data.
        Y(J)=Y(J)-RN1*(Y1*(N-J)+YN*(J-1))
11    CONTINUE
      IF(N+1.LE.M)THEN                 Zero pad.
        DO 12 J=N+1,M
          Y(J)=0.
12      CONTINUE
      ENDIF
      MO2=M/2
      CALL REALFT(Y,MO2,1)             Fourier transform.
      Y(1)=Y(1)/MO2
      FAC=1.                           Window function.
      DO 13 J=1,MO2-1                  Multiply the data by the window function.
```

```
    K=2*J+1
    IF(FAC.NE.0.)THEN
        FAC=AMAX1(0.,(1.-CONST*J**2)/MO2)
        Y(K)=FAC*Y(K)
        Y(K+1)=FAC*Y(K+1)
    ELSE                          Don't do unnecessary multiplies after window function is zero.
        Y(K)=0.
        Y(K+1)=0.
    ENDIF
 13 CONTINUE
FAC=AMAX1(0.,(1.-0.25*PTS**2)/MO2)  Last point.
Y(2)=FAC*Y(2)
CALL REALFT(Y,MO2,-1)             Inverse Fourier transform.
DO 14 J=1,N                       Restore the linear trend.
    Y(J)=RN1*(Y1*(N-J)+YN*(J-1))+Y(J)
 14 CONTINUE
RETURN
END
```

There is a different smoothing technique that we have found useful for data whose error distribution has very broad tails: At each data point, construct a "windowed median" of ordinates, that is the median of that point's ordinate and the ordinates of the $2M$ data points nearest in abscissa, M on each side. Near the edge of the graph, where there are fewer than M points available on one side, instead use more points on the other side, so that your medians are always of $2M$ points. Choose M to taste. The windowed median is not smooth (in the sense of differentiable), but it does fluctuate substantially less than the raw data, and it will often track otherwise noisy trends quite well. It is also genuinely nonparametric, i.e. invariant under reparametrizations of both abscissa and ordinate.

REFERENCES AND FURTHER READING:

Bevington, Philip R. 1969, *Data Reduction and Error Analysis for the Physical Sciences* (New York: McGraw-Hill), §13.1.

Chapter 14. Modeling of Data

14.0 Introduction

Given a set of observations, one often wants to condense and summarize the data by fitting it to a "model" that depends on adjustable parameters. Sometimes the model is simply a convenient class of functions, such as polynomials or Gaussians, and the fit supplies the appropriate coefficients. Other times, the model's parameters come from some underlying theory that the data are supposed to satisfy; examples are coefficients of rate equations in a complex network of chemical reactions, or orbital elements of a binary star. Modeling can also be used as a kind of constrained interpolation, where you want to extend a few data points into a continuous function, but with some underlying idea of what that function should look like.

The basic approach in all cases is usually the same: You choose or design a *figure-of-merit function* ("merit function," for short) that measures the agreement between the data and the model with a particular choice of parameters. The merit function is conventionally arranged so that small values represent close agreement. The parameters of the model are then adjusted to achieve a minimum in the merit function, yielding *best-fit parameters.* The adjustment process is thus a problem in minimization in many dimensions. This optimization was the subject of Chapter 10; however, there exist special, more efficient, methods that are specific to modeling, and we will discuss these in this chapter.

There are important issues that go beyond the mere finding of best-fit parameters. Data are generally not exact. They are subject to *measurement errors* (called *noise* in the context of signal-processing). Thus, typical data never exactly fit the model that is being used, even when that model is correct. We need the means to assess whether or not the model is appropriate, that is, we need to test the *goodness-of-fit* against some useful statistical standard.

We usually also need to know the accuracy with which parameters are determined by the data set. In other words, we need to know the likely errors of the best-fit parameters.

Finally, it is not uncommon in fitting data to discover that the merit function is not unimodal, with a single minimum. In some cases, we may be interested in global, rather than local questions. Not, "how good is this fit?", but rather, "how sure am I that there is not a *very much better* fit in some

corner of parameter space?" As we have seen in Chapter 10, especially §10.7, this kind of problem is generally quite difficult to solve.

The important message we want to deliver is that fitting of parameters is not the end-all of parameter estimation. To be genuinely useful, a fitting procedure should provide (i) parameters, (ii) error estimates on the parameters, and (iii) a statistical measure of goodness-of-fit. When the third item suggests that the model is an unlikely match to the data, then items (i) and (ii) are probably worthless. Unfortunately, many practitioners of parameter estimation never proceed beyond item (i)! They deem a fit acceptable if a graph of data and model "looks good." This approach is known as *chi-by-eye*. Luckily, its practitioners get what they deserve.

REFERENCES AND FURTHER READING:

Bevington, Philip R. 1969, *Data Reduction and Error Analysis for the Physical Sciences* (New York: McGraw-Hill).

Brownlee, K.A. 1965, *Statistical Theory and Methodology*, 2nd ed. (New York: Wiley).

Martin, B.R. 1971, *Statistics for Physicists* (New York: Academic Press).

von Mises, Richard. 1964, *Mathematical Theory of Probability and Statistics* (New York: Academic Press), Chapter X.

Korn, G.A., and Korn, T.M. 1968, *Mathematical Handbook for Scientists and Engineers*, 2nd ed. (New York: McGraw-Hill), Chapters 18–19.

14.1 Least Squares as a Maximum Likelihood Estimator

Suppose that we are fitting N data points (x_i, y_i) $i = 1, \ldots, N$, to a model which has M adjustable parameters a_j, $j = 1, \ldots, M$. The model predicts a functional relationship between the measured independent and dependent variables,

$$y(x) = y(x; a_1 \ldots a_M) \tag{14.1.1}$$

where the dependence on the parameters is indicated explicitly on the right-hand side.

What, exactly, do we want to minimize to get fitted values for the a_j's? The first thing that comes to mind is the familiar least-squares fit,

$$\text{minimize over } a_1 \ldots a_M : \qquad \sum_{i=1}^{N} \left[y_i - y(x_i; a_1 \ldots a_M)\right]^2 \tag{14.1.2}$$

But where does this come from? What general principles is it based on? The answer to these questions takes us into the subject of *maximum likelihood estimators*.

Given a particular data set of x_i's and y_i's, we have the intuitive feeling that some parameter sets $a_1 \dots a_M$ are very unlikely — those for which the model function $y(x)$ looks *nothing like* the data — while others may be very likely — those which closely resemble the data. How can we quantify this intuitive feeling? How can we select fitted parameters that are "most likely" to be correct? It is not meaningful to ask the question, "What is the probability that a particular set of fitted parameters $a_1 \dots a_M$ is correct?" The reason is that there is no statistical universe of models from which the parameters are drawn. There is just one model, the correct one, and a statistical universe of data sets that are drawn from it!

That being the case, we can, however, turn the question around, and ask, "*Given a particular set of parameters*, what is the probability that this data set could have occurred?" If the y_i's take on continuous values, the probability will always be zero unless we add the phrase, "...plus or minus some fixed Δy on each data point." So let's always take this phrase as understood. If the probability of obtaining the data set is infinitesimally small, then we can conclude that the parameters under consideration are "unlikely" to be right. Conversely, our intuition tells us that the data set should not be too improbable for the correct choice of parameters.

In other words, we identify the probability of the data given the parameters (which is a mathematically computable number), as the *likelihood* of the parameters given the data. This identification is entirely based on intuition. It has no formal mathematical basis in and of itself; as we already remarked, statistics is *not* a branch of mathematics!

Once we make this intuitive identification, however, it is only a small further step to decide to fit for the parameters $a_1 \dots a_M$ precisely by finding those values that *maximize* the likelihood defined in the above way. This form of parameter estimation is *maximum likelihood estimation.*

We are now ready to make the connection to (14.1.2). Suppose that each data point y_i has a measurement error that is independently random and distributed as a normal (Gaussian) distribution around the "true" model $y(x)$. And suppose that the standard deviations σ of these normal distributions are the same for all points. Then the probability of the data set is the product of the probabilities of each point,

$$P = \prod_{i=1}^{N} \left\{ \exp\left[-\frac{1}{2} \left(\frac{y_i - y(x_i)}{\sigma} \right)^2 \right] \Delta y \right\} \tag{14.1.3}$$

Notice that there is a factor Δy in each term in the product. Maximizing (14.1.3) is equivalent to maximizing its logarithm, or minimizing the negative of its logarithm, namely,

$$\left[\sum_{i=1}^{N} \frac{[y_i - y(x_i)]^2}{2\sigma^2} \right] - N \log \Delta y \tag{14.1.4}$$

Since N, σ and Δy are all constants, minimizing this equation is equivalent to minimizing (14.1.2).

What we see is that least-squares fitting *is* a maximum likelihood estimation of the fitted parameters *if* the measurement errors are independent and normally distributed with constant standard deviation. Notice that we made no assumption about the linearity or nonlinearity of the model $y(x; a_1 \dots)$ in its parameters $a_1 \dots a_M$. Just below, we will relax our assumption of constant standard deviations and obtain the very similar formulas for what is called "chi-square fitting" or "weighted least-squares fitting." First, however, let us discuss further our very stringent assumption of a normal distribution.

For a hundred years or so, mathematical statisticians have been in love with the fact that the probability distribution of the sum of a very large number of very small random deviations always converges to a normal distribution. (For precise statements of this *central limit theorem*, consult von Mises or other standard works on mathematical statistics.) This infatuation tended to focus interest away from the fact that, for real data, the normal distribution is often rather poorly realized, if it is realized at all. We are often taught, rather casually, that, on average, measurements will fall within $\pm\sigma$ of the true value 68 percent of the time, within $\pm 2\sigma$ 95 percent of the time, and within $\pm 3\sigma$ 99.7 percent of the time. Extending this, one would expect a measurement to be off by $\pm 20\sigma$ only one time out of 2×10^{88}. We all know that "glitches" are much more likely than *that*!

In some instances, the deviations from a normal distribution are easy to understand and quantify. For example, in measurements obtained by counting events, the measurement errors are usually distributed as a Poisson distribution, whose cumulative probability function was already discussed in §6.2. When the number of counts going into one data point is large, the Poisson distribution converges toward a Gaussian. However, the convergence is not uniform when measured in fractional accuracy. The more standard deviations out on the tail of the distribution, the larger the number of counts must be before a value close to the Gaussian is realized. The sign of the effect is always the same: the Gaussian predicts that "tail" events are much less likely than they actually (by Poisson) are. This causes such events, when they occur, to skew a least-squares fit much more than they ought.

Other times, the deviations from a normal distribution are not so easy to understand in detail. Experimental points are occasionally just *way off*. Perhaps the power flickered during a point's measurement, or someone kicked the apparatus, or someone wrote down a wrong number. Points like this are called *outliers*. They can easily turn a least-squares fit on otherwise adequate data into nonsense. Their probability of occurrence in the assumed Gaussian model is so small that the maximum likelihood estimator is willing to distort the whole curve to try to bring them, mistakenly, into line.

The subject of *robust statistics* deals with cases where the normal or Gaussian model is a bad approximation, or cases where outliers are important. We will discuss robust methods briefly in §14.7. All the sections between this one and that one assume, one way or the other, a Gaussian model for the measurement errors in the data. It it quite important that you keep the

limitations of that model in mind, even as you use the very useful methods which follow from assuming it.

Finally, note that our discussion of measurement errors has been limited to *statistical* errors, the kind that will average away if we only take enough data. Measurements are also susceptible to *systematic* errors that will not go away with any amount of averaging. For example, the calibration of a metal meter stick might depend on its temperature. If we take all our measurements at the same wrong temperature, then no amount of averaging or numerical processing will correct for this unrecognized systematic error.

Chi-Square Fitting

We considered the chi-square statistic once before, in §13.5. Here it arises in a slightly different context.

If each data point (x_i, y_i) has its own standard deviation σ_i, then equation (14.1.3) is modified only by putting a subscript i on the symbol σ. That subscript also propagates docilely into (14.1.4), so that the maximum likelihood estimate of the model parameters is obtained by minimizing the quantity

$$\chi^2 \equiv \sum_{i=1}^{N} \left(\frac{y_i - y(x_i; a_1 \dots a_M)}{\sigma_i} \right)^2 \tag{14.1.5}$$

called the "chi-square."

To whatever extent the measurement errors actually *are* normally distributed, the quantity χ^2 is correspondingly a sum of N squares of normally distributed quantities, each normalized to unit variance. Once we have adjusted the $a_1 \dots a_M$ to minimize the value of χ^2, the terms in the sum are not all statistically independent. However it turns out that the probability distribution for different values of χ^2 at its minimum can nevertheless be derived analytically, and is the *chi-square distribution for $N - M$ degrees of freedom.* We learned how to compute this probability function using the incomplete gamma function GAMMQ in §6.2. In particular, equation (6.2.18) gives the probability Q that the chi-square should exceed a particular value χ^2 by chance, where $\nu = N - M$ is the *number of degrees of freedom.* The quantity Q, or its complement $P \equiv 1 - Q$ is frequently tabulated in appendices to statistics books, but we generally find it easier to use GAMMQ and compute our own values: Q=GAMMQ(0.5*ν,0.5*χ^2).

This computed probability gives a quantitative measure for the goodness-of-fit of the model. If Q is a very small probability for some particular data set, then the apparent discrepancies are unlikely to be chance fluctuations. Much more probably either (i) the model is wrong — can be statistically rejected, or (ii) someone has lied to you about the size of the measurement errors σ_i — they are really larger than stated.

It is an important point that the chi-square probability Q does not directly measure the credibility of the assumption that the measurement errors

are normally distributed. It assumes they are. In most, but not all, cases, however, the effect of nonnormal errors is to create an abundance of outlier points. These decrease the probability Q, so that we can add another possible, though less definitive, conclusion to the above list: (iii) the measurement errors may not be normally distributed.

Possibility (iii) is fairly common, and also fairly benign. It is for this reason that reasonable experimenters are often rather tolerant of low probabilities Q. It is not uncommon to deem acceptable on equal terms any models with, say, $Q > 0.001$. This is not as sloppy as it sounds: truly *wrong* models will often be rejected with vastly smaller values of Q, 10^{-18}, say. However, if day-in and day-out you find yourself accepting models with $Q \sim 10^{-3}$, you really should track down the cause.

If you happen to know the actual distribution law of your measurement errors, then you might wish to *Monte Carlo simulate* some data sets drawn from a particular model, cf. §7.2–§7.3. You can then subject these synthetic data sets to your actual fitting procedure, so as to determine both the probability distribution of the χ^2 statistic, and also the accuracy with which your model parameters are reproduced by the fit. We discuss this further in §14.5. The technique is very general, but it can also be very expensive.

At the opposite extreme, it sometimes happens that the probability Q is too large, too near to 1, literally too good to be true! Nonnormal measurement errors cannot in general produce this disease, since the normal distribution is about as "compact" as a distribution can be. Almost always, the cause of too good a chi-square fit is that the experimenter, in a "fit" of conservativism, has *overestimated* his or her measurement errors. Very rarely, too good a chi-square signals actual fraud, data that has been "fudged" to fit the model.

A rule of thumb is that a "typical" value of χ^2 for a "moderately" good fit is $\chi^2 \approx \nu$. More precise is the statement that, asymptotically for large ν, the statistic χ^2 becomes normally distributed with a mean ν and a standard deviation $\sqrt{2\nu}$.

In some cases the uncertainties associated with a set of measurements are not known in advance, and considerations related to χ^2 fitting are used to derive a value for σ. If we assume that all measurements have the same standard deviation, $\sigma_i = \sigma$, and that the model does fit well, then we can proceed by first assigning an arbitrary constant σ to all points, next fitting for the model parameters by minimizing χ^2, and finally recomputing

$$\sigma^2 = \sum_{i=1}^{N} [y_i - y(x_i)]^2 / N \tag{14.1.6}$$

Obviously, this approach prohibits an independent assessment of goodness-of-fit, a fact occasionally missed by its adherents. When, however, the measurement error is not known, this approach at least allows *some* kind of error bar to be assigned to the points.

If we take the derivative of equation (14.1.5) with respect to the parameters a_k, we obtain equations which must hold at the chi-square minimum,

$$0 = \sum_{i=1}^{N} \left(\frac{y_i - y(x_i)}{\sigma_i^2} \right) \left(\frac{\partial y(x_i; \ldots a_k \ldots)}{\partial a_k} \right) \qquad k = 1, \ldots, M \tag{14.1.7}$$

Equation (14.1.7) is, in general, a set of M nonlinear equations for the M unknown a_k. Various of the procedures described subsequently in this chapter derive from (14.1.7) and its specializations.

REFERENCES AND FURTHER READING:

Bevington, Philip R. 1969, *Data Reduction and Error Analysis for the Physical Sciences* (New York: McGraw-Hill), Chapters 1-4.

von Mises, Richard. 1964, *Mathematical Theory of Probability and Statistics* (New York: Academic Press), §VI.C.

14.2 Fitting Data to a Straight Line

A concrete example will make the considerations of the previous section more meaningful. We consider the problem of fitting a set of N data points (x_i, y_i) to a straight-line model

$$y(x) = y(x; a, b) = a + bx \tag{14.2.1}$$

This problem is often called *linear regression*, a terminology that originated, long ago, in the social sciences. We assume that the uncertainty σ_i associated with each measurement y_i is known, and that the x_i's (values of the dependent variable) are known exactly.

To measure how well the model agrees with the data, we use the chi-square merit function (14.1.5), which in this case is

$$\chi^2(a, b) = \sum_{i=1}^{N} \left(\frac{y_i - a - bx_i}{\sigma_i} \right)^2 \tag{14.2.2}$$

If the measurement errors are normally distributed, then this merit function will give maximum likelihood parameter estimations of a and b; if the errors are not normally distributed, then the estimations are not maximum likelihood, but may still be useful in a practical sense. In §14.6, we will treat the case where outlier points are so numerous as to render the χ^2 merit function useless.

Equation (14.2.2) is minimized to determine a and b. At its minimum, derivatives of $\chi^2(a,b)$ with respect to a, b vanish.

$$
\begin{aligned}
0 = \frac{\partial \chi^2}{\partial a} &= -2 \sum_{i=1}^{N} \frac{y_i - a - b x_i}{\sigma_i^2} \\
0 = \frac{\partial \chi^2}{\partial b} &= -2 \sum_{i=1}^{N} \frac{x_i (y_i - a - b x_i)}{\sigma_i^2}
\end{aligned}
\tag{14.2.3}
$$

These conditions can be rewritten in a convenient form if we define the following sums,

$$
\begin{aligned}
S \equiv \sum_{i=1}^{N} \frac{1}{\sigma_i^2} \quad & S_x \equiv \sum_{i=1}^{N} \frac{x_i}{\sigma_i^2} \quad S_y \equiv \sum_{i=1}^{N} \frac{y_i}{\sigma_i^2} \\
S_{xx} \equiv \sum_{i=1}^{N} \frac{x_i^2}{\sigma_i^2} \quad & S_{xy} \equiv \sum_{i=1}^{N} \frac{x_i y_i}{\sigma_i^2}
\end{aligned}
\tag{14.2.4}
$$

With these definitions (14.2.3) becomes

$$
\begin{aligned}
aS + bS_x &= S_y \\
aS_x + bS_{xx} &= S_{xy}
\end{aligned}
\tag{14.2.5}
$$

The solution of these two equations in two unknowns is calculated as

$$
\begin{aligned}
\Delta &\equiv S S_{xx} - (S_x)^2 \\
a &= \frac{S_{xx} S_y - S_x S_{xy}}{\Delta} \\
b &= \frac{S S_{xy} - S_x S_y}{\Delta}
\end{aligned}
\tag{14.2.6}
$$

Equation (14.2.6) gives the solution for the best-fit model parameters a and b.

We are not done, however. We must estimate the probable uncertainties in the estimates of a and b, since obviously the measurement errors in the data must introduce some uncertainty in the determination of those parameters. If the data are independent, then each contributes its own bit of uncertainty to the parameters. Consideration of propagation of errors shows that the variance σ_f^2 in the value of any function will be

$$
\sigma_f^2 = \sum_{i=1}^{N} \sigma_i^2 \left(\frac{\partial f}{\partial y_i} \right)^2
\tag{14.2.7}
$$

For the straight line, the derivatives of a and b with respect to y_i can be directly evaluated from the solution:

$$\begin{aligned}\frac{\partial a}{\partial y_i} &= \frac{S_{xx} - S_x x_i}{\sigma_i^2 \Delta} \\ \frac{\partial b}{\partial y_i} &= \frac{S x_i - S_x}{\sigma_i^2 \Delta}\end{aligned} \tag{14.2.8}$$

Summing over the points as in (14.2.7), we get

$$\begin{aligned}\sigma_a^2 &= S_{xx}/\Delta \\ \sigma_b^2 &= S/\Delta\end{aligned} \tag{14.2.9}$$

which are the variances in the estimates of a and b, respectively. We will see in §14.5 that an additional number is also needed to characterize properly the probable uncertainty of the parameter estimation. That number is the *covariance* of a and b, and (as we will see below) is given by

$$\text{Cov}(a, b) = -S_x/\Delta \tag{14.2.10}$$

The coefficient of correlation between the uncertainty in a and the uncertainty in b, which is a number between -1 and 1, follows from (14.2.10) (compare equation 13.7.1),

$$r_{ab} = \frac{-S_x}{\sqrt{S S_{xx}}} \tag{14.2.11}$$

A positive value of r_{ab} indicates that the errors in a and b are likely to have the same sign, while a negative value indicates the errors are anticorrelated, likely to have opposite signs.

We are *still* not done. We must estimate the goodness-of-fit of the data to the model. Absent this estimate, we have not the slightest indication that the parameters a and b in the model have any meaning at all! The probability Q that a value of chi-square as *poor* as the value (14.2.2) should occur by chance is

$$Q = \text{GAMMQ}\left(\frac{N-2}{2}, \frac{\chi^2}{2}\right) \tag{14.2.12}$$

Here GAMMQ is our routine for the incomplete gamma function $Q(a, x)$, §6.2. If Q is larger than, say, 0.1, then the goodness-of-fit is believable. If it is larger than, say, 0.001, then the fit *may* be acceptable if the errors are nonnormal or

have been moderately underestimated. If Q is less than 0.001 then the model and/or estimation procedure can rightly be called into question. In this latter case, turn to §14.6 to proceed further.

If you do not know the individual measurement errors of the points σ_i, and are proceeding (dangerously) to use equation (14.1.6) for estimating these errors, then here is the procedure for estimating the probable uncertainties of the parameters a and b: Set $\sigma_i \equiv 1$ in all equations through (14.2.6), and multiply σ_a and σ_b, as obtained from equation (14.2.9), by the additional factor $\sqrt{\chi^2/(N-2)}$, where χ^2 is computed by (14.2.2) using the fitted parameters a and b. As discussed above, this procedure is equivalent to *assuming* a good fit, so you get no independent goodness-of-fit probability Q.

In §13.7 we promised a relation between the linear correlation coefficient r (equation 13.7.1) and a goodness-of-fit measure, χ^2 (equation 14.2.2). For unweighted data (all $\sigma_i = 1$), that relation is

$$\chi^2 = (1 - r^2)\mathrm{NVar}\,(y_1 \ldots y_N) \tag{14.2.13}$$

where

$$\mathrm{NVar}\,(y_1 \ldots y_N) \equiv \sum_{i=1}^{N} (y_i - \overline{y})^2 \tag{14.2.14}$$

For data with varying weights σ_i, the above equations remain valid if the sums in equation (13.7.1) are weighted by $1/\sigma_i^2$.

The following subroutine, `FIT`, carries out exactly the operations that we have discussed. When the weights σ are known in advance, the calculations exactly correspond to the formulas above. However, when weights σ are unavailable, the routine *assumes* equal values of σ for each point and *assumes* a good fit, as discussed in §14.1.

The formulas (14.2.6) are susceptible to roundoff error. Accordingly, we rewrite them as follows: Define

$$t_i = \frac{1}{\sigma_i}\left(x_i - \frac{S_x}{S}\right), \qquad i = 1, 2, \ldots, N \tag{14.2.15}$$

and

$$S_{tt} = \sum_{i=1}^{N} t_i^2 \tag{14.2.16}$$

Then, as you can verify by direct substitution,

$$b = \frac{1}{S_{tt}} \sum_{i=1}^{N} \frac{t_i y_i}{\sigma_i} \tag{14.2.17}$$

$$a = \frac{S_y - S_x b}{S} \tag{14.2.18}$$

$$\sigma_a^2 = \frac{1}{S}\left(1 + \frac{S_x^2}{SS_{tt}}\right) \tag{14.2.19}$$

$$\sigma_b^2 = \frac{1}{S_{tt}} \tag{14.2.20}$$

$$\text{Cov}(a,b) = -\frac{S_x}{SS_{tt}} \tag{14.2.21}$$

$$r_{ab} = \frac{\text{Cov}(a,b)}{\sigma_a \sigma_b} \tag{14.2.22}$$

```
SUBROUTINE FIT(X,Y,NDATA,SIG,MWT,A,B,SIGA,SIGB,CHI2,Q)
```

Given a set of NDATA points X(I),Y(I) with standard deviations SIG(I), fit them to a straight line $y = a + bx$ by minimizing χ^2. Returned are A,B and their respective probable uncertainties SIGA and SIGB, the chi-square CHI2, and the goodness-of-fit probability Q (that the fit would have χ^2 this large or larger). If MWT=0 on input, then the standard deviations are assumed to be unavailable: Q is returned as 1.0 and the normalization of CHI2 is to unit standard deviation on all points.

```
DIMENSION X(NDATA),Y(NDATA),SIG(NDATA)
SX=0.                           Initialize sums to zero.
SY=0.
ST2=0.
B=0.
IF(MWT.NE.0) THEN               Accumulate sums ...
    SS=0.
    DO 11 I=1,NDATA             ...with weights
        WT=1./(SIG(I)**2)
        SS=SS+WT
        SX=SX+X(I)*WT
        SY=SY+Y(I)*WT
    11 CONTINUE
ELSE
    DO 12 I=1,NDATA             ...or without weights.
        SX=SX+X(I)
        SY=SY+Y(I)
    12 CONTINUE
    SS=FLOAT(NDATA)
ENDIF
SXOSS=SX/SS
IF(MWT.NE.0) THEN
    DO 13 I=1,NDATA
        T=(X(I)-SXOSS)/SIG(I)
        ST2=ST2+T*T
        B=B+T*Y(I)/SIG(I)
    13 CONTINUE
ELSE
    DO 14 I=1,NDATA
        T=X(I)-SXOSS
        ST2=ST2+T*T
        B=B+T*Y(I)
    14 CONTINUE
ENDIF
B=B/ST2                         Solve for A, B, σa and σb.
A=(SY-SX*B)/SS
SIGA=SQRT((1.+SX*SX/(SS*ST2))/SS)
SIGB=SQRT(1./ST2)
CHI2=0.                         Calculate χ2
IF(MWT.EQ.0) THEN
    DO 15 I=1,NDATA
        CHI2=CHI2+(Y(I)-A-B*X(I))**2
```

```
         15 CONTINUE
      Q=1.
      SIGDAT=SQRT(CHI2/(NDATA-2))       For unweighted data evaluate typical SIG using CHI2, and
      SIGA=SIGA*SIGDAT                  adjust the standard deviations.
      SIGB=SIGB*SIGDAT
   ELSE
      DO 16 I=1,NDATA
         CHI2=CHI2+((Y(I)-A-B*X(I))/SIG(I))**2
16    CONTINUE
      Q=GAMMQ(0.5*(NDATA-2),0.5*CHI2)        §6.2
   ENDIF
   RETURN
   END
```

REFERENCES AND FURTHER READING:

Bevington, Philip R. 1969, *Data Reduction and Error Analysis for the Physical Sciences* (New York: McGraw-Hill), Chapter 6.

14.3 General Linear Least Squares

An immediate generalization of the previous section is to fit a set of data points (x_i, y_i) to a model which is not just a linear combination of 1 and x (namely $a + bx$), but rather a linear combination of *any* M specified functions of x. For example, the functions could be $1, x, x^2, \ldots, x^{M-1}$, in which case their general linear combination,

$$y(x) = a_1 + a_2 x + a_3 x^2 + \cdots + a_M x^{M-1} \tag{14.3.1}$$

is a polynomial of degree $M - 1$. Or, the functions could be sines and cosines, in which case their general linear combination can be a harmonic series.

The general form of this kind of model is

$$y(x) = \sum_{k=1}^{M} a_k X_k(x) \tag{14.3.2}$$

where $X_1(x), \ldots, X_M(x)$ are arbitrary fixed functions of x, called the *basis functions*.

Note that the functions $X_k(x)$ can be wildly nonlinear functions of x. In this discussion "linear" only refers to the model's dependence on its *parameters* a_k.

For these linear models we generalize the discussion of the previous section by defining a merit function

$$\chi^2 = \sum_{i=1}^{N} \left[\frac{y_i - \sum_{k=1}^{M} a_k X_k(x_i)}{\sigma_i} \right]^2 \tag{14.3.3}$$

As before σ_i is the measurement error (standard deviation) of the ith data point, presumed to be known. If the measurement errors are not known, they may all, as before, be set to the constant value $\sigma = 1$.

Once again, we will pick as best parameters those that minimize χ^2. There are several different techniques available for finding this minimum. Two are particularly useful, and we will discuss both in this section. To introduce them and elucidate their relationship, we need some notation.

Let $\mathbf{A}$ be a matrix whose $N \times M$ components are constructed from the M basis functions evaluated at the N abscissas x_i, and from the N measurement errors σ_i, by the prescription

$$A_{ij} = \frac{X_j(x_i)}{\sigma_i} \tag{14.3.4}$$

The matrix $\mathbf{A}$ is called the *design matrix* of the fitting problem. Notice that in general $\mathbf{A}$ has more rows than columns, $N \geq M$, since there must be more data points than model parameters to be solved for. (You can fit a straight line to two points, but not a very meaningful quintic!) The design matrix is shown schematically in Figure 14.3.1.

Also define a vector $\mathbf{b}$ of length N by

$$b_i = \frac{y_i}{\sigma_i} \tag{14.3.5}$$

and denote the M vector whose components are the parameters to be fitted, $a_1, \ldots, a_M$, by $\mathbf{a}$.

Solution by Use of the Normal Equations

The minimum of (14.3.3) occurs where the derivative of χ^2 with respect to all M parameters a_k vanishes. Specializing equation (14.1.7) to the case of the model (14.3.2), this condition yields the M equations

$$0 = \sum_{i=1}^{N} \frac{1}{\sigma_i^2} \left[y_i - \sum_{j=1}^{M} a_j X_j(x_i) \right] X_k(x_i) \qquad k = 1, \ldots, M \tag{14.3.6}$$

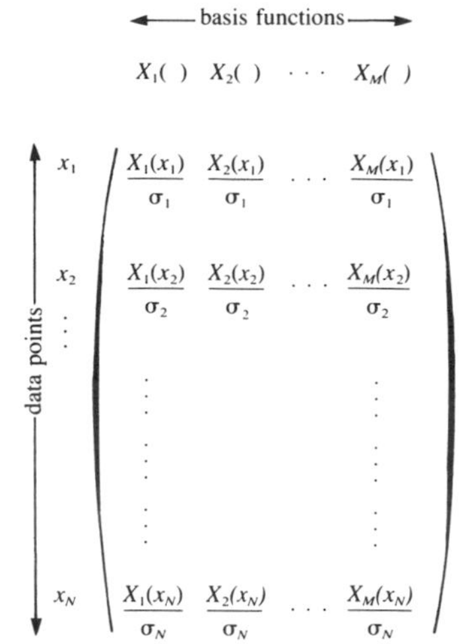

Figure 14.3.1. Design matrix for the least-squares fit of a linear combination of M basis functions to N data points. The matrix elements involve the basis functions evaluated at the values of the independent variable at which measurements are made, and the standard deviations of the measured dependent variable. The measured values of the dependent variable do not enter the design matrix.

Interchanging the order of summations, we can write (14.3.6) as the matrix equation

$$\sum_{j=1}^{M} \alpha_{kj} a_j = \beta_k \tag{14.3.7}$$

where

$$\alpha_{kj} = \sum_{i=1}^{N} \frac{X_j(x_i) X_k(x_i)}{\sigma_i^2} \qquad \text{or equivalently} \qquad [\alpha] = \mathbf{A}^T \cdot \mathbf{A} \tag{14.3.8}$$

an $M \times M$ matrix, and

$$\beta_k = \sum_{i=1}^{N} \frac{y_i X_k(x_i)}{\sigma_i^2} \qquad \text{or equivalently} \qquad [\beta] = \mathbf{A}^T \cdot \mathbf{b} \tag{14.3.9}$$

a vector of length M.

The equations (14.3.6) or (14.3.7) are called the *normal equations* of the least-squares problem. They can be solved for the vector of parameters $\mathbf{a}$ by the standard methods of Chapter 2, notably *LU* decomposition and backsubstitution or Gauss-Jordan elimination. In matrix form, the normal equations

can be written as either

$$[\alpha] \cdot \mathbf{a} = [\beta] \qquad \text{or as} \qquad \left(\mathbf{A}^T \cdot \mathbf{A}\right) \cdot \mathbf{a} = \mathbf{A}^T \cdot \mathbf{b} \tag{14.3.10}$$

The inverse matrix $C_{jk} \equiv [\alpha]_{jk}^{-1}$ is closely related to the probable (or, more precisely, *standard*) uncertainties of the estimated parameters $\mathbf{a}$. To estimate these uncertainties, consider that

$$a_j = \sum_{k=1}^{M} [\alpha]_{jk}^{-1} \beta_k = \sum_{k=1}^{M} C_{jk} \left[\sum_{i=1}^{N} \frac{y_i X_k(x_i)}{\sigma_i^2} \right] \tag{14.3.11}$$

and that the variance associated with the estimate a_j can be found as in (14.2.7) from

$$\sigma^2(a_j) = \sum_{i=1}^{N} \sigma_i^2 \left(\frac{\partial a_j}{\partial y_i} \right)^2 \tag{14.3.12}$$

Note that α_{jk} is independent of y_i, so that

$$\frac{\partial a_j}{\partial y_i} = \sum_{k=1}^{M} C_{jk} X_k(x_i)/\sigma_i^2 \tag{14.3.13}$$

Consequently, we find that

$$\sigma^2(a_j) = \sum_{k=1}^{M} \sum_{l=1}^{M} C_{jk} C_{jl} \left[\sum_{i=1}^{N} \frac{X_k(x_i) X_l(x_i)}{\sigma_i^2} \right] \tag{14.3.14}$$

The final term in brackets is just the matrix $[\alpha]$. Since this is the matrix inverse of $[C]$, (14.3.14) reduces immediately to

$$\sigma^2(a_j) = C_{jj} \tag{14.3.15}$$

In other words, the diagonal elements of $[C]$ are the variances (squared uncertainties) of the fitted parameters $\mathbf{a}$. It should not surprise you to learn that the off diagonal elements C_{jk} are the covariances between a_j and a_k (cf. 14.2.10); but we shall defer discussion of these to §14.5.

We will now give a routine that implements the above formulas for the general linear least-squares problem, by the method of normal equations. Since we wish to compute not only the solution vector $\mathbf{a}$ but also the covariance matrix $[C]$, it is most convenient to use Gauss-Jordan elimination

(routine GAUSSJ of §2.1) to perform the linear algebra. The operation count, in this application, is no larger than that for *LU* decomposition. If you have no need for the covariance matrix, however, you can save a factor of 3 on the linear algebra by switching to *LU* decomposition, without computation of the matrix inverse.

We need to warn you that the solution of a least-squares problem directly from the normal equations is rather susceptible to roundoff error. An alternative, and preferred, technique involves *QR* decomposition (§11.3 and §11.6) of the design matrix **A**. This is essentially what we did at the end of §14.2 for fitting data to a straight line, but without invoking all the machinery of *QR* to derive the necessary formulas. Later in this section, we will discuss other difficulties in the least-squares problem, for which the cure is *singular value decomposition* (SVD), of which we give an implementation. It turns out that SVD also fixes the roundoff problem, so it is our recommended technique for all but "easy" least-squares problems. It is for these easy problems that the following routine, which solves the normal equations, is intended.

The routine below introduces one bookkeeping complication that is quite useful in practical work. Frequently it is a matter of "art" to decide which parameters a_k in a model should be fit from the data set, and which should be held constant at fixed values, for example values predicted by a theory or measured in a previous experiment. One wants, therefore, to have a convenient means for "freezing" and "unfreezing" the parameters a_k. In the following routine the total number of parameters a_k is denoted MA (called M above), while MFIT is the number of parameters which are to be adjusted in minimizing the best fit. As input to the routine, you supply a list LISTA. The MFIT elements of LISTA contain the numbers of the parameters that are to be adjusted. The remaining MA−MFIT parameters will be held fixed at their input values. For example, if MA=8, MFIT=4, and LISTA contains the numbers 3,1,7,5, then the parameters a_3, a_1, a_7, a_5 will be adjusted. The other parameters (a_2, a_4, a_6, a_8) will be held fixed at their input values; notice that you *must* therefore initialize these input values before calling the program. On output, any frozen variable will have its variance and all its covariances returned as zero in the covariance matrix.

```
      SUBROUTINE LFIT(X,Y,SIG,NDATA,A,MA,LISTA,MFIT,COVAR,NCVM,CHISQ,FUNCS)
```

Given a set of NDATA points X(I),Y(I) with individual standard deviations SIG(I), use χ^2 minimization to determine MFIT of MA coefficients A of a function that depends linearly on A, $y = \sum_i \mathrm{A}_i \times \mathrm{AFUNC}_i(x)$. The array LISTA renumbers the parameters so that the first MFIT elements correspond to the parameters actually being determined; the remaining MA-MFIT elements are held fixed at their input value. The program returns values for the MA fit parameters A, χ^2, CHI2, and the covariance matrix COVAR(I,J). NCVM is the physical dimension of COVAR(NCVM,NCVM) in the calling routine. The user supplies a subroutine FUNCS(X,AFUNC,MA) that returns the MA basis functions evaluated at x =X in the array AFUNC.

```
      PARAMETER (MMAX=50)             Set to the maximum number of coefficients MA.
      DIMENSION X(NDATA),Y(NDATA),SIG(NDATA),A(MA),LISTA(MA),
     *    COVAR(NCVM,NCVM),BETA(MMAX),AFUNC(MMAX)
      KK=MFIT+1                       Check to see that LISTA contains a proper permutation of the coeffi-
      DO 12 J=1,MA                         cients and fill in any missing members.
        IHIT=0
        DO 11 K=1,MFIT
          IF (LISTA(K).EQ.J) IHIT=IHIT+1
11      CONTINUE
```

```
          IF (IHIT.EQ.0) THEN
              LISTA(KK)=J
              KK=KK+1
          ELSE IF (IHIT.GT.1) THEN
              PAUSE 'Improper set in LISTA'
          ENDIF
12      CONTINUE
      IF (KK.NE.(MA+1)) PAUSE 'Improper set in LISTA'
      DO 14 J=1,MFIT                     Initialize the (symmetric) matrix.
          DO 13 K=1,MFIT
              COVAR(J,K)=0.
13          CONTINUE
          BETA(J)=0.
14      CONTINUE
      DO 18 I=1,NDATA                    Loop over data to accumulate coefficients of the normal equations.
          CALL FUNCS(X(I),AFUNC,MA)
          YM=Y(I)
          IF(MFIT.LT.MA) THEN            Subtract off dependences on known pieces of the fitting function.
              DO 15 J=MFIT+1,MA
                  YM=YM-A(LISTA(J))*AFUNC(LISTA(J))
15              CONTINUE
          ENDIF
          SIG2I=1./SIG(I)**2
          DO 17 J=1,MFIT
              WT=AFUNC(LISTA(J))*SIG2I
              DO 16 K=1,J
                  COVAR(J,K)=COVAR(J,K)+WT*AFUNC(LISTA(K))
16              CONTINUE
              BETA(J)=BETA(J)+YM*WT
17          CONTINUE
18      CONTINUE
      IF (MFIT.GT.1) THEN
          DO 21 J=2,MFIT                 Fill in above the diagonal from symmetry.
              DO 19 K=1,J-1
                  COVAR(K,J)=COVAR(J,K)
19              CONTINUE
21          CONTINUE
      ENDIF
      CALL GAUSSJ(COVAR,MFIT,NCVM,BETA,1,1)   Matrix solution.
      DO 22 J=1,MFIT
          A(LISTA(J))=BETA(J)            Partition solution to appropriate coefficients A.
22      CONTINUE
      CHISQ=0.                           Evaluate CHI2 of the fit.
      DO 24 I=1,NDATA
          CALL FUNCS(X(I),AFUNC,MA)
          SUM=0.
          DO 23 J=1,MA
              SUM=SUM+A(J)*AFUNC(J)
23          CONTINUE
          CHISQ=CHISQ+((Y(I)-SUM)/SIG(I))**2
24      CONTINUE
      CALL COVSRT(COVAR,NCVM,MA,LISTA,MFIT)   Sort covariance matrix to true order of fitting
      RETURN                                  coefficients.
      END
```

That last call to a subroutine COVSRT is only for the purpose of spreading the MFIT×MFIT covariances back into the full MA×MA covariance matrix, sorted into the proper rows and columns and with zero variances and covariances set for variables which were held frozen. Thus, e.g., the variance of variable a_i will be in its natural place COVAR(i,i). If, instead, you are willing to look up

variances via the index LISTA, then you can omit the call. In that case, e.g., the variance of variable number LISTA(j) will be in COVAR(j,j).

The subroutine COVSRT is as follows.

```
SUBROUTINE COVSRT(COVAR,NCVM,MA,LISTA,MFIT)
   Given the covariance matrix COVAR of a fit for MFIT of MA total parameters, and their
   ordering LISTA(I), repack the covariance matrix to the true order of the parameters.
   Elements associated with fixed parameters will be zero. NCVM is the physical dimension
   of COVAR.
DIMENSION COVAR(NCVM,NCVM),LISTA(MFIT)
DO 12 J=1,MA-1                  Zero all elements below diagonal.
   DO 11 I=J+1,MA
      COVAR(I,J)=0.
11    CONTINUE
12 CONTINUE
DO 14 I=1,MFIT-1                Repack off-diagonal elements of fit into correct locations below diag-
   DO 13 J=I+1,MFIT                onal.
      IF(LISTA(J).GT.LISTA(I)) THEN
         COVAR(LISTA(J),LISTA(I))=COVAR(I,J)
      ELSE
         COVAR(LISTA(I),LISTA(J))=COVAR(I,J)
      ENDIF
13    CONTINUE
14 CONTINUE
SWAP=COVAR(1,1)                 Temporarily store original diagonal elements in top row, and zero the
DO 15 J=1,MA                       diagonal.
   COVAR(1,J)=COVAR(J,J)
   COVAR(J,J)=0.
15 CONTINUE
COVAR(LISTA(1),LISTA(1))=SWAP
DO 16 J=2,MFIT                  Now sort elements into proper order on diagonal.
   COVAR(LISTA(J),LISTA(J))=COVAR(1,J)
16 CONTINUE
DO 18 J=2,MA                    Finally, fill in above diagonal by symmetry.
   DO 17 I=1,J-1
      COVAR(I,J)=COVAR(J,I)
17    CONTINUE
18 CONTINUE
RETURN
END
```

Solution by Use of Singular Value Decomposition

In some applications, the normal equations are perfectly adequate for linear least-squares problems. However, in many cases the normal equations are very close to singular. A zero pivot element may be encountered during the solution of the linear equations (e.g. in GAUSSJ), in which case you get no solution at all. Or a very small pivot may occur, in which case you typically get fitted parameters a_k with very large magnitudes that are delicately (and unstably) balanced to cancel out almost precisely when the fitted function is evaluated.

Why does this commonly occur? The reason is that, more often than experimenters would like to admit, data do not clearly distinguish between two or more of the basis functions provided. If two such functions, or two different combinations of functions, happen to fit the data about equally well —

or equally badly — then the matrix $[\alpha]$, unable to distinguish between them, neatly folds up its tent and becomes singular. There is a certain mathematical irony in the fact that least-squares problems are *both* overdetermined (number of data points greater than number of parameters) *and* underdetermined (ambiguous combinations of parameters exist); but that is how it frequently is. The ambiguities can be extremely hard to notice *a priori* in complicated problems.

Enter singular value decomposition (SVD). This would be a good time for you to review the material in §2.9, which we will not repeat here. In the case of an overdetermined system, SVD produces a solution that is the best approximation in the least-squares sense, cf. equation (2.9.10). That is exactly what we want. In the case of an underdetermined system, SVD produces a solution whose values (for us, the a_k's) are smallest in the least-squares sense, cf. equation (2.9.8). That is also what we want: when some combination of basis functions is irrelevant to the fit, that combination will be driven down to a small, innocuous, value, rather than pushed up to delicately canceling infinities.

In terms of the design matrix $\mathbf{A}$ (equation 14.3.4) and the vector $\mathbf{b}$ (equation 14.3.5), minimization of χ^2 in (14.3.3) can be written as

$$\text{find} \quad \mathbf{a} \quad \text{which minimizes} \quad \chi^2 = |\mathbf{A} \cdot \mathbf{a} - \mathbf{b}|^2 \tag{14.3.16}$$

Comparing to equation (2.9.9), we see that this is precisely the problem which routines SVDCMP and SVBKSB are designed to solve. The solution, which is given by equation (2.9.12), can be rewritten as follows: If $\mathbf{U}$ and $\mathbf{V}$ enter the SVD decomposition of $\mathbf{A}$ according to equation (2.9.1), as computed by SVDCMP, then let the vectors $\mathbf{U}_{(i)}$ $i = 1, \ldots, M$ denote the *columns* of $\mathbf{U}$ (each one a vector of length N); and let the vectors $\mathbf{V}_{(i)}; i = 1, \ldots, M$ denote the *columns* of $\mathbf{V}$ (each one a vector of length M). Then the solution (2.9.12) of the least-squares problem (14.3.16) can be written as

$$\mathbf{a} = \sum_{i=1}^{M} \left(\frac{\mathbf{U}_{(i)} \cdot \mathbf{b}}{w_i} \right) \mathbf{V}_{(i)} \tag{14.3.17}$$

where the w_i are, as in §2.9, the singular values returned by SVDCMP.

Equation (14.3.17) says that the fitted parameters $\mathbf{a}$ are linear combinations of the columns of $\mathbf{V}$, with coefficients obtained by forming dot products of the columns of $\mathbf{U}$ with the weighted data vector (14.3.5). Though it is beyond our scope to prove here, it turns out that the standard (loosely, "probable") errors in the fitted parameters are also linear combinations of the columns of $\mathbf{V}$. In fact, equation (14.3.17) can be written in a form displaying these errors as

$$\mathbf{a} = \left[\sum_{i=1}^{M} \left(\frac{\mathbf{U}_{(i)} \cdot \mathbf{b}}{w_i} \right) \mathbf{V}_{(i)} \right] \pm \frac{1}{w_1} \mathbf{V}_{(1)} \pm \cdots \pm \frac{1}{w_M} \mathbf{V}_{(M)} \tag{14.3.18}$$

Here each $\pm$ is followed by a standard deviation. The amazing fact is that, decomposed in this fashion, the standard deviations are all mutually independent (uncorrelated). Therefore they can be added togther in root-mean-square fashion. What is going on is that the vectors $\mathbf{V}_{(i)}$ are the principal axes of the error ellipsoid of the fitted parameters $\mathbf{a}$ (see §14.5).

It follows that the variance in the estimate of a parameter a_j is given by

$$\sigma^2(a_j) = \sum_{i=1}^{M} \frac{1}{w_i^2}[\mathbf{V}_{(i)}]_j^2 = \sum_{i=1}^{M} \left(\frac{V_{ji}}{w_i}\right)^2 \tag{14.3.19}$$

whose result should be identical with (14.3.14). As before, you should not be surprised at the formula for the covariances, here given without proof,

$$\mathrm{Cov}(a_j, a_k) = \sum_{i=1}^{M} \left(\frac{V_{ji}V_{ki}}{w_i^2}\right) \tag{14.3.20}$$

We introduced this subsection by noting that the normal equations can fail by encountering a zero pivot. We have not yet, however, mentioned how SVD overcomes this problem. The answer is: If any singular value w_i is zero, its reciprocal in equation (14.3.18) should be set to zero, not infinity. (Compare the discussion preceding equation 2.9.7). This corresponds to adding to the fitted parameters $\mathbf{a}$ a *zero* multiple, rather than some random large multiple, of any linear combination of basis functions which are degenerate in the fit. It is a good thing to do!

Moreover, if a singular value w_i is nonzero but very small, you should also define *its* reciprocal to be zero, since its apparent value is probably an artifact of roundoff error, not a meaningful number. A plausible answer to the question "how small is small?", is to edit in this fashion all singular values whose ratio to the largest singular value is less than N times the machine precision ϵ. (You might argue for $\sqrt{N}$, or a constant, instead of N as the multiple; that starts getting into hardware-dependent questions.)

There is another reason for editing even *additional* singular values, ones large enough that roundoff error is not a question. Singular value decomposition allows you to identify linear combinations of variables which just happen not to contribute much to reducing the χ^2 of your data set. Editing these can sometimes reduce the probable error on your coefficients quite significantly, while increasing the minimum χ^2 only negligibly. We will learn more about identifying and treating such cases in §14.5. In the following routine, the point at which this kind of editing would occur is indicated.

Generally speaking, we recommend that you always use SVD techniques instead of using the normal equations. SVD's only significant disadvantage is that it requires an extra array of size $N \times M$ to store the whole design matrix. This storage is overwritten by the matrix $\mathbf{U}$. Storage is also required for the $M \times M$ matrix $\mathbf{V}$, but this is instead of the same-sized matrix for the coefficients of the normal equations. SVD can be significantly slower than

solving the normal equations; however its great advantage, that it (theoretically) *cannot fail* more than makes up for the speed disadvantage.

In the routine that follows, the matrices U,V and the vector W are input as working space. NP and MP are their various physical dimensions. The logical dimensions of the problem are NDATA data points by MA basis functions (and fitted parameters). If you care only about the values A of the fitted parameters, then U,V,W contain no useful information on output. If you want probable errors for the fitted parameters, read on.

```
SUBROUTINE SVDFIT(X,Y,SIG,NDATA,A,MA,U,V,W,MP,NP,CHISQ,FUNCS)
    Given a set of NDATA points X(I),Y(I) with individual standard deviations SIG(I), use χ²
    minimization to determine the MA coefficients A of the fitting function y = Σ_i A_i × AFUNC_i(x).
    Here we solve the fitting equations using singular value decomposition of the NDATA by MA
    matrix, as in §2.9. Arrays U,V,W provide workspace on input, on output they define the
    singular value decomposition, and can be used to obtain the covariance matrix. MP,NP are
    the physical dimensions of the matrices U,V,W, as indicated below. It is necessary that
    MP≥NDATA, NP≥MA. The program returns values for the MA fit parameters A, and χ², CHISQ.
    The user supplies a subroutine FUNCS(X,AFUNC,MA) that returns the MA basis functions
    evaluated at x =X in the array AFUNC.
PARAMETER(NMAX=1000,MMAX=50,TOL=1.E-5)       Max expected NDATA and MA.
DIMENSION X(NDATA),Y(NDATA),SIG(NDATA),A(MA),V(NP,NP),
*     U(MP,NP),W(NP),B(NMAX),AFUNC(MMAX)
DO 12 I=1,NDATA                 Accumulate coefficients of the fitting matrix.
    CALL FUNCS(X(I),AFUNC,MA)
    TMP=1./SIG(I)
    DO 11 J=1,MA
        U(I,J)=AFUNC(J)*TMP
11  CONTINUE
    B(I)=Y(I)*TMP
12 CONTINUE
CALL SVDCMP(U,NDATA,MA,MP,NP,W,V)   Singular value decomposition.
WMAX=0.                         Edit the singular values, given TOL from the parameter statement,
DO 13 J=1,MA                        between here ...
    IF(W(J).GT.WMAX)WMAX=W(J)
13 CONTINUE
THRESH=TOL*WMAX
DO 14 J=1,MA
    IF(W(J).LT.THRESH)W(J)=0.
14 CONTINUE                     ...and here.
CALL SVBKSB(U,W,V,NDATA,MA,MP,NP,B,A)
CHISQ=0.                        Evaluate chi-square.
DO 16 I=1,NDATA
    CALL FUNCS(X(I),AFUNC,MA)
    SUM=0.
    DO 15 J=1,MA
        SUM=SUM+A(J)*AFUNC(J)
15  CONTINUE
    CHISQ=CHISQ+((Y(I)-SUM)/SIG(I))**2
16 CONTINUE
RETURN
END
```

For covariances, you can easily modify the above routine to implement equation (14.3.20).

Feeding the matrix V and vector W output by the above program into the following short routine, you easily obtain variances of the fitted parameters A, whose square roots are standard deviations. The routine straightforwardly

implements equation (14.3.19) above, with the convention that singular values equal to zero are recognized as having been edited out of the fit.

```
SUBROUTINE SVDVAR(V,MA,NP,W,CVM,NCVM)
    To evaluate the covariance matrix CVM of the fit for MA parameters obtained by SVDFIT,
    call this routine with matrices V,W as returned from SVDFIT. NP,NCVM give the physical
    dimensions of V,W,CVM as indicated below.
PARAMETER (MMAX=20)          Set to the maximum number of fit parameters.
DIMENSION V(NP,NP),W(NP),CVM(NCVM,NCVM),WTI(MMAX)
DO 11 I=1,MA
    WTI(I)=0.
    IF(W(I).NE.0.)  WTI(I)=1./(W(I)*W(I))
11  CONTINUE
DO 14 I=1,MA                 Sum contributions to covariance matrix (14.3.20).
    DO 13 J=1,I
        SUM=0.
        DO 12 K=1,MA
            SUM=SUM+V(I,K)*V(J,K)*WTI(K)
12      CONTINUE
        CVM(I,J)=SUM
        CVM(J,I)=SUM
13  CONTINUE
14  CONTINUE
RETURN
END
```

Examples

Be aware that some apparently nonlinear problems can be expressed so that they are linear. For example, an exponential model with two parameters a and b,

$$y(x) = a\exp(-bx) \tag{14.3.21}$$

can be rewritten as

$$\log[y(x)] = c - bx \tag{14.3.22}$$

which is linear in its parameters c and b.

Also watch out for "non-parameters," as in

$$y(x) = a\exp(-bx + d) \tag{14.3.23}$$

Here the parameters a and d are, in fact, indistinguishable. This is a good example of where the normal equations will be exactly singular, and where SVD will find a zero singular value. SVD will then make a "least-squares" choice for setting a balance between a and d (or, rather, their equivalents the linear model derived by taking the logarithms). However — and this is true whenever SVD returns a zero singular value — you are better advised to figure out analytically where the degeneracy is among your basis functions, and then make appropriate deletions in the basis set.

Here are two examples for user-supplied routines FUNCS. The first one is trivial and fits a general polynomial to a set of data:

```
SUBROUTINE FPOLY(X,P,NP)
   Fitting routine for a polynomial of degree NP-1, with NP coefficients.
DIMENSION P(NP)
P(1)=1.
DO 11 J=2,NP
    P(J)=P(J-1)*X
 11 CONTINUE
RETURN
END
```

The second example is slightly less trivial. It is used to fit Legendre polynomials up to some order NL-1 through a data set.

```
SUBROUTINE FLEG(X,PL,NL)
   Fitting routine for an expansion with NL Legendre polynomials PL, evaluated using the
   recurrence relation as in §4.5.
DIMENSION PL(NL)
PL(1)=1.
PL(2)=X
IF(NL.GT.2) THEN
    TWOX=2.*X
    F2=X
    D=1.
    DO 11 J=3,NL
        F1=D
        F2=F2+TWOX
        D=D+1.
        PL(J)=(F2*PL(J-1)-F1*PL(J-2))/D
     11 CONTINUE
ENDIF
RETURN
END
```

REFERENCES AND FURTHER READING:

Bevington, Philip R. 1969, *Data Reduction and Error Analysis for the Physical Sciences* (New York: McGraw-Hill), Chapters 8,9.

Lawson, Charles L., and Hanson, Richard J. 1974, *Solving Least Squares Problems* (Englewood Cliffs, N.J.: Prentice-Hall).

Forsythe, George E., Malcolm, Michael A., and Moler, Cleve B. 1977, *Computer Methods for Mathematical Computations* (Englewood Cliffs, N.J.: Prentice-Hall), Chapter 9.

14.4 Nonlinear Models

We now consider fitting when the model depends *nonlinearly* on the set of M unknown parameters $a_k, k = 1, 2, \ldots, M$. We use the same approach as in previous sections, namely to define a χ^2 merit function and determine best-fit parameters by its minimization. With nonlinear dependences, however, the minimization must proceed iteratively. Given trial values for the parameters, we develop a procedure that improves the trial solution. The procedure is then repeated until χ^2 stops (or effectively stops) decreasing.

How is this problem different from the general nonlinear function minimization problem already dealt with in Chapter 10? Superficially, not at all: Sufficiently close to the minimum, we expect the χ^2 function to be well approximated by a quadratic form, which we can write as

$$\chi^2(\mathbf{a}) \approx \gamma - \mathbf{d} \cdot \mathbf{a} + \frac{1}{2}\mathbf{a} \cdot \mathbf{D} \cdot \mathbf{a} \tag{14.4.1}$$

where $\mathbf{d}$ is an M-vector and $\mathbf{D}$ is an $M \times M$ matrix. (Compare equation 10.6.1.) If the approximation is a good one, we know how to jump from the current trial parameters $\mathbf{a}_{cur}$ to the minimizing ones $\mathbf{a}_{min}$ in a single leap, namely

$$\mathbf{a}_{min} = \mathbf{a}_{cur} + \mathbf{D}^{-1} \cdot \left[-\nabla\chi^2(\mathbf{a}_{cur})\right] \tag{14.4.2}$$

(Compare equation 10.7.4, and reread the discussion leading up to it.)

On the other hand, (14.4.1) might be a poor local approximation to the shape of the function that we are trying to minimize at $\mathbf{a}_{cur}$. In that case, about all we can do is take a step down the gradient, as in the steepest descent method (§10.6). In other words,

$$\mathbf{a}_{next} = \mathbf{a}_{cur} - \text{constant} \times \nabla\chi^2(\mathbf{a}_{cur}) \tag{14.4.3}$$

where the constant is small enough not to exhaust the downhill direction.

To use (14.4.2) or (14.4.3), we must be able to compute the gradient of the χ^2 function at any set of parameters $\mathbf{a}$. To use (14.4.2) we also need the matrix $\mathbf{D}$, which is the second derivative matrix (Hessian matrix) of the χ^2 merit function, at any $\mathbf{a}$.

Now, this is the crucial difference from Chapter 10: There, we had no way of directly evaluating the Hessian matrix. We were only given the ability to evaluate the function to be minimized and (in some cases) its gradient. Therefore, we had to resort to iterative methods *not just* because our function was nonlinear, *but also* in order to build up information about the Hessian matrix. Sections 10.7 and 10.6 concerned themselves with two different techniques for building up this information.

Here, life is much simpler. We *know* exactly the form of χ^2, since it is based on a model function that we ourselves have specified. Therefore the Hessian matrix is known to us. Thus we are free to use (14.4.2) whenever we care to do so. The only reason to use (14.4.3) will be failure of (14.4.2) to improve the fit, signaling failure of (14.4.1) as a good local approximation.

Calculation of the Gradient and Hessian

The model to be fitted is

$$y = y(x; \mathbf{a}) \tag{14.4.4}$$

and the χ^2 merit function is

$$\chi^2(\mathbf{a}) = \sum_{i=1}^{N} \left[\frac{y_i - y(x_i; \mathbf{a})}{\sigma_i} \right]^2 \tag{14.4.5}$$

The gradient of χ^2 with respect to the parameters $\mathbf{a}$, which will be zero at the χ^2 minimum, has components

$$\frac{\partial \chi^2}{\partial a_k} = -2 \sum_{i=1}^{N} \frac{[y_i - y(x_i; \mathbf{a})]}{\sigma_i^2} \frac{\partial y(x_i; \mathbf{a})}{\partial a_k} \qquad k = 1, 2, \ldots, M \tag{14.4.6}$$

Taking an additional partial derivative gives

$$\frac{\partial^2 \chi^2}{\partial a_k \partial a_l} = 2 \sum_{i=1}^{N} \frac{1}{\sigma_i^2} \left[\frac{\partial y(x_i; \mathbf{a})}{\partial a_k} \frac{\partial y(x_i; \mathbf{a})}{\partial a_l} - [y_i - y(x_i; \mathbf{a})] \frac{\partial^2 y(x_i; \mathbf{a})}{\partial a_l \partial a_k} \right] \tag{14.4.7}$$

It is conventional to remove the factors of 2 by defining

$$\beta_k \equiv -\frac{1}{2} \frac{\partial \chi^2}{\partial a_k} \qquad \alpha_{kl} \equiv \frac{1}{2} \frac{\partial^2 \chi^2}{\partial a_k \partial a_l} \tag{14.4.8}$$

making $[\alpha] = \frac{1}{2}\mathbf{D}$ in equation (14.4.2), in terms of which that equation can be rewritten as the set of linear equations

$$\sum_{l=1}^{M} \alpha_{kl}\, \delta a_l = \beta_k \tag{14.4.9}$$

This set is solved for the increments δa_l that, added to the current approximation, give the next approximation. In the context of least-squares, the matrix $[\alpha]$, equal to one-half times the Hessian matrix, is usually called the *curvature matrix.*

Equation (14.4.3), the steepest descent formula, translates to

$$\delta a_l = \text{constant} \times \beta_l \tag{14.4.10}$$

Note that the components α_{kl} of the Hessian matrix (14.4.7) depend both on the first derivatives and on the second derivatives of the basis functions with respect to their parameters. Some treatments proceed to ignore the second derivative without comment. We will ignore it also, but only *after* a few comments.

Second derivatives occur because the gradient (14.4.6) already has a dependence on $\partial y_i/\partial a_k$, so the next derivative simply must contain terms involving $\partial^2 y_i/\partial a_l \partial a_k$. The second derivative term can be dismissed when it is zero (as in the linear case of equation 14.3.8), or small enough to be negligible when compared to the term involving the first derivative. It also has an additional possibility of being ignorably small in practice: The term multiplying the second derivative in equation (14.4.7) is $[y_i - y(x_i; \mathbf{a})]$. For a successful model, this term should just be the random measurement error of each point. This error can have either sign, and should in general be uncorrelated with the model. Therefore, the second derivative terms tend to cancel out when summed over i.

Inclusion of the second-derivative term can in fact be destabilizing if the model fits badly or is contaminated by outlier points that are unlikely to be offset by compensating points of opposite sign. From this point on, we will always use as the definition of α_{kl} the formula

$$\alpha_{kl} = \sum_{i=1}^{N} \frac{1}{\sigma_i^2} \left[\frac{\partial y(x_i; \mathbf{a})}{\partial a_k} \frac{\partial y(x_i; \mathbf{a})}{\partial a_l} \right] \tag{14.4.11}$$

This expression more closely resembles its linear cousin (14.3.8). You should understand that minor (or even major) fiddling with $[\alpha]$ has no effect at all on what final set of parameters $\mathbf{a}$ is reached, but only affects the iterative route that is taken in getting there. The condition at the χ^2 minimum, that $\beta_k = 0$ for all k, is independent of how $[\alpha]$ is defined.

Levenberg-Marquardt Method

Marquardt has put forth an elegant method, related to an earlier suggestion of Levenberg, for varying smoothly between the extremes of the inverse-Hessian method (14.4.9) and the steepest descent method (14.4.10). The latter method is used far from the minimum, switching continuously to the former as the minimum is approached. This *Levenberg-Marquardt method* (also called

Marquardt method) works very well in practice and has become the standard of nonlinear least-squares routines.

The method is based on two elementary, but important, insights. Consider the "constant" in equation (14.4.10). What should it be, even in order of magnitude? What sets its scale? There is no information about the answer in the gradient. That tells only the slope, not how far that slope extends. Marquardt's first insight is that the components of the Hessian matrix, even if they are not usable in any precise fashion, give *some* information about the order-of-magnitude scale of the problem.

The quantity χ^2 is nondimensional, i.e. is a pure number; this is evident from its definition (14.4.5). On the other hand, β_k has the dimensions of $1/a_k$, which may well be dimensional, i.e. have units like cm^{-1}, or kilowatt-hours, or whatever. (In fact, each component of β_k can have different dimensions!) The constant of proportionality between β_k and δa_k must therefore have the dimensions of a_k^2. Scan the components of $[\alpha]$ and you see that there is only one obvious quantity with these dimensions, and that is $1/\alpha_{kk}$, the reciprocal of the diagonal element. So that must set the scale of the constant. But that scale might itself be too big. So let's divide the constant by some (nondimensional) fudge factor λ, with the possibility of setting $\lambda \gg 1$ to cut down the step. In other words, replace equation (14.4.10) by

$$\delta a_l = \frac{1}{\lambda \alpha_{ll}} \beta_l \qquad \text{or} \qquad \lambda\, \alpha_{ll}\, \delta a_l = \beta_l \tag{14.4.12}$$

It is necessary that a_{ll} be positive, but this is guaranteed by definition (14.4.11) — another reason for adopting that equation.

Marquardt's second insight is that equations (14.4.12) and (14.4.9) can be combined if we define a new matrix α' by the following prescription

$$\begin{aligned} \alpha'_{jj} &\equiv \alpha_{jj}(1+\lambda) \\ \alpha'_{jk} &\equiv \alpha_{jk} \qquad (j \neq k) \end{aligned} \tag{14.4.13}$$

and then replace both (14.4.12) and (14.4.9) by

$$\sum_{l=1}^{M} \alpha'_{kl}\, \delta a_l = \beta_k \tag{14.4.14}$$

When λ is very large, the matrix α' is forced into being *diagonally dominant*, so equation (14.4.14) goes over to be identical to (14.4.12). On the other hand, as λ approaches zero, equation (14.4.14) goes over to (14.4.9).

Given an initial guess for the set of fitted parameters $\mathbf{a}$, the recommended Marquardt recipe is as follows:

- Compute $\chi^2(\mathbf{a})$.
- Pick a modest value for λ, say $\lambda = 0.001$.

- (†) Solve the linear equations (14.4.14) for $\delta\mathbf{a}$ and evaluate $\chi^2(\mathbf{a}+\delta\mathbf{a})$.
- If $\chi^2(\mathbf{a}+\delta\mathbf{a}) \geq \chi^2(\mathbf{a})$, *increase* λ by a factor of 10 (or any other substantial factor) and go back to (†).
- If $\chi^2(\mathbf{a}+\delta\mathbf{a}) < \chi^2(\mathbf{a})$, *decrease* λ by a factor of 10, update the trial solution $\mathbf{a} \leftarrow \mathbf{a}+\delta\mathbf{a}$, and go back to (†).

Also necessary is a condition for stopping. Iterating to convergence (to machine accuracy or to the roundoff limit) is generally wasteful and unnecessary since the minimum is at best only a statistical estimate of the parameters $\mathbf{a}$. As we will see in §14.5, a change in the parameters that changes χ^2 by an amount $\ll 1$ is *never* statistically meaningful.

Furthermore, it is not uncommon to find the parameters wandering around near the minimum in a flat valley of complicated topology. The reason is that Marquardt's method generalizes the method of normal equations (§14.3), hence has the same problem as that method with regard to near-degeneracy of the minimum. Outright failure by a zero pivot is possible, but unlikely. More often, a small pivot will generate a large correction which is then rejected, the value of λ being then increased. For sufficiently large λ the matrix $[\alpha']$ is positive definite and can have no small pivots. Thus the method does tend to stay away from zero pivots, but at the cost of a tendency to wander around doing steepest descent in very un-steep degenerate valleys.

These considerations suggest that, in practice, one might as well stop iterating on the first or second occasion that χ^2 decreases by a negligible amount, say either less than 0.1 absolutely or (in case roundoff prevents that being reached) some fractional amount like 10^{-3}. Don't stop after a step where χ^2 *increases*: that only shows that λ has not yet adjusted itself optimally.

Once the acceptable minimum has been found, one wants to set $\lambda = 0$ and compute the matrix

$$[C] \equiv [\alpha]^{-1} \tag{14.4.15}$$

which, as before, is the estimated covariance matrix of the standard errors in the fitted parameters $\mathbf{a}$ (see next section).

The following pair of subroutines encodes Marquardt's method for nonlinear parameter estimation. Much of the organization matches that used in `LFIT` of §14.3. In particular the array `LISTA` is on input a list of the `MFIT` parameters, out of `MA` total, that are desired to be fitted, the remaining parameters being held at their input values.

The routine `MRQMIN` performs one iteration of Marquardt's method. It is first called (once) with `ALAMDA` < 0, which signals the routine to initialize. `ALAMDA` is returned on the first and all subsequent calls as the suggested value of λ for the next iteration; `A` and `CHISQ` are always returned as the best parameters found so far and their χ^2. When convergence is deemed satisfactory, set `ALAMDA` to zero before a final call. The matrices `ALPHA` and `COVAR` (which were used as workspace in all previous calls) will then be set to the curvature and covariance matrices for the converged parameter values. The arguments `ALPHA`, `A`, and `CHISQ` must not be modified between calls, nor should `ALAMDA` be, except to set it to zero for the final call. When an uphill

step is taken, CHISQ and A are returned with their input (best) values, but ALAMDA is returned with an increased value.

The routine MRQMIN calls the routine MRQCOF for the computation of the matrix $[\alpha]$ (equation 14.4.11) and vector β (equations 14.4.6 and 14.4.8). In turn MRQCOF calls the user-supplied routine FUNCS(X,A,Y,DYDA) which for input values $X \equiv x_i$ and $A \equiv \mathbf{a}$ returns the model function $Y \equiv y(x_i;\mathbf{a})$ and the vector of derivatives $DYDA \equiv \partial y/\partial a_k$.

```
      SUBROUTINE MRQMIN(X,Y,SIG,NDATA,A,MA,LISTA,MFIT,
     *    COVAR,ALPHA,NCA,CHISQ,FUNCS,ALAMDA)
```

Levenberg-Marquardt method, attempting to reduce the value χ^2 of a fit between a set of NDATA points X(I),Y(I) with individual standard deviations SIG(I), and a nonlinear function dependent on MA coefficients A. The array LISTA numbers the parameters A such that the first MFIT elements correspond to values actually being adjusted; the remaining MA-MFIT parameters are held fixed at their input value. The program returns current best-fit values for the MA fit parameters A, and χ^2, CHISQ. The arrays COVAR(NCA,NCA), ALPHA(NCA,NCA) with physical dimension NCA ($\geq$ MFIT) are used as working space during most iterations. Supply a subroutine FUNCS(X,A,YFIT,DYDA,MA) that evaluates the fitting function YFIT, and its derivatives DYDA with respect to the fitting parameters A at X. On the first call provide an initial guess for the parameters A, and set ALAMDA<0 for initialization (which then sets ALAMDA=.001). If a step succeeds CHISQ becomes smaller and ALAMDA decreases by a factor of 10. If a step fails ALAMDA grows by a factor of 10. You must call this routine repeatedly until convergence is achieved. Then, make one final call with ALAMDA=0, so that COVAR(I,J) returns the covariance matrix, and ALPHA(I,J) the curvature matrix.

```
      PARAMETER (MMAX=20)            Set to largest number of fit parameters.
      DIMENSION X(NDATA),Y(NDATA),SIG(NDATA),A(MA),LISTA(MA),
     *  COVAR(NCA,NCA),ALPHA(NCA,NCA),ATRY(MMAX),BETA(MMAX),DA(MMAX)
      IF(ALAMDA.LT.0.)THEN           Initialization.
        KK=MFIT+1
        DO 12 J=1,MA                 Does LISTA contain a proper permutation of the coefficients?
          IHIT=0
          DO 11 K=1,MFIT
            IF(LISTA(K).EQ.J)IHIT=IHIT+1
11        CONTINUE
          IF (IHIT.EQ.0) THEN
            LISTA(KK)=J
            KK=KK+1
          ELSE IF (IHIT.GT.1) THEN
            PAUSE 'Improper permutation in LISTA'
          ENDIF
12      CONTINUE
        IF (KK.NE.(MA+1)) PAUSE 'Improper permutation in LISTA'
        ALAMDA=0.001
        CALL MRQCOF(X,Y,SIG,NDATA,A,MA,LISTA,MFIT,ALPHA,BETA,NCA,CHISQ,FUNCS)
        OCHISQ=CHISQ
        DO 13 J=1,MA
          ATRY(J)=A(J)
13      CONTINUE
      ENDIF
      DO 15 J=1,MFIT                 Alter linearized fitting matrix, by augmenting diagonal elements.
        DO 14 K=1,MFIT
          COVAR(J,K)=ALPHA(J,K)
14      CONTINUE
        COVAR(J,J)=ALPHA(J,J)*(1.+ALAMDA)
        DA(J)=BETA(J)
15    CONTINUE
      CALL GAUSSJ(COVAR,MFIT,NCA,DA,1,1)      Matrix solution.
      IF(ALAMDA.EQ.0.)THEN           Once converged evaluate covariance matrix with ALAMDA=0.
        CALL COVSRT(COVAR,NCA,MA,LISTA,MFIT)
        RETURN
```

```
      ENDIF
      DO 16 J=1,MFIT                    Did the trial succeed?
        ATRY(LISTA(J))=A(LISTA(J))+DA(J)
16    CONTINUE
      CALL MRQCOF(X,Y,SIG,NDATA,ATRY,MA,LISTA,MFIT,COVAR,DA,NCA,CHISQ,FUNCS)
      IF(CHISQ.LT.OCHISQ)THEN           Success, accept the new solution.
        ALAMDA=0.1*ALAMDA
        OCHISQ=CHISQ
        DO 18 J=1,MFIT
          DO 17 K=1,MFIT
            ALPHA(J,K)=COVAR(J,K)
17        CONTINUE
          BETA(J)=DA(J)
          A(LISTA(J))=ATRY(LISTA(J))
18      CONTINUE
      ELSE                              Failure, increase ALAMDA and return.
        ALAMDA=10.*ALAMDA
        CHISQ=OCHISQ
      ENDIF
      RETURN
      END
```

Notice the use of the routine COVSRT from §14.3. This is only for rearranging the covariance matrix COVAR into the order of all MA parameters. If you are willing to look up nonzero components corresponding to the MFIT fitted variables through the index LISTA, then you can omit all reference to COVSRT. The above routine also makes use of

```
      SUBROUTINE MRQCOF(X,Y,SIG,NDATA,A,MA,LISTA,MFIT,ALPHA,BETA,NALP,CHISQ,FUNCS)
        Used by MRQMIN to evaluate the linearized fitting matrix ALPHA, and vector BETA as in
        (14.4.8).
      PARAMETER (MMAX=20)
      DIMENSION X(NDATA),Y(NDATA),SIG(NDATA),ALPHA(NALP,NALP),BETA(MA),
     *    DYDA(MMAX),LISTA(MFIT),A(MA)
      DO 12 J=1,MFIT                    Initialize (symmetric) ALPHA, BETA.
        DO 11 K=1,J
          ALPHA(J,K)=0.
11      CONTINUE
        BETA(J)=0.
12    CONTINUE
      CHISQ=0.
      DO 15 I=1,NDATA                   Summation loop over all data.
        CALL FUNCS(X(I),A,YMOD,DYDA,MA)
        SIG2I=1./(SIG(I)*SIG(I))
        DY=Y(I)-YMOD
        DO 14 J=1,MFIT
          WT=DYDA(LISTA(J))*SIG2I
          DO 13 K=1,J
            ALPHA(J,K)=ALPHA(J,K)+WT*DYDA(LISTA(K))
13        CONTINUE
          BETA(J)=BETA(J)+DY*WT
14      CONTINUE
        CHISQ=CHISQ+DY*DY*SIG2I         And find χ².
15    CONTINUE
      DO 17 J=2,MFIT                    Fill in the symmetric side.
        DO 16 K=1,J-1
          ALPHA(K,J)=ALPHA(J,K)
16      CONTINUE
```

```
 17 CONTINUE
RETURN
END
```

Example

The following subroutine FGAUSS is an example of a user-supplied subroutine FUNCS. Used with the above routine MRQMIN (in turn using MRQCOF, COVSRT, and GAUSSJ) it fits for the model

$$y(x) = \sum_{k=1}^{K} B_k \exp\left[-\left(\frac{x - E_k}{G_k}\right)^2\right] \qquad (14.4.16)$$

which is a sum of K Gaussians, each having a variable position, amplitude, and width. We store the parameters in the order $B_1, E_1, G_1, B_2, E_2, G_2, \ldots, B_K, E_K, G_K$.

```
SUBROUTINE FGAUSS(X,A,Y,DYDA,NA)
    Y(X;A) is the sum of NA/3 Gaussians (14.4.16). The amplitude, center, and width of the
    Gaussians are stored in consecutive locations of A: A(I)= B_k, A(I+1)= E_k, A(I+2)= G_k,
    k = 1,...,NA/3.
DIMENSION A(NA),DYDA(NA)
Y=0.
DO 11 I=1,NA-1,3
    ARG=(X-A(I+1))/A(I+2)
    EX=EXP(-ARG**2)
    FAC=A(I)*EX*2.*ARG
    Y=Y+A(I)*EX
    DYDA(I)=EX
    DYDA(I+1)=FAC/A(I+2)
    DYDA(I+2)=FAC*ARG/A(I+2)
 11 CONTINUE
RETURN
END
```

REFERENCES AND FURTHER READING:

Bevington, Philip R. 1969, *Data Reduction and Error Analysis for the Physical Sciences* (New York: McGraw-Hill), Chapter 11.

Marquardt, D. W. 1963, *J. Soc. Ind. Appl. Math.*, vol. 11, pp. 431-441.

Jacobs, David A.H., ed. 1977, *The State of the Art in Numerical Analysis* (London: Academic Press), chapter III.2 (by J.E. Dennis).

14.5 Confidence Limits on Estimated Model Parameters

Several times already in this chapter we have made statements about the standard errors, or uncertainties, in a set of M estimated parameters $\mathbf{a}$. We have given some formulas for computing standard deviations or variances of individual parameters (equations 14.2.9, 14.3.15, 14.3.19), as well as some formulas for covariances between pairs of parameters (equation 14.2.10; remark following equation 14.3.15; equation 14.4.15).

In this section, we want to be more explicit regarding the precise meaning of these quantitative uncertainties, and to give further information about how quantitative confidence limits on fitted parameters can be estimated. The subject can get somewhat technical, and even somewhat confusing, so we will try to make precise statements, even when they must be offered without proof.

Figure 14.5.1 shows the conceptual scheme of an experiment which "measures" a set of parameters. There is some underlying true set of parameters $\mathbf{a}_{true}$ which are known to Mother Nature but hidden from the experimenter. These true parameters are statistically realized, along with random measurement errors, as a measured data set, which we will symbolize as $\mathcal{D}_{(0)}$. The data set $\mathcal{D}_{(0)}$ *is* known to the experimenter. He or she fits the data to a model by χ^2 minimization or some other technique, and obtains measured, i.e. fitted, values for the parameters, which we here denote $\mathbf{a}_{(0)}$.

Because measurement errors have a random component, $\mathcal{D}_{(0)}$ is not a unique realization of the true parameters $\mathbf{a}_{true}$. Rather, there are infinitely many other realizations of the true parameters as "hypothetical data sets" each of which *could* have been the one measured, but happened not to be. Let us symbolize these by $\mathcal{D}_{(1)}, \mathcal{D}_{(2)}, \ldots$. Each one, had it been realized, would have given a slightly different set of fitted parameters, $\mathbf{a}_{(1)}, \mathbf{a}_{(2)}, \ldots$, respectively. These parameter sets $\mathbf{a}_{(i)}$ therefore occur with some probability distribution in the M-dimensional space of all possible parameter sets $\mathbf{a}$. The actual measured set $\mathbf{a}_{(0)}$ is one member drawn from this distribution.

Even more interesting than the probability distribution of $\mathbf{a}_{(i)}$ would be the distribution of the difference $\mathbf{a}_{(i)} - \mathbf{a}_{true}$. This distribution differs from the former one by a translation that puts Mother Nature's true value at the origin. If we knew *this* distribution, we would know everything that there is to know about the quantitative uncertainties in our experimental measurement $\mathbf{a}_{(0)}$.

So the name of the game is to find some way of estimating or approximating the probability distribution of $\mathbf{a}_{(i)} - \mathbf{a}_{true}$ without knowing $\mathbf{a}_{true}$ and without having available to us an infinite universe of hypothetical data sets.

General Case: Confidence Limits by Monte Carlo Simulation

There is really only one way of making the desired estimation. That one way sometimes comes dressed up with fancy analytical formulas, or sometimes naked as a purely numerical procedure; but conceptually it is the same in both cases. When various extra mathematical assumptions are known to hold, the

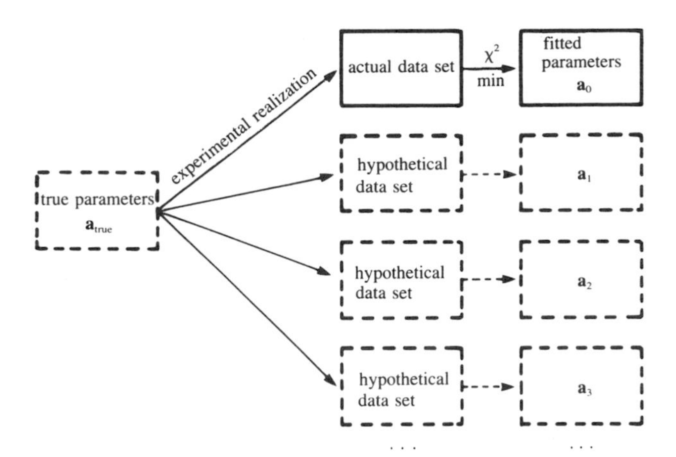

Figure 14.5.1. A statistical universe of data sets from an underlying model. True parameters $\mathbf{a}_{true}$ are realized in a data set, from which fitted (observed) parameters $\mathbf{a}_0$ are obtained. If the experiment were repeated many times, new data sets and new values of the fitted parameters would be obtained.

one way can be proved to give an "accurate" estimate; when they fail, its estimate may be crude. But in either case it is just about the only game in town. Here it is:

Although the measured parameter set $\mathbf{a}_{(0)}$ is not the true one, let us consider a fictitious world in which it *was* the true one. Since we hope that our measured parameters are not *too* wrong, we hope that that fictitious world is not too different from the actual world with parameters $\mathbf{a}_{true}$. In particular, let us hope — no, let us *assume* — that the shape of the probability distribution $\mathbf{a}_{(i)} - \mathbf{a}_{(0)}$ in the fictitious world is the same, or very nearly the same, as the shape of the probability distribution $\mathbf{a}_{(i)} - \mathbf{a}_{true}$ in the real world. Notice that we are not assuming that $\mathbf{a}_{(0)}$ and $\mathbf{a}_{true}$ are equal; they are certainly not. We are only assuming that the way in which random errors enter the experiment and data analysis does not vary rapidly as a function of $\mathbf{a}_{true}$, so that $\mathbf{a}_{(0)}$ can serve as a reasonable surrogate.

Now the distribution of $\mathbf{a}_{(i)} - \mathbf{a}_{(0)}$ in the fictitious world *is* within our power to calculate (see Figure 14.5.2). Starting with our parameters $\mathbf{a}_{(0)}$, we can *simulate* our own sets of "synthetic" realizations of these parameters as "synthetic data sets." The procedure is to draw random numbers from appropriate distributions (cf. §7.2–§7.3) so as to mimic our best understanding of the measurement errors in our apparatus. With such random draws, we construct data sets with exactly the same numbers of measured points, and precisely the same values of all control (independent) variables, as our actual data set $\mathcal{D}_{(0)}$. Let us call these simulated data sets $\mathcal{D}^S_{(1)}, \mathcal{D}^S_{(2)}, \ldots$. By construction these are supposed to have exactly the same statistical relationship to $\mathbf{a}_{(0)}$ as the $\mathcal{D}_{(i)}$'s have to $\mathbf{a}_{true}$.

Next, for each $\mathcal{D}^S_{(j)}$, perform exactly the same procedure for estimation of parameters, e.g. χ^2 minimization, as was performed on the actual data to get the parameters $\mathbf{a}_{(0)}$, giving simulated measured parameters $\mathbf{a}^S_{(1)}, \mathbf{a}^S_{(2)}, \ldots$.

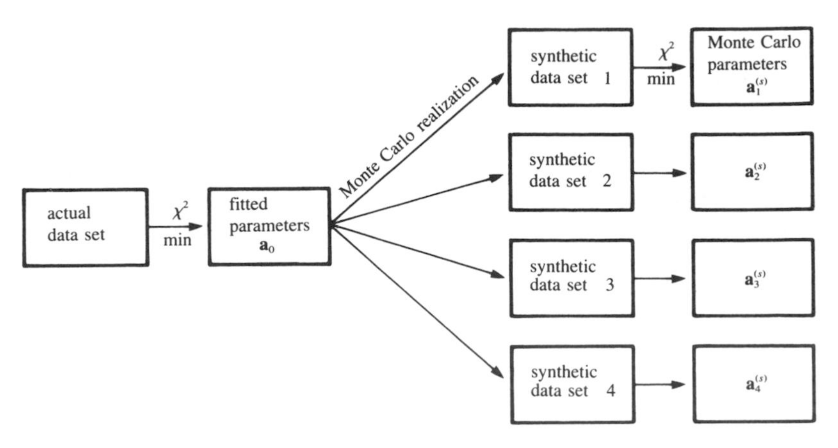

Figure 14.5.2. Monte Carlo simulation of an experiment. The fitted parameters from an actual experiment are used as surrogates for the true parameters. Computer-generated random numbers are used to simulate many synthetic data sets. Each of these is analyzed to obtain its fitted parameters. The distribution of these fitted parameters around the (known) surrogate true parameters is thus studied.

Each simulated measured parameter set yields a point $\mathbf{a}^S_{(i)} - \mathbf{a}_{(0)}$. Simulate enough data sets and enough derived simulated measured parameters, and you map out the desired probability distribution in M dimensions.

In fact, the ability to do *Monte Carlo simulations* in this fashion has revolutionized many fields of modern experimental science. Not only is one able to characterize the errors of parameter estimation in a very precise way. One can also try out on the computer different methods of parameter estimation, or different data reduction techniques, and seek to minimize the uncertainty of the result according to any desired criteria. Offered the choice between mastery of a five-foot shelf of analytical statistics books and middling ability at performing statistical Monte Carlo simulations, we would surely choose to have the latter skill.

Nevertheless, there are a few important analytic results which we will mention just below.

Rather than present all details of the probability distribution of errors in parameter estimation, it is common practice to summarize the distribution in the form of *confidence limits.* The full probability distribution is a function defined on the M-dimensional space of parameters $\mathbf{a}$. A *confidence region* (or *confidence interval*) is just a region of that M-dimensional space (hopefully a small region) that contains a certain (hopefully large) percentage of the total probability distribution. You point to a confidence region and say, e.g., "there is a 99 percent chance that the true parameter values fall within this region around the measured value."

It is worth emphasizing that you, the experimenter, get to pick both the *confidence level* (99 percent in the above example), and the shape of the confidence region. The only requirement is that your region does include the stated percentage of probability. Certain percentages are, however, customary in scientific usage: 68.3 percent (the lowest confidence worthy of quoting), 90 percent, 95.4 percent, 99 percent, and 99.73 percent. Higher confidence

levels are conventionally "ninety-nine point nine ... nine." As for shape, obviously you want a region that is compact and reasonably centered on your measurement $\mathbf{a}_{(0)}$, since the whole purpose of a confidence limit is to inspire confidence in that measured value. In one dimension, the convention is to use a line segment centered on the measured value; in higher dimensions, ellipses or ellipsoids are most frequently used.

You might suspect, correctly, that the numbers 68.3 percent, 95.4 percent, and 99.73 percent, and the use of ellipsoids, have some connection with a normal distribution. That is true historically, but not always relevant nowadays. In general, the probability distribution of the parameters will not be normal, and the above numbers, used as levels of confidence, are purely matters of convention.

Figure 14.5.3 sketches a possible probability distribution for the case $M = 2$. Shown are three different confidence regions which might usefully be given, all at the same confidence level. The two vertical lines enclose a band (horizontal inverval) which represents the 68 percent confidence interval for the variable a_1 without regard to the value of a_2. Similarly the horizontal lines enclose a 68 percent confidence interval for a_2. The ellipse shows a 68 percent confidence interval for a_1 and a_2 jointly. Notice that to enclose the same probability as the two bands, the ellipse must necessarily extend outside of both of them (a point we will return to below).

Use of Constant Chi-Square Boundaries as Confidence Limits

When the method used to estimate the parameters $\mathbf{a}_{(0)}$ is chi-square minimization, as in the previous sections of this chapter, then there is a natural choice for the shape of confidence intervals, whose use is almost universal. For the observed data set $\mathcal{D}_{(0)}$, the value of χ^2 is a minimum at $\mathbf{a}_{(0)}$. Call this minimum value χ^2_{min}. If the vector $\mathbf{a}$ of parameter values is perturbed away from $\mathbf{a}_{(0)}$, then χ^2 increases. The region within which χ^2 increases by no more than a set amount $\Delta\chi^2$ defines some M dimensional confidence region around $\mathbf{a}_{(0)}$. If $\Delta\chi^2$ is set to be a large number, this will be a big region; if it is small, it will be small. Somewhere in between there will be choices of $\Delta\chi^2$ which cause the region to contain, variously, 68 percent, 90 percent, etc. of probability distribution for $\mathbf{a}$'s, as defined above. These regions are taken as the confidence regions for the parameters $\mathbf{a}_{(0)}$.

Very frequently one is interested not in the full M-dimensional confidence region, but in individual confidence regions for some smaller number ν of parameters. For example, one might be interested in the confidence interval of each parameter taken separately (the bands in Figure 14.5.3), in which case $\nu = 1$. In that case, the natural confidence regions in the ν-dimensional subspace of the M-dimensional parameter space are the *projections* of the M-dimensional regions defined by fixed $\Delta\chi^2$ into the ν-dimensional spaces of interest. In Figure 14.5.4, for the case $M = 2$, we show regions corresponding to several values of $\Delta\chi^2$. The one-dimensional confidence interval in a_2 corresponding to the region bounded by $\Delta\chi^2 = 1$ lies between the lines A and A'.

Notice that the projection of the higher-dimensional region on the lower-dimension space is used, not the intersection. The intersection would be the

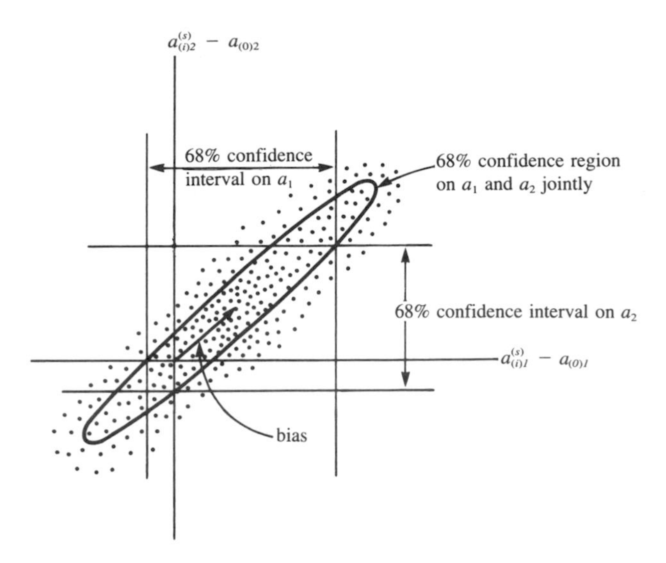

Figure 14.5.3. Confidence intervals in 1 and 2 dimensions. The same fraction of measured points (here 68%) lies (i) between the two vertical lines, (ii) between the two horizontal lines, (iii) within the ellipse.

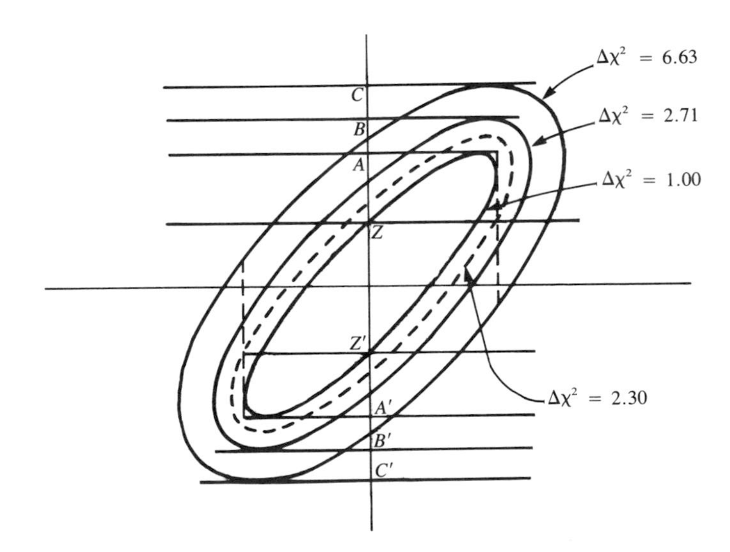

Figure 14.5.4. Confidence region ellipses corresponding to values of chi-square larger than the fitted minimum. The solid curves, with $\Delta\chi^2 = 1.00, 2.71, 6.63$ project onto one-dimensional intervals AA', BB', CC'. These intervals – not the ellipses themselves – contain 68.3%, 90%, and 99% of normally distributed data. The ellipse that contains 68.3% of normally distributed data is shown dashed, and has $\Delta\chi^2 = 2.30$. For additional numerical values, see accompanying table.

band between Z and Z'. It is *never* used. It is shown in the figure only for the purpose of making this cautionary point, that it should not be confused with the projection.

Probability Distribution of Parameters in the Normal Case

You may be wondering why we have, in this section up to now, made no connection at all with the error estimates that come out of the χ^2 fitting procedure, most notably the covariance matrix C_{ij}. The reason is this: χ^2 minimization is a useful means for estimating parameters even if the measurement errors are not normally distributed. While normally distributed errors are required if the χ^2 parameter estimate is to be a maximum likelihood estimator (§14.1), one is often willing to give up that property in return for the relative convenience of the χ^2 procedure. Only in extreme cases, measurement error distributions with very large "tails," is χ^2 minimization abandoned in favor of more robust techniques, as will be discussed in §14.6.

However, the formal covariance matrix that comes out of a χ^2 minimization has meaning *only* if (or to the extent that) the measurement errors actually are normally distributed. In the case of *non*normal errors, you are "allowed"

- to fit for parameters by minimizing χ^2
- to use a contour of constant $\Delta\chi^2$ as the boundary of your confidence region
- to use Monte Carlo simulation or detailed analytic calculation in determining *which* contour $\Delta\chi^2$ is the correct one for your desired confidence level
- to give the covariance matrix C_{ij} as the "formal covariance matrix of the fit on the assumption of normally distributed errors."

You are *not* allowed

- to interpret C_{ij} as the actual squared standard errors of the parameter estimation
- to use formulas that we now give for the case of normal errors, which establish quantitative relationships among $\Delta\chi^2$, C_{ij}, and the confidence level.

Here are the key theorems that hold when (i) the measurement errors are normally distributed, and either (ii) the model is linear in its parameters or (iii) the sample size is large enough that the uncertainties in the fitted parameters $\mathbf{a}$ do not extend outside a region in which the model could be replaced by a suitable linearized model. [Note that condition (iii) does not preclude your use of a nonlinear routine like `MQRFIT` to *find* the fitted parameters.]

Theorem A. χ^2_{min} is distributed as a chi-square distribution with $N-M$ degrees of freedom, where N is the number of data points and M is the number of fitted parameters. This is the basic theorem which lets you evaluate the goodness-of-fit of the model, as discussed above in §14.1. We list it first to remind you that unless the goodness-of-fit is credible, the whole estimation of parameters is suspect.

Theorem B. If $\mathbf{a}_{(j)}^S$ is drawn from the universe of simulated data sets with actual parameters $\mathbf{a}_{(0)}$, then the probability distribution of $\delta\mathbf{a} \equiv \mathbf{a}_{(j)}^S - \mathbf{a}_{(0)}$ is the multivariate normal distribution

$$P(\delta\mathbf{a})\, da_1 \ldots da_M = \text{const.} \times \exp\left(-\frac{1}{2}\delta\mathbf{a} \cdot [\alpha] \cdot \delta\mathbf{a}\right)\, da_1 \ldots da_M$$

where $[\alpha]$ is the curvature matrix defined in equation (14.4.8).

Theorem C. If $\mathbf{a}_{(j)}^S$ is drawn from the universe of simulated data sets with actual parameters $\mathbf{a}_{(0)}$, then the quantity $\Delta\chi^2 \equiv \chi^2(\mathbf{a}_{(j)}) - \chi^2(\mathbf{a}_{(0)})$ is distributed as a chi-square distribution with M degrees of freedom. Here the χ^2's are all evaluated using the fixed (actual) data set $\mathcal{D}_{(0)}$. This theorem makes the connection between particular values of $\Delta\chi^2$ and the fraction of the probability distribution that they enclose as an M-dimensional region, i.e., the confidence level of the M-dimensional confidence region.

Theorem D. Suppose that $\mathbf{a}_{(j)}^S$ is drawn from the universe of simulated data sets (as above), that its first ν components $a_1, \ldots, a_\nu$ are held fixed, and that its remaining $M - \nu$ components are varied so as to minimize χ^2. Call this minimum value χ_ν^2. Then $\Delta\chi_\nu^2 \equiv \chi_\nu^2 - \chi_{min}^2$ is distributed as a chi-square distribution with ν degrees of freedom. If you consult Figure 14.5.4, you will see that this theorem connects the *projected* $\Delta\chi^2$ region with a confidence level. In the figure, a point that is held fixed in a_2 and allowed to vary in a_1 minimizing χ^2 will seek out the ellipse whose top or bottom edge is tangent to the line of constant a_2, and is therefore the line that projects it onto the smaller dimensional space.

As a first example, let us consider the case $\nu = 1$, where we want to find the confidence interval of a single parameter, say a_1. Notice that the chi-square distribution with $\nu = 1$ degree of freedom is the same distribution as that of the square of a single normally distributed quantity. Thus $\Delta\chi_\nu^2 < 1$ occurs 68.3 percent of the time (1-σ for the normal distribution), $\Delta\chi_\nu^2 < 4$ occurs 95.4 percent of the time (2-σ for the normal distribution), $\Delta\chi_\nu^2 < 9$ occurs 99.73 percent of the time (3-σ for the normal distribution), etc. In this manner you find the $\Delta\chi_\nu^2$ which corresponds to your desired confidence level. (Additional values are given in the accompanying table.)

Let $\delta\mathbf{a}$ be a change in the parameters whose first component is arbitrary, δa_1, but the rest of whose components are chosen to minimize the $\Delta\chi^2$. Then Theorem D applies. The value of $\Delta\chi^2$ is given in general by

$$\Delta\chi^2 = \delta\mathbf{a} \cdot [\alpha] \cdot \delta\mathbf{a} \tag{14.5.1}$$

which follows from equation (14.4.8) applied at χ_{min}^2 where $\beta_k = 0$. Since $\delta\mathbf{a}$ by hypothesis minimizes χ^2 in all but its first component, the second through M^{th} components of the normal equations (14.4.9) continue to hold.

$\Delta\chi^2$ *as a Function of Confidence Level and Degrees of Freedom*						
			ν			
p	1	2	3	4	5	6
68.3%	1.00	2.30	3.53	4.72	5.89	7.04
90%	2.71	4.61	6.25	7.78	9.24	10.6
95.4%	4.00	6.17	8.02	9.70	11.3	12.8
99%	6.63	9.21	11.3	13.3	15.1	16.8
99.73%	9.00	11.8	14.2	16.3	18.2	20.1
99.99%	15.1	18.4	21.1	23.5	25.7	27.8

Therefore, the solution of (14.4.9) is

$$\delta\mathbf{a} = [\alpha]^{-1} \cdot \begin{pmatrix} c \\ 0 \\ \vdots \\ 0 \end{pmatrix} = [C] \cdot \begin{pmatrix} c \\ 0 \\ \vdots \\ 0 \end{pmatrix} \tag{14.5.2}$$

where c is one arbitrary constant that we get to adjust to make (14.5.1) give the desired left-hand value. Plugging (14.5.2) into (14.5.1) and using the fact that $[C]$ and $[\alpha]$ are inverse matrices of one another, we get

$$c = \delta a_1 / C_{11} \qquad \text{and} \qquad \Delta\chi^2_\nu = (\delta a_1)^2 / C_{11} \tag{14.5.3}$$

or

$$\delta a_1 = \pm\sqrt{\Delta\chi^2_\nu}\,\sqrt{C_{11}} \tag{14.5.4}$$

At last! A relation between the confidence interval $\pm\delta a_1$ and the formal standard error $\sigma_1 \equiv \sqrt{C_{11}}$. Not unreasonably, we find that the 68 percent confidence interval is $\pm\sigma_1$, the 95 percent confidence interval is $\pm 2\sigma_1$, etc.

These considerations hold not just for the individual parameters a_i, but also for any linear combination of them: If

$$b \equiv \sum_{k=1}^{M} c_i a_i = \mathbf{c} \cdot \mathbf{a} \tag{14.5.5}$$

then the 68 percent confidence interval on b is

$$\delta b = \pm\sqrt{\mathbf{c} \cdot [C] \cdot \mathbf{c}} \tag{14.5.6}$$

However, these simple, normal-sounding numerical relationships do *not* hold in the case $\nu > 1$. In particular, $\Delta\chi^2 = 1$ is not the boundary, nor does it project onto the boundary, of a 68.3 percent confidence region when $\nu > 1$. If you want to calculate not confidence intervals in one parameter, but

confidence ellipses in two parameters jointly, or ellipsoids in three, or higher, then you must follow the following prescription for implementing Theorems C and D above:

- Let ν be the number of fitted parameters whose joint confidence region you wish to display, $\nu \leq M$. Call these parameters the "parameters of interest."
- Let p be the confidence limit desired, e.g. $p = 0.68$ or $p = 0.95$.
- Find Δ (i.e. $\Delta\chi^2$) such that the probability of a chi-square variable with ν degrees of freedom being less than Δ is p. For some useful values of p and ν, Δ is given in the table. For other values, you can use the routine GAMMQ and a simple root-finding routine (e.g. bisection) to find Δ such that GAMMQ$(\nu/2, \Delta/2) = 1 - p$.
- Take the $M \times M$ covariance matrix $[C] = [\alpha]^{-1}$ of the chi-square fit. Copy the intersection of the ν rows and columns corresponding to the parameters of interest into a $\nu \times \nu$ matrix denoted $[C_{proj}]$.
- Invert the matrix $[C_{proj}]$. (In the one-dimensional case this was just taking the reciprocal of the element C_{11}.)
- The equation for the elliptical boundary of your desired confidence region in the ν-dimensional subspace of interest is

$$\Delta = \delta\mathbf{a}' \cdot [C_{proj}]^{-1} \cdot \delta\mathbf{a}' \tag{14.5.7}$$

  where $\delta\mathbf{a}'$ is the ν-dimensional vector of parameters of interest.

If you are confused at this point, you may find it helpful to compare Figure 14.5.4 and the accompanying table, considering the case $M = 2$ with $\nu = 1$ and $\nu = 2$. You should be able to verify the following statements: (i) The horizontal band between C and C' contains 99 percent of the probability distribution, so is a confidence limit on a_2 alone at this level of confidence. (ii) Ditto the band between B and B' at the 90 percent confidence level. (iii) The dashed ellipse, labeled by $\Delta\chi^2 = 2.30$, contains 68.3 percent of the probability distribution, so is a confidence region for a_1 and a_2 jointly, at this level of confidence.

Confidence Limits from Singular Value Decomposition

When you have obtained your χ^2 fit by singular value decomposition (§14.3), the information about the fit's formal errors comes packaged in a somewhat different, but generally more convenient, form. The columns of the matrix $\mathbf{V}$ are an orthonormal set of M vectors which are the principal axes of the $\Delta\chi^2 =$ constant ellipsoids. We denote the columns as $\mathbf{V}_{(1)} \ldots \mathbf{V}_{(M)}$. The lengths of those axes are inversely proportional to the corresponding singular values $w_1 \ldots w_M$; see Figure 14.5.5. The boundaries of the ellipsoids are thus given by

$$\Delta\chi^2 = w_1^2(\mathbf{V}_{(1)} \cdot \delta\mathbf{a})^2 + \cdots + w_M^2(\mathbf{V}_{(M)} \cdot \delta\mathbf{a})^2 \tag{14.5.8}$$

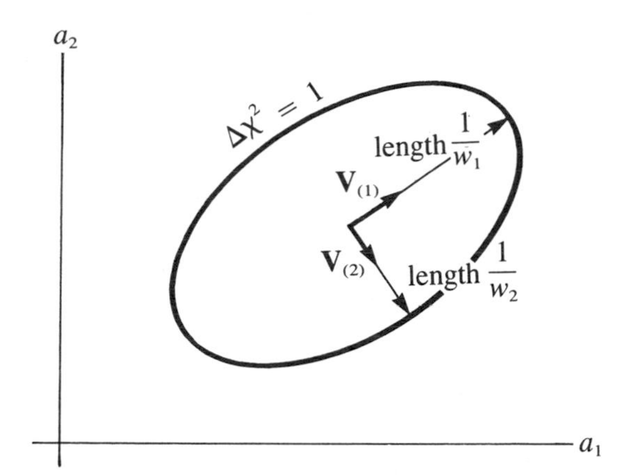

Figure 14.5.5. Relation of the confidence region ellipse $\Delta\chi^2 = 1$ to quantities computed by singular value decomposition. The vectors $\mathbf{V}_{(i)}$ are unit vectors along the principal axes of the confidence region. The semi-axes have lengths equal to the reciprocal of the singular values w_i. If the axes are all scaled by some constant factor α, $\Delta\chi^2$ is scaled by the factor α^2.

which is the justification for writing equation (14.3.18) above. Keep in mind that it is *much* easier to plot an ellipsoid given a list of its vector principal axes, than given its matrix quadratic form!

The formula for the covariance matrix $[C]$ in terms of the columns $\mathbf{V}_{(i)}$ is

$$[C] = \sum_{i=1}^{M} \frac{1}{w_i^2} \mathbf{V}_{(i)} \otimes \mathbf{V}_{(i)} \tag{14.5.9}$$

or, in components,

$$C_{jk} = \sum_{i=1}^{M} \frac{1}{w_i^2} V_{ji} V_{ki} \tag{14.5.10}$$

REFERENCES AND FURTHER READING:

Avni, Y. 1976, *Astrophysical Journal*, vol. 210, pp. 642–646.

Lampton, M., Margon, M., and Bowyer, S. 1976, *Astrophysical Journal*, vol. 208, pp. 177–190.

Brownlee, K.A. 1965, *Statistical Theory and Methodology*, 2nd ed. (New York: Wiley).

Martin, B.R. 1971, *Statistics for Physicists* (New York: Academic Press).

14.6 Robust Estimation

The concept of *robustness* has been mentioned in passing several times already. In §13.2 we noted that the median was a more robust estimator of central value than the mean; in §13.8 it was mentioned that rank correlation is more robust than linear correlation. The concept of outlier points as exceptions to a Gaussian model for experimental error was discussed in §14.1.

The term "robust" was coined in statistics by G.E.P. Box in 1953. Various definitions of greater or lesser mathematical rigor are possible for the term, but in general, referring to a statistical estimator, it means "insensitive to small departures from the idealized assumptions for which the estimator is optimized." The word "small" can have two different interpretations, both important: either fractionally small departures for all data points, or else fractionally large departures for a small number of data points. It is the latter interpretation, leading to the notion of outlier points, that is generally the most stressful for statistical procedures.

Statisticians have developed various sorts of robust statistical estimators. Many, if not most, can be grouped in one of three categories.

M-estimates follow from maximum-likelihood arguments very much as equations (14.1.5) and (14.1.7) followed from equation (14.1.3). M-estimates are usually the most relevant class for model-fitting, that is, estimation of parameters. We therefore consider these estimates in some detail below.

L-estimates are "linear combinations of order statistics." These are most applicable to estimations of central value and central tendency, though they can occasionally be applied to some problems in estimation of parameters. Two "typical" L-estimates will give you the general idea. They are (i) the median, and (ii) *Tukey's trimean*, defined as the weighted average of the first, second, and third quartile points in a distribution, with weights 1/4, 1/2, and 1/4 respectively.

R-estimates are estimates based on rank tests. For example, the equality or inequality of two distributions can be estimated by the *Wilcoxon test* of computing the mean rank of one distribution in a combined sample of both distributions. The Kolmogorov-Smirnov statistic (equation 13.5.4) and the Spearman rank-order correlation coefficient (13.8.1) are R-estimates in essence, if not always by formal definition.

Some other kinds of robust techniques, coming from the fields of optimal control and filtering rather than from the field of mathematical statistics, are mentioned at the end of this section. Some examples where robust statistical methods are desirable are shown in Figure 14.6.1.

Estimation of Parameters by Local M-estimates

Suppose we know that our measurement errors are not normally distributed. Then, in deriving a maximum-likelihood formula for the estimated

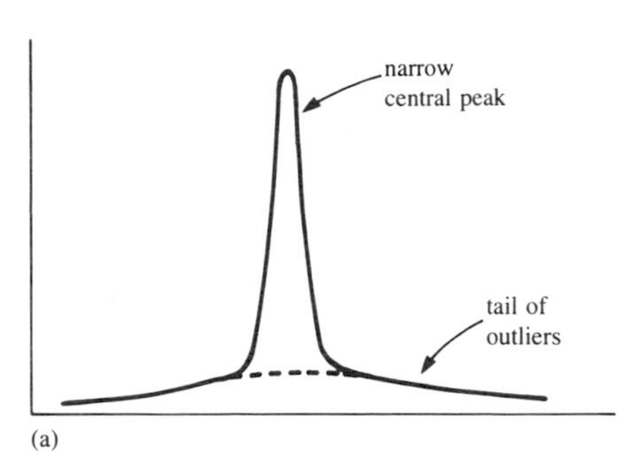

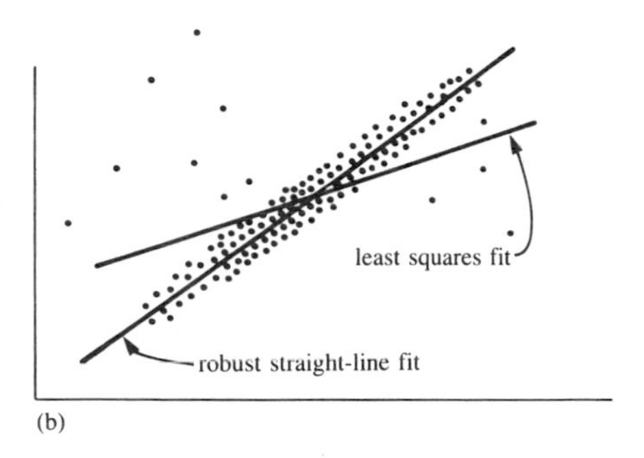

Figure 14.6.1. Examples where robust statistical methods are desirable: (a) A one-dimensional distribution with a tail of outliers; statistical fluctuations in these outliers can prevent accurate determination of the position of the central peak. (b) A distribution in two dimensions fitted to a straight line; non-robust techniques such as least-squares fitting can have undesired sensitivity to outlying points.

parameters $\mathbf{a}$ in a model $y(x;\mathbf{a})$, we would write instead of equation (14.1.3)

$$P = \prod_{i=1}^{N} \{\exp\left[-\rho(y_i, y\{x_i;\mathbf{a}\})\right]\Delta y\} \tag{14.6.1}$$

where the function ρ is the negative logarithm of the probability density. Taking the logarithm of (14.6.1) analogously with (14.1.4), we find that we want to minimize the expression

$$\sum_{i=1}^{N} \rho(y_i, y\{x_i;\mathbf{a}\}) \tag{14.6.2}$$

Very often, it is the case that the function ρ depends not independently on its two arguments, measured y_i and predicted $y(x_i)$, but only on their difference, at least if scaled by some weight factors σ_i which we are able to

assign to each point. In this case the M-estimate is said to be *local*, and we can replace (14.6.2) by the prescription

$$\text{minimize over } \mathbf{a} \qquad \sum_{i=1}^{N} \rho\left(\frac{y_i - y(x_i;\mathbf{a})}{\sigma_i}\right) \tag{14.6.3}$$

where the function $\rho(z)$ is a function of a single variable $z \equiv [y_i - y(x_i)]/\sigma_i$.

If we now define the derivative of $\rho(z)$ to be a function $\psi(z)$,

$$\psi(z) \equiv \frac{d\rho(z)}{dz} \tag{14.6.4}$$

then the generalization of (14.1.7) to the case of a general M-estimate is

$$0 = \sum_{i=1}^{N} \frac{1}{\sigma_i}\psi\left(\frac{y_i - y(x_i)}{\sigma_i}\right)\left(\frac{\partial y(x_i;\mathbf{a})}{\partial a_k}\right) \qquad k = 1,\ldots,M \tag{14.6.5}$$

If you compare (14.6.3) to (14.1.3), and (14.6.5) to (14.1.7), you see at once that the specialization for normally distributed errors is

$$\rho(z) = \frac{1}{2}z^2 \qquad \psi(z) = z \qquad \text{(normal)} \tag{14.6.6}$$

If the errors are distributed as a *double* or *two-sided exponential*, namely

$$\text{Prob}\,\{y_i - y(x_i)\} \sim \exp\left(-\left|\frac{y_i - y(x_i)}{\sigma_i}\right|\right) \tag{14.6.7}$$

then, by contrast,

$$\rho(x) = |z| \qquad \psi(z) = \text{sgn}(z) \qquad \text{(double exponential)} \tag{14.6.8}$$

Comparing to equation (14.6.3), we see that in this case the maximum likelihood estimator is obtained by minimizing the *mean absolute deviation*, rather than the mean square deviation. Here the tails of the distribution, although exponentially decreasing, are asymptotically much larger than any corresponding Gaussian.

A distribution with even more extensive — therefore sometimes even more realistic — tails is the *Cauchy* or *Lorentzian* distribution,

$$\text{Prob}\,\{y_i - y(x_i)\} \sim \frac{1}{1 + \frac{1}{2}\left(\frac{y_i - y(x_i)}{\sigma_i}\right)^2} \tag{14.6.9}$$

This implies

$$\rho(z) = \log\left(1 + \frac{1}{2}z^2\right) \qquad \psi(z) = \frac{z}{1 + \frac{1}{2}z^2} \qquad \text{(Lorentzian)} \tag{14.6.10}$$

Notice that the ψ function occurs as a weighting function in the generalized normal equations (14.6.5). For normally distributed errors, equation (14.6.6) says that the more deviant the points, the greater the weight. By contrast, when tails are somewhat more prominent, as in (14.6.7), then (14.6.8) says that all deviant points get the same relative weight, with only the sign information used. Finally, when the tails are even larger, (14.6.10) says the ψ increases with deviation, then starts *decreasing*, so that very deviant points — the true outliers — are not counted at all in the estimation of the parameters.

This general idea, that the weight given individual points should first increase with deviation, then decrease, motivates some additional prescriptions for ψ which do not especially correspond to standard, textbook probability distributions. Two examples are

Andrew's sine

$$\psi(z) = \begin{cases} \sin(z/c) & |z| < c\pi \\ 0 & |z| > c\pi \end{cases} \tag{14.6.11}$$

If the measurement errors happen to be normal after all, with standard deviations σ_i, then it can be shown that the optimal value for the constant c is $c = 2.1$.

Tukey's biweight

$$\psi(z) = \begin{cases} z(1 - z^2/c^2)^2 & |z| < c \\ 0 & |z| > c \end{cases} \tag{14.6.12}$$

where the optimal value of c for normal errors is $c = 6.0$.

Numerical Calculation of M-estimates

To fit a model by means of an M-estimate, you first decide which M-estimate you want, that is, which matching pair ρ, ψ you want to use. We rather like (14.6.8) or (14.6.10).

You then have to make Hobson's choice between two fairly difficult problems. Either find the solution of the nonlinear set of M equations (14.6.5), or else minimize the single function in M variables (14.6.3).

Notice that the function (14.6.8) has a discontinuous ψ, and a discontinuous derivative for ρ. Such discontinuities frequently wreak havoc on both general nonlinear equation solvers and general function minimizing routines. You might now think of rejecting (14.6.8) in favor of (14.6.10), which is smoother. However, you will find that the latter choice is also bad news for many general

equation solving or minimization routines: small changes in the fitted parameters can drive $\psi(z)$ off its peak into one or the other of its asymptotically small regimes. Therefore, different terms in the equation spring into or out of action (almost as bad as analytic discontinuities).

Don't despair. If your computer budget (or, for personal computers, patience) is up to it, this is an excellent application for the downhill simplex minimization algorithm exemplified in AMOEBA §10.4. That algorithm makes no assumptions about continuity, it just oozes downhill. It will work for virtually any sane choice of the function ρ.

It is very much to your (financial) advantage to find good starting values, however. Often this is done by first fitting the model by the standard χ^2 (nonrobust) techniques, e.g. as described in §14.3 or §14.4. The fitted parameters thus obtained are then used as starting values in AMOEBA, now using the robust choice of ρ and minimizing the expression (14.6.3).

Fitting a Line by Minimizing Absolute Deviation

Occasionally there is a special case that happens to be much easier than is suggested by the general strategy outlined above. The case of equations (14.6.7)–(14.6.8), when the model is a simple straight line

$$y(x;a,b) = a + bx \tag{14.6.13}$$

and where the weights σ_i are all equal, happens to be such a case. The problem is precisely the robust version of the problem posed in equation (14.2.1) above, namely fit a straight line through a set of data points. The merit function to be minimized is

$$\sum_{i=1}^{N} |y_i - a - bx_i| \tag{14.6.14}$$

rather than the χ^2 given by equation (14.2.2).

The key simplification is based on the following fact: The median c_M of a set of numbers c_i is also that value which minimizes the sum of the absolute deviations

$$\sum_i |c_i - c_M|$$

(Proof: Differentiate the above expression with respect to c_M and set it to zero.)

It follows that, for fixed b, the value of a which minimizes (14.6.14) is

$$a = \text{median}\,\{y_i - bx_i\} \tag{14.6.15}$$

Equation (14.6.5) for the parameter b is

$$0 = \sum_{i=1}^{N} x_i \operatorname{sgn}(y_i - a - bx_i) \tag{14.6.16}$$

If we replace a in this equation by the implied function $a(b)$ of (14.6.15), then we are left with an equation in a single variable which can be solved by bracketing and bisection, as described in §9.1. (In fact, it is dangerous to use any fancier method of root-finding, because of the discontinuities in equation 14.6.16.)

Here is a routine which does all this. It calls SORT (§8.2) to find the median by the sorting method, cf. §13.2. The bracketing and bisection are built in to the following routine, as is the χ^2 solution which generates the initial guesses for a and b. Notice that the evaluation of the right-hand side of (14.6.16) occurs in the function subroutine ROFUNC, with communication via a common block. To save memory, you could generate your data arrays directly into that common block, deleting them from this routine's calling sequence.

```
      SUBROUTINE MEDFIT(X,Y,NDATA,A,B,ABDEV)
         Fits y = a+bx by the criterion of least absolute deviations. The arrays X and Y, of length
         NDATA, are the input experimental points. The fitted parameters A and B are output, along
         with ABDEV which is the mean absolute deviation (in y) of the experimental points from the
         fitted line. This routine uses the routine ROFUNC, with communication via a common block.
      PARAMETER (NMAX=1000)
      EXTERNAL ROFUNC
      COMMON /ARRAYS/ NDATAT,XT(NMAX),YT(NMAX),ARR(NMAX),AA,ABDEVT
      DIMENSION X(NDATA),Y(NDATA)
      SX=0.
      SY=0.
      SXY=0.
      SXX=0.
      DO 11 J=1,NDATA               As a first guess for A and B, we will find the least-squares fitting line.
         XT(J)=X(J)
         YT(J)=Y(J)
         SX=SX+X(J)
         SY=SY+Y(J)
         SXY=SXY+X(J)*Y(J)
         SXX=SXX+X(J)**2
11    CONTINUE
      NDATAT=NDATA
      DEL=NDATA*SXX-SX**2
      AA=(SXX*SY-SX*SXY)/DEL        Least-squares solutions.
      BB=(NDATA*SXY-SX*SY)/DEL
      CHISQ=0.
      DO 12 J=1,NDATA
         CHISQ=CHISQ+(Y(J)-(AA+BB*X(J)))**2
12    CONTINUE
      SIGB=SQRT(CHISQ/DEL)          The standard deviation will give some idea of how big an iteration
      B1=BB                            step to take.
      F1=ROFUNC(B1)
      B2=BB+SIGN(3.*SIGB,F1)        Guess bracket as 3-σ away, in the downhill direction known from F1.
      F2=ROFUNC(B2)
1     IF(F1*F2.GT.0.)THEN           Bracketing.
         BB=2.*B2-B1
         B1=B2
```

```
          F1=F2
          B2=BB
          F2=ROFUNC(B2)
          GOTO 1
      ENDIF
      SIGB=0.01*SIGB                Refine until error a negligible number of standard deviations.
2     IF(ABS(B2-B1).GT.SIGB)THEN    Bisection.
          BB=0.5*(B1+B2)
          IF(BB.EQ.B1.OR.BB.EQ.B2)GOTO 3
          F=ROFUNC(BB)
          IF(F*F1.GE.0.)THEN
              F1=F
              B1=BB
          ELSE
              F2=F
              B2=BB
          ENDIF
          GOTO 2
      ENDIF
3     A=AA
      B=BB
      ABDEV=ABDEVT/NDATA
      RETURN
      END
```

```
      FUNCTION ROFUNC(B)
         Evaluates the right-hand side of equation (14.6.16) for a given value of B. Communication
         with the program MEDFIT is through a common block.
      PARAMETER (NMAX=1000)
      COMMON /ARRAYS/ NDATA,X(NMAX),Y(NMAX),ARR(NMAX),AA,ABDEV
      N1=NDATA+1
      NML=N1/2
      NMH=N1-NML
      DO 11 J=1,NDATA
          ARR(J)=Y(J)-B*X(J)
11    CONTINUE
      CALL SORT(NDATA,ARR)
      AA=0.5*(ARR(NML)+ARR(NMH))
      SUM=0.
      ABDEV=0.
      DO 12 J=1,NDATA
          D=Y(J)-(B*X(J)+AA)
          ABDEV=ABDEV+ABS(D)
          SUM=SUM+X(J)*SIGN(1.0,D)
12    CONTINUE
      ROFUNC=SUM
      RETURN
      END
```

Other Robust Techniques

Sometimes you may have *a priori* knowledge about the probable values and probable uncertainties of some parameters that you are trying to estimate from a data set. In such cases you may want to perform a fit that takes this advance information properly into account, neither completely freezing a parameter at a predetermined value (as in LFIT §14.3) nor completely leaving

it to be determined by the data set. The formalism for doing this is called "use of a priori covariances."

A related problem occurs in signal processing and control theory, where it is sometimes desired to "track" (i.e. maintain an estimate of) a time-varying signal in the presence of noise. If the signal is known to be characterized by some number of parameters that vary only slowly, then the formalism of *Kalman filtering* tells how the incoming, raw measurements of the signal should be processed to produce best parameter estimates as a function of time. For example, if the signal is a frequency-modulated sine wave, then the slowly varying parameter might be the instantaneous frequency. The Kalman filter for this case is called a *phase-locked loop* and is implemented in the circuitry of good radio receivers.

Consult Bryson and Ho, or Jazwinski for details on these and other techniques.

REFERENCES AND FURTHER READING:

Huber, P.J. 1981, *Robust Statistics* (New York: Wiley).

Launer, R.L., and Wilkinson, G.N., eds. 1979, *Robustness in Statistics* (New York: Academic Press).

Bryson, A. E., and Ho, Y.C. 1969, *Applied Optimal Control* (Waltham, Mass.: Ginn).

Jazwinski, A. H. 1970, *Stochastic Processes and Filtering Theory* (New York: Academic Press).

Chapter 15. Integration of Ordinary Differential Equations

15.0 Introduction

Problems involving ordinary differential equations (ODEs) can always be reduced to the study of sets of first order differential equations. For example the second order equation

$$\frac{d^2y}{dx^2} + q(x)\frac{dy}{dx} = r(x) \tag{15.0.1}$$

can be rewritten as two first order equations

$$\begin{aligned}\frac{dy}{dx} &= z(x)\\ \frac{dz}{dx} &= r(x) - q(x)z(x)\end{aligned} \tag{15.0.2}$$

where z is a new variable. This is exemplary of the procedure for an arbitrary ODE. The usual choice for the new variables is to let them be just derivatives of each other (and of the original variable). Occasionally, it is useful to incorporate into their definition some other factors in the equation, or some powers of the independent variable, for the purpose of mitigating singular behavior that could result in overflows or increased roundoff error. Let common sense be your guide: if you find that the original variables are smooth in a solution, while your auxiliary variables are doing crazy things, then figure out why and choose different auxiliary variables.

The generic problem in ordinary differential equations is thus reduced to the study of a set of N coupled *first order* differential equations for the functions y_i, $i = 1, 2, \ldots, N$, having the general form

$$\frac{dy_i(x)}{dx} = f_i'(x, y_1, \ldots, y_N). \qquad i = 1, \ldots, N \tag{15.0.3}$$

where the functions f_i' on the right-hand side are known. (The "prime" here does not mean to take a derivative; it is only a notational reminder that the f' functions *are* the derivatives of the y's.)

A problem involving ODEs is not completely specified by its equations. Even more crucial in determining how to attack the problem numerically is the nature of the problem's boundary conditions. Boundary conditions are algebraic conditions on the values of the functions y_i in (15.0.3). In general they can be satisfied at discrete specified points, but do not hold between those points, i.e. are not preserved automatically by the differential equations. Boundary conditions can be as simple as requiring that certain variables have certain numerical values, or as complicated as a set of nonlinear algebraic equations among the variables.

Usually, it is the nature of the boundary conditions that determines which numerical methods will be feasible. Boundary conditions divide into two broad categories.

- In *initial value problems* all the y_i are given at some starting value x_s, and it is desired to find the y_i's at some final point x_f, or at some discrete list of points (for example, at tabulated intervals).
- In *two-point boundary value problems*, on the other hand, boundary conditions are specified at more than one x. Typically, some of the conditions will be specified at x_s and the remainder at x_f.

This chapter will consider exclusively the initial value problem, deferring two-point boundary value problems, which are generally more difficult, to Chapter 16.

The underlying idea of any routine for solving the initial value problem is always this: Rewrite the dy's and dx's in (15.0.3) as finite steps Δy and Δx, and multiply the equations by Δx. This gives algebraic formulas for the change in the functions when the independent variable x is "stepped" by one "stepsize" Δx. In the limit of making the stepsize very small, a good approximation to the underlying differential equation is achieved. Literal implementation of this procedure results in *Euler's method* (15.1.1, below), which is, however, *not* recommended for any practical use. Euler's method is conceptually important, however; one way or another, practical methods all come down to this same idea: Add small increments to your functions corresponding to derivatives (right-hand sides of the equations) multiplied by stepsizes.

In this chapter we consider three major types of practical numerical methods for solving initial value problems for ODEs:

- Runge-Kutta methods
- Richardson extrapolation and its particular implementation as the Bulirsch-Stoer method
- predictor-corrector methods.

A brief description of each of these types follows.

1. *Runge-Kutta* methods propagate a solution over an interval by combining the information from several Euler-style steps (each involving one evaluation of the right-hand f's), and then using the information obtained to match a Taylor series expansion up to some higher order.

2. *Richardson extrapolation* uses the powerful idea of extrapolating a computed result to the value that *would* have been obtained if the stepsize

had been very much smaller than it actually was. In particular, extrapolation to zero stepsize is the desired goal. When combined with a particular way of taking individual steps (the *modified midpoint method*) and a particular kind of extrapolation (rational function extrapolation), Richardson extrapolation produces the *Bulirsch-Stoer method.*

3. *Predictor-corrector* methods store the solution along the way, and use those results to extrapolate the solution one step advanced; they then correct the extrapolation using derivative information at the new point. These are best for very smooth functions.

Runge-Kutta is what you use when (i) you don't know any better, or (ii) you have an intransigent problem where Bulirsch-Stoer is failing, or (iii) you have a trivial problem where computational efficiency is of no concern. Runge-Kutta succeeds virtually always; but it is not usually fastest. Predictor-corrector methods, since they use past information, are somewhat more difficult to start up, but, for many smooth problems, they are computationally more efficient than Runge-Kutta. In recent years Bulirsch-Stoer has been replacing predictor-corrector in many applications, but it is too soon to say that predictor-corrector is dominated in all cases. However, it appears that only rather sophisticated predictor-corrector routines are competitive. Accordingly, we have chosen *not* to give an implementation of predictor-corrector in this book. We discuss predictor-corrector further in §15.5, so that you can use a canned routine should you encounter a suitable problem. In our experience, the relatively simple Runge-Kutta and Bulirsch-Stoer routines we give are adequate for most problems.

Each of the three types of methods can be organized to monitor internal consistency. This allows numerical errors which are inevitably introduced into the solution to be controlled by automatic, (*adaptive*) changing of the fundamental stepsize. We always recommend that adaptive stepsize control be implemented, and we will do so below.

In general, all three types of methods can be applied to any initial value problem. Each comes with its own set of debits and credits that must be understood before it is used.

We have organized the routines in this chapter into three nested levels. The lowest or "nitty-gritty" level is the piece we call the *algorithm* routine. This implements the basic formulas of the method, starts with dependent variables y_i at x, and returns new values of the dependent variables at the value $x + h$. The algorithm routine also yields up some information about the quality of the solution after the step. The routine is dumb, however, and it is unable to make any adaptive decision about whether the solution is of acceptable quality or not.

That quality-control decision we encode in a *stepper* routine. The stepper routine calls the algorithm routine. It may reject the result, set a smaller stepsize, and call the algorithm routine again, until compatibility with a predetermined accuracy criterion has been achieved. The stepper's fundamental task is to take the largest stepsize consistent with specified performance. Only when this is accomplished does the true power of an algorithm come to light.

Above the stepper is the *driver* routine which starts and stops the integration, stores intermediate results, and generally acts as an interface with the user. There is nothing at all canonical about our driver routines. You should consider them to be examples, and you can customize them for your particular application.

Of the routines that follow, RK4 and MMID are algorithm routines; RKQC and BSSTEP are steppers; RKDUMB and ODEINT are drivers.

The final section of this chapter is a brief introduction to the subject of *stiff equations*, relevant both to ordinary differential equations and also to partial differential equations (Chapter 17).

REFERENCES AND FURTHER READING:

Gear, C. William. 1971, *Numerical Initial Value Problems in Ordinary Differential Equations* (Englewood Cliffs, N.J.: Prentice-Hall).

Acton, Forman S. 1970, *Numerical Methods That Work* (New York: Harper and Row), Chapter 5.

Stoer, J., and Bulirsch, R. 1980, *Introduction to Numerical Analysis* (New York: Springer-Verlag), Chapter 7.

15.1 Runge-Kutta Method

The formula for the Euler method is

$$y_{n+1} = y_n + hf'(x_n, y_n) \tag{15.1.1}$$

which advances a solution from x_n to $x_{n+1} \equiv x_n + h$. The formula is unsymmetrical: it advances the solution through an interval h, but uses derivative information only at the beginning of that interval (see Figure 15.1.1). That means (and you can verify by expansion in power series) that the step's error is only one power of h smaller than the correction, i.e $O(h^2)$ added to (15.1.1).

There are several reasons that Euler's method is not recommended for practical use, among them, (i) the method is not very accurate when compared to other, fancier, methods run at the equivalent stepsize, and (ii) neither is it very stable (see §15.6 below).

Consider, however, the use of a step like (15.1.1) to take a "trial" step to the midpoint of the interval. Then use the value of both x and y at that midpoint to compute the "real" step across the whole interval. Figure 15.1.2 illustrates the idea. In equations,

$$\begin{aligned} k_1 &= hf'(x_n, y_n) \\ k_2 &= hf'\left(x_n + \tfrac{1}{2}h, y_n + \tfrac{1}{2}k_1\right) \\ y_{n+1} &= y_n + k_2 + O(h^3) \end{aligned} \tag{15.1.2}$$

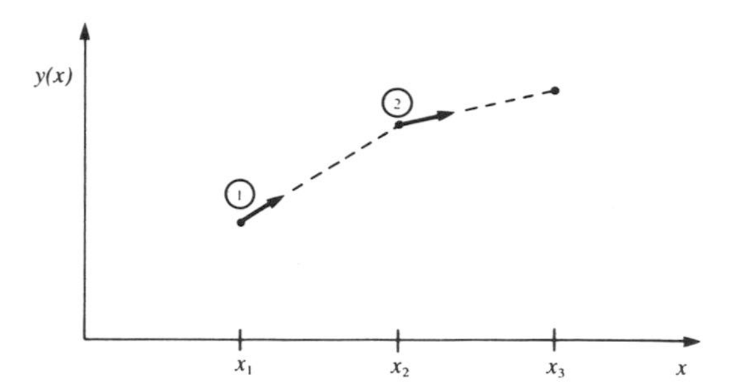

Figure 15.1.1. Euler's method. In this simplest (and least-accurate) method for integrating an ODE, the derivative at the starting point of each interval is extrapolated to find the next function value. The method has first-order accuracy.

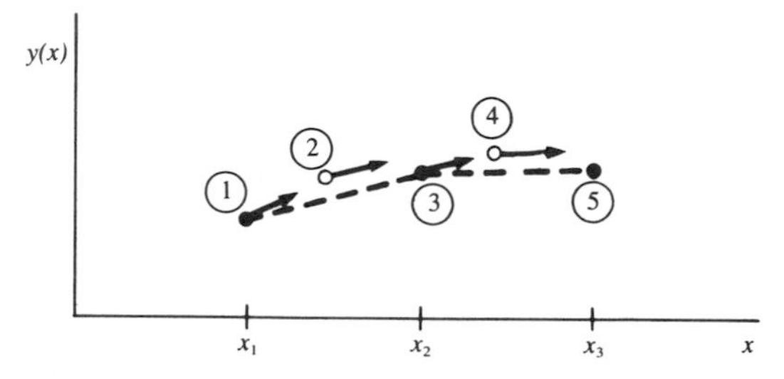

Figure 15.1.2. Midpoint method. Second order accuracy is obtained by using the initial derivative at each step to find a point halfway across the interval, then using the midpoint derivative across the full width of the interval. In the figure, filled dots represent final function values, while open dots represent function values that are discarded once their derivatives have been calculated and used.

As indicated in the error term, this symmetrization cancels out the first order error term, making the method *second order*. [A method is conventionally called n^{th} order if its error term is $O(h^{n+1})$.] In fact, (15.1.2) is called the *second-order Runge-Kutta* or *midpoint* method.

We needn't stop there. There are many ways to evaluate the right-hand side $f'(x, y)$ which all agree to first order, but which have different coefficients of higher-order error terms. Adding up the right combination of these, we can eliminate the error terms order by order. That is the basic idea of the Runge-Kutta method. Abramowitz and Stegun, and Gear, give various specific formulas which derive from this basic idea. By far the most often used, and arguably even most useful, is the *fourth-order Runge-Kutta formula*, which has a certain sleekness of organization about it:

$$\begin{aligned}
k_1 &= hf'(x_n, y_n) \\
k_2 &= hf'(x_n + \frac{h}{2}, y_n + \frac{k_1}{2}) \\
k_3 &= hf'(x_n + \frac{h}{2}, y_n + \frac{k_2}{2}) \\
k_4 &= hf'(x_n + h, y_n + k_3)
\end{aligned}$$

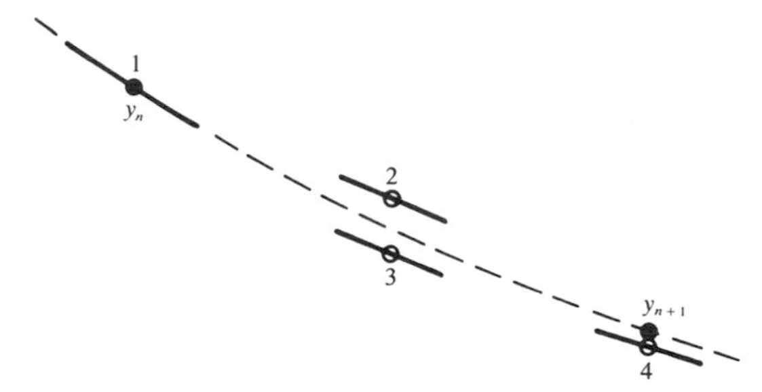

Figure 15.1.3. Fourth-order Runge-Kutta method. In each step the derivative is evaluated four times: once at the initial point, twice at trial midpoints, and once at a trial endpoint. From these derivatives the final function value (shown as a filled dot) is calculated. (See text for details.)

$$y_{n+1} = y_n + \frac{k_1}{6} + \frac{k_2}{3} + \frac{k_3}{3} + \frac{k_4}{6} + O(h^5) \tag{15.1.3}$$

The fourth-order Runge-Kutta method requires four evaluations of the right-hand side per step h (see Figure 15.1.3). This will be superior to the midpoint method (15.1.2) *if* at least twice as large a step is possible with (15.1.3) for the same accuracy. Is that so? The answer is: often, perhaps even usually, but surely not always! This takes us back to a central theme, namely that *high order* does not always mean *high accuracy.* The statement "fourth-order Runge-Kutta is generally superior to second-order" is a true one, but you should recognize it as a statement about the contemporary practice of science rather than as a statement about strict mathematics. That is, it reflects the nature of the problems that contemporary scientists like to solve.

By the same token, and with the same caveats, fourth-order Runge-Kutta is generally found superior to *higher*-order Runge-Kutta schemes, and that is why you rarely see those formulas written out. One interesting fact is that, for orders M higher than four, more than M function evaluations (though never more than $M+2$) are required. Thus fourth-order is a natural breakpoint.

For many scientific users, fourth-order Runge-Kutta is not just the first word on ODE integrators, but the last word as well. In fact, you can get pretty far on this old workhorse, especially if you combine it with an adaptive stepsize algorithm, as we shall do in the next section. Keep in mind, however, that the old workhorse's last trip may well be to take you to the poorhouse: Bulirsch-Stoer or predictor-corrector methods can be very much more efficient for problems where very high accuracy is a requirement. Those methods are the high-strung racehorses. Runge-Kutta is for ploughing the fields.

Here is the routine for carrying out one Runge-Kutta step on a set of `N` differential equations. You input the values of the independent variables, and you get out new values which are stepped by a stepsize `H` (which can be positive or negative). You will notice that the routine requires you to supply not only a routine `DERIVS` for calculating the right-hand side, but also values of the derivatives at the starting point. Why not let the routine call `DERIVS` for this first value? The answer will become clear only in the next section, but in brief is this: This call may not be your only one with these starting

conditions. You may have taken a previous step with too large a stepsize, and this is your replacement. In that case, you do not want to call DERIVS unnecessarily at the start. Note that the routine which follows has, therefore, only three calls to DERIVS.

```
SUBROUTINE RK4(Y,DYDX,N,X,H,YOUT,DERIVS)
    Given values for N variables Y and their derivatives DYDX known at X, use the fourth-order
    Runge-Kutta method to advance the solution over an interval H and return the incremented
    variables as YOUT, which need not be a distinct array from Y. The user supplies the
    subroutine DERIVS(X,Y,DYDX) which returns derivatives DYDX at X.
PARAMETER (NMAX=10)             Set to the maximum number of functions
DIMENSION Y(N),DYDX(N),YOUT(N),YT(NMAX),DYT(NMAX),DYM(NMAX)
HH=H*0.5
H6=H/6.
XH=X+HH
DO 11 I=1,N                     First step
    YT(I)=Y(I)+HH*DYDX(I)
 11 CONTINUE
CALL DERIVS(XH,YT,DYT)          Second step
DO 12 I=1,N
    YT(I)=Y(I)+HH*DYT(I)
 12 CONTINUE
CALL DERIVS(XH,YT,DYM)          Third step
DO 13 I=1,N
    YT(I)=Y(I)+H*DYM(I)
    DYM(I)=DYT(I)+DYM(I)
 13 CONTINUE
CALL DERIVS(X+H,YT,DYT)         Fourth step
DO 14 I=1,N                     Accumulate increments with proper weights.
    YOUT(I)=Y(I)+H6*(DYDX(I)+DYT(I)+2.*DYM(I))
 14 CONTINUE
RETURN
END
```

The Runge-Kutta method treats every step in a sequence of steps in identical manner. Prior behavior of a solution is not used in its propagation. This is mathematically proper, since any point along the trajectory of an ordinary differential equation can serve as an initial point. The fact that all steps are treated identically also makes it easy to incorporate Runge-Kutta into relatively simple "driver" schemes.

We consider adaptive stepsize control, discussed in the next section, an essential for serious computing. Occasionally, however, you just want to tabulate a function at equally spaced intervals, and without particularly high accuracy. In the most common case, you want to produce a graph of the function. Then all you need may be a simple driver program that goes from an initial x_s to a final x_f in a specified number of steps. To check accuracy, double the number of steps, repeat the integration, and compare results. This approach surely does not minimize computer time, and it can fail for problems whose nature *requires* a variable stepsize, but it may well minimize user effort. On small problems, this may be the paramount consideration.

Here is such a driver, self-explanatory, which tabulates the integrated functions in a common block PATH.

```
SUBROUTINE RKDUMB(VSTART,NVAR,X1,X2,NSTEP,DERIVS)
    Starting from initial values VSTART for NVAR functions, known at X1 use fourth-order
    Runge-Kutta to advance NSTEP equal increments to X2. The user supplied subroutine
    DERIVS(X,V,DVDX) evaluates derivatives. Results are stored in the common block PATH.
    Be sure to dimension the common block appropriately.
PARAMETER (NMAX=10)                    Set to the maximum number of functions
COMMON /PATH/ XX(200),Y(10,200)
DIMENSION VSTART(NVAR),V(NMAX),DV(NMAX)
DO 11 I=1,NVAR                         Load starting values.
    V(I)=VSTART(I)
    Y(I,1)=V(I)
11 CONTINUE
XX(1)=X1
X=X1
H=(X2-X1)/NSTEP
DO 13 K=1,NSTEP                        Take NSTEP steps.
    CALL DERIVS(X,V,DV)
    CALL RK4(V,DV,NVAR,X,H,V,DERIVS)
    IF(X+H.EQ.X)PAUSE 'Stepsize not significant in RKDUMB.'
    X=X+H
    XX(K+1)=X                          Store intermediate steps.
    DO 12 I=1,NVAR
        Y(I,K+1)=V(I)
12  CONTINUE
13 CONTINUE
RETURN
END
```

REFERENCES AND FURTHER READING:

Gear, C. William. 1971, *Numerical Initial Value Problems in Ordinary Differential Equations* (Englewood Cliffs, N.J.: Prentice-Hall), Chapter 2.

Abramowitz, Milton, and Stegun, Irene A. 1964, *Handbook of Mathematical Functions*, Applied Mathematics Series, vol. 55 (Washington: National Bureau of Standards; reprinted 1968 by Dover Publications, New York), §25.5.

Rice, J.R. 1983, *Numerical Methods, Software, and Analysis* (New York: McGraw-Hill), §9.2.

15.2 Adaptive Stepsize Control for Runge-Kutta

A good ODE integrator should exert some adaptive control over its own progress, making frequent changes in its stepsize. Usually the purpose of this adaptive stepsize control is to achieve some predetermined accuracy in the solution with minimum computational effort. Many small steps should tiptoe through treacherous terrain, while a few great strides should speed through smooth uninteresting countryside. The resulting gains in efficiency are not mere tens of percents or factors of two; they can sometimes be factors of ten, a hundred, or more. Sometimes accuracy may be demanded not directly in the solution itself, but in some related conserved quantity that can be monitored.

Implementation of adaptive stepsize control requires that the stepping algorithm return information about its performance, most important, an estimate of its truncation error. In this section we will learn how such information can be obtained. Obviously, the calculation of this information will add to the computational overhead, but the investment will generally be repaid handsomely.

With fourth-order Runge-Kutta, the most straightforward technique by far is *step doubling*. We take each step twice, once as a full step, then, independently, as two half steps (see Figure 15.2.1). How much overhead is this, say in terms of the number of evaluations of the right-hand sides? Each of the three separate Runge-Kutta steps in the procedure requires 4 evaluations, but the single and double sequences share a starting point, so the total is 11. This is to be compared not to 4, but to 8 (the two half-steps), since — stepsize control aside — we are achieving the accuracy of the smaller (half) stepsize. The overhead cost is therefore a factor 1.375. What does it buy us?

Let us denote the exact solution for an advance from x to $x + 2h$ by $y(x + 2h)$ and the two approximate solutions by y_1 (one step $2h$) and y_2 (2 steps each of size h). Since the basic method is fourth order, the true solution and the two numerical approximations are related by

$$
\begin{aligned}
y(x+2h) &= y_1 + (2h)^5\phi + O(h^6) + \dots \\
y(x+2h) &= y_2 + 2(h^5)\phi + O(h^6) + \dots
\end{aligned}
\qquad (15.2.1)
$$

where, to order h^5, the value ϕ remains constant over the step. [Taylor series expansion tells us the ϕ is a number whose order of magnitude is $y^{(5)}(x)/5!$]. The first expression in (15.2.1) involves $(2h)^5$ since the stepsize is $2h$, while the second expression involves $2(h^5)$ since the error on each step is $h^5\phi$. The difference between the two numerical estimates is a convenient indicator of truncation error

$$\Delta \equiv y_2 - y_1. \qquad (15.2.2)$$

It is this difference which we shall endeavor to keep to a desired degree of accuracy, neither too large nor too small. We do this by adjusting h.

It might also occur to you that, ignoring terms of order h^6 and higher, we can solve the two equations in (15.2.1) to improve our numerical estimate of the true solution $y(x + 2h)$, namely,

$$y(x+2h) = y_2 + \frac{\Delta}{15} + O(h^6). \qquad (15.2.3)$$

This estimate is accurate to *fifth order*, one order higher than the original Runge-Kutta steps. However, we can't have our cake and eat it: (15.2.3) may be fifth-order accurate, but we have no way of monitoring *its* truncation

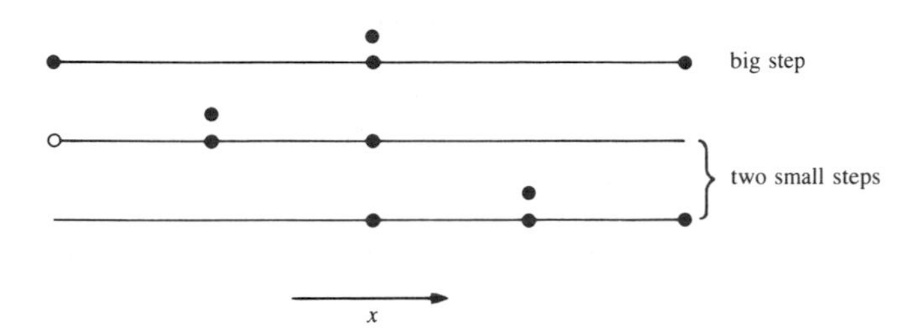

Figure 15.2.1. Step-doubling as a means for adaptive stepsize control in fourth-order Runge-Kutta. Points where the derivative is evaluated are shown as filled circles. The open circle represents the same derivatives as the filled circle immediately above it, so the total number of evaluations is 11 per two steps. Comparing the accuracy of the big step with the two small steps gives a criterion for adjusting the stepsize on the next step, or for rejecting the current step as inaccurate.

error. Higher order is not always higher accuracy! Use of (15.2.3) rarely does harm, but we have no way of directly knowing whether it is doing any good. Therefore we should use Δ as the error estimate and take as "gravy" any additional accuracy gain derived from (15.2.3).

Now that we know, at least approximately, what our error is, we need to consider how to keep it within desired bounds. What is the relation between Δ and h? According to (15.2.1)–(15.2.2), Δ scales as h^5. If we take a step h_1 and produce an error Δ_1, therefore, the step h_0 which *would have given* some other value Δ_0 is readily estimated as

$$h_0 = h_1 \left| \frac{\Delta_0}{\Delta_1} \right|^{0.2} \tag{15.2.4}$$

Henceforth we will let Δ_0 denote the *desired* accuracy. Then equation (15.2.4) is used in two ways: If Δ_1 is larger than Δ_0 in magnitude, the equation tells how much to decrease the stepsize *when we retry the present (failed) step.* If Δ_1 is smaller than Δ_0, on the other hand, then the equation tells how much we can safely increase the stepsize *for the next step.*

Our notation hides the fact that Δ_0 is actually a vector of desired accuracies, one for each equation in the set of ODEs. In general, our accuracy requirement will be that all equations are within their respective allowed errors. In other words, we will rescale the stepsize according to the needs of the "worst-offender" equation.

How is Δ_0, the desired accuracy, related to some looser prescription like "get a solution good to one part in 10^6"? That can be a subtle question, and it depends on exactly what your application is! You may be dealing with a set of equations whose dependent variables differ enormously in magnitude. In that case, you probably want to use fractional errors, $\Delta_0 = \epsilon y$, where ϵ is the number like 10^{-6} or whatever. On the other hand, you may have oscillatory functions that pass through zero but are bounded by some maximum values. In that case you probably want to set Δ_0 equal to ϵ times those maximum values.

A convenient way to fold these considerations into a generally useful stepper routine is this: One of the arguments of the routine will of course be the vector of dependent variables at the beginning of a proposed step. Call that Y. Let us require the user to specify for each step another, corresponding, vector argument YSCAL, and also an overall tolerance level EPS. Then the desired accuracy for the ith equation will be taken to be

$$\Delta_0 = \mathtt{EPS} \times \mathtt{YSCAL}_i \tag{15.2.5}$$

If you desire constant fractional errors, plug Y into the YSCAL calling slot (no need to copy the values into a different array). If you desire constant absolute errors relative to some maximum values, set the elements of YSCAL equal to those maximum values. A useful "trick" for getting constant fractional errors *except* "very" near zero crossings is to set $\mathtt{YSCAL}_i$ equal to $|\mathtt{Y}_i| + |h \times \mathtt{DYDX}_i|$. (The routine ODEINT, below, does this.)

Here is a more technical point. We have to consider one additional possibility for YSCAL. The error criteria mentioned thus far are "local," in that they bound the error of each step individually. In some applications you may be unusually sensitive about a "global" accumulation of errors, from beginning to end of the integration and in the worst possible case where the errors all are presumed to add with the same sign. Then, the smaller the stepsize h, the smaller the value Δ_0 that you will need to impose. Why? Because there will be *more steps* between your starting and ending values of x. In such cases you will want to set YSCAL proportional to h, typically to something like

$$\Delta_0 = \epsilon h \times \mathtt{DYDX}_i \tag{15.2.6}$$

This enforces fractional accuracy ϵ not on the values of Y but (much more stringently) on the *increments* to those values at each step. But now look back at (15.2.4). If Δ_0 has an implicit scaling with h, then the exponent 0.20 is no longer correct: when the stepsize is reduced from a too-large value, the new predicted value h_1 will fail to meet the desired accuracy when YSCAL is also altered to this new h_1 value. Instead of $0.20 = 1/5$, we must scale by the exponent $0.25 = 1/4$ for things to work out.

The exponents 0.20 and 0.25 are not really very different. This motivates us to adopt the following pragmatic approach, one which frees us from having to know in advance whether or not you, the user, plan to scale your YSCAL's with stepsize. Whenever we decrease a stepsize, let us use the larger value of the exponent (whether we need it or not!), and whenever we increase a stepsize, let us use the smaller exponent. Furthermore, because our estimates of error are not exact, but only accurate to the leading order in h, we are advised to put in a safety factor S which is a few percent smaller than unity.

Equation (15.2.4) is thus replaced by

$$h_0 = Sh_1 \left|\frac{\Delta_0}{\Delta_1}\right|^{0.20} \qquad \Delta_0 \geq \Delta_1$$
$$= Sh_1 \left|\frac{\Delta_0}{\Delta_1}\right|^{0.25} \qquad \Delta_0 < \Delta_1 \tag{15.2.7}$$

We have found this prescription to be a reliable one in practice.

Here, then, is a stepper program that takes one "quality-controlled" Runge-Kutta step.

```
      SUBROUTINE RKQC(Y,DYDX,N,X,HTRY,EPS,YSCAL,HDID,HNEXT,DERIVS)
         Fifth-order Runge-Kutta step with monitoring of local truncation error to ensure accuracy
         and adjust stepsize. Input are the dependent variable vector Y of length N and its derivative
         DYDX at the starting value of the independent variable X. Also input are the stepsize to
         be attempted HTRY, the required accuracy EPS, and the vector YSCAL against which the
         error is scaled. On output, Y and X are replaced by their new values, HDID is the stepsize
         which was actually accomplished, and HNEXT is the estimated next stepsize. DERIVS is the
         user-supplied subroutine that computes the right-hand side derivatives.
      PARAMETER (NMAX=10,PGROW=-0.20,PSHRNK=-0.25,FCOR=1./15.,
     *    ONE=1.,SAFETY=0.9,ERRCON=6.E-4)
         The value ERRCON equals (4/SAFETY)**(1/PGROW), see use below.
      EXTERNAL DERIVS
      DIMENSION Y(N),DYDX(N),YSCAL(N),YTEMP(NMAX),YSAV(NMAX),DYSAV(NMAX)
      XSAV=X                          Save initial values.
      DO 11 I=1,N
        YSAV(I)=Y(I)
        DYSAV(I)=DYDX(I)
11    CONTINUE
      H=HTRY                          Set stepsize to the initial trial value.
1     HH=0.5*H                        Take two half steps.
      CALL RK4(YSAV,DYSAV,N,XSAV,HH,YTEMP,DERIVS)
      X=XSAV+HH
      CALL DERIVS(X,YTEMP,DYDX)
      CALL RK4(YTEMP,DYDX,N,X,HH,Y,DERIVS)
      X=XSAV+H
      IF(X.EQ.XSAV)PAUSE 'Stepsize not significant in RKQC.'
      CALL RK4(YSAV,DYSAV,N,XSAV,H,YTEMP,DERIVS)     Take the large step.
      ERRMAX=0.                       Evaluate accuracy.
      DO 12 I=1,N
        YTEMP(I)=Y(I)-YTEMP(I)             YTEMP now contains the error estimate.
        ERRMAX=MAX(ERRMAX,ABS(YTEMP(I)/YSCAL(I)))
12    CONTINUE
      ERRMAX=ERRMAX/EPS               Scale relative to required tolerance.
      IF(ERRMAX.GT.ONE) THEN          Truncation error too large, reduce stepsize.
        H=SAFETY*H*(ERRMAX**PSHRNK)
        GOTO 1                        For another try.
      ELSE                            Step succeeded. Compute size of next step.
        HDID=H
        IF(ERRMAX.GT.ERRCON)THEN
          HNEXT=SAFETY*H*(ERRMAX**PGROW)
        ELSE
          HNEXT=4.*H
        ENDIF
      ENDIF
      DO 13 I=1,N                     Mop up fifth-order truncation error.
        Y(I)=Y(I)+YTEMP(I)*FCOR
```

```
  13 CONTINUE
  RETURN
  END
```

Noting that the above routines are all in single precision, don't be too greedy in specifying EPS. The punishment for excessive greediness is interesting and worthy of Gilbert and Sullivan's *Mikado*: the routine can always achieve an apparent *zero* error by making the stepsize so small that quantities of order hy' add to quantities of order y as if they were zero. Then the routine chugs happily along taking infinitely many infinitesimal steps and never changing the dependent variables one iota. (You guard against this catastrophic loss of your computer budget by signaling on abnormally small stepsizes or on the dependent variable vector remaining unchanged from step to step. On a personal computer you guard against it by not taking too long a lunch hour while your program is running.)

Here is a full-fledged "driver" for Runge-Kutta with adaptive stepsize control. We warmly recommend this routine, or one like it, for a variety of problems, notably including garden-variety ODEs or sets of ODEs, and definite integrals (augmenting the methods of Chapter 4). For storage of intermediate results (if you desire to inspect them) we assume a common block PATH, which can hold up to 200 steps. Because steps occur at unequal intervals results are only stored at intervals greater than DXSAV. Also in the block is KMAX, indicating the number of steps that can be stored. If KMAX=0 there is no intermediate storage, and the rest of the common block need not exist. Storage of steps stops if KMAX is exceeded, except that the ending values are always stored. Again, these controls are merely indicative of what you might need. ODEINT should be customized to the problem at hand.

```
SUBROUTINE ODEINT(YSTART,NVAR,X1,X2,EPS,H1,HMIN,NOK,NBAD,DERIVS,RKQC)
    Runge-Kutta driver with adaptive stepsize control. Integrate the NVAR starting values
    YSTART from X1 to X2 with accuracy EPS, storing intermediate results in the common block
    /PATH/. H1 should be set as a guessed first stepsize, HMIN as the minimum allowed stepsize
    (can be zero). On output NOK and NBAD are the number of good and bad (but retried and
    fixed) steps taken, and YSTART is replaced by values at the end of the integration interval.
    DERIVS is the user-supplied subroutine for calculating the right-hand side derivative, while
    RKQC is the name of the stepper routine to be used. PATH contains its own information
    about how often an intermediate value is to be stored.
PARAMETER (MAXSTP=10000,NMAX=10,TWO=2.0,ZERO=0.0,TINY=1.E-30)
COMMON /PATH/ KMAX,KOUNT,DXSAV,XP(200),YP(10,200)
    User storage for intermediate results. Preset DXSAV and KMAX.
DIMENSION YSTART(NVAR),YSCAL(NMAX),Y(NMAX),DYDX(NMAX)
X=X1
H=SIGN(H1,X2-X1)
NOK=0
NBAD=0
KOUNT=0
DO 11 I=1,NVAR
    Y(I)=YSTART(I)
  11 CONTINUE
IF (KMAX.GT.0) XSAV=X-DXSAV*TWO        Assures storage of first step.
DO 16 NSTP=1,MAXSTP                    Take at most MAXSTP steps.
    CALL DERIVS(X,Y,DYDX)
    DO 12 I=1,NVAR                     Scaling used to monitor accuracy. This general-purpose choice can
        YSCAL(I)=ABS(Y(I))+ABS(H*DYDX(I))+TINY      be modified if need be.
```

```
     12 CONTINUE
    IF(KMAX.GT.0)THEN
        IF(ABS(X-XSAV).GT.ABS(DXSAV)) THEN        Store intermediate results.
            IF(KOUNT.LT.KMAX-1)THEN
                KOUNT=KOUNT+1
                XP(KOUNT)=X
                DO 13 I=1,NVAR
                    YP(I,KOUNT)=Y(I)
13              CONTINUE
                XSAV=X
            ENDIF
        ENDIF
    ENDIF
    IF((X+H-X2)*(X+H-X1).GT.ZERO) H=X2-X    If step can overshoot end, cut down stepsize.
    CALL RKQC(Y,DYDX,NVAR,X,H,EPS,YSCAL,HDID,HNEXT,DERIVS)
    IF(HDID.EQ.H)THEN
        NOK=NOK+1
    ELSE
        NBAD=NBAD+1
    ENDIF
    IF((X-X2)*(X2-X1).GE.ZERO)THEN        Are we done?
        DO 14 I=1,NVAR
            YSTART(I)=Y(I)
14      CONTINUE
        IF(KMAX.NE.0)THEN
            KOUNT=KOUNT+1        Save final step.
            XP(KOUNT)=X
            DO 15 I=1,NVAR
                YP(I,KOUNT)=Y(I)
15          CONTINUE
        ENDIF
        RETURN                   Normal exit.
    ENDIF
    IF(ABS(HNEXT).LT.HMIN) PAUSE 'Stepsize smaller than minimum.'
    H=HNEXT
16 CONTINUE
PAUSE 'Too many steps.'
RETURN
END
```

REFERENCES AND FURTHER READING:

Gear, C. William. 1971, *Numerical Initial Value Problems in Ordinary Differential Equations* (Englewood Cliffs, N.J.: Prentice-Hall).

15.3 Modified Midpoint Method

This section discusses the *modified midpoint method*, which advances a vector of dependent variables $y(x)$ from a point x to a point $x + H$ by a sequence of n substeps each of size h,

$$h = H/n \tag{15.3.1}$$

In principle, one could use the modified midpoint method in its own right as an ODE integrator. In practice, the method finds its most important application as a part of the more powerful Bulirsch-Stoer technique, treated in §15.4. You can therefore consider this section as a preamble to §15.4.

The number of right-hand side evaluations required by the modified midpoint method is $n + 1$. The formulas for the method are

$$\begin{aligned} z_0 &\equiv y(x) \\ z_1 &= z_0 + hf'(x, z_0) \\ z_{m+1} &= z_{m-1} + 2hf'(x + mh, z_m) \qquad \text{for} \quad m = 1, 2, \ldots, n-1 \\ y(x + H) \approx y_n &\equiv \frac{1}{2}[z_n + z_{n-1} + hf'(x + H, z_n)] \end{aligned} \tag{15.3.2}$$

Here the z's are intermediate approximations which march along in steps of h, while y_n is the final approximation to $y(x+H)$. The method is basically a "centered difference" or "midpoint" method (compare equation 15.1.2), except at the first and last points. Those give the qualifier "modified."

The modified midpoint method is a second-order method, like (15.1.2), but with the advantage of requiring (asymptotically for large n) only one derivative evaluation per step h instead of the two required by second-order Runge-Kutta. Perhaps there are applications where the simplicity of (15.3.2), easily coded in-line in some other program, recommends it. In general, however, use of the modified midpoint method by itself will be dominated by fourth-order Runge-Kutta with adaptive stepsize control, as implemented in the preceding section.

The usefulness of the modified midpoint method to the Bulirsch-Stoer technique (§15.4) derives from a "deep" result about equations (15.3.2), due to Gragg. It turns out that the error of (15.3.2), expressed as a power series in h, the stepsize, contains only *even* powers of h,

$$y_n - y(x + H) = \sum_{i=1}^{\infty} \alpha_i h^{2i} \tag{15.3.3}$$

where H is held constant, but h changes by varying n in (15.3.1). The importance of this even power series is that, if we play our usual tricks of combining steps to knock out higher order error terms, we can gain *two* orders at a time!

For example, suppose n is even, and let $y_{n/2}$ denote the result of applying (15.3.1) and (15.3.2) with half as many steps, $n \rightarrow n/2$. Then the estimate

$$y(x + H) \approx \frac{4y_n - y_{n/2}}{3} \tag{15.3.4}$$

is *fourth-order* accurate, the same as fourth-order Runge-Kutta, but requires only about 1.5 derivative evaluations per step h instead of Runge-Kutta's 4 evaluations. Don't be too anxious to implement (15.3.4), since we will soon do even better.

Now would be a good time to look back at the routine QSIMP in §4.2, and especially to compare equation (4.2.4) with equation (15.3.4) above. You will see that the transition in Chapter 4 to the idea of Richardson extrapolation, as embodied in Romberg integration of §4.3, is exactly analogous to the transition in going from this section to the next one.

Here is the routine that implements the modified midpoint method, which will be used below.

```
SUBROUTINE MMID(Y,DYDX,NVAR,XS,HTOT,NSTEP,YOUT,DERIVS)
    Modified midpoint step. Dependent variable vector Y of length NVAR and its derivative
    vector DYDX are input at XS. Also input is HTOT, the total step to be made, and NSTEP,
    the number of substeps to be used. The output is returned as YOUT, which need not be a
    distinct array from Y; if it is distinct, however, then Y and DYDX are returned undamaged.
PARAMETER (NMAX=10)
DIMENSION Y(NVAR),DYDX(NVAR),YOUT(NVAR),YM(NMAX),YN(NMAX)
H=HTOT/NSTEP                        Stepsize this trip.
DO 11 I=1,NVAR
    YM(I)=Y(I)
    YN(I)=Y(I)+H*DYDX(I)            First step.
  11 CONTINUE
X=XS+H
CALL DERIVS(X,YN,YOUT)              Will use YOUT for temporary storage of derivatives.
H2=2.*H
DO 13 N=2,NSTEP                     General step.
    DO 12 I=1,NVAR
        SWAP=YM(I)+H2*YOUT(I)
        YM(I)=YN(I)
        YN(I)=SWAP
      12 CONTINUE
    X=X+H
    CALL DERIVS(X,YN,YOUT)
  13 CONTINUE
DO 14 I=1,NVAR                      Last step.
    YOUT(I)=0.5*(YM(I)+YN(I)+H*YOUT(I))
  14 CONTINUE
RETURN
END
```

REFERENCES AND FURTHER READING:

Gear, C. William. 1971, *Numerical Initial Value Problems in Ordinary Differential Equations* (Englewood Cliffs, N.J.: Prentice-Hall), §6.1.4.

Stoer, J., and Bulirsch, R. 1980, *Introduction to Numerical Analysis* (New York: Springer-Verlag), §7.2.12.

15.4 Richardson Extrapolation and the Bulirsch-Stoer Method

The techniques described in this section are not for differential equations containing nonsmooth functions. For example, you might have a differential equation whose right-hand side involves a function which is evaluated by table look-up and interpolation. If so, go back to Runge-Kutta with adaptive stepsize choice: That method does an excellent job of feeling its way through rocky or discontinuous terrain. It is also an excellent choice for quick-and-dirty, low-accuracy solution of a set of equations. A second warning is that the techniques in this section are not particularly good for differential equations which have singular points *inside* the interval of integration. A regular solution must tiptoe very carefully across such points. Runge-Kutta with adaptive stepsize can sometimes effect this; more generally, there are special techniques available for such problems, beyond our scope here.

Apart from those two caveats, we believe that the Bulirsch-Stoer method, discussed in this section, is the best known way to obtain high-accuracy solutions to ordinary differential equations with minimal computational effort. (A possible exception, rarely encountered in practice, is discussed in the next section.)

Three key ideas are involved. The first is *Richardson's deferred approach to the limit*, which we already met in §4.3 on Romberg integration. The idea is to consider the final answer of a numerical calculation as itself being an analytic function (if a complicated one) of an adjustable parameter like the stepsize h. That analytic function can be probed by performing the calculation with various values of h, *none* of them being necessarily small enough to yield the accuracy that we desire. When we know enough about the function, we *fit* it to some analytic form, and then *evaluate* it at that mythical and golden point $h = 0$ (see Figure 15.4.1). Richardson extrapolation is a method for turning straw into gold! (Lead into gold for alchemist readers.)

The second idea has to do with what kind of fitting function is used. Bulirsch and Stoer first recognized the strength of *rational function extrapolation* in Richardson-type applications. That strength is to break the shackles of the power series and its limited radius of convergence, out only to the distance of the first pole in the complex plane. Rational function fits can remain good approximations to analytic functions even after the various terms in powers of h all have comparable magnitudes. In other words, h can be so large as to make the whole notion of the "order" of the method meaningless — and the method can still work superbly. You might wish at this point to review §3.1–§3.2, where rational function extrapolation was already discussed.

The third idea was discussed in the section before this one, namely to use a method whose error function is strictly even, allowing the rational function approximation to be in terms of the variable h^2 instead of just h.

Put these ideas together and you have the *Bulirsch-Stoer method.* A single Bulirsch-Stoer step takes us from x to $x+H$, where H is supposed to be quite a large — not at all infinitesimal — distance. That single step is a grand leap

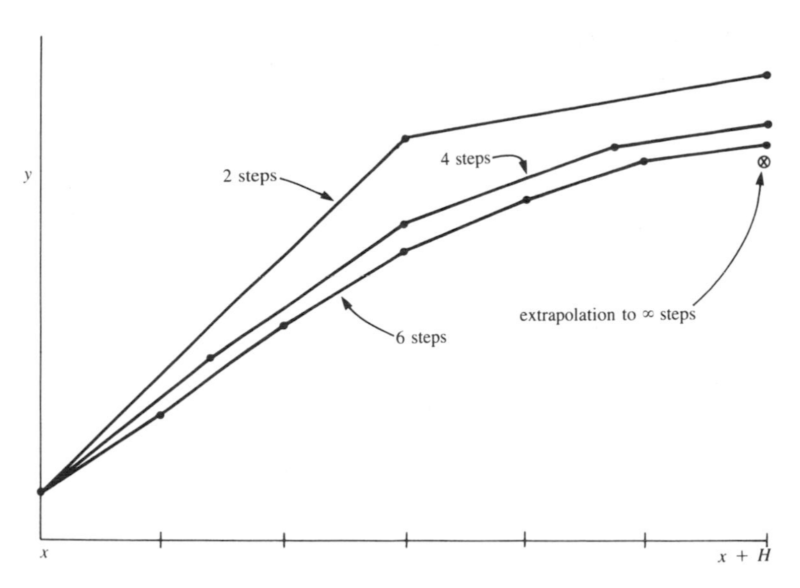

Figure 15.4.1. Richardson extrapolation as used in the Bulirsch-Stoer method. A large interval H is spanned by different sequences of finer and finer substeps. Their results are extrapolated to an answer which is supposed to correspond to infinitely fine substeps. In the Bulirsch-Stoer method, the integrations are done by the modified midpoint method, and the extrapolation technique is rational function extrapolation.

consisting of many (e.g. dozens to hundreds) substeps of modified midpoint method, which are then rational-function extrapolated to zero stepsize.

The sequence of separate attempts to cross the interval H is made with increasing values of n, the number of substeps. A conventional sequence of n's is

$$n = 2, 4, 6, 8, 12, 16, 24, 32, 48, 64, 96, \ldots, [n_j = 2n_{j-2}], \ldots \tag{15.4.1}$$

(The more obvious choice $n = 2, 4, 8, 16, \ldots$ makes h too small too rapidly.) For each step, we do not know in advance how far up this sequence we will go. After each successive n is tried, a rational function extrapolation is attempted. That extrapolation returns both extrapolated values and error estimates. If the errors are not satisfactory, we go higher in n. If they are satisfactory, we go on to the next step and begin anew with $n = 2$.

Of course there must be some upper limit, beyond which we conclude that there is some obstacle in our path in the interval H, so that we must reduce H rather than just subdivide it more finely. In the implementations below, the maximum number of n's to be tried is called `IMAX`. We usually take this equal to 11; the $11th$ value of the sequence (15.4.1) is 96, so this is the maximum number of subdivisions of H that we allow.

Another adjustable parameter is the number of previous estimates of the functions at $x + H$ (different values of n) to incorporate into the rational function fit. One possibility would be to use all the estimates available,

$n = 2, 4, 6, \ldots$ etc. Experience shows that, by the time n gets moderately large, the early values are not very relevant. Therefore we usually take the parameter NUSE, the number of estimates to use, equal to 7. (These are values suggested by Gear.)

We enforce error control, as in the Runge-Kutta method, by monitoring internal consistency, and adapting stepsize to match a prescribed bound on the local truncation error. Each new result from the sequence of modified midpoint integrations allows a tableau like that in §3.1 (or more precisely, like that of equations 3.2.6–3.2.7) to be extended by one additional set of diagonals. The size of the new correction added at each stage is taken as the (conservative) error estimate.

Different from Runge-Kutta, and not a fully solved problem in the literature or in practice, is the question of just *how* to increase or decrease the big stepsize H. The problem is not a lack of acceptable ways. The problem is that almost *any* sensible scheme works well, so it is difficult to show that any particular scheme accomplishes the goal of minimizing computation over some universe of hypothetical problems. A key point to keep in mind is that *each* Bulirsch-Stoer step effectively ranges over a factor of, say, up to 96/2 in stepsize (cf. 15.4.1 above). Furthermore, each step can be effectively of very high order in h^2 — if "order" means anything at all for values of h as large as the method often takes! Therefore the range of accuracy already accessible to a step, 48^2 to some high power, is immense. It is a fairly rare event for a Bulirsch-Stoer step to fail at all, and such failure is usually associated with starting or ending transients or internal singularities in the solution.

One desideratum is to keep the number of sequences tried below or at the value NUSE, since after this point early information is thrown away without affecting the solution. As each step succeeds, we learn the value of n at which success occurs. We can then try the next step with H scaled so that the same value h will be reached on, say, sequence number NUSE − 1. When NUSE − 1 succeeds, we might tweak the stepsize up a little bit; while when NUSE succeeds, we might tweak it down a bit.

Of course, if the step truly fails — goes all the way up to IMAX without finding an acceptable error — then we have to take more drastic action, decreasing H substantially and trying the step over.

The following implementation of a Bulirsch-Stoer step has exactly the same calling sequence as the quality-controlled Runge-Kutta stepper RKQC. This means that the driver ODEINT in §15.2 can be used for Bulirsch-Stoer as well as Runge-Kutta: just substitute BSSTEP for RKQC in ODEINT's argument list *and be sure to make the routines consistently either single- or double-precision.* The routine BSSTEP calls MMID to take the modified midpoint sequences, and calls RZEXTR, given below, to do the rational function extrapolation.

```
SUBROUTINE BSSTEP(Y,DYDX,NV,X,HTRY,EPS,YSCAL,HDID,HNEXT,DERIVS)
```

Bulirsch-Stoer step with monitoring of local truncation error to ensure accuracy and adjust stepsize. Input are the dependent variable vector Y of length NV and its derivative DYDX at the starting value of the independent variable X. Also input are the stepsize to be attempted HTRY, the required accuracy EPS, and the vector YSCAL against which the error is scaled. On output, Y and X are replaced by their new values, HDID is the stepsize

```
      which was actually accomplished, and HNEXT is the estimated next stepsize. DERIVS is the
      user-supplied subroutine that computes the right-hand side derivatives.
      PARAMETER (NMAX=10,IMAX=11,NUSE=7,ONE=1.E0,SHRINK=.95E0,GROW=1.2E0)
      DIMENSION Y(NV),DYDX(NV),YSCAL(NV),YERR(NMAX),
     *    YSAV(NMAX),DYSAV(NMAX),YSEQ(NMAX),NSEQ(IMAX)
      DATA NSEQ /2,4,6,8,12,16,24,32,48,64,96/
      H=HTRY
      XSAV=X
      DO 11 I=1,NV                      Save the starting values.
        YSAV(I)=Y(I)
        DYSAV(I)=DYDX(I)
11    CONTINUE
1     DO 10 I=1,IMAX                    Evaluate the sequence of modified midpoint integrations.
        CALL MMID(YSAV,DYSAV,NV,XSAV,H,NSEQ(I),YSEQ,DERIVS)
        XEST=(H/NSEQ(I))**2             Squared, since error series is even.
        CALL RZEXTR(I,XEST,YSEQ,Y,YERR,NV,NUSE) Perform rational function extrapolation.
        ERRMAX=0.                       Check local truncation error.
        DO 12 J=1,NV
          ERRMAX=MAX(ERRMAX,ABS(YERR(J)/YSCAL(J)))
12      CONTINUE
        ERRMAX=ERRMAX/EPS               Scale accuracy relative to tolerance.
        IF(ERRMAX.LT.ONE) THEN          Step converged.
          X=X+H
          HDID=H
          IF(I.EQ.NUSE)THEN
            HNEXT=H*SHRINK
          ELSE IF(I.EQ.NUSE-1)THEN
            HNEXT=H*GROW
          ELSE
            HNEXT=(H*NSEQ(NUSE-1))/NSEQ(I)
          ENDIF
          RETURN                        Normal return.
        ENDIF
10    CONTINUE
      If here, then step failed, quite unusual for this method. We reduce the stepsize and try
      again.
      H=0.25*H/2**((IMAX-NUSE)/2)
      IF(X+H.EQ.X)PAUSE 'Step size underflow.'
      GOTO 1
      END
```

The rational function extrapolation routine is based on the same algorithm as RATINT §3.2. It is simpler in that it is always extrapolating to zero, rather than to an arbitrary value. However it is more complicated in that it must individually extrapolate each component of a vector of quantities.

```
      SUBROUTINE RZEXTR(IEST,XEST,YEST,YZ,DY,NV,NUSE)
      Use diagonal rational function extrapolation to evaluate NV functions at X = 0 by fitting
      a diagonal rational function to a sequence of estimates with progressively smaller values
      X = XEST, and corresponding function vectors YEST. This call is number IEST in the
      sequence of calls. The extrapolation uses at most the last NUSE estimates. Extrapolated
      function values are output as YZ, and their estimated error is output as DY.
      PARAMETER (IMAX=11,NMAX=10,NCOL=7)
      Maximum expected value of NUSE is NCOL; of NV is NMAX; of IEST is IMAX.
      DIMENSION X(IMAX),YEST(NV),YZ(NV),DY(NV),D(NMAX,NCOL),FX(NCOL)
      X(IEST)=XEST                      Save current independent variable.
      IF(IEST.EQ.1) THEN
        DO 11 J=1,NV
          YZ(J)=YEST(J)
          D(J,1)=YEST(J)
          DY(J)=YEST(J)
```

```
        11 CONTINUE
ELSE
    M1=MIN(IEST,NUSE)                Use at most NUSE previous members.
    DO 12 K=1,M1-1
        FX(K+1)=X(IEST-K)/XEST
      12 CONTINUE
    DO 14 J=1,NV                     Evaluate next diagonal in tableau.
        YY=YEST(J)
        V=D(J,1)
        C=YY
        D(J,1)=YY
        DO 13 K=2,M1
            B1=FX(K)*V
            B=B1-C
            IF(B.NE.0.) THEN
                B=(C-V)/B
                DDY=C*B
                C=B1*B
            ELSE                     Care needed to avoid division by 0.
                DDY=V
            ENDIF
            IF (K.NE.M1) V=D(J,K)
            D(J,K)=DDY
            YY=YY+DDY
          13 CONTINUE
        DY(J)=DDY
        YZ(J)=YY
      14 CONTINUE
ENDIF
RETURN
END
```

Rational function extrapolation can fail. The extrapolated function might have a pole at the desired evaluation point. Or, more commmonly, there might be two poles that very nearly cancel, so that roundoff becomes a problem. In the above routine, the test for division by zero prevents program crash, but disguises the fact that the quantity computed as B1-C may have lost all significance.

You can use Bulirsch-Stoer for years without encountering a failure, or you can be stopped dead on your first attempt. It all depends on the nature of your ODEs, the precision of your machine, and (we sometimes think) whether the fates are smiling on you. The rewards of mastering Bulirsch-Stoer are so great that we urge persistence.

Keep a watch for failed steps, where HDID is returned with a value less than HTRY. If these are not rare, then Bulirsch-Stoer is in trouble. You can try to save it with either of two options:

- Take a couple of quality-controlled Runge-Kutta steps to get over the rough spot. A call to RKQC above is exactly substitutable for a call to BSSTEP. (You might, however, reduce the suggested stepsize HTRY by a factor of 16 or 32, in recognition of Runge-Kutta's necessarily smaller steps. If you don't do this, RKQC will have to seek out the smaller stepsize by itself, with additional effort.)
- If the problem is not a rough spot in your solutions, but purely an artifact in the rational function extrapolation, then a less drastic therapy is to use polynomial extrapolation instead of rational

function extrapolation for a step or two. Polynomial extrapolation does not involve any divisions, and it can be less finicky than rational function extrapolation — also less powerful!

If you are trying the second of the above options, you will want the following polynomial extrapolation routine, which is an exact substitution for RZEXTR above.

```
SUBROUTINE PZEXTR(IEST,XEST,YEST,YZ,DY,NV,NUSE)
PARAMETER (IMAX=11,NCOL=7,NMAX=10)
DIMENSION X(IMAX),YEST(NV),YZ(NV),DY(NV),QCOL(NMAX,NCOL),D(NMAX)
X(IEST)=XEST                     Store current dependent value.
DO 11 J=1,NV
    DY(J)=YEST(J)
    YZ(J)=YEST(J)
  11 CONTINUE
IF(IEST.EQ.1) THEN               Store first estimate in first column.
    DO 12 J=1,NV
        QCOL(J,1)=YEST(J)
      12 CONTINUE
ELSE
    M1=MIN(IEST,NUSE)            Use at most NUSE previous estimates.
    DO 13 J=1,NV
        D(J)=YEST(J)
      13 CONTINUE
    DO 15 K1=1,M1-1
        DELTA=1./(X(IEST-K1)-XEST)
        F1=XEST*DELTA
        F2=X(IEST-K1)*DELTA
        DO 14 J=1,NV             Propagate tableau 1 diagonal more.
            Q=QCOL(J,K1)
            QCOL(J,K1)=DY(J)
            DELTA=D(J)-Q
            DY(J)=F1*DELTA
            D(J)=F2*DELTA
            YZ(J)=YZ(J)+DY(J)
          14 CONTINUE
      15 CONTINUE
    DO 16 J=1,NV
        QCOL(J,M1)=DY(J)
      16 CONTINUE
ENDIF
RETURN
END
```

REFERENCES AND FURTHER READING:

Stoer, J., and Bulirsch, R. 1980, *Introduction to Numerical Analysis* (New York: Springer-Verlag), §7.2.14.

Gear, C. William. 1971, *Numerical Initial Value Problems in Ordinary Differential Equations* (Englewood Cliffs, N.J.: Prentice-Hall), §6.2.

15.5 Predictor-Corrector Methods

We suspect that predictor-corrector integrators have had their day, and that they are no longer the method of choice for most problems in ODEs. For high-precision applications, or applications where evaluations of the right-hand sides are expensive, Bulirsch-Stoer dominates. For convenience, or for low-precision, adaptive-stepsize Runge-Kutta dominates. Predictor-corrector methods have been, we think, squeezed out in the middle. There is possibly only one exceptional case: high-precision solution of very smooth equations with very complicated right-hand sides, as we will describe later.

Nevertheless, these methods have had a long historical run. Textbooks are full of information on them, and there are a lot of standard ODE programs around that are based on predictor-corrector methods. Many capable researchers have a lot of experience with predictor-corrector routines, and they see no reason to make a precipitous change of habit. You, the knowledgeable practitioner, had better be familiar with the principles involved, and even with the sorts of bookkeeping details that are the bane of these methods. Otherwise there will be a big surprise in store when you first have to fix a problem in a predictor-corrector routine.

Unlike the approaches discussed thus far, *predictor-corrector* methods do not step forward as though each new point were another initial value. Instead these methods record past function values and extrapolate them (via polynomial extrapolation) to *predict* what the next step is going to yield. This is called the *predictor step*. The predictor step does not involve any evaluations of the right-hand side of the equations. Why do it? The answer becomes clear if you think about how integrating an ODE is different from finding the integral of a function: For a function, the integrand has a known dependence on the independent variable x, and can be evaluated at will. For an ODE, the "integrand" is the right-hand side, which depends both on x and on the dependent variables y.

If by some magic we knew the value of the y's that enter into the right-hand side calculation, then integrating an ODE would be exactly the same as integrating a function. We could do the integration by Simpson's rule, or any of the other quadrature rules in §4.1. At whatever value x our chosen rule requires us to evaluate the derivative function f', we do so by evaluating $f'(x, y)$ with the "magically" obtained value of y.

In a predictor-corrector method, the predictor step *is* the desired "magic." Its extrapolated value may not be exact, but if it is accurate to as high an order as the integration will be, then that is good enough. The Simpson-like integration, using the prediction step's value of y in evaluating the right-hand side, is called the *corrector step*. Figure 15.5.1 illustrates the general idea.

Let us here dispose of two silly ideas that might occur to you in an idle moment: (1) Why do any corrector steps at all; why not do predictor step after predictor step? Don't predictor steps alone have the same order as the whole method? Answer: Yes, they do. But order is not the same as accuracy! Repeated extrapolation, with no evaluations of the right-hand

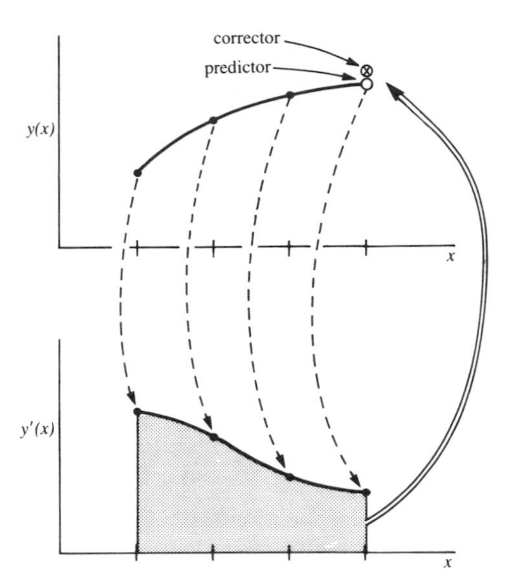

Figure 15.5.1. A predictor-corrector method illustrated schematically. In the upper figure, three already-known points are shown as filled dots. These are polynomial-extrapolated to obtain the predictor, shown as an open dot. The lower figure shows the derivatives as evaluated at each of the upper points, including the predictor. A polynomial is passed through these derivative points and the area under the curve is evaluated. That area, being the integral of y', is the desired corrector point, which becomes a filled dot in the upper figure on the next step. Practical predictor-corrector methods use more complicated combinations of function and derivatives on both predictor and corrector steps.

sides and thus no influx of new information, is wildly unstable and hence wholly inaccurate. (2) If one corrector step is good, aren't many better? Why not use each corrector as an improved predictor and iterate to convergence on each step? Answer: Even if you had a *perfect* predictor, the step would still be accurate only to the finite order of the corrector. This incurable error term is on the same order as that which your iteration is supposed to cure, so you are at best only changing the coefficient in front of the error term by a fractional amount. So dubious an improvement is certainly not worth the effort. Your extra effort would be better spent in taking a smaller stepsize.

As described so far, you might think it desirable or necessary to predict several intervals ahead at each step, then to use all these intervals, with various weights, in a Simpson-like corrector step. That is not a good idea. Extrapolation is the least stable part of the procedure, and it is desirable to minimize its effect. Therefore, the integration steps of a predictor-corrector method are overlapping, each one involving several stepsize intervals h, but extending just one such interval farther than the previous ones. Only that one extended interval is extrapolated by each predictor step.

The most popular predictor-corrector methods are the so-called Adams-Bashforth-Moulton schemes, which have good stability properties. The

Adams-Bashforth part is the predictor. For example, the 4th order case is

$$\text{predictor:} \qquad y_1 = y_0 + \frac{h}{24}(55y'_0 - 59y'_{-1} + 37y'_{-2} - 9y'_{-3}) + O(h^5) \quad (15.5.1)$$

Here information at the current point x_0, together with the three previous points x_{-1}, x_{-2} and x_{-3} (assumed equally spaced), is used to predict the value y_1 at the next point, x_1. The prime means calculate the derivative using as the dependent variable the quantity that is primed.

The Adams-Moulton part is the corrector. For the same order as the predictor, the corrector has the same number of terms. The 4th order case is

$$\text{corrector:} \qquad y_1 = y_0 + \frac{h}{24}(9y'_1 + 19y'_0 - 5y'_{-1} + y'_{-2}) + O(h^5) \qquad (15.5.2)$$

Without the trial value of y_1 from the predictor step to insert on the right-hand side, the corrector would be a nasty implicit equation for y_1.

There are actually three separate processes occurring in this method: the predictor step, which we call P, the evaluation of the derivative y'_1 from the latest value of y, which we call E, and the corrector step which we call C. In this notation, iterating m times with the corrector (a practice we inveighed against earlier) would be written $\mathrm{P(EC)}^m$. One also has the choice of finishing with a C or an E step. The lore is that a final E is superior, so the strategy usually recommended is PECE.

The difference between the predicted and corrected function values supplies information on the local truncation error that can be used to control accuracy and to adjust stepsize. Unfortunately, predictor-corrector methods are not very flexible as far as stepsize adjustment is concerned.

Since the formulas require results from four equally spaced steps, halving and doubling adjustments to stepsize are the most simply realized. Suppose that we attempt to take a step, but the error indicated by the difference "predictor minus corrector" exceeds the required tolerance. We can use interpolation to generate "old" results at half the current stepsize, and then try again with $h/2$ as the new step.

If the routine is working particularly well, on the other hand, then we can double the stepsize simply by discarding every other previous point and using $2h$. However, we can only do this if we had the foresight to save the necessary information at the last *seven* points for a 4th order method. Thus, the predictor-corrector requires some considerable bookkeeping on prior information, and some checks to allow doubling only when enough steps are available. Also, it is wasteful to double prematurely and then be forced immediately to halve again. For this reason it is best to set a fairly strict accuracy criterion for doubling.

Starting and stopping pose obvious problems for predictor-corrector methods. For starting, we need the initial values plus three previous steps to prime the pump. Stopping is a problem because equal steps are unlikely

to land directly on the desired termination point. The solution is to use Runge-Kutta (or any other available procedure) to start and stop.

The reason we have chosen not to give an implementation of a PC method is that, because of the bookkeeping complexity, the code is about twice as long as the comparable Runge-Kutta or Bulirsch-Stoer routines given earlier. Moreover, such a routine is no better than Bulirsch-Stoer, and in fact is often less efficient.

There do exist even more complicated PC methods that adaptively change the *order* of the method as the integration proceeds, as well as the stepsize. They also use more complicated integration schemes with unequal spacing of steps. For very smooth functions, very high order methods get invoked. If the right-hand side of the equation is relatively complicated, so that the expense of evaluating it outweighs the bookkeeping expense, then the best PC packages can outperform Bulirsch-Stoer on such problems. As you can imagine, however, such a variable-stepsize, variable-order method is a nightmare to program. If you suspect that your problem is suitable for this treatment, we recommend use of a canned PC package. For further details consult Gear or Shampine and Gordon.

Our prediction is that, as extrapolation methods like Bulirsch-Stoer continue to gain sophistication, they will eventually beat out PC methods in all applications. We are willing, however, to be corrected.

REFERENCES AND FURTHER READING:

Acton, Forman S. 1970, *Numerical Methods That Work* (New York: Harper and Row), Chapter 5.

Gear, C. William. 1971, *Numerical Initial Value Problems in Ordinary Differential Equations* (Englewood Cliffs, N.J.: Prentice-Hall), Chapter 9.

Hamming, R.W. 1962, *Numerical Methods for Engineers and Scientists* (New York: McGraw-Hill), Chapters 14, 15.

Stoer, J., and Bulirsch, R. 1980, *Introduction to Numerical Analysis* (New York: Springer-Verlag), Chapter 7.

Shampine, L.F., and Gordon, M.K. 1975, *Computer Solution of Ordinary Differential Equations. The Initial Value Problem.* (San Francisco: W.H Freeman).

15.6 Stiff Sets of Equations

As soon as one deals with more than one first-order differential equation, the possibility of a *stiff* set of equations arises. Stiffness occurs in a problem where there are two or more very different scales of the independent variable on which the dependent variables are changing. For example, consider the second-order equation

$$y'' = f(x)y \tag{15.6.1}$$

where $f >> 0$. (This is equivalent to a set of two first-order equations by the usual substitution $y_1 = y$, $y_2 = y'$.) For large values of $f(x)$, y is approximately the sum of two exponentials. For example, if we approximate f as a constant, say $f = 100$, then

$$y = Ae^{-10x} + Be^{10x}, \qquad A, B = \text{constant} \tag{15.6.2}$$

Often the desired solution is the dying exponential. Thus if we integrated equation (15.6.1) with the boundary conditions

$$y(0) = 1 \qquad y'(0) = -10 \tag{15.6.3}$$

the true solution is $y = e^{-10x}$. However, the integration methods given so far in this chapter would give a numerical solution that would start off decaying as e^{-10x}, but would then "explode" as e^{10x} as x becomes large. The reason is clear: any roundoff or truncation error as we start the integration, no matter how small, is equivalent to a small admixture near the origin of the unwanted other solution, e^{10x}. Thus

$$y_{numerical} \approx e^{-10x} + \epsilon e^{10x} \tag{15.6.4}$$

No matter how small ϵ is made (e.g. by taking a very small stepsize), sooner or later the second term in (15.6.4) dominates.

"Simple" stiff equations like (15.6.1) can be handled by a change of variable called the Riccati transformation (see Acton, p. 148). Before considering a general strategy for stiff equations, let us consider an example which shows that stiff systems need not have divergent solutions. Gear gives the following set of equations:

$$\begin{aligned} u' &= 998u + 1998v \\ v' &= -999u - 1999v \end{aligned} \tag{15.6.5}$$

with boundary conditions

$$u(0) = 1 \qquad v(0) = 0 \tag{15.6.6}$$

By means of the transformation

$$u = 2y - z \qquad v = -y + z \tag{15.6.7}$$

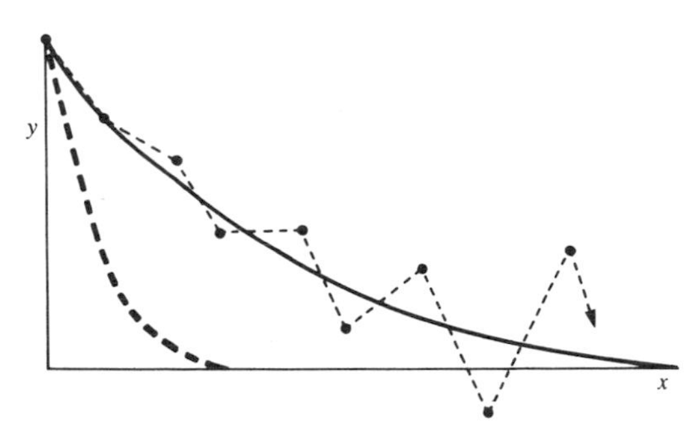

Figure 15.6.1. Example of an instability encountered in integrating a stiff equation (schematic). Here it is supposed that the equation has two solutions, shown as solid and dashed lines. Although the initial conditions are such as to give the solid solution, the stability of the integration (shown as the unstable dotted sequence of segments) is determined by the more rapidly-varying dashed solution, even after that solution has effectively died away to zero. Implicit integration methods are the cure.

we find the solution

$$\begin{aligned} u &= 2e^{-x} - e^{-1000x} \\ v &= -e^{-x} + e^{-1000x} \end{aligned} \tag{15.6.8}$$

If we integrated the system (15.6.5) with any of the methods given so far in this chapter, the presence of the e^{-1000x} term would require a stepsize $h \ll 1/1000$ for the method to be stable (the reason for this is explained below). This is so even though the e^{-1000x} term is completely negligible in determining the values of u and v as soon as one is away from the origin (see Figure 15.6.1).

This is the generic disease of stiff equations: we are required to follow the variation in the solution on the shortest length scale to maintain stability of the integration, even though accuracy requirements allow a much larger stepsize.

To see how we might cure this problem, consider the single equation

$$y' = -cy \tag{15.6.9}$$

where $c > 0$ is a constant. The explicit (or *forward*) Euler scheme for integrating this equation with stepsize h is

$$y_{n+1} = y_n + hy'_n = (1 - ch)y_n \tag{15.6.10}$$

The method is called explicit because the new value y_{n+1} is given explicitly in terms of the old value y_n. Clearly the method is unstable if $h > 2/c$, for then $y_n \to \infty$ as $n \to \infty$.

The simplest cure is to resort to *implicit* differencing, where the right-hand side is evaluated at the *new* y location. In this case, we get the *backward Euler* scheme:

$$y_{n+1} = y_n + hy'_{n+1} \tag{15.6.11}$$

or

$$y_{n+1} = \frac{y_n}{1+ch} \tag{15.6.12}$$

The method is absolutely stable: even as $h \to \infty$, $y_{n+1} \to 0$, which is in fact the correct solution of the differential equation. If we think of x as representing time, then the implicit method converges to the true equilibrium solution (i.e. the solution at late times) for large stepsizes. This nice feature of implicit methods holds only for linear systems, but even in the general case implicit methods give better stability. Of course, we give up *accuracy* in following the evolution towards equilibrium if we use large stepsizes, but we maintain *stability.*

These considerations can easily be generalized to sets of linear equations with constant coefficients:

$$\mathbf{y}' = -\mathbf{C} \cdot \mathbf{y} \tag{15.6.13}$$

where $\mathbf{C}$ is a positive definite matrix. Explicit differencing gives

$$\mathbf{y}_{n+1} = (\mathbf{1} - \mathbf{C}h) \cdot \mathbf{y}_n \tag{15.6.14}$$

Now a matrix $\mathbf{A}^n$ tends to zero as $n \to \infty$ only if the largest eigenvalue of $\mathbf{A}$ has magnitude less than unity. Thus $\mathbf{y}_n$ is bounded as $n \to \infty$ only if the largest eigenvalue of $\mathbf{1} - \mathbf{C}h$ is less than 1, or in other words

$$h < \frac{2}{\lambda_{max}} \tag{15.6.15}$$

where λ_{max} is the largest eigenvalue of $\mathbf{C}$.

On the other hand, implicit differencing gives

$$\mathbf{y}_{n+1} = \mathbf{y}_n + h\mathbf{y}'_{n+1} \tag{15.6.16}$$

or

$$\mathbf{y}_{n+1} = (\mathbf{1} + \mathbf{C}h)^{-1} \cdot \mathbf{y}_n \tag{15.6.17}$$

If the eigenvalues of $\mathbf{C}$ are λ, then the eigenvalues of $(\mathbf{1}+\mathbf{C}h)^{-1}$ are $(1+\lambda h)^{-1}$, which has magnitude less than one for all h. (Recall that all the eigenvalues of a positive definite matrix are nonnegative.) Thus the method is stable for

all stepsizes h. The penalty we pay for this stability is that we are required to invert a matrix at each step.

Not all equations are linear with constant coefficients, unfortunately! For the system

$$\mathbf{y}' = \mathbf{f}(x, \mathbf{y}) \tag{15.6.18}$$

(note that the notation differs from previous sections by omission of the reminder "prime" on the right-hand side) implicit differencing gives

$$\mathbf{y}_{n+1} = \mathbf{y}_n + h\mathbf{f}(x_{n+1}, \mathbf{y}_{n+1}) \tag{15.6.19}$$

In general this is some nasty set of nonlinear equations that has to be solved iteratively at each step. Usually we can get away with linearizing the equations:

$$\mathbf{y}_{n+1} = \mathbf{y}_n + h \left[\mathbf{f}(x_{n+1}, \mathbf{y}_n) + \left. \frac{\partial \mathbf{f}}{\partial \mathbf{y}} \right|_{\mathbf{y}_n} \cdot (\mathbf{y}_{n+1} - \mathbf{y}_n) \right] \tag{15.6.20}$$

Here $\partial \mathbf{f} / \partial \mathbf{y}$ is the matrix of the partial derivatives of the right-hand side, and so at each step we have to invert the matrix

$$\mathbf{1} - h \frac{\partial \mathbf{f}}{\partial \mathbf{y}} \tag{15.6.21}$$

to find $\mathbf{y}_{n+1}$. This procedure is called a "semi-implicit" method. It is not guaranteed to be stable, but it usually is, because the behavior is locally similar to the case of a constant matrix $\mathbf{C}$ described above.

So far we have dealt only with implicit methods that are first-order accurate. While these are very robust, most problems will tolerate higher order methods. We can easily construct a second-order method by taking the average of the explicit and implicit first-order methods:

$$y_{n+1} = y_n + \frac{h}{2}(y'_{n+1} + y'_n) \tag{15.6.22}$$

This is simply the trapezoidal rule if y' does not depend on y. For our model equation (15.6.9), we find

$$y_{n+1} = \frac{1 - ch/2}{1 + ch/2} y_n \tag{15.6.23}$$

showing stability. For the system (15.6.18), we can construct a semi-implicit scheme analogous to (15.6.20):

$$\mathbf{y}_{n+1} = \mathbf{y}_n + \frac{h}{2}\left[\mathbf{f}(x_{n+1}, \mathbf{y}_n) + \left.\frac{\partial \mathbf{f}}{\partial \mathbf{y}}\right|_{\mathbf{y}_n} \cdot (\mathbf{y}_{n+1} - \mathbf{y}_n) + \mathbf{f}(x_n, \mathbf{y}_n)\right] \quad (15.6.24)$$

It is quite complicated to design higher order implicit methods, especially with a good scheme for automatic stepsize adjustment. Higher order methods analogous to Runge-Kutta and predictor-corrector methods have been developed. Gear gives a well tested routine in his book, and references to more recent developments can be found in Stoer and Bulirsch.

REFERENCES AND FURTHER READING:

Acton, Forman S. 1970, *Numerical Methods That Work* (New York: Harper and Row).

Gear, C. William. 1971, *Numerical Initial Value Problems in Ordinary Differential Equations* (Englewood Cliffs, N.J.: Prentice-Hall).

Stoer, J., and Bulirsch, R. 1980, *Introduction to Numerical Analysis* (New York: Springer-Verlag).

Chapter 16. Two Point Boundary Value Problems

16.0 Introduction

When ordinary differential equations are required to satisfy boundary conditions at more than one value of the independent variable, the resulting problem is called a *two point boundary value problem.* As the terminology indicates, the most common case by far is where boundary conditions are supposed to be satisfied at two points — usually the starting and ending values of the integration. However the phrase "two point boundary value problem" is also used loosely to include more complicated cases, e.g., where some conditions are specified at endpoints, others at interior (usually singular) points.

The crucial distinction between initial value problems (Chapter 15) and two point boundary value problems (this chapter) is that in the former case we are able to start an acceptable solution at its beginning (initial values) and just march it along by numerical integration to its end (final values); while in the present case, the boundary conditions at the starting point do not determine a unique solution to start with — and a "random" choice among the solutions which satisfy these (incomplete) starting boundary conditions is almost certain *not* to satisfy the boundary conditions at the other specified point(s).

It should not surprise you that iteration is in general required to meld these spatially scattered boundary conditions into a single global solution of the differential equations. For this reason, two point boundary value problems require considerably more effort to solve than do initial value problems. You have to integrate your differential equations over the interval of interest, or perform an analogous "relaxation" procedure (see below), at least several, and sometimes very many, times. Only in the special case of linear differential equations can you say in advance just how many such iterations will be required.

The "standard" two point boundary value problem has the following form: We desire the solution to a set of N coupled first-order ordinary differential equations, satisfying n_1 boundary conditions at the starting point x_1, and a remaining set of $n_2 = N - n_1$ boundary conditions at the final point x_2. (Recall that all differential equations of order higher than first can be written as coupled sets of first-order equations, cf. §15.0.)

The differential equations are

$$\frac{dy_i(x)}{dx} = g_i(x, y_1, y_2, \ldots, y_N) \qquad i = 1, 2, \ldots, N \tag{16.0.1}$$

At x_1, the solution is supposed to satisfy

$$B_{1j}(x_1, y_1, y_2, \ldots, y_N) = 0 \qquad j = 1, \ldots, n_1 \tag{16.0.2}$$

while at x_2, it is supposed to satisfy

$$B_{2k}(x_2, y_1, y_2, \ldots, y_N) = 0 \qquad k = 1, \ldots, n_2 \tag{16.0.3}$$

There are two distinct classes of numerical methods for solving two point boundary value problems. In the *shooting method* (§16.1) we choose values for all of the dependent variables at one boundary. These values must be consistent with any boundary conditions for *that* boundary, but otherwise are arranged to depend on arbitrary free parameters whose values we initially "randomly" guess. We then integrate the ODEs by initial value methods, arriving at the other boundary (and/or any interior points with boundary conditions specified). In general, we find discrepancies from the desired boundary values there. Now we have a multidimensional root-finding problem, as was treated in Chapter 9: find the adjustment of the free parameters at the starting point that zeros the discrepancies at the other boundary point(s). If we liken integrating the differential equations to following the trajectory of a shot from gun to target, then picking the initial conditions corresponds to aiming (see Figure 16.0.1). The shooting method provides a systematic approach to taking a set of "ranging" shots that allow us to improve our "aim" systematically.

As another variant of the shooting method (§16.2), we can guess unknown free parameters at both ends of the domain, integrate the equations to a common midpoint, and seek to adjust the guessed parameters so that the solution joins "smoothly" at the fitting point. In all shooting methods, trial solutions satisfy the differential equations "exactly" (or as exactly as we care to make our numerical integration), but the trial solutions come to satisfy the required boundary conditions only after the iterations are finished.

Relaxation methods use a different approach. The differential equations are replaced by finite difference equations on a mesh of points that cover the range of the integration. A trial solution consists of values for the dependent variables at each mesh point, *not* satisfying the desired finite difference equations, nor necessarily even satisfying the required boundary conditions. The iteration, now called *relaxation*, consists of adjusting all the values on the mesh so as to bring them into successively closer agreement with the finite difference equations and, simultaneously, with the boundary conditions (see Figure 16.0.2). For example, if the problem involves three coupled equations

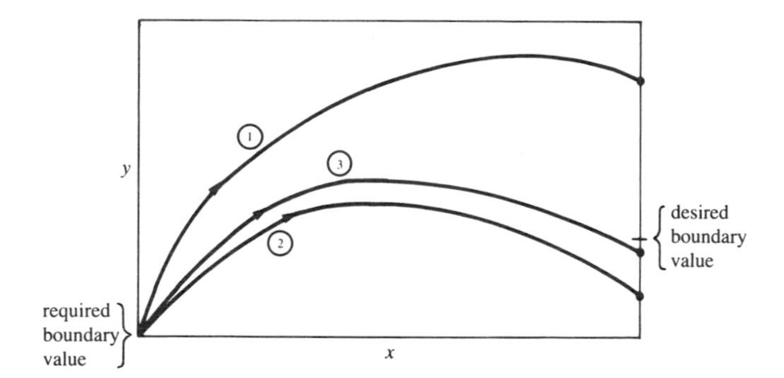

Figure 16.0.1. Shooting method (schematic). Trial integrations that satisfy the boundary condition at one endpoint are "launched." The discrepancies from the desired boundary condition at the other endpoint are used to adjust the starting conditions, until boundary conditions at both endpoints are ultimately satisfied.

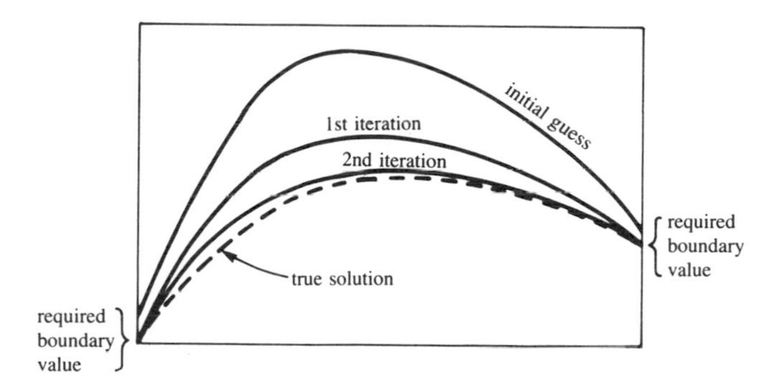

Figure 16.0.2. Relaxation method (schematic). An initial solution is guessed that approximately satisfies the differential equation and boundary conditions. An iterative process adjusts the function to bring it into close agreement with the true solution.

and a mesh of one hundred points, we must guess and improve three hundred variables representing the solution.

With all this adjustment, you may be surprised that relaxation is ever an efficient method, but (for the right problems) it really is! Relaxation works better than shooting when the boundary conditions are especially delicate or subtle, or where they involve complicated algebraic relations that cannot easily be solved in closed form. Relaxation works best when the solution is smooth and not highly oscillatory. Such oscillations would require many grid points for accurate representation. The number and position of required points may not be known *a priori*. Shooting methods are usually preferred in such cases, because their variable stepsize integrations adjust naturally to a solution's peculiarities.

Relaxation methods are often preferred when the ODEs have extraneous solutions which, while not appearing in the final solution satisfying all boundary conditions, may wreak havoc on the initial value integrations required by

shooting. The typical case is that of trying to maintain a dying exponential in the presence of growing exponentials.

Good initial guesses are the secret of efficient relaxation methods. Often one has to solve a problem many times, each time with a slightly different value of some parameter. In that case, the previous solution is usually a good initial guess when the parameter is changed, and relaxation will work well.

Until you have enough experience to make your own judgment between the two methods, you might wish to follow the advice of your authors as notorious computer gunslingers: We always shoot first, and only then relax.

Problems Reducible to the Standard Boundary Value Problem

There are two important problems that can be reduced to the standard boundary value problem described by equations (16.0.1) – (16.0.3). The first is the *eigenvalue problem for differential equations.* Here the right-hand side of the system of differential equations depends on a parameter λ,

$$\frac{dy_i(x)}{dx} = g_i(x, y_1, \ldots, y_N, \lambda) \tag{16.0.4}$$

and one has to satisfy $N+1$ boundary conditions instead of just N. The problem is overdetermined and in general there is no solution for arbitrary values of λ. For certain special values of λ, the eigenvalues, equation (16.0.4) does have a solution.

We reduce this problem to the standard case by introducing a new dependent variable

$$y_{N+1} \equiv \lambda \tag{16.0.5}$$

and another differential equation

$$\frac{dy_{N+1}}{dx} = 0 \tag{16.0.6}$$

An example of this trick is given in §16.4.

The other case that can be put in the standard form is a *free boundary problem.* Here only one boundary abscissa x_1 is specified, while the other boundary x_2 is to be determined so that the system (16.0.1) has a solution satisfying a total of $N+1$ boundary conditions. Here we again add an extra constant dependent variable:

$$y_{N+1} \equiv x_2 - x_1 \tag{16.0.7}$$

$$\frac{dy_{N+1}}{dx} = 0 \tag{16.0.8}$$

We also define a new *independent* variable t by setting

$$x - x_1 \equiv t\, y_{N+1}, \qquad 0 \leq t \leq 1 \tag{16.0.9}$$

The system of $N+1$ differential equations for dy_i/dt is now in the standard form, with t varying between the known limits 0 and 1.

REFERENCES AND FURTHER READING:

Keller, Herbert B. 1968, *Numerical Methods for Two-Point Boundary-Value Problems* (Waltham, Mass.: Blaisdell).

Kippenhan, R., Weigert, A., and Hofmeister, E. 1968, *Methods of Computational Physics*, vol. 7, pp. 129 ff.

Eggleton, P.P. 1971, *Monthly Notices of the Royal Astronomical Society*, vol. 151, pp. 351 ff.

London, R.A., and Flannery, B.P. 1982, *The Astrophysical Journal*, vol. 258, pp. 260–269.

16.1 The Shooting Method

In this section we discuss "pure" shooting, where the integration proceeds from x_1 to x_2, and we try to match boundary conditions at the end of the integration. In the next section, we describe shooting to an intermediate fitting point, where the solution to the equations and boundary conditions is found by launching "shots" from both sides of the interval and trying to match continuity conditions at some intermediate point.

The shooting method exactly implements multidimensional Newton-Raphson (§9.6). It seeks to zero n_2 functions of n_2 variables. The functions are obtained by integrating N differential equations from x_1 to x_2. Let us see how this works:

At the starting point x_1 there are N starting values y_i to be specified, but subject to n_1 conditions. Therefore there are $n_2 = N - n_1$ *freely specifiable* starting values. Let us imagine that these freely specifiable values are the components of a vector $\mathbf{V}$ that lives in a vector space of dimension n_2. Then you, the user, knowing the functional form of the boundary conditions (16.0.2), can write a subroutine which generates a complete set of N starting values $\mathbf{y}$, satisfying the boundary conditions at x_1, from an arbitrary vector value of $\mathbf{V}$ in which there are no restrictions on the n_2 component values. In other words, (16.0.2) converts to a prescription

$$y_i(x_1) = y_i(x_1; V_1 \ldots V_{n_2}) \qquad i = 1 \ldots N \tag{16.1.1}$$

Below, the subroutine which implements (16.1.1) will be called LOAD.

Notice that the components of $\mathbf{V}$ might be exactly the values of certain "free" components of $\mathbf{y}$, with the other components of $\mathbf{y}$ determined by the boundary conditions. Alternatively, the components of $\mathbf{V}$ might parametrize the solutions which satisfy the starting boundary conditions in some other convenient way. Boundary conditions often impose algebraic relations among the y_i, rather than specific values for each of them. Using some auxiliary set of parameters often makes it easier to "solve" the boundary relations for a consistent set of y_i's. It makes no difference which way you go, as long as your vector space of $\mathbf{V}$'s generates (through 16.1.1) all allowed starting vectors $\mathbf{y}$.

Given a particular $\mathbf{V}$, a particular $\mathbf{y}(x_1)$ is thus generated. It can then be turned into a $\mathbf{y}(x_2)$ by integrating the ODEs to x_2 as an initial value problem (e.g. using Chapter 15's ODEINT). Now, at x_2, let us define a *discrepancy vector* $\mathbf{F}$, also of dimension n_2, whose components measure how far we are from satisfying the n_2 boundary conditions at x_2 (16.0.3). Simplest of all is just to use the right-hand sides of (16.0.3),

$$F_k = B_{2k}(x_2, \mathbf{y}) \qquad k = 1 \ldots n_2 \tag{16.1.2}$$

As in the case of $\mathbf{V}$, however, you can use any other convenient parametrization, as long as your space of $\mathbf{F}$'s spans the space of possible discrepancies from the desired boundary conditions, with all components of $\mathbf{F}$ equal to zero if and only if the boundary conditions at x_2 are satisfied. Below, you will be asked to supply a user-written subroutine SCORE which uses (16.0.3) to convert an N-vector of ending values $\mathbf{y}(x_2)$ into an n_2-vector of discrepancies $\mathbf{F}$.

Now, as far as Newton-Raphson is concerned, we are nearly in business. We want to find a vector value of $\mathbf{V}$ which zeros the vector value of $\mathbf{F}$. We do this by computing (iteratively, as many times as required) the solution of a set of n_2 linear equations

$$[\alpha] \cdot \delta\mathbf{V} = -\mathbf{F} \tag{16.1.3}$$

and then adding the correction back,

$$\mathbf{V}^{new} = \mathbf{V}^{old} + \delta\mathbf{V} \tag{16.1.4}$$

In (16.1.3), the matrix $[\alpha]$ has components given by

$$[\alpha]_{ij} = \frac{\partial F_i}{\partial V_j} \tag{16.1.5}$$

It is not feasible to compute these partial derivatives analytically. Rather, each requires a *separate* integration of the N ODEs, followed by the evaluation of

$$\frac{\partial F_i}{\partial V_j} \approx \frac{F_i(V_1, \ldots, V_j + \Delta V_j, \ldots) - F_i(V_1, \ldots, V_j, \ldots)}{\Delta V_j} \tag{16.1.6}$$

This is done in the program below. You, the user, will have to supply a vector $\mathbf{\Delta V}$ of appropriate size steps for the finite differencing of derivatives.

For some problems the initial stepsize might depend sensitively upon the initial conditions. It is straightforward to alter `LOAD` to include a suggested stepsize `H1` as another returned argument.

A complete cycle of the shooting method thus requires $n_2 + 1$ integrations of the N coupled ODEs: One integration to evaluate the current degree of mismatch, and n_2 for the partial derivatives. Each new cycle requires a new round of $n_2 + 1$ integrations. This illustrates the enormous extra effort involved in solving two point boundary value problems compared with intial value problems.

If the differential equations are *linear*, then only one complete cycle is required, since (16.1.3)–(16.1.4) should take us right to the solution. A second round can be useful, however, in mopping up some (never all) of the roundoff error.

Obviously, choosing the increments ΔV_j for taking numerical derivatives is crucial to the success of the shooting method. If we pick ΔV_j too small, the derivative will be meaningless because we lose accuracy by roundoff, or because the integration of the ODEs in any case maintains only limited accuracy. If we pick ΔV_j too large, then values $y_i(x_2)$ may "whiplash" all over the place, so that equation (16.1.6) does not well approximate the derivative. However, in practice, *either* it is not difficult to distingush between large and small, *or* shooting won't work anyway. That is to say, if the increment required for an accurate approximation of the numerical derivative is "small," then the solution is probably very sensitive to initial conditions. In such cases relaxation often is the preferred technique.

Like any Newton-Raphson method, shooting will work if the the functions vary smoothly in the neighborhood of a solution, and if the initial guess is "close enough" to the actual solution. The method converges quadratically in the vicinity of a solution, with the remaining error ϵ in the root becoming ϵ^2 after each cycle of the method.

The subroutine `SHOOT` carries out one cycle of the shooting method for `NVAR` ODEs with `N2` adjustable starting values at `X1` (equal to the number of boundary conditions at `X2`). `SHOOT` performs *one cycle* of shooting, and returns updated values for parameters `V` describing the initial conditions at `X1`, and also `N2` values of the increments `DV`. Also returned is `F`, the discrepancy vector at `X2` that should be zero for a perfect solution.

Two point boundary value problems are usually sufficiently complex that the user will want to encode procedures to examine details of the solution after each cycle of the method. Consequently, we have not built `SHOOT` with an iteration loop that cycles to convergence. If you have a simple problem it is

trivial to add such a loop by supplying tolerances on DV,F and iterating until they are met. As given here, SHOOT uses the quality controlled Runge-Kutta method of §15.2 to integrate the ODEs, but any of the other methods could just as well be used.

You, the user, must supply SHOOT with: (i) a subroutine LOAD(X1,V,Y) which returns the N-vector Y (satisfying the starting boundary conditions, of course), given the freely-specifiable variables of V at the initial point X1; (ii) a subroutine SCORE(X2,Y,F) which returns the discrepancy vector F of the ending boundary conditions, given the vector Y at the end point X2; (iii) a vector of suggested increments DELV to be used in the finite differences (16.1.6); (iv) a starting vector V; (v) a subroutine DERIVS for the ODE integration; and other obvious parameters as described in the header comment below.

```
SUBROUTINE SHOOT(NVAR,V,DELV,N2,X1,X2,EPS,H1,HMIN,F,DV)
   Improve the trial solution of a two point boundary value problem for NVAR coupled ODEs
   shooting from X1 to X2. Initial values for the NVAR ODEs at X1 are generated from
   the N2 coefficients V, using the user-supplied routine LOAD. The routine integrates the
   ODEs to X2 using the Runge-Kutta method with tolerance EPS, initial step size H1, and
   minimum step size HMIN. At X2 it calls the user-supplied subroutine SCORE to evaluate
   the N2 functions F that ought to be zero to satisfy the boundary conditions at X2. Multi-
   dimensional Newton-Raphson is then used to develop a linear matrix equation for the
   N2 increments DV to the adjustable parameters V. These increments are solved for and
   added before return. The user-supplied subroutine DERIVS(X,Y,DYDX) supplies derivative
   information to the ODE integrator (see Chapter 15).
PARAMETER (NP=20)              At most NP coupled ODEs, or change as desired.
COMMON /PATH/ KMAX             For compatibility with ODEINT.
DIMENSION V(N2),DELV(N2),F(N2),DV(N2),Y(NP),DFDV(NP,NP),INDX(NP)
EXTERNAL DERIVS,RKQC
KMAX=0
CALL LOAD(X1,V,Y)              Integrate from X1 with best trial values.
CALL ODEINT(Y,NVAR,X1,X2,EPS,H1,HMIN,NOK,NBAD,DERIVS,RKQC)
CALL SCORE(X2,Y,F)
DO 12 IV=1,N2                  Vary boundary conditions at X1.
    SAV=V(IV)
    V(IV)=V(IV)+DELV(IV)       Increment parameter IV.
    CALL LOAD(X1,V,Y)
    CALL ODEINT(Y,NVAR,X1,X2,EPS,H1,HMIN,NOK,NBAD,DERIVS,RKQC)
    CALL SCORE(X2,Y,DV)
    DO 11 I=1,N2               Evaluate numerical derivatives of N2 matching conditions.
        DFDV(I,IV)=(DV(I)-F(I))/DELV(IV)
11  CONTINUE
    V(IV)=SAV                  Restore incremented parameter.
12 CONTINUE
DO 13 IV=1,N2
    DV(IV)=-F(IV)
13 CONTINUE
CALL LUDCMP(DFDV,N2,NP,INDX,DET)     Solve linear equations.
CALL LUBKSB(DFDV,N2,NP,INDX,DV)
DO 14 IV=1,N2                  Increment boundary parameters.
    V(IV)=V(IV)+DV(IV)
14 CONTINUE
RETURN
END
```

REFERENCES AND FURTHER READING:

Acton, Forman S. 1970, *Numerical Methods That Work* (New York: Harper and Row).

Keller, Herbert B. 1968, *Numerical Methods for Two-Point Boundary-Value Problems* (Waltham, Mass.: Blaisdell).

16.2 Shooting to a Fitting Point

The shooting method described in §16.1 tacitly assumed that the "shots" would be able to traverse the entire domain of integration, even at the early stages of convergence to a correct solution. In some problems it can happen that, for very wrong starting conditions, an initial solution can't even get from x_1 to x_2 without encountering some incalculable, or catastrophic, result. For example, the argument of a square root might go negative, causing the numerical code to crash. Simple shooting would be stymied.

A different, but related, case is where the endpoints are both singular points of the set of ODEs. One frequently needs to use special methods to integrate near the singular points, analytic asymptotic expansions, for example. In such cases it is feasible to integrate in the direction *away* from a singular point, using the special method to get through the first little bit and then reading off "initial" values for further numerical integration. However it is usually not feasible to integrate *into* a singular point, if only because one has not usually expended the same analytic effort to obtain expansions of "wrong" solutions near the singular point (those not satisfying the desired boundary condition).

The solution to the above mentioned difficulties is *shooting to a fitting point.* Instead of integrating from x_1 to x_2, we integrate first from x_1 to some point x_f that is *between* x_1 and x_2; and second from x_2 (in the opposite direction) to x_f.

If (as before) the number of boundary conditions imposed at x_1 is n_1, and the number imposed at x_2 is n_2, then there are n_2 freely specifiable starting values at x_1 and n_1 freely specifiable starting values at x_2. (If you are confused by this, go back to §16.1.) We can therefore define an n_2-vector $\mathbf{V}_{(1)}$ of starting parameters at x_1, and a prescription `LOAD1(X1,V1,Y)` for mapping $\mathbf{V}_{(1)}$ into a $\mathbf{y}$ which satisfies the boundary conditions at x_1,

$$y_i(x_1) = y_i(x_1; V_{(1)1} \dots V_{(1)n_2}) \qquad i = 1 \dots N \tag{16.2.1}$$

Likewise we can define an n_1-vector $\mathbf{V}_{(2)}$ of starting parameters at x_2, and a prescription `LOAD2(X2, V2, Y)` for mapping $\mathbf{V}_{(2)}$ into a $\mathbf{y}$ which satisfies the boundary conditions at x_2,

$$y_i(x_2) = y_i(x_2; V_{(2)1} \dots V_{(2)n_1}) \qquad i = 1 \dots N \tag{16.2.2}$$

We thus have a total of N freely adjustable parameters in the combination of $\mathbf{V}_{(1)}$ and $\mathbf{V}_{(2)}$. The N conditions that must be satisfied are that there be agreement in N components of $\mathbf{y}$ at x_f between the values obtained integrating from one side and from the other,

$$y_i(x_f; \mathbf{V}_{(1)}) = y_i(x_f; \mathbf{V}_{(2)}) \qquad i = 1, \dots, N \tag{16.2.3}$$

In some problems, the N matching conditions can be better described (physically, mathematically, or numerically) by using N different functions F_i, $i = 1 \ldots N$, each possibly depending on the N components y_i. In those cases, (16.2.3) is replaced by

$$F_i[\mathbf{y}(x_f; \mathbf{V}_{(1)})] = F_i[\mathbf{y}(x_f; \mathbf{V}_{(2)})] \qquad i = 1, \ldots, N \tag{16.2.4}$$

In the program below, the user-supplied subroutine SCORE(XF,Y,F) is supposed to map an input N-vector $\mathbf{y}$ into an output N-vector $\mathbf{F}$. In most cases, you can dummy this subroutine as the identity mapping.

Shooting to a fitting point uses Newton-Raphson exactly as in §16.1. Comparing closely with the routine SHOOT of the previous section, you should have no difficulty in understanding the following routine SHOOTF. The main differences in use are that you have to supply both V1 and V2; DELV1 and DELV2; LOAD1 and LOAD2.

In fitting problems it is unlikely that the magnitude of the initial stepsize will be identical for integrations from each boundary. In the routine below, the subroutine ODEINT will automatically adjust the step to achieve the proper accuracy. Alternatively, you can modify the routine to accept individual suggested stepsizes for the two integrations.

```
SUBROUTINE SHOOTF(NVAR,V1,V2,DELV1,DELV2,N1,N2,X1,X2,XF,EPS,H1,HMIN,F,DV1,DV2)
```

Improve the trial solution of a two point boundary value problem for NVAR coupled ODEs shooting from X1 and X2 to a fitting point XF. Initial values for the NVAR ODEs at X1 (X2) are generated from the N2 (N1) coefficients V1 (V2), using the user-supplied routine LOAD1 (LOAD2). The routine integrates the ODEs to XF using the Runge-Kutta method with tolerance EPS, initial stepsize H1, and minimum stepsize HMIN. At XF it calls the user-supplied subroutine SCORE to evaluate the NVAR functions F that ought to match at XF. Multi-dimensional Newton-Raphson is then used to develop a linear matrix equation for the N2 (N1) increments DV1 (DV2) to the adjustable parameters V1 (V2). These increments are solved for and added before return. The user-supplied subroutine DERIVS(X,Y,DYDX) supplies derivative information to the ODE integrator (see Chapter 15).

```
      PARAMETER (NP=20)                At most NP equations, or change as desired.
      COMMON /PATH/ KMAX               For compatibility with ODEINT.
      DIMENSION V1(N2),DELV1(N2),V2(N1),DELV2(N1),F(NVAR),DV1(N2),
     *    DV2(N1),Y(NP),F1(NP),F2(NP),DFDV(NP,NP),INDX(NP)
      EXTERNAL DERIVS,RKQC
      KMAX=0
      CALL LOAD1(X1,V1,Y)              Path from X1 to XF with best trial values V1.
      CALL ODEINT(Y,NVAR,X1,XF,EPS,H1,HMIN,NOK,NBAD,DERIVS,RKQC)
      CALL SCORE(XF,Y,F1)
      CALL LOAD2(X2,V2,Y)              Path from X2 to XF with best trial values V2.
      CALL ODEINT(Y,NVAR,X2,XF,EPS,H1,HMIN,NOK,NBAD,DERIVS,RKQC)
      CALL SCORE(XF,Y,F2)
      J=0
      DO 12 IV=1,N2                    Vary boundary conditions at X1.
          J=J+1
          SAV=V1(IV)
          V1(IV)=V1(IV)+DELV1(IV)
          CALL LOAD1(X1,V1,Y)
          CALL ODEINT(Y,NVAR,X1,XF,EPS,H1,HMIN,NOK,NBAD,DERIVS,RKQC)
          CALL SCORE(XF,Y,F)
          DO 11 I=1,NVAR               Evaluate numerical derivatives of NVAR fitting conditions.
              DFDV(I,J)=(F(I)-F1(I))/DELV1(IV)
11        CONTINUE
          V1(IV)=SAV                   Restore boundary parameter.
```

```
   12 CONTINUE
DO 14 IV=1,N1                         Next vary boundary conditions at X2.
    J=J+1
    SAV=V2(IV)
    V2(IV)=V2(IV)+DELV2(IV)
    CALL LOAD2(X2,V2,Y)
    CALL ODEINT(Y,NVAR,X2,XF,EPS,H1,HMIN,NOK,NBAD,DERIVS,RKQC)
    CALL SCORE(XF,Y,F)
    DO 13 I=1,NVAR
        DFDV(I,J)=(F2(I)-F(I))/DELV2(IV)
      13 CONTINUE
    V2(IV)=SAV
   14 CONTINUE
DO 15 I=1,NVAR
    F(I)=F1(I)-F2(I)
    F1(I)=-F(I)                       F1 used as handy temporary.
   15 CONTINUE
CALL LUDCMP(DFDV,NVAR,NP,INDX,DET)    Solve to find increments to free parameters.
CALL LUBKSB(DFDV,NVAR,NP,INDX,F1)
J=0
DO 16 IV=1,N2                         Increment adjustable boundary parameters at X1.
    J=J+1
    V1(IV)=V1(IV)+F1(J)
    DV1(IV)=F1(J)
   16 CONTINUE
DO 17 IV=1,N1                         Increment adjustable boundary parameters at X2.
    J=J+1
    V2(IV)=V2(IV)+F1(J)
    DV2(IV)=F1(J)
   17 CONTINUE
RETURN
END
```

REFERENCES AND FURTHER READING:

Acton, Forman S. 1970, *Numerical Methods That Work* (New York: Harper and Row).

Keller, Herbert B. 1968, *Numerical Methods for Two-Point Boundary-Value Problems* (Waltham, Mass.: Blaisdell).

16.3 Relaxation Methods

In *relaxation methods* we replace ODEs by approximate *finite difference equations* (FDEs) on a grid or mesh of points that spans the domain of interest. As a typical example, we could replace a general first-order differential equation

$$\frac{dy}{dx} = g(x, y) \tag{16.3.1}$$

with an algebraic equation relating function values at two points $k, k-1$:

$$y_k - y_{k-1} - (x_k - x_{k-1})\, g\left[\tfrac{1}{2}(x_k + x_{k-1}), \tfrac{1}{2}(y_k + y_{k-1})\right] = 0 \tag{16.3.2}$$

The form of the FDE in (16.3.2) illustrates the idea, but not uniquely: there are many ways to turn the ODE into an FDE. When the problem involves N coupled first-order ODEs represented by FDEs on a mesh of M points, a solution consists of values for N dependent functions given at each of the M mesh points, or $N \times M$ variables in all. The relaxation method, like many of the methods described before, determines the solution by starting with a guess and improving it, iteratively. As the iterations improve the solution, the result is said to *relax* to the true solution.

While several iteration schemes are possible, for most problems our old standby, multi-dimensional Newton's method, works well. The method produces a matrix equation that must be solved, but the matrix takes a special, "block diagonal" form, that allows it to be inverted far more economically both in time and storage than would be possible for a general matrix of size $(MN) \times (MN)$. Since MN can easily be several thousand, this is crucial for the feasibility of the method.

Our implementation couples at most pairs of points, as in equation (16.3.2). More points can be coupled, but then the method becomes more complex. We will provide enough background so that you can write a more general scheme if you have the patience to do so.

Let us develop a general set of algebraic equations that represent the ODEs by FDEs. The ODE problem is exactly identical to that expressed in equations (16.0.1)–(16.0.3) where we had N coupled first-order equations that satisfy n_1 boundary conditions at x_1 and $n_2 = N - n_1$ boundary conditions at x_2. We first define a mesh or grid by a set of $k = 1, 2, ..., M$ points at which we supply values for the dependent variable x_k. In particular, x_1 is the initial boundary, and x_M is the final boundary. We use the notation $\mathbf{y}_k$ to refer to the entire set of dependent variables $y_1, y_2, \ldots, y_N$ at point x_k. At an arbitrary point k in the middle of the mesh, we approximate the set of N first-order ODEs by algebraic relations of the form

$$0 = \mathbf{E}_k = \mathbf{y}_k - \mathbf{y}_{k-1} - (x_k - x_{k-1})\mathbf{g}_k(x_k, x_{k-1}, \mathbf{y}_k, \mathbf{y}_{k-1}), \quad k = 2, 3, \ldots, M \tag{16.3.3}$$

The notation signifies that $\mathbf{g}_k$ can be evaluated using information from both points $k, k-1$. The FDEs labeled by $\mathbf{E}_k$ provide N equations coupling $2N$ variables at points $k, k-1$. There are $M-1$ points, $k = 2, 3, \ldots, M$, at which difference equations of the form (16.3.3) apply. Thus the FDEs provide a total of $(M-1)N$ equations for the MN unknowns. The remaining N equations come from the boundary conditions.

At the first boundary we have

$$0 = \mathbf{E}_1 = \mathbf{B}(x_1, \mathbf{y}_1) \tag{16.3.4}$$

while at the second boundary

$$0 = \mathbf{E}_{M+1} = \mathbf{C}(x_{M+1}, \mathbf{y}_{M+1}) \tag{16.3.5}$$

The vectors $\mathbf{E}_1$ and $\mathbf{B}$ have only n_1 nonzero components, corresponding to the n_1 boundary conditions at x_1. It will turn out to be useful to take these nonzero components to be the *last* n_1 components. In other words, $E_{j,1} \neq 0$ only for $j = n_2 + 1, n_2 + 2, \ldots, N$. At the other boundary, only the first n_2 components of $\mathbf{E}_{M+1}$ and $\mathbf{C}$ are nonzero: $E_{j,M+1} \neq 0$ only for $j = 1, 2, \ldots, n_2$.

The "solution" of the FDE problem in (16.3.3)–(16.3.5) consists of a set of variables $y_{j,k}$, the values of the N variables y_j at the M points x_k. The algorithm we describe below requires an initial guess for the $y_{j,k}$. We then determine increments $\Delta y_{j,k}$ such that $y_{j,k} + \Delta y_{j,k}$ is an improved approximation to the solution.

Equations for the increments are developed by expanding the FDEs in first-order Taylor series with respect to small changes $\Delta \mathbf{y}_k$. At an interior point, $k = 2, 3, \ldots, M$ this gives:

$$\begin{aligned}\mathbf{E}_k(\mathbf{y}_k + \Delta\mathbf{y}_k, \mathbf{y}_{k-1} + \Delta\mathbf{y}_{k-1}) &\approx \mathbf{E}_k(\mathbf{y}_k, \mathbf{y}_{k-1}) \\ &+ \sum_{n=1}^{N} \frac{\partial \mathbf{E}_k}{\partial y_{n,k-1}} \Delta y_{n,k-1} + \sum_{n=1}^{N} \frac{\partial \mathbf{E}_k}{\partial y_{n,k}} \Delta y_{n,k}\end{aligned} \tag{16.3.6}$$

For a solution we want the updated value $\mathbf{E}(\mathbf{y}+\Delta\mathbf{y})$ to be zero, so the general set of equations at an interior point can be written in matrix form as

$$\sum_{n=1}^{N} S_{j,n} \Delta y_{n,k-1} + \sum_{n=N+1}^{2N} S_{j,n} \Delta y_{n-N,k} = -E_{j,k}, \quad j = 1, 2, \ldots, N \tag{16.3.7}$$

where

$$S_{j,n} = \frac{\partial E_{j,k}}{\partial y_{n,k-1}}, \quad S_{j,n+N} = \frac{\partial E_{j,k}}{\partial y_{n,k}}, \quad n = 1, 2, \ldots, N \tag{16.3.8}$$

The quantity $S_{j,n}$ is an $N \times 2N$ matrix at each point k. Each interior point thus supplies a block of N equations coupling $2N$ corrections to the solution variables at the points $k, k-1$.

Similarly, the algebraic relations at the boundaries can be expanded in a first-order Taylor series for increments that improve the solution. Since $\mathbf{E}_1$ depends only on $\mathbf{y}_1$, we find at the first boundary:

$$\sum_{n=1}^{N} S_{j,n} \Delta y_{n,1} = -E_{j,1}, \quad j = n_2 + 1, n_2 + 2, \ldots, N \tag{16.3.9}$$

where

$$S_{j,n} = \frac{\partial E_{j,1}}{\partial y_{n,1}}, \quad n = 1, 2, \ldots, N \tag{16.3.10}$$

```
X X X X X                                V   B
X X X X X                                V   B
X X X X X                                V   B
X X X X X X X X X X                      V   B
X X X X X X X X X X                      V   B
X X X X X X X X X X                      V   B
X X X X X X X X X X                      V   B
X X X X X X X X X X                      V   B
          X X X X X X X X X X            V   B
          X X X X X X X X X X            V   B
          X X X X X X X X X X            V   B
          X X X X X X X X X X            V   B
          X X X X X X X X X X            V   B
                    X X X X X X X X X X  V   B
                    X X X X X X X X X X  V   B
                    X X X X X X X X X X  V   B
                    X X X X X X X X X X  V   B
                    X X X X X X X X X X  V   B
                              X X X X X  V   B
                              X X X X X  V   B
```

Figure 16.3.1. Matrix structure of a set of linear finite difference equations (FDEs) with boundary conditions imposed at both endpoints. Here X represents a coefficient of the FDEs, V represents a component of the unknown solution vector, B is a component of the known right-hand side. Empty spaces represent zeros. The matrix equation is to be solved by a special form of Gaussian elimination. (See text for details.)

At the second boundary,

$$\sum_{n=1}^{N} S_{j,n}\Delta y_{n,M} = -E_{j,M+1}, \quad j = 1,2,\ldots,n_2 \tag{16.3.11}$$

where

$$S_{j,n} = \frac{\partial E_{j,M+1}}{\partial y_{n,M}}, \quad n = 1,2,\ldots,N \tag{16.3.12}$$

We thus have in equations (16.3.7)–(16.3.12) a set of linear equations to be solved for the corrections $\Delta\mathbf{y}$, iterating until the corrections are sufficiently small. The equations have a special structure, because each $S_{j,n}$ couples only points $k, k-1$. Figure 16.3.1 illustrates the typical structure of the complete matrix equation for the case of 5 variables and 4 mesh points, with 3 boundary conditions at the first boundary and 2 at the second. The 3×5 block of nonzero entries in the top left-hand corner of the matrix comes from the boundary condition $S_{j,n}$ at point $k = 1$. The next three 5×10 blocks are the $S_{j,n}$ at the interior points, coupling variables at mesh points (2,1), (3,2) and (4,3). Finally we have the block corresponding to the second boundary condition.

We can solve equations (16.3.7)–(16.3.12) for the increments $\Delta\mathbf{y}$ using a form of Gaussian elimination which exploits the special structure of the matrix to minimize the total number of operations, and which minimizes storage of matrix coefficients by packing the elements in a special blocked structure. (You might wish to review Chapter 2, especially §2.2, if you are unfamiliar

```
1   X X                                         V  B
  1 X X                                         V  B
    1 X X                                       V  B
      1        X X                              V  B
        1      X X                              V  B
          1    X X                              V  B
            1  X X                              V  B
              1 X X                             V  B
                1          X X                  V  B
                  1        X X                  V  B
                    1      X X                  V  B
                      1    X X                  V  B
                        1  X X                  V  B
                          1           X X       V  B
                            1         X X       V  B
                              1       X X       V  B
                                1     X X       V  B
                                  1   X X       V  B
                                    1           V  B
                                      1         V  B
```

Figure 16.3.2. Target structure of the Gaussian elimination. Once the matrix of Figure 16.3.1 has been reduced to this form, the solution follows quickly by backsubstitution.

with the steps involved in Gaussian elimination.) Recall that Gaussian elimination consists of manipulating the equations by elementary operations such as dividing rows of coefficients by a common factor to produce unity in diagonal elements, and adding appropriate multiples of other rows to produce zeros below the diagonal. Here we take advantage of the block structure by performing a bit more reduction than in pure Gaussian elimination, so that the storage of coefficients is minimized. Figure 16.3.2 shows the form that we wish to achieve by elimination, just prior to the backsubstitution step. Only a small subset of the reduced $MN \times MN$ matrix elements needs to be stored as the elimination progresses. Once the matrix elements reach the stage in Figure 16.3.2, the solution follows quickly by a backsubstitution procedure.

Furthermore, the entire procedure, except the backsubstitution step, operates only on one block of the matrix at a time. The procedure contains four types of operations: (1) partial reduction to zero of certain elements of a block using results from a previous step, (2) elimination of the square structure of the remaining block elements such that the square section contains unity along the diagonal, and zero in off-diagonal elements, (3) storage of the remaining nonzero coefficients for use in later steps, and (4) backsubstitution. We illustrate the steps schematically by figures.

Consider the block of equations describing corrections available from the initial boundary conditions. We have n_1 equations for N unknown corrections. We wish to transform the first block so that its left-hand $n_1 \times n_1$ square section becomes unity along the diagonal, and zero in off-diagonal elements. Figure 16.3.3 shows the original and final form of the first block of the matrix. In the figure we designate matrix elements that are subject to diagonalization by "D," and elements that will be altered by "A"; in the final block, elements that are stored are labeled by "S." We get from start to finish by selecting in turn n_1 "pivot" elements from among the first n_1 columns, normalizing the pivot row so that the value of the "pivot" element is unity, and adding appropriate multiples of this row to the remaining rows so that they contain

```
(a)  D D D A A      V   A
     D D D A A      V   A
     D D D A A      V   A

(b)  1 0 0 S S      V   S
     0 1 0 S S      V   S
     0 0 1 S S      V   S
```

Figure 16.3.3. Reduction process for the first (upper left) block of the matrix in Figure 16.3.1. (a) Original form of the block, (b) final form. (See text for explanation.)

zeros in the pivot column. In its final form, the reduced block expresses values for the corrections to the first n_1 variables at mesh point 1 in terms of values for the remaining n_2 unknown corrections at point 1, i.e. we now know what the first n_1 elements are in terms of the remaining n_2 elements. We store only the the final set of n_2 nonzero columns from the initial block, plus the column for the altered right-hand side of the matrix equation.

We must emphasize here an important detail of the method. To exploit the reduced storage allowed by operating on blocks, it is essential that the ordering of columns in the `S` matrix of derivatives be such that pivot elements can be found among the first n_1 rows of the matrix. This means that the n_1 boundary conditions at the first point must contain some dependence on the first `J=1,2,...,`n_1 dependent variables, `Y(J,1)`. If not, then the original square $n_1 \times n_1$ subsection of the first block will appear to be singular, and the method will fail. Alternatively, we would have to allow the search for pivot elements to involve all N columns of the block, and this would require column swapping and far more bookkeeping. The code provides a simple method of reordering the variables, i.e., the columns of the `S` matrix, so that this can be done easily. End of important detail.

Next consider the block of N equations representing the FDEs that describe the relation between the $2N$ corrections at points 2 and 1. The elements of that block, together with results from the previous step are illustrated in Figure 16.3.4. Note that by adding suitable multiples of rows from the first block we can reduce to zero the first n_1 columns of the block (labeled by "Z"), and, to do so, we will need to alter only the columns from $n_1 + 1$ to N and the vector element on the right-hand side. Of the remaining columns we can diagonalize a square subsection of $N \times N$ elements, labeled by "D" in the figure. In the process we alter the final set of $n_2 + 1$ columns, denoted "A" in the figure. The second half of the figure shows the block when we finish operating on it, with the stored $(n_2 + 1) \times N$ elements labeled by "S".

If we operate on the next set of equations corresponding to the FDEs coupling corrections at points 3 and 2, we see that the state of available results and new equations exactly reproduces the situation described in the previous paragraph. Thus, we can carry out those steps again for each block in turn through block M. Finally on block $M+1$ we encounter the remaining boundary conditions.

Figure 16.3.5 shows the final block of n_2 FDEs relating the N corrections for variables at mesh point M, together with the result of reducing the previous block. Again, we can first use the prior results to zero the first n_1 columns of the block. Now, when we diagonalize the remaining square section, we strike gold: we get values for the final n_2 corrections at mesh point M.

```
(a)  1 0 0 S S                 V  S
     0 1 0 S S                 V  S
     0 0 1 S S                 V  S
     Z Z Z D D D D D A A       V  A
     Z Z Z D D D D D A A       V  A
     Z Z Z D D D D D A A       V  A
     Z Z Z D D D D D A A       V  A
     Z Z Z D D D D D A A       V  A

(b)  1 0 0 S S                 V  S
     0 1 0 S S                 V  S
     0 0 1 S S                 V  S
     0 0 0 1 0 0 0 0 S S       V  S
     0 0 0 0 1 0 0 0 S S       V  S
     0 0 0 0 0 1 0 0 S S       V  S
     0 0 0 0 0 0 1 0 S S       V  S
     0 0 0 0 0 0 0 1 S S       V  S
```

Figure 16.3.4. Reduction process for intermediate blocks of the matrix in Figure 16.3.1. (a) Original form. (b) final form. (See text for explanation.)

```
(a)  0 0 0 1 0 0 0 0 S S V  S
     0 0 0 0 1 0 0 0 S S V  S
     0 0 0 0 0 1 0 0 S S V  S
     0 0 0 0 0 0 1 0 S S V  S
     0 0 0 0 0 0 0 1 S S V  S
                 Z Z Z D D V  A
                 Z Z Z D D V  A

(b)  0 0 0 1 0 0 0 0 S S V  S
     0 0 0 0 1 0 0 0 S S V  S
     0 0 0 0 0 1 0 0 S S V  S
     0 0 0 0 0 0 1 0 S S V  S
     0 0 0 0 0 0 0 1 S S V  S
                 0 0 0 1 0 V  S
                 0 0 0 0 1 V  S
```

Figure 16.3.5. Reduction process for the last (lower right) block of the matrix in Figure 16.3.1. (a) Original form, (b) final form. (See text for explanation.)

With the final block reduced, the matrix has the desired form shown previously in Figure 16.3.2, and the matrix is ripe for backsubstitution. Starting with the bottom row and working up towards the top, at each stage we can simply determine one unknown correction in terms of known quantities.

The subroutine `SOLVDE` organizes the steps described above. The principal procedures used in the algorithm are performed by subroutines called internally by `SOLVDE`. The subroutine `RED` eliminates leading columns of the `S` matrix using results from prior blocks. `PINVS` diagonalizes the square subsection of `S` and stores unreduced coefficients. `BKSUB` carries out the backsubstitution step. The user of `SOLVDE` must understand the calling arguments, as described below, and supply a subroutine `DIFEQ`, called by `SOLVDE`, that evaluates the `S` matrix for each block.

Most of the arguments in the call to `SOLVDE` have already been described, but some require discussion. Array `Y(J,K)` contains the initial guess for the solution, with `J` labeling the dependent variables at mesh points `K`. The problem involves `NE` FDEs spanning points `K=1,..., M`. `NB` boundary conditions apply at the first point `K=1`. The array `INDEXV(J)` establishes the correspon-

dence between columns of the S matrix, equations (16.3.8), (16.3.10), and (16.3.12), and the dependent variables. As described above it is essential that the NB boundary conditions at K=1 involve the dependent variables referenced by the first NB columns of the S matrix. Thus, columns J of the S matrix can be ordered by the user in DIFEQ to refer to derivatives with respect to the dependent variable INDEXV(J).

The subroutine only attempts ITMAX correction cycles before returning, even if the solution has not converged. The parameters CONV, SLOWC, SCALV relate to convergence. Each inversion of the matrix produces corrections for NE variables at M mesh points. We want these to become vanishingly small as the iterations proceed, but we must define a measure for the size of corrections. This error "norm" is very problem specific, so the user might wish to rewrite this section of the code as appropriate. In the program below we compute a value for the average correction ERR by summing the absolute value of all corrections, weighted by a scale factor appropriate to each type of variable:

$$\mathtt{ERR} = \frac{1}{\mathtt{M} \times \mathtt{NE}} \sum_{\mathtt{K=1}}^{\mathtt{M}} \sum_{\mathtt{J=1}}^{\mathtt{NE}} \frac{|\Delta Y(\mathtt{J,K})|}{\mathtt{SCALV(J)}} \tag{16.3.13}$$

When ERR$\leq$CONV, the method has converged. Note that the user gets to supply an array SCALV which measures the typical size of each variable.

Obviously, if ERR is large, we are far from a solution, and perhaps it is a bad idea to believe that the corrections generated from a first-order Taylor series are accurate. The number SLOWC modulates application of corrections. After each iteration we only apply a fraction of the corrections found by matrix inversion:

$$Y(\mathtt{J,K}) \to Y(\mathtt{J,K}) + \frac{\mathtt{SLOWC}}{\max(\mathtt{SLOWC,ERR})} \Delta Y(\mathtt{J,K}) \tag{16.3.14}$$

Thus, when ERR>SLOWC only a fraction of the corrections are used, but when ERR$\leq$SLOWC the entire correction gets applied.

The call statement also supplies SOLVDE with the array Y(NYJ,NYK) containing the initial trial solution, and workspace arrays C(NCI,NCJ,NCK), S(NSI,NSJ). The array C is the blockbuster: it stores the unreduced elements of the matrix built up for the backsubstitution step. If there are M mesh points, then there will be NCK=M+1 blocks, each requiring NCI=NE rows and NCJ=NE-NB+1 columns. Although large, this is small compared with $(\mathtt{NE}\times\mathtt{M})^2$ elements required for the whole matrix if we did not break it into blocks.

We now describe the workings of the user supplied subroutine DIFEQ. The parameters of the subroutine are given by

```
SUBROUTINE DIFEQ(K,K1,K2,JSF,IS1,ISF,INDEXV,NE,S,NSI,NSJ,Y,NYJ,NYK)
```

The only information returned from DIFEQ to SOLVDE is the matrix of derivatives S(I,J); all other arguments are input to DIFEQ and should not

be altered. K indicates the current mesh point, or block number. K1,K2 label the first and last point in the mesh. If K=K1 or K>K2, the block involves the boundary conditions at the first or final points; otherwise the block acts on FDEs coupling variables at points K-1, K.

The convention on storing information into the array S(I,J) follows that used in equations (16.3.8), (16.3.10) and (16.3.12): rows I label equations, columns J refer to derivatives with respect to dependent variables in the solution. Recall that each equation will depend on the NE dependent variables at either one or two points. Thus, J runs from 1 to either NE or 2*NE. The column ordering for dependent variables at each point must agree with the list supplied in INDEXV(J). Thus, for a block not at a boundary, the first column multiplies ΔY(L=INDEXV(1),K-1), and the column NE+1 multiplies ΔY(L=INDEXV(1),K). IS1,ISF give the numbers of the starting and final *rows* that need to be filled in the S matrix for this block. JSF labels the column in which the difference equations $E_{j,k}$ of equations (16.3.3)–(16.3.5) are stored. Thus, −S(I,JSF) is the vector on the right-hand side of the matrix. The reason for the minus sign is that DIFEQ supplies the actual difference equation, $E_{j,k}$, not its negative. Note that SOLVDE supplies a value for JSF such that the difference equation is put in the column *just after* all derivatives in the S matrix. Thus, DIFEQ expects to find values entered into S(I,J) for rows IS1$\leq$ I $\leq$ ISF and 1$\leq$ J $\leq$ JSF.

Finally, S (dimensioned NSI,NSJ) and Y (dimensioned NYJ,NYK) supply DIFEQ with storage for S and the solution variables Y for this iteration. An example of how to use this routine is given in the next section.

Many ideas in the following code are due to Peter Eggleton.

```
      SUBROUTINE SOLVDE(ITMAX,CONV,SLOWC,SCALV,INDEXV,NE,NB,M,
     *    Y,NYJ,NYK,C,NCI,NCJ,NCK,S,NSI,NSJ)
C        Driver routine for solution of two point boundary value problems by relaxation. ITMAX is
C        the maximum number of iterations. CONV is the convergence criterion (see text). SLOWC
C        controls the fraction of corrections actually used after each iteration. SCALV contains
C        typical sizes for each dependent variable, used to weight errors. INDEXV lists the column
C        ordering of variables used to construct the matrix S of derivatives. (The NB boundary
C        conditions at the first mesh point must contain some dependence on the first NB variables
C        listed in INDEXV.) The problem involves NE equations for NE adjustable dependent variables
C        at each point. At the first mesh point there are NB boundary conditions. There are a
C        total of M mesh points. Y is the two-dimensional array of size (NYJ,NYK) that contains
C        the initial guess for all the dependent variables at each mesh point. On each iteration,
C        it is updated by the calculated correction. The arrays C(NCI,NCJ,NCK), S(NSI,NSJ)
C        supply dummy storage used by the relaxation code; the minimum dimensions must satisfy:
C        NCI=NE, NCJ=NE-NB+1, NCK=M+1, NSI=NE, NSJ=2*NE+1.
      PARAMETER (NMAX=10)           Largest expected value of NE.
      DIMENSION Y(NYJ,NYK),C(NCI,NCJ,NCK),S(NSI,NSJ),SCALV(NYJ),INDEXV(NYJ)
      DIMENSION ERMAX(NMAX),KMAX(NMAX)
      K1=1                          Set up row and column markers.
      K2=M
      NVARS=NE*M
      J1=1
      J2=NB
      J3=NB+1
      J4=NE
      J5=J4+J1
      J6=J4+J2
      J7=J4+J3
      J8=J4+J4
```

```
      J9=J8+J1
      IC1=1
      IC2=NE-NB
      IC3=IC2+1
      IC4=NE
      JC1=1
      JCF=IC3
      DO 16 IT=1,ITMAX                 Primary iteration loop.
        K=K1                           Boundary conditions at first point.
        CALL DIFEQ(K,K1,K2,J9,IC3,IC4,INDEXV,NE,S,NSI,NSJ,Y,NYJ,NYK)
        CALL PINVS(IC3,IC4,J5,J9,JC1,K1,C,NCI,NCJ,NCK,S,NSI,NSJ)
        DO 11 K=K1+1,K2                Finite difference equations at all point pairs.
          KP=K-1
          CALL DIFEQ(K,K1,K2,J9,IC1,IC4,INDEXV,NE,S,NSI,NSJ,Y,NYJ,NYK)
          CALL RED(IC1,IC4,J1,J2,J3,J4,J9,IC3,JC1,JCF,KP,
     *         C,NCI,NCJ,NCK,S,NSI,NSJ)
          CALL PINVS(IC1,IC4,J3,J9,JC1,K,C,NCI,NCJ,NCK,S,NSI,NSJ)
11      CONTINUE
        K=K2+1                         Final boundary conditions.
        CALL DIFEQ(K,K1,K2,J9,IC1,IC2,INDEXV,NE,S,NSI,NSJ,Y,NYJ,NYK)
        CALL RED(IC1,IC2,J5,J6,J7,J8,J9,IC3,JC1,JCF,K2,
     *       C,NCI,NCJ,NCK,S,NSI,NSJ)
        CALL PINVS(IC1,IC2,J7,J9,JCF,K2+1,C,NCI,NCJ,NCK,S,NSI,NSJ)
        CALL BKSUB(NE,NB,JCF,K1,K2,C,NCI,NCJ,NCK)    Back substitution
        ERR=0.
        DO 13 J=1,NE                   Convergence check, accumulate average error.
          JV=INDEXV(J)
          ERRJ=0.
          KM=0
          VMAX=0.
          DO 12 K=K1,K2                Find point with largest error, for each dependent variable.
            VZ=ABS(C(J,1,K))
            IF(VZ.GT.VMAX) THEN
                 VMAX=VZ
                 KM=K
            ENDIF
            ERRJ=ERRJ+VZ
12        CONTINUE
          ERR=ERR+ERRJ/SCALV(JV)          Note weighting for each dependent variable.
          ERMAX(J)=C(J,1,KM)/SCALV(JV)
          KMAX(J)=KM
13      CONTINUE
        ERR=ERR/NVARS
        FAC=SLOWC/MAX(SLOWC,ERR)          Reduce correction applied when error is large.
        DO 15 JV=1,NE                  Apply corrections.
          J=INDEXV(JV)
          DO 14 K=K1,K2
            Y(J,K)=Y(J,K)-FAC*C(JV,1,K)
14        CONTINUE
15      CONTINUE
        WRITE(*,100) IT,ERR,FAC,(KMAX(J),ERMAX(J),J=1,NE)    Summary of corrections for this
        IF(ERR.LT.CONV) RETURN                                step.
16    CONTINUE
      PAUSE 'ITMAX exceeded'           Convergence failed.
100   FORMAT(1X,I4,2F12.6,(/5X,I5,F12.6))
      RETURN
      END
```

```
SUBROUTINE BKSUB(NE,NB,JF,K1,K2,C,NCI,NCJ,NCK)
   Back substitution, used internally by SOLVDE.
DIMENSION C(NCI,NCJ,NCK)
NBF=NE-NB
IM=1
DO 13 K=K2,K1,-1               Use recurrence relations to eliminate remaining dependences.
   IF (K.EQ.K1) IM=NBF+1         Special handling of first point.
   KP=K+1
   DO 12 J=1,NBF
      XX=C(J,JF,KP)
      DO 11 I=IM,NE
         C(I,JF,K)=C(I,JF,K)-C(I,J,K)*XX
11       CONTINUE
12     CONTINUE
13   CONTINUE
DO 16 K=K1,K2                  Reorder corrections to be in column 1.
   KP=K+1
   DO 14 I=1,NB
      C(I,1,K)=C(I+NBF,JF,K)
14     CONTINUE
   DO 15 I=1,NBF
      C(I+NB,1,K)=C(I,JF,KP)
15     CONTINUE
16   CONTINUE
RETURN
END


SUBROUTINE PINVS(IE1,IE2,JE1,JSF,JC1,K,C,NCI,NCJ,NCK,S,NSI,NSJ)
   Diagonalize the square subsection of the S matrix, and store the recursion coefficients in
   C; used internally by SOLVDE.
PARAMETER (ZERO=0.,ONE=1.,NMAX=10)
DIMENSION C(NCI,NCJ,NCK),S(NSI,NSJ),PSCL(NMAX),INDXR(NMAX)
JE2=JE1+IE2-IE1
JS1=JE2+1
DO 12 I=IE1,IE2                Implicit pivoting, as in §2.1.
   BIG=ZERO
   DO 11 J=JE1,JE2
      IF(ABS(S(I,J)).GT.BIG) BIG=ABS(S(I,J))
11     CONTINUE
   IF(BIG.EQ.ZERO) PAUSE 'Singular matrix, row all 0'
   PSCL(I)=ONE/BIG
   INDXR(I)=0
12   CONTINUE
DO 18 ID=IE1,IE2
   PIV=ZERO
   DO 14 I=IE1,IE2             Find pivot element.
      IF(INDXR(I).EQ.0) THEN
         BIG=ZERO
         DO 13 J=JE1,JE2
            IF(ABS(S(I,J)).GT.BIG) THEN
               JP=J
               BIG=ABS(S(I,J))
            ENDIF
13           CONTINUE
         IF(BIG*PSCL(I).GT.PIV) THEN
            IPIV=I
            JPIV=JP
            PIV=BIG*PSCL(I)
         ENDIF
      ENDIF
14     CONTINUE
   IF(S(IPIV,JPIV).EQ.ZERO) PAUSE 'Singular matrix'
   INDXR(IPIV)=JPIV            In place reduction. Save column ordering.
```

```
    PIVINV=ONE/S(IPIV,JPIV)
    DO 15 J=JE1,JSF                 Normalize pivot row.
        S(IPIV,J)=S(IPIV,J)*PIVINV
15  CONTINUE
    S(IPIV,JPIV)=ONE
    DO 17 I=IE1,IE2                 Reduce nonpivot elements in column.
        IF(INDXR(I).NE.JPIV) THEN
            IF(S(I,JPIV).NE.ZERO) THEN
                DUM=S(I,JPIV)
                DO 16 J=JE1,JSF
                    S(I,J)=S(I,J)-DUM*S(IPIV,J)
16              CONTINUE
                S(I,JPIV)=ZERO
            ENDIF
        ENDIF
17  CONTINUE
18 CONTINUE
JCOFF=JC1-JS1                       Sort and store unreduced coefficients.
ICOFF=IE1-JE1
DO 21 I=IE1,IE2
    IROW=INDXR(I)+ICOFF
    DO 19 J=JS1,JSF
        C(IROW,J+JCOFF,K)=S(I,J)
19  CONTINUE
21 CONTINUE
RETURN
END
```

```
      SUBROUTINE RED(IZ1,IZ2,JZ1,JZ2,JM1,JM2,JMF,IC1,JC1,JCF,KC,
     *    C,NCI,NCJ,NCK,S,NSI,NSJ)
```

Reduce columns JZ1-JZ2 of the S matrix, using previous results as stored in the C matrix. Only columns JM1-JM2,JMF are affected by the prior results. RED is used internally by SOLVDE.

```
DIMENSION C(NCI,NCJ,NCK),S(NSI,NSJ)
LOFF=JC1-JM1
IC=IC1
DO 14 J=JZ1,JZ2                     Loop over columns to be zeroed.
    DO 12 L=JM1,JM2                 Loop over columns altered.
        VX=C(IC,L+LOFF,KC)
        DO 11 I=IZ1,IZ2             Loop over rows.
            S(I,L)=S(I,L)-S(I,J)*VX
11      CONTINUE
12  CONTINUE
    VX=C(IC,JCF,KC)
    DO 13 I=IZ1,IZ2                 Plus final element.
        S(I,JMF)=S(I,JMF)-S(I,J)*VX
13  CONTINUE
    IC=IC+1
14 CONTINUE
RETURN
END
```

"Algebraically Difficult" Sets of Differential Equations

Relaxation methods allow you to take advantage of an additional opportunity that, while not obvious, can speed up some calculations enormously.

It is not necessary that the set of variables $y_{j,k}$ correspond exactly with the dependent variables of the original differential equations. They can be related to those variables through algebraic equations. Obviously, it is only necessary that the solution variables allow us to *evaluate* the functions $y, g, \mathbf{B}, \mathbf{C}$ that are used to construct the FDEs from the ODEs. In some problems g depends on functions of y that are known only implicitly, so that iterative solutions are necessary to evaluate functions in the ODEs. Often one can dispense with this "internal" nonlinear problem by defining a new set of variables from which both y, g and the boundary conditions can be obtained directly. A typical example occurs in physical problems where the equations require solution of a complex equation of state that can be expressed in more convenient terms using variables other than the original dependent variables in the ODE. While this approach is analogous to performing an *analytic* change of variables directly on the original ODEs, such an analytic transformation might be prohibitively complicated. The change of variables in the relaxation method is easy and requires no analytic manipulations.

REFERENCES AND FURTHER READING:

Eggleton, P. P. 1971, *Monthly Notices of the Royal Astronomical Society*, vol. 151, pp. 351 ff.

Keller, Herbert B. 1968, *Numerical Methods for Two-Point Boundary-Value Problems* (Waltham, Mass.: Blaisdell).

Kippenhan, R., Weigert, A., and Hofmeister, E. 1968, *Methods of Computational Physics*, vol. 7, pp. 129 ff.

16.4 A Worked Example: Spheroidal Harmonics

The best way to understand the algorithms of the previous sections is to see them employed to solve an actual problem. As a sample problem, we have selected the computation of spheroidal harmonics. (The more common name is spheroidal angle functions, but we prefer the explicit reminder of the kinship with spherical harmonics.)

Spheroidal harmonics typically arise when certain partial differential equations are solved by separation of variables in spheroidal coordinates. They satisfy the following differential equation on the interval $-1 \le x \le 1$:

$$\frac{d}{dx}\left[(1-x^2)\frac{dS}{dx}\right] + \left(\lambda - c^2x^2 - \frac{m^2}{1-x^2}\right)S = 0 \tag{16.4.1}$$

Here m is an integer, c is the "oblateness parameter," and λ is the eigenvalue. Despite the notation, c^2 can be positive or negative. For $c^2 > 0$ the functions are called "prolate," while if $c^2 < 0$ they are called "oblate." The equation has singular points at $x = \pm 1$ and is to be solved subject to the boundary conditions that the solution be regular at $x = \pm 1$. Only for certain values of λ, the eigenvalue, will this be possible.

If we consider first the spherical case, where $c = 0$, we recognize the differential equation for Legendre functions $P_n^m(x)$. In this case the eigenvalues are $\lambda_{mn} = n(n+1)$, $n = m, m+1, \ldots$. The integer n labels successive eigenvalues for fixed m: when $n = m$ we have the lowest eigenvalue, and the corresponding eigenfunction has no nodes in the interval $-1 < x < 1$; when $n = m+1$ we have the next eigenvalue, and the eigenfunction has one node inside $(-1, 1)$; and so on.

A similar situation holds for the general case $c^2 \neq 0$. We write the eigenvalues of (16.4.1) as $\lambda_{mn}(c)$ and the eigenfunctions as $S_{mn}(x;c)$. For fixed m, $n = m, m+1, \ldots$ labels the successive eigenvalues.

The computation of $\lambda_{mn}(c)$ and $S_{mn}(x;c)$ traditionally has been quite difficult. Complicated recurrence relations, power series expansions, etc., can be found in the references for this section. Cheap computing makes evaluation by direct solution of the differential equation quite feasible.

The first step is to investigate the behavior of the solution near the singular points $x = \pm 1$. Substituting a power series expansion of the form

$$S = (1 \pm x)^\alpha \sum_{k=0}^{\infty} a_k (1 \pm x)^k \tag{16.4.2}$$

in equation (16.4.1), we find that the regular solution has $\alpha = m/2$. (Without loss of generality we can take $m \geq 0$ since $m \to -m$ is a symmetry of the equation.) We get an equation that is numerically more tractable if we factor out this behavior. Accordingly we set

$$S = (1 - x^2)^{m/2} y \tag{16.4.3}$$

We then find from (16.4.1) that y satisfies the equation

$$(1 - x^2)\frac{d^2 y}{dx^2} - 2(m+1)x\frac{dy}{dx} + (\mu - c^2 x^2)y = 0 \tag{16.4.4}$$

where

$$\mu \equiv \lambda - m(m+1) \tag{16.4.5}$$

Both equations (16.4.1) and (16.4.5) are invariant under the replacement $x \to -x$. Thus the functions S and y must also be invariant, except possibly for an overall scale factor. (Since the equations are linear, a constant multiple of a solution is also a solution.) Because the solutions will be normalized, the scale factor can only be ± 1. If $n - m$ is odd, there are an odd number of zeros in the interval $(-1, 1)$. Thus we must choose the antisymmetric solution

$y(-x) = -y(x)$ which has a zero at $x = 0$. Conversely, if $n - m$ is even we must have the symmetric solution. Thus

$$y_{mn}(-x) = (-1)^{n-m} y_{mn}(x) \tag{16.4.6}$$

and similarly for S_{mn}.

The boundary conditions on (16.4.4) require that y be regular at $x = \pm 1$. In other words, near the endpoints the solution takes the form

$$y = a_0 + a_1(1 - x^2) + a_2(1 - x^2)^2 + \dots \tag{16.4.7}$$

Substituting this expansion in equation (16.4.4) and letting $x \to 1$, we find that

$$a_1 = -\frac{\mu - c^2}{4(m+1)} a_0 \tag{16.4.8}$$

Equivalently,

$$y'(1) = \frac{\mu - c^2}{2(m+1)} y(1) \tag{16.4.9}$$

A similar equation holds at $x = -1$ with a minus sign on the right-hand side. The irregular solution has a different relation between function and derivative at the endpoints.

Instead of integrating the equation from -1 to 1, we can exploit the symmetry (16.4.6) to integrate from 0 to 1. The boundary condition at $x = 0$ is

$$\begin{aligned} y(0) &= 0, \quad n - m \text{ odd} \\ y'(0) &= 0, \quad n - m \text{ even} \end{aligned} \tag{16.4.10}$$

A third boundary condition comes from the fact that any constant multiple of a solution y is a solution. We can thus *normalize* the solution. We adopt the normalization that the function S_{mn} has the same limiting behavior as P_n^m at $x = 1$:

$$\lim_{x \to 1} (1 - x^2)^{-m/2} S_{mn}(x; c) = \lim_{x \to 1} (1 - x^2)^{-m/2} P_n^m(x) \tag{16.4.11}$$

Various normalization conventions in the literature are tabulated by Flammer.

The imposition of three boundary conditions for the second-order equation (16.4.4) makes it an eigenvalue problem for λ or equivalently for μ. We write it in the standard form by setting

$$y_1 = y \tag{16.4.12}$$
$$y_2 = y' \tag{16.4.13}$$
$$y_3 = \mu \tag{16.4.14}$$

Then

$$y_1' = y_2 \tag{16.4.15}$$
$$y_2' = \frac{1}{1-x^2}\left[2x(m+1)y_2 - (y_3 - c^2x^2)y_1\right] \tag{16.4.16}$$
$$y_3' = 0 \tag{16.4.17}$$

The boundary condition at $x = 0$ in this notation is

$$\begin{aligned} y_1 &= 0, \quad n-m \text{ odd} \\ y_2 &= 0, \quad n-m \text{ even} \end{aligned} \tag{16.4.18}$$

At $x = 1$ we have two conditions:

$$y_2 = \frac{y_3 - c^2}{2(m+1)} y_1 \tag{16.4.19}$$
$$y_1 = \lim_{x\to 1}(1-x^2)^{-m/2} P_n^m(x) = \frac{(-1)^m (n+m)!}{2^m m!(n-m)!} \equiv \gamma \tag{16.4.20}$$

We now have to decide what numerical method to use. If we just want a few isolated values of λ or S, shooting is probably the quickest method. However, if we want values for a large sequence of values of c, relaxation is good. Relaxation rewards a good initial guess with rapid convergence, and the previous solution should be a good initial guess if c is changed only slightly. Since we want to show how to use the relaxation routines of the previous section anyway, this is a good excuse.

For simplicity, we choose a uniform grid on the interval $0 \le x \le 1$. For a total of M mesh points, we have

$$h = \frac{1}{M-1} \tag{16.4.21}$$
$$x_k = (k-1)h, \qquad k = 1, 2, \ldots, M \tag{16.4.22}$$

At interior points $k = 2, 3, \dots, M$, equation (16.4.15) gives

$$E_{1,k} = y_{1,k} - y_{1,k-1} - \frac{h}{2}(y_{2,k} + y_{2,k-1}) \tag{16.4.23}$$

Equation (16.4.16) gives

$$\begin{aligned} E_{2,k} &= y_{2,k} - y_{2,k-1} - \beta_k \\ &\times \left[\frac{(x_k + x_{k-1})(m+1)(y_{2,k} + y_{2,k-1})}{2} - \alpha_k \frac{(y_{1,k} + y_{1,k-1})}{2} \right] \end{aligned} \tag{16.4.24}$$

where

$$\alpha_k = \frac{y_{3,k} + y_{3,k-1}}{2} - \frac{c^2(x_k + x_{k-1})^2}{4} \tag{16.4.25}$$

$$\beta_k = \frac{h}{1 - \frac{1}{4}(x_k + x_{k-1})^2} \tag{16.4.26}$$

Finally, equation (16.4.17) gives

$$E_{3,k} = y_{3,k} - y_{3,k-1} \tag{16.4.27}$$

Now recall that the matrix of partial derivatives $S_{i,j}$ of equation (16.3.8) is defined so that i labels the equation and j the variable. In our case, j runs from 1 to 3 for y_j at $k-1$ and from 4 to 6 for y_j at k. Thus equation (16.4.23) gives

$$\begin{aligned} S_{1,1} &= -1, \quad S_{1,2} = -\frac{h}{2}, \quad S_{1,3} = 0 \\ S_{1,4} &= 1, \quad S_{1,5} = -\frac{h}{2}, \quad S_{1,6} = 0 \end{aligned} \tag{16.4.28}$$

Similarly equation (16.4.24) yields

$$\begin{aligned} &S_{2,1} = \alpha_k \beta_k / 2, \quad S_{2,2} = -1 - \beta_k (x_k + x_{k-1})(m+1)/2, \\ &S_{2,3} = \beta_k (y_{1,k} + y_{1,k-1})/4 \quad S_{2,4} = S_{2,1}, \\ &S_{2,5} = 2 + S_{2,2}, \quad S_{2,6} = S_{2,3} \end{aligned} \tag{16.4.29}$$

while from equation (16.4.27) we find

$$\begin{aligned} S_{3,1} &= 0, \quad S_{3,2} = 0, \quad S_{3,3} = -1 \\ S_{3,4} &= 0, \quad S_{3,5} = 0, \quad S_{3,6} = 1 \end{aligned} \tag{16.4.30}$$

At $x = 0$ we have the boundary condition

$$E_{3,1} = \begin{cases} y_{1,1} & n-m \text{ odd} \\ y_{2,1} & n-m \text{ even} \end{cases} \tag{16.4.31}$$

Recall the convention adopted in the `SOLVDE` routine that for one boundary condition at $k = 1$ only $S_{3,j}$ can be nonzero. Also, j takes on the values 4 to 6 since the boundary condition involves only y_k, not y_{k-1}. Accordingly, the only nonzero values of $S_{3,j}$ at $x = 0$ are

$$\begin{aligned} S_{3,4} &= 1, \qquad n-m \text{ odd} \\ S_{3,5} &= 1, \qquad n-m \text{ even} \end{aligned} \tag{16.4.32}$$

At $x = 1$ we have

$$E_{1,M+1} = y_{2,M} - \frac{y_{3,M} - c^2}{2(m+1)} y_{1,M} \tag{16.4.33}$$

$$E_{2,M+1} = y_{1,M} - \gamma \tag{16.4.34}$$

Thus

$$S_{1,4} = -\frac{y_{3,M} - c^2}{2(m+1)}, \quad S_{1,5} = 1, \quad S_{1,6} = -\frac{y_{1,M}}{2(m+1)} \tag{16.4.35}$$

$$S_{2,4} = 1, \quad S_{2,5} = 0, \quad S_{2,6} = 0 \tag{16.4.36}$$

We now give a sample program that implements the above algorithm. We need a main program, `SFROID`, that calls the routine `SOLVDE`, and we must supply the subroutine `DIFEQ` called by `SOLVDE`. For simplicity we choose an equally spaced mesh of `M` $= 41$ points, that is, $h = .025$. As we shall see, this gives good accuracy for the eigenvalues up to moderate values of $n - m$.

Since the boundary condition at $x = 0$ does not involve y_1 if $n - m$ is even, we have to use the `INDEXV` feature of `SOLVDE`. Recall that the value of `INDEXV(J)` describes which column of `S(I,J)` the variable `Y(J)` has been put

in. If $n-m$ is even, we need to interchange the columns for y_1 and y_2 so that there is not a zero pivot element in S(I,J).

The program prompts for values of m and n. It then computes an initial guess for y based on the Legendre function P_n^m. It next prompts for c^2, solves for y, prompts for c^2, solves for y using the previous values as an initial guess, and so on.

```
      PROGRAM SFROID
C     Sample program using SOLVDE. Computes eigenvalues of spheroidal harmonics Smn(x;c)
C     for m >= 0 and n >= m. In the program, m is MM, c^2 is C2, and gamma of equation (16.4.20)
C     is ANORM.
      PARAMETER(NE=3,M=41,NB=1,NCI=NE,NCJ=NE-NB+1,NCK=M+1,NSI=NE,
     *    NSJ=2*NE+1,NYJ=NE,NYK=M)
      COMMON X(M),H,MM,N,C2,ANORM             Communicates with DIFEQ.
      DIMENSION SCALV(NE),INDEXV(NE),Y(NE,M),C(NCI,NCJ,NCK),S(NSI,NSJ)
      ITMAX=100
      CONV=5.E-6
      SLOWC=1.
      H=1./(M-1)
      C2=0.
      WRITE(*,*)'ENTER M,N'
      READ(*,*)MM,N
      IF(MOD(N+MM,2).EQ.1)THEN        No interchanges necessary.
          INDEXV(1)=1
          INDEXV(2)=2
          INDEXV(3)=3
      ELSE                            Interchange y1 and y2.
          INDEXV(1)=2
          INDEXV(2)=1
          INDEXV(3)=3
      ENDIF
      ANORM=1.                        Compute gamma.
      IF(MM.NE.0)THEN
        Q1=N
        DO 11 I=1,MM
          ANORM=-.5*ANORM*(N+I)*(Q1/I)
          Q1=Q1-1.
11      CONTINUE
      ENDIF
      DO 12 K=1,M-1                   Initial guess.
        X(K)=(K-1)*H
        FAC1=1.-X(K)**2
        FAC2=FAC1**(-MM/2.)
        Y(1,K)=PLGNDR(N,MM,X(K))*FAC2      Pnm from §6.6.
        DERIV=-((N-MM+1)*PLGNDR(N+1,MM,X(K))-(N+1)*
     *      X(K)*PLGNDR(N,MM,X(K)))/FAC1   Derivative of Pnm from a recurrence relation.
        Y(2,K)=MM*X(K)*Y(1,K)/FAC1+DERIV*FAC2
        Y(3,K)=N*(N+1)-MM*(MM+1)
12    CONTINUE
      X(M)=1.                         Initial guess at x = 1 done separately.
      Y(1,M)=ANORM
      Y(3,M)=N*(N+1)-MM*(MM+1)
      Y(2,M)=(Y(3,M)-C2)*Y(1,M)/(2.*(MM+1.))
      SCALV(1)=ABS(ANORM)
      SCALV(2)=MAX(ABS(ANORM),Y(2,M))
      SCALV(3)=MAX(1.,Y(3,M))
1     CONTINUE
      WRITE (*,*) 'ENTER C**2 OR 999 TO END'
      READ (*,*) C2
      IF(C2.EQ.999.)STOP
      CALL SOLVDE(ITMAX,CONV,SLOWC,SCALV,INDEXV,NE,NB,M,Y,NYJ,NYK,
     *    C,NCI,NCJ,NCK,S,NSI,NSJ)
      WRITE (*,*) ' M = ',MM,' N = ',N,
```

```
*      ' C**2 = ',C2,' LAMBDA = ',Y(3,1)+MM*(MM+1)
   GO TO 1                              for another value of c^2.
   END


   SUBROUTINE DIFEQ(K,K1,K2,JSF,IS1,ISF,INDEXV,NE,S,NSI,NSJ,Y,NYJ,NYK)
      Returns matrix S(I,J) for SOLVDE.
   PARAMETER(M=41)
   COMMON X(M),H,MM,N,C2,ANORM
   DIMENSION Y(NYJ,NYK),S(NSI,NSJ),INDEXV(NYJ)
   IF(K.EQ.K1) THEN                    Boundary condition at first point.
       IF(MOD(N+MM,2).EQ.1)THEN
           S(3,3+INDEXV(1))=1.  Equation (16.4.32).
           S(3,3+INDEXV(2))=0.
           S(3,3+INDEXV(3))=0.
           S(3,JSF)=Y(1,1)      Equation (16.4.31).
       ELSE
           S(3,3+INDEXV(1))=0.  Equation (16.4.32).
           S(3,3+INDEXV(2))=1.
           S(3,3+INDEXV(3))=0.
           S(3,JSF)=Y(2,1)      Equation (16.4.31).
       ENDIF
   ELSE IF(K.GT.K2) THEN               Boundary conditions at last point.
       S(1,3+INDEXV(1))=-(Y(3,M)-C2)/(2.*(MM+1.))  Equation (16.4.35).
       S(1,3+INDEXV(2))=1.
       S(1,3+INDEXV(3))=-Y(1,M)/(2.*(MM+1.))
       S(1,JSF)=Y(2,M)-(Y(3,M)-C2)*Y(1,M)/(2.*(MM+1.)) Equation (16.4.33).
       S(2,3+INDEXV(1))=1.      Equation (16.4.36).
       S(2,3+INDEXV(2))=0.
       S(2,3+INDEXV(3))=0.
       S(2,JSF)=Y(1,M)-ANORM    Equation (16.4.34).
   ELSE                                Interior point.
       S(1,INDEXV(1))=-1.       Equation (16.4.28).
       S(1,INDEXV(2))=-.5*H
       S(1,INDEXV(3))=0.
       S(1,3+INDEXV(1))=1.
       S(1,3+INDEXV(2))=-.5*H
       S(1,3+INDEXV(3))=0.
       TEMP=H/(1.-(X(K)+X(K-1))**2*.25)
       TEMP2=.5*(Y(3,K)+Y(3,K-1))-C2*.25*(X(K)+X(K-1))**2
       S(2,INDEXV(1))=TEMP*TEMP2*.5           Equation (16.4.29).
       S(2,INDEXV(2))=-1.-.5*TEMP*(MM+1.)*(X(K)+X(K-1))
       S(2,INDEXV(3))=.25*TEMP*(Y(1,K)+Y(1,K-1))
       S(2,3+INDEXV(1))=S(2,INDEXV(1))
       S(2,3+INDEXV(2))=2.+S(2,INDEXV(2))
       S(2,3+INDEXV(3))=S(2,INDEXV(3))
       S(3,INDEXV(1))=0.        Equation (16.4.30).
       S(3,INDEXV(2))=0.
       S(3,INDEXV(3))=-1.
       S(3,3+INDEXV(1))=0.
       S(3,3+INDEXV(2))=0.
       S(3,3+INDEXV(3))=1.
       S(1,JSF)=Y(1,K)-Y(1,K-1)-.5*H*(Y(2,K)+Y(2,K-1)) Equation (16.4.23).
       S(2,JSF)=Y(2,K)-Y(2,K-1)-TEMP*((X(K)+X(K-1))     Equation (16.4.24).
*          *.5*(MM+1.)*(Y(2,K)+Y(2,K-1))-TEMP2*
*          .5*(Y(1,K)+Y(1,K-1)))
       S(3,JSF)=Y(3,K)-Y(3,K-1) Equation (16.4.27).
   ENDIF
   RETURN
   END
```

You can run the program and check it against values of $\lambda_{mn}(c)$ given in the tables at the back of Flammer's book or in Table 21.1 of Abramowitz and Stegun. Typically it converges in about 3 iterations. The table below gives a few comparisons.

Selected Output of SFROID				
m	n	c^2	λ_{exact}	$\lambda_{\mathtt{SFROID}}$
2	2	0.1	6.01427	6.01427
		1.0	6.14095	6.14095
		4.0	6.54250	6.54253
2	5	1.0	30.4361	30.4372
		16.0	36.9963	37.0135
4	11	−1.0	131.560	131.554

REFERENCES AND FURTHER READING:

Abramowitz, Milton, and Stegun, Irene A. 1964, *Handbook of Mathematical Functions*, Applied Mathematics Series, vol. 55 (Washington: National Bureau of Standards; reprinted 1968 by Dover Publications, New York), §21.

Flammer, C. 1957, *Spheroidal Wave Functions* (Stanford, Calif.: Stanford University Press).

Morse, P.M., and Feshbach, H. 1953, *Methods of Theoretical Physics*, Part II (New York: McGraw-Hill), pp. 1502 ff.

16.5 Automated Allocation of Mesh Points

In relaxation problems, you have to choose values for the independent variable at the mesh points. This is called *allocating* the grid or mesh. The usual procedure is to pick a plausible set of values and, if it works, to be content. If it doesn't work, increasing the number of points usually cures the problem.

If we know ahead of time where our solutions will be rapidly varying, we can put more grid points there and less elsewhere. Alternatively, we can solve the problem first on a uniform mesh and then examine the solution to see where we should add more points. We then repeat the solution with the improved grid. The object of the exercise is to allocate points in such a way as to represent the solution accurately.

It is also possible to automate the allocation of mesh points, so that it is done "dynamically" during the relaxation process. This powerful technique not only improves the accuracy of the relaxation method, but also (as we will

see in the next section) allows internal singularities to be handled in quite a neat way. Here we learn how to accomplish the automatic allocation.

We want to focus attention on the dependent variable x, and consider two alternative reparametrizations of it. The first, we term q; this is just the coordinate corresponding to the mesh points themselves, so that $q = 1$ at K= 1, $q = 2$ at K= 2, and so on. Between any two mesh points we have $\Delta q = 1$. In the change of independent variable in the ODEs from x to q,

$$\frac{d\mathbf{y}}{dx} = \mathbf{g} \tag{16.5.1}$$

becomes

$$\frac{d\mathbf{y}}{dq} = \mathbf{g}\frac{dx}{dq} \tag{16.5.2}$$

In terms of q, equation (16.5.2) as an FDE might be written

$$\mathbf{y}_k - \mathbf{y}_{k-1} - \tfrac{1}{2}[(\mathbf{g}\frac{dx}{dq})_k + (\mathbf{g}\frac{dx}{dq})_{k-1}] = 0 \tag{16.5.3}$$

or some related version. Note that dx/dq should accompany $\mathbf{g}$. The transformation between x and q depends only on the *Jacobian* dx/dq. Its reciprocal dq/dx is proportional to the density of mesh points.

Now, given the function $\mathbf{y}(x)$, or its approximation at the current stage of relaxation, we are supposed to have some idea of how we want to specify the density of mesh points. For example, we might want dq/dx to be larger where $\mathbf{y}$ is changing rapidly, or near to the boundaries, or both. In fact, we can probably make up a formula for what we would like dq/dx to be proportional to. The problem is that we do not know the proportionality constant. That is, the formula which we might invent would not have the correct integral over the whole range of x so as to make q vary from 1 to M, according to its definition. To solve this problem we introduce a second reparametrization $Q(q)$, where Q is a new independent variable. The relation between Q and q is taken to be *linear*, so that a mesh spacing formula for dQ/dx differs only in its unknown proportionality constant. A linear relation implies

$$\frac{d^2Q}{dq^2} = 0 \tag{16.5.4}$$

or, expressed in the usual manner as coupled first order equations,

$$\frac{dQ(x)}{dq} = \psi \qquad \frac{d\psi}{dq} = 0 \tag{16.5.5}$$

where ψ is a new intermediate variable. We add these two equations to the set of ODEs being solved.

Completing the prescription, we add a third ODE that is just our desired mesh-density function, namely

$$\phi(x) = \frac{dQ}{dx} = \frac{dQ}{dq}\frac{dq}{dx} \tag{16.5.6}$$

where $\phi(x)$ is chosen by us. Written in terms of the mesh variable q, this equation is

$$\frac{dx}{dq} = \frac{\psi}{\phi(x)} \tag{16.5.7}$$

Notice that $\phi(x)$ should be chosen to be positive definite, so that the density of mesh points is everywhere positive. Otherwise (16.5.7) can have a zero in its denominator.

To use automated mesh spacing, you add the three ODEs (16.5.5) and (16.5.7) to your set of equations, i.e. to the array `Y(J,K)`. Now x becomes a dependent variable! Q and ψ also become new dependent variables. Normally, evaluating ϕ requires little extra work since it will be composed from pieces of the g's that exist anyway. The automated procedure allows one to investigate quickly how the numerical results might be affected by various strategies for mesh spacing. (A special case occurs if the desired mesh spacing function Q can be found analytically, i.e., dQ/dx is directly integrable. Then, you only need to add two additional equations, those in 16.5.5, and two new variables x, ψ.)

As an example of a typical strategy for implementing this scheme, consider a system with one dependent variable $y(x)$. We could set

$$dQ = \frac{dx}{\Delta} + \frac{|d\ln y|}{\delta} \tag{16.5.8}$$

or

$$\phi(x) = \frac{dQ}{dx} = \frac{1}{\Delta} + \left|\frac{dy/dx}{y\delta}\right| \tag{16.5.9}$$

where Δ and δ are constants that we choose. The first term would give a uniform spacing in x if it alone were present. The second term forces more grid points to be used where y is changing rapidly. The constants act to make every logarithmic change in y of an amount δ about as "attractive" to a grid point as a change in x of amount Δ. You adjust the constants according to taste. Other strategies are possible, such as a logarithmic spacing in x, replacing dx in the first term with $d\ln x$.

REFERENCES AND FURTHER READING:

Eggleton, P. P. 1971, *Monthly Notices of the Royal Astronomical Society*, vol. 151, pp. 351 ff.

Kippenhan, R., Weigert, A., and Hofmeister, E. 1968, *Methods of Computational Physics*, vol. 7, pp. 129 ff.

16.6 Handling Internal Boundary Conditions or Singular Points

Singularities can occur in the interiors of two point boundary value problems. Typically, there is a point x_s at which a derivative must be evaluated by an expression of the form

$$S(x_s) = \frac{N(x_s, \mathbf{y})}{D(x_s, \mathbf{y})} \tag{16.6.1}$$

where the denominator $D(x_s, \mathbf{y}) = 0$. In physical problems with finite answers, singular points usually come with their own cure: where $D \to 0$, then the physical solution $\mathbf{y}$ must be such as to make $N \to 0$ simultaneously, in such a way that the ratio takes on a meaningful value. This constraint on the solution $\mathbf{y}$ is often called a *regularity condition.* The condition that $D(x_s, \mathbf{y})$ satisfy some special constraint at x_s is entirely analogous to an extra boundary condition, an algebraic relation among the dependent variables that must hold at a point.

We discussed a related situation earlier, in §16.2, when we described the "fitting point method" to handle the task of integrating equations with singular behavior at the boundaries. In those problems you are unable to integrate from one side of the domain to the other. However, the ODEs do have well behaved derivatives and solutions in the neighborhood of the singularity, so it is readily possible to integrate away from the point. Both the relaxation method and the method of "shooting" to a fitting point handle such problems easily. Also, in those problems the presence of singular behavior served to isolate some special boundary values that had to be satisfied to solve the equations.

The difference here is that we are concerned with singularities arising at intermediate points, where the location of the singular point depends on the solution, so is not known *a priori.* Consequently, we face a circular task: The singularity prevents us from finding a numerical solution, but we need a numerical solution to find its location. Such singularities are also associated with selecting a special value for some variable which allows the solution to satisfy the regularity condition at the singular point. Thus, internal singularities take on aspects of being internal boundary conditions.

One way of handling internal singularities is to treat the problem as a free boundary problem, as discussed at the end of §16.0. Suppose, as a simple example, we consider the equation

$$\frac{dy}{dx} = \frac{N(x,y)}{D(x,y)} \tag{16.6.2}$$

where N and D are required to pass through zero at some unknown point x_s. We add the equation

$$z \equiv x_s - x_1 \qquad \frac{dz}{dx} = 0 \tag{16.6.3}$$

where x_s is the unknown location of the singularity, and change the independent variable to t by setting

$$x - x_1 = tz, \qquad 0 \le t \le 1 \tag{16.6.4}$$

The boundary conditions at $t = 1$ become

$$N(x,y) = 0, \qquad D(x,y) = 0 \tag{16.6.5}$$

Use of an adaptive mesh as discussed in the previous section is another way to overcome the difficulties of an internal singularity. For the problem (16.6.2), we add the mesh spacing equations

$$\frac{dQ}{dq} = \psi \tag{16.6.6}$$

$$\frac{d\psi}{dq} = 0 \tag{16.6.7}$$

with a simple mesh spacing function that maps x uniformly into q, where q runs from 1 to M, the number of mesh points:

$$Q(x) = x, \quad \frac{dQ}{dx} = 1 \tag{16.6.8}$$

Having added three first-order differential equations, we must also add their corresponding boundary conditions. If there were no singularity, these could simply be

$$\text{at} \quad q = 1: \qquad x = x_1, \quad Q = 0 \tag{16.6.9}$$

$$\text{at} \quad q = M: \qquad x = x_2 \tag{16.6.10}$$

and a total of N values y_i specified at $q = 1$. In this case the problem is essentially an initial-value problem with all boundary conditions specified at x_1 and the mesh spacing function is superfluous.

However, in the actual case at hand we impose the conditions

$$\text{at} \quad q = 1: \qquad x = x_1, \quad Q = 0 \tag{16.6.11}$$

$$\text{at} \quad q = M: \quad N(x,y) = 0, \quad D(x,y) = 0 \tag{16.6.12}$$

and $N - 1$ values y_i at $q = 1$. The "missing" y_i is to be adjusted, in other words, so as to make the solution go through the singular point in a regular (zero-over-zero) rather than irregular (finite-over-zero) manner. Notice also that these boundary conditions do not directly impose a value for x_2, which becomes an adjustable parameter that the code varies in an attempt to match the regularity condition.

In this example the singularity occurred at a boundary, and the complication arose because the location of the boundary was unknown. In other problems we might wish to continue the integration beyond the internal singularity. For the example given above, we could simply integrate the ODEs to the singular point, then as a separate problem recommence the integration from the singular point on as far we care to go. However, in other cases the singularity occurs internally, but does not completely determine the problem: There are still some more boundary conditions to be satisfied further along in the mesh. Such cases present no difficulty in principle, but do require some adaptation of the relaxation code given in §16.3. In effect all you need to do is to add a "special" block of equations at the mesh point where the internal boundary conditions occur, and do the proper bookkeeping.

Figure 16.6.1 illustrates a concrete example where the overall problem contains 5 equations with 2 boundary conditions at the first point, one "internal" boundary condition, and two final boundary conditions. The figure shows the structure of the overall matrix equations along the diagonal in the vicinity of the special block. In the middle of the domain, blocks typically involve 5 equation (rows) in 10 unknowns (columns). For each block prior to the special block, the initial boundary conditions provided enough information to zero the first two columns of the blocks. The five FDEs eliminate five more columns, and the final three columns need to be stored for the back-substitution step (as described in §16.3). To handle the extra condition we break the normal cycle and add a special block with only one equation: the internal boundary condition. This effectively reduces the required storage of unreduced coefficients by one column for the rest of the grid, and allows us to reduce to zero the first three columns of subsequent blocks. The subroutines `RED, PINVS, BKSUB` can readily handle these cases with minor recoding, but each problem makes for a special case, and you will have to make the modifications as required.

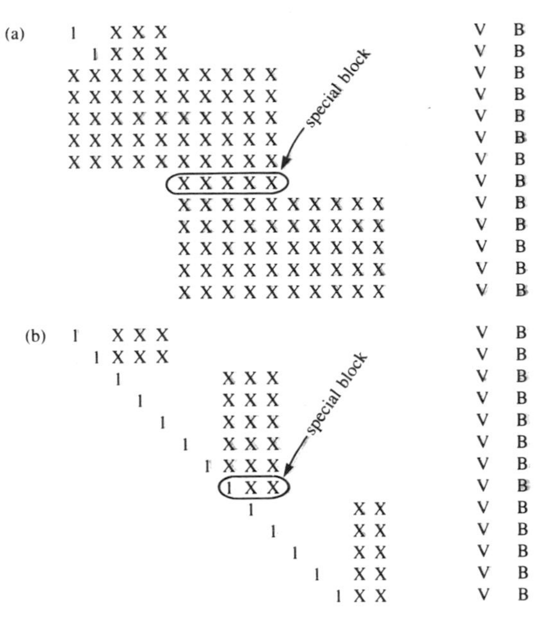

Figure 16.6.1. FDE matrix structure with an internal boundary condition. The internal condition introduces a special block. (a) Original form, compare with Figure 16.3.1; (b) final form, compare with Figure 16.3.2.

REFERENCES AND FURTHER READING:

London, R.A., and Flannery, B.P. 1982, *The Astrophysical Journal*, vol. 258, pp. 260–269.

Chapter 17. Partial Differential Equations

17.0 Introduction

The numerical treatment of partial differential equations is, by itself, a vast subject. Partial differential equations are at the heart of many, if not most, computer analyses or simulations of continuous physical systems, such as fluids, electromagnetic fields, the human body, etc. The intent of this chapter is to give the briefest possible useful introduction. Ideally, there would be an entire second volume of *Numerical Recipes* dealing with partial differential equations alone. (The references below provide, of course, available alternatives.)

In most mathematics books, partial differential equations (PDEs) are classified into the three categories, *hyperbolic, parabolic* and *elliptic*, on the basis of their *characteristics*, or curves of information propagation. The prototypical example of a hyperbolic equation is the one-dimensional *wave* equation

$$\frac{\partial^2 u}{\partial t^2} = v^2 \frac{\partial^2 u}{\partial x^2} \tag{17.0.1}$$

where $v =$ constant is the velocity of wave propagation. The prototypical parabolic equation is the *diffusion* equation

$$\frac{\partial u}{\partial t} = -\frac{\partial}{\partial x}\left(D\frac{\partial u}{\partial x}\right) \tag{17.0.2}$$

where D is the diffusion coefficient. The prototypical elliptic equation is the *Poisson* equation

$$\frac{\partial^2 u}{\partial x^2} + \frac{\partial^2 u}{\partial y^2} = \rho(x, y) \tag{17.0.3}$$

where the source term ρ is given. If the source term is equal to zero, the equation is *Laplace's equation.*

From a computational point of view, the classification into these three canonical types is not very meaningful — or at least not as important as some other essential distinctions. Equations (17.0.1) and (17.0.2) both define *initial value* or *Cauchy* problems: If information on u (perhaps including time derivative information) is given at some initial time t_0 for all x, then the equations describe how $u(x, t)$ propagates itself forward in time. In other words, equations (17.0.1) and (17.0.2) describe time evolution. The goal of a numerical code should be to track that time evolution with some desired accuracy.

By contrast, equation (17.0.3) directs us to find a single "static" function $u(x, y)$ which satisfies the equation within some (x, y) region of interest, and which – one must also specify – has some desired behavior on the boundary of the region of interest. These problems are called *boundary value problems.* In general it is not possible stably to just "integrate in from the boundary" in the same sense that an initial value problem can be "integrated forward in time." Therefore, the goal of a numerical code is somehow to converge on the correct solution everywhere at once.

This, then, is the most important classification from a computational point of view: Is the problem at hand an *initial value* (time evolution) problem? or is it a *boundary value* (static solution) problem? Figure 17.0.1 emphasizes the distinction. Notice that while the italicized terminology is standard, the terminology in parentheses is a much better description of the dichotomy from a computational perspective. The subclassification of initial value problems into parabolic and hyperbolic is much less important because (i) many actual problems are of a mixed type, and (ii) as we will see, most hyperbolic problems get parabolic pieces mixed into them by the time one is discussing practical computational schemes.

Initial Value Problems

An initial value problem is defined by answers to the following questions:

- What are the dependent variables to be propagated forward in time?
- What is the evolution equation for each variable? Usually the evolution equations will all be coupled, with more than one dependent variable appearing on the right-hand side of each equation.
- What is the highest time derivative that occurs in each variable's evolution equation? If possible, this time derivative should be put alone on the equation's left-hand side. Not only the value of a variable, but also the value of all its time derivatives — up to the highest one — must be specified to define the evolution.
- What special equations (boundary conditions) govern the evolution in time of points on the boundary of the spatial region of interest? Examples: *Dirichlet conditions* specify the values of the boundary points as a function of time; *Neumann conditions* specify the values of the normal gradients on the boundary; *outgoing-wave boundary conditions* are just what they say.

Sections 17.1–17.3 of this chapter deal with initial value problems of several different forms. We make no pretence of completeness, but rather hope

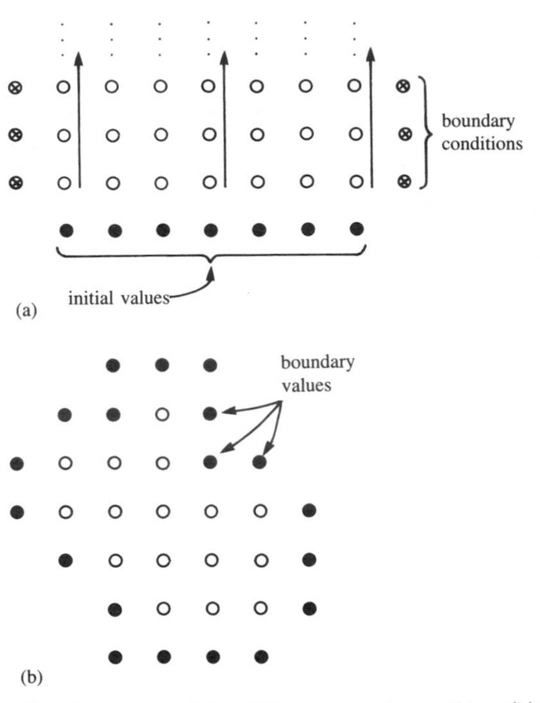

Figure 17.0.1. Initial value problem (a) and boundary value problem (b) are contrasted. In (a) initial values are given on one "time slice", and it is desired to advance the solution in time, computing successive rows of open dots in the direction shown by the arrows. Boundary conditions at the left and right edges of each row (⊗) must also be supplied, but only one row at a time. Only one, or a few, previous rows need be maintained in memory. In (b), boundary values are specified around the edge of a grid, and an iterative process is employed to find the values of all the internal points (open circles). All gridpoints must be maintained in memory.

to convey a certain amount of generalizable information through a few carefully chosen model examples. These examples will illustrate an important point: one's principal *computational* concern must be the *stability* of the algorithm. Many reasonable-looking algorithms for initial value problems just don't work — they are numerically unstable.

Boundary Value Problems

The questions that define a boundary value problem are

- What are the variables?
- What equations are satisfied in the interior of the region of interest?
- What equations are satisfied by points on the boundary of the region of interest. (Here Dirichlet and Neumann conditions are possible choices for elliptic second-order equations, but more complicated boundary conditions can also be encountered.)

In contrast to initial value problems, stability is relatively easy to achieve for boundary value problems. Thus, the *efficiency* of the algorithms, both in computational load and storage requirements, becomes the principal concern.

Because all the conditions on a boundary value problem must be satisfied "simultaneously," these problems usually boil down, at least conceptually, to the solution of large numbers of simultaneous algebraic equations. When such equations are nonlinear, they are usually solved by linearization and iteration; so without much loss of generality we can view the problem as being the solution of special, large linear sets of equations.

As an example, one which we will refer to in §17.4 – §17.6 as our "model problem," let us consider the solution of equation (17.0.3) by the *finite difference method.* We represent the function $u(x, y)$ by its values at the discrete set of points

$$\begin{aligned} x_j &= x_0 + j\Delta, \qquad & j &= 0, 1, ..., J, \\ y_l &= y_0 + l\Delta, \qquad & l &= 0, 1, ..., L \end{aligned} \tag{17.0.4}$$

where Δ is the *grid spacing.* From now on, we will write $u_{j,l}$ for $u(x_j, y_l)$, and $\rho_{j,l}$ for $\rho(x_j, y_l)$. For (17.0.3) we substitute a finite-difference representation (see Figure 17.0.2),

$$\frac{u_{j+1,l} - 2u_{j,l} + u_{j-1,l}}{\Delta^2} + \frac{u_{j,l+1} - 2u_{j,l} + u_{j,l-1}}{\Delta^2} = \rho_{j,l} \tag{17.0.5}$$

or equivalently

$$u_{j+1,l} + u_{j-1,l} + u_{j,l+1} + u_{j,l-1} - 4u_{j,l} = \Delta^2 \rho_{j,l} \tag{17.0.6}$$

To write this system of linear equations in matrix form we need to make a vector out of u. Let us number the two dimensions of grid points in a single one-dimensional sequence by defining

$$i \equiv j(L+1) + l \qquad \text{for} \qquad j = 0, 1, ..., J, \qquad l = 0, 1, ..., L \tag{17.0.7}$$

In other words, i increases most rapidly along the columns representing y values. Equation (17.0.6) now becomes

$$u_{i+L+1} + u_{i-(L+1)} + u_{i+1} + u_{i-1} - 4u_i = \Delta^2 \rho_i \tag{17.0.8}$$

This equation only holds at the interior points $j = 1, 2, ..., J-1; l = 1, 2, ..., L-1$.

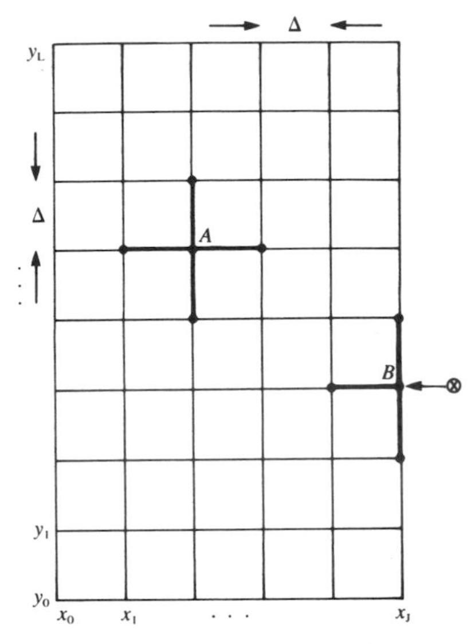

Figure 17.0.2. Finite difference representation of a second-order elliptic equation on a two-dimensional grid. The second derivatives at the point A are evaluated using the points to which A is shown connected. The second derivatives at point B are evaluated using the connected points and also using "right-hand side" boundary information, shown schematically as ⊗.

The points where

$$
\begin{aligned}
j &= 0 \qquad [\text{i.e., } i = 0, ..., L] \\
j &= J \qquad [\text{i.e., } i = J(L+1), ..., J(L+1)+L] \\
l &= 0 \qquad [\text{i.e., } i = 0, L+1, ..., J(L+1)] \\
l &= L \qquad [\text{i.e., } i = L, L+1+L, ..., J(L+1)+L]
\end{aligned}
\tag{17.0.9}
$$

are boundary points where either u or its derivative has been specified. If we pull all this "known" information over to the right-hand side of equation (17.0.8), then the equation takes the form

$$\mathbf{A} \cdot \mathbf{u} = \mathbf{b} \tag{17.0.10}$$

where $\mathbf{A}$ has the form shown in Figure 17.0.3. The matrix $\mathbf{A}$ is called "tridi-

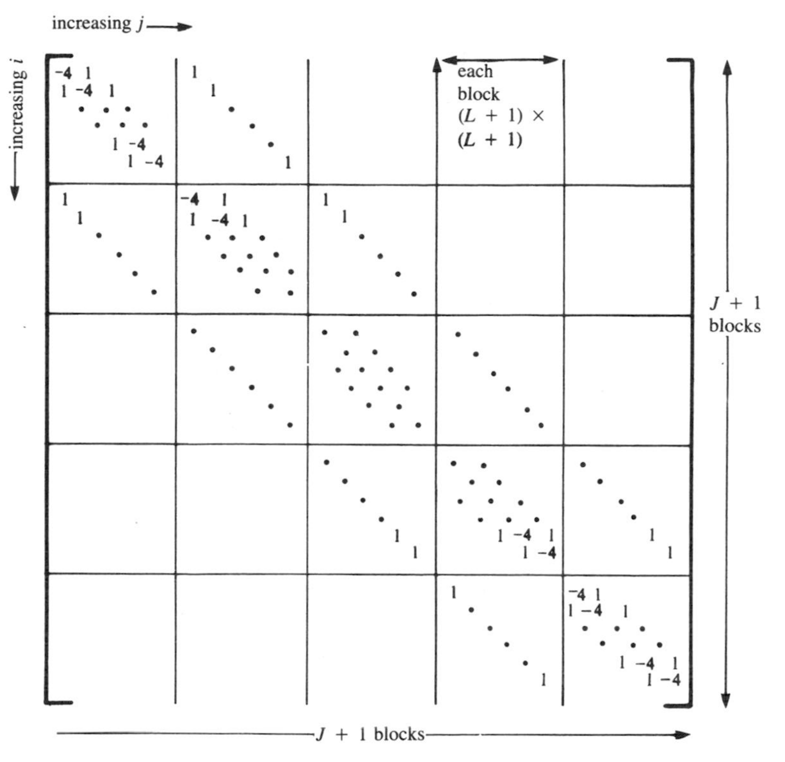

Figure 17.0.3. Matrix structure derived from a second-order elliptic equation (here equation 17.0.6). All elements not shown are zero. The matrix has diagonal blocks that are themselves tridiagonal, and sub- and super-diagonal blocks that are diagonal. This form is called "tridiagonal with fringes." A matrix this sparse would never be stored in its full form as shown here.

agonal with fringes." A general linear second-order elliptic equation

$$\begin{aligned} a(x,y)\frac{\partial^2 u}{\partial x^2} &+ b(x,y)\frac{\partial u}{\partial x} \\ &+ c(x,y)\frac{\partial^2 u}{\partial y^2} + d(x,y)\frac{\partial u}{\partial y} \\ &+ e(x,y)\frac{\partial^2 u}{\partial x \partial y} + f(x,y)u = g(x,y) \end{aligned} \tag{17.0.11}$$

will lead to a matrix of similar structure except that the nonzero entries will not be constants.

As a rough classification, there are three different approaches to the solution of equation (17.0.10), not all applicable in all cases: relaxation methods, "rapid" methods (e.g. Fourier methods), and direct matrix methods.

Relaxation methods make immediate use of the structure of the sparse

matrix **A**. The matrix is split into two parts

$$\mathbf{A} = \mathbf{E} - \mathbf{F} \tag{17.0.12}$$

where **E** is easily invertible and **F** is the remainder. Then (17.0.10) becomes

$$\mathbf{E} \cdot \mathbf{u} = \mathbf{F} \cdot \mathbf{u} + \mathbf{b} \tag{17.0.13}$$

The relaxation method involves choosing an initial guess $\mathbf{u}^{(0)}$ and then solving successively for iterates $\mathbf{u}^{(r)}$ from

$$\mathbf{E} \cdot \mathbf{u}^{(r)} = \mathbf{F} \cdot \mathbf{u}^{(r-1)} + \mathbf{b} \tag{17.0.14}$$

Since **E** is chosen to be easily invertible, each iteration is fast. We will discuss relaxation methods in some detail in §17.5, and from a slightly different point of view ("operator splitting") in §17.6.

Rapid methods apply for only a rather special class of equations: those with constant coefficients, or, more generally, those that are separable in the chosen coordinates. In addition, the boundaries must coincide with coordinate lines. This special class of equations is met quite often in practice. We defer detailed discussion to §17.4.

Matrix methods attempt to solve the equation

$$\mathbf{A} \cdot \mathbf{x} = \mathbf{b} \tag{17.0.15}$$

directly. The degree to which this is practical depends very strongly on the exact structure of the matrix **A** for the problem at hand, so our discussion can go no farther than a few remarks and references at this point.

Sparseness of the matrix *must* be the guiding force. Otherwise the matrix problem is prohibitively large. For example, the simplest problem on a 100×100 spatial grid would involve 10000 unknown $u_{j,l}$'s, implying a 10000×10000 matrix **A**, containing 10^8 elements!

As we discussed at the end of §10.6, if **A** is symmetric and positive definite (as it usually is in elliptic problems), the conjugate-gradient algorithm can be used. In practice, rounding error often spoils the effectiveness of the conjugate gradient algorithm for solving finite-difference equations. However, it is useful when incorporated in methods that first rewrite the equations so that **A** is transformed to a matrix $\mathbf{A}'$ that is close to the identity matrix. The quadratic surface defined by the equations then has almost spherical contours, and the conjugate gradient algorithm works very well. An example of such an algorithm is the *incomplete Choleski conjugate gradient method (ICCG)* (see Meijerink and van der Vorst; van der Vorst; Kershaw).

Another method that relies on a transformation approach is the *strongly implicit procedure* of Stone. A program called SIPSOL that implements this routine has been published by Jesshope.

A third class of matrix methods is the Analyze-Factorize-Operate approach as described in §2.10.

Generally speaking, when you have the storage available to implement these methods — not nearly as much as the 10^8 above, but usually much more than is required by relaxation methods — then you should consider doing so. Their execution times can be superior to the best relaxation methods by factors of, e.g., five. For grids larger than, say, 100×100, however, it is generally found that only relaxation methods, or "rapid" methods when they are applicable, are possible.

There is More to Life than Finite Differencing

Besides finite differencing, there are other methods for solving PDEs. Most important are finite element, Monte Carlo, spectral, and variational methods. Unfortunately, we shall barely be able to do justice to finite differencing in this chapter, and so shall not be able to discuss these other methods in this book. You can consult the references below for more details of the other techniques. Finite element methods are often preferred by practitioners in solid mechanics and structural engineering; these methods allow considerable freedom in putting computational elements where you want them, important when dealing with highly irregular geometries. Spectral methods are preferred for very regular geometries and smooth functions; they converge more rapidly than finite-difference methods (cf. §17.4), but they do not work well for problems with discontinuities.

REFERENCES AND FURTHER READING:

Ames, W.F. 1977, *Numerical Methods for Partial Differential Equations*, 2nd ed. (New York: Academic Press).

Richtmyer, R.D., and Morton, K.W. 1967, *Difference Methods for Initial Value Problems*, 2nd ed. (New York: Wiley-Interscience).

Roache, P.J. 1976, *Computational Fluid Dynamics* (Albuquerque: Hermosa).

Strang, G., and Fix, G. 1973, *An Analysis of the Finite Element Method* (Englewood Cliffs, N.J.: Prentice-Hall).

Mitchell, A.R., and Griffiths, D.F. 1980, *The Finite Difference Method in Partial Differential Equations* (New York: Wiley) [includes discussion of finite element methods].

Dorr, F.W. 1970, *S.I.A.M. Review*, vol. 12, pp. 248–263 [review of rapid Poisson equation methods].

Jesshope, C.R. 1979, *Computer Physics Communications*, vol. 17, pp. 383–391.

Kershaw, D.S. 1970, *Journal of Computational Physics*, vol. 26, pp. 43–65.

Meijerink, J.A., and van der Vorst, H.A. 1977, *Mathematics of Computation*, vol. 31, pp. 148–162.

Stone, H.J. 1968, *S.I.A.M. Journal of Numerical Analysis*, vol. 5, pp. 530–558.

van der Vorst, H.A. 1981, *Journal of Computational Physics*, vol. 44, pp. 1–19 [review of sparse iterative methods].

17.1 Flux-Conservative Initial Value Problems

A large class of initial value (time-evolution) PDEs in one space dimension can be cast into the form of a *flux-conservative equation*,

$$\frac{\partial \mathbf{u}}{\partial t} = -\frac{\partial \mathbf{F}(\mathbf{u})}{\partial x} \tag{17.1.1}$$

where $\mathbf{u}$ and $\mathbf{F}$ are vectors, and where (in some cases) $\mathbf{F}$ may depend not only on $\mathbf{u}$ but also on spatial derivatives of $\mathbf{u}$. The vector $\mathbf{F}$ is called the *conserved flux*.

For example, the prototypical hyperbolic equation, the one-dimensional wave equation with constant velocity of propagation v

$$\frac{\partial^2 u}{\partial t^2} = v^2 \frac{\partial^2 u}{\partial x^2} \tag{17.1.2}$$

can be rewritten as a set of two first-order equations

$$\begin{aligned} \frac{\partial r}{\partial t} &= v\frac{\partial s}{\partial x} \\ \frac{\partial s}{\partial t} &= v\frac{\partial r}{\partial x} \end{aligned} \tag{17.1.3}$$

where

$$\begin{aligned} r &\equiv v\frac{\partial u}{\partial x} \\ s &\equiv \frac{\partial u}{\partial t} \end{aligned} \tag{17.1.4}$$

In this case r and s become the two components of $\mathbf{u}$, and the flux is given by the linear matrix relation

$$\mathbf{F}(\mathbf{u}) = \begin{pmatrix} 0 & -v \\ -v & 0 \end{pmatrix} \cdot \mathbf{u} \tag{17.1.5}$$

[The physicist-reader may recognize equations (17.1.3) as analogous to Maxwell's equations for one-dimensional propagation of electromagnetic waves.]

We will consider, in this section, a prototypical example of the general flux-conservative equation (17.1.1), namely the equation for a scalar u,

$$\frac{\partial u}{\partial t} = -v\frac{\partial u}{\partial x} \tag{17.1.6}$$

with v a constant. As it happens, we already know analytically that the general solution of this equation is a wave propagating in the positive x-direction,

$$u = f(x - vt) \tag{17.1.7}$$

where f is an arbitrary function. However, the numerical strategies that we develop will be equally applicable to the more general equations represented by (17.1.1). In some contexts, equation (17.1.6) is called an *advective* equation, because the quantity u is transported by a "fluid flow" with a velocity v.

How do we go about finite differencing equation (17.1.6) (or, analogously, 17.1.1)? The straightforward approach is to choose equally spaced points along both the t- and x-axes. Thus denote

$$\begin{aligned} x_j &= x_0 + j\Delta x, \qquad & j &= 0, 1, \ldots, J, \\ t_n &= t_0 + n\Delta t, \qquad & n &= 0, 1, \ldots, N \end{aligned} \tag{17.1.8}$$

Let u_j^n denote $u(t_n, x_j)$. We have several choices for representing the time derivative term. The obvious way is to set

$$\left.\frac{\partial u}{\partial t}\right|_{j,n} = \frac{u_j^{n+1} - u_j^n}{\Delta t} + O(\Delta t) \tag{17.1.9}$$

This is called *forward Euler* differencing (cf. equation 15.1.1). While forward Euler is only first-order accurate in Δt, it has the advantage that one is able to calculate quantities at timestep $n+1$ in terms of only quantities known at timestep n. For the space derivative, we can use a second-order representation still using only quantities known at timestep n:

$$\left.\frac{\partial u}{\partial x}\right|_{j,n} = \frac{u_{j+1}^n - u_{j-1}^n}{2\Delta x} + O(\Delta x^2) \tag{17.1.10}$$

The resulting finite-difference approximation to equation (17.1.6) is called the FTCS representation (Forward Time Centered Space),

$$\frac{u_j^{n+1} - u_j^n}{\Delta t} = -v\left(\frac{u_{j+1}^n - u_{j-1}^n}{2\Delta x}\right) \tag{17.1.11}$$

which can easily be rearranged to be a formula for u_j^{n+1} in terms of the other quantities. The FTCS scheme is illustrated in Figure 17.1.1. It's a

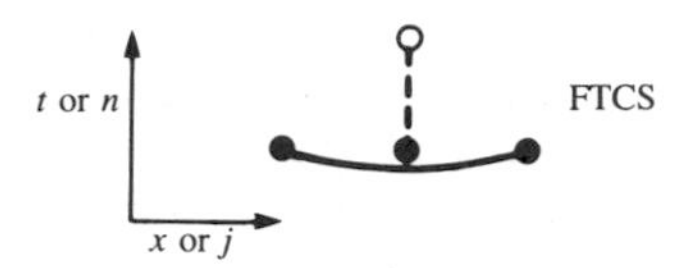

Figure 17.1.1. Representation of the Forward Time Centered Space (FTCS) differencing scheme. In this and subsequent figures, the open circle is the new point at which the solution is desired; filled circles are known points whose function values are used in calculating the new point; the solid lines connect points that are used to calculate spatial derivatives; the dashed lines connect points that are used to calculate time derivatives. The FTCS scheme is generally unstable for hyperbolic problems and cannot usually be used.

fine example of an algorithm that is easy to derive, takes little storage, and executes quickly. Too bad it doesn't work! (See below.)

The FTCS representation is an *explicit* scheme. This means that u_j^{n+1} for each j can be calculated explicitly from the quantities that are already known. Later we shall meet *implicit* schemes, which require us to solve implicit equations coupling the u_j^{n+1} for various j. (Explicit and implicit methods for ordinary differential equations were discussed in §15.6.) The FTCS algorithm is also an example of a *single-level* scheme, since only values at time level n have to be stored to find values at time level $n+1$.

von Neumann Stability Analysis

Unfortunately, equation (17.1.11) is of very limited usefulness. It is an *unstable* method, which can only be used (if at all) to study waves for a short fraction of one oscillation period. To find alternative methods with more general applicability, we must introduce the *von Neumann stability analysis.*

The von Neumann analysis is local: we imagine that the coefficients of the difference equations are so slowly varying as to be considered constant in space and time. In that case, the independent solutions, or *eigenmodes*, of the difference equations are all of the form

$$u_j^n = \xi^n e^{ikj\Delta x} \tag{17.1.12}$$

where k is a real spatial wave number (which can have any value) and $\xi = \xi(k)$ is a complex number that depends on k. The key fact is that the time dependence of a single eigenmode is nothing more than successive integer powers of the complex number ξ. Therefore, the difference equations are unstable (have exponentially growing modes) if $|\xi(k)| > 1$ for *some* k. The number ξ is called the *amplification factor* at a given wave number k.

To find $\xi(k)$, we simply substitute (17.1.12) back into (17.1.11). Dividing by ξ^n, we get

$$\xi(k) = 1 - i\frac{v\Delta t}{\Delta x}\sin k\Delta x \tag{17.1.13}$$

whose modulus is > 1 for *all* k; so the FTCS scheme is unconditionally unstable.

If the velocity v were a function of t and x, then we would write v_j^n in equation (17.1.11). In the von Neumann stability analysis we would still treat v as a constant, the idea being that for v slowly varying the analysis is local. In fact, even in the case of strictly constant v, the von Neumann analysis does not rigorously treat the end effects at $j = 0$ and $j = N$.

More generally, if the equation's right-hand side were nonlinear in u, then a von Neumann analysis would linearize by writing $u = u_0 + \delta u$, expanding to linear order in δu. Assuming that the u_0 quantities already satisfy the difference equation exactly, the analysis would look for an unstable eigenmode of δu.

Despite its lack of rigor, the von Neumann method generally gives valid answers and is much easier to apply than more careful methods. We accordingly adopt it exclusively. (See, for example, Richtmyer and Morton for a discussion of other methods of stability analysis.)

Lax Method

The instability in the FTCS method can be cured by a simple change due to Lax. One replaces the term u_j^n in the time derivative term by its average (Figure 17.1.2):

$$u_j^n \to \frac{1}{2}\left(u_{j+1}^n + u_{j-1}^n\right) \tag{17.1.14}$$

This turns (17.1.11) into

$$u_j^{n+1} = \frac{1}{2}\left(u_{j+1}^n + u_{j-1}^n\right) - \frac{v\Delta t}{2\Delta x}\left(u_{j+1}^n - u_{j-1}^n\right) \tag{17.1.15}$$

Substituting equation (17.1.12), we find for the amplification factor

$$\xi = \cos k\Delta x - i\frac{v\Delta t}{\Delta x}\sin k\Delta x \tag{17.1.16}$$

The stability condition $|\xi|^2 \le 1$ leads to the requirement

$$\frac{|v|\Delta t}{\Delta x} \le 1 \tag{17.1.17}$$

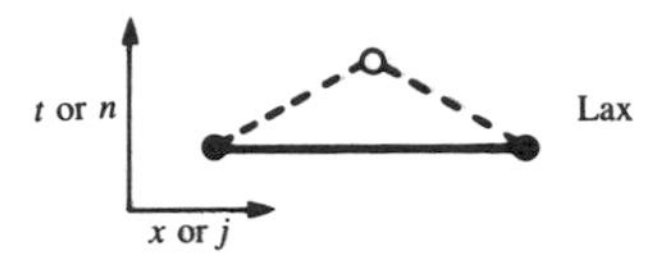

Figure 17.1.2. Representation of the Lax differencing scheme, as in the previous figure. The stability criterion for this scheme is the Courant condition.

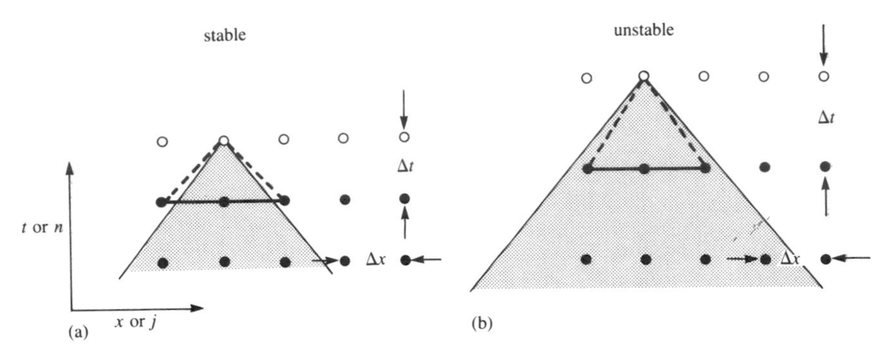

Figure 17.1.3. Courant condition for stability of a differencing scheme. The PDEs of an initial value problem imply that the value at a point depends on information within some domain of dependency to the past, shown here shaded. A differencing scheme has its own domain of dependency determined by the choice of points on one time slice (shown as connected solid dots) whose values are used in determining a new point (shown connected by dashed lines). A differencing scheme is Courant stable if the differencing domain of dependency is larger than that of the PDE's, as in (a), and unstable if the relationship is the reverse, as in (b).

This is the famous Courant-Friedrichs-Lewy stability criterion, often called simply the *Courant condition.* Intuitively, the stability condition can be understood as follows (Figure 17.1.3): The quantity u_j^{n+1}in equation (17.1.15) is computed from information at points $j-1$ and $j+1$ at time n. In other words, x_{j-1} and x_{j+1} are the boundaries of the spatial region that is allowed to communicate information to u_j^{n+1}. Now recall that in the continuum wave equation, information actually propagates with a maximum velocity v. If the point u_j^{n+1} is outside of the shaded region in Figure 17.1.3, then it requires information from points more distant than the differencing scheme allows. Lack of that information gives rise to an instability. Therefore, Δt cannot be made too large.

The surprising result, that the simple replacement (17.1.14) stabilizes the FTCS scheme, is our first encounter with the fact that differencing PDEs is an art as much as a science. To see if we can demystify the art somewhat, let us compare the FTCS and Lax schemes by rewriting equation (17.1.15) so that it is in the form of equation (17.1.11) with a remainder term:

$$\frac{u_j^{n+1}-u_j^n}{\Delta t} = -v\left(\frac{u_{j+1}^n - u_{j-1}^n}{2\Delta x}\right) + \frac{1}{2}\left(\frac{u_{j+1}^n - 2u_j^n + u_{j-1}^n}{\Delta t}\right) \qquad (17.1.18)$$

But this is exactly the FTCS representation of the equation

$$\frac{\partial u}{\partial t} = -v\frac{\partial u}{\partial x} + \frac{(\Delta x)^2}{2\Delta t}\nabla^2 u \tag{17.1.19}$$

where $\nabla^2 = \partial^2/\partial x^2$ in one dimension. We have, in effect, added a diffusion term to the equation, or, if you recall the form of the Navier-Stokes equation for viscous fluid flow, a dissipative term. The Lax scheme is thus said to have *numerical dissipation*, or *numerical viscosity*. We can see this also in the amplification factor. Unless $|v|\Delta t$ is exactly equal to Δx, $|\xi| < 1$ and the amplitude of the wave decreases spuriously.

Isn't a spurious decrease as bad as a spurious increase? No. The scales that we hope to study accurately are those that encompass many gridpoints, so that they have $k\Delta x \ll 1$. (The spatial wave number k is defined by equation 17.1.12.) For these scales, the amplification factor can be seen to be very close to one, in both the stable and unstable scheme. The stable and unstable schemes are therefore about equally accurate. For the unstable scheme, however, short scales with $k\Delta x \sim 1$, *which we are not interested in*, will blow up and swamp the interesting part of the solution. Much better to have a stable scheme in which these short wavelengths die away innocuously. Both the stable and the unstable schemes are *inaccurate* for these short wavelengths, but the inaccuracy is of a tolerable character when the scheme is stable.

When the independent variable $\mathbf{u}$ is a vector, then the von Neumann analysis is slightly more complicated. For example, we can consider equation (17.1.3), rewritten as

$$\frac{\partial}{\partial t}\begin{bmatrix} r \\ s \end{bmatrix} = -\frac{\partial}{\partial x}\begin{bmatrix} vs \\ vr \end{bmatrix} \tag{17.1.20}$$

The Lax method for this equation is

$$\begin{aligned} r_j^{n+1} &= \frac{1}{2}(r_{j+1}^n + r_{j-1}^n) + \frac{v\Delta t}{2\Delta x}(s_{j+1}^n - s_{j-1}^n) \\ s_j^{n+1} &= \frac{1}{2}(s_{j+1}^n + s_{j-1}^n) + \frac{v\Delta t}{2\Delta x}(r_{j+1}^n - r_{j-1}^n) \end{aligned} \tag{17.1.21}$$

The von Neumann stability analysis now proceeds by assuming that the eigenmode is of the following (vector) form,

$$\begin{bmatrix} r_j^n \\ s_j^n \end{bmatrix} = \xi^n e^{ikj\Delta x}\begin{bmatrix} r^0 \\ s^0 \end{bmatrix} \tag{17.1.22}$$

Here the vector on the right-hand side is a constant (both in space and in time) eigenvector, and ξ is a complex number, as before. Substituting (17.1.22)

into (17.1.21), and dividing by the power ξ^n, gives the homogeneous vector equation

$$\begin{bmatrix} (\cos k\Delta x) - \xi & i\dfrac{v\Delta t}{\Delta x}\sin k\Delta x \\ i\dfrac{v\Delta t}{\Delta x}\sin k\Delta x & (\cos k\Delta x) - \xi \end{bmatrix} \cdot \begin{bmatrix} r^0 \\ s^0 \end{bmatrix} = \begin{bmatrix} 0 \\ 0 \end{bmatrix} \tag{17.1.23}$$

This admits a solution only if the determinant of the matrix on the left vanishes, a condition easily shown to yield the two roots ξ

$$\xi = \cos k\Delta x \pm i\frac{v\Delta t}{\Delta x}\sin k\Delta x \tag{17.1.24}$$

The stability condition is that both roots satisfy $|\xi| \leq 1$. This again turns out to be simply the Courant condition (17.1.17).

Other Varieties of Error

Thus far we have been concerned with *amplitude error*, because of its intimate connection with the stability or instability of a differencing scheme. Other varieties of error are relevant when we shift our concern to accuracy, rather than stability.

Finite-difference schemes for hyperbolic equations can exhibit dispersion, or *phase errors*. For example, even if we set $v\Delta t/\Delta x = 1$ in equation (17.1.16), we find

$$\xi = e^{-ik\Delta x} \tag{17.1.25}$$

An arbitrary initial wave packet is a superposition of modes with different k's. At each timestep the modes get multiplied by different phase factors (17.1.25), depending on their value of k. After a while, the phase relations of the modes can become hopelessly garbled and the wave packet disperses. Note from (17.1.25) that the dispersion becomes large as soon as the wavelength becomes comparable to the grid spacing Δx.

A third type of error is one associated with nonlinear hyperbolic equations and is therefore sometimes called *nonlinear instability*. For example, a piece of the Euler or Navier-Stokes equations for fluid flow looks like

$$\frac{\partial v}{\partial t} = -v\frac{\partial v}{\partial x} + \dots \tag{17.1.26}$$

The nonlinear term in v can cause a transfer of energy in Fourier space from long wavelengths to short wavelengths. This results in a wave profile steepening until a vertical profile or "shock" develops. Since the von Neumann analysis suggests that the stability can depend on $k\Delta x$, a scheme that was

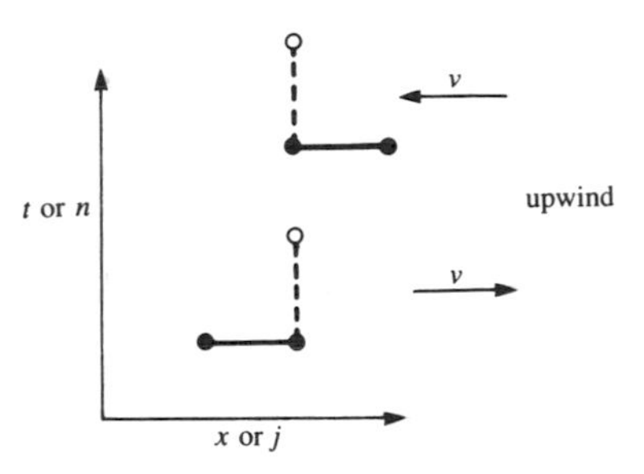

Figure 17.1.4. Representation of upwind differencing schemes. The upper scheme is stable when the advection constant v is negative, as shown; the lower scheme is stable when the advection constant v is positive, also as shown. The Courant condition must, of course, also be satisfied.

stable for shallow profiles can become unstable for steep profiles. This kind of difficulty arises in a differencing scheme where the cascade in Fourier space is halted at the shortest wavelength representable on the grid, that is, at $k \sim 1/\Delta x$. If energy simply accumulates in these modes, it eventually swamps the energy in the long wavelength modes of interest.

Nonlinear instability and shock formation is thus somewhat controlled by numerical viscosity such as that discussed in connection with equation (17.1.18) above. In some fluid problems, however, shock formation is not merely an annoyance, but an actual physical behavior of the fluid whose detailed study is a goal. Then, numerical viscosity alone may not be adequate or sufficiently controllable. While there are some alternative methods, the standard way of handling physical shocks is by adding *artificial viscosity* to the difference equations. For a discussion of this specialized technique, see, for example, Richtmyer and Morton or Roache.

For wave equations, propagation errors (amplitude or phase) are usually most worrisome. For advective equations, on the other hand, *transport errors* are usually of greater concern. In the Lax scheme, equation (17.1.15), a disturbance in the advected quantity u at mesh point j propagates to mesh points $j+1$ and $j-1$ at the next time step. In reality, however, if the velocity v is positive then only mesh point $j+1$ should be affected.

The simplest way to model the transport properties "better" is to use *upwind differencing* (see Figure 17.1.4):

$$\frac{u_j^{n+1}-u_j^n}{\Delta t} = -v_j^n \begin{cases} \dfrac{u_j^n - u_{j-1}^n}{\Delta x}, & v_j^n > 0 \\ \dfrac{u_{j+1}^n - u_j^n}{\Delta x}, & v_j^n < 0 \end{cases} \tag{17.1.27}$$

Note that this scheme is only first-order, not second-order, accurate in the calculation of the spatial derivatives. How can it be "better"? The answer is one that annoys the mathematicians: The goal of numerical simulations is not always "accuracy" in a strictly mathematical sense, but sometimes "fidelity"

to the underlying physics in a sense that is looser and more pragmatic. In such contexts, some kinds of error are much more tolerable than others. Upwind differencing generally adds fidelity to problems where the advected variables are liable to undergo sudden changes of state, e.g., as they pass through shocks or other discontinuities. You will have to be guided by the specific nature of your own problem.

For the differencing scheme (17.1.27), the amplification factor (for constant v) is

$$\xi = 1 - \left|\frac{v\Delta t}{\Delta x}\right|(1-\cos k\Delta x) - i\frac{v\Delta t}{\Delta x}\sin k\Delta x \tag{17.1.28}$$

$$|\xi|^2 = 1 - 2\left|\frac{v\Delta t}{\Delta x}\right|\left(1-\left|\frac{v\Delta t}{\Delta x}\right|\right)(1-\cos k\Delta x) \tag{17.1.29}$$

So the stability criterion $|\xi|^2 \leq 1$ is (again) simply the Courant condition (17.1.17).

There are various ways of improving the accuracy of first-order upwind differencing. In the continuum equation, material originally a distance $v\Delta t$ away arrives at a given point after a time interval Δt. In the first-order method, the material always arrives from Δx away. If $v\Delta t \ll \Delta x$ (to insure accuracy), this can cause a large error. One way of reducing this error is to interpolate u between $j-1$ and j before transporting it. This gives effectively a second-order method. Various schemes for second-order upwind differencing are discussed and compared by Centrella and Wilson and by Hawley et al.

Second-Order Accuracy in Time

When using a method that is first-order accurate in time but second-order accurate in space, one generally has to take $v\Delta t$ significantly smaller than Δx to achieve desired accuracy, say, by at least a factor of 5. Thus the Courant condition is not actually the limiting factor with such schemes in practice. However, there are schemes that are second-order accurate in both space and time, and these can often be pushed right to their stability limit, with correspondingly smaller computation times.

For example, the *staggered leapfrog* method for the conservation equation (17.1.1) is defined as follows (Figure 17.1.5): Using the values of u^n at time t^n, compute the fluxes F_j^n. Then compute new values u^{n+1} using the time-centered values of the fluxes:

$$u_j^{n+1} - u_j^{n-1} = -\frac{\Delta t}{\Delta x}(F_{j+1}^n - F_{j-1}^n) \tag{17.1.30}$$

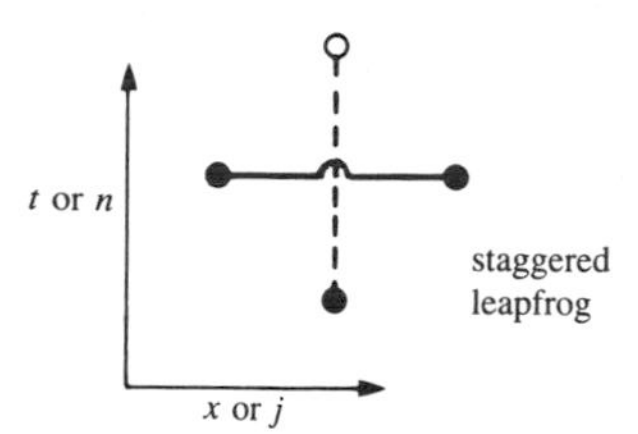

Figure 17.1.5. Representation of the staggered leapfrog differencing scheme. Note that information from two previous time slices is used in obtaining the desired point. This scheme is second-order accurate in both space and time.

The name comes from the fact that the time levels in the time derivative term "leapfrog" over the time levels in the space derivative term. The method requires that u^{n-1} and u^n be stored to compute u^{n+1}.

For our simple model equation (17.1.6), staggered leapfrog takes the form

$$u_j^{n+1} - u_j^{n-1} = -\frac{v\Delta t}{\Delta x}(u_{j+1}^n - u_{j-1}^n) \tag{17.1.31}$$

The von Neumann stability analysis now gives a quadratic equation for ξ, rather than a linear one, because of the occurrence of three consecutive powers of ξ when the form (17.1.12) for an eigenmode is substituted into equation (17.1.31),

$$\xi^2 - 1 = -2i\xi\frac{v\Delta t}{\Delta x}\sin k\Delta x \tag{17.1.32}$$

whose solution is

$$\xi = -i\frac{v\Delta t}{\Delta x}\sin k\Delta x \pm \sqrt{1 - \left(\frac{v\Delta t}{\Delta x}\sin k\Delta x\right)^2} \tag{17.1.33}$$

Thus the Courant condition is again required for stability.

[If u were a vector $\mathbf{u}$, then an eigenmode like (17.1.22) would, in the von Neumann analysis, lead to a determinant whose components were themselves polynomials in ξ. Setting this determinant to zero would give a higher order polynomial equation for the amplification factor ξ.]

In fact, in equation (17.1.33), $|\xi|^2 = 1$ for any $v\Delta t \leq \Delta x$. This is the great advantage of the staggered leapfrog method: There is no amplitude dissipation. For equations more complicated than our simple model equation, especially nonlinear equations, the method usually becomes unstable when the gradients get large. The instability is related to the fact that odd and even mesh points are completely decoupled, like the black and white squares of a chessboard, as shown in Figure 17.1.6. This mesh drifting instability is cured

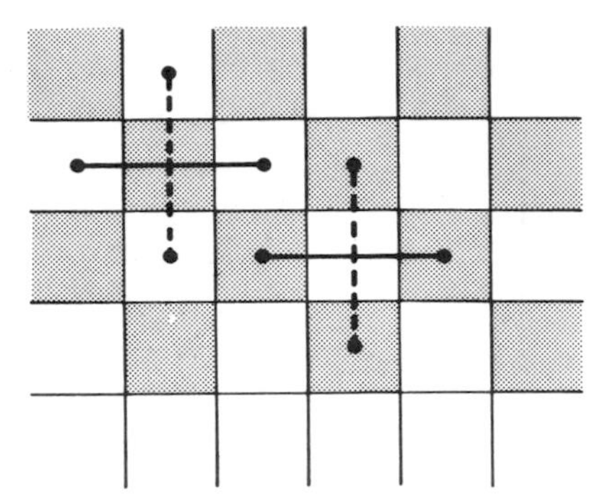

Figure 17.1.6. Origin of mesh-drift instabilities in a staggered leapfrog scheme. If the mesh points are imagined to lie in the squares of a chess board, then white squares couple to themselves, black to themselves, but there is no coupling between white and black. The fix is to introduce a small diffusive mesh-coupling piece.

by coupling the two meshes through a numerical viscosity term, e.g. adding to the right side of (17.1.31) a small coefficient ($\ll 1$) times $u^n_{j+1} - 2u^n_j + u^n_{j-1}$.

The *Two-Step Lax-Wendroff* scheme is a second-order in time method that avoids large numerical dissipation and mesh drifting. One defines intermediate values $u_{j+1/2}$ at the half timesteps $t_{n+1/2}$ and the half mesh points $x_{j+1/2}$. These are calculated by the Lax scheme:

$$u^{n+1/2}_{j+1/2} = \frac{1}{2}(u^n_{j+1} + u^n_j) - \frac{\Delta t}{2\Delta x}(F^n_{j+1} - F^n_j) \tag{17.1.34}$$

Using these variables, one calculates the fluxes $F^{n+1/2}_{j+1/2}$. Then the updated values u^{n+1}_j are calculated by the properly-centered expression

$$u^{n+1}_j = u^n_j - \frac{\Delta t}{\Delta x}(F^{n+1/2}_{j+1/2} - F^{n+1/2}_{j-1/2}) \tag{17.1.35}$$

The provisional values $u^{n+1/2}_{j+1/2}$ are now discarded. (See Figure 17.1.7.)

Let us investigate the stability of this method for our model advective equation, where $F = vu$. Substitute (17.1.34) in (17.1.35) to get

$$\begin{aligned} u^{n+1}_j = u^n_j - \alpha \biggl[&\frac{1}{2}(u^n_{j+1} + u^n_j) - \frac{1}{2}\alpha(u^n_{j+1} - u^n_j) \\ &-\frac{1}{2}(u^n_j + u^n_{j-1}) + \frac{1}{2}\alpha(u^n_j - u^n_{j-1}) \biggr] \end{aligned} \tag{17.1.36}$$

where

$$\alpha \equiv \frac{v\Delta t}{\Delta x} \tag{17.1.37}$$

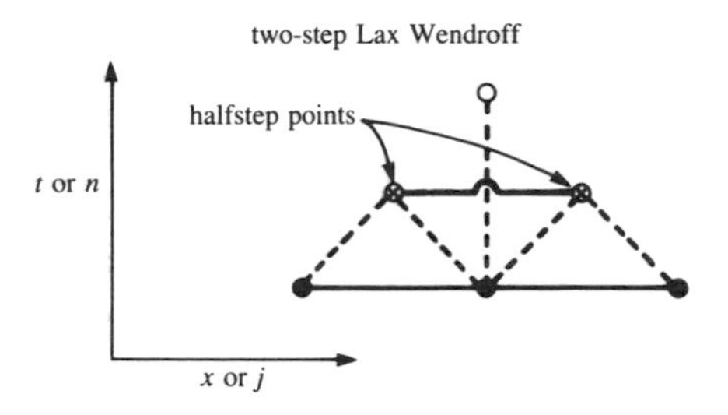

Figure 17.1.7. Representation of the two-step Lax-Wendroff differencing scheme. Two halfstep points (⊗) are calculated by the Lax method. These, plus one of the original points, produce the new point via staggered leapfrog. Halfstep points are used only temporarily and do not require storage allocation on the grid. This scheme is second-order accurate in both space and time.

Then

$$\xi = 1 - i\alpha \sin k\Delta x - \alpha^2(1 - \cos k\Delta x) \tag{17.1.38}$$

so

$$|\xi|^2 = 1 - \alpha^2(1 - \alpha^2)(1 - \cos k\Delta x)^2 \tag{17.1.39}$$

The stability criterion $|\xi|^2 \le 1$ is therefore $\alpha^2 \le 1$, or $v\Delta t \le \Delta x$ as usual. Incidentally, you should not think that the Courant condition is the only stability requirement that ever turns up in PDEs. It keeps doing so in our model examples just because those examples are so simple in form. The method of analysis is, however, general.

Except when $\alpha = 1$, $|\xi|^2 < 1$ in (17.1.39), so some amplitude damping does occur. The effect is relatively small, however, for wavelengths large compared with the mesh size Δx. If we expand (17.1.39) for small $k\Delta x$, we find

$$|\xi|^2 = 1 - \alpha^2(1 - \alpha^2)\frac{(k\Delta x)^4}{4} + \dots \tag{17.1.40}$$

The departure from unity occurs only at fourth order in k. This should be contrasted with equation (17.1.16) for the Lax method, which shows that

$$|\xi|^2 = 1 - (1 - \alpha^2)(k\Delta x)^2 + \dots \tag{17.1.41}$$

for small $k\Delta x$.

In summary, our recommendation for initial value problems that can be cast in flux-conservative form is to use either the staggered leapfrog or the

Two-Step Lax-Wendroff method. For problems sensitive to transport errors, upwind differencing or one of its refinements should be considered.

REFERENCES AND FURTHER READING:

Ames, W.F. 1977, *Numerical Methods for Partial Differential Equations*, 2nd ed. (New York: Academic Press), Chapter 4.

Richtmyer, R.D., and Morton, K.W. 1967, *Difference Methods for Initial Value Problems*, 2nd ed. (New York: Wiley-Interscience).

Roache, P.J. 1976, *Computational Fluid Dynamics* (Albuquerque: Hermosa).

Centrella, J., and Wilson, J. R. 1984, *Astrophysical Journal Supplement*, vol. 54, pp. 229–249, Appendix B.

Hawley, J. F., Smarr. L. L., and Wilson, J. R. 1984, *Astrophysical Journal Supplement*, vol. 55, pp. 211–246, §2c.

17.2 Diffusive Initial Value Problems

Recall the model parabolic equation, the diffusion equation in one space dimension,

$$\frac{\partial u}{\partial t} = \frac{\partial}{\partial x}\left(D\frac{\partial u}{\partial x}\right) \tag{17.2.1}$$

where D is the diffusion coefficient. Actually, this equation is a flux-conservative equation of the form considered in the previous section, with

$$F = -D\frac{\partial u}{\partial x} \tag{17.2.2}$$

the flux in the x-direction. We will assume $D \geq 0$, otherwise equation (17.2.1) has physically unstable solutions: A small disturbance evolves to become more and more concentrated instead of dispersing. (Don't make the mistake of trying to find a stable differencing scheme for a problem whose underlying PDEs are themselves unstable!)

Even though (17.2.1) is of the form already considered, it is useful to consider it as a model in its own right. The particular form of flux (17.2.2), and its direct generalizations, occur quite frequently in practice. Moreover, we have already seen that numerical viscosity and artificial viscosity can introduce diffusive pieces like the right-hand side of (17.2.1) in many other situations.

Consider first the case when D is a constant. Then the equation

$$\frac{\partial u}{\partial t} = D\frac{\partial^2 u}{\partial x^2} \tag{17.2.3}$$

can be differenced in the obvious way:

$$\frac{u_j^{n+1} - u_j^n}{\Delta t} = D\left[\frac{u_{j+1}^n - 2u_j^n + u_{j-1}^n}{(\Delta x)^2}\right] \tag{17.2.4}$$

This is the FTCS scheme again, except that it is a second derivative that has been differenced on the right-hand side. But this makes a world of difference! The FTCS scheme was unstable for the hyperbolic equation; however, a quick calculation shows that the amplification factor for equation (17.2.4) is

$$\xi = 1 - \frac{4D\Delta t}{(\Delta x)^2}\sin^2\left(\frac{k\Delta x}{2}\right) \tag{17.2.5}$$

The requirement $|\xi| \leq 1$ leads to the stability criterion

$$\frac{2D\Delta t}{(\Delta x)^2} \leq 1 \tag{17.2.6}$$

The physical interpretation of the restriction (17.2.6) is that the maximum allowed timestep is, up to a numerical factor, the diffusion time across a cell of width Δx.

More generally, the diffusion time τ across a spatial scale of size λ is of order

$$\tau \sim \frac{\lambda^2}{D} \tag{17.2.7}$$

Usually we are interested in modeling accurately the evolution of features with spatial scales $\lambda \gg \Delta x$. If we are limited to timesteps satisfying (17.2.6), we will need to evolve through of order $\lambda^2/(\Delta x)^2$ steps before things start to happen on the scale of interest. This number of steps is usually prohibitive. We must therefore find a stable way of taking timesteps comparable to, or perhaps — for accuracy — somewhat smaller than, the time scale of (17.2.7).

This goal poses an immediate "philosophical" question. Obviously the large timesteps that we propose to take are going to be woefully inaccurate for the small scales that we have decided not to be interested in. We want those scales to do something stable, "innocuous," and perhaps not too physically unreasonable. We want to build this innocuous behavior into our differencing scheme. What should it be?

There are two different answers, each of which has its pros and cons. The first answer is to seek a differencing scheme that drives small scale features to their *equilibrium* forms, e.g. satisfying equation (17.2.3) with the left-hand side set to zero. This answer generally makes the best physical sense; but, as we will see, it leads to a differencing scheme ("fully implicit") that is only *first-order* accurate in time for the scales that we are interested in. The second

answer is to let small scale features *maintain* their initial amplitudes, so that the evolution of the larger scale features of interest takes place superposed with a kind of "frozen in" (though fluctuating) background of small scale stuff. This answer gives a differencing scheme ("Crank-Nicholson") that is *second-order* accurate in time. Toward the end of an evolution calculation, however, one might want to switch over to some steps of the other kind, to drive the small-scale stuff into equilibrium. Let us now see where these distinct differencing schemes come from:

Consider the following differencing of (17.2.3),

$$\frac{u_j^{n+1} - u_j^n}{\Delta t} = D\left[\frac{u_{j+1}^{n+1} - 2u_j^{n+1} + u_{j-1}^{n+1}}{(\Delta x)^2}\right] \tag{17.2.8}$$

This is exactly like the FTCS scheme (17.2.4), except that the spatial derivatives on the right-hand side are evaluated at timestep $n+1$. Schemes with this character are called *fully implicit* or *backward time*, by contrast with FTCS (which is called *fully explicit*). To solve equation (17.2.8) one has to solve a set of simultaneous linear equations at each timestep for the u_j^{n+1}. Fortunately, this is a simple problem because the system is tridiagonal: Just group the terms in equation (17.2.8) appropriately:

$$-\alpha u_{j-1}^{n+1} + (1+2\alpha)u_j^{n+1} - \alpha u_{j+1}^{n+1} = u_j^n, \qquad j = 1, 2 \ldots J-1 \tag{17.2.9}$$

where

$$\alpha \equiv \frac{D\Delta t}{(\Delta x)^2} \tag{17.2.10}$$

Supplemented by Dirichlet or Neumann boundary conditions at $j = 0$ and $j = J$, equation (17.2.9) is clearly a tridiagonal system, which can easily be solved at each timestep by the method of §2.6.

What is the behavior of (17.2.8) for very large timesteps? The answer is seen most clearly in (17.2.9), in the limit $\alpha \to \infty$ ($\Delta t \to \infty$). Dividing by α, we see that the difference equations are just the finite-difference form of the equilibrium equation

$$\frac{\partial^2 u}{\partial x^2} = 0 \tag{17.2.11}$$

What about stability? The amplification factor for equation (17.2.8) is

$$\xi = \frac{1}{1 + 4\alpha \sin^2\left(\dfrac{k\Delta x}{2}\right)} \tag{17.2.12}$$

Clearly $|\xi| < 1$ for any stepsize Δt. The scheme is unconditionally stable. The details of the small-scale evolution from the initial conditions are obviously inaccurate for large Δt. But, as advertised, the correct equilibrium solution is obtained. This is the characteristic feature of implicit methods.

Here, on the other hand, is how one gets to the second of our above philosophical answers, combining the stability of an implicit method with the accuracy of a method that is second-order in both space and time. Simply form the average of the explicit and implicit FTCS schemes:

$$\frac{u_j^{n+1} - u_j^n}{\Delta t} = \frac{D}{2}\left[\frac{(u_{j+1}^{n+1} - 2u_j^{n+1} + u_{j-1}^{n+1}) + (u_{j+1}^n - 2u_j^n + u_{j-1}^n)}{(\Delta x)^2}\right] \tag{17.2.13}$$

Here both the left- and right-hand sides are centered at timestep $n + \frac{1}{2}$, so the method is second-order accurate in time as claimed. The amplification factor is

$$\xi = \frac{1 - 2\alpha \sin^2\left(\frac{k\Delta x}{2}\right)}{1 + 2\alpha \sin^2\left(\frac{k\Delta x}{2}\right)} \tag{17.2.14}$$

so the method is stable for any size Δt. This scheme is called the *Crank-Nicholson* scheme, and is our recommended method for any simple diffusion problem (perhaps supplemented by a few fully implicit steps at the end). (See Figure 17.2.1.)

Now turn to some generalizations of the simple diffusion equation (17.2.3). Suppose first that the diffusion coefficient D is not constant, say $D = D(x)$. We can adopt either of two strategies. First, we can make an analytic change of variable

$$y = \int \frac{dx}{D(x)} \tag{17.2.15}$$

Then

$$\frac{\partial u}{\partial t} = \frac{\partial}{\partial x} D(x) \frac{\partial u}{\partial x} \tag{17.2.16}$$

becomes

$$\frac{\partial u}{\partial t} = \frac{1}{D(y)} \frac{\partial^2 u}{\partial y^2} \tag{17.2.17}$$

and we evaluate D at the appropriate y_j. Heuristically, the stability criterion

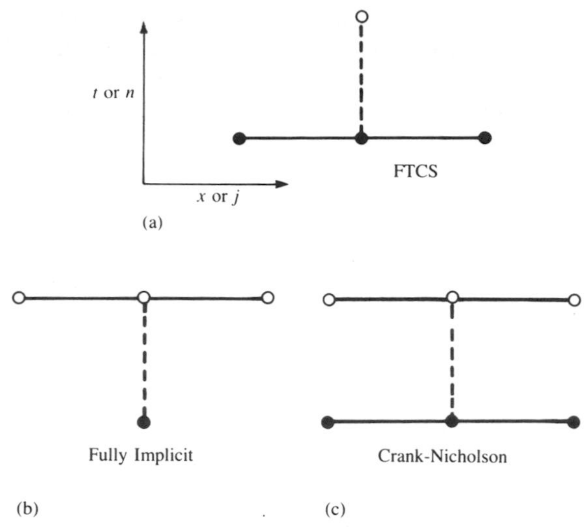

Figure 17.2.1. Three differencing schemes for diffusive problems (shown as in Figure 17.1.2). (a) Forward Time Center Space is first-order accurate, but stable only for sufficiently small timesteps. (b) Fully Implicit is stable for arbitrarily large timesteps, but is still only first-order accurate. (c) Crank-Nicholson is second-order accurate, and is usually stable for large timesteps.

(17.2.6) in an explicit scheme becomes

$$\Delta t \le \min_j \left[\frac{(\Delta y)^2}{2D_j^{-1}} \right] \tag{17.2.18}$$

Note that constant spacing Δy in y does not imply constant spacing in x.

An alternative method that does not require analytically tractable forms for D is simply to difference equation (17.2.16) as it stands, centering everything appropriately. Thus the FTCS method becomes

$$\frac{u_j^{n+1} - u_j^n}{\Delta t} = \frac{D_{j+1/2}(u_{j+1}^n - u_j^n) - D_{j-1/2}(u_j^n - u_{j-1}^n)}{(\Delta x)^2} \tag{17.2.19}$$

where

$$D_{j+1/2} \equiv D(x_{j+1/2}) \tag{17.2.20}$$

and the heuristic stability criterion is

$$\Delta t \le \min_j \left[\frac{(\Delta x)^2}{2D_{j+1/2}} \right] \tag{17.2.21}$$

The Crank-Nicholson method can be generalized similarly.

The second complication one can consider is a nonlinear diffusion problem, for example where $D = D(u)$. Explicit schemes can be generalized in the obvious way. For example, in equation (17.2.19) write

$$D_{j+1/2} = \frac{1}{2}\left[D(u_{j+1}^n) + D(u_j^n)\right] \tag{17.2.22}$$

Implicit schemes are not as easy. The replacement (17.2.22) with $n \to n+1$ leaves us with a nasty set of coupled nonlinear equations to solve at each timestep. Instead, we linearize them by writing

$$D(u_j^{n+1}) = D(u_j^n) + (u_j^{n+1} - u_j^n)\left.\frac{\partial D}{\partial u}\right|_{j,n} \tag{17.2.23}$$

This reduces the problem to tridiagonal form again and in practice usually retains the stability advantages of implicit differencing.

Schrödinger Equation

Sometimes the physical problem being solved imposes constraints on the differencing scheme that we have not yet taken into account. For example, consider the time-dependent Schrödinger equation of quantum mechanics. This is basically a parabolic equation for the evolution of a complex quantity ψ. For the scattering of a wavepacket by a one-dimensional potential $V(x)$, the equation has the form

$$i\frac{\partial \psi}{\partial t} = -\frac{\partial^2 \psi}{\partial x^2} + V(x)\psi \tag{17.2.24}$$

(Here we have chosen units so that Planck's constant $\hbar = 1$ and the particle mass $m = 1/2$.) One is given the initial wavepacket, $\psi(x, t = 0)$, together with boundary conditions that $\psi \to 0$ at $x \to \pm\infty$. Suppose we content ourselves with first-order accuracy in time, but want to use an implicit scheme, for stability. A slight generalization of (17.2.8) leads to

$$i\left[\frac{\psi_j^{n+1} - \psi_j^n}{\Delta t}\right] = -\left[\frac{\psi_{j+1}^{n+1} - 2\psi_j^{n+1} + \psi_{j-1}^{n+1}}{(\Delta x)^2}\right] + V_j \psi_j^{n+1} \tag{17.2.25}$$

for which

$$\xi = \frac{1}{1 + i\left[\dfrac{4\Delta t}{(\Delta x)^2}\sin^2\left(\dfrac{k\Delta x}{2}\right) + V_j \Delta t\right]} \tag{17.2.26}$$

This is unconditionally stable, but unfortunately is not *unitary*. The underlying physical problem requires that the total probability of finding the particle somewhere remains unity. This is represented formally by the modulus-square norm of ψ remaining unity:

$$\int_{-\infty}^{\infty} |\psi|^2 dx = 1 \tag{17.2.27}$$

The initial wave function $\psi(x,0)$ is normalized to satisfy (17.2.27). The Schrödinger equation (17.2.24) then guarantees that this condition is satisfied at all later times.

Let us write equation (17.2.24) in the form

$$i\frac{\partial \psi}{\partial t} = H\psi \tag{17.2.28}$$

where the operator H is

$$H = -\frac{\partial^2}{\partial x^2} + V(x) \tag{17.2.29}$$

The formal solution of equation (17.2.28) is

$$\psi(x,t) = e^{-iHt}\psi(x,0) \tag{17.2.30}$$

where the exponential of the operator is defined by its power series expansion.

The explicit FTCS scheme approximates (17.2.30) as

$$\psi_j^{n+1} = (1 - iH\Delta t)\psi_j^n \tag{17.2.31}$$

where H is represented by a centered finite-difference approximation in x. The stable implicit scheme (17.2.25) is, by contrast,

$$\psi_j^{n+1} = (1 + iH\Delta t)^{-1}\psi_j^n \tag{17.2.32}$$

These are both first-order accurate in time, as can be seen by expanding equation (17.2.30). However, neither operator in (17.2.31) or (17.2.32) is unitary.

The correct way to difference Schrödinger's equation is to use *Cayley's form* for the finite-difference representation of e^{-iHt}, which is second-order accurate *and* unitary:

$$e^{-iHt} \simeq \frac{1 - \frac{1}{2}iH\Delta t}{1 + \frac{1}{2}iH\Delta t} \tag{17.2.33}$$

In other words,

$$(1 + \frac{1}{2}iH\Delta t)\psi_j^{n+1} = (1 - \frac{1}{2}iH\Delta t)\psi_j^n \tag{17.2.34}$$

On replacing H by its finite-difference approximation in x, we have a complex tridiagonal system to solve. The method is stable, unitary, and second-order accurate in space and time. In fact, it is simply the Crank-Nicholson method once again!

REFERENCES AND FURTHER READING:

Ames, W.F. 1977, *Numerical Methods for Partial Differential Equations*, 2nd ed. (New York: Academic Press), Chapter 2.

Goldberg, A., Schey, H.M., and Schwartz, J.L. 1967, *American Journal of Physics*, vol. 35, pp. 177–186. [Schrödinger equation].

Galbraith, I., Ching, Y.S., and Abraham, E. 1984, *American Journal of Physics*, vol. 52, pp. 60–68. [Schrödinger equation].

17.3 Initial Value Problems in Multidimensions

The methods described in §17.1 and §17.2 for problems in $1 + 1$ dimension (one space and one time dimension) can easily be generalized to $N + 1$ dimensions. However, the computing power necessary to solve the resulting equations is enormous. If you have solved a one-dimensional problem with 100 spatial grid points, solving the two-dimensional version with 100×100 mesh points requires *at least* 100 times as much computing. You generally have to be content with very modest spatial resolution in multidimensional problems.

Indulge us in offering a bit of advice about the development and testing of multidimensional PDE codes: You should always develop and test your programs on *very small* grids, e.g., 8×8, even though the resulting accuracy is so poor as to be useless. When your program is all debugged and demonstrably stable, *then* you can increase the grid size to a reasonable one and start looking at the results. We have actually heard someone protest, "my program would be unstable for a crude grid, but I am sure the instability will go away on a larger grid." That is nonsense of a most pernicious sort, evidencing total confusion between accuracy and stability. In fact, new instabilities sometimes do show up on *larger* grids; but old instabilities never (in our experience) just go away.

Forced to live with modest grid sizes, some people recommend going to higher order methods in an attempt to improve accuracy. This is very dangerous. Unless the solution you are looking for is known to be smooth, and the high order method you are using is known to be extremely stable, we do not recommend anything higher than second-order in time (for sets of first-order equations). For spatial differencing, we recommend the order of the

underlying PDEs, perhaps allowing second-order spatial differencing for first-order-in-space PDEs. When you increase the order of a differencing method to greater than the order of the original PDEs, you introduce spurious solutions to the difference equations. This does not create a problem if they all happen to decay exponentially; otherwise you are going to see all hell break loose!

Lax Method for a Flux-Conservative Equation

As an example, we show how to generalize the Lax method (17.1.15) to two dimensions for the conservation equation

$$\frac{\partial u}{\partial t} = -\nabla \cdot \mathbf{F} = -\left(\frac{\partial F_x}{\partial x} + \frac{\partial F_y}{\partial y}\right) \tag{17.3.1}$$

Use a spatial grid with

$$\begin{aligned} x_j &= x_0 + j\Delta \\ y_l &= y_0 + l\Delta \end{aligned} \tag{17.3.2}$$

We have chosen $\Delta x = \Delta y \equiv \Delta$ for simplicity. Then the Lax scheme is

$$\begin{aligned} u_{j,l}^{n+1} = &\frac{1}{4}(u_{j+1,l}^n + u_{j-1,l}^n + u_{j,l+1}^n + u_{j,l-1}^n) \\ &- \frac{\Delta t}{2\Delta}(F_{j+1,l}^n - F_{j-1,l}^n + F_{j,l+1}^n - F_{j,l-1}^n) \end{aligned} \tag{17.3.3}$$

Note that as an abbreviated notation F_{j+1} and F_{j-1} refer to F_x, while F_{l+1} and F_{l-1} refer to F_y.

Let us carry out a stability analysis for the model advective equation (analog of 17.1.6) with

$$F_x = v_x u, \qquad F_y = v_y u \tag{17.3.4}$$

This requires an eigenmode with two dimensions in space, though still only a simple dependence on powers of ξ in time,

$$u_{j,i}^n = \xi^n e^{ik_x j\Delta} e^{ik_y l\Delta} \tag{17.3.5}$$

Substituting in equation (17.3.3), we find

$$\xi = \frac{1}{2}(\cos k_x\Delta + \cos k_y\Delta) - i\alpha_x \sin k_x\Delta - i\alpha_y \sin k_y\Delta \tag{17.3.6}$$

where

$$\alpha_x = \frac{v_x \Delta t}{\Delta}, \qquad \alpha_y = \frac{v_y \Delta t}{\Delta} \tag{17.3.7}$$

The expression for $|\xi|^2$ can be manipulated into the form

$$\begin{aligned} |\xi|^2 = 1 &- (\sin^2 k_x\Delta + \sin^2 k_y\Delta)\left[\frac{1}{2} - (\alpha_x^2 + \alpha_y^2)\right] \\ &- \frac{1}{4}(\cos k_x\Delta - \cos k_y\Delta)^2 \\ &- (\alpha_y \sin k_x\Delta - \alpha_x \sin k_y\Delta)^2 \end{aligned} \tag{17.3.8}$$

The last two terms are negative, and so the stability requirement $|\xi|^2 \leq 1$ becomes

$$\frac{1}{2} - (\alpha_x^2 + \alpha_y^2) \geq 0 \tag{17.3.9}$$

or

$$\Delta t \leq \frac{\Delta}{\sqrt{2}(v_x^2 + v_y^2)^{1/2}} \tag{17.3.10}$$

This is an example of the general result for the N-dimensional Courant condition: If $|v|$ is the maximum propagation velocity in the problem, then

$$\Delta t \leq \frac{\Delta}{\sqrt{N}|v|} \tag{17.3.11}$$

is the Courant condition.

Diffusion Equation in Multidimensions

Let us consider the two-dimensional diffusion equation,

$$\frac{\partial u}{\partial t} = D\left(\frac{\partial^2 u}{\partial x^2} + \frac{\partial^2 u}{\partial y^2}\right) \tag{17.3.12}$$

An explicit method, such as FTCS, can be generalized from the one-dimensional case in the obvious way. However, we have seen that diffusive problems are usually best treated implicitly. Suppose we try to implement the Crank-Nicholson scheme in two dimensions. This would give us

$$u_{j,l}^{n+1} = u_{j,l}^n + \frac{1}{2}\alpha\left(\delta_x^2 u_{j,l}^{n+1} + \delta_x^2 u_{j,l}^n + \delta_y^2 u_{j,l}^{n+1} + \delta_y^2 u_{j,l}^n\right) \tag{17.3.13}$$

Here

$$\alpha \equiv \frac{D\Delta t}{\Delta^2} \qquad \Delta \equiv \Delta x = \Delta y \tag{17.3.14}$$

$$\delta_x^2 u_{j,l}^n \equiv u_{j+1,l}^n - 2u_{j,l}^n + u_{j-1,l}^n \tag{17.3.15}$$

and similarly for $\delta_y^2 u_{j,l}^n$. This is certainly a viable scheme; the problem arises in solving the coupled linear equations. Whereas in one space dimension the system was tridiagonal, that is no longer true, though the matrix is still very sparse. One possibility is to use a suitable sparse matrix technique (see §2.10 and §17.0).

Another possibility, which we generally prefer, is a slightly different way of generalizing the Crank-Nicholson algorithm. It is still second-order acccurate in time and space, and unconditionally stable, but the equations are easier to solve than (17.3.13). Called the *alternating-direction implicit method (ADI)*, this scheme is our first mention of the powerful concept of *operator splitting* or *time splitting*. We will discuss that concept in more detail in §17.6. As a preview here, the idea is to divide each timestep into two steps of size $\Delta t/2$. In each substep, a different dimension is treated implicitly:

$$\begin{aligned} u_{j,l}^{n+1/2} &= u_{j,l}^n + \frac{1}{2}\alpha\left(\delta_x^2 u_{j,l}^{n+1/2} + \delta_y^2 u_{j,l}^n\right) \\ u_{j,l}^{n+1} &= u_{j,l}^{n+1/2} + \frac{1}{2}\alpha\left(\delta_x^2 u_{j,l}^{n+1/2} + \delta_y^2 u_{j,l}^{n+1}\right) \end{aligned} \tag{17.3.16}$$

The advantage of this method is that each substep requires only the solution of a simple tridiagonal system.

It is at this point that we turn our attention from initial value problems to boundary value problems. These will occupy us for the remainder of the chapter.

REFERENCES AND FURTHER READING:

Ames, W.F. 1977, *Numerical Methods for Partial Differential Equations*, 2nd ed. (New York: Academic Press).

17.4 Fourier and Cyclic Reduction Methods for Boundary Value Problems

As discussed in §17.0, most boundary value problems (elliptic equations, for example) reduce to solving large sparse linear systems of the form

$$\mathbf{A} \cdot \mathbf{u} = \mathbf{b} \tag{17.4.1}$$

either once, for boundary value equations that are linear, or iteratively, for boundary value equations that are nonlinear.

Two important techniques lead to "rapid" solution of equation (17.4.1) when the sparse matrix is of certain frequently occurring forms. The *Fourier transform method* is directly applicable when the equations have coefficients that are constant in space. The *cyclic reduction* method is somewhat more general; its applicability is related to the question of whether the equations are separable (in the sense of "separation of variables"). Both methods require the boundaries to coincide with the coordinate lines. Finally, for some problems, there is a powerful combination of these two methods called *FACR (Fourier Analysis and Cyclic Reduction)*. We now consider each method in turn, using equation (17.0.3), with finite difference representation (17.0.6), as a model example.

Fourier Transform Method

The discrete inverse Fourier transform in both x and y is

$$u_{jl} = \frac{1}{JL} \sum_{m=0}^{J-1} \sum_{n=0}^{L-1} \widehat{u}_{mn} e^{-2\pi ijm/J} e^{-2\pi iln/L} \tag{17.4.2}$$

This can be computed using the FFT independently in each dimension, or else all at once via the routine FOURN of §12.11. Similarly,

$$\rho_{jl} = \frac{1}{JL} \sum_{m=0}^{J-1} \sum_{n=0}^{L-1} \widehat{\rho}_{mn} e^{-2\pi ijm/J} e^{-2\pi iln/L} \tag{17.4.3}$$

If we substitute expressions (17.4.2) and (17.4.3) in our model problem (17.0.6), we find

$$\widehat{u}_{mn} \left(e^{2\pi im/J} + e^{-2\pi im/J} + e^{2\pi in/L} + e^{-2\pi in/L} - 4 \right) = \widehat{\rho}_{mn} \Delta^2 \tag{17.4.4}$$

or

$$\widehat{u}_{mn} = \frac{\widehat{\rho}_{mn}\Delta^2}{2\left(\cos\dfrac{2\pi m}{J} + \cos\dfrac{2\pi n}{L} - 2\right)} \tag{17.4.5}$$

Thus the strategy for solving equation (17.0.6) by FFT techniques is:

• Compute $\widehat{\rho}_{mn}$ as the Fourier transform

$$\widehat{\rho}_{mn} = \sum_{j=0}^{J-1}\sum_{l=0}^{L-1} \rho_{jl}\, e^{2\pi imj/J} e^{2\pi inl/L} \tag{17.4.6}$$

• Compute $\widehat{u}_{mn}$ from equation (17.4.5).
• Compute u_{jl} by the inverse Fourier transform (17.4.2).

The above procedure is valid for periodic boundary conditions. In other words, the solution satisfies

$$u_{jl} = u_{j+J,l} = u_{j,l+L} \tag{17.4.7}$$

Consider now a Dirichlet boundary condition $u = 0$ on the rectangular boundary. Instead of the expansion (17.4.2), we now need an expansion in sine waves:

$$u_{jl} = \frac{2}{J}\frac{2}{L}\sum_{m=1}^{J-1}\sum_{n=1}^{L-1} \widehat{u}_{mn} \sin\frac{\pi jm}{J} \sin\frac{\pi ln}{L} \tag{17.4.8}$$

This satisfies the boundary conditions that $u = 0$ at $j = 0, J$ and at $l = 0, L$. If we substitute this expansion and the analogous one for ρ_{jl} into equation (17.0.6), we find that the solution procedure parallels that for periodic boundary conditions:

• Compute $\widehat{\rho}_{mn}$ by the sine transform

$$\widehat{\rho}_{mn} = \sum_{j=1}^{J-1}\sum_{l=1}^{L-1} \rho_{jl} \sin\frac{\pi jm}{J} \sin\frac{\pi ln}{L} \tag{17.4.9}$$

(A fast sine transform algorithm was given in §12.3.)

• Compute $\widehat{u}_{mn}$ from the expression analogous to (17.4.5),

$$\widehat{u}_{mn} = \frac{\Delta^2\widehat{\rho}_{mn}}{2\left(\cos\dfrac{\pi m}{J} + \cos\dfrac{\pi n}{L} - 2\right)} \tag{17.4.10}$$

- Compute u_{jl} by the inverse sine transform (17.4.8).

If we have inhomogeneous boundary conditions, for example $u = 0$ on all boundaries except $u = f(y)$ on the boundary $x = J\Delta$, we have to add to the above solution a solution u^H of the homogeneous equation

$$\frac{\partial^2 u}{\partial x^2} + \frac{\partial^2 u}{\partial y^2} = 0 \tag{17.4.11}$$

that satisfies the required boundary conditions. In the continuum case, this would be an expression of the form

$$u^H = \sum_n A_n \sinh \frac{n\pi x}{J\Delta} \sin \frac{n\pi y}{L\Delta} \tag{17.4.12}$$

where A_n would be found by requiring that $u = f(y)$ at $x = J\Delta$. In the discrete case, we have

$$u_{jl}^H = \frac{2}{L} \sum_{n=1}^{L-1} A_n \sinh \frac{\pi n j}{J} \sin \frac{\pi n l}{L} \tag{17.4.13}$$

If $f(y = l\Delta) \equiv f_l$, then we get A_n from the inverse formula

$$A_n = \frac{1}{\sinh \pi n} \sum_{l=1}^{L-1} f_l \sin \frac{\pi n l}{L} \tag{17.4.14}$$

The complete solution to the problem is

$$u = u_{jl} + u_{jl}^H \tag{17.4.15}$$

By adding appropriate terms of the form (17.4.12), we can handle inhomogeneous terms on any boundary surface.

A much simpler procedure for handling inhomogeneous terms is to note that whenever boundary terms appear on the left-hand side of (17.0.6), they can be taken over to the right-hand side since they are known. The effective source term is therefore ρ_{jl} plus a contribution from the boundary terms. For example, instead of equation (17.4.13), we write down equation (17.0.6) at $j = J - 1$:

$$u_{J,l} + u_{J-2,l} + u_{J-1,l+1} + u_{J-1,l-1} - 4u_{J-1,l} = \Delta^2 \rho_{J-1,l} \tag{17.4.16}$$

The term $u_{J,l}$ is just the boundary value f_l, so we take it over to the right-hand side. The problem is now equivalent to the case of zero boundary conditions, except that one row of the source term is modified by the replacement

$$\Delta^2 \rho_{J-1,l} \to \Delta^2 \rho_{J-1,l} - f_l \tag{17.4.17}$$

Note that it is crucial that we use the sine transform to automatically impose the zero boundary condition for this method to work.

The case of Neumann boundary conditions $\nabla u = 0$ is handled by the cosine expansion (see §12.3)

$$u_{jl} = \frac{1}{2}\widehat{u}_{00} + \frac{2}{J}\frac{2}{L}\sum_{m=1}^{J-1}\sum_{n=1}^{L-1} \widehat{u}_{mn} \cos\frac{\pi jm}{J} \cos\frac{\pi ln}{L} \tag{17.4.18}$$

Inhomogeneous terms $\nabla u = g$ can be again included by adding a suitable solution of the homogeneous equation, or more simply by taking boundary terms over to the right-hand side. For example, the condition

$$\frac{\partial u}{\partial x} = g(y) \qquad \text{at} \quad x = 0 \tag{17.4.19}$$

becomes

$$\frac{u_{1,l} - u_{-1,l}}{2\Delta} = g_l \tag{17.4.20}$$

where $g_l \equiv g(y = l\Delta)$. This equation must be satisfied in conjunction with equation (17.0.6) at $j = 0$:

$$u_{1,l} + u_{-1,l} + u_{0,l+1} + u_{0,l-1} - 4u_{0,l} = \Delta^2 \rho_{0,l} \tag{17.4.21}$$

Define a new variable

$$u^*_{-1,l} = u_{-1,l} + 2\Delta g_l \tag{17.4.22}$$

Then equations (17.4.20) and (17.4.21) can be rewritten as

$$u_{1,l} - u^*_{-1,l} = 0 \tag{17.4.23}$$
$$u_{1,l} + u^*_{-1,l} + u_{0,l+1} + u_{0,l-1} - 4u_{0,l} = \Delta^2 \rho_{0,l} + 2\Delta g_l \tag{17.4.24}$$

Equations (17.4.23) and (17.4.24) are the equations for a zero gradient problem, with the source term modified by the replacement

$$\Delta^2 \rho_{0,l} \to \Delta^2 \rho_{0,l} + 2\Delta g_l \tag{17.4.25}$$

Again it is crucial that we use the cosine transform here to automatically impose the zero gradient boundary condition.

Cyclic Reduction

Evidently the FFT method works only when the original PDE has constant coefficients, and boundaries that coincide with the coordinate lines. An alternative algorithm, which can be used on somewhat more general equations, is called *cyclic reduction (CR)*.

We illustrate cyclic reduction on the equation

$$\frac{\partial^2 u}{\partial x^2} + \frac{\partial^2 u}{\partial y^2} + b(y)\frac{\partial u}{\partial y} + c(y)u = g(x,y) \tag{17.4.26}$$

This form arises very often in practice from the Helmholtz or Poisson equations in polar, cylindrical or spherical coordinate systems. More general separable equations are treated by Swartzrauber.

The finite-difference form of equation (17.4.26) can be written as a set of vector equations

$$\mathbf{u}_{j-1} + \mathbf{T}\cdot\mathbf{u}_j + \mathbf{u}_{j+1} = \mathbf{g}_j\Delta^2 \tag{17.4.27}$$

Here the index j comes from differencing in the x-direction, while the y-differencing (denoted by the index l previously) has been left in vector form. The matrix $\mathbf{T}$ has the form

$$\mathbf{T} = \mathbf{B} - 2\mathbf{1} \tag{17.4.28}$$

where the $\mathbf{21}$ comes from the x-differencing and the matrix $\mathbf{B}$ from the y-differencing. The matrix $\mathbf{B}$, and hence $\mathbf{T}$, is tridiagonal with variable coefficients.

The CR method is derived by writing down three successive equations like (17.4.27):

$$\begin{aligned}
\mathbf{u}_{j-2} + \mathbf{T}\cdot\mathbf{u}_{j-1} + \mathbf{u}_j &= \mathbf{g}_{j-1}\Delta^2 \\
\mathbf{u}_{j-1} + \mathbf{T}\cdot\mathbf{u}_j + \mathbf{u}_{j+1} &= \mathbf{g}_j\Delta^2 \\
\mathbf{u}_j + \mathbf{T}\cdot\mathbf{u}_{j+1} + \mathbf{u}_{j+2} &= \mathbf{g}_{j+1}\Delta^2
\end{aligned} \tag{17.4.29}$$

Matrix-multiplying the middle equation by $-\mathbf{T}$ and then adding the three equations, we get

$$\mathbf{u}_{j-2} + \mathbf{T}^{(1)}\cdot\mathbf{u}_j + \mathbf{u}_{j+2} = \mathbf{g}_j^{(1)}\Delta^2 \tag{17.4.30}$$

This is an equation of the same form as (17.4.27), with

$$\begin{aligned} \mathbf{T}^{(1)} &= 2\mathbf{1} - \mathbf{T}^2 \\ \mathbf{g}_j^{(1)} &= \Delta^2(\mathbf{g}_{j-1} - \mathbf{T}\cdot\mathbf{g}_j + \mathbf{g}_{j+1}) \end{aligned} \tag{17.4.31}$$

After one level of CR, we have reduced the number of equations by a factor of two. Since the resulting equations are of the same form as the original equation, we can repeat the process. Taking the number of mesh points to be a power of 2 for simplicity, we finally end up with a single equation for the central line of variables:

$$\mathbf{T}^{(f)}\cdot\mathbf{u}_{J/2} = \Delta^2\mathbf{g}_{J/2}^{(f)} - \mathbf{u}_0 - \mathbf{u}_J \tag{17.4.32}$$

Here we have written $\mathbf{u}_0$ and $\mathbf{u}_J$ on the right-hand side because they are known boundary values. Equation (17.4.32) can be solved for $\mathbf{u}_{J/2}$ by the standard tridiagonal algorithm. The two equations at level $f-1$ involve $\mathbf{u}_{J/4}$ and $\mathbf{u}_{3J/4}$. The equation for $\mathbf{u}_{J/4}$ involves $\mathbf{u}_0$ and $\mathbf{u}_{J/2}$, both of which are known, and hence can be solved by the usual tridiagonal routine. A similar result holds true at every stage, so we end up solving $J-1$ tridiagonal systems.

In practice, equations (17.4.31) should be rewritten to avoid numerical instability. For these and other practical details, we refer you to the review paper of Buzbee et al.

FACR Method

The *best* way to solve equations of the form (17.4.26), including the constant coefficient problem (17.0.3), is a combination of Fourier analysis and cyclic reduction, the FACR method (Hockney 1965, 1971; Temperton). If at the r'th stage of CR we Fourier analyze the equations of the form (17.4.30) along y, that is, with respect to the suppressed vector index, we will have a tridiagonal system in the x-direction for each y-Fourier mode:

$$\widehat{u}_{j-2^r}^k + \lambda_k^{(r)}\widehat{u}_j^k + \widehat{u}_{j+2^r}^k = \Delta^2 g_j^{(r)k} \tag{17.4.33}$$

Here $\lambda_k^{(r)}$ is the eigenvalue of $\mathbf{T}^{(r)}$ corresponding to the k'th Fourier mode. For the equation (17.0.3), equation (17.4.5) shows that $\lambda_k^{(r)}$ will involve terms like $\cos\frac{2\pi k}{L} - 2$ raised to a power. Solve the tridiagonal systems for $\widehat{u}_j^k$ at the levels $j = 2^r, 2\times 2^r, 4\times 2^r, ..., J-2^r$. Fourier synthesize to get the y-values on these x-lines. Then fill in the intermediate x-lines as in the original CR algorithm.

The trick is to choose the number of levels of CR so as to minimize the total number of arithmetic operations. One can show that for a typical case of a 128×128 mesh, the optimal level is $r = 2$; asymptotically, $r \to \log_2(\log_2 J)$.

A rough estimate of running times for these algorithms for equation (17.0.3) is as follows: the FFT method (in both x and y) and the CR method are roughly comparable. FACR with $r = 0$ (that is, FFT in one dimension and solve the tridiagonal equations by the usual algorithm in the other dimension) gives about a factor of two gain in speed. The optimal FACR with $r = 2$ gives another factor of two gain in speed.

REFERENCES AND FURTHER READING:

Swartzrauber, P.N. 1977, *S.I.A.M Review*, vol. 19, pp. 490 – 501.

Buzbee, B.L, Golub, G.H., and Nielson, C.W. 1970, *S.I.A.M. Journal of Numerical Analysis*, vol. 7, pp. 627–656; see also *op. cit.* vol. 11, pp. 753–763.

Hockney, R.W. 1965, *Journal Assn. Comp. Mach.*, vol. 12, p. 95 ff.

Hockney, R.W. 1970, in *Methods of Computational Physics*, vol. 9 (New York: Academic Press), pp. 135–211.

Hockney, R.W., and Eastwood, J.W. 1981, *Computer Simulation Using Particles* (New York: McGraw Hill), Chapter 6.

Temperton, C. 1980, *Journal of Computational Physics*, vol. 34, pp. 314–329.

17.5 Relaxation Methods for Boundary Value Problems

As we mentioned in §17.0, relaxation methods involve splitting the sparse matrix that arises from finite differencing and then iterating until a solution is found.

There is another way of thinking about relaxation methods that is somewhat more physical. Suppose we wish to solve the elliptic equation

$$\mathcal{L}u = \rho \tag{17.5.1}$$

where $\mathcal{L}$ represents some elliptic operator and ρ is the source term. Rewrite the equation as a diffusion equation,

$$\frac{\partial u}{\partial t} = \mathcal{L}u - \rho \tag{17.5.2}$$

An initial distribution u *relaxes* to an equilibrium solution as $t \to \infty$. This equilibrium has all time derivatives vanishing. Therefore it is the solution of the original elliptic problem (17.5.1). We see that all the machinery of §17.2, on diffusive initial value equations, can be brought to bear on the solution of boundary value problems by relaxation methods.

Let us apply this idea to our model problem (17.0.3). The diffusion equation is

$$\frac{\partial u}{\partial t} = \frac{\partial^2 u}{\partial x^2} + \frac{\partial^2 u}{\partial y^2} - \rho \tag{17.5.3}$$

If we use FTCS differencing (cf. equation 17.2.4), we get

$$u_{j,l}^{n+1} = u_{j,l}^n + \frac{\Delta t}{\Delta^2}\left(u_{j+1,l}^n + u_{j-1,l}^n + u_{j,l+1}^n + u_{j,l-1}^n - 4u_{j,l}^n\right) - \rho_{j,l}\Delta t \tag{17.5.4}$$

Recall from (17.2.6) that FTCS differencing is stable in one spatial dimension only if $\Delta t/\Delta^2 \leq \frac{1}{2}$. In two dimensions this becomes $\Delta t/\Delta^2 \leq \frac{1}{4}$. Suppose we try to take the largest possible timestep, and set $\Delta t = \Delta^2/4$. Then equation (17.5.4) becomes

$$u_{j,l}^{n+1} = \frac{1}{4}\left(u_{j+1,l}^n + u_{j-1,l}^n + u_{j,l+1}^n + u_{j,l-1}^n\right) - \frac{\Delta^2}{4}\rho_{j,l} \tag{17.5.5}$$

Thus the algorithm consists of using the average of u at its four nearest-neighbor points on the grid (plus the contribution from the source). This procedure is then iterated until convergence.

This method is in fact a classical method with origins dating back to the last century, called *Jacobi's method* (not to be confused with the Jacobi method for eigenvalues). The method is not practical because it converges too slowly. However, it is the basis for understanding the modern methods, which are always compared with it.

Another classical method is the *Gauss-Seidel* method. Here we make use of updated values of u on the right-hand side of (17.5.5) as soon as they become available. In other words, the averaging is done "in place" instead of being "copied" from an earlier timestep to a later one. If we are proceeding along the rows, incrementing j for fixed l, we have

$$u_{j,l}^{n+1} = \frac{1}{4}\left(u_{j+1,l}^n + u_{j-1,l}^{n+1} + u_{j,l+1}^n + u_{j,l-1}^{n+1}\right) - \frac{\Delta^2}{4}\rho_{j,l} \tag{17.5.6}$$

This method is also slowly converging and only of theoretical interest, but some analysis of it will be instructive.

Let us look at the Jacobi and Gauss-Seidel methods in terms of the matrix splitting concept. We change notation and call $\mathbf{u}$ "$\mathbf{x}$", to conform to standard matrix notation. To solve

$$\mathbf{A} \cdot \mathbf{x} = \mathbf{b} \tag{17.5.7}$$

we can consider splitting $\mathbf{A}$ as

$$\mathbf{A} = \mathbf{L} + \mathbf{D} + \mathbf{U} \tag{17.5.8}$$

where $\mathbf{D}$ is the diagonal part of $\mathbf{A}$, $\mathbf{L}$ is the lower triangle of $\mathbf{A}$ with zeros on the diagonal, and $\mathbf{U}$ is the upper triangle of $\mathbf{A}$ with zeros on the diagonal.

In the Jacobi method we write for the r'th step of iteration

$$\mathbf{D} \cdot \mathbf{x}^{(r)} = -(\mathbf{L} + \mathbf{U}) \cdot \mathbf{x}^{(r-1)} + \mathbf{b} \tag{17.5.9}$$

For our model problem (17.5.5), $\mathbf{D}$ is simply the identity matrix. The Jacobi method converges for matrices $\mathbf{A}$ that are "diagonally dominant" in a sense that can be made mathematically precise. For matrices arising from finite differencing, this condition is usually met.

What is the rate of convergence of the Jacobi method? A detailed analysis is beyond our scope, but here is some of the flavor: The matrix $-\mathbf{D}^{-1} \cdot (\mathbf{L}+\mathbf{U})$ is the *iteration matrix* which, apart from an additive term, maps one set of $\mathbf{x}$'s into the next. The iteration matrix has eigenvalues, each one of which reflects the factor by which the amplitude of a particular eigenmode of undesired residual is suppressed during one iteration. Evidently those factors had better all have modulus < 1 for the relaxation to work at all! The rate of convergence of the method is set by the rate for the slowest-decaying eigenmode, i.e., the factor with largest modulus. The modulus of this largest factor, therefore lying between 0 and 1, is called the *spectral radius* of the relaxation operator, denoted ρ_s.

The number of iterations r required to reduce the overall error by a factor 10^{-p} is thus estimated by

$$r \approx \frac{p \ln 10}{(-\ln \rho_s)} \tag{17.5.10}$$

In general, the spectral radius ρ_s goes asymptotically to the value 1 as the grid size J is increased, so that more iterations are required. For any given equation, grid geometry, *and boundary condition*, the spectral radius can, in principle, be computed analytically. For example, for equation (17.5.5) on a $J \times J$ grid with Dirichlet boundary conditions on all four sides, the asymptotic formula for large J turns out to be

$$\rho_s \simeq 1 - \frac{\pi^2}{2J^2} \tag{17.5.11}$$

The number of iterations r required to reduce the error by a factor of 10^{-p} is thus

$$r \simeq \frac{2pJ^2 \ln 10}{\pi^2} \simeq \frac{1}{2} pJ^2 \tag{17.5.12}$$

In other words, the number of iterations is proportional to the number of mesh points, J^2. Since 100×100 and larger problems are common, it is clear that the Jacobi method is only of academic interest.

The Gauss-Seidel method, equation (17.5.6), corresponds to the matrix decomposition

$$(\mathbf{L} + \mathbf{D}) \cdot \mathbf{x}^{(r)} = -\mathbf{U} \cdot \mathbf{x}^{(r-1)} + \mathbf{b} \tag{17.5.13}$$

The fact that $\mathbf{L}$ is on the left-hand side of the equation follows from the updating in place, as you can easily check if you write out (17.5.13) in components. One can show (Varga; Young; Stoer and Bulirsch) that the spectral radius is just the square of the spectral radius of the Jacobi method. For our model problem, therefore,

$$\rho_s \simeq 1 - \frac{\pi^2}{J^2} \tag{17.5.14}$$

$$r \simeq \frac{pJ^2 \ln 10}{\pi^2} \simeq \frac{1}{4} p J^2 \tag{17.5.15}$$

The factor of two improvement in the number of iterations over the Jacobi method still leaves the method impractical.

Simultaneous Over-Relaxation (SOR)

We get a practical algorithm if we make an *overcorrection* to the value of $\mathbf{x}^{(r)}$ at the r'th stage of Gauss-Seidel iteration, thus anticipating future corrections. Add and subtract $\mathbf{x}^{(r-1)}$ on the right-hand side of equation (17.5.13), and write the Gauss-Seidel method as

$$\mathbf{x}^{(r)} = \mathbf{x}^{(r-1)} - (\mathbf{L} + \mathbf{D})^{-1} \cdot [(\mathbf{L} + \mathbf{D} + \mathbf{U}) \cdot \mathbf{x}^{(r-1)} - \mathbf{b}] \tag{17.5.16}$$

The term in square brackets is just the residual vector $\xi^{(r-1)}$, so

$$\mathbf{x}^{(r)} = \mathbf{x}^{(r-1)} - (\mathbf{L} + \mathbf{D})^{-1} \cdot \xi^{(r-1)} \tag{17.5.17}$$

Now *overcorrect*, defining

$$\mathbf{x}^{(r)} = \mathbf{x}^{(r-1)} - \omega(\mathbf{L} + \mathbf{D})^{-1} \cdot \xi^{(r-1)} \tag{17.5.18}$$

Here ω is called the *overrelaxation parameter*, and the method is called *simultaneous overrelaxation* (SOR).

The following theorems can be proved (Varga; Young; Stoer and Bulirsch):

- The method is convergent only for $0 < \omega < 2$. If $0 < \omega < 1$, we speak of *underrelaxation.*
- Under certain mathematical restrictions generally satisfied by matrices arising from finite differencing, only overrelaxation ($1 < \omega < 2$) can give faster convergence than the Gauss-Seidel method.
- If ρ_{Jacobi} is the spectral radius of the Jacobi iteration (so that the square of it is the spectral radius of the Gauss-Seidel iteration), then the *optimal* choice for ω is given by

$$\omega = \frac{2}{1+\sqrt{1-\rho_{Jacobi}^2}} \tag{17.5.19}$$

- For this optimal choice, the spectral radius for SOR is

$$\rho_{SOR} = \left(\frac{\rho_{Jacobi}}{1+\sqrt{1-\rho_{Jacobi}^2}}\right)^2 \tag{17.5.20}$$

As an application of the above results, consider our model problem for which ρ_{Jacobi} is given by equation (17.5.11). Then equations (17.5.19) and (17.5.20) give

$$\omega \simeq \frac{2}{1+\pi/J} \tag{17.5.21}$$

$$\rho_{SOR} \simeq 1 - \frac{2\pi}{J} \qquad \text{for large} \quad J \tag{17.5.22}$$

Equation (17.5.10) gives for the number of iterations to reduce the initial error by a factor of 10^{-p},

$$r \simeq \frac{pJ\ln 10}{2\pi} \simeq \frac{1}{3}pJ \tag{17.5.23}$$

Comparing with equations (17.5.12) or (17.5.15), we see that optimal SOR requires of order J iterations, as opposed to of order J^2. Since J is typically 100 or even larger, this makes a tremendous difference! Equation (17.5.23) leads to the mnemonic that 3-figure accuracy ($p = 3$) requires a number of iterations equal to the number of mesh points along a side of the grid. For 6-figure accuracy, we require about twice as many iterations.

How do we choose ω for a problem for which the answer is not known analytically? That is just the weak point of SOR! The advantages of SOR obtain only in a fairly narrow window around the correct value of ω. It is better to take ω slightly too large, rather than slightly too small, but best to get it right.

One way to choose ω is to map your problem approximately onto a known problem, replacing the coefficients in the equation by average values. Note, however, that the known problem must have the same grid size and boundary conditions as the actual problem. We give for reference purposes the value of ρ_{Jacobi} for our model problem on a rectangular $J \times L$ grid, allowing for the possibility that $\Delta x \neq \Delta y$:

$$\rho_{Jacobi} = \frac{\cos\frac{\pi}{J} + \left(\frac{\Delta x}{\Delta y}\right)^2 \cos\frac{\pi}{L}}{1 + \left(\frac{\Delta x}{\Delta y}\right)^2} \tag{17.5.24}$$

Equation (17.5.24) holds for homogeneous Dirichlet or Neumann boundary conditions. For periodic boundary conditions, make the replacement $\pi \to 2\pi$.

A second way, which is especially useful if you plan to solve many similar elliptic equations each time with slightly different coefficients, is to determine the optimum value ω empirically on the first equation and then use that value for the remaining equations. Various automated schemes for doing this and for "seeking out" the best values of ω are described in the literature.

While the matrix notation introduced earlier is useful for theoretical analyses, for practical implementation of the SOR algorithm we need explicit formulas. Consider a general second-order elliptic equation in x and y, finite differenced on a square as for our model equation. Corresponding to each row of the matrix $\mathbf{A}$ is an equation of the form

$$a_{j,l}u_{j+1,l} + b_{j,l}u_{j-1,l} + c_{j,l}u_{j,l+1} + d_{j,l}u_{j,l-1} + e_{j,l}u_{j,l} = f_{j,l} \tag{17.5.25}$$

For our model equation, we had $a = b = c = d = 1, e = -4$. The quantity f is proportional to the source term. The iterative procedure is defined by solving (17.5.25) for $u_{j,l}$:

$$u^*{}_{j,l} = \frac{1}{e_{j,l}}\left(f_{j,l} - a_{j,l}u_{j+1,l} - b_{j,l}u_{j-1,l} - c_{j,l}u_{j,l+1} - d_{j,l}u_{j,l-1}\right) \tag{17.5.26}$$

Then $u_{j,l}^{\text{new}}$ is a weighted average

$$u_{j,l}^{\text{new}} = \omega u_{j,l}{}^* + (1-\omega)u_{j,l}^{\text{old}} \tag{17.5.27}$$

We calculate it as follows: The residual at any stage is

$$\xi_{j,l} = a_{j,l}u_{j+1,l} + b_{j,l}u_{j-1,l} + c_{j,l}u_{j,l+1} + d_{j,l}u_{j,l-1} + e_{j,l}u_{j,l} - f_{j,l} \tag{17.5.28}$$

and the SOR algorithm (17.5.18) or (17.5.27) is

$$u_{j,l}^{\text{new}} = u_{j,l}^{\text{old}} - \omega \frac{\xi_{j,l}}{e_{j,l}} \tag{17.5.29}$$

This formulation is very easy to program, and the norm of the residual vector $\xi_{j,l}$ can be used as a criterion for terminating the iteration.

Another practical point concerns the order in which mesh points are processed. The obvious strategy is simply to proceed in order down the rows (or columns). Alternatively, suppose we divide the mesh into "odd" and "even" meshes, like the black and white squares of a chessboard. Then equation (17.5.26) shows that the odd points depend only on the even mesh values and vice versa. Accordingly, we can carry out one half-sweep updating the odd points, say, and then another half-sweep updating the even points with the new odd values. For the version of SOR implemented below, we shall adopt odd-even ordering.

The last practical point is that in practice the asymptotic rate of convergence in SOR is not attained until of order J iterations. The error often grows by a factor of 20 before convergence sets in. A trivial modification to SOR resolves this problem. It is based on the observation that, while ω is the optimum *asymptotic* relaxation parameter, it is not necessarily a good initial choice. In SOR with *Chebyshev acceleration*, one uses odd-even ordering and changes ω at each half-sweep according to the following prescription:

$$\begin{aligned}
\omega^{(0)} &= 1 \\
\omega^{(1/2)} &= 1/(1 - \rho_{Jacobi}^2/2) \\
\omega^{(n+1/2)} &= 1/(1 - \rho_{Jacobi}^2 \omega^{(n)}/4), \qquad n = 1/2, 1, ..., \infty \\
\omega^{(\infty)} &\rightarrow \omega_{optimal}
\end{aligned} \tag{17.5.30}$$

The beauty of Chebyshev acceleration is that the norm of the error always decreases with each iteration. (This is the norm of the actual error in $u_{j,l}$. The norm of the residual $\xi_{j,l}$ need not decrease monotonically.) While the asymptotic rate of convergence is the same as ordinary SOR, there is never any excuse for not using Chebyshev acceleration to reduce the total number of iterations required.

We should also mention the existence of so-called block methods of SOR. These methods can improve convergence by factors like $\sqrt{2}$, but are more complicated to program.

Here we give a routine for SOR with Chebyshev acceleration.

```
      SUBROUTINE SOR(A,B,C,D,E,F,U,JMAX,RJAC)
```

Simultaneous overrelaxation solution of equation (17.5.25) with Chebyshev acceleration. **A**, **B**, **C**, **D**, **E** and **F** are input as the coefficients of the equation, each dimensioned to the grid size **JMAX** × **JMAX**. **U** is input as the initial guess to the solution, usually zero, and returns with the final value. **RJAC** is input as the spectral radius of the Jacobi iteration, or an estimate of it.

```
      IMPLICIT REAL*8(A-H,O-Z)      Double precision is a good idea for JMAX bigger than about 25.
      DIMENSION A(JMAX,JMAX),B(JMAX,JMAX),C(JMAX,JMAX),
     *     D(JMAX,JMAX),E(JMAX,JMAX),F(JMAX,JMAX),U(JMAX,JMAX)
      PARAMETER(MAXITS=1000,EPS=1.D-5,ZERO=0.D0,HALF=.5D0,QTR=.25D0,ONE=1.D0)
      ANORMF=ZERO                   Compute initial norm of residual and terminate iteration when norm
      DO 12 J=2,JMAX-1                  has been reduced by a factor EPS.
        DO 11 L=2,JMAX-1
          ANORMF=ANORMF+ABS(F(J,L))       Assumes initial U is zero.
11      CONTINUE
12    CONTINUE
      OMEGA=ONE
      DO 15 N=1,MAXITS
        ANORM=ZERO
        DO 14 J=2,JMAX-1
          DO 13 L=2,JMAX-1
            IF(MOD(J+L,2).EQ.MOD(N,2))THEN       Odd-even ordering.
              RESID=A(J,L)*U(J+1,L)+B(J,L)*U(J-1,L)+
     *             C(J,L)*U(J,L+1)+D(J,L)*U(J,L-1)+
     *             E(J,L)*U(J,L)-F(J,L)
              ANORM=ANORM+ABS(RESID)
              U(J,L)=U(J,L)-OMEGA*RESID/E(J,L)
            ENDIF
13        CONTINUE
14      CONTINUE
        IF(N.EQ.1) THEN
          OMEGA=ONE/(ONE-HALF*RJAC**2)
        ELSE
          OMEGA=ONE/(ONE-QTR*RJAC**2*OMEGA)
        ENDIF
        IF((N.GT.1).AND.(ANORM.LT.EPS*ANORMF))RETURN
15    CONTINUE
      PAUSE 'MAXITS exceeded'
      END
```

REFERENCES AND FURTHER READING:

Hockney, R.W., and Eastwood, J.W. 1981, *Computer Simulation Using Particles* (New York: McGraw-Hill) Chapter 6.

Stoer, J., and Bulirsch, R. 1980, *Introduction to Numerical Analysis* (New York: Springer-Verlag), §8.3–§8.5.

Varga, R.S. 1962, *Matrix Iterative Analysis* (Englewood Cliffs, N.J.: Prentice-Hall).

Young, D.M. 1971, *Iterative Solution of Large Linear Systems* (New York: Academic Press).

Rigal, A. 1979, *Journal of Computational Physics*, vol. 32, pp. 10–23.

17.6 Operator Splitting Methods and ADI

The basic idea of operator splitting, which is also called *time splitting* or *the method of fractional steps*, is this: Suppose you have an initial value equation of the form

$$\frac{\partial u}{\partial t} = \mathcal{L}u \tag{17.6.1}$$

where $\mathcal{L}$ is some operator. While $\mathcal{L}$ is not necessarily linear, suppose that it can at least be written as a linear sum of m pieces, which act additively on u,

$$\mathcal{L}u = \mathcal{L}_1 u + \mathcal{L}_2 u + \cdots + \mathcal{L}_m u \qquad (17.6.1)$$

Finally, suppose that for *each* of the pieces, you already know a differencing scheme for updating the variable u from timestep n to timestep $n+1$, valid if that piece of the operator were the *only* one on the right-hand side. We will write these updatings symbolically as

$$\begin{aligned} u^{n+1} &= \mathcal{U}_1(u^n, \Delta t) \\ u^{n+1} &= \mathcal{U}_2(u^n, \Delta t) \\ &\cdots \\ u^{n+1} &= \mathcal{U}_m(u^n, \Delta t) \end{aligned} \tag{17.6.3}$$

Now, one form of operator splitting would be to get from n to $n+1$ by the following sequence of updatings:

$$\begin{aligned} u^{n+(1/m)} &= \mathcal{U}_1(u^n, \Delta t) \\ u^{n+(2/m)} &= \mathcal{U}_2(u^{n+(1/m)}, \Delta t) \\ &\cdots \\ u^{n+1} &= \mathcal{U}_m(u^{n+(m-1)/m}, \Delta t) \end{aligned} \tag{17.6.4}$$

For example, a combined advective-diffusion equation, such as

$$\frac{\partial u}{\partial t} = -v\frac{\partial u}{\partial x} + D\frac{\partial^2 u}{\partial x^2} \tag{17.6.5}$$

might profitably use an explicit scheme for the advective term combined with a Crank-Nicholson or other implicit scheme for the diffusion term.

We already introduced the *alternating direction implicit (ADI)* method in §17.3. That is a slightly different kind of operator splitting. Let us reinterpret (17.6.3) to have a different meaning: Let $\mathcal{U}_1$ now denote an updating method that includes algebraically *all* the pieces of the total operator $\mathcal{L}$, but which is desirably *stable* only for the $\mathcal{L}_1$ piece; likewise $\mathcal{U}_2, \dots \mathcal{U}_m$. Then a method of getting from u^n to u^{n+1} is

$$\begin{aligned} u^{n+1/m} &= \mathcal{U}_1(u^n, \Delta t/m) \\ u^{n+2/m} &= \mathcal{U}_2(u^{n+1/m}, \Delta t/m) \\ &\cdots \\ u^{n+1} &= \mathcal{U}_m(u^{n+(m-1)/m}, \Delta t/m) \end{aligned} \tag{17.6.6}$$

The timestep for each fractional step in (17.6.6) is now only $1/m$ of the full timestep, because each partial operation acts with all the terms of the original operator.

Equation (17.6.6) is usually, though not always, stable as a differencing scheme for the operator $\mathcal{L}$. In fact, as a rule of thumb, it is often sufficient to have stable $\mathcal{U}_i$'s only for the operator pieces having the highest number of spatial derivatives – the other $\mathcal{U}_i$'s can be *unstable* – to make the overall scheme stable!

We previously discussed ADI as a method for solving the time-dependent heat-flow equation

$$\frac{\partial u}{\partial t} = \nabla^2 u - \rho \tag{17.6.7}$$

By letting $t \to \infty$ one also gets an iterative method for solving the elliptic equation

$$\nabla^2 u = \rho \tag{17.6.8}$$

In either case, the operator splitting is of the form

$$\mathcal{L} = \mathcal{L}_x + \mathcal{L}_y \tag{17.6.9}$$

where $\mathcal{L}_x$ represents the differencing in x and $\mathcal{L}_y$ that in y.

For example, in our model problem (17.0.6) with $\Delta x = \Delta y = \Delta$, we have

$$\begin{aligned} \mathcal{L}_x u &= 2u_{j,l} - u_{j+1,l} - u_{j-1,l} \\ \mathcal{L}_y u &= 2u_{j,l} - u_{j,l+1} - u_{j,l-1} \end{aligned} \tag{17.6.10}$$

More complicated operators may be similarly split, but there is some art involved. A bad choice of splitting can lead to an algorithm that fails to converge. Usually one tries to base the splitting on the physical nature of the problem. We know for our model problem that an initial transient diffuses away, and we set up the x and y splitting to mimic diffusion in each dimension.

Having chosen a splitting, we difference the time-dependent equation (17.6.7) implicitly in two half-steps:

$$\begin{aligned}\frac{u^{n+1/2}-u^n}{\Delta t/2} &= -\frac{\mathcal{L}_x u^{n+1/2}+\mathcal{L}_y u^n}{\Delta^2}-\rho\\ \frac{u^{n+1}-u^{n+1/2}}{\Delta t/2} &= -\frac{\mathcal{L}_x u^{n+1/2}+\mathcal{L}_y u^{n+1}}{\Delta^2}-\rho\end{aligned} \tag{17.6.11}$$

(cf. equation 17.3.16). Here we have suppressed the spatial indices (j, l). In matrix notation, equations (17.6.11) are

$$(\mathbf{L}_x + r\mathbf{1})\cdot\mathbf{u}^{n+1/2} = (r\mathbf{1}-\mathbf{L}_y)\cdot\mathbf{u}^n - \Delta^2\rho \tag{17.6.12}$$

$$(\mathbf{L}_y + r\mathbf{1})\cdot\mathbf{u}^{n+1} = (r\mathbf{1}-\mathbf{L}_x)\cdot\mathbf{u}^{n+1/2} - \Delta^2\rho \tag{17.6.13}$$

where

$$r \equiv \frac{2\Delta^2}{\Delta t} \tag{17.6.14}$$

The matrices on the left-hand sides of equations (17.6.12) and (17.6.13) are tridiagonal (and usually positive definite), so the equations can be solved by the standard tridiagonal algorithm. Given $\mathbf{u}^n$, one solves (17.6.12) for $\mathbf{u}^{n+1/2}$, substitutes on the right-hand side of (17.6.13), and then solves for $\mathbf{u}^{n+1}$. The key question is how to choose the iteration parameter r, the analog of a choice of timestep for an initial value problem.

As usual, we want to minimize the spectral radius of the iteration matrix. Although it is beyond our scope to go into details here, it turns out that, for the optimal choice of r, the ADI method has the same rate of convergence as does SOR (§17.5). Since the individual iteration steps in the ADI method are much more complicated than in SOR, the ADI method would appear to be inferior. This is certainly true if we choose the same parameter r for every iteration step. The trick is to choose a *different* r for each step.

The determination of an optimal sequence of r's for an arbitrary elliptic problem is an unsolved problem. The answer is known for a set of N iterations if the operators $\mathcal{L}_x$ and $\mathcal{L}_y$ have a common set of eigenvectors, and if the eigenvalues can be bounded by some $\lambda_{\min}$ and $\lambda_{\max}$. Even though this is a restrictive set of assumptions, in practice the same prescription often works satisfactorily on more general problems, so we present it here. Although there are general formulas (in terms of elliptic functions) for arbitrary N, usually we don't know N in advance. We therefore choose a sequence of N r_n's, and then repeat the cycle after N iterations. The formulas for r_n are simplest when N is a power of 2, $N = 2^k$ say, so we might as well choose N in this way.

Suppose the eigenvalues of $\mathcal{L}_x$ and $\mathcal{L}_y$ are bounded by the interval $[\alpha, \beta]$, where $0 < \alpha < \beta$ since the operators are positive definite. The prescription is: Define

$$\alpha_0 = \alpha \quad , \quad \beta_0 = \beta \tag{17.6.15}$$

and recursively compute

$$\alpha_{j+1} = \sqrt{\alpha_j \beta_j}, \qquad \beta_{j+1} = \frac{\alpha_j + \beta_j}{2}, \qquad j = 0, 1, ..., k-1 \tag{17.6.16}$$

Now set

$$s_1^{(0)} = \sqrt{\alpha_k \beta_k} \tag{17.6.17}$$

For each $j = 0, 1, ..., k-1$, recursively compute $s_n^{(j+1)}$ for $n = 1, 2, ..., 2^{j+1}$ as the solutions of the following quadratic equation in x:

$$s_n^{(j)} = \frac{1}{2}\left(x + \frac{\alpha_{k-1-j}\beta_{k-1-j}}{x}\right), \qquad n = 1, 2, ..., 2^j \tag{17.6.18}$$

Here n labels the 2^{j+1} solutions x of equation (17.6.18) for each j. Finally, put

$$r_n = s_n^{(k)}, \qquad n = 1, 2, ..., N = 2^k \tag{17.6.19}$$

The astute reader will recognize that this procedure is related to the arithmetic-geometric mean method for computing elliptic functions.

For the model problem (Poisson equation on a square with Dirichlet boundary conditions), bounds on the eigenvalues turn out to be

$$\begin{aligned} \alpha &= 2\left(1 - \cos\frac{\pi}{J}\right), \\ \beta &= 2\left(1 - \cos\frac{(J-1)\pi}{J}\right) \end{aligned} \tag{17.6.20}$$

One can show using induction that the above prescription leads to a spectral radius of the iteration matrix for a cycle of N iterations

$$\rho^{(N)}(ADI) \simeq 1 - 4\left(\frac{\pi}{4J}\right)^{1/k} \quad , \quad J \to \infty \tag{17.6.21}$$

This should be contrasted with equation (17.5.22). The k'th root in equation (17.6.21) leads to a dramatic increase in the rate of convergence.

For a more complicated problem, one can usually estimate bounds α and β by mapping the problem onto a known model problem, although as before the mapping must be onto a problem with the same types of boundary conditions. The result (17.6.20) holds for homogeneous Dirichlet or Neumann boundary conditions. For periodic boundary conditions, make the replacement $\pi \rightarrow 2\pi$. For an example where the eigenvalues of $\mathcal{L}_x$ are not the same as the eigenvalues of $\mathcal{L}_y$, see Spanier.

To minimize the total number of iterations, one should choose N to be a power of 2 close to $\approx \ln(4J/\pi)$, but the exact value is not crucial. Very roughly, the ADI method requires about $J/(8\log_{10}J)$ less operations than optimal SOR. Reductions in execution time of 6 are common for $J = 100$, while for $J = 1000$ a factor of 40 is possible.

In summary, the method of choice among relaxation methods is ADI. Occasionally there are problems for which ADI does not converge, in which case SOR with Chebyshev acceleration should be tried. SOR is much easier to program than ADI, and this accounts in part for its continued popularity. In an attempt to persuade you that ADI is not that much more complicated, we give below an ADI routine with $\mathcal{L}_x$ and $\mathcal{L}_y$ defined as the following slight generalization of equation (17.6.10):

$$
\begin{aligned}
(\mathcal{L}_x u)_{jl} &= a_{j,l}u_{j-1,l} + b_{j,l}u_{j,l} + c_{j,l}u_{j+1,l} \\
(\mathcal{L}_y u)_{jl} &= d_{j,l}u_{j,l-1} + e_{j,l}u_{j,l} + f_{j,l}u_{j,l+1}
\end{aligned}
\tag{17.6.22}
$$

To cut down on the number of arithmetic operations per iteration, we organize the computation as follows:

- Initialize a vector ψ by

$$
\psi = (r\mathbf{1} - \mathbf{L}_y) \cdot \mathbf{u}^0 \tag{17.6.23}
$$

- For $n = 0, 1, \ldots$, solve

$$
(\mathbf{L}_x + r\mathbf{1}) \cdot \mathbf{u}^{n+1/2} = \psi - \rho\Delta^2 \tag{17.6.24}
$$

  for $\mathbf{u}^{n+1/2}$.
- Set

$$
\phi = -\psi + 2r\mathbf{u}^{n+1/2} \tag{17.6.25}
$$

  (The quantity ϕ can overwrite ψ in storage.)

- Solve

$$(\mathbf{L}_y + r\mathbf{1}) \cdot \mathbf{u}^{n+1} = \phi \qquad (17.6.26)$$

  for $\mathbf{u}^{n+1}$
- If you have not yet converged, set

$$\psi = -\phi + 2r\mathbf{u}^{n+1} \qquad (17.6.27)$$

  and continue from (17.6.24) again.

The ADI routine given below could be made even more efficient, but at the expense of readability. For typical compilers, the SOR routine given earlier is about the same speed as the ADI routine on the model problem for $J = 10$, a factor of 4 slower for $J = 40$, and a factor of 6.5 slower for $J = 100$. These numbers are in agreement with the theoretical estimates quoted earlier.

```
SUBROUTINE ADI(A,B,C,D,E,F,G,U,JMAX,K,ALPHA,BETA,EPS)
    ADI solution of equations (17.6.12) and (17.6.13), with the operators as defined in equa-
    tion (17.6.22). On input, A, B, C, D, E and F contain the coefficients of the equation.
    G contains the right-hand side, while U is input as the initial guess, usually zero. All
    these arrays are dimensioned to the grid size, JMAX × JMAX. The routine carries out 2**K
    iterations with different values of r, and then repeats. ALPHA and BETA are user-supplied
    bounds for the eigenvalues of ℒx and ℒy, while EPS is the desired reduction in the norm of
    the residual. Note that the routine as given requires a double precision version of TRIDAG
    from §2.6.
IMPLICIT REAL*8(A-H,O-Z)      Double precision is a good idea for JMAX bigger than about 25.
PARAMETER(JJ=100,KK=6,NRR=2**(KK-1),MAXITS=100,ZERO=0.D0,TWO=2.D0,HALF=.5D0)
DIMENSION A(JMAX,JMAX),B(JMAX,JMAX),C(JMAX,JMAX),D(JMAX,JMAX),
*      E(JMAX,JMAX),F(JMAX,JMAX),G(JMAX,JMAX),U(JMAX,JMAX),
*      AA(JJ),BB(JJ),CC(JJ),RR(JJ),UU(JJ),PSI(JJ,JJ),
*      ALPH(KK),BET(KK),R(NRR),S(NRR,KK)
IF(JMAX.GT.JJ)PAUSE 'Increase JJ'
IF(K.GT.KK-1)PAUSE 'Increase KK'
K1=K+1
NR=2**K
ALPH(1)=ALPHA                 Determine r's from (17.6.15)-(17.6.19).
BET(1)=BETA
DO 11 J=1,K
    ALPH(J+1)=SQRT(ALPH(J)*BET(J))
    BET(J+1)=HALF*(ALPH(J)+BET(J))
11  CONTINUE
S(1,1)=SQRT(ALPH(K1)*BET(K1))
DO 13 J=1,K
    AB=ALPH(K1-J)*BET(K1-J)
    DO 12 N=1,2**(J-1)
        DISC=SQRT(S(N,J)**2-AB)
        S(2*N,J+1)=S(N,J)+DISC
        S(2*N-1,J+1)=AB/S(2*N,J+1)
12      CONTINUE
13  CONTINUE
DO 14 N=1,NR
    R(N)=S(N,K1)
14  CONTINUE
ANORMG=ZERO                   Compute initial residual, assuming U is zero.
DO 16 J=2,JMAX-1
    DO 15 L=2,JMAX-1
```

```
          ANORMG=ANORMG+ABS(G(J,L))
          PSI(J,L)=-D(J,L)*U(J,L-1)+(R(1)-E(J,L))*U(J,L)
*             -F(J,L)*U(J,L+1)          Equation (17.6.23).
15      CONTINUE
16    CONTINUE
      NITS=MAXITS/NR
      DO 27 KITS=1,NITS
        DO 24 N=1,NR                    Start cycle of 2**K iterations.
          IF(N.EQ.NR)THEN
            NEXT=1
          ELSE
            NEXT=N+1
          ENDIF
          RFACT=R(N)+R(NEXT)            This is "2r" in (17.6.27).
          DO 19 L=2,JMAX-1
            DO 17 J=2,JMAX-1            Solve (17.6.24).
              AA(J-1)=A(J,L)
              BB(J-1)=B(J,L)+R(N)
              CC(J-1)=C(J,L)
              RR(J-1)=PSI(J,L)-G(J,L)
17          CONTINUE
            CALL TRIDAG(AA,BB,CC,RR,UU,JMAX-2)
            DO 18 J=2,JMAX-1
              PSI(J,L)=-PSI(J,L)+TWO*R(N)*UU(J-1)    Equation (17.6.25).
18          CONTINUE
19        CONTINUE
          DO 23 J=2,JMAX-1
            DO 21 L=2,JMAX-1            Solve (17.6.26).
              AA(L-1)=D(J,L)
              BB(L-1)=E(J,L)+R(N)
              CC(L-1)=F(J,L)
              RR(L-1)=PSI(J,L)
21          CONTINUE
            CALL TRIDAG(AA,BB,CC,RR,UU,JMAX-2)
            DO 22 L=2,JMAX-1
              U(J,L)=UU(L-1)            Store current value of solution.
              PSI(J,L)=-PSI(J,L)+RFACT*UU(L-1)     Equation (17.6.27).
22          CONTINUE
23        CONTINUE
24      CONTINUE
        ANORM=ZERO                      Check residual for convergence every 2**K iterations.
        DO 26 J=2,JMAX-1
          DO 25 L=2,JMAX-1
            RESID=A(J,L)*U(J-1,L)+(B(J,L)+E(J,L))*U(J,L)
*                +C(J,L)*U(J+1,L)+D(J,L)*U(J,L-1)
*                +F(J,L)*U(J,L+1)+G(J,L)
            ANORM=ANORM+ABS(RESID)
25        CONTINUE
26      CONTINUE
        IF(ANORM.LT.EPS*ANORMG)RETURN
27    CONTINUE
      PAUSE 'MAXITS exceeded'
      END
```

REFERENCES AND FURTHER READING:

Spanier, J. 1967, in *Mathematical Methods for Digital Computers, Volume 2* (New York: John Wiley), Chapter 11.

Stoer, J., and Bulirsch, R. 1980, *Introduction to Numerical Analysis* (New York: Springer-Verlag), §8.6.

Varga, R.S. 1962, *Matrix Iterative Analysis* (Englewood Cliffs, N.J.: Prentice-Hall).

Young, D.M. 1971, *Iterative Solution of Large Linear Systems* (New York: Academic Press).

References

The references collected here are those of general usefulness, cited in more than one section of this book. More specialized sources, cited in a single section, are not repeated here.

We first list a small number of books which, in our view, are the nucleus of a useful personal reference collection on numerical methods and numerical analysis. These are the books that we like to have within easy reach.

Acton, Forman S. 1970, *Numerical Methods That Work* (New York: Harper and Row)

Ames, W.F. 1977, *Numerical Methods for Partial Differential Equations*, 2nd ed. (New York: Academic Press)

Brigham, E. Oran. 1974, *The Fast Fourier Transform* (Englewood Cliffs, N.J.: Prentice-Hall)

Dahlquist, Germund, and Bjorck, Ake. 1974, *Numerical Methods* (Englewood Cliffs, N.J.: Prentice-Hall)

Forsythe, George E., Malcolm, Michael A., and Moler, Cleve B. 1977, *Computer Methods for Mathematical Computations* (Englewood Cliffs, N.J.: Prentice-Hall)

Gear, C. William. 1971, *Numerical Initial Value Problems in Ordinary Differential Equations* (Englewood Cliffs, N.J.: Prentice-Hall)

Ralston, Anthony, and Rabinowitz, Philip. 1978, *A First Course in Numerical Analysis*, 2nd ed. (New York: McGraw-Hill)

Stoer, J., and Bulirsch, R. 1980, *Introduction to Numerical Analysis* (New York: Springer-Verlag)

Wilkinson, J.H., and Reinsch, C. 1971, *Linear Algebra*, vol. II of *Handbook for Automatic Computation* (New York: Springer-Verlag)

We next list the much larger collection of books, which, in our view, should be included in any serious research library on computing, numerical methods, or analysis.

Abramowitz, Milton, and Stegun, Irene A. 1964, *Handbook of Mathematical Functions*, Applied Mathematics Series, vol. 55 (Washington: National Bureau of Standards; reprinted 1968 by Dover Publications, New York)

Acton, Forman S. 1970, *Numerical Methods That Work* (New York: Harper and Row)

Ames, W.F. 1977, *Numerical Methods for Partial Differential Equations*, 2nd ed. (New York: Academic Press)

Bevington, Philip R. 1969, *Data Reduction and Error Analysis for the Physical Sciences* (New York: McGraw-Hill)

Bloomfield, P. 1976, *Fourier Analysis of Time Series – An Introduction* (New York: Wiley)

Brent, Richard P. 1973, *Algorithms for Minimization without Derivatives* (Englewood Cliffs, N.J.: Prentice-Hall)

Brigham, E. Oran. 1974, *The Fast Fourier Transform* (Englewood Cliffs, N.J.: Prentice-Hall)

Brownlee, K.A. 1965, *Statistical Theory and Methodology*, 2nd ed. (New York: Wiley)

Bunch, J.R., and Rose, D.J. (eds.) 1976, *Sparse Matrix Computations* (New York: Academic Press)

Carnahan, Brice, Luther, H.A., and Wilkes, James O. 1969, *Applied Numerical Methods* (New York: Wiley)

Champeney, D.C. 1973, *Fourier Transforms and Their Physical Applications* (New York: Academic Press)

Childers, Donald G. (ed.). 1978, *Modern Spectrum Analysis* (New York: IEEE Press)

Cooper, L., and Steinberg, D. 1970, *Introduction to Methods of Optimization* (Philadelphia: Saunders)

Dahlquist, Germund, and Bjorck, Ake. 1974, *Numerical Methods* (Englewood Cliffs, N.J.: Prentice-Hall)

Dantzig, G.B. 1963, *Linear Programming and Extensions* (Princeton, N.J.: Princeton University Press)

Dongarra, J.J., et al. 1979, *LINPACK User's Guide* (Philadelphia: Society for Industrial and Applied Mathematics)

Downie, N.M., and Heath, R.W. 1965, *Basic Statistical Methods*, 2nd ed. (New York: Harper and Row)

Duff, I.S., and Stewart, G.W. (eds.) 1979, *Sparse Matrix Proceedings 1978* (Philadelphia: S.I.A.M.)

Elliott, D.F., and Rao, K.R. 1982, *Fast Transforms: Algorithms, Analyses, Applications* (New York: Academic Press)

Fike, C.T. 1968, *Computer Evaluation of Mathematical Functions* (Englewood Cliffs, N.J.: Prentice-Hall)

Forsythe, George E., Malcolm, Michael A., and Moler, Cleve B. 1977, *Computer Methods for Mathematical Computations* (Englewood Cliffs, N.J.: Prentice-Hall)

Forsythe, George E., and Moler, Cleve B. 1967, *Computer Solution of Linear Algebraic Systems* (Englewood Cliffs, N.J.: Prentice-Hall)

Gass, S.T. 1969, *Linear Programming*, 3rd ed. (New York: McGraw-Hill)

Gear, C. William. 1971, *Numerical Initial Value Problems in Ordinary Differential Equations* (Englewood Cliffs, N.J.: Prentice-Hall)

Golub, Gene H., and Van Loan, Charles F. 1983, *Matrix Computations* (Baltimore: Johns Hopkins University Press)

Goodwin, E.T. (ed.) 1961, *Modern Computing Methods*, 2nd ed. (New York: Philosophical Library)

Hamming, R.W. 1962, *Numerical Methods for Engineers and Scientists* (New York: McGraw-Hill)

Hart, John F., et al. 1968, *Computer Approximations* (New York: Wiley)

Hastings, Cecil. 1955, *Approximations for Digital Computers* (Princeton: Princeton University Press)

Hoel, P.G. 1971, *Introduction to Mathematical Statistics*, 4th ed. (New York: Wiley)

Householder, A.S. 1970, *The Numerical Treatment of a Single Nonlinear Equation* (New York: McGraw-Hill)

Huber, P.J. 1981, *Robust Statistics* (New York: Wiley)

IMSL Library Reference Manual, 1980, ed. 8 (IMSL Inc., 7500 Bellaire Boulevard, Houston TX 77036)

Isaacson, Eugene, and Keller, Herbert B. 1966, *Analysis of Numerical Methods* (New York: Wiley)

Jacobs, David A.H., ed. 1977, *The State of the Art in Numerical Analysis* (London: Academic Press)

Johnson, Lee W., and Riess, R. Dean. 1982, *Numerical Analysis*, 2nd ed. (Reading, Mass.: Addison-Wesley)

Keller, Herbert B. 1968, *Numerical Methods for Two-Point Boundary-Value Problems* (Waltham, Mass.: Blaisdell)

Kendall, Maurice, and Stuart, Alan. 1977, *The Advanced Theory of Statistics*, 4th ed. (London: Griffin and Co.)

Knuth, Donald E. 1968, *Fundamental Algorithms*, vol. 1 of *The Art of Computer Programming* (Reading, Mass.: Addison-Wesley)

Knuth, Donald E. 1981, *Seminumerical Algorithms*, 2nd ed., vol. 2 of *The Art of Computer Programming* (Reading, Mass.: Addison-Wesley)

Knuth, Donald E. 1973, *Sorting and Searching*, vol. 3 of *The Art of Computer Programming* (Reading, Mass.: Addison-Wesley)

Korn, G.A., and Korn, T.M. 1968, *Mathematical Handbook for Scientists and Engineers*, 2nd ed. (New York: McGraw-Hill)

Kuenzi, H.P., Tzschach, H.G., and Zehnder, C.A. 1971 *Numerical Methods of Mathematical Optimization* (New York: Academic Press)

Lanczos, Cornelius. 1956, *Applied Analysis* (Englewood Cliffs, N.J.: Prentice-Hall)

Land, A.H., and Powell, S. 1973, *Fortran Codes for Mathematical Programming* (London: Wiley-Interscience)

Launer, R.L., and Wilkinson, G.N., eds. 1979, *Robustness in Statistics* (New York: Academic Press)

Lawson, Charles L., and Hanson, Richard J. 1974, *Solving Least Squares Problems* (Englewood Cliffs, N.J.: Prentice-Hall)

Lehmann, E.L. 1975, *Nonparametrics: Statistical Methods Based on Ranks* (San Francisco: Holden-Day)

Luke, Yudell L. 1975, *Mathematical Functions and Their Approximations* (New York: Academic Press)

Magnus, Wilhelm, and Oberhettinger, Fritz. 1949, *Formulas and Theorems for the Functions of Mathematical Physics* (New York: Chelsea)

Martin, B.R. 1971, *Statistics for Physicists* (New York: Academic Press)

Mathews, Jon, and Walker, R.L. 1970, *Mathematical Methods of Physics*, 2nd ed. (Reading, Mass.: W.A. Benjamin/Addison-Wesley)

von Mises, Richard. 1964, *Mathematical Theory of Probability and Statistics* (New York: Academic Press)

Murty, K.G. 1976, *Linear and Combinatorial Programming* (New York: Wiley)

NAG Fortran Library Manual Mark 8, 1980 (NAG Central Office, 7 Banbury Road, Oxford OX26NN, U.K.)

Nussbaumer, H.J. 1982, *Fast Fourier Transform and Convolution Algorithms* (New York: Springer-Verlag)

Ortega, J., and Rheinboldt, W. 1970, *Iterative Solution of Nonlinear Equations in Several Variables* (New York: Academic Press)

Ostrowski, A.M. 1966, *Solutions of Equations and Systems of Equations*, 2nd ed. (New York: Academic Press)

Polak, E. 1971, *Computational Methods in Optimization* (New York: Academic Press)

Ralston, Anthony, and Rabinowitz, Philip. 1978, *A First Course in Numerical Analysis*, 2nd ed. (New York: McGraw-Hill)

Rice, J.R. 1983, *Numerical Methods, Software, and Analysis* (New York: McGraw-Hill)

Richtmyer, R.D., and Morton, K.W. 1967, *Difference Methods for Initial Value Problems*, 2nd ed. (New York: Wiley-Interscience)

Roache, P.J. 1976, *Computational Fluid Dynamics* (Albuquerque: Hermosa)

Robinson, E.A., and Treitel, S. 1980, *Geophysical Signal Analysis* (Englewood Cliffs, N.J.: Prentice-Hall)

Smith, B.T., et al. 1976, *Matrix Eigensystem Routines — EISPACK Guide*, 2nd ed., vol. 6 of Lecture Notes in Computer Science (New York: Springer-Verlag)

SPSS: Statistical Package for the Social Sciences, 2nd ed., by Norman H. Nie, et al. (New York: McGraw-Hill)

Stoer, J., and Bulirsch, R. 1980, *Introduction to Numerical Analysis* (New York: Springer-Verlag)

Tewarson, R.P. 1973, *Sparse Matrices* (New York: Academic Press)

Westlake, Joan R. 1968, *A Handbook of Numerical Matrix Inversion and Solution of Linear Equations* (New York: Wiley)

Wilkinson, J.H. 1965, *The Algebraic Eigenvalue Problem* (New York: Oxford University Press)

Wilkinson, J.H., and Reinsch, C. 1971, *Linear Algebra*, vol. II of *Handbook for Automatic Computation* (New York: Springer-Verlag)

Numerical Recipes in Pascal

Introduction

This last section of *Numerical Recipes* gives translations into Pascal of all the FORTRAN routines from the main body of the text, Chapters 1–17. We emphasize that the procedures given are translations of the FORTRAN, not Pascal programs written from scratch. This means that, stylistically, a lot of the original language shows through. Pascal programmers with a strong sense of style may feel some discomfort. However, there are two separate reasons for our proceeding in this way: First, since the Pascal versions are uncommented (for reasons of economy in the production of this book), we want the comments and descriptions in the main body of the book to apply to both languages equally. Second, we want to decrease the probability of errors (or, rather, different errors!) in one or the other language.

In fact, much of the translation from FORTRAN to Pascal was done automatically by a computer program written in the language SNOBOL4, a powerful (if old-fashioned) pattern-matching and string processing language. The mechanical "flavor" of the translations will also be evident to a Pascal programmer, for example in the redundant BEGIN...END constructions which appear. A good exercise for the Pascal student is to clean up and "de-FORTRANify" the programs that follow, giving special attention to the use of Pascal's richer set of control structures.

While Pascal has better control structures, it is *not* at present (early 1985) a better language for serious numerical work overall. In some crucial areas, as we will discuss below, "standard" Pascal is not very standardized; so we have had to go to rather elaborate lengths to prepare routines which run correctly on a variety of different machines and compilers. In general, subject to some caveats detailed below, the routines that follow will run on the following systems:

- IBM PC or compatibles running TURBO Pascal
- IBM PC or compatibles running UCSD p-system Pascal
- DEC VAX systems running VMS with VAX-11 Pascal
- IBM mainframes running IBM Pascal/VS

We expect that the programs can be run on many other combinations of computer, operating system, and compiler; but we have not ourselves done so. See the section on customizing, below.

Differences and Deficiencies in Pascal

Certain differences between Pascal and FORTRAN affect how the programs that follow are written and can be used.

In Pascal, a procedure can make use of variable or type identifiers which are neither declared in the procedure nor passed as an argument, *if* such identifiers are declared in the *calling* program or procedure. These identifiers are thus *global* from the point of view of the called procedure, and they can be put to good use. We will use the convention that identifiers beginning with "gl" are global; that is, they must be defined in the calling routine. Whenever we do this, however, we include the required specification statements as comments in the listing of the called procedure. Virtually all of the procedures that follow have such comments.

According to the Pascal standard, local variables become undefined on exit from a procedure — there is no analog of the FORTRAN SAVE statement. In several places, such as in the random number generators, the comment thus instructs you to define certain variables as global that were local in the FORTRAN versions.

By far the most important difference, and notable deficiency, in Pascal concerns so-called *conformant arrays.* In FORTRAN, a subroutine whose argument list includes a matrix A_{ij} is able to pass the dimension of the matrix dynamically,

```
SUBROUTINE DUMMY(A,N)
DIMENSION A(N,N)
```

Here storage for the array A is not set aside in the subroutine. The subroutine uses the storage location of A in the calling program ("call by reference"). If the compiler encounters a reference to A(6,4), say, it can be replaced by the address of A(1,1) plus an offset equal to $5 + 3\text{N}$ (recall that in FORTRAN the first index increases most rapidly). The key benefit is that the subroutine DUMMY can be called at different times with arguments A of different dimensions, a concept as natural to FORTRAN programmers as the air they breathe.

Pascal provides for a standard with two "Levels." The Level 1 standard does in fact provide for conformant arrays by a construction like

```
PROCEDURE dummy(a : ARRAY [1..n,1..n] OF real);
```

whereupon the matrix a and the integer n both become available inside the procedure.

Unfortunately, the Level 1 standard exists, at the time of writing, mostly only on paper. Almost all implementations are of the Level 0 standard, which does not include conformant arrays. In fact, the U.S. domestic (ANSI/IEEE) standard not only fails to go beyond Level 0, it *requires* that any use of conformant arrays be treated "in a manner similar to that specified for errors"!

In the programs following, we have no choice but to assume that the *calling routine* has declared the exact physical dimensions of any array or matrix to be passed, and defined those dimensions in a TYPE identifier, as in

```
TYPE
    glnbyn = ARRAY [1..n,1..n] OF real;
```

Then, and only then, are we able to pass both the array and its size to a procedure. The procedure declaration becomes

```
PROCEDURE dummy (VAR a: glnbyn; n: integer);
```

Things get tricky when you need to use several different procedures together. For example, suppose you are going to call DUMMY and also

```
PROCEDURE dummy2 (VAR b: glmbym; m: integer);
```

where the comment for DUMMY2 instructs you to declare a global type

```
TYPE
    glmbym = ARRAY [1..m,1..m] OF real;
```

If you now declare

```
CONST
    n = 10; m = 10;
VAR
    x : glnbyn;
```

and execute the statements

```
dummy(x,n);
dummy2(x,m);
```

you will get an error message on many Pascal compilers. This is so even though the types glnbyn and glmbym are effectively the same! The solution is to declare

```
TYPE
    glnbyn = ARRAY [1..n,1..n] OF real;
    glmbym = glnbyn;
```

If you plan to use several of the procedures from this book within a single program, you will have to figure out a minimal set of distinct TYPE declarations, and make all other required type declarations equivalent to one or another of these in the above fashion. Don't blame us, blame Pascal!

Things can get even more complicated: For example, the routines SPLIN2 and SPLIE2 call SPLINE, once with a one-dimensional array of length n and another time with an array of length m *where* m *is not equal to* n. In this case we must declare only one type, whose dimension is equal to the *maximum* of n and m, and then pass the value of n or m with each call to SPLINE. Clearly the universal implementation of conformant arrays will be a big plus for numerical computations in Pascal.

Hardly any current Pascal compilers allow you to pass a procedure or function name as an argument of another procedure. Usually we have been able to work around this restriction, and where necessary you will find comments instructing you on what to name certain functions that you must supply. This restriction, however, can be quite an annoyance if you want, e.g., to use the same integration routine to integrate several different functions. Since the ANSI/IEEE standard *does* allow function names as arguments, we expect that it is only a matter of time before implementations fall into line.

The next problem with Pascal, from our point of view, is that the standard does not provide for any mechanism to initialize variables, like the DATA statement in FORTRAN. For a program with a small number of variables, one can just use assignment statements — compare for example the FORTRAN and Pascal versions of QGAUS. In this case, the assignments are executed each time the program is called; the resulting inefficiency is inelegant, but tolerable. If the procedure is to be called many times, efficiency can be improved by making the variables global and initializing them only once in the main program.

This strategy is awkward for initializing large numbers of variables. For example, the procedure CYFUN would require 656 assignment statements! Our make-do solution is to declare the variables global and to have a global flag that tells if the procedure is being called for the first time. If so, we read the variables in from a text file.

There is a general conceptual weakness in Pascal's CONST declaration. While, philosophically, the language insists on explicit declarations and typings, CONSTs are, in effect, implicitly typed. In many implementations, only reals and integers are allowed, with perhaps boolean variables and character strings also settable. We think that, as a minimum, array constants should be provided for.

TURBO Pascal in fact provides for "typed constants" which are as good or better than FORTRAN's DATA statement, and which allow arrays. We approve; but to remain within the present standard, we have not made use of these typed constants in the routines that follow.

This brings us to another deplorable aspect of the Pascal standard: There is no uniform way to associate an external disk file with an internal file and open it for reading. Only a few of the procedures in this book read data files. However many of the test driver routines (available separately, see front of this book) do so. If we had adopted one particular method or another, you would have to make editing changes in a very large number of files.

We have saved you this work by instead adopting *no* particular method. When we want to open an external file and perform a reset on it, we call an (undefined) procedure

```
glopen(infile,'filename');
```

It is up to you to define such a procedure for your system, but you have to do this only once in your main program, not many times in our procedures. Some examples of suitable glopen procedures appear below.

Another area of Pascal that could use some standardization is that of variable types. The types integer and real are always defined, but *what* they define varies widely from system to system.

Many Pascal compilers, especially those for microcomputers, support only 2-byte integers and not 4-byte integers. The standard requires that a predefined identifier maxint be the largest allowed integer; by printing it, you can always determine what kinds of integers you have. (For 2-byte integers, maxint is generally 32768 or 32767.) The calendar routines in Chapter 1 will not work with 2-byte integers, nor will the random number generators RAN1 and RAN2. However, RAN3 can be implemented in real arithmetic, and so we have made this the "default" portable random number generator for our Pascal routines. Accordingly, routines like EXPDEV, POIDEV, and so on call RAN3 in the Pascal versions instead of RAN1. We have also changed the constants in IRBIT1 and IRBIT2 so that the routines will not overflow 2-byte integers. This renders them little more than "demonstration models"; if you have 4-byte integers, change the constants to agree with the FORTRAN versions.

A similar lack of standardization plagues the type real. Most compilers provide only a single level of precision, either with about 11 decimal significant figures (intermediate between FORTRAN's REAL*4 and REAL*8), or else with about 14 decimal significant figures (equivalent to REAL*8). A few compilers provide two levels of precision, with distinct predefined types real (REAL*4)and double (REAL*8). We do not know of a compiler with *only* the equivalent of REAL*4, not quite 8 decimal significant figures, but would not be too surprised to learn of one.

Our FORTRAN routines are generally written with REAL*4 (single precision) real variables. We have used REAL*8 variables as indicators of those relatively rare situations where there is particular sensitivity to roundoff error. To preserve this indication in the Pascal programs, we have used both of the types real and double. If, as is likely, your compiler does not recognize this latter type, it is a simple matter to "define it away" by embedding

```
TYPE
   double = real;
```

in your main program. Likewise, you must "define away" the predefined function for converting a double precision value to single precision, by embedding

```
FUNCTION sngl(x : real): real;
BEGIN sngl := x END;
```

in your main program.

A definite deficiency in Pascal is its lack of a complex data type, along with corresponding complex arithmetic. In the routines below, we do all complex arithmetic "by hand," carrying real and imaginary parts separately.

Customization: Use of MODFILE.PAS

You can implement the customizations recommended thus far in a single, short file, which we will call MODFILE.PAS. Each PROGRAM listed below, and

all the demonstration PROGRAMs in the separately available *Numerical Recipes Example Book*, already contain a statement which includes MODFILE.PAS. You should also include the file in any PROGRAMs that you write, if they reference any of the procedures whose listings follow. Note that you do *not* include MODFILE.PAS in procedures or functions, but only in their embedding main PROGRAMs.

The file MODFILE.PAS should resolve the following two issues: (i) definition of glopen; and (ii) definition of the type double and of the function sngl. Here are some examples of MODFILE.PAS.

For TURBO Pascal on the IBM PC:

```
TYPE
    double = real;
    string10 = string[10];
FUNCTION sngl(x : real): real;
    BEGIN sngl := x END;
PROCEDURE glopen(VAR infile : text; filename : string10);
    BEGIN
        assign(infile,filename);
        reset(infile)
    END;
```

For IBM Pascal/VS on IBM mainframe machines:

```
TYPE
    double = real;
FUNCTION sngl(x : real): real;
    BEGIN sngl := x END;
PROCEDURE glopen(VAR infile : text; filename : string);
    BEGIN
        reset(infile,'NAME='||filename||'.*')
    END;
```

For UCSD Pascal on a machine running UCSD p-system:

```
TYPE
    double = real;
FUNCTION sngl(x : real): real;
    BEGIN sngl := x END;
PROCEDURE glopen(VAR infile : text; filename : string);
    BEGIN
        reset(infile,filename)
    END;
```

For VAX-11 Pascal on a VAX system running VMS:

```
TYPE
    char13 = PACKED ARRAY [1..13] OF char;
PROCEDURE glopen(VAR infile : text; filename : char13);
(* Filename must be exactly 13 characters long, may include trailing blanks. *)
    BEGIN
```

```
        open(infile,filename,old);
        reset(infile)
    END;
```

Format of the Procedure Listings

Pascal allows more than one program statement on an input line. In fact, it pays no attention to how a program is organized on the page, considering "carriage return" as exactly the same as a simple space. However, programs are most readable when there is no more than one statement on a line, and when keywords that denote both the beginning and end of a control structure are prominently displayed. If the economics of book production were not a consideration, we would print all the Pascal routines in a format like the following:

```
PROCEDURE eigsrt(VAR d: glnp; VAR v: glnpnp; n: integer);
(* Programs using routine EIGSRT must define the types
TYPE
    glnp = ARRAY [1..np] OF real;
    glnpnp = ARRAY [1..np,1..np] OF real;
where np is the physical dimension of the arrays to be used (v and d) *)
VAR
    k,j,i: integer;
    p: real;
BEGIN
    FOR i := 1 to n-1 DO BEGIN
        k := i;
        p := d[i];
        FOR j := i+1 to n DO BEGIN
            IF (d[j] >= p) THEN BEGIN
                k := j;
                p := d[j]
            END
        END;
        IF (k <> i) THEN BEGIN
            d[k] := d[i];
            d[i] := p;
            FOR j := 1 to n DO BEGIN
                p := v[j,i];
                v[j,i] := v[j,k];
                v[j,k] := p
            END
        END
    END
END;
```

Unfortunately, there is a close connection between the length of a book and the price that you, as reader, must pay for it. We have therefore adopted a more compact format for the listings that follow. *Within the body of a control block*, we put as many statements on a program line as will fit. (This turns out to save about 15% in the length of the listings.) Also, lines containing only the statement END are wrapped to the preceding line. (This turns out to save about another 15% in length.) The above program thus becomes

```
PROCEDURE eigsrt(VAR d: glnp; VAR v: glnpnp; n: integer);
(* Programs using routine EIGSRT must define the types
TYPE
   glnp = ARRAY [1..np] OF real;
   glnpnp = ARRAY [1..np,1..np] OF real;
where np is the physical dimension of the arrays to be used (v and d) *)
VAR
   k,j,i: integer; p: real;
BEGIN
   FOR i := 1 TO n-1 DO BEGIN
      k := i; p := d[i];
      FOR j := i+1 TO n DO BEGIN
         IF (d[j] >= p) THEN BEGIN
            k := j; p := d[j] END END;
      IF (k <> i) THEN BEGIN
         d[k] := d[i]; d[i] := p;
         FOR j := 1 TO n DO BEGIN
            p := v[j,i]; v[j,i] := v[j,k]; v[j,k] := p END END END
END;
```

In this format, a control structure acts on everything indented more than its first line. Reading along, when you encounter a line at the same level of indentation as the beginning of the preceding control structure, you know that you are out of its block. If you encounter an "outdented" line, you know that you are also out of additional enclosing blocks, and you look up the page to see which ones. Of course, in case of any doubt, you can actually count the ENDs. It takes only a little practice to get used to this convention.

The machine-readable `Pascal` program listings, separately available on diskette, are in the longer format. If you are studying a routine closely, or modifying it, you can print out that format as desired.

We intend to produce a subsequent edition of *Numerical Recipes* with `Pascal` as its principal language. There, the programs will be represented in the longer, clearer, format; they will be in the main body of the text; they will be commented. If you would like to see this done sooner, rather than later, we invite you to make your views known to us, c/o Numerical Recipes Software, at the address listed in the front of this book. Note that an edition of this book in the `C` language is already available.

Chapter 1. Preliminaries

Introduction

```
PROCEDURE flmoon(n,nph: integer; VAR jd: integer; VAR frac: real);
VAR
   i: integer; rad,xtra,t2,t,c,as,am: real;
BEGIN
   rad := 3.14159265/180.0; c := n+nph/4.0; t := c/1236.85; t2 := sqr(t);
   as := 359.2242+29.105356*c; am := 306.0253+385.816918*c+0.010730*t2;
   jd := 2415020+28*n+7*nph;
   xtra := 0.75933+1.53058868*c+(1.178e-4-1.55e-7*t)*t2;
   IF ((nph = 0) OR (nph = 2)) THEN BEGIN
      xtra := xtra+(0.1734-3.93e-4*t)*sin(rad*as)-0.4068*sin(rad*am) END
   ELSE IF ((nph = 1) OR (nph = 3)) THEN BEGIN
      xtra := xtra+(0.1721-4.0e-4*t)*sin(rad*as)-0.6280*sin(rad*am) END
   ELSE BEGIN
      writeln('pause in FLMOON - nph is unknown.'); readln
   END;
   IF (xtra >= 0.0) THEN i := trunc(xtra) ELSE i := trunc(xtra-1.0);
   jd := jd+i; frac := xtra-i
END;
```

Program Organization and Control Structures

```
FUNCTION julday(mm,id,iyyy: integer): integer;
CONST
   igreg=588829;
VAR
   ja,jm,jy,jul: integer;
BEGIN
   IF (iyyy = 0) THEN BEGIN
      writeln('there is no year zero.'); readln; END;
   IF (iyyy < 0) THEN iyyy := iyyy+1;
   IF (mm > 2) THEN BEGIN
      jy := iyyy; jm := mm+1 END
   ELSE BEGIN
      jy := iyyy-1; jm := mm+13 END;
   jul := trunc(365.25*jy)+trunc(30.6001*jm)+id+1720995;
   IF (id+31*(mm+12*iyyy) >= igreg) THEN BEGIN
      ja := trunc(0.01*jy); jul := jul+2-ja+trunc(0.25*ja) END;
   julday := jul
END;
```

```
PROGRAM badluk(input,output);
LABEL 1,2;
CONST
   zon=-5.0; iybeg=1900; iyend=2000;
VAR
   timzon,frac: real; ic,icon,idwk,im: integer; iyyy,jd,jday,n: integer;
(*$I MODFILE.PAS *)
(*$I JULDAY.PAS *)
(*$I FLMOON.PAS *)
BEGIN
   timzon := zon/24.0;
   writeln('Full moons on Friday the 13th from',iybeg:5,' to',iyend:5);
   FOR iyyy := iybeg TO iyend DO BEGIN
      FOR im := 1 TO 12 DO BEGIN
         jday := julday(im,13,iyyy);
         idwk := (jday+1) MOD 7;
         IF (idwk = 5) THEN BEGIN
```

```
                  n := trunc(12.37*(iyyy-1900+(im-0.5)/12.0));
                  icon := 0;
1:                flmoon(n,2,jd,frac);
                  frac := 24.0*(frac+timzon);
                  IF (frac < 0.0) THEN BEGIN
                     jd := jd-1; frac := frac+24.0 END;
                  IF (frac > 12) THEN BEGIN
                     jd := jd+1; frac := frac-12.0 END
                  ELSE BEGIN
                     frac := frac+12.0 END;
                  IF (jd = jday) THEN BEGIN
                     writeln; writeln(im:2,'/',13:2,'/',iyyy:4);
                     writeln('Full moon ',frac:5:1,
                        ' hrs after midnight (EST).');
                     GOTO 2 END
                  ELSE BEGIN
                     IF (jday >= jd) THEN ic := 1 ELSE ic := -1;
                     IF (ic = -icon) THEN GOTO 2;
                     icon := ic; n := n+ic END;
                  GOTO 1;
2:             END
            END END
END.

PROCEDURE caldat(julian: integer; VAR mm,id,iyyy: integer);
CONST
   igreg=2299161;
VAR
   je,jd,jc,jb,jalpha,ja: integer;
BEGIN
   IF (julian >= igreg) THEN BEGIN
      jalpha := trunc(((julian-1867216)-0.25)/36524.25);
      ja := julian+1+jalpha-trunc(0.25*jalpha) END
   ELSE BEGIN
      ja := julian END;
   jb := ja+1524; jc := trunc(6680.0+((jb-2439870)-122.1)/365.25);
   jd := 365*jc+trunc(0.25*jc); je := trunc((jb-jd)/30.6001);
   id := jb-jd-trunc(30.6001*je); mm := je-1;
   IF (mm > 12) THEN mm := mm-12;
   iyyy := jc-4715;
   IF (mm > 2) THEN iyyy := iyyy-1;
   IF (iyyy <= 0) THEN iyyy := iyyy-1
END;
```

Chapter 2. Solution of Linear Algebraic Equations

Gauss-Jordan Elimination

```
PROCEDURE gaussj(VAR a: glnpbynp; n,np: integer;
         VAR b: glnpbymp; m,mp: integer);
(* Programs using GAUSSJ must define the types
TYPE
   glnpbynp = ARRAY [1..np,1..np] OF real;
   glnpbymp = ARRAY [1..np,1..mp] OF real;
   glnp = ARRAY [1..np] OF integer;
in the main routine. *)
VAR
   big,dum,pivinv: real; i,icol,irow,j,k,l,ll: integer;
   indxc,indxr,ipiv: glnp;
BEGIN
```

```
   FOR j := 1 TO n DO BEGIN
      ipiv[j] := 0 END;
   FOR i := 1 TO n DO BEGIN
      big := 0.0;
      FOR j := 1 TO n DO BEGIN
         IF (ipiv[j] <> 1) THEN BEGIN
            FOR k := 1 TO n DO BEGIN
               IF (ipiv[k] = 0) THEN BEGIN
                  IF (abs(a[j,k]) >= big) THEN BEGIN
                     big := abs(a[j,k]); irow := j; icol := k END END
               ELSE IF (ipiv[k] > 1) THEN BEGIN
                  writeln('pause 1 in GAUSSJ - singular matrix'); readln
               END END END END;
      ipiv[icol] := ipiv[icol]+1;
      IF (irow <> icol) THEN BEGIN
         FOR l := 1 TO n DO BEGIN
            dum := a[irow,l]; a[irow,l] := a[icol,l]; a[icol,l] := dum
         END;
         FOR l := 1 TO m DO BEGIN
            dum := b[irow,l]; b[irow,l] := b[icol,l]; b[icol,l] := dum
         END END;
      indxr[i] := irow; indxc[i] := icol;
      IF (a[icol,icol] = 0.0) THEN BEGIN
         writeln('pause 2 in GAUSSJ - singular matrix'); readln
      END;
      pivinv := 1.0/a[icol,icol]; a[icol,icol] := 1.0;
      FOR l := 1 TO n DO BEGIN
         a[icol,l] := a[icol,l]*pivinv END;
      FOR l := 1 TO m DO BEGIN
         b[icol,l] := b[icol,l]*pivinv END;
      FOR ll := 1 TO n DO BEGIN
         IF (ll <> icol) THEN BEGIN
            dum := a[ll,icol]; a[ll,icol] := 0.0;
            FOR l := 1 TO n DO BEGIN
               a[ll,l] := a[ll,l]-a[icol,l]*dum END;
            FOR l := 1 TO m DO BEGIN
               b[ll,l] := b[ll,l]-b[icol,l]*dum END END END END;
   FOR l := n DOWNTO 1 DO BEGIN
      IF (indxr[l] <> indxc[l]) THEN BEGIN
         FOR k := 1 TO n DO BEGIN
            dum := a[k,indxr[l]]; a[k,indxr[l]] := a[k,indxc[l]];
            a[k,indxc[l]] := dum END END END
END;
```

LU Decomposition

```
PROCEDURE ludcmp(VAR a: glnpbynp; n,np: integer;
      VAR indx: glindx; VAR d: real);
(* Programs using LUDCMP must define the types
TYPE
   glnpbynp = ARRAY [1..np,1..np] OF real;
   glnarray = ARRAY [1..n] OF real;
   glindx = ARRAY [1..n] OF integer;
in the main routine. *)
CONST
   tiny=1.0e-20;
VAR
   k,j,imax,i: integer; sum,dum,big: real; vv: glnarray;
BEGIN
   d := 1.0;
   FOR i := 1 TO n DO BEGIN
      big := 0.0;
```

```
        FOR j := 1 TO n DO IF (abs(a[i,j]) > big) THEN big := abs(a[i,j]);
        IF (big = 0.0) THEN BEGIN
            writeln('pause in LUDCMP - singular matrix'); readln
        END;
        vv[i] := 1.0/big END;
    FOR j := 1 TO n DO BEGIN
        FOR i := 1 TO j-1 DO BEGIN
            sum := a[i,j];
            FOR k := 1 TO i-1 DO BEGIN
                sum := sum-a[i,k]*a[k,j] END;
            a[i,j] := sum END;
        big := 0.0;
        FOR i := j TO n DO BEGIN
            sum := a[i,j];
            FOR k := 1 TO j-1 DO BEGIN
                sum := sum-a[i,k]*a[k,j] END;
            a[i,j] := sum;
            dum := vv[i]*abs(sum);
            IF (dum >= big) THEN BEGIN
                big := dum; imax := i END END;
        IF (j <> imax) THEN BEGIN
            FOR k := 1 TO n DO BEGIN
                dum := a[imax,k]; a[imax,k] := a[j,k]; a[j,k] := dum END;
            d := -d; vv[imax] := vv[j] END;
        indx[j] := imax;
        IF (a[j,j] = 0.0) THEN a[j,j] := tiny;
        IF (j <> n) THEN BEGIN
            dum := 1.0/a[j,j];
            FOR i := j+1 TO n DO BEGIN
                a[i,j] := a[i,j]*dum END END END;
END;
```

```
PROCEDURE lubksb(a: glnpbynp; n,np: integer; indx: glindx; VAR b: glnarray);
(* Programs using LUBKSB must define the types
TYPE
    glnarray = ARRAY [1..n] OF real;
    glindx = ARRAY [1..n] OF integer;
    glnpbynp = ARRAY [1..np,1..np] OF real;
in the main routine *)
VAR
    j,ip,ii,i: integer; sum: real;
BEGIN
    ii := 0;
    FOR i := 1 TO n DO BEGIN
        ip := indx[i]; sum := b[ip]; b[ip] := b[i];
        IF (ii <> 0) THEN BEGIN
            FOR j := ii TO i-1 DO BEGIN
                sum := sum-a[i,j]*b[j] END END
        ELSE IF (sum <> 0.0) THEN BEGIN
            ii := i END;
        b[i] := sum END;
    FOR i := n DOWNTO 1 DO BEGIN
        sum := b[i];
        IF (i < n) THEN BEGIN
            FOR j := i+1 TO n DO BEGIN
                sum := sum-a[i,j]*b[j] END END;
        b[i] := sum/a[i,i] END
END;
```

Tridiagonal Systems of Equations

```
PROCEDURE tridag(a,b,c,r: glnarray; VAR u: glnarray; n: integer);
(* Programs using routine TRIDAG should define the type
TYPE
   glnarray = ARRAY [1..n] OF real;
in the main routine. *)
VAR
   j: integer; bet: real; gam: glnarray;
BEGIN
   IF (b[1] = 0.0) THEN BEGIN writeln('pause 1 in TRIDAG'); readln END;
   bet := b[1]; u[1] := r[1]/bet;
   FOR j := 2 TO n DO BEGIN
      gam[j] := c[j-1]/bet; bet := b[j]-a[j]*gam[j];
      IF (bet = 0.0) THEN BEGIN writeln('pause 2 in TRIDAG'); readln END;
      u[j] := (r[j]-a[j]*u[j-1])/bet END;
   FOR j := n-1 DOWNTO 1 DO BEGIN
      u[j] := u[j]-gam[j+1]*u[j+1] END
END;
```

Iterative Improvement of a Solution to Linear Equations

```
PROCEDURE mprove(a,alud: glnpbynp; n,np: integer; indx: glindx;
       b: glnarray; VAR x: glnarray);
(* Programs using routine MPROVE must define the types
TYPE
   glnarray = ARRAY [1..n] OF real;
   glindx = ARRAY [1..n] OF integer;
   glnpbynp = ARRAY [1..np,1..np] OF real;
in the main routine. *)
VAR
   j,i: integer; sdp: double; r: glnarray;
BEGIN
   FOR i := 1 TO n DO BEGIN
      sdp := -b[i];
      FOR j := 1 TO n DO BEGIN
         sdp := sdp+a[i,j]*x[j] END;
      r[i] := sngl(sdp) END;
   lubksb(alud,n,np,indx,r);
   FOR i := 1 TO n DO BEGIN
      x[i] := x[i]-r[i] END
END;
```

Vandermonde Matrices and Toeplitz Matrices

```
PROCEDURE vander(x: glnarray; VAR w: glnarray; q: glnarray; n: integer);
(* Programs using the routine VANDER must define the type
TYPE
   glnarray = ARRAY [1..n] OF real;
in the main routine. *)
CONST
   zero=0.0; one=1.0;
VAR
   k1,k,j,i: integer; xx,t,s,b: real; c: glnarray;
BEGIN
   IF (n = 1) THEN BEGIN
      w[1] := q[1] END
   ELSE BEGIN
      FOR i := 1 TO n DO BEGIN
         c[i] := zero END;
```

```
        c[n] := -x[1];
        FOR i := 2 TO n DO BEGIN
            xx := -x[i];
            FOR j := (n+1-i) TO (n-1) DO BEGIN
                c[j] := c[j]+xx*c[j+1] END;
            c[n] := c[n]+xx END;
        FOR i := 1 TO n DO BEGIN
            xx := x[i]; t := one; b := one; s := q[n]; k := n;
            FOR j := 2 TO n DO BEGIN
                k1 := k-1; b := c[k]+xx*b; s := s+q[k1]*b; t := xx*t+b;
                k := k1 END;
            w[i] := s/t END END
END;

PROCEDURE toeplz(r: gltwon; VAR x: glnarray; y: glnarray; n: integer);
(* Programs using routine TOEPLZ must define the types
    glnarray = ARRAY [1..n] OF real;
    gltwon = ARRAY [1..2*n] OF real;
in the main routine. *)
LABEL 98,99;
VAR
    m2,m1,m,k,j: integer; sxn,shn,sgn,sgd,sd,qt2,qt1,qq,pt2,pt1,pp: real;
    g,h: glnarray;
BEGIN
    IF (r[n] = 0.0) THEN GOTO 99;
    x[1] := y[1]/r[n];
    IF (n = 1) THEN GOTO 99;
    g[1] := r[n-1]/r[n]; h[1] := r[n+1]/r[n];
    FOR m := 1 TO n DO BEGIN
        m1 := m+1; sxn := -y[m1]; sd := -r[n];
        FOR j := 1 TO m DO BEGIN
            sxn := sxn+r[n+m1-j]*x[j]; sd := sd+r[n+m1-j]*g[m-j+1] END;
        IF (sd = 0.0) THEN GOTO 98;
        x[m1] := sxn/sd;
        FOR j := 1 TO m DO BEGIN
            x[j] := x[j]-x[m1]*g[m-j+1] END;
        IF (m1 = n) THEN GOTO 99;
        sgn := -r[n-m1]; shn := -r[n+m1]; sgd := -r[n];
        FOR j := 1 TO m DO BEGIN
            sgn := sgn+r[n+j-m1]*g[j]; shn := shn+r[n+m1-j]*h[j];
            sgd := sgd+r[n+j-m1]*h[m-j+1] END;
        IF ((sd = 0.0) OR (sgd = 0.0)) THEN GOTO 98;
        g[m1] := sgn/sgd; h[m1] := shn/sd; k := m;
        m2 := (m+1) DIV 2;
        pp := g[m1]; qq := h[m1];
        FOR j := 1 TO m2 DO BEGIN
            pt1 := g[j]; pt2 := g[k]; qt1 := h[j]; qt2 := h[k];
            g[j] := pt1-pp*qt2; g[k] := pt2-pp*qt1; h[j] := qt1-qq*pt2;
            h[k] := qt2-qq*pt1; k := k-1 END END;
    writeln('pause in TOEPLZ - should not arrive here!'); readln;
    GOTO 99;
98: writeln('pause in TOEPLZ - Levinson method fails');
    writeln('matrix has a singular principal minor'); readln;
99: END;
```

Singular Value Decomposition

```
PROCEDURE svdcmp(VAR a: glmpbynp; m,n,mp,np: integer;
        VAR w: glnparray; VAR v: glnpbynp);
(* Programs using routine SVDCMP must define the types
TYPE
    glnparray = ARRAY [1..np] OF real;
```

```
    glmpbynp = ARRAY [1..mp,1..np] OF real;
    glnpbynp = ARRAY [1..np,1..np] OF real;
in the main routine. *)
LABEL 1,2,3;
CONST
    nmax=100;
VAR
    nm,l,k,j,jj,its,i: integer; z,y,x,scale,s,h,g,f,c,anorm: real;
    rv1: ARRAY [1..nmax] OF real;
FUNCTION sign(a,b: real): real;
    BEGIN
        IF (b >= 0.0) THEN sign := abs(a) ELSE sign := -abs(a)
    END;
FUNCTION max(a,b: real): real;
    BEGIN
        IF (a > b) THEN max := a ELSE max := b
    END;
BEGIN
    g := 0.0; scale := 0.0; anorm := 0.0;
    FOR i := 1 TO n DO BEGIN
        l := i+1; rv1[i] := scale*g; g := 0.0; s := 0.0; scale := 0.0;
        IF (i <= m) THEN BEGIN
            FOR k := i TO m DO BEGIN
                scale := scale+abs(a[k,i]) END;
            IF (scale <> 0.0) THEN BEGIN
                FOR k := i TO m DO BEGIN
                    a[k,i] := a[k,i]/scale; s := s+a[k,i]*a[k,i] END;
                f := a[i,i]; g := -sign(sqrt(s),f); h := f*g-s;
                a[i,i] := f-g;
                IF (i <> n) THEN BEGIN
                    FOR j := l TO n DO BEGIN
                        s := 0.0;
                        FOR k := i TO m DO BEGIN
                            s := s+a[k,i]*a[k,j] END;
                        f := s/h;
                        FOR k := i TO m DO BEGIN
                            a[k,j] := a[k,j]+ f*a[k,i] END END END;
                FOR k := i TO m DO BEGIN
                    a[k,i] := scale*a[k,i] END END END;
        w[i] := scale*g; g := 0.0; s := 0.0; scale := 0.0;
        IF ((i <= m) AND (i <> n)) THEN BEGIN
            FOR k := l TO n DO BEGIN
                scale := scale+abs(a[i,k]) END;
            IF (scale <> 0.0) THEN BEGIN
                FOR k := l TO n DO BEGIN
                    a[i,k] := a[i,k]/scale; s := s+a[i,k]*a[i,k] END;
                f := a[i,l]; g := -sign(sqrt(s),f); h := f*g-s;
                a[i,l] := f-g;
                FOR k := l TO n DO BEGIN
                    rv1[k] := a[i,k]/h END;
                IF (i <> m) THEN BEGIN
                    FOR j := l TO m DO BEGIN
                        s := 0.0;
                        FOR k := l TO n DO BEGIN
                            s := s+a[j,k]*a[i,k] END;
                        FOR k := l TO n DO BEGIN
                            a[j,k] := a[j,k] +s*rv1[k] END END END;
                FOR k := l TO n DO BEGIN
                    a[i,k] := scale*a[i,k] END END END;
        anorm := max(anorm,(abs(w[i])+abs(rv1[i]))) END;
    FOR i := n DOWNTO 1 DO BEGIN
        IF (i < n) THEN BEGIN
            IF (g <> 0.0) THEN BEGIN
                FOR j := l TO n DO BEGIN
```

```
                    v[j,i] := (a[i,j]/a[i,l])/g END;
                FOR j := 1 TO n DO BEGIN
                    s := 0.0;
                    FOR k := 1 TO n DO BEGIN
                        s := s+a[i,k]*v[k,j] END;
                    FOR k := 1 TO n DO BEGIN
                        v[k,j] := v[k,j]+s*v[k,i] END END END;
            FOR j := l TO n DO BEGIN
                v[i,j] := 0.0; v[j,i] := 0.0 END END;
        v[i,i] := 1.0; g := rv1[i]; l := i END;
    FOR i := n DOWNTO 1 DO BEGIN
        l := i+1; g := w[i];
        IF (i < n) THEN BEGIN
            FOR j := l TO n DO BEGIN
                a[i,j] := 0.0 END END;
        IF (g <> 0.0) THEN BEGIN
            g := 1.0/g;
            IF (i <> n) THEN BEGIN
                FOR j := l TO n DO BEGIN
                    s := 0.0;
                    FOR k := l TO m DO BEGIN
                        s := s+a[k,i]*a[k,j] END;
                    f := (s/a[i,i])*g;
                    FOR k := i TO m DO BEGIN
                        a[k,j] := a[k,j]+f*a[k,i] END END END;
            FOR j := i TO m DO BEGIN
                a[j,i] := a[j,i]*g END END
        ELSE BEGIN
            FOR j := i TO m DO BEGIN
                a[j,i] := 0.0 END END;
        a[i,i] := a[i,i]+1.0 END;
    FOR k := n DOWNTO 1 DO BEGIN
        FOR its := 1 TO 30 DO BEGIN
            FOR l := k DOWNTO 1 DO BEGIN
                nm := l-1;
                IF ((abs(rv1[l])+anorm) = anorm) THEN GOTO 2;
                IF ((abs(w[nm])+anorm) = anorm) THEN GOTO 1
            END;
1:          c := 0.0;
            s := 1.0;
            FOR i := l TO k DO BEGIN
                f := s*rv1[i];
                IF ((abs(f)+anorm) <> anorm) THEN BEGIN
                    g := w[i]; h := sqrt(f*f+g*g); w[i] := h; h := 1.0/h;
                    c := (g*h); s := -(f*h);
                    FOR j := 1 TO m DO BEGIN
                        y := a[j,nm]; z := a[j,i]; a[j,nm] := (y*c)+(z*s);
                        a[j,i] := -(y*s)+(z*c) END END END;
2:          z := w[k];
            IF (l = k) THEN BEGIN
                IF (z < 0.0) THEN BEGIN
                    w[k] := -z;
                    FOR j := 1 TO n DO BEGIN
                    v[j,k] := -v[j,k] END END;
            GOTO 3
            END;
            IF (its = 30) THEN BEGIN
                writeln ('no convergence in 30 SVDCMP iterations'); readln
            END;
            x := w[l]; nm := k-1; y := w[nm]; g := rv1[nm]; h := rv1[k];
            f := ((y-z)*(y+z)+(g-h)*(g+h))/(2.0*h*y); g := sqrt(f*f+1.0);
            f := ((x-z)*(x+z)+h*((y/(f+sign(g,f)))-h))/x; c := 1.0; s := 1.0;
            FOR j := l TO nm DO BEGIN
                i := j+1; g := rv1[i]; y := w[i]; h := s*g; g := c*g;
```

```
                z := sqrt(f*f+h*h); rv1[j] := z; c := f/z; s := h/z;
                f := (x*c)+(g*s); g := -(x*s)+(g*c); h := y*s; y := y*c;
                FOR jj := 1 TO n DO BEGIN
                    x := v[jj,j]; z := v[jj,i]; v[jj,j] := (x*c)+(z*s);
                    v[jj,i] := -(x*s)+(z*c) END;
                z := sqrt(f*f+h*h); w[j] := z;
                IF (z <> 0.0) THEN BEGIN
                    z := 1.0/z; c := f*z; s := h*z END;
                f := (c*g)+(s*y); x := -(s*g)+(c*y);
                FOR jj := 1 TO m DO BEGIN
                    y := a[jj,j]; z := a[jj,i]; a[jj,j] := (y*c)+(z*s);
                    a[jj,i] := -(y*s)+(z*c) END END;
            rv1[l] := 0.0; rv1[k] := f; w[k] := x END;
3:  END
END;

PROCEDURE svbksb(u: glmpbynp; w: glnparray; v: glnpbynp;
        m,n,mp,np: integer; b: glmparray; VAR x: glnparray);
(* Programs using SVBKSB must define the types
TYPE
    glnparray = ARRAY [1..np] OF real;
    glmparray = ARRAY [1..mp] OF real;
    glnpbynp = ARRAY [1..np,1..np] OF real;
    glmpbynp = ARRAY [1..mp,1..np] OF real;
in the main routine. *)
VAR
    jj,j,i: integer; s: real; tmp: glnparray;
BEGIN
    FOR j := 1 TO n DO BEGIN
        s := 0.0;
        IF (w[j] <> 0.0) THEN BEGIN
            FOR i := 1 TO m DO BEGIN
                s := s+u[i,j]*b[i] END;
            s := s/w[j] END;
        tmp[j] := s END;
    FOR j := 1 TO n DO BEGIN
        s := 0.0;
        FOR jj := 1 TO n DO BEGIN
            s := s+v[j,jj]*tmp[jj]; END;
        x[j] := s END
END;
```

Sparse Linear Systems

```
PROCEDURE sparse(b: glnarray; n: integer; VAR x: glnarray; VAR rsq: real);
(* Programs using routine SPARSE must define the type
TYPE
    glnarray = ARRAY [1..n] OF real;
in the main routine. They must also provide two routines,
PROCEDURE asub(x: glnarray; VAR y: glnarray; n: integer);
and
PROCEDURE atsub(x: glnarray; VAR z: glnarray; n: integer);
which calculate A*x and (A transpose)*x *)
LABEL 1,99;
CONST
    eps=1.0e-6;
VAR
    j,iter,irst: integer; rp,gg,gam,eps2,dgg,bsq,anum,aden: real;
    g,h,xi,xj: glnarray;
BEGIN
    eps2 := n*sqr(eps); irst := 0;
1:  irst := irst+1;
```

```
    asub(x,xi,n); rp := 0.0; bsq := 0.0;
    FOR j := 1 TO n DO BEGIN
        bsq := bsq+sqr(b[j]); xi[j] := xi[j]-b[j]; rp := rp+sqr(xi[j]) END;
    atsub(xi,g,n);
    FOR j := 1 TO n DO BEGIN
        g[j] := -g[j]; h[j] := g[j] END;
    FOR iter := 1 TO 10*n DO BEGIN
        asub(h,xi,n); anum := 0.0; aden := 0.0;
        FOR j := 1 TO n DO BEGIN
            anum := anum+g[j]*h[j]; aden := aden+sqr(xi[j]) END;
        IF (aden = 0.0) THEN BEGIN
            writeln('pause in routine SPARSE');
            writeln('very singular matrix'); readln END;
        anum := anum/aden;
        FOR j := 1 TO n DO BEGIN
            xi[j] := x[j]; x[j] := x[j]+anum*h[j] END;
        asub(x,xj,n); rsq := 0.0;
        FOR j := 1 TO n DO BEGIN
            xj[j] := xj[j]-b[j]; rsq := rsq+sqr(xj[j]) END;
        IF ((rsq = rp) OR (rsq <= bsq*eps2)) THEN GOTO 99;
        IF (rsq > rp) THEN BEGIN
            FOR j := 1 TO n DO BEGIN
                x[j] := xi[j] END;
            IF (irst >= 3) THEN GOTO 99;
            GOTO 1
        END;
        rp := rsq; atsub(xj,xi,n); gg := 0.0; dgg := 0.0;
        FOR j := 1 TO n DO BEGIN
            gg := gg+sqr(g[j]); dgg := dgg+(xi[j]+g[j])*xi[j] END;
        IF (gg = 0.0) THEN GOTO 99;
        gam := dgg/gg;
        FOR j := 1 TO n DO BEGIN
            g[j] := -xi[j]; h[j] := g[j]+gam*h[j] END END;
    writeln('pause in routine SPARSE');
    writeln('too many iterations'); readln;
99: END;
```

Chapter 3. Interpolation and Extrapolation

Polynomial Interpolation and Extrapolation

```
PROCEDURE polint(xa,ya: glnarray; n: integer;
        x: real; VAR y,dy: real);
(* Programs using routine POLINT must define the type
TYPE
    glnarray = ARRAY [1..n] OF real;
in the main routine. *)
VAR
    ns,m,i: integer; w,hp,ho,dift,dif,den: real; c,d: glnarray;
BEGIN
    ns := 1; dif := abs(x-xa[1]);
    FOR i := 1 TO n DO BEGIN
        dift := abs(x-xa[i]);
        IF (dift < dif) THEN BEGIN
            ns := i; dif := dift END;
        c[i] := ya[i]; d[i] := ya[i] END;
    y := ya[ns]; ns := ns-1;
    FOR m := 1 TO n-1 DO BEGIN
        FOR i := 1 TO n-m DO BEGIN
            ho := xa[i]-x; hp := xa[i+m]-x; w := c[i+1]-d[i]; den := ho-hp;
            IF (den = 0.0) THEN BEGIN
                writeln ('pause in routine POLINT'); readln END;
```

```
            den := w/den; d[i] := hp*den; c[i] := ho*den END;
        IF ((2*ns) < (n-m)) THEN BEGIN
            dy := c[ns+1] END
        ELSE BEGIN
            dy := d[ns]; ns := ns-1 END;
        y := y+dy END
END;
```

Rational Function Interpolation and Extrapolation

```
PROCEDURE ratint(xa,ya: glnarray; n: integer; x: real; VAR y,dy: real);
(* Programs using routine RATINT must define the type
TYPE
    glnarray = ARRAY [1..n] OF real;
in the main routine. *)
LABEL 99;
CONST
    tiny=1.0e-25;
VAR
    ns,m,i: integer; w,t,hh,h,dd: real; c,d: glnarray;
BEGIN
    ns := 1; hh := abs(x-xa[1]);
    FOR i := 1 TO n DO BEGIN
        h := abs(x-xa[i]);
        IF (h = 0.0) THEN BEGIN
            y := ya[i]; dy := 0.0;
            GOTO 99 END
        ELSE IF (h < hh) THEN BEGIN
            ns := i; hh := h END;
        c[i] := ya[i]; d[i] := ya[i]+tiny END;
    y := ya[ns]; ns := ns-1;
    FOR m := 1 TO n-1 DO BEGIN
        FOR i := 1 TO n-m DO BEGIN
            w := c[i+1]-d[i]; h := xa[i+m]-x; t := (xa[i]-x)*d[i]/h;
            dd := t-c[i+1];
            IF (dd = 0.0) THEN BEGIN
                writeln('pause in routine RATINT'); readln
            END;
            dd := w/dd; d[i] := c[i+1]*dd; c[i] := t*dd END;
        IF (2*ns < n-m) THEN BEGIN
            dy := c[ns+1] END
        ELSE BEGIN
            dy := d[ns]; ns := ns-1 END;
        y := y+dy END;
99: END;
```

Cubic Spline Interpolation

```
PROCEDURE spline(x,y: glnarray; n: integer; yp1,ypn: real;
        VAR y2: glnarray);
(* Programs using routine SPLINE must define the type
TYPE
    glnarray = ARRAY [1..n] OF real;
in the main routine. *)
VAR
    i,k: integer; p,qn,sig,un: real; u: glnarray;
BEGIN
    IF (yp1 > 0.99e30) THEN BEGIN
        y2[1] := 0.0; u[1] := 0.0 END
    ELSE BEGIN
```

```
        y2[1] := -0.5; u[1] := (3.0/(x[2]-x[1]))*((y[2]-y[1])/(x[2]-x[1])-yp1)
    END;
    FOR i := 2 TO n-1 DO BEGIN
        sig := (x[i]-x[i-1])/(x[i+1]-x[i-1]); p := sig*y2[i-1]+2.0;
        y2[i] := (sig-1.0)/p; u[i] := (y[i+1]-y[i])/(x[i+1]-x[i])
            -(y[i]-y[i-1])/(x[i]-x[i-1]);
        u[i] := (6.0*u[i]/(x[i+1]-x[i-1])-sig*u[i-1])/p END;
    IF (ypn > 0.99e30) THEN BEGIN
        qn := 0.0; un := 0.0 END
    ELSE BEGIN
        qn := 0.5; un := (3.0/(x[n]-x[n-1]))*(ypn-(y[n]-y[n-1])/(x[n]-x[n-1]))
    END;
    y2[n] := (un-qn*u[n-1])/(qn*y2[n-1]+1.0);
    FOR k := n-1 DOWNTO 1 DO BEGIN
        y2[k] := y2[k]*y2[k+1]+u[k] END
END;


PROCEDURE splint(xa,ya,y2a: glnarray; n: integer;
         x: real; VAR y: real);
(* Programs using routine SPLINT must define the type
TYPE
    glnarray = ARRAY [1..n] OF real;
in the main routine. *)
VAR
    klo,khi,k: integer; h,b,a: real;
BEGIN
    klo := 1; khi := n;
    WHILE (khi-klo > 1) DO BEGIN
        k := (khi+klo) DIV 2;
        IF (xa[k] > x) THEN khi := k ELSE klo := k
    END;
    h := xa[khi]-xa[klo];
    IF (h = 0.0) THEN BEGIN
        writeln ('pause in routine SPLINT');
        writeln (' ... bad XA input'); readln END;
    a := (xa[khi]-x)/h; b := (x-xa[klo])/h; y := a*ya[klo]+b*ya[khi]+
        ((a*a*a-a)*y2a[klo]+(b*b*b-b)*y2a[khi])*(h*h)/6.0
END;
```

How to Search an Ordered Table

```
PROCEDURE locate(xx: glnarray; n: integer;
           x: real; VAR j: integer);
(* Programs which use routine LOCATE must define the type
TYPE
    glnarray = ARRAY [1..n] OF real;
in the main routine. *)
VAR
    ju,jm,jl: integer;
BEGIN
    jl := 0; ju := n+1;
    WHILE (ju-jl > 1) DO BEGIN
        jm := (ju+jl) DIV 2;
        IF ((xx[n] > xx[1]) = (x > xx[jm])) THEN jl := jm
        ELSE ju := jm
    END;
    j := jl
END;
```

```
PROCEDURE hunt(xx: glnarray; n: integer;
      x: real; VAR jlo: integer);
(* Programs using routine HUNT must define the type
TYPE
   glnarray = ARRAY [1..n] OF real;
in the main routine. *)
LABEL 1,2,3,4;
VAR
   jm,jhi,inc: integer; ascnd: boolean;
BEGIN
   ascnd := xx[n] > xx[1];
   IF ((jlo <= 0) OR (jlo > n)) THEN BEGIN
      jlo := 0; jhi := n+1;
      GOTO 3
   END;
   inc := 1;
   IF ((x >= xx[jlo]) = ascnd ) THEN BEGIN
1:    jhi := jlo+inc;
      IF (jhi > n) THEN BEGIN
         jhi := n+1 END
      ELSE IF ((x >= xx[jhi]) = ascnd ) THEN BEGIN
         jlo := jhi; inc := inc+inc;
         GOTO 1
      END END
   ELSE BEGIN
      jhi := jlo;
2:    jlo := jhi-inc;
      IF (jlo < 1) THEN BEGIN
         jlo := 0 END
      ELSE IF ((x < xx[jlo]) = ascnd ) THEN BEGIN
         jhi := jlo; inc := inc+inc;
         GOTO 2
      END END;
3: IF ((jhi-jlo) = 1) THEN GOTO 4;
   jm := (jhi+jlo) DIV 2;
   IF ((x > xx[jm]) = ascnd ) THEN BEGIN
      jlo := jm END
   ELSE BEGIN
      jhi := jm END;
   GOTO 3;
4:
END;
```

Coefficients of the Interpolating Polynomial

```
PROCEDURE polcoe(x,y: glnarray; n: integer; VAR cof: glnarray);
(* Programs using routine POLCOE must define the type
TYPE
   glnarray = ARRAY [1..narray] OF real;
in the main routine. *)
VAR
   k,j,i: integer; phi,ff,b: real; s: glnarray;
BEGIN
   FOR i := 1 TO n DO BEGIN
      s[i] := 0.0; cof[i] := 0.0 END;
   s[n] := -x[1];
   FOR i := 2 TO n DO BEGIN
      FOR j := n+1-i TO n-1 DO BEGIN
         s[j] := s[j]-x[i]*s[j+1] END;
      s[n] := s[n]-x[i] END;
   FOR j := 1 TO n DO BEGIN
      phi := n;
```

```
        FOR k := n-1 DOWNTO 1 DO BEGIN
            phi := k*s[k+1]+x[j]*phi END;
        ff := y[j]/phi; b := 1.0;
        FOR k := n DOWNTO 1 DO BEGIN
            cof[k] := cof[k]+b*ff; b := s[k]+x[j]*b END END
END;
```

```
PROCEDURE polcof(xa,ya: glnarray; n: integer; VAR cof: glnarray);
(* Programs using routine POLCOF must define the type
TYPE
    glnarray = ARRAY [1..n] OF real;
in the main routine. *)
VAR
    k,j,i: integer; xmin,dy: real; x,y: glnarray;
BEGIN
    FOR j := 1 TO n DO BEGIN
        x[j] := xa[j]; y[j] := ya[j] END;
    FOR j := 1 TO n DO BEGIN
        polint(x,y,n+1-j,0.0,cof[j],dy); xmin := 1.0E38; k := 0;
        FOR i := 1 TO n+1-j DO BEGIN
            IF (abs(x[i]) < xmin) THEN BEGIN
                xmin := abs(x[i]); k := i END;
            IF (x[i] <> 0.0) THEN y[i] := (y[i]-cof[j])/x[i]
        END;
        IF (k < (n+1-j)) THEN BEGIN
            FOR i := k+1 TO n+1-j DO BEGIN
                y[i-1] := y[i]; x[i-1] := x[i] END END END
END;
```

Interpolation in Two or More Dimensions

```
PROCEDURE polin2(x1a: glmarray; x2a: glnarray; ya: glmbyn;
        m,n: integer; x1,x2: real; VAR y,dy: real);
(* Programs using POLIN2 must define the types
TYPE
    glmarray = ARRAY [1..m] OF real;
    glnarray = ARRAY [1..n] OF real;
    glmbyn = ARRAY [1..m,1..n] OF real;
in the main routine. *)
VAR
    k,j: integer; ymtmp: glmarray; yntmp: glnarray;
BEGIN
    FOR j := 1 TO m DO BEGIN
        FOR k := 1 TO n DO BEGIN
            yntmp[k] := ya[j,k] END;
        polint(x2a,yntmp,n,x2,ymtmp[j],dy) END;
    polint(x1a,ymtmp,m,x1,y,dy)
END;
```

```
PROCEDURE bcucof(y,y1,y2,y12: gl4array; d1,d2: real; VAR c: gl4by4);
(* Programs using routine BCUCOF must define the types
TYPE
    gl4array = ARRAY [1..4] OF real;
    gl4by4 = ARRAY [1..4,1..4] OF real;
in the main routine. They must also declare the variables
VAR
    glflag: boolean;
    wt: ARRAY [1..16,1..16] OF real;
and initialize glflag to true. The values of wt are read from the file
bcucof.dat whose contents are listed at the end of this routine. The procedure
GLOPEN assigns bcucof.dat to infile and opens the file for reading. *)
```

```
VAR
    l,k,j,i: integer; xx,d1d2: real;
    cl,x: ARRAY[1..16] OF real;
    infile: text;
BEGIN
    IF glflag THEN BEGIN
        glflag := FALSE;
        glopen(infile,'bcucof.dat');
        FOR i := 1 TO 16 DO FOR k := 1 to 16 DO read(infile,wt[k,i]);
        close(infile) END;
    d1d2 := d1*d2;
    FOR i := 1 TO 4 DO BEGIN
        x[i] := y[i]; x[i+4] := y1[i]*d1; x[i+8] := y2[i]*d2;
        x[i+12] := y12[i]*d1d2 END;
    FOR i := 1 TO 16 DO BEGIN
        xx := 0.0;
        FOR k := 1 TO 16 DO xx := xx+wt[i,k]*x[k];
        cl[i] := xx END;
    l := 0;
    FOR i := 1 TO 4 DO
        FOR j := 1 TO 4 DO BEGIN
            l := l+1; c[i,j] := cl[l] END
END;
(* Contents of the file bcucof.dat
1 0 -3 2 0 0 0 0 -3 0 9 -6 2 0 -6 4 0 0 0 0 0 0 0 0 3 0 -9 6 -2 0 6 -4
0 0 0 0 0 0 0 0 0 0 9 -6 0 0 -6 4 0 0 3 -2 0 0 0 0 0 0 -9 6 0 0 6 -4
0 0 0 0 1 0 -3 2 -2 0 6 -4 1 0 -3 2 0 0 0 0 0 0 0 0 -1 0 3 -2 1 0 -3 2
0 0 0 0 0 0 0 0 0 0 -3 2 0 0 3 -2 0 0 0 0 0 0 3 -2 0 0 -6 4 0 0 3 -2
0 1 -2 1 0 0 0 0 0 -3 6 -3 0 2 -4 2 0 0 0 0 0 0 0 0 0 3 -6 3 0 -2 4 -2
0 0 0 0 0 0 0 0 0 0 -3 3 0 0 2 -2 0 0 -1 1 0 0 0 0 0 0 3 -3 0 0 -2 2
0 0 0 0 0 1 -2 1 0 -2 4 -2 0 1 -2 1 0 0 0 0 0 0 0 0 0 -1 2 -1 0 1 -2 1
0 0 0 0 0 0 0 0 0 0 1 -1 0 0 -1 1 0 0 0 0 0 0 -1 1 0 0 2 -2 0 0 -1 1 *)


PROCEDURE bcuint(y,y1,y2,y12: gl4array; x1l,x1u,x2l,x2u,x1,x2: real;
        VAR ansy,ansy1,ansy2: real);
(* Programs using procedure BCUINT must define the types
TYPE
    gl4array = ARRAY [1..4] OF real;
    gl4by4 = ARRAY [1..4,1..4] OF real;
in the main routine. *)
VAR
    i: integer; t,u,d1,d2: real; c: gl4by4;
BEGIN
    d1 := x1u-x1l; d2 := x2u-x2l; bcucof(y,y1,y2,y12,d1,d2,c);
    IF ((x1u = x1l) OR (x2u = x2l)) THEN BEGIN
        writeln('pause in routine BCUINT - bad input'); readln
    END;
    t := (x1-x1l)/d1; u := (x2-x2l)/d2; ansy := 0.0; ansy2 := 0.0;
    ansy1 := 0.0;
    FOR i := 4 DOWNTO 1 DO BEGIN
        ansy := t*ansy+((c[i,4]*u+c[i,3])*u+c[i,2])*u+c[i,1];
        ansy2 := t*ansy2+(3.0*c[i,4]*u+2.0*c[i,3])*u+c[i,2];
        ansy1 := u*ansy1+(3.0*c[4,i]*t+2.0*c[3,i])*t+c[2,i] END;
    ansy1 := ansy1/d1; ansy2 := ansy2/d2
END;


PROCEDURE splie2(x1a,x2a: glnarray; ya: glmbyn;
        m,n: integer; VAR y2a: glmbyn);
(* Programs using routine SPLIE2 must define the types
TYPE
    glnarray = ARRAY [1..nn] OF real;
    glmbyn = ARRAY [1..m,1..n] OF real;
in the main routine. The dimension nn of glnarray must be set to the larger
```

```
of n and m. *)
VAR
    k,j: integer; ytmp,y2tmp: glnarray;
BEGIN
    FOR j := 1 TO m DO BEGIN
        FOR k := 1 TO n DO BEGIN
            ytmp[k] := ya[j,k] END;
        spline(x2a,ytmp,n,1.0e30,1.0e30,y2tmp);
        FOR k := 1 TO n DO BEGIN
            y2a[j,k] := y2tmp[k] END END
END;
```

```
PROCEDURE splin2(x1a,x2a: glnarray; ya,y2a: glmbyn;
         m,n: integer; x1,x2: real; VAR y: real);
(* Programs using routine SPLIN2 must define the types
TYPE
    glnarray = ARRAY [1..nn] OF real;
    glmbyn = ARRAY [1..m,1..n] OF real;
in the main routine. The dimension nn of glnarray must be set
to the larger of n and m *)
VAR
    k,j: integer; ytmp,y2tmp,yytmp: glnarray;
BEGIN
    FOR j := 1 TO m DO BEGIN
        FOR k := 1 TO n DO BEGIN
            ytmp[k] := ya[j,k]; y2tmp[k] := y2a[j,k] END;
        splint(x2a,ytmp,y2tmp,n,x2,yytmp[j]) END;
    spline(x1a,yytmp,m,1.0e30,1.0e30,y2tmp); splint(x1a,yytmp,y2tmp,m,x1,y)
END;
```

Chapter 4. Integration of Functions

Elementary Algorithms

```
PROCEDURE trapzd(a,b: real; VAR s: real; n: integer);
(* Programs calling TRAPZD must provide a function
func(x:real):real which is to be integrated. They must
also define the variable
VAR
    glit: integer;
in the main routine. *)
VAR
    j: integer; x,tnm,sum,del: real;
BEGIN
    IF (n = 1) THEN BEGIN
        s := 0.5*(b-a)*(func(a)+func(b)); glit := 1 END
    ELSE BEGIN
        tnm := glit; del := (b-a)/tnm; x := a+0.5*del; sum := 0.0;
        FOR j := 1 TO glit DO BEGIN
            sum := sum+func(x); x := x+del END;
        s := 0.5*(s+(b-a)*sum/tnm); glit := 2*glit END
END;
```

```
PROCEDURE qtrap(a,b: real;VAR s: real);
LABEL 99;
CONST
    eps=1.0e-6; jmax=20;
VAR
    j: integer; olds: real;
BEGIN
    olds := -1.0e30;
    FOR j := 1 TO jmax DO BEGIN
        trapzd(a,b,s,j);
        IF (abs(s-olds) < eps*abs(olds)) THEN GOTO 99;
        olds := s END;
    writeln ('pause in QTRAP - too many steps'); readln;
99: END;

PROCEDURE qsimp(a,b: real; VAR s: real);
LABEL 99;
CONST
    eps=1.0e-6; jmax=20;
VAR
    j: integer; st,ost,os: real;
BEGIN
    ost := -1.0e30; os := -1.0e30;
    FOR j := 1 TO jmax DO BEGIN
        trapzd(a,b,st,j); s := (4.0*st-ost)/3.0;
        IF (abs(s-os) < eps*abs(os)) THEN GOTO 99;
        os := s; ost := st END;
    writeln ('pause in QSIMP - too many steps'); readln;
99: END;
```

Romberg Integration

```
PROCEDURE qromb(a,b: real; VAR ss: real);
(* Programs using routine QROMB must define type
TYPE
    glnarray = ARRAY [1..n] OF real;
just as for routine POLINT which it calls. In this case
n should have a value no smaller than the constant k below. *)
LABEL 99;
CONST
    eps=1.0e-6; jmax=20;
    jmaxp=21;    (* jmax+1 *)
    k=5;
VAR
    i,j: integer; dss: real;
    h,s: ARRAY[1..jmaxp] OF real;
    c,d: glnarray;
BEGIN
    h[1] := 1.0;
    FOR j := 1 TO jmax DO BEGIN
        trapzd(a,b,s[j],j);
        IF (j >= k) THEN BEGIN
            FOR i := 1 TO k DO BEGIN
                c[i] := h[j-k+i]; d[i] := s[j-k+i] END;
            polint(c,d,k,0.0,ss,dss);
            IF (abs(dss) < eps*abs(ss)) THEN GOTO 99
        END;
        s[j+1] := s[j]; h[j+1] := 0.25*h[j] END;
    writeln('pause in QROMB - too many steps'); readln;
99: END;
```

Improper Integrals

```
PROCEDURE midpnt(a,b: real; VAR s: real; n: integer);
(* Programs using routine MIDPNT must supply a function
func(x:real):real which is to be integrated. They must also
declare an iteration counter
VAR
    glit: integer; *)
VAR
    j: integer; x,tnm,sum,del,ddel: real;
BEGIN
    IF (n = 1) THEN BEGIN
        s := (b-a)*func(0.5*(a+b)); glit := 1 END
    ELSE BEGIN
        tnm := glit; del := (b-a)/(3.0*tnm); ddel := del+del;
        x := a+0.5*del; sum := 0.0;
        FOR j := 1 TO glit DO BEGIN
            sum := sum+func(x); x := x+ddel; sum := sum+func(x); x := x+del
        END;
        s := (s+(b-a)*sum/tnm)/3.0; glit := 3*glit END
END;
```

```
PROCEDURE qromo(a,b: real; VAR ss: real);
(* Programs using routine QROMO must define the type
TYPE
    glnarray = ARRAY [1..n] OF real;
in the main routine, where n is equal to the constant k below. The routine
func(x:real):real in the calling routine must return the value of the function
to be integrated. You must choose MIDPNT, MIDSQL, MIDSQU or MIDINF at the
indicated point below *)
LABEL 99;
CONST
    eps=1.0e-6; jmax=14;
    jmaxp=15;    (* jmaxp=jmax+1 *)
    k=5;
    km=4;        (* km=k-1 *)
VAR
    i,j: integer; dss: real;
    h,s: ARRAY [1..jmaxp] OF real;
    c,d: glnarray;
BEGIN
    h[1] := 1.0;
    FOR j := 1 TO jmax DO BEGIN
(* Here you must choose the appropriate integration method *)
        midsql(a,b,s[j],j);
        IF (j >= k) THEN BEGIN
            FOR i := 1 TO k DO BEGIN
                c[i] := h[j-k+i]; d[i] := s[j-k+i] END;
            polint(c,d,k,0.0,ss,dss);
            IF (abs(dss) < (eps*abs(ss))) THEN GOTO 99
        END;
        s[j+1] := s[j]; h[j+1] := h[j]/9.0 END;
    writeln('pause in QROMO - too many steps');
    readln;
99: END;
```

```
PROCEDURE midinf(aa,bb: real; VAR s: real; n: integer);
(* Programs using MIDINF must define the function to be integrated
with the declaration FUNCTION func(x: real): real; They must also declare
the global variable
VAR
    glit: integer; *)
VAR
    j: integer; x,tnm,sum,del,ddel,b,a: real;
FUNCTION funk(x: real): real;
    BEGIN
        funk := func(1.0/x)/sqr(x) END;
BEGIN
    b := 1.0/aa; a := 1.0/bb;
    IF (n = 1) THEN BEGIN
        s := (b-a)*funk(0.5*(a+b)); glit := 1 END
    ELSE BEGIN
        tnm := glit; del := (b-a)/(3.0*tnm); ddel := del+del;
        x := a+0.5*del; sum := 0.0;
        FOR j := 1 TO glit DO BEGIN
            sum := sum+funk(x); x := x+ddel; sum := sum+funk(x); x := x+del
        END;
        s := (s+(b-a)*sum/tnm)/3.0; glit := 3*glit END
END;
```

```
PROCEDURE midsql(aa,bb: real; VAR s: real; n: integer);
(* Programs using MIDSQL must define the function to be integrated
with the declaration FUNCTION func(x: real): real; They must also declare
the global variable
VAR
    glit: integer; *)
VAR
    j: integer; x,tnm,sum,del,ddel,b,a: real;
FUNCTION funk(x: real): real;
    BEGIN
        funk := 2.0*x*func(aa+sqr(x)) END;
BEGIN
    b := sqrt(bb-aa); a := 0.0;
    IF (n = 1) THEN BEGIN
        s := (b-a)*funk(0.5*(a+b)); glit := 1 END
    ELSE BEGIN
        tnm := glit; del := (b-a)/(3.0*tnm); ddel := del+del;
        x := a+0.5*del; sum := 0.0;
        FOR j := 1 TO glit DO BEGIN
            sum := sum+funk(x); x := x+ddel; sum := sum+funk(x); x := x+del
        END;
        s := (s+(b-a)*sum/tnm)/3.0; glit := 3*glit END
END;
```

```
PROCEDURE midsqu(aa,bb: real; VAR s: real; n: integer);
(* Programs using MIDSQU must define the function to be integrated
with the declaration FUNCTION func(x: real): real; They must also declare
the global variable
VAR
    glit: integer; *)
VAR
    j: integer; x,tnm,sum,del,ddel,b,a: real;
FUNCTION funk(x: real): real;
    BEGIN
        funk := 2.0*x*func(bb-sqr(x)) END;
BEGIN
    b := sqrt(bb-aa); a := 0.0;
```

```
    IF (n = 1) THEN BEGIN
        s := (b-a)*funk(0.5*(a+b)); glit := 1 END
    ELSE BEGIN
        tnm := glit; del := (b-a)/(3.0*tnm); ddel := del+del;
        x := a+0.5*del; sum := 0.0;
        FOR j := 1 TO glit DO BEGIN
            sum := sum+funk(x); x := x+ddel; sum := sum+funk(x); x := x+del
        END;
        s := (s+(b-a)*sum/tnm)/3.0; glit := 3*glit END
END;
```

Gaussian Quadratures

```
PROCEDURE qgaus(a,b: real; VAR ss: real);
(* Programs using routine QGAUS must externally define
a function func(x:real):real which is to be integrated *)
VAR
    j: integer; xr,xm,dx: real;
    w,x: ARRAY[1..5] OF real;
BEGIN
    x[1] := 0.1488743389; x[2] := 0.4333953941; x[3] := 0.6794095682;
    x[4] := 0.8650633666; x[5] := 0.97390652; w[1] := 0.2955242247;
    w[2] := 0.2692667193; w[3] := 0.2190863625; w[4] := 0.1494513491;
    w[5] := 0.06667134; xm := 0.5*(b+a); xr := 0.5*(b-a); ss := 0;
    FOR j := 1 TO 5 DO BEGIN
        dx := xr*x[j]; ss := ss+w[j]*(func(xm+dx)+func(xm-dx)) END;
    ss := xr*ss
END;
```

```
PROCEDURE gauleg(x1,x2: double; VAR x,w: darray; n: integer);
(* Programs using routine GAULEG must define the type
TYPE
    darray=ARRAY [1..n] OF double;
in the calling program *)
CONST
    eps = 3.0e-11; (* adjust to your floating precision *)
VAR
      m,j,i: integer; z1,z,xm,xl,pp,p3,p2,p1: double;
BEGIN
    m := (n+1) DIV 2;
    xm := 0.5*(x2+x1); xl := 0.5*(x2-x1);
    FOR i := 1 TO m DO BEGIN
        z := cos(3.141592654*(i-0.25)/(n+0.5));
        REPEAT
                  p1 := 1.0; p2 := 0.0;
                  FOR j := 1 TO n DO BEGIN
                     p3 := p2; p2 := p1; p1 := ((2.0*j-1.0)*z*p2-(j-1.0)*p3)/j
                  END;
                  pp := n*(z*p1-p2)/(z*z-1.0); z1 := z; z := z1-p1/pp;
        UNTIL (abs(z-z1) <= eps);
        x[i] := xm-xl*z; x[n+1-i] := xm+xl*z;
        w[i] := 2.0*xl/((1.0-z*z)*pp*pp); w[n+1-i] := w[i] END
END;
```

Multidimensional Integrals

```
PROCEDURE quad3d(x1,x2: real; VAR ss: real);
(* Evaluates 3-dimensional integral with z integration innermost, then
y integration, and finally x integration. Unlike FORTRAN version, calls QGAUS
(here called QGAUS3) recursively. Programs using routine QUAD3D must define
the integrand by
FUNCTION func(x,y,z: real): real;
and functions for the limits of integration by
FUNCTION y1(x: real): real;
FUNCTION y2(x: real): real;
FUNCTION z1(x,y: real): real;
FUNCTION z2(x,y: real): real;
Also global variables
VAR
    glx,gly: real;
are required. *)
    PROCEDURE qgaus3(a,b: real; VAR ss: real; n: integer);
    VAR
        j: integer; xr,xm,dx: real;
        w,x: ARRAY [1..5] OF real;
    FUNCTION f(x: real; n: integer): real;
        VAR
            ss: real;
        BEGIN
            IF n=1 THEN BEGIN
                glx := x; qgaus3(y1(glx),y2(glx),ss,2); f := ss END
            ELSE IF n=2 THEN BEGIN
                gly := x; qgaus3(z1(glx,gly),z2(glx,gly),ss,3); f := ss END
            ELSE f := func(glx,gly,x)
        END;
    BEGIN
        x[1] := 0.1488743389; x[2] := 0.4333953941; x[3] := 0.6794095682;
        x[4] := 0.8650633666; x[5] := 0.97390652; w[1] := 0.2955242247;
        w[2] := 0.2692667193; w[3] := 0.2190863625; w[4] := 0.1494513491;
        w[5] := 0.06667134; xm := 0.5*(b+a); xr := 0.5*(b-a); ss := 0;
        FOR j := 1 TO 5 DO BEGIN
            dx := xr*x[j]; ss := ss+w[j]*(f(xm+dx,n)+f(xm-dx,n)) END;
        ss := xr*ss END;
BEGIN
    qgaus3(x1,x2,ss,1)
END;
```

Chapter 5. Evaluation of Functions

Series and Their Convergence

```
PROCEDURE eulsum(VAR sum: real; term: real; jterm: integer);
(* Programs using routine EULSUM must declare the variable array
    glwksp: ARRAY [1..np] OF real;
where np is a physical dimension larger than any value of jterm
to be used. Also declare
VAR
    glnterm: integer; *)
VAR
    j: integer; tmp,dum: real;
BEGIN
    IF (jterm = 1) THEN BEGIN
        glnterm := 1; glwksp[1] := term; sum := 0.5*term END
    ELSE BEGIN
        tmp := glwksp[1]; glwksp[1] := term;
```

```
        FOR j := 1 TO glnterm-1 DO BEGIN
            dum := glwksp[j+1]; glwksp[j+1] := 0.5*(glwksp[j]+tmp);
            tmp := dum END;
        glwksp[glnterm+1] := 0.5*(glwksp[glnterm]+tmp);
        IF (abs(glwksp[glnterm+1]) <= abs(glwksp[glnterm])) THEN BEGIN
            sum := sum+0.5*glwksp[glnterm+1]; glnterm := glnterm+1 END
        ELSE BEGIN
            sum := sum+glwksp[glnterm+1] END END
END;
```

Polynomials and Rational Functions

```
PROCEDURE ddpoly(c: glcarray; nc: integer; x: real;
         VAR pd: glpdarray; nd: integer);
(* Programs using routine DDPOLY must define the types
TYPE
    glcarray = ARRAY [1..nc] OF integer;
    glpdarray = ARRAY [1..nd] OF integer;
in the main routine. *)
VAR
    nnd,j,i: integer; cnst: real;
BEGIN
    pd[1] := c[nc];
    FOR j := 2 TO nd DO BEGIN
        pd[j] := 0.0 END;
    FOR i := nc-1 DOWNTO 1 DO BEGIN
        IF (nd < (nc+1-i)) THEN nnd := nd ELSE nnd := nc+1-i;
        FOR j := nnd DOWNTO 2 DO BEGIN
            pd[j] := pd[j]*x+pd[j-1] END;
        pd[1] := pd[1]*x+c[i] END;
    cnst := 2.0;
    FOR i := 3 TO nd DO BEGIN
        pd[i] := cnst*pd[i]; cnst := cnst*i END
END;
```

```
PROCEDURE poldiv(u: glnarray; n: integer; v: glnvarray; nv: integer;
         VAR q,r: glnarray);
(* Programs using routine POLDIV must define the types
TYPE
    glnarray = ARRAY [1..n] OF real;
    glnvarray = ARRAY [1..nv] OF real;
in the main routine. *)
VAR
    k,j: integer;
BEGIN
    FOR j := 1 TO n DO BEGIN
        r[j] := u[j]; q[j] := 0.0 END;
    FOR k := n-nv DOWNTO 0 DO BEGIN
        q[k+1] := r[nv+k]/v[nv];
        FOR j := nv+k-1 DOWNTO k+1 DO BEGIN
            r[j] := r[j]-q[k+1]*v[j-k] END END;
    r[nv] := 0.0
END;
```

Chebyshev Approximation

```
PROCEDURE chebft(a,b: real; VAR c: glcarray; n: integer);
(* Programs using the routine CHEBFT must define the type
TYPE
    glcarray=ARRAY [1..n] OF real;
in the calling routine, and must externally define a function
```

```
func(y:real):real which is to be analyzed. *)
CONST
    pi=3.141592653589793;
VAR
    k,j: integer; y,fac,bpa,bma: real; sum: double; f: glcarray;
BEGIN
    bma := 0.5*(b-a); bpa := 0.5*(b+a);
    FOR k := 1 TO n DO BEGIN
        y := cos(pi*(k-0.5)/n); f[k] := func(y*bma+bpa) END;
    fac := 2.0/n;
    FOR j := 1 TO n DO BEGIN
        sum := 0.0;
        FOR k := 1 TO n DO BEGIN
            sum := sum+f[k]*cos((pi*(j-1))*((k-0.5)/n)) END;
        c[j] := sngl(fac*sum) END
END;

FUNCTION chebev(a,b: real; c: glcarray; m: integer; x:real): real;
(* Programs using routine CHEBEV must define the type
glcarray as in routine CHEBFT. *)
VAR
    d,dd,sv,y,y2: real; j: integer;
BEGIN
    IF (((x-a)*(x-b)) > 0.0) THEN BEGIN
        writeln('pause in CHEBEV - x not in range.'); readln
    END;
    d := 0.0; dd := 0.0; y := (2.0*x-a-b)/(b-a); y2 := 2.0*y;
    FOR j := m DOWNTO 2 DO BEGIN
        sv := d; d := y2*d-dd+c[j]; dd := sv END;
    chebev := y*d-dd+0.5*c[1]
END;
```

Derivatives or Integrals of Chebyshev-approximated Functions

```
PROCEDURE chint(a,b: real; c: glcarray; VAR cint: glcarray; n: integer);
(* Programs using routine CHINT must define the type glcarray as in
routine CHEBFT. *)
VAR
    j: integer; sum,fac,con: real;
BEGIN
    con := 0.25*(b-a); sum := 0.0; fac := 1.0;
    FOR j := 2 TO n-1 DO BEGIN
        cint[j] := con*(c[j-1]-c[j+1])/(j-1); sum := sum+fac*cint[j];
        fac := -fac END;
    cint[n] := con*c[n-1]/(n-1); sum := sum+fac*cint[n]; cint[1] := 2.0*sum
END;

PROCEDURE chder(a,b: real; c: glcarray; VAR cder: glcarray; n: integer);
(* Programs using routine CHDER must define the type glcarray
as in routine CHEBFT. *)
VAR
    j: integer; con: real;
BEGIN
    cder[n] := 0.0; cder[n-1] := 2*(n-1)*c[n];
    IF (n >= 3) THEN BEGIN
        FOR j := n-2 DOWNTO 1 DO BEGIN
            cder[j] := cder[j+2]+2*j*c[j+1] END END;
    con := 2.0/(b-a);
    FOR j := 1 TO n DO BEGIN
        cder[j] := cder[j]*con END
END;
```

Polynomial Approximation from Chebyshev Coefficients

```
PROCEDURE chebpc(c: glcarray;VAR d: glcarray; n: integer);
(* Programs using routine CHEBPC must define the type
glcarray as in routine CHEBFT. *)
VAR
    k,j: integer; sv: real; dd: glcarray;
BEGIN
    FOR j := 1 TO n DO BEGIN
        d[j] := 0.0; dd[j] := 0.0 END;
    d[1] := c[n];
    FOR j := n-1 DOWNTO 2 DO BEGIN
        FOR k := n-j+1 DOWNTO 2 DO BEGIN
            sv := d[k]; d[k] := 2.0*d[k-1]-dd[k]; dd[k] := sv END;
        sv := d[1]; d[1] := -dd[1]+c[j]; dd[1] := sv END;
    FOR j := n DOWNTO 2 DO BEGIN
        d[j] := d[j-1]-dd[j] END;
    d[1] := -dd[1]+0.5*c[1]
END;
```

```
PROCEDURE pcshft(a,b: real; VAR d: glcarray; n: integer);
(* Programs using routine PCSHFT must define the type
glcarray as in routine CHEBFT. *)
VAR
    k,j: integer; fac,cnst: real;
BEGIN
    cnst := 2.0/(b-a); fac := cnst;
    FOR j := 2 TO n DO BEGIN
        d[j] := d[j]*fac; fac := fac*cnst END;
    cnst := 0.5*(a+b);
    FOR j := 1 TO n-1 DO BEGIN
        FOR k := n-1 DOWNTO j DO BEGIN
            d[k] := d[k]-cnst*d[k+1] END END
END;
```

Chapter 6. Special Functions

Gamma Function and Related Functions

```
FUNCTION gammln(xx: real): real;
CONST
   stp = 2.50662827465; half = 0.5; one = 1.0; fpf = 5.5;
VAR
   x,tmp,ser: double; j: integer;
   cof: ARRAY [1..6] OF double;
BEGIN
   cof[1] := 76.18009173; cof[2] := -86.50532033; cof[3] := 24.01409822;
   cof[4] := -1.231739516; cof[5] := 0.120858003e-2; cof[6] := -0.536382e-5;
   x := xx-one; tmp := x+fpf; tmp := (x+half)*ln(tmp)-tmp; ser := one;
   FOR j := 1 TO 6 DO BEGIN
      x := x+one; ser := ser+cof[j]/x END;
   gammln := sngl(tmp+ln(stp*ser))
END;
```

```
FUNCTION factrl(n: integer): real;
(* Programs using routing FACTRL must declare the variables
VAR
    glntop: integer;
    gla: ARRAY [1..33] OF real;
and initialize the values
    glntop := 0; gla[1] := 1.0;
in the main routine. *)
VAR
    j: integer;
BEGIN
    IF (n < 0) THEN BEGIN
        writeln('pause in FACTRL - negative factorial'); readln END
    ELSE IF (n <= glntop) THEN BEGIN
        factrl := gla[n+1] END
    ELSE IF (n <= 32) THEN BEGIN
        FOR j := glntop+1 TO n DO BEGIN
            gla[j+1] := j*gla[j] END;
        glntop := n; factrl := gla[n+1] END
    ELSE BEGIN
        factrl := exp(gammln(n+1.0)) END
END;

FUNCTION bico(n,k: integer): real;
BEGIN
    bico := round(exp(factln(n)-factln(k)-factln(n-k)));
END;

FUNCTION factln(n: integer): real;
(* Programs using routine FACTLN must declare the array
VAR
    gla: ARRAY [1..100] OF real;
and must initialize the array to the values
    FOR i := 1 TO 100 DO gla[i] := -1.0;
in the main routine. *)
BEGIN
    IF (n < 0) THEN BEGIN
        writeln ('pause in FACTLN - negative factorial'); readln END
    ELSE IF (n <= 99) THEN BEGIN
        IF (gla[n+1] < 0.0) THEN gla[n+1] := gammln(n+1.0);
        factln := gla[n+1] END
    ELSE BEGIN
        factln := gammln(n+1.0) END
END;

FUNCTION beta(VAR z,w: real): real;
BEGIN
    beta := exp(gammln(z)+gammln(w)-gammln(z+w))
END;
```

Incomplete Gamma Function and Related Functions

```
FUNCTION gammp(a,x: real): real;
VAR
    gammcf,gln: real;
BEGIN
    IF ((x < 0.0) OR (a <= 0.0)) THEN BEGIN
        writeln('pause in GAMMP - invalid arguments'); readln
    END;
    IF (x < (a+1.0)) THEN BEGIN
```

```
        gser(a,x,gammcf,gln); gammp := gammcf END
    ELSE BEGIN
        gcf(a,x,gammcf,gln); gammp := 1.0-gammcf END
END;

FUNCTION gammq(a,x: real): real;
VAR
    gamser,gln: real;
BEGIN
    IF ((x < 0.0) OR (a <= 0.0)) THEN BEGIN
        writeln('pause in GAMMQ - invalid arguments'); readln
    END;
    IF (x < a+1.0) THEN BEGIN
        gser(a,x,gamser,gln); gammq := 1.0-gamser END
    ELSE BEGIN
        gcf(a,x,gamser,gln); gammq := gamser END
END;

PROCEDURE gser(a,x: real; VAR gamser,gln: real);
LABEL 1;
CONST
    itmax=100; eps=3.0e-7;
VAR
    n: integer; sum,del,ap: real;
BEGIN
    gln := gammln(a);
    IF (x <= 0.0) THEN BEGIN
        IF (x < 0.0) THEN BEGIN
            writeln('pause in GSER - x less than 0'); readln
        END;
        gamser := 0.0 END
    ELSE BEGIN
        ap := a; sum := 1.0/a; del := sum;
        FOR n := 1 TO itmax DO BEGIN
            ap := ap+1.0; del := del*x/ap; sum := sum+del;
            IF (abs(del) < abs(sum)*eps) THEN GOTO 1
        END;
        writeln('pause in GSER - a too large, itmax too small'); readln;
1:      gamser := sum*exp(-x+a*ln(x)-gln)
    END
END;

PROCEDURE gcf(a,x: real; VAR gammcf,gln: real);
LABEL 1;
CONST
    itmax=100; eps=3.0e-7;
VAR
    n: integer; gold,g,fac,b1,b0,anf,ana,an,a1,a0: real;
BEGIN
    gln := gammln(a); gold := 0.0; a0 := 1.0; a1 := x; b0 := 0.0;
    b1 := 1.0; fac := 1.0;
    FOR n := 1 TO itmax DO BEGIN
        an := 1.0*n; ana := an-a; a0 := (a1+a0*ana)*fac;
        b0 := (b1+b0*ana)*fac; anf := an*fac; a1 := x*a0+anf*a1;
        b1 := x*b0+anf*b1;
        IF (a1 <> 0.0) THEN BEGIN
            fac := 1.0/a1; g := b1*fac;
            IF (abs((g-gold)/g) < eps) THEN GOTO 1;
            gold := g END END;
    writeln('pause in GCF - a too large, itmax too small'); readln;
1:  gammcf := exp(-x+a*ln(x)-gln)*g
END;
```

```
FUNCTION erf(x: real): real;
BEGIN
   IF (x < 0.0) THEN BEGIN
      erf := -gammp(0.5,sqr(x)) END
   ELSE BEGIN
      erf := gammp(0.5,sqr(x)) END
END;

FUNCTION erfc(VAR x: real): real;
BEGIN
   IF (x < 0.0) THEN BEGIN
      erfc := 1.0+gammp(0.5,sqr(x)) END
   ELSE BEGIN
      erfc := gammq(0.5,sqr(x)) END
END;

FUNCTION erfcc(x: real): real;
VAR
   t,z,ans: real;
BEGIN
   z := abs(x); t := 1.0/(1.0+0.5*z);
   ans := t*exp(-z*z-1.26551223+t*(1.00002368+
      t*(0.37409196+t*(0.09678418+t*(-0.18628806+
      t*(0.27886807+t*(-1.13520398+t*(1.48851587+
      t*(-0.82215223+t*0.17087277)))))))));
   IF (x >= 0.0) THEN erfcc := ans
   ELSE erfcc := 2.0-ans
END;
```

Incomplete Beta Function and Related Functions

```
FUNCTION betai(a,b,x: real): real;
VAR
   bt: real;
BEGIN
   IF ((x < 0.0) OR (x > 1.0)) THEN BEGIN
      writeln('pause in routine BETAI'); readln
   END;
   IF ((x = 0.0) OR (x = 1.0)) THEN bt := 0.0
   ELSE bt := exp(gammln(a+b)-gammln(a)-gammln(b)
           +a*ln(x)+b*ln(1.0-x));
   IF (x < ((a+1.0)/(a+b+2.0))) THEN
      betai := bt*betacf(a,b,x)/a
   ELSE betai := 1.0-bt*betacf(b,a,1.0-x)/b
END;

FUNCTION betacf(a,b,x: real): real;
LABEL 1;
CONST
   itmax=100; eps=3.0e-7;
VAR
   tem,qap,qam,qab,em,d: real; bz,bpp,bp,bm,az,app: real; am,aold,ap: real;
   m: integer;
BEGIN
   am := 1.0; bm := 1.0; az := 1.0; qab := a+b; qap := a+1.0;
   qam := a-1.0; bz := 1.0-qab*x/qap;
   FOR m := 1 TO itmax DO BEGIN
      em := m; tem := em+em; d := em*(b-m)*x/((qam+tem)*(a+tem));
      ap := az+d*am; bp := bz+d*bm;
      d := -(a+em)*(qab+em)*x/((a+tem)*(qap+tem)); app := ap+d*az;
      bpp := bp+d*bz; aold := az; am := ap/bpp; bm := bp/bpp;
```

```
        az := app/bpp; bz := 1.0;
        IF ((abs(az-aold)) < (eps*abs(az))) THEN GOTO 1
    END;
    writeln('pause in BETACF');
    writeln('a or b too big, or itmax too small'); readln;
1:  betacf := az
END;
```

Bessel Functions of Integer Order

```
FUNCTION bessj0(x: real): real;
VAR
    ax,xx,z: real; y,ans,ans1,ans2: double;
BEGIN
    IF (abs(x) < 8.0) THEN BEGIN
        y := sqr(x); ans1 := 57568490574.0+y*(-13362590354.0+y*(651619640.7
            +y*(-11214424.18+y*(77392.33017+y*(-184.9052456)))));
        ans2 := 57568490411.0+y*(1029532985.0+y*(9494680.718
            +y*(59272.64853+y*(267.8532712+y*1.0))));
        bessj0 := sngl(ans1/ans2) END
    ELSE BEGIN
        ax := abs(x); z := 8.0/ax; y := sqr(z); xx := ax-0.785398164;
        ans1 := 1.0+y*(-0.1098628627e-2+y*(0.2734510407e-4
            +y*(-0.2073370639e-5+y*0.2093887211e-6)));
        ans2 := -0.1562499995e-1+y*(0.1430488765e-3
            +y*(-0.6911147651e-5+y*(0.7621095161e-6 -y*0.934945152e-7)));
        ans := sqrt(0.636619772/ax)*(cos(xx)*ans1-z*sin(xx)*ans2);
        bessj0 := sngl(ans) END
END;

FUNCTION bessy0(x: real): real;
VAR
    xx,z: real; y,ans,ans1,ans2: double;
BEGIN
    IF (x < 8.0) THEN BEGIN
        y := sqr(x); ans1 := -2957821389.0+y*(7062834065.0+y*(-512359803.6
            +y*(10879881.29+y*(-86327.92757+y*228.4622733))));
        ans2 := 40076544269.0+y*(745249964.8+y*(7189466.438
            +y*(47447.26470+y*(226.1030244+y*1.0))));
        ans := (ans1/ans2)+0.636619772*bessj0(x)*ln(x);
        bessy0 := sngl(ans) END
    ELSE BEGIN
        z := 8.0/x; y := sqr(z); xx := x-0.785398164;
        ans1 := 1.0+y*(-0.1098628627e-2+y*(0.2734510407e-4
            +y*(-0.2073370639e-5+y*0.2093887211e-6)));
        ans2 := -0.1562499995e-1+y*(0.1430488765e-3
            +y*(-0.6911147651e-5+y*(0.7621095161e-6+y*(-0.934945152e-7))));
        ans := sin(xx)*ans1+z*cos(xx)*ans2; ans := sqrt(0.636619772/x)*ans;
        bessy0 := sngl(ans) END
END;

FUNCTION bessj1(x: real): real;
VAR
    ax,xx,z: real; y,ans,ans1,ans2: double;
FUNCTION sign(x: real): real;
    BEGIN
        IF x >= 0.0 THEN sign := 1.0
        ELSE sign := -1.0;
    END;
BEGIN
    IF (abs(x) < 8.0) THEN BEGIN
```

```
        y := sqr(x); ans1 := x*(72362614232.0+y*(-7895059235.0+y*(242396853.1
            +y*(-2972611.439+y*(15704.48260+y*(-30.16036606))))));
        ans2 := 144725228442.0+y*(2300535178.0+y*(18583304.74
            +y*(99447.43394+y*(376.9991397+y*1.0))));
        bessj1 := sngl(ans1/ans2) END
    ELSE BEGIN
        ax := abs(x); z := 8.0/ax; y := sqr(z); xx := ax-2.356194491;
        ans1 := 1.0+y*(0.183105e-2+y*(-0.3516396496e-4
            +y*(0.2457520174e-5+y*(-0.240337019e-6))));
        ans2 := 0.04687499995+y*(-0.2002690873e-3
            +y*(0.8449199096e-5+y*(-0.88228987e-6+y*0.105787412e-6)));
        ans := sqrt(0.636619772/ax)*(cos(xx)*ans1 -z*sin(xx)*ans2)*sign(x);
        bessj1 := sngl(ans) END
END;


FUNCTION bessy1(x: real): real;
VAR
    xx,z: real; y,ans,ans1,ans2: double;
BEGIN
    IF (x < 8.0) THEN BEGIN
        y := sqr(x); ans1 := x*(-0.4900604943e13+y*(0.1275274390e13
            +y*(-0.5153438139e11+y*(0.7349264551e9
            +y*(-0.4237922726e7+y*0.8511937935e4)))));
        ans2 := 0.2499580570e14+y*(0.4244419664e12
            +y*(0.3733650367e10+y*(0.2245904002e8
            +y*(0.1020426050e6+y*(0.3549632885e3+y*1.0)))));
        ans := (ans1/ans2)+0.636619772*(bessj1(x)*ln(x)-1.0/x);
        bessy1 := sngl(ans) END
    ELSE BEGIN
        z := 8.0/x; y := sqr(z); xx := x-2.356194491;
        ans1 := 1.0+y*(0.183105e-2+y*(-0.3516396496e-4
            +y*(0.2457520174e-5+y*(-0.240337019e-6))));
        ans2 := 0.04687499995+y*(-0.2002690873e-3
            +y*(0.8449199096e-5+y*(-0.88228987e-6+y*0.105787412e-6)));
        ans := sqrt(0.636619772/x)*(sin(xx)*ans1+z*cos(xx)*ans2);
        bessy1 := sngl(ans) END
END;


FUNCTION bessj(n: integer; x: real): real;
CONST
    iacc=40; bigno=1.0e10; bigni=1.0e-10;
VAR
    bj,bjm,bjp,sum,tox,ans: real; j,jsum,m: integer;
BEGIN
    IF (n < 2) THEN BEGIN
        writeln('pause in BESSJ'); readln END;
    IF (x=0.0) THEN ans := 0.0
    ELSE IF (abs(x) > 1.0*n) THEN BEGIN
        tox := 2.0/abs(x);
        bjm := bessj0(abs(x)); bj := bessj1(abs(x));
        FOR j := 1 TO n-1 DO BEGIN
            bjp := j*tox*bj-bjm; bjm := bj; bj := bjp END;
        ans := bj END
    ELSE BEGIN
        tox := 2.0/abs(x);
        m := 2*((n+trunc(sqrt(1.0*(iacc*n)))) DIV 2);
        ans := 0.0; jsum := 0; sum := 0.0; bjp := 0.0; bj := 1.0;
        FOR j := m DOWNTO 1 DO BEGIN
            bjm := j*tox*bj-bjp; bjp := bj; bj := bjm;
            IF (abs(bj) > bigno) THEN BEGIN
                bj := bj*bigni; bjp := bjp*bigni; ans := ans*bigni;
                sum := sum*bigni END;
            IF (jsum <> 0) THEN sum := sum+bj;
```

```
            jsum := 1-jsum;
            IF (j = n) THEN ans := bjp
        END;
        sum := 2.0*sum-bj; ans := ans/sum END;
    IF (x<0.0) AND ((n MOD 2)=1) THEN ans := -ans;
    bessj := ans
END;
```

```
FUNCTION bessy(n: integer; x: real): real;
VAR
    by,bym,byp,tox: real; j: integer;
BEGIN
    IF (n < 2) THEN BEGIN
        writeln('pause in BESSY - index n less than 2'); readln
    END;
    tox := 2.0/x; by := bessy1(x); bym := bessy0(x);
    FOR j := 1 TO n-1 DO BEGIN
        byp := j*tox*by-bym; bym := by; by := byp END;
    bessy := by
END;
```

Modified Bessel Functions of Integer Order

```
FUNCTION bessi0(x: real): real;
VAR
    ax: real; y,ans: double;
BEGIN
    IF (abs(x) < 3.75) THEN BEGIN
        y := sqr(x/3.75); ans := 1.0+y*(3.5156229+y*(3.0899424+y*(1.2067492+y*
            (0.2659732+y*(0.360768e-1+y*0.45813e-2))))) END
    ELSE BEGIN
        ax := abs(x); y := 3.75/ax;
        ans := (exp(ax)/sqrt(ax))*(0.39894228+y*(0.1328592e-1
            +y*(0.225319e-2+y*(-0.157565e-2+y*(0.916281e-2
            +y*(-0.2057706e-1+y*(0.2635537e-1+y*(-0.1647633e-1
            +y*0.392377e-2)))))))) END;
    bessi0 := sngl(ans)
END;
```

```
FUNCTION bessk0(x: real): real;
VAR
    y,ans: double;
BEGIN
    IF (x <= 2.0) THEN BEGIN
        y := x*x/4.0; ans := (-ln(x/2.0)*bessi0(x))+(-0.57721566+y*(0.42278420
            +y*(0.23069756+y*(0.3488590e-1+y*(0.262698e-2
            +y*(0.10750e-3+y*0.74e-5)))))) END
    ELSE BEGIN
        y := (2.0/x); ans := (exp(-x)/sqrt(x))*(1.25331414+y*(-0.7832358e-1
            +y*(0.2189568e-1+y*(-0.1062446e-1+y*(0.587872e-2
            +y*(-0.251540e-2+y*0.53208e-3)))))) END;
    bessk0 := sngl(ans)
END;
```

```
FUNCTION bessi1(x: real): real;
VAR
   ax: real; y,ans: double;
BEGIN
   IF (abs(x) < 3.75) THEN BEGIN
      y := sqr(x/3.75);
      ans := x*(0.5+y*(0.87890594+y*(0.51498869+y*(0.15084934
         +y*(0.2658733e-1+y*(0.301532e-2+y*0.32411e-3)))))) END
   ELSE BEGIN
      ax := abs(x); y := 3.75/ax;
      ans := 0.2282967e-1+y*(-0.2895312e-1+y*(0.1787654e-1-y*0.420059e-2));
      ans := 0.39894228+y*(-0.3988024e-1+y*(-0.362018e-2
         +y*(0.163801e-2+y*(-0.1031555e-1+y*ans))));
      ans := (exp(ax)/sqrt(ax))*ans;
      IF (x<0.0) THEN ans := -ans END;
   bessi1 := sngl(ans)
END;

FUNCTION bessk1(x: real): real;
VAR
   y,ans: double;
BEGIN
   IF (x <= 2.0) THEN BEGIN
      y := x*x/4.0; ans := (ln(x/2.0)*bessi1(x))+(1.0/x)*(1.0+y*(0.15443144
         +y*(-0.67278579+y*(-0.18156897+y*(-0.1919402e-1
         +y*(-0.110404e-2+y*(-0.4686e-4))))))) END
   ELSE BEGIN
      y := 2.0/x; ans := (exp(-x)/sqrt(x))*(1.25331414+y*(0.23498619
         +y*(-0.3655620e-1+y*(0.1504268e-1+y*(-0.780353e-2
         +y*(0.325614e-2+y*(-0.68245e-3))))))) END;
   bessk1 := sngl(ans)
END;

FUNCTION bessi(n: integer; x: real): real;
CONST
   iacc=40; bigno=1.0e10; bigni=1.0e-10;
VAR
   bi,bim,bip,tox,ans: real; j,m: integer;
BEGIN
   IF (n < 2) THEN BEGIN
      writeln('pause in routine BESSI');
      writeln('index n is less than 2'); readln END;
   IF (x=0.0) THEN bessi := 0.0 ELSE BEGIN
      ans := 0.0; tox := 2.0/abs(x); bip := 0.0; bi := 1.0;
      m := 2*(n+trunc(sqrt(iacc*n)));
      FOR j := m DOWNTO 1 DO BEGIN
         bim := bip+j*tox*bi; bip := bi; bi := bim;
         IF (abs(bi) > bigno) THEN BEGIN
            ans := ans*bigni; bi := bi*bigni; bip := bip*bigni END;
         IF (j=n) THEN ans := bip
      END;
      IF (x<0.0) AND ((n MOD 2)=1) THEN ans := -ans;
      bessi := ans*bessi0(x)/bi END
END;

FUNCTION bessk(n: integer; x: real): real;
VAR
   tox,bkp,bkm,bk: real; j: integer;
BEGIN
   IF (n < 2) THEN BEGIN
      writeln('pause in routine BESSK');
      writeln('index n less than 2'); readln END;
   tox := 2.0/x; bkm := bessk0(x); bk := bessk1(x);
   FOR j := 1 TO n-1 DO BEGIN
      bkp := bkm+j*tox*bk; bkm := bk; bk := bkp END;
```

```
    bessk := bk
END;
```

Spherical Harmonics

```
FUNCTION plgndr(l,m: integer; x: real): real;
VAR
    fact,pll,pmm,pmmp1,somx2: real; i,ll: integer;
BEGIN
    IF ((m < 0) OR (m > l) OR (abs(x) > 1.0)) THEN BEGIN
        writeln('Pause in routine PLGNDR');
        writeln('bad arguments'); readln END;
    pmm := 1.0;
    IF (m > 0) THEN BEGIN
        somx2 := sqrt((1.0-x)*(1.0+x)); fact := 1.0;
        FOR i := 1 TO m DO BEGIN
            pmm := -pmm*fact*somx2; fact := fact+2.0 END END;
    IF (l = m) THEN BEGIN
        plgndr := pmm END
    ELSE BEGIN
        pmmp1 := x*(2*m+1)*pmm;
        IF (l = m+1) THEN BEGIN
            plgndr := pmmp1 END
        ELSE BEGIN
            FOR ll := m+2 TO l DO BEGIN
                pll := (x*(2*ll-1)*pmmp1-(ll+m-1)*pmm)/(ll-m); pmm := pmmp1;
                pmmp1 := pll END;
            plgndr := pll END END
END;
```

Elliptic Integrals and Jacobian Elliptic Functions

```
FUNCTION el2(x,qqc,aa,bb: real): real;
LABEL 1;
CONST
    pi=3.14159265; ca=0.0003; cb=1.0e-9;
VAR
    a,b,c,d,e,f,g: real; em,eye,p,qc,y,z: real; l: integer;
BEGIN
    IF (x = 0.0) THEN el2 := 0.0
    ELSE IF (qqc <> 0.0) THEN BEGIN
        qc := qqc; a := aa; b := bb; c := sqr(x); d := 1.0+c;
        p := sqrt((1.0+c*sqr(qc))/d); d := x/d; c := d/(2.0*p); z := a-b;
        eye := a; a := 0.5*(b+a); y := abs(1.0/x); f := 0.0; l := 0;
        em := 1.0; qc := abs(qc);
1:      b := eye*qc+b;
        e := em*qc; g := e/p; d := f*g+d; f := c; eye := a; p := g+p;
        c := 0.5*(d/p+c); g := em; em := qc+em; a := 0.5*(b/em+a);
        y := -e/y+y;
        IF (y = 0.0) THEN y := sqrt(e)*cb;
        IF (abs(g-qc) > ca*g) THEN BEGIN
            qc := sqrt(e)*2.0; l := l+l;
            IF (y < 0.0) THEN l := l+1;
            GOTO 1
        END;
        IF (y < 0.0) THEN l := l+1;
        e := (arctan(em/y)+pi*l)*a/em;
        IF (x < 0.0) THEN e := -e;
        el2 := e+c*z END
    ELSE BEGIN
        writeln('pause in routine EL2'); readln
    END
END;
```

```
FUNCTION cel(qqc,pp,aa,bb: real): real;
LABEL 1;
CONST
    ca=0.0003; pio2=1.5707963268;
VAR
    a,b,e,f,g: real; em,p,q,qc: real;
BEGIN
    IF (qqc = 0.0) THEN BEGIN
        writeln('pause in routine CEL'); readln END;
    qc := abs(qqc); a := aa; b := bb; p := pp; e := qc; em := 1.0;
    IF (p > 0.0) THEN BEGIN
        p := sqrt(p); b := b/p END
    ELSE BEGIN
        f := qc*qc; q := 1.0-f; g := 1.0-p; f := f-p; q := q*(b-a*p);
        p := sqrt(f/g); a := (a-b)/g; b := -q/(g*g*p)+a*p END;
1:  f := a;
    a := a+b/p; g := e/p; b := b+f*g; b := b+b; p := g+p; g := em;
    em := qc+em;
    IF (abs(g-qc) > (g*ca)) THEN BEGIN
        qc := sqrt(e); qc := qc+qc; e := qc*em;
        GOTO 1
    END;
    cel := pio2*(b+a*em)/(em*(em+p))
END;

PROCEDURE sncndn(uu,emmc: real; VAR sn,cn,dn: real);
LABEL 1;
CONST
    ca=0.0003;
VAR
    a,b,c,d,emc,u: real; i,ii,l: integer; bo: boolean;
    em,en: ARRAY [1..13] OF real;
FUNCTION cosh(u: real): real;
    BEGIN cosh := 0.5*(exp(u)+exp(-u)) END;
FUNCTION tanh(u: real): real;
    VAR
        u2,epu,emu: real;
    BEGIN
        epu := exp(u); emu := 1.0/epu;
        IF (abs(u)<0.3) THEN BEGIN
            u2 := u*u;
            tanh := 2*u*(1+u2/6*(1+u2/20*(1+u2/42*(1+u2/72))))/(epu+emu) END
        ELSE BEGIN tanh := (epu-emu)/(epu+emu) END
    END;
BEGIN
    emc := emmc; u := uu;
    IF (emc <> 0.0) THEN BEGIN
        bo := (emc < 0.0);
        IF (bo) THEN BEGIN
            d := 1.0-emc; emc := -emc/d; d := sqrt(d); u := d*u END;
        a := 1.0; dn := 1.0;
        FOR i := 1 TO 13 DO BEGIN
            l := i; em[i] := a; emc := sqrt(emc); en[i] := emc;
            c := 0.5*(a+emc);
            IF (abs(a-emc) <= ca*a) THEN GOTO 1;
            emc := a*emc; a := c END;
1:      u := c*u;
        sn := sin(u); cn := cos(u);
        IF (sn <> 0.0) THEN BEGIN
            a := cn/sn; c := a*c;
            FOR ii := l DOWNTO 1 DO BEGIN
                b := em[ii]; a := c*a; c := dn*c; dn := (en[ii]+a)/(b+a);
                a := c/b END;
            a := 1.0/sqrt(sqr(c)+1.0);
```

```
            IF (sn < 0.0) THEN sn := -a
            ELSE sn := a;
            cn := c*sn END;
         IF (bo) THEN BEGIN
            a := dn; dn := cn; cn := a; sn := sn/d END; END
      ELSE BEGIN
         cn := 1.0/cosh(u); dn := cn; sn := tanh(u) END
END;
```

Chapter 7. Random Numbers

Uniform Deviates

```
FUNCTION ran0(VAR idum: integer): real;
(* RANO assumes that the system random number generator is called RANDOM,
with no arguments. Programs using RANO must declare the following
variables
VAR
   gly: real;
   glv: ARRAY [1..97] OF real;
in the main routine. *)
VAR
   dum: real; j: integer;
BEGIN
   IF (idum < 0) THEN BEGIN
      idum := 1;
      FOR j := 1 TO 97 DO dum := random;
      FOR j := 1 TO 97 DO glv[j] := random;
      gly := random END;
   j := 1+trunc(97.0*gly);
   IF ((j > 97) OR (j < 1)) THEN BEGIN
      writeln('pause in routine RANO'); readln END;
   gly := glv[j]; ran0 := gly; glv[j] := random
END;
```

```
FUNCTION ran1(VAR idum: integer): real;
(* Programs using RAN1 must declare the following variables
VAR
   glix1,glix2,glix3: integer;
   glr: ARRAY [1..97] OF real;
in the main program. *)
CONST
   m1=259200; ia1=7141; ic1=54773;
   rm1=3.8580247e-6;     (* 1.0/m1 *)
   m2=134456; ia2=8121; ic2=28411;
   rm2=7.4373773e-6;     (* 1.0/m2 *)
   m3=243000; ia3=4561; ic3=51349;
VAR
   j: integer;
BEGIN
   IF (idum < 0) THEN BEGIN
      glix1 := (ic1-idum) MOD m1;
      glix1 := (ia1*glix1+ic1) MOD m1;
      glix2 := glix1 MOD m2;
      glix1 := (ia1*glix1+ic1) MOD m1;
      glix3 := glix1 MOD m3;
      FOR j := 1 TO 97 DO BEGIN
         glix1 := (ia1*glix1+ic1) MOD m1;
         glix2 := (ia2*glix2+ic2) MOD m2;
         glr[j] := (glix1+glix2*rm2)*rm1 END;
      idum := 1 END;
```

```
    glix1 := (ia1*glix1+ic1) MOD m1;
    glix2 := (ia2*glix2+ic2) MOD m2;
    glix3 := (ia3*glix3+ic3) MOD m3;
    j := 1 + (97*glix3) DIV m3;
    IF ((j > 97) OR (j < 1)) THEN BEGIN
        writeln('pause in routine RAN1'); readln
    END;
    ran1 := glr[j]; glr[j] := (glix1+glix2*rm2)*rm1
END;


FUNCTION ran2(VAR idum: integer): real;
(* Programs using RAN2 must declare the following variables
VAR
    gliy: integer;
    glir: ARRAY [1..97] OF integer;
in the main program. *)
CONST
    m=714025; ia=1366; ic=150889;
    rm=1.400512e-6;        (* 1.0/m *)
VAR
    j: integer;
BEGIN
    IF (idum < 0) THEN BEGIN
        idum := (ic-idum) MOD m;
        FOR j := 1 TO 97 DO BEGIN
            idum := (ia*idum+ic) MOD m;
            glir[j] := idum END;
        idum := (ia*idum+ic) MOD m;
        gliy := idum END;
    j := 1 + (97*gliy) DIV m;
    IF ((j > 97) OR (j < 1)) THEN BEGIN
        writeln('pause in routine RAN2'); readln
    END;
    gliy := glir[j]; ran2 := gliy*rm;
    idum := (ia*idum+ic) MOD m;
    glir[j] := idum
END;


FUNCTION ran3(VAR idum: integer): real;
(* Programs using RAN3 must declare the following variables
VAR
    glinext,glinextp: integer;
    glma: ARRAY [1..55] OF real;
in the main routine. Machines with 4-byte integers can use the integer
implementation of this routine, substituting glma of type integer, the
commented CONST and VAR declarations, and the MOD function in the third
line after the BEGIN. *)
(* CONST
    mbig=1000000000; mseed=161803398; mz=0; fac=1.0e-9;
VAR
    i,ii,k,mj,mk: integer; *)
CONST
    mbig=4.0e6; mseed=1618033.0; mz=0.0;
    fac=2.5e-7; (* 1/mbig *)
VAR
    i,ii,k: integer; mj,mk: real;
BEGIN
    IF (idum < 0) THEN BEGIN
        mj := mseed+idum;
        (* The following IF block is mj := mj MOD mbig; for real variables. *)
        IF mj>=0.0 THEN mj := mj-mbig*trunc(mj/mbig)
            ELSE mj := mbig-abs(mj)+mbig*trunc(abs(mj)/mbig);
        glma[55] := mj; mk := 1;
```

```
        FOR i := 1 TO 54 DO BEGIN
            ii := 21*i MOD 55;
            glma[ii] := mk; mk := mj-mk;
            IF (mk < mz) THEN mk := mk+mbig;
            mj := glma[ii] END;
        FOR k := 1 TO 4 DO BEGIN
            FOR i := 1 TO 55 DO BEGIN
                glma[i] := glma[i]-glma[1+((i+30) MOD 55)];
                IF (glma[i] < mz) THEN glma[i] := glma[i]+mbig
            END END;
        glinext := 0; glinextp := 31; idum := 1 END;
    glinext := glinext+1;
    IF (glinext = 56) THEN glinext := 1;
    glinextp := glinextp+1;
    IF (glinextp = 56) THEN glinextp := 1;
    mj := glma[glinext]-glma[glinextp];
    IF (mj < mz) THEN mj := mj+mbig;
    glma[glinext] := mj; ran3 := mj*fac
END;
```

Transformation Method: Exponential and Normal Deviates

```
FUNCTION expdev(VAR idum: integer): real;
BEGIN
    expdev := -ln(ran3(idum))
END;
```

```
FUNCTION gasdev(VAR idum: integer): real;
(* Programs using GASDEV must declare the variables
VAR
    gliset: integer; glgset: real;
in the main routine and must intialize gliset to
    gliset := 0;    *)
VAR
    fac,r,v1,v2: real;
BEGIN
    IF (gliset = 0) THEN BEGIN
        REPEAT
            v1 := 2.0*ran3(idum)-1.0; v2 := 2.0*ran3(idum)-1.0;
            r := sqr(v1)+sqr(v2);
        UNTIL (r < 1.0);
        fac := sqrt(-2.0*ln(r)/r); glgset := v1*fac; gasdev := v2*fac;
        gliset := 1 END
    ELSE BEGIN
        gasdev := glgset; gliset := 0 END
END;
```

Rejection Method: Gamma, Poisson, Binomial Deviates

```
FUNCTION gamdev(VAR ia,idum: integer): real;
VAR
    am,e,s,v1,v2,x,y: real; j: integer;
BEGIN
    IF (ia < 1) THEN BEGIN
        writeln('pause in routine GAMDEV'); readln END;
    IF (ia < 6) THEN BEGIN
        x := 1.0;
        FOR j := 1 TO ia DO x := x*ran3(idum);
```

```
        x := -ln(x); END
    ELSE BEGIN
        REPEAT
            REPEAT
                REPEAT
                    v1 := 2.0*ran3(idum)-1.0; v2 := 2.0*ran3(idum)-1.0;
                UNTIL ((sqr(v1)+sqr(v2)) <= 1.0);
                y := v2/v1; am := ia-1; s := sqrt(2.0*am+1.0); x := s*y+am;
            UNTIL (x > 0.0);
            e := (1.0+sqr(y))*exp(am*ln(x/am)-s*y);
        UNTIL (ran3(idum) <= e)
    END;
    gamdev := x
END;


FUNCTION poidev(xm: real; VAR idum: integer): real;
(* Programs using POIDEV must declare the variables
VAR
    gloldm,glsq,glalxm,glg: real;
in the main program and should intialize gloldm to
    gloldm := -1.0;    *)
CONST
    pi=3.141592654;
VAR
    em,t,y: real;
BEGIN
    IF (xm < 12.0) THEN BEGIN
        IF (xm <> gloldm) THEN BEGIN
            gloldm := xm; glg := exp(-xm) END;
        em := -1; t := 1.0;
        REPEAT
            em := em+1.0; t := t*ran3(idum);
        UNTIL (t <= glg)
    END ELSE BEGIN
        IF (xm <> gloldm) THEN BEGIN
            gloldm := xm; glsq := sqrt(2.0*xm); glalxm := ln(xm);
            glg := xm*glalxm-gammln(xm+1.0) END;
        REPEAT
            REPEAT
                y := pi*ran3(idum); y := sin(y)/cos(y); em := glsq*y+xm;
            UNTIL (em >= 0.0);
            em := trunc(em);
            t := 0.9*(1.0+sqr(y))*exp(em*glalxm-gammln(em+1.0)-glg);
        UNTIL (ran3(idum) <= t)
    END;
    poidev := em
END;



FUNCTION bnldev(pp: real; n: integer; VAR idum: integer): real;
LABEL 1;
CONST
    pi=3.141592654;
VAR
    am,em,en,g,angle: real; oldg,p,pc,bnl: real;
    pclog,plog,pold,sq,t,y: real; j,nold: integer;
BEGIN
    nold := -1; pold := -1.0;
    IF (pp <= 0.5) THEN p := pp ELSE p := 1.0-pp;
    am := n*p;
    IF (n < 25) THEN BEGIN
        bnl := 0.0;
        FOR j := 1 TO n DO BEGIN
```

```
            IF (ran3(idum) < p) THEN bnl := bnl+1.0
        END END
    ELSE IF (am < 1.0) THEN BEGIN
        g := exp(-am); t := 1.0;
        FOR j := 0 TO n DO BEGIN
            t := t*ran3(idum);
            IF (t < g) THEN GOTO 1
        END;
        j := n;
1:      bnl := j
    END ELSE BEGIN
        IF (n <> nold) THEN BEGIN
            en := n; oldg := gammln(en+1.0); nold := n END;
        IF (p <> pold) THEN BEGIN
            pc := 1.0-p; plog := ln(p); pclog := ln(pc); pold := p END;
        sq := sqrt(2.0*am*pc);
        REPEAT
            REPEAT
                angle := pi*ran3(idum); y := sin(angle)/cos(angle);
                em := sq*y+am;
            UNTIL ((em >= 0.0) AND (em < en+1.0));
            em := trunc(em); t := 1.2*sq*(1.0+sqr(y))*exp(oldg-gammln(em+1.0)
                -gammln(en-em+1.0)+em*plog+(en-em)*pclog);
        UNTIL (ran3(idum) <= t);
        bnl := em END;
    IF (p <> pp) THEN bnl := n-bnl;
    bnldev := bnl
END;
```

Generation of Random Bits

```
FUNCTION irbit1(VAR iseed: integer): integer;
(* This routine runs much faster if you can perform bitwise logical operations
on integers. For example, here is a TURBO Pascal version:
CONST
    ib1=1; ib3=4; ib5=16; ib14=8192;
VAR
    newbit: boolean;
BEGIN
    newbit := (iseed AND ib14) <> 0;
    IF ((iseed AND ib5) <> 0) THEN newbit := NOT newbit;
    IF ((iseed AND ib3) <> 0) THEN newbit := NOT newbit;
    IF ((iseed AND ib1) <> 0) THEN newbit := NOT newbit;
    iseed := iseed SHL 1;
    IF (newbit) THEN BEGIN
        irbit1 := 1; iseed := iseed OR ib1
    END ELSE BEGIN
        irbit1 := 0; iseed := iseed AND (NOT ib1)
    END
END; *)

(* Here is the slower version for other Pascal systems: *)
CONST
    ib1=1; ib3=4; ib5=16;
    ib14=8192;     (* Values chosen not to overflow 2-byte integers *)
VAR
    mask: integer; newbit: boolean;
FUNCTION iand(i1,i2: integer): integer;
    VAR
        i: integer;
    BEGIN
```

```
        IF ((i1=0) OR (i2=0)) THEN iand := 0
        ELSE BEGIN
            i := ord(odd(i1) AND odd(i2));
            i1 := i1 DIV 2; i2 := i2 DIV 2;
            iand := 2*iand(i1,i2) + i END END;
FUNCTION inot(ib: integer): integer;
    BEGIN inot := maxint-ib END;
FUNCTION ior(i1,i2: integer): integer;
    VAR
        i: integer;
    BEGIN
        IF ((i1=0) AND (i2=0)) THEN ior := 0
        ELSE BEGIN
            i := ord(odd(i1) OR odd(i2));
            i1 := i1 DIV 2; i2 := i2 DIV 2;
            ior := 2*ior(i1,i2) + i END END;
BEGIN
    mask := maxint DIV 2;
    newbit := iand(iseed,ib14) <> 0;
    IF (iand(iseed,ib5) <> 0) THEN newbit := NOT newbit;
    IF (iand(iseed,ib3) <> 0) THEN newbit := NOT newbit;
    IF (iand(iseed,ib1) <> 0) THEN newbit := NOT newbit;
    irbit1 := 0; iseed := iand(2*iand(mask,iseed),inot(ib1));
    IF (newbit) THEN BEGIN
        irbit1 := 1; iseed := ior(iseed,ib1); END
END;


FUNCTION irbit2(VAR iseed: integer): integer;
(* This routine runs much faster if you can perform bitwise logical operations
on integers. For example, here is a TURBO Pascal version:
CONST
    ib1=1; ib3=4; ib5=16; ib14=8192; mask = 21;
BEGIN
    IF (iseed AND ib14) <> 0 THEN BEGIN
        iseed := ((iseed XOR mask) SHL 1) OR ib1;
        irbit2 := 1 END
    ELSE BEGIN
        iseed := (iseed SHL 1) AND (NOT ib1);
        irbit2 := 0 END
END; *)
(* Here is the slower version for other Pascal systems: *)
CONST
    ib1=1; ib3=4; ib5=16;
    ib14=8192;    (* Values chosen not to overflow 2-byte integers *)
    mask = 21;    (* ib1+ib3+ib5 *)
VAR
    mask1: integer;
FUNCTION iand(i1,i2: integer): integer;
    VAR
        i: integer;
    BEGIN
        IF ((i1=0) OR (i2=0)) THEN iand := 0
        ELSE BEGIN
            i := ord(odd(i1) AND odd(i2));
            i1 := i1 DIV 2; i2 := i2 DIV 2;
            iand := 2*iand(i1,i2) + i END END;
FUNCTION inot(i: integer): integer;
    BEGIN inot := maxint-i END;
FUNCTION ior(i1,i2: integer): integer;
    VAR
        i: integer;
    BEGIN
        IF ((i1=0) AND (i2=0)) THEN ior := 0
        ELSE BEGIN
```

```
            i := ord(odd(i1) OR odd(i2));
            i1 := i1 DIV 2; i2 := i2 DIV 2;
            ior := 2*ior(i1,i2) + i END END;
FUNCTION ieor(i1,i2: integer): integer;
    VAR
        i: integer;
    BEGIN
        IF ((i1=0) AND (i2=0)) THEN ieor := 0
        ELSE BEGIN
            i := ord((odd(i1) AND NOT odd(i2))
                    OR (odd(i2) AND NOT odd(i1)));
            i1 := i1 DIV 2; i2 := i2 DIV 2;
            ieor := 2*ieor(i1,i2) + i END END;
BEGIN
    mask1 := maxint DIV 2;
    IF (iand(iseed,ib14) <> 0) THEN BEGIN
        iseed := ior(2*iand(ieor(iseed,mask),mask1),ib1); irbit2 := 1 END
    ELSE BEGIN
        iseed := iand(2*iand(iseed,mask1),inot(ib1)); irbit2 := 0 END
END;
```

The Data Encryption Standard

```
FUNCTION ran4(VAR idum: integer): real;
(* Programs using routine RAN4 must define the global variables
TYPE
    gl64array = ARRAY [1..64] OF integer;
    gl65reals = ARRAY [1..65] OF real;
VAR
    glnewkey: integer; glinp,glkey: gl64array; glpow: gl65reals;
in the main routine. The initialization block, IF (idum < 0), has been
written in real arithmetic to avoid overflow on machines with 2-byte
integers. With 4-byte integers, this block can be simplified with MOD
and DIV. *)
CONST
    im=11979; rm=11979.0; a=430.0; c=2531.0; nacc=24;
VAR
    isav,j: integer; jot: gl64array; r4,dum: real;
BEGIN
    IF (idum < 0) THEN BEGIN
        dum := idum MOD im;
        IF (dum < 0.0) THEN dum := dum+rm;
        glpow[1] := 0.5;
        FOR j := 1 TO 64 DO BEGIN
            dum := dum*a+c; dum := dum-rm*trunc(dum/rm);
            glkey[j] := trunc(2.0*dum/rm);
            glinp[j] := trunc(4.0*dum/rm) MOD 2;
            glpow[j+1] := 0.5*glpow[j] END;
        idum := round(dum); glnewkey := 1 END;
    isav := glinp[64];
    IF (isav <> 0) THEN BEGIN
        glinp[4] := 1-glinp[4]; glinp[3] := 1-glinp[3];
        glinp[1] := 1-glinp[1] END;
    FOR j := 64 DOWNTO 2 DO BEGIN
        glinp[j] := glinp[j-1] END;
    glinp[1] := isav; des(glinp,glkey,glnewkey,0,jot); r4 := 0;
    FOR j := 1 TO nacc DO BEGIN
        IF (jot[j] <> 0) THEN r4 := r4+glpow[j]
    END;
    ran4 := r4
END;
```

```
PROCEDURE des(input,key: gl64array; VAR newkey: integer;
          isw: integer; VAR jotput: gl64array);
(* Programs using routine DES must define the type
TYPE
    gl32array = ARRAY [1..32] OF integer;
    gl64array = ARRAY [1..64] OF integer;
    gl48array = ARRAY [1..48] OF real;
in the main routine. They must also declare the variables
VAR
    ip,ipm: gl64array; desflg: boolean;
and initialize desflg to true. The values of ip and ipm are read from
the file desinp.dat whose contents are listed at the end of this routine.
The procedure GLOPEN is used to assign desinp.dat to infile and open the file
for reading. *)
VAR
    j,ii,ic,i: integer; titmp,icf: gl32array; itmp: gl64array;
    kns: ARRAY[1..48,1..16] OF integer;
    tkns: gl48array; infile: text;
BEGIN
    IF desflg THEN BEGIN
        desflg := false; glopen(infile,'desinp.dat');
        FOR i := 1 TO 64 DO read(infile,ip[i]);
        FOR i := 1 TO 64 DO read(infile,ipm[i]);
        close(infile) END;
    IF (newkey <> 0) THEN BEGIN
        newkey := 0;
        FOR i := 1 TO 16 DO BEGIN
            ks(key,i,tkns);
            FOR j := 1 TO 48 DO kns[j,i] := tkns[j]
        END END;
    FOR j := 1 TO 64 DO itmp[j] := input[ip[j]];
    FOR i := 1 TO 16 DO BEGIN
        ii := i;
        IF (isw = 1) THEN ii := 17-i;
        FOR j := 1 TO 48 DO tkns[j] := kns[j,ii];
        FOR j := 1 TO 32 DO titmp[j] := itmp[32+j];
        cyfun(titmp,tkns,icf);
        FOR j := 1 TO 32 DO BEGIN
            ic := icf[j]+itmp[j]; itmp[j] := itmp[j+32];
            itmp[j+32] := abs(ic MOD 2)
        END END;
    FOR j := 1 TO 32 DO BEGIN
        ic := itmp[j]; itmp[j] := itmp[j+32]; itmp[j+32] := ic END;
    FOR j := 1 TO 64 DO jotput[j] := itmp[ipm[j]]
END;
(* Contents of the file desinp.dat
58 50 42 34 26 18 10 2 60 52 44 36 28 20 12 4 62 54 46
38 30 22 14 6 64 56 48 40 32 24 16 8 57 49 41 33 25 17 9 1 59 51
43 35 27 19 11 3 61 53 45 37 29 21 13 5 63 55 47 39 31 23 15 7
40 8 48 16 56 24 64 32 39 7 47 15 55 23 63 31 38 6 46 14
54 22 62 30 37 5 45 13 53 21 61 29 36 4 44 12 52 20 60 28 35 3
43 11 51 19 59 27 34 2 42 10 50 18 58 26 33 1 41 9 49 17 57 25 *)


PROCEDURE ks(key: gl64array; n: integer; VAR kn: gl48array);
(* Programs using routine KS must define the types
TYPE
    gl48array = ARRAY [1..48] OF integer;
    gl56array = ARRAY [1..56] OF integer;
    gl64array = ARRAY [1..64] OF integer;
in the main routine. They must also declare the variables
VAR
    glicd,ipc1: gl56array; ipc2: gl48array; ksflg: boolean;
and initialize ksflg to true. The values of ipc1 and ipc2 are read from
the file ksinpu.dat whose contents are given at the end of this routine. The
```

```
procedure GLOPEN assigns ksinpu.dat to infile and opens the file for reading. *)
VAR
   j,it,id,ic,i: integer; infile: text;
BEGIN
IF ksflg THEN BEGIN
      ksflg := false; glopen(infile,'ksinpu.dat');
      FOR i := 1 TO 56 DO read(infile,ipc1[i]);
      FOR i := 1 TO 48 DO read(infile,ipc2[i]);
      close(infile) END;
   IF (n = 1) THEN BEGIN
      FOR j := 1 TO 56 DO BEGIN
         glicd[j] := key[ipc1[j]] END END;
   it := 2;
   IF ((n = 1) OR (n = 2) OR (n = 9) OR (n = 16)) THEN it := 1;
   FOR i := 1 TO it DO BEGIN
      ic := glicd[1]; id := glicd[29];
      FOR j := 1 TO 27 DO BEGIN
         glicd[j] := glicd[j+1]; glicd[j+28] := glicd[j+29] END;
      glicd[28] := ic; glicd[56] := id END;
   FOR j := 1 TO 48 DO kn[j] := glicd[ipc2[j]]
END;
(* Contents of the file ksinpu.dat
57 49 41 33 25 17 9 1 58 50 42 34 26 18 10 2 59 51 43 35 27 19 11 3 60 52 44 36
63 55 47 39 31 23 15 7 62 54 46 38 30 22 14 6 61 53 45 37 29 21 13 5 28 20 12 4
14 17 11 24 1 5 3 28 15 6 21 10 23 19 12 4 26 8 16 7 27 20 13 2
41 52 31 37 47 55 30 40 51 45 33 48 44 49 39 56 34 53 46 42 50 36 29 32 *)
PROCEDURE cyfun(ir: gl32array; k: gl48array; VAR iout: gl32array);
(* Programs using routine CYFUN must define the types
TYPE
   gl32array = ARRAY [1..32] OF integer;
   gl48array = ARRAY [1..48] OF integer;
in the main routine. They must also declare the variables
VAR
   iet: gl48array; ipp: gl32array;
   is: ARRAY[1..16,1..4,1..8] OF integer;
   ibin: ARRAY[1..4,1..16] OF integer;
   cyflg: boolean;
and initialize cyflg to true. The values of iet, ipp, is and ibin are read
from the file cyfuni.dat. The procedure GLOPEN assigns cyfuni.dat to
infile and opens the file for reading. *)
VAR
   jj,irow,icol,kk,iss,ki,j: integer; ie: gl48array; itmp: gl32array;
   infile: text;
BEGIN
   IF cyflg THEN BEGIN
      cyflg := false; glopen(infile,'cyfuni.dat');
      FOR j := 1 TO 48 DO read(infile,iet[j]);
      FOR j := 1 TO 32 DO read(infile,ipp[j]);
      FOR jj := 1 TO 8 DO
         FOR ki := 1 TO 4 DO
            FOR j := 1 TO 16 DO read(infile,is[j,ki,jj]);
      FOR j := 1 TO 16 DO
         FOR ki := 1 TO 4 DO read(infile,ibin[ki,j]);
      close(infile) END;
   FOR j := 1 TO 48 DO ie[j] := (((ir[iet[j]]+k[j]) MOD 2)+2) MOD 2;
   FOR jj := 1 TO 8 DO BEGIN
      j := 6*jj-5; irow := 2*ie[j]+ie[j+5];
      icol := 8*ie[j+1]+4*ie[j+2]+2*ie[j+3]+ie[j+4];
      iss := is[icol+1,irow+1,jj]; kk := 4*(jj-1);
      FOR ki := 1 TO 4 DO itmp[kk+ki] := ibin[ki,iss+1]
   END;
   FOR j := 1 TO 32 DO iout[j] := itmp[ipp[j]]
END;
(* Contents of the file cyfuni.dat
```

```
32 1 2 3 4 5 4 5 6 7 8 9 8 9 10 11 12 13 12 13 14 15 16 17
16 17 18 19 20 21 20 21 22 23 24 25 24 25 26 27 28 29 28 29 30 31 32 1
16 7 20 21 29 12 28 17 1 15 23 26 5 18 31 10
2 8 24 14 32 27 3 9 19 13 30 6 22 11 4 25
14 4 13 1 2 15 11 8 3 10 6 12 5 9 0 7 0 15 7 4 14 2 13 1 10 6 12 11 9 5 3 8
4 1 14 8 13 6 2 11 15 12 9 7 3 10 5 0 15 12 8 2 4 9 1 7 5 11 3 14 10 0 6 13
15 1 8 14 6 11 3 4 9 7 2 13 12 0 5 10 3 13 4 7 15 2 8 14 12 0 1 10 6 9 11 5
0 14 7 11 10 4 13 1 5 8 12 6 9 3 2 15 13 8 10 1 3 15 4 2 11 6 7 12 0 5 14 9
10 0 9 14 6 3 15 5 1 13 12 7 11 4 2 8 13 7 0 9 3 4 6 10 2 8 5 14 12 11 15 1
13 6 4 9 8 15 3 0 11 1 2 12 5 10 14 7 1 10 13 0 6 9 8 7 4 15 14 3 11 5 2 12
7 13 14 3 0 6 9 10 1 2 8 5 11 12 4 15 13 8 11 5 6 15 0 3 4 7 2 12 1 10 14 9
10 6 9 0 12 11 7 13 15 1 3 14 5 2 8 4 3 15 0 6 10 1 13 8 9 4 5 11 12 7 2 14
2 12 4 1 7 10 11 6 8 5 3 15 13 0 14 9 14 11 2 12 4 7 13 1 5 0 15 10 3 9 8 6
4 2 1 11 10 13 7 8 15 9 12 5 6 3 0 14 11 8 12 7 1 14 2 13 6 15 0 9 10 4 5 3
12 1 10 15 9 2 6 8 0 13 3 4 14 7 5 11 10 15 4 2 7 12 9 5 6 1 13 14 0 11 3 8
9 14 15 5 2 8 12 3 7 0 4 10 1 13 11 6 4 3 2 12 9 5 15 10 11 14 1 7 6 0 8 13
4 11 2 14 15 0 8 13 3 12 9 7 5 10 6 1 13 0 11 7 4 9 1 10 14 3 5 12 2 15 8 6
1 4 11 13 12 3 7 14 10 15 6 8 0 5 9 2 6 11 13 8 1 4 10 7 9 5 0 15 14 2 3 12
13 2 8 4 6 15 11 1 10 9 3 14 5 0 12 7 1 15 13 8 10 3 7 4 12 5 6 11 0 14 9 2
7 11 4 1 9 12 14 2 0 6 10 13 15 3 5 8 2 1 14 7 4 10 8 13 15 12 9 0 3 5 6 11
0 0 0 0 0 0 0 1 0 0 1 0 0 0 1 1 0 1 0 0 0 1 0 1 0 1 1 0 0 1 1 1
1 0 0 0 1 0 0 1 1 0 1 0 1 0 1 1 1 1 0 0 1 1 0 1 1 1 1 0 1 1 1 1 *)
```

Chapter 8. Sorting

Straight Insertion and Shell's Method

```
PROCEDURE piksrt(n: integer; VAR arr: glsarray);
(* Programs using routine PIKSRT must define the type
TYPE
   glsarray = ARRAY [1..np] OF real;
in the main routine, with np >= n.    *)
LABEL 10;
VAR
   j,i: integer; a: real;
BEGIN
   FOR j := 2 TO n DO BEGIN
      a := arr[j];
      FOR i := j-1 DOWNTO 1 DO BEGIN
         IF (arr[i] <= a) THEN GOTO 10;
         arr[i+1] := arr[i] END;
      i := 0;
10:   arr[i+1] := a
   END
END;
```

```
PROCEDURE piksr2(n: integer; VAR arr,brr: glsarray);
(* Programs using routine PIKSR2 must define the type
TYPE
   glsarray = ARRAY [1..np] OF real;
in the main routine, with np >= n.    *)
LABEL 10;
VAR
   j,i: integer; b,a: real;
BEGIN
   FOR j := 2 TO n DO BEGIN
      a := arr[j]; b := brr[j];
      FOR i := j-1 DOWNTO 1 DO BEGIN
         IF (arr[i] <= a) THEN GOTO 10;
         arr[i+1] := arr[i]; brr[i+1] := brr[i] END;
```

```
        i := 0;
10:     arr[i+1] := a;
        brr[i+1] := b END
END;

PROCEDURE shell(n: integer; VAR arr: glnarray);
(* Programs using routine SHELL must define the type
TYPE
    glnarray = ARRAY [1..np] OF real;
in the main routine, with np >= n.    *)
LABEL 3;
CONST
    aln2i=1.442695022; tiny=1.0e-5;
VAR
    nn,m,lognb2,l,k,j,i: integer; t: real;
BEGIN
    lognb2 := trunc(ln(n)*aln2i+tiny); m := n;
    FOR nn := 1 TO lognb2 DO BEGIN
        m := m DIV 2;
        k := n-m;
        FOR j := 1 TO k DO BEGIN
            i := j;
3:          l := i+m;
            IF (arr[l] < arr[i]) THEN BEGIN
                t := arr[i]; arr[i] := arr[l]; arr[l] := t; i := i-m;
                IF (i >= 1) THEN GOTO 3
            END END END
END;
```

Heapsort

```
PROCEDURE sort(n: integer; VAR ra: glsarray);
(* Program using routine SORT must define the type
TYPE
    glsarray = ARRAY [1..np] OF real;
in the main routine, with np >= n.    *)
LABEL 99;
VAR
    l,j,ir,i: integer; rra: real;
BEGIN
    l := (n DIV 2)+1;
    ir := n;
    WHILE true DO BEGIN
        IF (l > 1) THEN BEGIN
            l := l-1; rra := ra[l] END
        ELSE BEGIN
            rra := ra[ir]; ra[ir] := ra[1]; ir := ir-1;
            IF (ir = 1) THEN BEGIN
                ra[1] := rra;
                GOTO 99
            END END;
        i := l; j := l+l;
        WHILE (j <= ir) DO BEGIN
            IF (j < ir) THEN
                IF (ra[j] < ra[j+1]) THEN j := j+1;
            IF (rra < ra[j]) THEN BEGIN
                ra[i] := ra[j]; i := j; j := j+j END
            ELSE
                j := ir+1 END;
        ra[i] := rra END;
99: END;
```

```
PROCEDURE sort2(n: integer;VAR ra,rb: glsarray);
(* Programs using routine SORT2 must define type
TYPE
    glsarray = ARRAY [1..np] OF real;
in the main routine, with np >= n.    *)
LABEL 99;
VAR
    l,j,ir,i: integer; rrb,rra: real;
BEGIN
    l := (n DIV 2)+1;
    ir := n;
    WHILE true DO BEGIN
        IF (l > 1) THEN BEGIN
            l := l-1; rra := ra[l]; rrb := rb[l] END
        ELSE BEGIN
            rra := ra[ir]; rrb := rb[ir]; ra[ir] := ra[1]; rb[ir] := rb[1];
            ir := ir-1;
            IF (ir = 1) THEN BEGIN
                ra[1] := rra; rb[1] := rrb;
                GOTO 99
            END END;
        i := l; j := l+l;
        WHILE (j <= ir) DO BEGIN
            IF (j < ir) THEN
                IF (ra[j] < ra[j+1]) THEN j := j+1;
            IF (rra < ra[j]) THEN BEGIN
                ra[i] := ra[j]; rb[i] := rb[j]; i := j; j := j+j END
            ELSE j := ir+1
        END;
        ra[i] := rra; rb[i] := rrb END;
99: END;
```

Indexing and Ranking

```
PROCEDURE indexx(n: integer; arrin: glsarray; VAR indx: gliarray);
(* Programs using routine INDEXX must define the types
TYPE
    glsarray = ARRAY [1..np] OF real;
    gliarray = ARRAY [1..np] OF integer;
in the main routine, with np >= n.    *)
LABEL 99;
VAR
    l,j,ir,indxt,i: integer; q: real;
BEGIN
    FOR j := 1 TO n DO BEGIN
        indx[j] := j END;
    l := (n DIV 2) + 1;
    ir := n;
    WHILE true DO BEGIN
        IF (l > 1) THEN BEGIN
                l := l-1; indxt := indx[l]; q := arrin[indxt] END
        ELSE BEGIN
            indxt := indx[ir]; q := arrin[indxt]; indx[ir] := indx[1];
            ir := ir-1;
            IF (ir = 1) THEN BEGIN
                indx[1] := indxt;
                GOTO 99
            END END;
        i := l; j := l+l;
        WHILE (j <= ir) DO BEGIN
            IF (j < ir) THEN BEGIN
                IF (arrin[indx[j]] < arrin[indx[j+1]]) THEN j := j+1
```

```
            END;
            IF (q < arrin[indx[j]]) THEN BEGIN
                indx[i] := indx[j]; i := j; j := j+j END
            ELSE
                j := ir+1 END;
        indx[i] := indxt END;
99: END;

PROCEDURE sort3(n: integer; VAR ra,rb,rc,wksp: glsarray;
        VAR iwksp: gliarray);
(* Programs using routine SORT3 must define the types
TYPE
    glsarray = ARRAY [1..np] OF real;
    gliarray = ARRAY [1..np] OF integer;
in the main routine, with np >= n.    *)
VAR
    j: integer;
BEGIN
    indexx(n,ra,iwksp);
    FOR j := 1 TO n DO BEGIN
        wksp[j] := ra[j] END;
    FOR j := 1 TO n DO BEGIN
        ra[j] := wksp[iwksp[j]] END;
    FOR j := 1 TO n DO BEGIN
        wksp[j] := rb[j] END;
    FOR j := 1 TO n DO BEGIN
        rb[j] := wksp[iwksp[j]] END;
    FOR j := 1 TO n DO BEGIN
        wksp[j] := rc[j] END;
    FOR j := 1 TO n DO BEGIN
        rc[j] := wksp[iwksp[j]] END
END;

PROCEDURE rank(n: integer; indx: gliarray; VAR irank: gliarray);
(* Programs using routine RANK must define type
TYPE
    gliarray = ARRAY [1..np] OF real;
in the main routine, with np >= n.    *)
VAR
    j: integer;
BEGIN
    FOR j := 1 TO n DO BEGIN
        irank[indx[j]] := j END
END;
```

Quicksort

```
PROCEDURE qcksrt(n: integer; VAR arr: glarray);
(* Programs using routine QCKSRT must define the type
TYPE
    glarray = ARRAY [1..np] OF real;
in the main routine, with np >= n.    *)
LABEL 11,21,22,30,99;
CONST
    m=7; nstack=50; fm=7875; fa=211.0; fc=1663.0;
VAR
    l,jstack,j,ir,iq,i: integer; fx,fmi,a: real;
    istack: ARRAY[1..nstack] OF integer;
BEGIN
    fmi := 1.0/fm; jstack := 0; l := 1; ir := n; fx := 0.0;
```

```
    WHILE true DO BEGIN
        IF ((ir-l) < m) THEN BEGIN
            FOR j := l+1 TO ir DO BEGIN
                a := arr[j];
                FOR i := j-1 DOWNTO 1 DO BEGIN
                    IF (arr[i] <= a) THEN GOTO 11;
                    arr[i+1] := arr[i] END;
                i := 0;
11:             arr[i+1] := a
            END;
            IF (jstack = 0) THEN GOTO 99;
            ir := istack[jstack]; l := istack[jstack-1]; jstack := jstack-2 END
        ELSE BEGIN
            i := l; j := ir; fx := (fx*fa+fc)/fm; fx := fx-trunc(fx);
            iq := l+(ir-l+1)*trunc(fx*fmi); a := arr[iq]; arr[iq] := arr[l];
21:         IF (j > 0) THEN BEGIN
                IF (a < arr[j]) THEN BEGIN
                    j := j-1;
                    GOTO 21
                END END;
            IF (j <= i) THEN BEGIN
                arr[i] := a;
                GOTO 30
            END;
            arr[i] := arr[j]; i := i+1;
22:         IF (i <= n) THEN IF (a > arr[i]) THEN BEGIN
                i := i+1;
                GOTO 22
            END;
            IF (j <= i) THEN BEGIN
                arr[j] := a; i := j;
                GOTO 30
            END;
            arr[j] := arr[i]; j := j-1;
            GOTO 21;
30:         jstack := jstack+2;
            IF (jstack > nstack) THEN BEGIN
                writeln('pause in QCKSRT - NSTACK must be made larger'); readln
            END;
            IF ((ir-i) >= (i-l)) THEN BEGIN
                istack[jstack] := ir; istack[jstack-1] := i+1; ir := i-1 END
            ELSE BEGIN
                istack[jstack] := i-1; istack[jstack-1] := l; l := i+1 END
        END END;
99: END;
```

Determination of Equivalence Classes

```
PROCEDURE eclass(VAR nf: glnarray; n: integer;
        lista,listb: glmarray; m: integer);
(* Programs using routine ECLASS must define the types
TYPE
    glnarray = ARRAY [1..n] OF integer;
    glmarray = ARRAY [1..m] OF integer;
in the main routine. *)
VAR
    l,k,j: integer;
BEGIN
    FOR k := 1 TO n DO BEGIN
        nf[k] := k END;
    FOR l := 1 TO m DO BEGIN
```

```
        j := lista[l];
        WHILE (nf[j] <> j) DO j := nf[j];
        k := listb[l];
        WHILE (nf[k] <> k) DO k := nf[k];
        IF (j <> k) THEN nf[j] := k
    END;
    FOR j := 1 TO n DO WHILE (nf[j] <> nf[nf[j]]) DO nf[j] := nf[nf[j]];
END;

PROCEDURE eclazz(VAR nf: glnarray; n: integer);
(* Programs using routine ECLAZZ must supply a boolean function
equiv(i,j:integer):boolean which indicates (TRUE or FALSE) whether
i and j belong to the same equivalence class. They must also
define the type
TYPE
    glnarray = ARRAY [1..n] OF integer;
in the main routine. *)
VAR
    kk,jj: integer;
BEGIN
    nf[1] := 1;
    FOR jj := 2 TO n DO BEGIN
        nf[jj] := jj;
        FOR kk := 1 TO jj-1 DO BEGIN
             nf[kk] := nf[nf[kk]];
             IF (equiv(jj,kk)) THEN nf[nf[nf[kk]]] := jj
        END END;
    FOR jj := 1 TO n DO nf[jj] := nf[nf[jj]]
END;
```

Chapter 9. Root Finding and Nonlinear Sets of Equations

Introduction

```
PROCEDURE scrsho;
(* Programs using routine SCRSHO must externally define a
function fx(x:real):real which is to be plotted. *)
LABEL 1,99;
CONST
   iscr=60; jscr=21; blank=' '; zero='-'; yy='l'; xx='-'; ff='x';
VAR
   jz,j,i: integer; ysml,ybig,x2,x1,x,dyj,dx: real;
   y: ARRAY [1..iscr] OF real;
   scr: ARRAY [1..iscr,1..jscr] OF char;
BEGIN
1: writeln('Enter x1 x2 (x1=x2 to stop): '); readln(x1,x2);
   IF (x1 = x2) THEN GOTO 99;
   FOR j := 1 TO jscr DO BEGIN
       scr[1,j] := yy; scr[iscr,j] := yy END;
   FOR i := 2 TO iscr-1 DO BEGIN
       scr[i,1] := xx; scr[i,jscr] := xx;
       FOR j := 2 TO jscr-1 DO BEGIN
           scr[i,j] := blank END END;
   dx := (x2-x1)/(iscr-1); x := x1; ybig := 0.0; ysml := ybig;
   FOR i := 1 TO iscr DO BEGIN
       y[i] := fx(x);
       IF (y[i] < ysml) THEN ysml := y[i];
       IF (y[i] > ybig) THEN ybig := y[i];
       x := x+dx END;
   IF (ybig = ysml) THEN ybig := ysml+1.0;
```

```
    dyj := (jscr-1)/(ybig-ysml); jz := 1-trunc(ysml*dyj);
    FOR i := 1 TO iscr DO BEGIN
        scr[i,jz] := zero; j := 1+trunc((y[i]-ysml)*dyj); scr[i,j] := ff
    END;
    write(' ',ybig:10:3,' ');
    FOR i := 1 TO iscr DO write(scr[i,jscr]);
    writeln;
    FOR j := jscr-1 DOWNTO 2 DO BEGIN
        write(' ':12);
        FOR i := 1 TO iscr DO write(scr[i,j]);
        writeln END;
    write(' ',ysml:10:3,' ');
    FOR i := 1 TO iscr DO write(scr[i,1]);
    writeln; writeln(' ':8,x1:10:3,' ':44,x2:10:3);
    GOTO 1;
99: END;
```

Bracketing and Bisection

```
PROCEDURE zbrac(VAR x1,x2: real; VAR succes: boolean);
(* Programs using routine ZBRAC must externally define a
function fx(x:real):real which is to be bracketed. *)
LABEL 99;
CONST
    factor=1.6; ntry=50;
VAR
    j: integer; f2,f1: real;
BEGIN
    IF (x1 = x2) THEN BEGIN
        writeln('pause in routine ZBRAC');
        writeln('you have to guess an initial range'); readln END;
    f1 := fx(x1); f2 := fx(x2); succes := true;
    FOR j := 1 TO ntry DO BEGIN
        IF (f1*f2 < 0.0) THEN GOTO 99;
        IF (abs(f1) < abs(f2)) THEN BEGIN
            x1 := x1+factor*(x1-x2); f1 := fx(x1) END
        ELSE BEGIN
            x2 := x2+factor*(x2-x1); f2 := fx(x2) END END;
    succes := false;
99: END;

PROCEDURE zbrak(x1,x2: real; n: integer;
        VAR xb1,xb2: glnbmax; VAR nb: integer);
(* Programs using routine ZBRAK must externally define a function
fx(x:real):real which is to be analyzed for roots. Also, the
calling program must define type
TYPE
    glnbmax = ARRAY [1..nbmax] OF real;
where nbmax is the maximum number to be required for 'nb'. *)
LABEL 99;
VAR
    nbb,i: integer; x,fp,fc,dx: real;
BEGIN
    nbb := nb; nb := 0; x := x1; dx := (x2-x1)/n; fp := fx(x);
    FOR i := 1 TO n DO BEGIN
        x := x+dx; fc := fx(x);
        IF ((fc*fp) < 0.0) THEN BEGIN
            nb := nb+1; xb1[nb] := x-dx; xb2[nb] := x END;
        fp := fc;
        IF (nbb = nb) THEN GOTO 99;
    END;
99: END;
```

```
FUNCTION rtbis(x1,x2,xacc: real): real;
(* Programs using routine RTBIS must externally define a function
fx(x:real):real which is to be analyzed for roots. *)
LABEL 99;
CONST
    jmax=40;
VAR
    dx,f,fmid,xmid,rtb: real; j: integer;
BEGIN
    fmid := fx(x2); f := fx(x1);
    IF ((f*fmid) >= 0.0) THEN BEGIN
        writeln('pause in RTBIS');
        writeln('Root must be bracketed for bisection.'); readln
    END;
    IF (f < 0.0) THEN BEGIN
        rtb := x1; dx := x2-x1 END
    ELSE BEGIN
        rtb := x2; dx := x1-x2 END;
    FOR j := 1 TO jmax DO BEGIN
        dx := dx*0.5; xmid := rtb+dx; fmid := fx(xmid);
        IF (fmid <= 0.0) THEN rtb := xmid;
        IF ((abs(dx) < xacc) OR (fmid = 0.0)) THEN GOTO 99
    END;
    writeln('pause in RTBIS - too many bisections'); readln;
99: rtbis := rtb
END;
```

Secant Method and False Position Method

```
FUNCTION rtflsp(x1,x2,xacc: real): real;
(* Programs using routine RTFLSP must externally define a function
fx(x:real):real which is to be analyzed for roots. *)
LABEL 99;
CONST
    maxit=30;
VAR
    xl,xh,swap,fl: real; dx,del,f,fh,rtf: real; j: integer;
BEGIN
    fl := fx(x1); fh := fx(x2);
    IF (fl*fh > 0.0) THEN BEGIN
        writeln('pause in routine RTFLSP');
        writeln('Root must be bracketed for false position'); readln
    END;
    IF (fl < 0.0) THEN BEGIN
        xl := x1; xh := x2 END
    ELSE BEGIN
        xl := x2; xh := x1; swap := fl; fl := fh; fh := swap END;
    dx := xh-xl;
    FOR j := 1 TO maxit DO BEGIN
        rtf := xl+dx*fl/(fl-fh); f := fx(rtf);
        IF (f < 0.0) THEN BEGIN
            del := xl-rtf; xl := rtf; fl := f END
        ELSE BEGIN
            del := xh-rtf; xh := rtf; fh := f END;
        dx := xh-xl;
        IF ((abs(del) < xacc) OR (f = 0.0)) THEN GOTO 99
    END;
    writeln('pause in routine RTFLSP');
    writeln('maximum number of iterations exceeded'); readln;
99: rtflsp := rtf
END;
```

```
FUNCTION rtsec(x1,x2,xacc: real): real;
(* Programs using routine RTSEC must externally define a
function fx(x:real):real which is to be analyzed for roots. *)
LABEL 99;
CONST
    maxit=30;
VAR
    dx,f,fl,swap,xl,rts: real; j: integer;
BEGIN
    fl := fx(x1); f := fx(x2);
    IF (abs(fl) < abs(f)) THEN BEGIN
        rts := x1; xl := x2; swap := fl; fl := f; f := swap END
    ELSE BEGIN
        xl := x1; rts := x2 END;
    FOR j := 1 TO maxit DO BEGIN
        dx := (xl-rts)*f/(f-fl); xl := rts; fl := f; rts := rts+dx;
        f := fx(rts);
        IF ((abs(dx) < xacc) OR (f = 0.0)) THEN GOTO 99
    END;
    writeln('pause in routine RTSEC');
    writeln('maximum number of iterations exceeded'); readln;
99: rtsec := rts
END;
```

Van Wijngaarden–Dekker–Brent Method

```
FUNCTION zbrent(x1,x2,tol: real): real;
(* Programs using routine ZBRENT must externally define a function
fx(x:real):real a root of which is to be found. *)
LABEL 99;
CONST
    itmax=100; eps=3.0e-8;
VAR
    a,b,c,d,e: real; min1,min2,min: real; fa,fb,fc,p,q,r: real;
    s,tol1,xm: real; iter: integer;
BEGIN
    a := x1; b := x2; fa := fx(a); fb := fx(b);
    IF (fb*fa > 0.0) THEN BEGIN
        writeln('pause in routine ZBRENT');
        writeln('root must be bracketed'); readln END;
    fc := fb;
    FOR iter := 1 TO itmax DO BEGIN
        IF (fb*fc > 0.0) THEN BEGIN
            c := a; fc := fa; d := b-a; e := d END;
        IF (abs(fc) < abs(fb)) THEN BEGIN
            a := b; b := c; c := a; fa := fb; fb := fc; fc := fa END;
        tol1 := 2.0*eps*abs(b)+0.5*tol; xm := 0.5*(c-b);
        IF ((abs(xm) <= tol1) OR (fb = 0.0)) THEN BEGIN
            zbrent := b; GOTO 99 END;
        IF ((abs(e) >= tol1) AND (abs(fa) > abs(fb))) THEN BEGIN
            s := fb/fa;
            IF (a = c) THEN BEGIN
                p := 2.0*xm*s; q := 1.0-s END
            ELSE BEGIN
                q := fa/fc; r := fb/fc;
                p := s*(2.0*xm*q*(q-r)-(b-a)*(r-1.0));
                q := (q-1.0)*(r-1.0)*(s-1.0) END;
            IF (p > 0.0) THEN q := -q;
            p := abs(p); min1 := 3.0*xm*q-abs(tol1*q); min2 := abs(e*q);
            IF (min1 < min2) THEN min := min1 ELSE min := min2;
            IF (2.0*p < min) THEN BEGIN
```

```
               e := d; d := p/q END
            ELSE BEGIN
               d := xm; e := d END END
         ELSE BEGIN
            d := xm; e := d END;
         a := b; fa := fb;
         IF (abs(d) > tol1) THEN BEGIN
            b := b+d END
         ELSE BEGIN
            IF (xm >= 0) THEN BEGIN
               b := b+abs(tol1) END
            ELSE BEGIN
               b := b-abs(tol1) END END;
         fb := fx(b) END;
   writeln('pause in routine ZBRENT');
   writeln('maximum number of iterations exceeded'); readln; zbrent := b;
99: END;
```

Newton-Raphson Method Using Derivative

```
FUNCTION rtnewt(x1,x2,xacc: real): real;
(* Programs using routine RTNEWT must externally define procedure
funcd(x,f,df:real) which returns the function value f and its
derivative df at the point x. *)
LABEL 99;
CONST
   jmax=20;
VAR
   df,dx,f,rtn: real; j: integer;
BEGIN
   rtn := 0.5*(x1+x2);
   FOR j := 1 TO jmax DO BEGIN
      funcd(rtn,f,df); dx := f/df; rtn := rtn-dx;
      IF ((x1-rtn)*(rtn-x2) < 0.0) THEN BEGIN
         writeln('pause in routine RTNEWT');
         writeln('jumped out of brackets'); readln END;
      IF (abs(dx) < xacc) THEN GOTO 99
   END;
   writeln('pause in routine RTNEWT');
   writeln('maximum number of iterations exceeded'); readln;
99: rtnewt := rtn
END;
```

```
FUNCTION rtsafe(x1,x2,xacc: real): real;
(* Programs using routine RTSAFE must externally define procedure
funcd(x,f,df:real) which returns the function value f and its
derivative df at the point x. *)
LABEL 99;
CONST
   maxit=100;
VAR
   df,dx,dxold,f,fh,fl: real; temp,xh,xl,rts: real; j: integer;
BEGIN
   funcd(x1,fl,df); funcd(x2,fh,df);
   IF (fl*fh >= 0.0) THEN BEGIN
      writeln('pause in routine RTSAFE');
      writeln('root must be bracketed'); readln END;
   IF (fl < 0.0) THEN BEGIN
      xl := x1; xh := x2 END
   ELSE BEGIN
      xh := x1; xl := x2 END;
   rts := 0.5*(x1+x2); dxold := abs(x2-x1); dx := dxold; funcd(rts,f,df);
```

```
    FOR j := 1 TO maxit DO BEGIN
        IF((((rts-xh)*df-f)*((rts-xl)*df-f) >= 0.0)
        OR (abs(2.0*f) > abs(dxold*df))) THEN BEGIN
            dxold := dx; dx := 0.5*(xh-xl); rts := xl+dx;
            IF (xl = rts) THEN GOTO 99 END
        ELSE BEGIN
            dxold := dx; dx := f/df; temp := rts; rts := rts-dx;
            IF (temp = rts) THEN GOTO 99
        END;
        IF (abs(dx) < xacc) THEN GOTO 99;
        funcd(rts,f,df);
        IF (f < 0.0) THEN BEGIN
            xl := rts END
        ELSE BEGIN
            xh := rts END END;
    writeln('pause in RTSAFE');
    writeln('maximum number of iterations exceeded'); readln;
99: rtsafe := rts
END;
```

Roots of Polynomials

```
PROCEDURE laguer(a: glcarray; m: integer; VAR x: gl2array;
       eps: real; polish: boolean);
(* Programs using routine LAGUER must define in the main routine the types
TYPE
    glcarray = ARRAY [1..2*m+2] OF real; gl2array = ARRAY [1..2] OF real; *)
LABEL 99; CONST epss=6.0e-8; mxit=100;
VAR
    j,iter: integer; err,dxold,cdx,abx,dum: real;
    sq,h,gp,gm,g2,g: gl2array; b,d,dx,f,x1,cdum: gl2array;
PROCEDURE cdiv(a,b: gl2array; VAR c:gl2array);
(* Complex division of a by b, answer in c *)
VAR    r,den: real;
BEGIN
    IF (abs(b[1]) >= abs(b[2])) THEN BEGIN
        r := b[2]/b[1]; den := b[1]+r*b[2];
        c[1] := (a[1]+a[2]*r)/den; c[2] := (a[2]-a[1]*r)/den END
    ELSE BEGIN
        r := b[1]/b[2]; den := b[2]+r*b[1];
        c[1] := (a[1]*r+a[2])/den; c[2] := (a[2]*r-a[1])/den END END;
FUNCTION cabs(a: gl2array): real; (* Complex absolute value of a *)
VAR    x,y : real;
BEGIN
    x := abs(a[1]); y := abs(a[2]);
    IF (x = 0.0) THEN cabs := y ELSE IF (y = 0.0) THEN cabs := x
    ELSE IF (x > y) THEN cabs := x*sqrt(1.0+sqr(y/x))
    ELSE cabs := y*sqrt(1.0+sqr(x/y)) END;
PROCEDURE csqrt(a: gl2array; VAR b: gl2array);
(* Returns complex square root of a in b *)
VAR    x,y,u,v,w,r : real;
BEGIN
    IF ((a[1] = 0.0) AND (a[2] = 0.0)) THEN BEGIN
        u := 0.0; v := 0.0 END
    ELSE BEGIN
        x := abs(a[1]); y := abs(a[2]);
        IF (x >= y) THEN w := sqrt(x)*sqrt(0.5*(1.0+sqrt(1.0+sqr(y/x))))
        ELSE BEGIN
            r := x/y; w := sqrt(y)*sqrt(0.5*(r+sqrt(1.0+sqr(r)))) END;
        IF (a[1] >= 0.0) THEN BEGIN
            u := w; v := a[2]/(2.0*u) END
        ELSE BEGIN
```

```
          IF (a[2] >= 0.0) THEN v := w ELSE v := -w; u := a[2]/(2.0*v) END END;
   b[1] := u; b[2] := v; END;
BEGIN
   dxold := cabs(x);
   FOR iter := 1 to mxit DO BEGIN
      b[1] := a[2*m+1]; b[2] := a[2*m+2]; err := cabs(b); d[1] := 0.0;
      d[2] := 0.0; f[1] := 0.0; f[2] := 0.0; abx := cabs(x);
      FOR j := m DOWNTO 1 DO BEGIN
         dum := f[1]; f[1] := x[1]*f[1]-x[2]*f[2]+d[1];
         f[2] := x[1]*f[2]+x[2]*dum+d[2]; dum := d[1];
         d[1] := x[1]*d[1]-x[2]*d[2]+b[1]; d[2] := x[1]*d[2]+x[2]*dum+b[2];
         dum := b[1]; b[1] := x[1]*b[1]-x[2]*b[2]+a[2*j-1];
         b[2] := x[1]*b[2]+x[2]*dum+a[2*j]; err := cabs(b)+abx*err END;
      err := epss*err;
      IF (cabs(b) <= err) THEN BEGIN
         dx[1] := 0.0; dx[2] := 0.0; GOTO 99 END
      ELSE BEGIN
         cdiv(d,b,g); g2[1] := sqr(g[1])-sqr(g[2]); g2[2] := 2.0*g[1]*g[2];
         cdiv(f,b,cdum); h[1] := g2[1]-2.0*cdum[1]; h[2] := g2[2]-2.0*cdum[2];
         cdum[1] := (m-1)*(m*h[1]-g2[1]); cdum[2] := (m-1)*(m*h[2]-g2[2]);
         csqrt(cdum,sq); gp[1] := g[1]+sq[1]; gp[2] := g[2]+sq[2];
         gm[1] := g[1]-sq[1]; gm[2] := g[2]-sq[2];
         IF(cabs(gp) < cabs(gm)) THEN BEGIN
            gp[1] := gm[1]; gp[2] := gm[2] END;
         cdum[1] := m; cdum[2] := 0.0; cdiv(cdum,gp,dx); END;
      x1[1] := x[1]-dx[1]; x1[2] := x[2]-dx[2];
      IF ((x[1] = x1[1]) AND (x[2] = x1[2])) THEN GOTO 99;
      x[1] := x1[1]; x[2] := x1[2]; cdx := cabs(dx);
      IF ((iter > 6) AND (cdx >= dxold)) THEN GOTO 99; dxold := cdx;
      IF (not polish) THEN
         IF (cabs(dx) <= eps*cabs(x)) THEN GOTO 99 END;
   writeln('pause in routine LAGUER - too many iterations'); readln;
99: END;


PROCEDURE zroots(a: glcarray; m: integer; VAR roots: glcarray;
         polish: boolean);
(* Programs using routine ZROOTS must define the types
TYPE
   glcarray = ARRAY [1..2*m+2] OF real;
   gl2array = ARRAY [1..2] OF real;
in the main routine. *)
LABEL 10;
CONST
   eps=2.0e-6;
VAR
   jj,j,i: integer; dum: real; b,c,x: gl2array; ad: glcarray;
BEGIN
   FOR j := 1 TO 2*(m+1) DO BEGIN
      ad[j] := a[j] END;
   FOR j := m DOWNTO 1 DO BEGIN
      x[1] := 0.0; x[2] := 0.0; laguer(ad,j,x,eps,false);
      IF (abs(x[2]) <= (2.0*sqr(eps)*abs(x[1]))) THEN BEGIN
         x[2] := 0.0 END;
      roots[2*j-1] := x[1]; roots[2*j] := x[2]; b[1] := ad[2*j+1];
      b[2] := ad[2*j+2];
      FOR jj := j DOWNTO 1 DO BEGIN
         c[1] := ad[2*jj-1]; c[2] := ad[2*jj]; ad[2*jj-1] := b[1];
         ad[2*jj] := b[2]; dum := b[1]; b[1] := b[1]*x[1]-b[2]*x[2]+c[1];
         b[2] := dum*x[2]+b[2]*x[1]+c[2] END END;
   IF (polish) THEN BEGIN
      FOR j := 1 TO m DO BEGIN
         x[1] := roots[2*j-1]; x[2] := roots[2*j]; laguer(a,m,x,eps,true);
         roots[2*j-1] := x[1]; roots[2*j] := x[2] END END;
   FOR j := 2 TO m DO BEGIN
```

```
        x[1] := roots[2*j-1]; x[2] := roots[2*j];
        FOR i := j-1 DOWNTO 1 DO BEGIN
            IF (roots[2*i-1] <= x[1]) THEN GOTO 10;
            roots[2*i+1] := roots[2*i-1]; roots[2*i+2] := roots[2*i] END;
        i := 0;
10:     roots[2*i+1] := x[1];
        roots[2*i+2] := x[2] END
END;
```

```
PROCEDURE qroot(p: glnarray; n: integer; VAR b,c: real; eps: real);
(* Programs using procedure QROOT must define the types
TYPE
    glnarray = ARRAY [1..n] OF real;
    glnvarray = ARRAY [1..3] OF real;
in the main routine *)
LABEL 99;
CONST
    itmax=20; tiny=1.0e-6;
VAR
    iter,i: integer; sc,sb,s,rc,rb,r,dv,delc,delb: real;
    q,qq,rem: glnarray; d: glnvarray;
BEGIN
    d[3] := 1.0;
    FOR iter := 1 TO itmax DO BEGIN
        d[2] := b; d[1] := c; poldiv(p,n,d,3,q,rem); s := rem[1];
        r := rem[2]; poldiv(q,n-1,d,3,qq,rem); sc := -rem[1]; rc := -rem[2];
        FOR i := n-1 DOWNTO 1 DO BEGIN
            q[i+1] := q[i] END;
        q[1] := 0.0; poldiv(q,n,d,3,qq,rem); sb := -rem[1]; rb := -rem[2];
        dv := 1.0/(sb*rc-sc*rb); delb := (r*sc-s*rc)*dv;
        delc := (-r*sb+s*rb)*dv; b := b+delb; c := c+delc;
        IF(((abs(delb) <= eps*abs(b)) OR (abs(b) < tiny)) AND
         ((abs(delc) <= eps*abs(c)) OR (abs(c) < tiny))) THEN GOTO 99
    END;
    writeln('pause in routine QROOT - too many iterations');
99: END;
```

Newton-Raphson Method for Nonlinear Systems of Equations

```
PROCEDURE mnewt(ntrial: integer; VAR x: glnarray; n: integer;
        tolx,tolf: real);
(* Programs using routine MNEWT must define the types
TYPE
    glnarray = ARRAY [1..n] OF real;
    glnbyn = ARRAY [1..n,1..n] OF real;
    glindx = ARRAY [1..n] OF integer;
in the main routine. They must also supply a routine
PROCEDURE usrfun(x: glnarray; n: integer; VAR alpha: glnbyn;
         VAR beta: glnarray);    *)
LABEL 99;
VAR
    k,i: integer; errx,errf,d: real; beta: glnarray; alpha: glnbyn;
    indx: glindx;
BEGIN
    FOR k := 1 TO ntrial DO BEGIN
        usrfun(x,n,alpha,beta); errf := 0.0;
        FOR i := 1 TO n DO errf := errf+abs(beta[i]);
        IF (errf <= tolf) THEN GOTO 99;
        ludcmp(alpha,n,np,indx,d); lubksb(alpha,n,np,indx,beta); errx := 0.0;
        FOR i := 1 TO n DO BEGIN
```

```
            errx := errx+abs(beta[i]); x[i] := x[i]+beta[i] END;
         IF (errx <= tolx) THEN GOTO 99
      END;
 99: END;
```

Chapter 10. Minimization or Maximization of Functions

Golden Section Search in One Dimension

```
PROCEDURE mnbrak(VAR ax,bx,cx,fa,fb,fc: real);
(* Programs using routine MNBRAK must supply an external
function func(x:real):real for which a minimum is to be found *)
LABEL 1;
CONST
   gold=1.618034; glimit=100.0; tiny=1.0e-20;
VAR
   ulim,u,r,q,fu,dum: real;
FUNCTION max(a,b: real): real;
   BEGIN
      IF (a > b) THEN max := a ELSE max := b
   END;
FUNCTION sign(a,b: real): real;
   BEGIN
      IF (b >= 0.0) THEN sign := abs(a) ELSE sign := -abs(a)
   END;
BEGIN
   fa := func(ax); fb := func(bx);
   IF (fb > fa) THEN BEGIN
      dum := ax; ax := bx; bx := dum; dum := fb; fb := fa; fa := dum
   END;
   cx := bx+gold*(bx-ax); fc := func(cx);
1: IF (fb >= fc) THEN BEGIN
      r := (bx-ax)*(fb-fc); q := (bx-cx)*(fb-fa);
      u := bx-((bx-cx)*q-(bx-ax)*r)/ (2.0*sign(max(abs(q-r),tiny),q-r));
      ulim := bx+glimit*(cx-bx);
      IF ((bx-u)*(u-cx) > 0.0) THEN BEGIN
         fu := func(u);
         IF (fu < fc) THEN BEGIN
            ax := bx; fa := fb; bx := u; fb := fu;
            GOTO 1 END
         ELSE IF (fu > fb) THEN BEGIN
            cx := u; fc := fu;
            GOTO 1
         END;
         u := cx+gold*(cx-bx); fu := func(u) END
      ELSE IF ((cx-u)*(u-ulim) > 0.0) THEN BEGIN
         fu := func(u);
         IF (fu < fc) THEN BEGIN
            bx := cx; cx := u; u := cx+gold*(cx-bx); fb := fc;
            fc := fu; fu := func(u) END END
      ELSE IF ((u-ulim)*(ulim-cx) >= 0.0) THEN BEGIN
         u := ulim; fu := func(u) END
      ELSE BEGIN
         u := cx+gold*(cx-bx); fu := func(u) END;
      ax := bx; bx := cx; cx := u; fa := fb; fb := fc; fc := fu;
      GOTO 1
   END
END;
```

```
FUNCTION golden(ax,bx,cx,tol: real; VAR xmin: real): real;
(* Programs using routine GOLDEN must supply an external
function func(x:real):real whose minimum is to be found. *)
CONST
   r=0.61803399;
VAR
   f0,f1,f2,f3,c: real; x0,x1,x2,x3: real;
BEGIN
   c := 1.0-r; x0 := ax; x3 := cx;
   IF (abs(cx-bx) > abs(bx-ax)) THEN BEGIN
      x1 := bx; x2 := bx+c*(cx-bx) END
   ELSE BEGIN
      x2 := bx; x1 := bx-c*(bx-ax) END;
   f1 := func(x1); f2 := func(x2);
   WHILE (abs(x3-x0) > tol*(abs(x1)+abs(x2))) DO BEGIN
      IF (f2 < f1) THEN BEGIN
         x0 := x1; x1 := x2; x2 := r*x1+c*x3; f0 := f1; f1 := f2;
         f2 := func(x2) END
      ELSE BEGIN
         x3 := x2; x2 := x1; x1 := r*x2+c*x0; f3 := f2; f2 := f1;
         f1 := func(x1) END END;
   IF (f1 < f2) THEN BEGIN
      golden := f1; xmin := x1 END
   ELSE BEGIN
      golden := f2; xmin := x2 END
END;
```

Parabolic Interpolation and Brent's Method in One Dimension

```
FUNCTION brent(ax,bx,cx,tol: real; VAR xmin: real): real;
(* Programs using routine BRENT must supply an external function
func(x:real):real whose minimum is to be found. *)
LABEL 1,2,3;
CONST
   itmax=100; cgold=0.3819660; zeps=1.0e-10;
VAR
   a,b,d,e,etemp: real; fu,fv,fw,fx: real; iter: integer;
   p,q,r,tol1,tol2: real; u,v,w,x,xm: real;
FUNCTION sign(a,b: real): real;
   BEGIN
      IF (b >= 0.0) THEN sign := abs(a) ELSE sign := -abs(a)
   END;
BEGIN
   IF ax < cx THEN a := ax ELSE a := cx;
   IF ax > cx THEN b := ax ELSE b := cx;
   v := bx; w := v; x := v; e := 0.0; fx := func(x); fv := fx; fw := fx;
   FOR iter := 1 TO itmax DO BEGIN
      xm := 0.5*(a+b); tol1 := tol*abs(x)+zeps; tol2 := 2.0*tol1;
      IF (abs(x-xm) <= (tol2-0.5*(b-a))) THEN GOTO 3;
      IF (abs(e) > tol1) THEN BEGIN
         r := (x-w)*(fx-fv); q := (x-v)*(fx-fw); p := (x-v)*q-(x-w)*r;
         q := 2.0*(q-r);
         IF (q > 0.0) THEN p := -p;
         q := abs(q); etemp := e; e := d;
         IF((abs(p) >= abs(0.5*q*etemp)) OR (p <= q*(a-x))
            OR (p >= q*(b-x))) THEN GOTO 1;
         d := p/q; u := x+d;
         IF (((u-a) < tol2) OR ((b-u) < tol2)) THEN d := sign(tol1,xm-x);
         GOTO 2
      END;
1:    IF (x >= xm) THEN e := a-x ELSE e := b-x;
```

```
        d := cgold*e;
2:      IF (abs(d) >= tol1) THEN u := x+d ELSE u := x+sign(tol1,d);
        fu := func(u);
        IF (fu <= fx) THEN BEGIN
            IF (u >= x) THEN a := x ELSE b := x;
            v := w; fv := fw; w := x; fw := fx; x := u; fx := fu END
        ELSE BEGIN
            IF (u < x) THEN a := u ELSE b := u;
            IF ((fu <= fw) OR (w = x)) THEN BEGIN
                v := w; fv := fw; w := u; fw := fu END
            ELSE IF ((fu <= fv) OR (v = x) OR (v = 2)) THEN BEGIN
                v := u; fv := fu END END END;
    writeln('pause in routine BRENT - too many iterations');
3:  xmin := x;
    brent := fx
END;
```

One-Dimensional Search with First Derivatives

```
FUNCTION dbrent(ax,bx,cx,tol: real; VAR xmin: real): real;
(* Programs using routine DBRENT must define the external
functions func(x:real):real and dfunc(x:real):real which are,
respectively, the function whose minimum is to be found,
and its derivative. *)
LABEL 1,2,3;
CONST
    itmax=100; zeps=1.0e-10;
VAR
    a,b,d,d1,d2: real; du,dv,dw,dx: real; e,fu,fv,fw,fx: real;
    iter: integer; olde,tol1,tol2: real; u,u1,u2,v,w,x,xm: real;
    ok1,ok2: boolean;
FUNCTION sign(a,b: real): real;
    BEGIN
        IF (b >= 0.0) THEN sign := abs(a) ELSE sign := -abs(a)
    END;
BEGIN
    IF ax < cx THEN a := ax ELSE a := cx;
    IF ax > cx THEN b := ax ELSE b := cx;
    v := bx; w := v; x := v; e := 0.0; fx := func(x); fv := fx; fw := fx;
    dx := dfunc(x); dv := dx; dw := dx;
    FOR iter := 1 TO itmax DO BEGIN
        xm := 0.5*(a+b); tol1 := tol*abs(x)+zeps; tol2 := 2.0*tol1;
        IF (abs(x-xm) <= (tol2-0.5*(b-a))) THEN GOTO 3;
        IF (abs(e) > tol1) THEN BEGIN
            d1 := 2.0*(b-a); d2 := d1;
            IF (dw <> dx) THEN d1 := (w-x)*dx/(dx-dw);
            IF (dv <> dx) THEN d2 := (v-x)*dx/(dx-dv);
            u1 := x+d1; u2 := x+d2;
            ok1 := ((a-u1)*(u1-b) > 0.0) AND (dx*d1 <= 0.0);
            ok2 := ((a-u2)*(u2-b) > 0.0) AND (dx*d2 <= 0.0);
            olde := e; e := d;
            IF (NOT (ok1 OR ok2)) THEN GOTO 1
            ELSE IF (ok1 AND ok2) THEN BEGIN
                IF (abs(d1) < abs(d2)) THEN BEGIN
                    d := d1 END
                ELSE BEGIN
                    d := d2 END END
            ELSE IF (ok1) THEN BEGIN
                d := d1 END
            ELSE BEGIN
                d := d2 END;
```

```
            IF (abs(d) > abs(0.5*olde)) THEN GOTO 1;
            u := x+d;
            IF (((u-a) < tol2) OR ((b-u) < tol2)) THEN BEGIN
                d := sign(tol1,xm-x) END;
            GOTO 2
        END;
1:      IF (dx >= 0.0) THEN e := a-x ELSE e := b-x;
        d := 0.5*e;
2:      IF (abs(d) >= tol1) THEN BEGIN
            u := x+d; fu := func(u) END
        ELSE BEGIN
            u := x+sign(tol1,d); fu := func(u);
            IF (fu > fx) THEN GOTO 3
        END;
        du := dfunc(u);
        IF (fu <= fx) THEN BEGIN
            IF (u >= x) THEN a := x ELSE b := x;
            v := w; fv := fw; dv := dw; w := x; fw := fx; dw := dx;
            x := u; fx := fu; dx := du END
        ELSE BEGIN
            IF (u < x) THEN a := u ELSE b := u;
            IF ((fu <= fw) OR (w = x)) THEN BEGIN
                v := w; fv := fw; dv := dw; w := u; fw := fu; dw := du END
            ELSE IF ((fu < fv) OR (v = x) OR (v = w)) THEN BEGIN
                v := u; fv := fu; dv := du END END END;
    writeln('pause in routine DBRENT - too many iterations');
3:  xmin := x;
    dbrent := fx
END;
```

Downhill Simplex Method in Multidimensions

```
PROCEDURE amoeba(VAR p: glmpnp; VAR y: glmp; ndim: integer;
         ftol: real; VAR iter: integer);
(* Programs using routine AMOEBA must supply an external function
func(pr:glnp):real whose minimum is to be found. They must
also define types
TYPE
    glmpnp = ARRAY [1..mp,1..np] OF real;
    glmp = ARRAY [1..mp] OF real;
    glnp = ARRAY [1..np] OF real;
where mp and np are physical dimensions *)
LABEL 99;
CONST
    alpha=1.0; beta=0.5; gamma=2.0; itmax=500;
VAR
    mpts,j,inhi,ilo,ihi,i: integer; yprr,ypr,rtol: real; pr,prr,pbar: glnp;
BEGIN
    mpts := ndim+1; iter := 0;
    WHILE true DO BEGIN
        ilo := 1;
        IF (y[1] > y[2]) THEN BEGIN
            ihi := 1; inhi := 2 END
        ELSE BEGIN
            ihi := 2; inhi := 1 END;
        FOR i := 1 TO mpts DO BEGIN
            IF (y[i] < y[ilo]) THEN ilo := i;
            IF (y[i] > y[ihi]) THEN BEGIN
                inhi := ihi; ihi := i END
            ELSE IF (y[i] > y[inhi]) THEN
                IF (i <> ihi) THEN inhi := i
        END;
```

```
        rtol := 2.0*abs(y[ihi]-y[ilo])/(abs(y[ihi])+abs(y[ilo]));
        IF (rtol < ftol) THEN GOTO 99;
        IF (iter = itmax) THEN BEGIN
            writeln('pause in AMOEBA - too many iterations'); readln
        END;
        iter := iter+1;
        FOR j := 1 TO ndim DO pbar[j] := 0.0;
        FOR i := 1 TO mpts DO
            IF (i <> ihi) THEN FOR j := 1 TO ndim DO pbar[j] := pbar[j]+p[i,j];
        FOR j := 1 TO ndim DO BEGIN
            pbar[j] := pbar[j]/ndim;
            pr[j] := (1.0+alpha)*pbar[j]-alpha*p[ihi,j] END;
        ypr := func(pr);
        IF (ypr <= y[ilo]) THEN BEGIN
            FOR j := 1 TO ndim DO prr[j] := gamma*pr[j]+(1.0-gamma)*pbar[j];
            yprr := func(prr);
            IF (yprr < y[ilo]) THEN BEGIN
                FOR j := 1 TO ndim DO p[ihi,j] := prr[j];
                y[ihi] := yprr END
            ELSE BEGIN
                FOR j := 1 TO ndim DO p[ihi,j] := pr[j];
                y[ihi] := ypr END END
        ELSE IF (ypr >= y[inhi]) THEN BEGIN
            IF (ypr < y[ihi]) THEN BEGIN
                FOR j := 1 TO ndim DO p[ihi,j] := pr[j];
                y[ihi] := ypr END;
            FOR j := 1 TO ndim DO prr[j] := beta*p[ihi,j]+(1.0-beta)*pbar[j];
            yprr := func(prr);
            IF (yprr < y[ihi]) THEN BEGIN
                FOR j := 1 TO ndim DO p[ihi,j] := prr[j];
                y[ihi] := yprr END
            ELSE BEGIN
                FOR i := 1 TO mpts DO
                    IF (i <> ilo) THEN BEGIN
                        FOR j := 1 TO ndim DO BEGIN
                            pr[j] := 0.5*(p[i,j]+p[ilo,j]); p[i,j] := pr[j]
                        END;
                        y[i] := func(pr) END END END
        ELSE BEGIN
            FOR j := 1 TO ndim DO p[ihi,j] := pr[j];
            y[ihi] := ypr END END;
99: END;
```

Conjugate Gradient Methods in Multidimensions

```
PROCEDURE powell(VAR p: glnarray; VAR xi: glnpbynp; n,np: integer;
        ftol: real; VAR iter: integer; VAR fret: real);
(* Programs using routine POWELL must define the types
TYPE
    glnarray = ARRAY [1..n] OF real;
    glnpbynp = ARRAY [1..np,1..np] OF real;
in the main routine. *)
LABEL 1,99;
CONST
    itmax=200;
VAR
    j,ibig,i: integer; t,fptt,fp,del: real; pt,ptt,xit: glnarray;
BEGIN
    fret := fnc(p);
    FOR j := 1 TO n DO BEGIN
        pt[j] := p[j] END;
    iter := 0;
```

```
1:  iter := iter+1;
    fp := fret; ibig := 0; del := 0.0;
    FOR i := 1 TO n DO BEGIN
        FOR j := 1 TO n DO BEGIN
            xit[j] := xi[j,i] END;
        fptt := fret; linmin(p,xit,n,fret);
        IF (abs(fptt-fret) > del) THEN BEGIN
            del := abs(fptt-fret); ibig := i END END;
    IF (2.0*abs(fp-fret) <= ftol*(abs(fp)+abs(fret))) THEN GOTO 99;
    IF (iter = itmax) THEN BEGIN
        writeln('pause in routine POWELL');
        writeln('too many interations'); readln END;
    FOR j := 1 TO n DO BEGIN
        ptt[j] := 2.0*p[j]-pt[j]; xit[j] := p[j]-pt[j]; pt[j] := p[j] END;
    fptt := fnc(ptt);
    IF (fptt >= fp) THEN GOTO 1;
    t := 2.0*(fp-2.0*fret+fptt)*sqr(fp-fret-del)-del*sqr(fp-fptt);
    IF (t >= 0.0) THEN GOTO 1;
    linmin(p,xit,n,fret);
    FOR j := 1 TO n DO BEGIN
        xi[j,ibig] := xit[j] END;
    GOTO 1;
99: END;


PROCEDURE linmin(VAR p,xi: glnarray; n: integer; VAR fret: real);
(* Programs using routine LINMIN must define the type
TYPE
    glnarray = ARRAY [1..n] OF real;
They must also declare the variables
VAR
    ncom: integer; pcom,xicom: glnarray;
in the main routine. Also the function FUNC referenced by BRENT
and MNBRAK must be set to return the function F1DIM. *)
CONST
    tol=1.0e-4;
VAR
    j: integer; xx,xmin,fx,fb,fa,bx,ax: real;
BEGIN
    ncom := n;
    FOR j := 1 TO n DO BEGIN
        pcom[j] := p[j]; xicom[j] := xi[j] END;
    ax := 0.0; xx := 1.0; bx := 2.0; mnbrak(ax,xx,bx,fa,fx,fb);
    fret := brent(ax,xx,bx,tol,xmin);
    FOR j := 1 TO n DO BEGIN
        xi[j] := xmin*xi[j]; p[j] := p[j]+xi[j] END
END;


FUNCTION f1dim(x: real): real;
(* Programs using F1DIM must declare the variables
TYPE
    glndim = ARRAY [1..ndim] OF real;
VAR
    ncom: integer; pcom,xicom: glndim;
and assign values to them externally. *)
VAR
    j: integer; xt: glndim;
BEGIN
    FOR j := 1 TO ncom DO BEGIN
        xt[j] := pcom[j]+x*xicom[j] END;
    f1dim := fnc(xt)
END;
```

Variable Metric Methods in Multidimensions

```
PROCEDURE frprmn(VAR p: glnarray; n: integer; ftol: real;
        VAR iter: integer; VAR fret: real);
(* Programs using routine FRPRMN must supply a
FUNCTION fnc(p: glnarray):real; and a
PROCEDURE dfnc(p: glnarray; VAR g: glnarray);
which evaluate a function and its gradient. They must
also define the type
TYPE
    glnarray = ARRAY [1..n] OF real;
in the main routine. *)
LABEL 99;
CONST
    itmax=200; eps=1.0e-10;
VAR
    j,its: integer; gg,gam,fp,dgg: real; g,h,xi: glnarray;
BEGIN
    fp := fnc(p); dfnc(p,xi);
     FOR j := 1 TO n DO BEGIN
        g[j] := -xi[j]; h[j] := g[j]; xi[j] := h[j] END;
    FOR its := 1 TO itmax DO BEGIN
        iter := its; linmin(p,xi,n,fret);
        IF ((2.0*abs(fret-fp)) <= (ftol*(abs(fret)+abs(fp)+eps)))
            THEN GOTO 99;
        fp := fnc(p); dfnc(p,xi); gg := 0.0; dgg := 0.0;
        FOR j := 1 TO n DO BEGIN
            gg := gg+sqr(g[j]);
(*            dgg := dgg+sqr(xi[j])    *)
            dgg := dgg+(xi[j]+g[j])*xi[j] END;
        IF (gg = 0.0) THEN GOTO 99;
        gam := dgg/gg;
        FOR j := 1 TO n DO BEGIN
            g[j] := -xi[j]; h[j] := g[j]+gam*h[j]; xi[j] := h[j] END END;
    writeln('pause in routine FRPRMN');
    writeln('too many iterations'); readln;
99: END;
```

```
FUNCTION df1dim(x: real): real;
(* Programs using routine DF1DIM must define the type
TYPE
    glnarray = ARRAY [1..n] OF real;
They must also define the variables
VAR
    ncom: integer; pcom,xicom: glnarray
in the main routine, and externally assign them values. *)
VAR
    df1: real; j: integer; xt,df: glnarray;
BEGIN
    FOR j := 1 TO ncom DO BEGIN
        xt[j] := pcom[j]+x*xicom[j] END;
    dfunc(xt,df); df1 := 0.0;
    FOR j := 1 TO ncom DO BEGIN
        df1 := df1+df[j]*xicom[j] END;
    df1dim := df1
END;
```

```
PROCEDURE dfpmin(VAR p: glnarray; n: integer; ftol: real;
        VAR iter: integer; VAR fret: real);
(* Programs using routine DFPMIN must supply a
FUNCTION fnc(p: glnarray):real; and a
PROCEDURE dfnc(p: glnarray; VAR g: glnarray);
```

```
which evaluate a function and its gradient. They must
also define the types
TYPE
    glnarray = ARRAY [1..n] OF real;
    glnbyn = ARRAY [1..n,1..n] OF real;
in the main routine. *)
LABEL 99;
CONST
    itmax=200; eps=1.0e-10;
VAR
    j,i,its: integer; fp,fae,fad,fac: real; xi,g,dg: glnarray;
    hdg: glnarray; hessin: glnbyn;
BEGIN
    fp := fnc(p); dfnc(p,g);
    FOR i := 1 TO n DO BEGIN
        FOR j := 1 TO n DO BEGIN
            hessin[i,j] := 0.0 END;
        hessin[i,i] := 1.0; xi[i] := -g[i] END;
    FOR its := 1 TO itmax DO BEGIN
        iter := its; linmin(p,xi,n,fret);
        IF ((2.0*abs(fret-fp)) <= (ftol*(abs(fret)+abs(fp)+eps)))
            THEN GOTO 99;
        fp := fret;
        FOR i := 1 TO n DO BEGIN
            dg[i] := g[i] END;
        fret := fnc(p); dfnc(p,g);
        FOR i := 1 TO n DO BEGIN
            dg[i] := g[i]-dg[i] END;
        FOR i := 1 TO n DO BEGIN
            hdg[i] := 0.0;
            FOR j := 1 TO n DO BEGIN
                hdg[i] := hdg[i]+hessin[i,j]*dg[j] END END;
        fac := 0.0; fae := 0.0;
        FOR i := 1 TO n DO BEGIN
            fac := fac+dg[i]*xi[i]; fae := fae+dg[i]*hdg[i] END;
        fac := 1.0/fac; fad := 1.0/fae;
        FOR i := 1 TO n DO BEGIN
            dg[i] := fac*xi[i]-fad*hdg[i] END;
        FOR i := 1 TO n DO BEGIN
            FOR j := 1 TO n DO BEGIN
                hessin[i,j] := hessin[i,j]+fac*xi[i]*xi[j]
                -fad*hdg[i]*hdg[j]+fae*dg[i]*dg[j]; END END;
        FOR i := 1 TO n DO BEGIN
            xi[i] := 0.0;
            FOR j := 1 TO n DO BEGIN
                xi[i] := xi[i]-hessin[i,j]*g[j] END END END;
    writeln('pause in routine DFPMIN');
    writeln('too many iterations'); readln;
99: END;
```

Linear Programming and the Simplex Method

```
PROCEDURE simplx(VAR a: glmpbynp; m,n,mp,np,m1,m2,m3: integer;
         VAR icase: integer; VAR izrov: glnarray;
         VAR iposv: glmarray);
(* Programs using routine SIMPLX must define the types
TYPE
    glmpbynp = ARRAY [1..mp,1..np] OF real;
    glnarray = ARRAY [1..n] OF integer;
    glmarray = ARRAY [1..m] OF integer;
    glmparray = ARRAY [1..mp] OF integer;
    glnparray = ARRAY [1..np] OF integer;
```

```
in the main routine. *)
LABEL 1,2,10,20,30,99;
CONST eps=1.0e-6;
VAR
   nl2,nl1,m12,kp,kh,k,is,ir,ip,i: integer; q1,bmax: real; l1: glnparray;
   l2,l3: glmparray;
BEGIN
   IF (m <> (m1+m2+m3)) THEN BEGIN
      writeln('pause in routine SIMPLX');
      writeln('bad input constraint counts'); readln END;
   nl1 := n;
   FOR k := 1 TO n DO BEGIN
      l1[k] := k; izrov[k] := k END;
   nl2 := m;
   FOR i := 1 TO m DO BEGIN
      IF (a[i+1,1] < 0.0) THEN BEGIN
         writeln('pause in routine SIMPLX');
         writeln('bad input tableau'); readln END;
      l2[i] := i; iposv[i] := n+i END;
   FOR i := 1 TO m2 DO BEGIN
      l3[i] := 1 END;
   ir := 0;
   IF ((m2+m3) = 0) THEN GOTO 30;
   ir := 1;
   FOR k := 1 TO n+1 DO BEGIN
      q1 := 0.0;
      FOR i := m1+1 TO m DO BEGIN
         q1 := q1+a[i+1,k] END;
      a[m+2,k] := -q1 END;
10: simp1(a,mp,np,m+1,l1,nl1,0,kp,bmax);
   IF ((bmax <= eps) AND (a[m+2,1] < -eps)) THEN BEGIN
      icase := -1; GOTO 99 END
   ELSE IF ((bmax <= eps) AND (a[m+2,1] <= eps)) THEN BEGIN
      m12 := m1+m2+1;
      IF (m12 <= m) THEN BEGIN
         FOR ip := m12 TO m DO BEGIN
            IF (iposv[ip] = (ip+n)) THEN BEGIN
               simp1(a,mp,np,ip,l1,nl1,1,kp,bmax);
               IF (bmax > 0.0) THEN GOTO 1
            END END END;
      ir := 0; m12 := m12-1;
      IF ((m1+1) > m12) THEN GOTO 30;
      FOR i := m1+1 TO m12 DO BEGIN
         IF (l3[i-m1] = 1) THEN BEGIN
            FOR k := 1 TO n+1 DO BEGIN
               a[i+1,k] := -a[i+1,k] END END END;
      GOTO 30
   END;
   simp2(a,m,n,mp,np,l2,nl2,ip,kp,q1);
   IF (ip = 0) THEN BEGIN
      icase := -1; GOTO 99
   END;
1: simp3(a,mp,np,m+1,n,ip,kp);
   IF (iposv[ip] >= (n+m1+m2+1)) THEN BEGIN
      FOR k := 1 TO nl1 DO BEGIN
         IF (l1[k] = kp) THEN GOTO 2
      END;
2:    nl1 := nl1-1;
      FOR is := k TO nl1 DO BEGIN
         l1[is] := l1[is+1] END END
   ELSE BEGIN
      IF (iposv[ip] < (n+m1+1)) THEN GOTO 20;
      kh := iposv[ip]-m1-n;
      IF (l3[kh] = 0) THEN GOTO 20;
```

```
        l3[kh] := 0 END;
    a[m+2,kp+1] := a[m+2,kp+1]+1.0;
    FOR i := 1 TO m+2 DO BEGIN
        a[i,kp+1] := -a[i,kp+1] END;
20: is := izrov[kp];
    izrov[kp] := iposv[ip]; iposv[ip] := is;
    IF (ir <> 0) THEN GOTO 10;
30: simp1(a,mp,np,0,l1,nl1,0,kp,bmax);
    IF (bmax <= 0.0) THEN BEGIN
        icase := 0; GOTO 99
    END;
    simp2(a,m,n,mp,np,l2,nl2,ip,kp,q1);
    IF (ip = 0) THEN BEGIN
        icase := 1; GOTO 99
    END;
    simp3(a,mp,np,m,n,ip,kp);
    GOTO 20;
99: END;


PROCEDURE simp1(a: glmpbynp; mp,np,mm: integer;
        l1: glnparray; nl1,iabf: integer;
        VAR kp: integer; VAR bmax: real);
(* Programs using routine SIMP1 must define the types
TYPE
    glmpbynp = ARRAY [1..mp,1..np] OF real;
    glnparray = ARRAY [1..np] OF integer;
in the main routine. *)
LABEL 99;
VAR
    k: integer; test: real;
BEGIN
    kp := l1[1]; bmax := a[mm+1,kp+1];
    IF (nl1 < 2) THEN GOTO 99;
    FOR k := 2 TO nl1 DO BEGIN
        IF (iabf = 0) THEN BEGIN
            test := a[mm+1,l1[k]+1]-bmax END
        ELSE BEGIN
            test := abs(a[mm+1,l1[k]+1])-abs(bmax) END;
        IF (test > 0.0) THEN BEGIN
            bmax := a[mm+1,l1[k]+1]; kp := l1[k] END END;
99: END;


PROCEDURE simp2(a: glmpbynp; m,n,mp,np: integer;
        l2: glmparray; nl2: integer; VAR ip: integer;
        kp: integer; VAR q1: real);
(* Programs using routine SIMP2 must define the types
TYPE
    glmpbynp = ARRAY [1..mp,1..np] OF real;
    glmparray = ARRAY [1..mp] OF integer;
in the main routine. *)
LABEL 2,6,99;
VAR
    k,ii,i: integer; qp,q0,q: real;
BEGIN
    ip := 0;
    IF (nl2 < 1) THEN GOTO 99;
    FOR i := 1 TO nl2 DO BEGIN
        IF (a[l2[i]+1,kp+1] < 0.0) THEN GOTO 2
    END;
    GOTO 99;
2:  q1 := -a[l2[i]+1,1]/a[l2[i]+1,kp+1];
    ip := l2[i];
    IF ((i+1) > nl2) THEN GOTO 99;
```

```
    FOR i := i+1 TO nl2 DO BEGIN
        ii := l2[i];
        IF (a[ii+1,kp+1] < 0.0) THEN BEGIN
            q := -a[ii+1,1]/a[ii+1,kp+1];
            IF (q < q1) THEN BEGIN
                ip := ii; q1 := q END
            ELSE IF (q = q1) THEN BEGIN
                FOR k := 1 TO n DO BEGIN
                    qp := -a[ip+1,k+1]/a[ip+1,kp+1];
                    q0 := -a[ii+1,k+1]/a[ii+1,kp+1];
                    IF (q0 <> qp) THEN GOTO 6
                END;
6:                  IF (q0 < qp) THEN ip := ii
            END END END;
99: END;

PROCEDURE simp3(VAR a: glmpbynp; mp,np,i1,k1,ip,kp: integer);
(* Programs using routine SIMP3 must define the type
TYPE
    glmpbynp = ARRAY [1..mp,1..np] OF real;
in the main routine. *)
VAR
    kk,ii: integer; piv: real;
BEGIN
    piv := 1.0/a[ip+1,kp+1];
    IF (i1 >= 0) THEN BEGIN
        FOR ii := 1 TO (i1+1) DO BEGIN
            IF ((ii-1) <> ip) THEN BEGIN
                a[ii,kp+1] := a[ii,kp+1]*piv;
                FOR kk := 1 TO k1+1 DO BEGIN
                    IF ((kk-1) <> kp) THEN BEGIN
                        a[ii,kk] := a[ii,kk] -a[ip+1,kk]*a[ii,kp+1] END END
            END END END;
    FOR kk := 1 TO k1+1 DO BEGIN
        IF ((kk-1) <> kp) THEN a[ip+1,kk] := -a[ip+1,kk]*piv
    END;
    a[ip+1,kp+1] := piv
END;
```

Combinatorial Minimization: Method of Simulated Annealing

```
PROCEDURE anneal(x,y : cityarray; VAR iorder: iarray; ncity: integer);
(* Programs using routine ANNEAL must define types
    cityarray : ARRAY [1..ncity] OF real;
    iarray : ARRAY [1..ncity] OF integer;
in the main routine. *)
LABEL 10,20,99;
CONST
    tfactr = 0.9;
TYPE
    nsix = ARRAY [1..6] OF integer;
VAR
    ans : boolean; path,de,t : real; nover,nlimit,i1,i2,idum,iseed: integer;
    i,j,k,nsucc,nn,idec : integer; n : nsix;
FUNCTION alen(x1,x2,y1,y2: real): real;
BEGIN
    alen := sqrt(sqr(x2-x1)+sqr(y2-y1))
END;
PROCEDURE revcst(x,y: cityarray; iorder: iarray; ncity: integer;
        VAR n: nsix; VAR de: real);
VAR
    xx,yy : ARRAY [1..6] OF real;
```

```
    j,ii : integer;
BEGIN
    n[3] := 1 + ((n[1]+ncity-2) MOD ncity);
    n[4] := 1 + (n[2] MOD ncity);
    FOR j := 1 to 4 DO BEGIN
        ii := iorder[n[j]]; xx[j] := x[ii]; yy[j] := y[ii] END;
    de := -alen(xx[1],xx[3],yy[1],yy[3])-alen(xx[2],xx[4],yy[2],yy[4])
        +alen(xx[1],xx[4],yy[1],yy[4])+alen(xx[2],xx[3],yy[2],yy[3])
END;
PROCEDURE reverse(VAR iorder: iarray; ncity: integer; n: nsix);
VAR
    nn,j,k,l,itmp : integer;
BEGIN
    nn := (1+((n[2]-n[1]+ncity) MOD ncity)) DIV 2;
    FOR j := 1 to nn DO BEGIN
        k := 1 + ((n[1]+j-2) MOD ncity);
        l := 1 + ((n[2]-j+ncity) MOD ncity);
        itmp := iorder[k]; iorder[k] := iorder[l]; iorder[l] := itmp END
END;
PROCEDURE trncst(x,y: cityarray; iorder: iarray; ncity: integer;
        VAR n: nsix; VAR de: real);
VAR
    xx,yy : ARRAY [1..6] OF real;
    j,ii : integer;
BEGIN
    n[4] := 1 + (n[3] MOD ncity);
    n[5] := 1 + ((n[1]+ncity-2) MOD ncity);
    n[6] := 1 + (n[2] MOD ncity);
    FOR j := 1 to 6 DO BEGIN
        ii := iorder[n[j]]; xx[j] := x[ii]; yy[j] := y[ii] END;
    de := -alen(xx[2],xx[6],yy[2],yy[6])-alen(xx[1],xx[5],yy[1],yy[5])
        -alen(xx[3],xx[4],yy[3],yy[4])+alen(xx[1],xx[3],yy[1],yy[3])
        +alen(xx[2],xx[4],yy[2],yy[4])+alen(xx[5],xx[6],yy[5],yy[6])
END;
PROCEDURE trnspt(VAR iorder: iarray; ncity: integer; n: nsix);
CONST
    maxcity=1000;
VAR
    jorder : ARRAY [1..maxcity] OF integer;
    m1,m2,m3,nn,j,jj : integer;
BEGIN
    m1 := 1 + ((n[2]-n[1]+ncity) MOD ncity);
    m2 := 1 + ((n[5]-n[4]+ncity) MOD ncity);
    m3 := 1 + ((n[3]-n[6]+ncity) MOD ncity);
    nn := 1;
    FOR j := 1 to m1 DO BEGIN
        jj := 1 + ((j+n[1]-2) MOD ncity);
        jorder[nn] := iorder[jj]; nn := nn+1 END;
    IF (m2>0) THEN BEGIN
        FOR j := 1 to m2 DO BEGIN
            jj := 1+((j+n[4]-2) MOD ncity);
            jorder[nn] := iorder[jj]; nn := nn+1 END END;
    IF (m3>0) THEN BEGIN
        FOR j := 1 to m3 DO BEGIN
            jj := 1 + ((j+n[6]-2) MOD ncity);
            jorder[nn] := iorder[jj]; nn := nn+1 END END;
    FOR j := 1 to ncity DO BEGIN
        iorder[j] := jorder[j] END
END;
PROCEDURE metrop(de,t: real; VAR ans: boolean);
(* Programs using routine METROP must declare the variable
VAR
    gljdum : integer;
and initialize its value to
```

```
    gljdum := 1;
 in the main routine. *)
 BEGIN
    ans := (de<0.0) OR (ran3(gljdum)<exp(-de/t))
 END;
 BEGIN
    nover := 100*ncity; nlimit := 10*ncity; path := 0.0; t := 0.5;
    FOR i := 1 to (ncity-1) DO BEGIN
       i1 := iorder[i]; i2 := iorder[i+1];
       path := path+alen(x[i1],x[i2],y[i1],y[i2]) END;
    i1 := iorder[ncity]; i2 := iorder[1];
    path := path+alen(x[i1],x[i2],y[i1],y[i2]); idum := -1; iseed := 111;
    FOR j := 1 to 100 DO BEGIN
       nsucc := 0;
       FOR k := 1 to nover DO BEGIN
10:            n[1] := 1+trunc(ncity*ran3(idum));
          n[2] := 1+trunc((ncity-1)*ran3(idum));
          IF (n[2]>=n[1]) THEN n[2] := n[2]+1;
          nn := 1+((n[1]-n[2]+ncity-1) MOD ncity);
          IF (nn<3) THEN goto 10;
          idec := irbit1(iseed);
          IF (idec=0) THEN BEGIN
             n[3] := n[2]+trunc(abs(nn-2)*ran3(idum))+1;
             n[3] := 1+((n[3]-1) MOD ncity);
             trncst(x,y,iorder,ncity,n,de); metrop(de,t,ans);
             IF ans THEN BEGIN
                nsucc := nsucc+1; path := path+de; trnspt(iorder,ncity,n)
             END END
          ELSE BEGIN
             revcst(x,y,iorder,ncity,n,de); metrop(de,t,ans);
             IF ans THEN BEGIN
                nsucc := nsucc+1; path := path+de;
                reverse(iorder,ncity,n) END END;
          IF (nsucc>=nlimit) THEN goto 20
       END;
20:       writeln;
       writeln('T =',t:10:6,'     Path Length =',path:12:6);
       writeln('Successful Moves: ',nsucc:6);
       t := t*tfactr;
       IF (nsucc=0) THEN goto 99
    END;
99:
 END;
```

Chapter 11. Eigensystems

Jacobi Transformations of a Symmetric Matrix

```
PROCEDURE jacobi(VAR a: glnpnp; n: integer;VAR d: glnp;
      VAR v: glnpnp; VAR nrot: integer);
(* Programs using routine JACOBI must define the types
TYPE
   glnpnp = ARRAY [1..np,1..np] OF real;
   glnp = ARRAY [1..np] OF real;
where 'np by np' is the physical dimension of the array
a into which all arrays are loaded for analysis. *)
LABEL 99;
CONST
   nmax=100;
VAR
   j,iq,ip,i: integer; tresh,theta,tau,t,sm,s,h,g,c: real;
   b,z: ARRAY [1..nmax] OF real;
```

```
BEGIN
   FOR ip := 1 TO n DO BEGIN
      FOR iq := 1 TO n DO BEGIN
         v[ip,iq] := 0.0 END;
      v[ip,ip] := 1.0 END;
   FOR ip := 1 TO n DO BEGIN
      b[ip] := a[ip,ip]; d[ip] := b[ip]; z[ip] := 0.0 END;
   nrot := 0;
   FOR i := 1 TO 50 DO BEGIN
      sm := 0.0;
      FOR ip := 1 TO n-1 DO BEGIN
         FOR iq := ip+1 TO n DO BEGIN
            sm := sm+abs(a[ip,iq]) END END;
      IF (sm = 0.0) THEN GOTO 99;
      IF (i < 4) THEN tresh := 0.2*sm/sqr(n)
      ELSE tresh := 0.0;
      FOR ip := 1 TO n-1 DO BEGIN
         FOR iq := ip+1 TO n DO BEGIN
            g := 100.0*abs(a[ip,iq]);
            IF ((i > 4) AND ((abs(d[ip])+g) = abs(d[ip]))
               AND ((abs(d[iq])+g) = abs(d[iq]))) THEN
               a[ip,iq] := 0.0
            ELSE IF (abs(a[ip,iq]) > tresh) THEN BEGIN
               h := d[iq]-d[ip];
               IF ((abs(h)+g) = abs(h)) THEN BEGIN
                  t := a[ip,iq]/h END
               ELSE BEGIN
                  theta := 0.5*h/a[ip,iq];
                  t := 1.0/(abs(theta)+sqrt(1.0+sqr(theta)));
                  IF (theta < 0.0) THEN t := -t
               END;
               c := 1.0/sqrt(1+sqr(t)); s := t*c; tau := s/(1.0+c);
               h := t*a[ip,iq]; z[ip] := z[ip]-h; z[iq] := z[iq]+h;
               d[ip] := d[ip]-h; d[iq] := d[iq]+h; a[ip,iq] := 0.0;
               FOR j := 1 TO ip-1 DO BEGIN
                  g := a[j,ip]; h := a[j,iq]; a[j,ip] := g-s*(h+g*tau);
                  a[j,iq] := h+s*(g-h*tau) END;
               FOR j := ip+1 TO iq-1 DO BEGIN
                  g := a[ip,j]; h := a[j,iq]; a[ip,j] := g-s*(h+g*tau);
                  a[j,iq] := h+s*(g-h*tau) END;
               FOR j := iq+1 TO n DO BEGIN
                  g := a[ip,j]; h := a[iq,j]; a[ip,j] := g-s*(h+g*tau);
                  a[iq,j] := h+s*(g-h*tau) END;
               FOR j := 1 TO n DO BEGIN
                  g := v[j,ip]; h := v[j,iq]; v[j,ip] := g-s*(h+g*tau);
                  v[j,iq] := h+s*(g-h*tau) END;
               nrot := nrot+1 END END END;
      FOR ip := 1 TO n DO BEGIN
         b[ip] := b[ip]+z[ip]; d[ip] := b[ip]; z[ip] := 0.0 END END;
   writeln('pause in routine JACOBI');
   writeln('50 iterations should not happen'); readln;
99: END;
```

```
PROCEDURE eigsrt(VAR d: glnp; VAR v: glnpnp; n: integer);
(* Programs using routine EIGSRT must define the types
TYPE
   glnp = ARRAY [1..np] OF real;
   glnpnp = ARRAY [1..np,1..np] OF real;
where np is the physical dimension of the arrays to be used (v and d) *)
VAR
   k,j,i: integer; p: real;
BEGIN
   FOR i := 1 TO n-1 DO BEGIN
      k := i; p := d[i];
```

```
        FOR j := i+1 TO n DO BEGIN
            IF (d[j] >= p) THEN BEGIN
                k := j; p := d[j] END END;
        IF (k <> i) THEN BEGIN
            d[k] := d[i]; d[i] := p;
            FOR j := 1 TO n DO BEGIN
                p := v[j,i]; v[j,i] := v[j,k]; v[j,k] := p END END END
END;
```

Reduction of a Symmetric Matrix to Tridiagonal Form

```
PROCEDURE tred2(VAR a: glnpnp; n: integer; VAR d,e: glnp);
(* Programs using routine TRED2 must define the types
TYPE
    glnp = ARRAY [1..np] OF real;
    glnpnp = ARRAY [1..np,1..np] OF real;
where 'np by np' is the physical dimension of the matrix to be analyzed. *)
VAR
    l,k,j,i: integer; scale,hh,h,g,f: real;
FUNCTION sign(a,b: real): real;
    BEGIN
        IF (b < 0) THEN sign := -abs(a) ELSE sign := abs(a)
    END;
BEGIN
    IF (n > 1) THEN BEGIN
        FOR i := n DOWNTO 2 DO BEGIN
            l := i-1; h := 0.0; scale := 0.0;
            IF (l > 1) THEN BEGIN
                FOR k := 1 TO l DO BEGIN
                    scale := scale+abs(a[i,k]) END;
                IF (scale = 0.0) THEN BEGIN
                    e[i] := a[i,l] END
                ELSE BEGIN
                    FOR k := 1 TO l DO BEGIN
                        a[i,k] := a[i,k]/scale; h := h+sqr(a[i,k]) END;
                    f := a[i,l]; g := -sign(sqrt(h),f); e[i] := scale*g;
                    h := h-f*g; a[i,l] := f-g; f := 0.0;
                    FOR j := 1 TO l DO BEGIN
                (* Next statement can be omitted if eigenvectors not wanted *)
                        a[j,i] := a[i,j]/h; g := 0.0;
                        FOR k := 1 TO j DO BEGIN
                            g := g+a[j,k]*a[i,k] END;
                        IF (l > j) THEN FOR k := j+1 TO l DO g := g+a[k,j]*a[i,k];
                        e[j] := g/h; f := f+e[j]*a[i,j] END;
                    hh := f/(h+h);
                    FOR j := 1 TO l DO BEGIN
                        f := a[i,j]; g := e[j]-hh*f; e[j] := g;
                        FOR k := 1 TO j DO a[j,k] := a[j,k]-f*e[k]-g*a[i,k]
                    END END END
            ELSE BEGIN
                e[i] := a[i,l] END;
            d[i] := h END END;
    (* Next statement can be omitted if eigenvectors not wanted *)
    d[1] := 0.0; e[1] := 0.0;
    FOR i := 1 TO n DO BEGIN
    (* Contents of this loop can be omitted if eigenvectors not wanted,
        except for statement d[i] := a[i,i]; *)
        l := i-1;
        IF (d[i] <> 0.0) THEN BEGIN
            FOR j := 1 TO l DO BEGIN
                g := 0.0;
                FOR k := 1 TO l DO BEGIN
```

```
                g := g+a[i,k]*a[k,j] END;
            FOR k := 1 TO l DO BEGIN
                a[k,j] := a[k,j]-g*a[k,i] END END END;
        d[i] := a[i,i]; a[i,i] := 1.0;
        IF (l >= 1) THEN BEGIN
            FOR j := 1 TO l DO BEGIN
                a[i,j] := 0.0; a[j,i] := 0.0 END END END
END;
```

Eigenvalues and Eigenvectors of a Tridiagonal Matrix

```
PROCEDURE tqli(VAR d,e: glnp; n: integer; VAR z: glnpnp);
(* Programs using routine TQLI must define the types
TYPE
    glnp = ARRAY [1..np] OF real;
    glnpnp = ARRAY [1..np,1..np] OF real;
where np is the physical dimension of the matrix to be analyzed. *)
LABEL 1,2;
VAR
    m,l,iter,i,k: integer; s,r,p,g,f,dd,c,b: real;
FUNCTION sign(a,b: real): real;
    BEGIN
        IF (b < 0) THEN sign := -abs(a) ELSE sign := abs(a)
    END;
BEGIN
    IF (n > 1) THEN BEGIN
        FOR i := 2 TO n DO BEGIN
            e[i-1] := e[i] END;
        e[n] := 0.0;
        FOR l := 1 TO n DO BEGIN
            iter := 0;
1:          FOR m := l TO n-1 DO BEGIN
                dd := abs(d[m])+abs(d[m+1]);
                IF (abs(e[m])+dd = dd) THEN GOTO 2
            END;
            m := n;
2:          IF (m <> l) THEN BEGIN
                IF (iter = 30) THEN BEGIN
                    writeln('pause in routine TQLI');
                    writeln('too many iterations'); readln END;
                iter := iter+1; g := (d[l+1]-d[l])/(2.0*e[l]);
                r := sqrt(sqr(g)+1.0); g := d[m]-d[l]+e[l]/(g+sign(r,g));
                s := 1.0; c := 1.0; p := 0.0;
                FOR i := m-1 DOWNTO l DO BEGIN
                    f := s*e[i]; b := c*e[i];
                    IF (abs(f) >= abs(g)) THEN BEGIN
                        c := g/f; r := sqrt(sqr(c)+1.0); e[i+1] := f*r;
                        s := 1.0/r; c := c*s END
                    ELSE BEGIN
                        s := f/g; r := sqrt(sqr(s)+1.0); e[i+1] := g*r;
                        c := 1.0/r; s := s*c END;
                    g := d[i+1]-p; r := (d[i]-g)*s+2.0*c*b; p := s*r;
                    d[i+1] := g+p; g := c*r-b;
                (* Next loop can be omitted if eigenvectors not wanted *)
                    FOR k := 1 TO n DO BEGIN
                        f := z[k,i+1]; z[k,i+1] := s*z[k,i]+c*f;
                        z[k,i] := c*z[k,i]-s*f END END;
                d[l] := d[l]-p; e[l] := g; e[m] := 0.0;
                GOTO 1
            END END END
END;
```

Reduction of a General Matrix to Hessenberg Form

```
PROCEDURE balanc(VAR a: glnpnp; n: integer);
(* Programs using routine BALANC should define the type
TYPE
    glnpnp = ARRAY [1..np,1..np] OF real;
where 'np by np' is the physical dimension of the array to be analyzed. *)
CONST
    radix=2.0;
VAR
    last,j,i: integer; s,r,g,f,c,sqrdx: real;
BEGIN
    sqrdx := sqr(radix);
    REPEAT
        last := 1;
        FOR i := 1 TO n DO BEGIN
            c := 0.0; r := 0.0;
            FOR j := 1 TO n DO
                IF (j <> i) THEN BEGIN
                    c := c+abs(a[j,i]); r := r+abs(a[i,j]) END;
            IF ((c <> 0.0) AND (r <> 0.0)) THEN BEGIN
                g := r/radix; f := 1.0; s := c+r;
                WHILE (c < g) DO BEGIN
                    f := f*radix; c := c*sqrdx END;
                g := r*radix;
                WHILE (c > g) DO BEGIN
                    f := f/radix; c := c/sqrdx END;
                IF ((c+r)/f < 0.95*s) THEN BEGIN
                    last := 0; g := 1.0/f;
                    FOR j := 1 TO n DO a[i,j] := a[i,j]*g;
                    FOR j := 1 TO n DO a[j,i] := a[j,i]*f
                END END END;
    UNTIL (last <> 0)
END;


PROCEDURE elmhes(VAR a: glnpnp; n: integer);
(* Programs using routine ELMHES must define the type
TYPE
    glnpnp = ARRAY [1..np,1..np]
where 'np by np' is the physical dimension of the matrix to be reduced. *)
VAR
    m,j,i: integer; y,x: real;
BEGIN
    IF (n > 2) THEN BEGIN
        FOR m := 2 TO n-1 DO BEGIN
            x := 0.0; i := m;
            FOR j := m TO n DO BEGIN
                IF (abs(a[j,m-1]) > abs(x)) THEN BEGIN
                    x := a[j,m-1]; i := j END END;
            IF (i <> m) THEN BEGIN
                FOR j := m-1 TO n DO BEGIN
                    y := a[i,j]; a[i,j] := a[m,j]; a[m,j] := y END;
                FOR j := 1 TO n DO BEGIN
                    y := a[j,i]; a[j,i] := a[j,m]; a[j,m] := y END END;
            IF (x <> 0.0) THEN BEGIN
                FOR i := m+1 TO n DO BEGIN
                    y := a[i,m-1];
                    IF (y <> 0.0) THEN BEGIN
                        y := y/x; a[i,m-1] := y;
                        FOR j := m TO n DO a[i,j] := a[i,j]-y*a[m,j];
                        FOR j := 1 TO n DO a[j,m] := a[j,m]+y*a[j,i]
                    END END END END END
END;
```

The QR Algorithm for Real Hessenberg Matrices

```
PROCEDURE hqr(VAR a: glnpnp; n: integer; VAR wr,wi: glnp);
(* Programs using routine HQR must define the type
TYPE
    glnpnp = ARRAY [1..np,1..np] OF real;
    glnp = ARRAY [1..np] OF real;
where 'np by np' is the physical dimension of the matrix whose
eigenvalues are to be found. *)
LABEL 2,3,4;
VAR
    nn,m,l,k,j,its,i,mmin: integer; z,y,x,w,v,u,t,s,r,q,p,anorm: real;
FUNCTION sign(a,b: real): real;
    BEGIN
        IF (b < 0.0) THEN sign := -abs(a) ELSE sign := abs(a)
    END;
FUNCTION min(a,b: integer): integer;
    BEGIN
        IF (a < b) THEN min := a ELSE min := b
    END;
BEGIN
    anorm := abs(a[1,1]);
    FOR i := 2 TO n DO BEGIN
        FOR j := i-1 TO n DO BEGIN
            anorm := anorm+abs(a[i,j]) END END;
    nn := n; t := 0.0;
    WHILE (nn >= 1) DO BEGIN
        its := 0;
2:      FOR l := nn DOWNTO 2 DO BEGIN
            s := abs(a[l-1,l-1])+abs(a[l,l]);
            IF (s = 0.0) THEN s := anorm;
            IF ((abs(a[l,l-1])+s) = s) THEN GOTO 3
        END;
        l := 1;
3:      x := a[nn,nn];
        IF (l = nn) THEN BEGIN
            wr[nn] := x+t; wi[nn] := 0.0; nn := nn-1 END
        ELSE BEGIN
            y := a[nn-1,nn-1]; w := a[nn,nn-1]*a[nn-1,nn];
            IF (l = nn-1) THEN BEGIN
                p := 0.5*(y-x); q := sqr(p)+w; z := sqrt(abs(q)); x := x+t;
                IF (q >= 0.0) THEN BEGIN
                    z := p+sign(z,p); wr[nn] := x+z; wr[nn-1] := wr[nn];
                    IF (z <> 0.0) THEN wr[nn] := x-w/z;
                    wi[nn] := 0.0; wi[nn-1] := 0.0 END
                ELSE BEGIN
                    wr[nn] := x+p; wr[nn-1] := wr[nn]; wi[nn] := z;
                    wi[nn-1] := -z END;
                nn := nn-2 END
            ELSE BEGIN
                IF (its = 30) THEN BEGIN
                    writeln('pause in routine HQR');
                    writeln('too many iterations'); readln END;
                IF ((its = 10) OR (its = 20)) THEN BEGIN
                    t := t+x;
                    FOR i := 1 TO nn DO BEGIN
                        a[i,i] := a[i,i]-x END;
                    s := abs(a[nn,nn-1])+abs(a[nn-1,nn-2]); x := 0.75*s;
                    y := x; w := -0.4375*sqr(s) END;
                its := its+1;
                FOR m := nn-2 DOWNTO 1 DO BEGIN
                    z := a[m,m]; r := x-z; s := y-z;
                    p := (r*s-w)/a[m+1,m]+a[m,m+1]; q := a[m+1,m+1]-z-r-s;
                    r := a[m+2,m+1]; s := abs(p)+abs(q)+abs(r); p := p/s;
```

```
                    q := q/s; r := r/s;
                    IF (m = 1) THEN GOTO 4;
                    u := abs(a[m,m-1])*(abs(q)+abs(r));
                    v := abs(p)*(abs(a[m-1,m-1])+abs(z) +abs(a[m+1,m+1]));
                    IF ((u+v) = v) THEN GOTO 4
                END;
4:              FOR i := m+2 TO nn DO BEGIN
                    a[i,i-2] := 0.0;
                    IF (i <> (m+2)) THEN a[i,i-3] := 0.0
                END;
                FOR k := m TO nn-1 DO BEGIN
                    IF (k <> m) THEN BEGIN
                        p := a[k,k-1]; q := a[k+1,k-1]; r := 0.0;
                        IF (k <> (nn-1)) THEN
                            r := a[k+2,k-1]; x := abs(p)+abs(q)+abs(r);
                        IF (x <> 0.0) THEN BEGIN
                            p := p/x; q := q/x; r := r/x END END;
                    s := sign(sqrt(sqr(p)+sqr(q)+sqr(r)),p);
                    IF (s <> 0.0) THEN BEGIN
                        IF (k = m) THEN BEGIN
                            IF (l <> m) THEN
                             a[k,k-1] := -a[k,k-1]; END
                        ELSE BEGIN
                            a[k,k-1] := -s*x END;
                        p := p+s; x := p/s; y := q/s; z := r/s; q := q/p;
                        r := r/p;
                        FOR j := k TO nn DO BEGIN
                            p := a[k,j]+q*a[k+1,j];
                            IF (k <> (nn-1)) THEN BEGIN
                                p := p+r*a[k+2,j]; a[k+2,j] := a[k+2,j]-p*z
                            END;
                            a[k+1,j] := a[k+1,j]-p*y; a[k,j] := a[k,j]-p*x
                        END;
                        mmin := min(nn,k+3);
                        FOR i := 1 TO mmin DO BEGIN
                            p := x*a[i,k]+y*a[i,k+1];
                            IF (k <> (nn-1)) THEN BEGIN
                                p := p+z*a[i,k+2]; a[i,k+2] := a[i,k+2]-p*r
                            END;
                            a[i,k+1] := a[i,k+1]-p*q; a[i,k] := a[i,k]-p END
                    END END;
                GOTO 2
            END END END
END;
```

Chapter 12. Fourier Transform Spectral Methods

Fast Fourier Transform (FFT)

```
PROCEDURE four1(VAR data: gldarray; nn,isign: integer);
(* Programs using routine FOUR1 must define type
TYPE
    gldarray = ARRAY [1..nn2] OF real;
in the calling routine, where nn2=nn+nn. *)
VAR
    ii,jj,n,mmax,m,j,istep,i: integer; wtemp,wr,wpr,wpi,wi,theta: double;
    tempr,tempi: real;
BEGIN
    n := 2*nn; j := 1;
    FOR ii := 1 TO nn DO BEGIN
```

```
        i := 2*ii-1;
        IF (j > i) THEN BEGIN
            tempr := data[j]; tempi := data[j+1]; data[j] := data[i];
            data[j+1] := data[i+1]; data[i] := tempr; data[i+1] := tempi
        END;
        m := n DIV 2;
        WHILE ((m >= 2) AND (j > m)) DO BEGIN
            j := j-m;
            m := m DIV 2
        END;
        j := j+m END;
    mmax := 2;
    WHILE (n > mmax) DO BEGIN
        istep := 2*mmax; theta := 6.28318530717959/(isign*mmax);
        wpr := -2.0*sqr(sin(0.5*theta)); wpi := sin(theta); wr := 1.0;
        wi := 0.0;
        FOR ii := 1 TO (mmax DIV 2) DO BEGIN
            m := 2*ii-1;
            FOR jj := 0 TO ((n-m) DIV istep) DO BEGIN
                i := m + jj*istep; j := i+mmax;
                tempr := sngl(wr)*data[j]-sngl(wi)*data[j+1];
                tempi := sngl(wr)*data[j+1]+sngl(wi)*data[j];
                data[j] := data[i]-tempr; data[j+1] := data[i+1]-tempi;
                data[i] := data[i]+tempr; data[i+1] := data[i+1]+tempi END;
            wtemp := wr; wr := wr*wpr-wi*wpi+wr; wi := wi*wpr+wtemp*wpi+wi
        END;
        mmax := istep END
END;
```

FFT of Real Functions, Sine and Cosine Transforms

```
PROCEDURE twofft(data1,data2: glnarray;
        VAR fft1,fft2: gl2narray; n: integer);
(* Programs using routine TWOFFT must define types
TYPE
    glnarray = ARRAY [1..n] OF real;
    gl2narray = ARRAY [1..2*n] OF real;
where n is the dimension of the real-valued data arrays. *)
VAR
    nn3,nn2,nn,jj,j: integer; rep,rem,aip,aim: real;
BEGIN
    nn := n+n; nn2 := nn+2; nn3 := nn+3;
    FOR j := 1 TO n DO BEGIN
        jj := j+j; fft1[jj-1] := data1[j]; fft1[jj] := data2[j] END;
    four1(fft1,n,1); fft2[1] := fft1[2]; fft1[2] := 0.0; fft2[2] := 0.0;
    FOR jj := 1 TO (n DIV 2) DO BEGIN
        j := 2*jj+1; rep := 0.5*(fft1[j]+fft1[nn2-j]);
        rem := 0.5*(fft1[j]-fft1[nn2-j]); aip := 0.5*(fft1[j+1]+fft1[nn3-j]);
        aim := 0.5*(fft1[j+1]-fft1[nn3-j]); fft1[j] := rep; fft1[j+1] := aim;
        fft1[nn2-j] := rep; fft1[nn3-j] := -aim; fft2[j] := aip;
        fft2[j+1] := -rem; fft2[nn2-j] := aip; fft2[nn3-j] := rem END
END;

PROCEDURE realft(VAR data: gldarray; n,isign: integer);
(* Programs using routine REALFT must define the type
TYPE
    gldarray = ARRAY [1..2*n] OF real;
where 2*n is the dimension of the input data array. When
routine FOUR1 is used with REALFT, its data type 'gldarray'
should be set as in this program. *)
VAR
    wr,wi,wpr,wpi,wtemp,theta: double; i,i1,i2,i3,i4: integer;
```

```
    c1,c2,h1r,h1i,h2r,h2i,wrs,wis: real;
BEGIN
    theta := 6.28318530717959/(2.0*n); c1 := 0.5;
    IF (isign = 1) THEN BEGIN
        c2 := -0.5; four1(data,n,1); END
    ELSE BEGIN
        c2 := 0.5; theta := -theta; END;
    wpr := -2.0*sqr(sin(0.5*theta)); wpi := sin(theta);
    wr := 1.0+wpr; wi := wpi;
    FOR i := 2 TO (n DIV 2) DO BEGIN
        i1 := i+i-1; i2 := i1+1; i3 := n+n+3-i2; i4 := i3+1;
        wrs := sngl(wr); wis := sngl(wi); h1r := c1*(data[i1]+data[i3]);
        h1i := c1*(data[i2]-data[i4]); h2r := -c2*(data[i2]+data[i4]);
        h2i := c2*(data[i1]-data[i3]); data[i1] := h1r+wrs*h2r-wis*h2i;
        data[i2] := h1i+wrs*h2i+wis*h2r; data[i3] := h1r-wrs*h2r+wis*h2i;
        data[i4] := -h1i+wrs*h2i+wis*h2r; wtemp := wr;
        wr := wr*wpr-wi*wpi+wr; wi := wi*wpr+wtemp*wpi+wi END;
    IF (isign = 1) THEN BEGIN
        h1r := data[1]; data[1] := h1r+data[2]; data[2] := h1r-data[2] END
    ELSE BEGIN
        h1r := data[1]; data[1] := c1*(h1r+data[2]);
        data[2] := c1*(h1r-data[2]); four1(data,n,-1) END
END;


PROCEDURE sinft(VAR y: glyarray; n: integer);
(* Programs using routine SINFT must define the type
TYPE
    glyarray = ARRAY [1..n] OF real;
where n is the dimension of the input data. *)
VAR
    jj,j,m,n2: integer; sum,y1,y2: real; theta,wi,wr,wpi,wpr,wtemp: double;
BEGIN
    theta := 3.14159265358979/n; wr := 1.0; wi := 0.0;
    wpr := -2.0*sqr(sin(0.5*theta)); wpi := sin(theta); y[1] := 0.0;
    m := n DIV 2;
    n2 := n+2;
    FOR j := 2 TO (m+1) DO BEGIN
        wtemp := wr; wr := wr*wpr-wi*wpi+wr; wi := wi*wpr+wtemp*wpi+wi;
        y1 := sngl(wi)*(y[j]+y[n2-j]); y2 := 0.5*(y[j]-y[n2-j]);
        y[j] := y1+y2; y[n2-j] := y1-y2 END;
    realft(y,m,+1); sum := 0.0; y[1] := 0.5*y[1]; y[2] := 0.0;
    FOR jj := 0 TO (m-1) DO BEGIN
        j := 2*jj+1; sum := sum+y[j]; y[j] := y[j+1]; y[j+1] := sum END
END;


PROCEDURE cosft(VAR y: glyarray; n,isign: integer);
(* Programs using routine COSFT must define the type
TYPE
    glyarray = ARRAY [1..n] OF real;
where n is the dimension of the input data array. *)
VAR
    enf0,even,odd,sum,sume,sumo,y1,y2: real;
    theta,wi,wr,wpi,wpr,wtemp: double; jj,j,m,n2: integer;
BEGIN
    theta := 3.14159265358979/n; wr := 1.0; wi := 0.0;
    wpr := -2.0*sqr(sin(0.5*theta)); wpi := sin(theta); sum := y[1];
    m := n DIV 2;
    n2 := n+2;
    FOR j := 2 TO (m+1) DO BEGIN
        wtemp := wr; wr := wr*wpr-wi*wpi+wr; wi := wi*wpr+wtemp*wpi+wi;
        y1 := 0.5*(y[j]+y[n2-j]); y2 := (y[j]-y[n2-j]);
        y[j] := y1-sngl(wi)*y2; y[n2-j] := y1+sngl(wi)*y2;
        sum := sum+sngl(wr)*y2 END;
```

```
    realft(y,m,+1); y[2] := sum;
    FOR jj := 2 TO m DO BEGIN
        j := 2*jj; sum := sum+y[j]; y[j] := sum END;
    IF (isign = -1) THEN BEGIN
        even := y[1]; odd := y[2];
        FOR jj := 1 TO (m-1) DO BEGIN
            j := 2*jj+1; even := even+y[j]; odd := odd+y[j+1] END;
        enf0 := 2.0*(even-odd); sumo := y[1]-enf0; sume := (2.0*odd/n)-sumo;
        y[1] := 0.5*enf0; y[2] := y[2]-sume;
        FOR jj := 1 TO (m-1) DO BEGIN
            j := 2*jj+1; y[j] := y[j]-sumo; y[j+1] := y[j+1]-sume END END
END;
```

Convolution and Deconvolution Using the FFT

```
PROCEDURE convlv(data: glnarray; n: integer; respns: glnarray; m: integer;
        isign: integer; VAR ans: gln2array);
(* Programs using routine CONVLV must define the types
TYPE
    glnarray = ARRAY [1..n] OF real;
    gln2array = ARRAY [1..n2] OF real;
where n is the dimension of the data and n2=2*n. NOTE: when used with CONVLV,
the data dimension in FOUR1 and in TWOFFT must be the same as gln2array here.
i.e. TYPE gldarray = gln2array; gl2narray = gln2array *)
VAR
    no2,i,ii: integer; dum,mag2: real; fft: gln2array;
BEGIN
    FOR i := 1 TO ((m-1) DIV 2) DO BEGIN
        respns[n+1-i] := respns[m+1-i] END;
    FOR i := ((m+3) DIV 2) TO (n-((m-1) DIV 2)) DO BEGIN
        respns[i] := 0.0 END;
    twofft(data,respns,fft,ans,n);
    no2 := n DIV 2;
    FOR i := 1 TO (no2+1) DO BEGIN
        ii := 2*i;
        IF (isign = 1) THEN BEGIN
            dum := ans[ii-1];
            ans[ii-1] := (fft[ii-1]*ans[ii-1]-fft[ii]*ans[ii])/no2;
            ans[ii] := (fft[ii]*dum+fft[ii-1]*ans[ii])/no2 END
        ELSE IF (isign = -1) THEN BEGIN
            IF ((sqr(ans[ii-1])+sqr(ans[ii])) = 0.0) THEN BEGIN
                writeln('pause in routine CONVLV');
                writeln('deconvolving at response zero'); readln END;
            dum := ans[ii-1]; mag2 := sqr(ans[ii-1])+sqr(ans[ii]);
            ans[ii-1] := (fft[ii-1]*ans[ii-1]+fft[ii]*ans[ii])/mag2/no2;
            ans[ii] := (fft[ii]*dum-fft[ii-1]*ans[ii])/mag2/no2 END
        ELSE BEGIN
            writeln('pause in routine CONVLV');
            writeln('no meaning for ISIGN'); readln
        END END;
    ans[2] := ans[n+1]; realft(ans,no2,-1)
END;
```

Correlation and Autocorrelation Using the FFT

```
PROCEDURE correl(data1,data2: glnarray; n: integer; VAR ans: gl2narray);
(* Programs using routine CORREL must define the type
TYPE
    glnarray = ARRAY [1..n] OF real;
```

```
    gl2narray = ARRAY [1..2*n] OF real;
in the main routine. *)
VAR
    no2,i,ii: integer; dum: real; fft: gl2narray;
BEGIN
    twofft(data1,data2,fft,ans,n);
    no2 := n DIV 2;
    FOR i := 1 TO (no2+1) DO BEGIN
        ii := 2*i; dum := ans[ii-1];
        ans[ii-1] := (fft[ii-1]*ans[ii-1]+fft[ii]*ans[ii])/no2;
        ans[ii] := (fft[ii]*dum-fft[ii-1]*ans[ii])/no2 END;
    ans[2] := ans[n+1]; realft(ans,no2,-1)
END;
```

Power Spectrum Estimation Using the FFT

```
PROCEDURE spctrm(VAR p: glmarray; m,k: integer; ovrlap: boolean;
        VAR w1: gl4marray; VAR w2: glmarray);
(* Programs using routine SPCTRM must include 'dfile' as
a main routine parameter, and declare
VAR
    dfile: text;
They must open and close this file appropriately.
Also they must define data types
TYPE
    glmarray = ARRAY [1..m] OF real;
    gl4marray = ARRAY [1..4*m] OF real;
and must declare two workspace matrices w1,w2 with types as shown
in the arguments above. *)
VAR
    mm,m44,m43,m4,kk,joffn,joff,j2,j,jj: integer; w,sumw,facp,facm,den: real;
FUNCTION window(j: integer; facm,facp: real): real;
    BEGIN
        window := (1.0-abs(((j-1)-facm)*facp))        (* Parzen *)
(*        window := 1.0                     *)    (* Square *)
(*        window := (1.0-sqr(((j-1)-facm)*facp))     *)    (* Welch *)
    END;
BEGIN
    mm := m+m; m4 := mm+mm; m44 := m4+4; m43 := m4+3; den := 0.0;
    facm := m-0.5; facp := 1.0/(m+0.5); sumw := 0.0;
    FOR j := 1 TO mm DO sumw := sumw+sqr(window(j,facm,facp));
    FOR j := 1 TO m DO p[j] := 0.0;
    IF (ovrlap) THEN BEGIN
        FOR j := 1 TO m DO read(dfile,w2[j])
    END;
    FOR kk := 1 TO k DO BEGIN
        FOR joff := -1 TO 0 DO BEGIN
            IF (ovrlap) THEN BEGIN
                FOR j := 1 TO m DO w1[joff+j+j] := w2[j];
                FOR j := 1 TO m DO read(dfile,w2[j]);
                joffn := joff+mm;
                FOR j := 1 TO m DO w1[joffn+j+j] := w2[j] END
            ELSE BEGIN
                FOR jj := 0 TO ((m4-joff-2) DIV 2) DO BEGIN
                    j := joff+2+2*jj; read(dfile,w1[j]) END END END;
        FOR j := 1 TO mm DO BEGIN
            j2 := j+j; w := window(j,facm,facp); w1[j2] := w1[j2]*w;
            w1[j2-1] := w1[j2-1]*w END;
        four1(w1,mm,1); p[1] := p[1]+sqr(w1[1])+sqr(w1[2]);
        FOR j := 2 TO m DO BEGIN
            j2 := j+j; p[j] := p[j]+sqr(w1[j2])+sqr(w1[j2-1])
```

```
            +sqr(w1[m44-j2])+sqr(w1[m43-j2]) END;
        den := den+sumw END;
    den := m4*den;
    FOR j := 1 TO m DO p[j] := p[j]/den
END;
```

Linear Prediction and Linear Predictive Coding

```
PROCEDURE memcof(data: glnarray; n,m: integer; VAR pm: real;
        VAR cof: glmarray; wk1,wk2: glnarray; wkm: glmarray);
(* Programs using routine MEMCOF must define the data types
TYPE
    glnarray = ARRAY [1..n] OF real;
    glmarray = ARRAY [1..m] OF real;
and must provide workspace arrays wk1,wk2,wkm with the dimensions
shown in the arguments above. *)
LABEL 99;
VAR
    k,j,i: integer; pneum,p,denom: real;
BEGIN
    p := 0.0;
    FOR j := 1 TO n DO BEGIN
        p := p+sqr(data[j]) END;
    pm := p/n; wk1[1] := data[1]; wk2[n-1] := data[n];
    FOR j := 2 TO n-1 DO BEGIN
        wk1[j] := data[j]; wk2[j-1] := data[j] END;
    FOR k := 1 TO m DO BEGIN
        pneum := 0.0; denom := 0.0;
        FOR j := 1 TO n-k DO BEGIN
            pneum := pneum+wk1[j]*wk2[j];
            denom := denom+sqr(wk1[j])+sqr(wk2[j]) END;
        cof[k] := 2.0*pneum/denom; pm := pm*(1.0-sqr(cof[k]));
        IF (k <> 1) THEN BEGIN
            FOR i := 1 TO k-1 DO BEGIN
                cof[i] := wkm[i]-cof[k]*wkm[k-i] END END;
        IF (k = m) THEN GOTO 99;
        FOR i := 1 TO k DO BEGIN
            wkm[i] := cof[i] END;
        FOR j := 1 TO n-k-1 DO BEGIN
            wk1[j] := wk1[j]-wkm[k]*wk2[j]; wk2[j] := wk2[j+1]-wkm[k]*wk1[j+1]
        END END;
99: END;
```

```
FUNCTION evlmem(fdt: real; cof: glmarray; m: integer; pm: real): real;
(* Programs using routine EVLMEM must define the types
TYPE
    glmarray = ARRAY [1..m] OF real;
where m is the dimension of the array of coefficients. *)
VAR
    wr,wi,wpr,wpi,wtemp,theta: double; sumi,sumr: real; i: integer;
BEGIN
    theta := 6.28318530717959*fdt; wpr := cos(theta); wpi := sin(theta);
    wr := 1.0; wi := 0.0; sumr := 1.0; sumi := 0.0;
    FOR i := 1 TO m DO BEGIN
        wtemp := wr; wr := wr*wpr-wi*wpi; wi := wi*wpr+wtemp*wpi;
        sumr := sumr-cof[i]*sngl(wr); sumi := sumi-cof[i]*sngl(wi) END;
    evlmem := pm/(sqr(sumr)+sqr(sumi))
END;
```

```
PROCEDURE fixrts(VAR d: glnparray; npoles: integer);
(* Programs using routine FIXRTS must define the type
TYPE
    glnparray = ARRAY [1..npoles] OF real;
    glcarray = ARRAY [1..2*npoles+2] OF real;
in the main routine. *)
VAR
    j,i: integer; size,dum: real; polish: boolean; a,roots: glcarray;
BEGIN
    a[2*npoles+1] := 1.0; a[2*npoles+2] := 0.0;
    FOR j := npoles DOWNTO 1 DO BEGIN
        a[2*j-1] := -d[npoles+1-j]; a[2*j] := 0.0 END;
    polish := true; zroots(a,npoles,roots,polish);
    FOR j := 1 TO npoles DO BEGIN
        size := sqr(roots[2*j-1])+sqr(roots[2*j]);
        IF (size > 1.0) THEN BEGIN
            roots[2*j-1] := roots[2*j-1]/size; roots[2*j] := roots[2*j]/size
        END END;
    a[1] := -roots[1]; a[2] := -roots[2]; a[3] := 1.0; a[4] := 0.0;
    FOR j := 2 TO npoles DO BEGIN
        a[2*j+1] := 1.0; a[2*j+2] := 0.0;
        FOR i := j DOWNTO 2 DO BEGIN
            dum := a[2*i-1]; a[2*i-1] := a[2*i-3]-a[2*i-1]*roots[2*j-1]
                    +a[2*i]*roots[2*j]; a[2*i] := a[2*i-2]-dum*roots[2*j]
                    -a[2*i]*roots[2*j-1] END;
        dum := a[1]; a[1] := -a[1]*roots[2*j-1]+a[2]*roots[2*j];
        a[2] := -dum*roots[2*j]-a[2]*roots[2*j-1] END;
    FOR j := 1 TO npoles DO BEGIN
        d[npoles+1-j] := -a[2*j-1] END
END;
```

```
PROCEDURE predic(data: gldarray; ndata: integer; d: glnparray;
        npoles: integer; VAR future: glnfarray; nfut: integer);
(* Programs using routine PREDIC must define the types
TYPE
    gldarray = ARRAY [1..ndata] OF real;
    glnparray = ARRAY [1..npoles] OF real;
    glnfarray = ARRAY [1..nfut] OF real;
in the main routine. *)
VAR
    k,j: integer; sum,discrp: real; reg: glnparray;
BEGIN
    FOR j := 1 TO npoles DO BEGIN
        reg[j] := data[ndata+1-j] END;
    FOR j := 1 TO nfut DO BEGIN
        discrp := 0.0; sum := discrp;
        FOR k := 1 TO npoles DO BEGIN
            sum := sum+d[k]*reg[k] END;
        FOR k := npoles DOWNTO 2 DO BEGIN
            reg[k] := reg[k-1] END;
        reg[1] := sum; future[j] := sum END
END;
```

FFT in Two or More Dimensions

```
PROCEDURE fourn(VAR data: gldarray; nn: glnnarray; ndim,isign: integer);
(* Programs using routine FOURN must define the types
TYPE
    gldarray = ARRAY [1..ndat2] OF real;
    glnnarray = ARRAY [1..ndim] OF integer;
where ndat2 is twice the product of the transform lengths nn(i). *)
VAR
```

```
    i1,i2,i3,i2rev,i3rev,ibit,idim: integer;
    ip1,ip2,ip3,ifp1,ifp2,k1,k2,n: integer; ii1,ii2,ii3: integer;
    nprev,nrem,ntot: integer; tempi,tempr: real;
    theta,wi,wpi,wpr,wr,wtemp: double;
BEGIN
    ntot := 1;
    FOR idim := 1 TO ndim DO BEGIN
        ntot := ntot*nn[idim] END;
    nprev := 1;
    FOR idim := 1 TO ndim DO BEGIN
        n := nn[idim];
        nrem := ntot DIV (n*nprev);
        ip1 := 2*nprev; ip2 := ip1*n; ip3 := ip2*nrem; i2rev := 1;
        FOR ii2 := 0 TO ((ip2-1) DIV ip1) DO BEGIN
            i2 := 1+ii2*ip1;
            IF (i2 < i2rev) THEN BEGIN
                FOR ii1 := 0 TO ((ip1-2) DIV 2) DO BEGIN
                    i1 := i2+ii1*2;
                    FOR ii3 := 0 TO ((ip3-i1) DIV ip2) DO BEGIN
                        i3 := i1+ii3*ip2; i3rev := i2rev+i3-i2;
                        tempr := data[i3]; tempi := data[i3+1];
                        data[i3] := data[i3rev]; data[i3+1] := data[i3rev+1];
                        data[i3rev] := tempr; data[i3rev+1] := tempi END END
            END;
            ibit := ip2 DIV 2;
            WHILE ((ibit >= ip1) AND (i2rev > ibit)) DO BEGIN
                i2rev := i2rev-ibit;
                ibit := ibit DIV 2
            END;
            i2rev := i2rev+ibit END;
        ifp1 := ip1;
        WHILE (ifp1 < ip2) DO BEGIN
            ifp2 := 2*ifp1;
            theta := isign*6.28318530717959/(ifp2 DIV ip1);
            wpr := -2.0*sqr(sin(0.5*theta)); wpi := sin(theta); wr := 1.0;
            wi := 0.0;
            FOR ii3 := 0 TO ((ifp1-1) DIV ip1) DO BEGIN
                i3 := 1+ii3*ip1;
                FOR ii1 := 0 TO ((ip1-2) DIV 2) DO BEGIN
                    i1 := i3+ii1*2;
                    FOR ii2 := 0 TO ((ip3-i1) DIV ifp2) DO BEGIN
                        i2 := i1+ii2*ifp2; k1 := i2; k2 := k1+ifp1;
                        tempr := sngl(wr)*data[k2] -sngl(wi)*data[k2+1];
                        tempi := sngl(wr)*data[k2+1] +sngl(wi)*data[k2];
                        data[k2] := data[k1]-tempr;
                        data[k2+1] := data[k1+1]-tempi;
                        data[k1] := data[k1]+tempr;
                        data[k1+1] := data[k1+1]+tempi END END;
                wtemp := wr; wr := wr*wpr-wi*wpi+wr;
                wi := wi*wpr+wtemp*wpi+wi END;
            ifp1 := ifp2 END;
        nprev := n*nprev END
END;
```

Chapter 13. Statistical Description of Data

Moments of a Distribution

```
PROCEDURE moment(data: narray; n: integer;
        VAR ave,adev,sdev,svar,skew,curt: real);
(* Programs using routine MOMENT must define the type
TYPE
```

```
    narray = ARRAY [1..n] OF real;
in the calling routine *)
VAR
    j: integer; s,p: real;
BEGIN
    IF (n <= 1) THEN BEGIN
        writeln('pause in MOMENT - n must be at least 2'); readln
    END;
    s := 0.0;
    FOR j := 1 TO n DO s := s+data[j];
    ave := s/n; adev := 0.0; svar := 0.0; skew := 0.0; curt := 0.0;
    FOR j := 1 TO n DO BEGIN
        s := data[j]-ave; adev := adev+abs(s); p := s*s; svar := svar+p;
        p := p*s; skew := skew+p; p := p*s; curt := curt+p END;
    adev := adev/n; svar := svar/(n-1); sdev := sqrt(svar);
    IF (svar <> 0.0) THEN BEGIN
        skew := skew/(n*sdev*sdev*sdev); curt := curt/(n*sqr(svar))-3.0 END
    ELSE BEGIN
        writeln('pause in MOMENT - no skew/kurtosis when variance = 0'); readln
    END
END;
```

Efficient Search for the Median

```
PROCEDURE mdian1(VAR x: narray; n: integer; VAR xmed: real);
(* Program using routine MDIAN1 must define the type
TYPE
    narray = ARRAY [1..n] OF real;
in the calling routine *)
VAR
    n2: integer;
BEGIN
    sort(n,x);
    n2 := n DIV 2;
    IF (2*n2 = n) THEN xmed := 0.5*(x[n2]+x[n2+1])
    ELSE xmed := x[n2+1]
END;
```

```
PROCEDURE mdian2(x: narray; n: integer; VAR xmed: real);
(* Programs using routine MDIAN2 must define the type
TYPE
    narray = ARRAY [1..n] OF real;
in the calling routine *)
LABEL 1;
CONST
    big=1.0e30; afac=1.5; amp=1.5;
VAR
    np,nm,j: integer; xx,xp,xm,sumx,sum,eps: real;
    stemp,dum,ap,am,aa,a: real;
BEGIN
    a := 0.5*(x[1]+x[n]); eps := abs(x[n]-x[1]); ap := big; am := -big;
1:  sum := 0.0; sumx := 0.0; np := 0; nm := 0; xp := big; xm := -big;
    FOR j := 1 TO n DO BEGIN
        xx := x[j];
        IF (xx <> a) THEN BEGIN
            IF (xx > a) THEN BEGIN
                np := np+1;
                IF (xx < xp) THEN xp := xx END
            ELSE IF (xx < a) THEN BEGIN
                nm := nm+1;
```

```
                IF (xx > xm) THEN xm := xx
            END;
            dum := 1.0/(eps+abs(xx-a)); sum := sum+dum; sumx := sumx+xx*dum
        END END;
    stemp := (sumx/sum)-a;
    IF ((np-nm) >= 2) THEN BEGIN
        am := a;
        IF (stemp < 0.0) THEN aa := xp
        ELSE aa := xp+stemp*amp;
        IF (aa > ap) THEN aa := 0.5*(a+ap);
        eps := afac*abs(aa-a); a := aa;
        GOTO 1 END
    ELSE IF ((nm-np) >= 2) THEN BEGIN
        ap := a;
        IF (stemp > 0.0) THEN aa := xm
        ELSE aa := xm+stemp*amp;
        IF (aa < am) THEN aa := 0.5*(a+am);
        eps := afac*abs(aa-a); a := aa;
        GOTO 1 END
    ELSE IF (n MOD 2) = 0 THEN BEGIN
        IF (np = nm) THEN xmed := 0.5*(xp+xm)
        ELSE IF (np > nm) THEN xmed := 0.5*(a+xp)
        ELSE xmed := 0.5*(xm+a) END
    ELSE BEGIN
        IF (np = nm) THEN xmed := a
        ELSE IF (np > nm) THEN xmed := xp
        ELSE xmed := xm
    END
END;
```

Do Two Distributions Have the Same Means or Variances?

```
PROCEDURE ttest(data1: glnarray; n1: integer;
       data2: glmarray; n2: integer; VAR t,prob: real);
(* Programs using routine TTEST must define the types
TYPE
    glnarray = ARRAY [1..n1] OF real;
    glmarray = ARRAY [1..n2] OF real;
in the main routine, with n2 less than or equal to n1. *)
VAR
    i: integer; var2,var1,svar,df,ave2,ave1: real;
BEGIN
    avevar(data1,n1,ave1,var1);
    IF (n2 > n1) THEN BEGIN
        writeln('pause in routine TTEST');
        writeln('The first array should be the larger'); readln
    END;
    FOR i := 1 TO n2 DO data1[i] := data2[i];
    avevar(data1,n2,ave2,var2); df := n1+n2-2;
    svar := ((n1-1)*var1+(n2-1)*var2)/df;
    t := (ave1-ave2)/sqrt(svar*(1.0/n1+1.0/n2));
    prob := betai(0.5*df,0.5,df/(df+sqr(t)))
END;
```

```
PROCEDURE avevar(data: glnparray; n: integer; VAR ave,svar: real);
(* Programs using routine AVEVAR must define type
TYPE
    glnparray = ARRAY [1..np] OF real;
in the main routine, with np >= n.    *)
VAR
    j: integer; s: real;
BEGIN
    ave := 0.0; svar := 0.0;
    FOR j := 1 TO n DO BEGIN
        ave := ave+data[j]; END;
    ave := ave/n;
    FOR j := 1 TO n DO BEGIN
        s := data[j]-ave; svar := svar+s*s END;
    svar := svar/(n-1)
END;

PROCEDURE tutest(data1: glnarray; n1: integer; data2: glmarray;
        n2: integer; VAR t,prob: real);
(* Programs using routine TUTEST must define the types
TYPE
    glnarray = ARRAY [1..n1] OF real;
    glmarray = ARRAY [1..n2] OF real;
in the main routine, with n2 less than or equal to n1.    *)
VAR
    var2,var1,df,ave2,ave1: real; i: integer;
BEGIN
    IF (n2 > n1) THEN BEGIN
        writeln('pause in routine TUTEST');
        writeln('first array must be the larger'); readln END;
    avevar(data1,n1,ave1,var1);
    FOR i := 1 TO n2 DO data1[i] := data2[i];
    avevar(data1,n2,ave2,var2); t := (ave1-ave2)/sqrt(var1/n1+var2/n2);
    df := sqr(var1/n1+var2/n2)/ (sqr(var1/n1)/(n1-1)+sqr(var2/n2)/(n2-1));
    prob := betai(0.5*df,0.5,df/(df+sqr(t)))
END;

PROCEDURE tptest(data1,data2: glnparray; n: integer; VAR t,prob: real);
(* Programs using routine TPTEST must define type
TYPE
    glnparray := ARRAY [1..np];
in the main routine, with np >= n. *)
VAR
    j: integer; var2,var1,sd,df,cov,ave2,ave1: real;
BEGIN
    avevar(data1,n,ave1,var1); avevar(data2,n,ave2,var2); cov := 0.0;
    FOR j := 1 TO n DO BEGIN
        cov := cov+(data1[j]-ave1)*(data2[j]-ave2); END;
    df := n-1; cov := cov/df; sd := sqrt((var1+var2-2.0*cov)/n);
    t := (ave1-ave2)/sd; prob := betai(0.5*df,0.5,df/(df+sqr(t)))
END;

PROCEDURE ftest(data1: gln1array; n1: integer;
        data2: gln2array; n2: integer; VAR f,prob: real);
(* Programs using routine FTEST must define the types
TYPE
    gln1array = ARRAY [1..n1] OF real;
    gln2array = ARRAY [1..n2] OF real;
in the main routine, with n2 less than or equal to n1. *)
VAR
    i: integer; var2,var1,df2,df1,ave2,ave1: real;
BEGIN
    IF (n2 > n1) THEN BEGIN
```

```
        writeln('pause in routine FTEST');
        writeln('first array must be the larger'); readln END;
    avevar(data1,n1,ave1,var1);
    FOR i := 1 TO n2 DO data1[i] := data2[i];
    avevar(data1,n2,ave2,var2);
    IF (var1 > var2) THEN BEGIN
        f := var1/var2; df1 := n1-1; df2 := n2-1 END
    ELSE BEGIN
        f := var2/var1; df1 := n2-1; df2 := n1-1 END;
    prob := 2.0*betai(0.5*df2,0.5*df1,df2/(df2+df1*f));
    IF (prob > 1.0) THEN prob := 2.0-prob;
END;
```

Are Two Distributions Different?

```
PROCEDURE chsone(bins,ebins: barray; nbins,knstrn: integer;
         VAR df,chsq,prob: real);
(* Programs using routine CHSONE must define the types
TYPE
    barray = ARRAY [1..nbins] OF real;
in the main routine.    *)
VAR
    j: integer;
BEGIN
    df := nbins-1-knstrn; chsq := 0.0;
    FOR j := 1 TO nbins DO BEGIN
        IF (ebins[j] <= 0.0) THEN BEGIN
            writeln('pause in CHSONE - bad expected number')
        END;
        chsq := chsq+sqr(bins[j]-ebins[j])/ebins[j] END;
    prob := gammq(0.5*df,0.5*chsq)
END;

PROCEDURE chstwo(bins1,bins2: barray; nbins,knstrn: integer;
         VAR df,chsq,prob: real);
(* Programs using routine CHSTWO must define type
TYPE
    barray = ARRAY [1..nbins] OF real;
in the main routine.    *)
VAR
    j: integer;
BEGIN
    df := nbins-1-knstrn; chsq := 0.0;
    FOR j := 1 TO nbins DO BEGIN
        IF ((bins1[j] = 0.0) AND (bins2[j] = 0.0)) THEN BEGIN
            df := df-1.0 END
        ELSE BEGIN
            chsq := chsq+sqr(bins1[j]-bins2[j])/ (bins1[j]+bins2[j]) END
    END;
    prob := gammq(0.5*df,0.5*chsq)
END;

PROCEDURE ksone(VAR data: gldarray; n: integer; VAR d,prob: real);
(* Programs using routine KSONE must define the type
TYPE
    gldarray = ARRAY [1..n] OF real;
in the main routine. *)
VAR
    j: integer; fo,fn,ff,en,dt: real;
BEGIN
    sort(n,data); en := n; d := 0.0; fo := 0.0;
```

```
    FOR j := 1 TO n DO BEGIN
        fn := j/en; ff := func(data[j]);
        IF (abs(fo-ff) > abs(fn-ff)) THEN BEGIN
            dt := abs(fo-ff) END
        ELSE BEGIN
            dt := abs(fn-ff) END;
        IF (dt > d) THEN d := dt;
        fo := fn END;
    prob := probks(sqrt(en)*d)
END;


PROCEDURE kstwo(VAR data1: glarray1; n1: integer;
        VAR data2: glarray2; n2: integer;
        VAR d,prob: real);
(* Programs using routine KSTWO must define the types
TYPE
    glarray1 = ARRAY [1..n1] OF real;
    glarray2 = ARRAY [1..n2] OF real;
in the main routine. *)
VAR
    i,j1,j2: integer;
    dum,en1,en2,fn1,fn2,dt,d1,d2: real;
BEGIN
    IF (n2 > n1) THEN BEGIN
        writeln('pause in routine KSTWO');
        writeln('first input array must be the larger'); readln
    END;
    sort(n1,data1);
    FOR i := 1 TO n2 DO BEGIN
        dum := data1[i];
        data1[i] := data2[i];
        data2[i] := dum
    END;
    sort(n2,data1);
    FOR i := 1 TO n2 DO BEGIN
        dum := data1[i];
        data1[i] := data2[i];
        data2[i] := dum
    END;
    en1 := n1; en2 := n2; j1 := 1; j2 := 1; fn1 := 0.0; fn2 := 0.0;
    d := 0.0;
    WHILE ((j1 <= n1) AND (j2 <= n2)) DO BEGIN
        d1 := data1[j1]; d2 := data2[j2];
        IF (d1 <= d2) THEN BEGIN
            fn1 := j1/en1; j1 := j1+1 END;
        IF (d2 <= d1) THEN BEGIN
            fn2 := j2/en2; j2 := j2+1 END;
        dt := abs(fn2-fn1);
        IF (dt > d) THEN d := dt
    END;
    prob := probks(sqrt(en1*en2/(en1+en2))*d)
END;


FUNCTION probks(alam: real): real;
LABEL 1; CONST eps1=0.001; eps2=1.0e-8;
VAR
    a2,fac,sum,term,termbf: real; j: integer;
BEGIN
    a2 := -2.0*alam*alam; fac := 2.0; sum := 0.0; termbf := 0.0;
    FOR j := 1 TO 100 DO BEGIN
        term := fac*exp(a2*sqr(j)); sum := sum+term;
        IF ((abs(term) <= (eps1*termbf)) OR (abs(term) <= (eps2*sum))) THEN BEGIN
            probks := sum;
```

```
            GOTO 1 END
        ELSE BEGIN
            fac := -fac; termbf := abs(term) END END;
    probks := 1.0;
1:  END;
```

Contingency Table Analysis of Two Distributions

```
PROCEDURE cntab1(nn: narray; ni,nj: integer;
        VAR chisq,df,prob,cramrv,ccc: real);
(* Programs using routine CNTAB1 must define type
TYPE
    narray = ARRAY [1..ni,1..nj] OF integer;
in the calling routine. *)
CONST
    maxi=100; maxj=100; tiny=1.0e-30;
VAR
    nnj,nni,j,i,min: integer; sum,expctd: real;
    sumi: ARRAY[1..maxi] OF real;
    sumj: ARRAY[1..maxj] OF real;
BEGIN
    sum := 0; nni := ni; nnj := nj;
    FOR i := 1 TO ni DO BEGIN
        sumi[i] := 0.0;
        FOR j := 1 TO nj DO BEGIN
            sumi[i] := sumi[i]+nn[i,j]; sum := sum+nn[i,j]; END;
        IF (sumi[i] = 0.0) THEN nni := nni-1;
    END;
    FOR j := 1 TO nj DO BEGIN
        sumj[j] := 0.0;
        FOR i := 1 TO ni DO BEGIN
            sumj[j] := sumj[j]+nn[i,j]; END;
        IF (sumj[j] = 0.0) THEN nnj := nnj-1;
    END;
    df := nni*nnj-nni-nnj+1; chisq := 0.0;
    FOR i := 1 TO ni DO BEGIN
        FOR j := 1 TO nj DO BEGIN
            expctd := sumj[j]*sumi[i]/sum;
            chisq := chisq+sqr(nn[i,j]-expctd)/(expctd+tiny) END END;
    prob := gammq(0.5*df,0.5*chisq);
    IF ((nni-1) < (nnj-1)) THEN BEGIN
        min := nni-1 END
    ELSE BEGIN
        min := nnj-1 END;
    cramrv := sqrt(chisq/(sum*min)); ccc := sqrt(chisq/(chisq+sum))
END;

PROCEDURE cntab2(nn: narray; ni,nj: integer;
        VAR h,hx,hy,hygx,hxgy,uygx,uxgy,uxy: real);
(* Programs using routine CNTAB2 must define type
TYPE
    narray = ARRAY [1..ni,1..nj] OF integer;
in the calling routine. *)
CONST
    maxi=100; maxj=100; tiny=1.0e-30;
VAR
    j,i: integer; sum,p: real;
    sumi: ARRAY[1..maxi] OF real;
    sumj: ARRAY[1..maxj] OF real;
BEGIN
    sum := 0;
    FOR i := 1 TO ni DO BEGIN
```

```
        sumi[i] := 0.0;
        FOR j := 1 TO nj DO BEGIN
            sumi[i] := sumi[i]+nn[i,j]; sum := sum+nn[i,j] END END;
    FOR j := 1 TO nj DO BEGIN
        sumj[j] := 0.0;
        FOR i := 1 TO ni DO BEGIN
            sumj[j] := sumj[j]+nn[i,j] END END;
    hx := 0.0;
    FOR i := 1 TO ni DO BEGIN
        IF (sumi[i] <> 0.0) THEN BEGIN
            p := sumi[i]/sum; hx := hx-p*ln(p) END END;
    hy := 0.0;
    FOR j := 1 TO nj DO BEGIN
        IF (sumj[j] <> 0.0) THEN BEGIN
            p := sumj[j]/sum; hy := hy-p*ln(p) END END;
    h := 0.0;
    FOR i := 1 TO ni DO BEGIN
        FOR j := 1 TO nj DO BEGIN
            IF (nn[i,j] <> 0) THEN BEGIN
                p := nn[i,j]/sum; h := h-p*ln(p) END END END;
    hygx := h-hx; hxgy := h-hy; uygx := (hy-hygx)/(hy+tiny);
    uxgy := (hx-hxgy)/(hx+tiny); uxy := 2.0*(hx+hy-h)/(hx+hy+tiny)
END;
```

Linear Correlation

```
PROCEDURE pearsn(x,y: nparray; n: integer; VAR r,prob,z: real);
(* Programs using routine PEARSN must define type
TYPE
    nparray = ARRAY [1..n] OF real;
in the main routine. *)
CONST
    tiny=1.0e-20;
VAR
    j: integer; yt,xt,t,syy,sxy,sxx,df,ay,ax: real;
BEGIN
    ax := 0.0; ay := 0.0;
    FOR j := 1 TO n DO BEGIN
        ax := ax+x[j]; ay := ay+y[j] END;
    ax := ax/n; ay := ay/n; sxx := 0.0; syy := 0.0; sxy := 0.0;
    FOR j := 1 TO n DO BEGIN
        xt := x[j]-ax; yt := y[j]-ay; sxx := sxx+sqr(xt);
        syy := syy+sqr(yt); sxy := sxy+xt*yt; END;
    r := sxy/sqrt(sxx*syy); z := 0.5*ln(((1.0+r)+tiny)/((1.0-r)+tiny));
    df := n-2; t := r*sqrt(df/(((1.0-r)+tiny)*((1.0+r)+tiny)));
    prob := betai(0.5*df,0.5,df/(df+sqr(t)))
(*    prob := erfcc(abs(z*sqrt(n-1.0))/1.4142136)    *)
END;
```

Nonparametric or Rank Correlation

```
PROCEDURE spear(data1,data2: narray; n: integer;
        VAR wksp1,wksp2: narray;
        VAR d,zd,probd,rs,probrs: real);
(* Programs using routine SPEAR must define types
TYPE
    narray = ARRAY [1..n] OF real;
    glsarray = narray;
in the calling routine *)
```

```
VAR
    j: integer; vard,t,sg,sf,fac,en3n,en,df,aved: real;
BEGIN
    FOR j := 1 TO n DO BEGIN
        wksp1[j] := data1[j]; wksp2[j] := data2[j] END;
    sort2(n,wksp1,wksp2); crank(n,wksp1,sf); sort2(n,wksp2,wksp1);
    crank(n,wksp2,sg); d := 0.0;
    FOR j := 1 TO n DO d := d+sqr(wksp1[j]-wksp2[j]);
    en := n; en3n := en*en*en-en; aved := en3n/6.0-(sf+sg)/12.0;
    fac := (1.0-sf/en3n)*(1.0-sg/en3n);
    vard := ((en-1.0)*sqr(en)*sqr(en+1.0)/36.0)*fac;
    zd := (d-aved)/sqrt(vard); probd := erfcc(abs(zd)/1.4142136);
    rs := (1.0-(6.0/en3n)*(d+0.5*(sf+sg)))/fac;
    t := rs*sqrt((en-2.0)/((1.0+rs)*(1.0-rs))); df := en-2.0;
    probrs := betai(0.5*df,0.5,df/(df+sqr(t)))
END;


PROCEDURE crank(n: integer; VAR w: narray; VAR s: real);
(* Programs using routine CRANK must define type
TYPE
    narray = ARRAY [1..n] OF real;
in the calling routine *)
LABEL 2;
VAR
    j,ji,jt,lbl1,lbl2: integer; t,rank: real;
BEGIN
    s := 0.0; j := 1;
    WHILE (j < n) DO BEGIN
        IF (w[j+1] <> w[j]) THEN BEGIN
            w[j] := j; j := j+1 END
        ELSE BEGIN
            FOR jt := j+1 TO n DO BEGIN
                IF (w[jt] <> w[J]) THEN GOTO 2;
            END;
            jt := n+1;
2:          rank := 0.5*(j+jt-1);
            FOR ji := j TO jt-1 DO W[ji] := rank;
            t := jt-j; s := s+t*t*t-t; j := jt END END;
    IF (j = n) THEN w[n] := n
END;


PROCEDURE kendl1(data1,data2: glnarray; n: integer; VAR tau,z,prob: real);
(* Programs using routine KENDL1 must define type
TYPE
    glnarray = ARRAY [1..n] OF real;
in the calling routine *)
VAR
    n2,n1,k,j,is: integer; svar,aa,a2,a1: real;
BEGIN
    n1 := 0; n2 := 0; is := 0;
    FOR j := 1 TO n-1 DO BEGIN
        FOR k := j+1 TO n DO BEGIN
            a1 := data1[j]-data1[k]; a2 := data2[j]-data2[k]; aa := a1*a2;
            IF (aa <> 0.0) THEN BEGIN
                n1 := n1+1; n2 := n2+1;
                IF (aa > 0.0) THEN BEGIN
                    is := is+1 END
                ELSE BEGIN
                    is := is-1 END END
            ELSE BEGIN
                IF (a1 <> 0.0) THEN n1 := n1+1;
                IF (a2 <> 0.0) THEN n2 := n2+1
            END END END;
```

```
    tau := is/(sqrt(n1)*sqrt(n2)); svar := (4.0*n+10.0)/(9.0*n*(n-1.0));
    z := tau/sqrt(svar); prob := erfcc(abs(z)/1.4142136)
END;

PROCEDURE kendl2(tab: gldarray; i,j,ip,jp: integer; VAR tau,z,prob: real);
(* Programs using routine KENDL2 must define type
TYPE
    gldarray = ARRAY [1..ip,1..jp] OF real;
in the calling program, where ip and jp are physical array dimensions
large enough to encompass the logical dimensions i and j.    *)
VAR
    nn,mm,m2,m1,lj,li,l,kj,ki,k: integer; svar,s,points,pairs,en2,en1: real;
BEGIN
    en1 := 0.0; en2 := 0.0; s := 0.0; nn := i*j; points := tab[i,j];
    FOR k := 0 TO nn-2 DO BEGIN
        ki := k DIV j;
        kj := k-j*ki; points := points+tab[ki+1,kj+1];
        FOR l := k+1 TO nn-1 DO BEGIN
            li := l DIV j;
            lj := l-j*li; m1 := li-ki; m2 := lj-kj; mm := m1*m2;
            pairs := tab[ki+1,kj+1]*tab[li+1,lj+1];
            IF (mm <> 0) THEN BEGIN
                en1 := en1+pairs; en2 := en2+pairs;
                IF (mm > 0) THEN BEGIN
                    s := s+pairs END
                ELSE BEGIN
                    s := s-pairs END END
            ELSE BEGIN
                IF (m1 <> 0) THEN en1 := en1+pairs;
                IF (m2 <> 0) THEN en2 := en2+pairs
            END END END;
    tau := s/sqrt(en1*en2);
    svar := (4.0*points+10.0)/(9.0*points*(points-1.0)); z := tau/sqrt(svar);
    prob := erfcc(abs(z)/1.4142136)
END;
```

Smoothing of Data

```
PROCEDURE smooft(VAR y: glyarray; n: integer; pts: real);
(* Programs using routine SMOOFT must define the type
TYPE
    glyarray = ARRAY [1..mp] OF real;
in the main routine, with mp >= (integral power of 2) >= n+2*pts *)
VAR
    nmin,m,mo2,k,j: integer; yn,y1,rn1,fac,cnst: real;
BEGIN
    m := 2; nmin := n+round(2.0*pts);
    WHILE (m < nmin) DO m := 2*m;
    cnst := sqr(pts/m); y1 := y[1]; yn := y[n]; rn1 := 1.0/(n-1.0);
    FOR j := 1 TO n DO BEGIN
        y[j] := y[j]-rn1*(y1*(n-j)+yn*(j-1)) END;
    IF (n+1 <= m) THEN BEGIN
        FOR j := n+1 TO m DO BEGIN
            y[j] := 0.0 END END;
    mo2 := m DIV 2;
    realft(y,mo2,1); y[1] := y[1]/mo2; fac := 1.0;
    FOR j := 1 TO (mo2-1) DO BEGIN
        k := 2*j+1;
        IF (fac <> 0.0) THEN BEGIN
            fac := (1.0-cnst*j*j)/mo2;
            IF (fac < 0.0) THEN fac := 0.0;
            y[k] := fac*y[k]; y[k+1] := fac*y[k+1] END
```

```
        ELSE BEGIN
            y[k] := 0.0; y[k+1] := 0.0 END END;
    fac := (1.0-0.25*pts*pts)/mo2;
    IF (fac < 0.0) THEN fac := 0.0;
    y[2] := fac*y[2]; realft(y,mo2,-1);
    FOR j := 1 TO n DO BEGIN
        y[j] := rn1*(y1*(n-j)+yn*(j-1))+y[j] END
END;
```

Chapter 14. Modeling of Data

Fitting Data to a Straight Line

```
PROCEDURE fit(x,y: glndata; ndata: integer; sig: glndata; mwt: integer;
           VAR a,b,siga,sigb,chi2,q: real);
(* Programs using routine FIT must define the type
TYPE
    glndata = ARRAY [1..ndata] OF real;
in the main routine.    *)
VAR
    i: integer; wt,t,sy,sxoss,sx,st2,ss,sigdat: real;
BEGIN
    sx := 0.0; sy := 0.0; st2 := 0.0; b := 0.0;
    IF (mwt <> 0)THEN BEGIN
        ss := 0.0;
        FOR i := 1 TO ndata DO BEGIN
            wt := 1.0/sqr(sig[i]); ss := ss+wt; sx := sx+x[i]*wt;
            sy := sy+y[i]*wt END END
    ELSE BEGIN
        FOR i := 1 TO ndata DO BEGIN
            sx := sx+x[i]; sy := sy+y[i] END;
        ss := ndata END;
    sxoss := sx/ss;
    IF (mwt <> 0)THEN BEGIN
        FOR i := 1 TO ndata DO BEGIN
            t := (x[i]-sxoss)/sig[i]; st2 := st2+t*t; b := b+t*y[i]/sig[i]
        END END
    ELSE BEGIN
        FOR i := 1 TO ndata DO BEGIN
            t := x[i]-sxoss; st2 := st2+t*t; b := b+t*y[i] END END;
    b := b/st2; a := (sy-sx*b)/ss; siga := sqrt((1.0+sx*sx/(ss*st2))/ss);
    sigb := sqrt(1.0/st2); chi2 := 0.0;
    IF (mwt = 0)THEN BEGIN
        FOR i := 1 TO ndata DO BEGIN
            chi2 := chi2+sqr(y[i]-a-b*x[i]) END;
        q := 1.0; sigdat := sqrt(chi2/(ndata-2)); siga := siga*sigdat;
        sigb := sigb*sigdat END
    ELSE BEGIN
        FOR i := 1 TO ndata DO BEGIN
            chi2 := chi2+sqr((y[i]-a-b*x[i])/sig[i]) END;
        q := gammq(0.5*(ndata-2),0.5*chi2) END;
END;
```

General Linear Least Squares

```
PROCEDURE lfit(x,y,sig: glndata; ndata: integer; VAR a: glmma; mma: integer;
        lista: gllista; mfit: integer; VAR covar: glcovar;
        ncvm: integer; VAR chisq: real);
(* Programs using routine LFIT must define the types
TYPE
```

```
    glndata = ARRAY [1..ndata] OF real;
    glmma = ARRAY [1..mma] OF real;
    gllista = ARRAY [1..mma] OF integer;
    glcovar = ARRAY [1..ncvm,1..ncvm] OF real;
    glnpbymp = ARRAY [1..ncvm,1..1] OF real;
in the main routine. *)
VAR
    k,kk,j,ihit,i: integer; ym,wt,sum,sig2i: real; beta: glnpbymp;
    afunc: glmma;
BEGIN
    kk := mfit+1;
    FOR j := 1 TO mma DO BEGIN
        ihit := 0;
        FOR k := 1 TO mfit DO BEGIN
            IF (lista[k] = j) THEN ihit := ihit+1
        END;
        IF (ihit = 0) THEN BEGIN
            lista[kk] := j; kk := kk+1 END
        ELSE IF (ihit > 1) THEN BEGIN
            writeln('pause in routine LFIT');
            writeln('improper permutation in LISTA'); readln
        END END;
    IF (kk <> (mma+1)) THEN BEGIN
        writeln('pause in routine LFIT');
        writeln('improper permutation in LISTA'); readln
    END;
    FOR j := 1 TO mfit DO BEGIN
        FOR k := 1 TO mfit DO BEGIN
            covar[j,k] := 0.0 END;
        beta[j,1] := 0.0 END;
    FOR i := 1 TO ndata DO BEGIN
        funcs(x[i],afunc,mma); ym := y[i];
        IF (mfit < mma) THEN BEGIN
            FOR j := (mfit+1) TO mma DO BEGIN
                ym := ym-a[lista[j]]*afunc[lista[j]] END END;
        sig2i := 1.0/sqr(sig[i]);
        FOR j := 1 TO mfit DO BEGIN
            wt := afunc[lista[j]]*sig2i;
            FOR k := 1 TO j DO BEGIN
                covar[j,k] := covar[j,k]+wt*afunc[lista[k]] END;
            beta[j,1] := beta[j,1]+ym*wt END END;
    IF (mfit > 1) THEN BEGIN
        FOR j := 2 TO mfit DO BEGIN
            FOR k := 1 TO j-1 DO BEGIN
                covar[k,j] := covar[j,k] END END END;
    gaussj(covar,mfit,ncvm,beta,1,1);
    FOR j := 1 TO mfit DO BEGIN
        a[lista[j]] := beta[j,1] END;
    chisq := 0.0;
    FOR i := 1 TO ndata DO BEGIN
        funcs(x[i],afunc,mma); sum := 0.0;
        FOR j := 1 TO mma DO BEGIN
            sum := sum+a[j]*afunc[j] END;
        chisq := chisq+sqr((y[i]-sum)/sig[i]) END;
    covsrt(covar,ncvm,mma,lista,mfit)
END;

PROCEDURE covsrt(VAR covar: glcovar; ncvm: integer; ma: integer;
        lista: gllista; mfit: integer);
(* Programs using routine COVSRT must define the types
TYPE
    glcovar = ARRAY [1..ncvm,1..ncvm] OF real;
    gllista = ARRAY [1..mfit] OF integer;
in the calling program. *)
```

```
VAR
    j,i: integer; swap: real;
BEGIN
    FOR j := 1 TO ma-1 DO BEGIN
        FOR i := j+1 TO ma DO BEGIN
            covar[i,j] := 0.0 END END;
    FOR i := 1 TO mfit-1 DO BEGIN
        FOR j := i+1 TO mfit DO BEGIN
            IF (lista[j] > lista[i]) THEN BEGIN
                covar[lista[j],lista[i]] := covar[i,j] END
            ELSE BEGIN
                covar[lista[i],lista[j]] := covar[i,j] END END END;
    swap := covar[1,1];
    FOR j := 1 TO ma DO BEGIN
        covar[1,j] := covar[j,j]; covar[j,j] := 0.0 END;
    covar[lista[1],lista[1]] := swap;
    FOR j := 2 TO mfit DO BEGIN
        covar[lista[j],lista[j]] := covar[1,j] END;
    FOR j := 2 TO ma DO BEGIN
        FOR i := 1 TO j-1 DO BEGIN
            covar[i,j] := covar[j,i] END END
END;


PROCEDURE svdfit(x,y,sig: glndata; ndata: integer; VAR a: glmma; mma: integer;
        VAR u: glmpbynp; VAR v: glnpbynp; VAR w: glnparray; mp,np:
        integer; VAR chisq: real);
(* Programs using routine SVDFIT must define the types
TYPE
    glndata = ARRAY [1..ndata] OF real;
    glmma = ARRAY [1..mma] OF real;
    glmpbynp = ARRAY [1..mp,1..np] OF real;
    glnpbynp = ARRAY [1..np,1..np] OF real;
    glnparray = ARRAY [1..np] OF real;
in the main routine. Implementations without conformant arrays require mma=np
and the declaration
    glnparray = glmma;
and also mp=ndata, with the declaration
    glmparray = glndata;
in the main routine for use by the procedure SVBKSB. All user programs must also
include a routine that computes the desired basis functions with the declaration
PROCEDURE func(x: real: VAR p: glmma; mma: integer);
which should return the values of first mma basis functions at x in p.
FPOLY and FLEG (renamed to FUNC) are possible choices. *)
CONST
    tol=1.0e-5;
VAR
    j,i: integer; wmax,tmp,thresh,sum: real; b: glndata; afunc: glmma;
BEGIN
    FOR i := 1 TO ndata DO BEGIN
        func(x[i],afunc,mma); tmp := 1.0/sig[i];
        FOR j := 1 TO mma DO u[i,j] := afunc[j]*tmp;
        b[i] := y[i]*tmp END;
    svdcmp(u,ndata,mma,mp,np,w,v); wmax := 0.0;
    FOR j := 1 TO mma DO IF (w[j] > wmax) THEN wmax := w[j];
    thresh := tol*wmax;
    FOR j := 1 TO mma DO IF (w[j] < thresh) THEN w[j] := 0.0;
    svbksb(u,w,v,ndata,mma,mp,np,b,a); chisq := 0.0;
    FOR i := 1 TO ndata DO BEGIN
        func(x[i],afunc,mma); sum := 0.0;
        FOR j := 1 TO mma DO sum := sum+a[j]*afunc[j];
        chisq := chisq+sqr((y[i]-sum)/sig[i]) END
END;
```

```
PROCEDURE svdvar(v: glnpbynp; ma,np: integer; w: glnparray;
        VAR cvm: glcvm; ncvm: integer);
(* Programs using routine SVDVAR must define the types
TYPE
    glnpbynp = ARRAY [1..np,1..np] OF real;
    glnparray = ARRAY [1..np] OF real;
    glcvm = ARRAY [1..ncvm,1..ncvm] OF real;
in the calling program. *)
VAR
    k,j,i: integer; sum: real; wti: glnparray;
BEGIN
    FOR i := 1 TO ma DO BEGIN
        wti[i] := 0.0;
        IF (w[i] <> 0.0) THEN wti[i] := 1.0/(w[i]*w[i])
    END;
    FOR i := 1 TO ma DO BEGIN
        FOR j := 1 TO i DO BEGIN
            sum := 0.0;
            FOR k := 1 TO ma DO BEGIN
                sum := sum+v[i,k]*v[j,k]*wti[k] END;
            cvm[i,j] := sum; cvm[j,i] := sum END END
END;


PROCEDURE fpoly(x: real; VAR p: glnparray; np: integer);
(* Programs using routine FPOLY must define the type
TYPE
    glnparray = ARRAY [1..np] OF real;
in the main routine.    *)
VAR
    j: integer;
BEGIN
    p[1] := 1.0;
    FOR j := 2 TO np DO BEGIN
        p[j] := p[j-1]*x END
END;


PROCEDURE fleg(x: real; VAR pl: glnlarray; nl: integer);
(* Programs using routine FLEG must define the type
TYPE
    glnlarray = ARRAY [1..nl] OF real;
in the main routine. *)
VAR
    j: integer; twox,f2,f1,d: real;
BEGIN
    pl[1] := 1.0; pl[2] := x;
    IF (nl > 2) THEN BEGIN
        twox := 2.0*x; f2 := x; d := 1.0;
        FOR j := 3 TO nl DO BEGIN
            f1 := d; f2 := f2+twox; d := d+1.0;
            pl[j] := (f2*pl[j-1]-f1*pl[j-2])/d END END
END;
```

Nonlinear Models

```
PROCEDURE mrqmin(x,y,sig: glndata; ndata: integer;
        VAR a: glmma; mma: integer; lista: gllista;
        mfit: integer; VAR covar,alpha: glncabynca;
        nca: integer; VAR chisq,alamda: real);
(* Programs using routine MRQMIN must define the types
TYPE
    glndata = ARRAY [1..ndata] OF real;
```

```
    glmma = ARRAY [1..mma] OF real;
    gllista = ARRAY [1..mma] OF integer;
    glncabynca = ARRAY [1..nca,1..nca] OF real;
and the variables
VAR
    glochisq: real; glbeta: glmma;
in the main routine. Also note that this routine calls MRQCOF, which
requires a user-defined procedure FUNCS, described in that routine.    *)
LABEL 99;
VAR
    k,kk,j,ihit: integer; atry,da: glmma; oneda: glncabynca;
BEGIN
    IF (alamda < 0.0) THEN BEGIN
        kk := mfit+1;
        FOR j := 1 TO mma DO BEGIN
            ihit := 0;
            FOR k := 1 TO mfit DO BEGIN
                IF (lista[k] = j) THEN ihit := ihit+1
            END;
            IF (ihit = 0) THEN BEGIN
                lista[kk] := j; kk := kk+1 END
            ELSE IF (ihit > 1) THEN BEGIN
                writeln('pause 1 in routine MRQMIN');
                writeln('Improper permutation in LISTA'); readln
            END END;
        IF (kk <> (mma+1)) THEN BEGIN
            writeln('pause 2 in routine MRQMIN');
            writeln('Improper permutation in LISTA'); readln
        END;
        alamda := 0.001;
        mrqcof(x,y,sig,ndata,a,mma,lista,mfit,alpha,glbeta,nca,chisq);
        glochisq := chisq;
        FOR j := 1 TO mma DO BEGIN
            atry[j] := a[j] END END;
    FOR j := 1 TO mfit DO BEGIN
        FOR k := 1 TO mfit DO BEGIN
            covar[j,k] := alpha[j,k] END;
        covar[j,j] := alpha[j,j]*(1.0+alamda); oneda[j,1] := glbeta[j] END;
    gaussj(covar,mfit,nca,oneda,1,1);
    FOR j := 1 TO mfit DO da[j] := oneda[j,1];
    IF (alamda = 0.0) THEN BEGIN
        covsrt(covar,nca,mma,lista,mfit);
        GOTO 99
    END;
    FOR j := 1 TO mfit DO BEGIN
        atry[lista[j]] := a[lista[j]]+da[j] END;
    mrqcof(x,y,sig,ndata,atry,mma,lista,mfit,covar,da,nca,chisq);
    IF (chisq < glochisq) THEN BEGIN
        alamda := 0.1*alamda; glochisq := chisq;
        FOR j := 1 TO mfit DO BEGIN
            FOR k := 1 TO mfit DO BEGIN
                alpha[j,k] := covar[j,k] END;
            glbeta[j] := da[j]; a[lista[j]] := atry[lista[j]] END END
    ELSE BEGIN
        alamda := 10.0*alamda; chisq := glochisq END;
99: END;

PROCEDURE mrqcof(x,y,sig: glndata; ndata: integer;
        VAR a: glmma; mma: integer; lista: gllista;
        mfit: integer; VAR alpha: glnalbynal;
        VAR beta: glmma; nalp: integer; VAR chisq: real);
(* Programs using routine MRQMIN must provide a
PROCEDURE funcs(xx:real; a:glmma; yfit:real; dyda:glmma; mma:integer);
that evaluates the fitting function yfit and its derivatives dyda
```

```
with respect to the parameters a at point xx. Also they
must define the types
TYPE
    glndata = ARRAY [1..ndata] OF real;
    glmma = ARRAY [1..mma] OF real;
    gllista = ARRAY [1..mma] OF integer;
    glnalbynal = ARRAY [1..nalp,1..nalp] OF real;
in the main routine *)
VAR
    k,j,i: integer; ymod,wt,sig2i,dy: real; dyda: glmma;
BEGIN
    FOR j := 1 TO mfit DO BEGIN
        FOR k := 1 TO j DO BEGIN
            alpha[j,k] := 0.0 END;
        beta[j] := 0.0 END;
    chisq := 0.0;
    FOR i := 1 TO ndata DO BEGIN
        funcs(x[i],a,ymod,dyda,mma); sig2i := 1.0/(sig[i]*sig[i]);
        dy := y[i]-ymod;
        FOR j := 1 TO mfit DO BEGIN
            wt := dyda[lista[j]]*sig2i;
            FOR k := 1 TO j DO BEGIN
                alpha[j,k] := alpha[j,k]+wt*dyda[lista[k]] END;
            beta[j] := beta[j]+dy*wt END;
        chisq := chisq+dy*dy*sig2i END;
    FOR j := 2 TO mfit DO BEGIN
        FOR k := 1 TO j-1 DO BEGIN
            alpha[k,j] := alpha[j,k] END END
END;

PROCEDURE fgauss(x: real; a: glnparam; VAR y: real;
        VAR dyda: glnparam; na: integer);
(* Programs using routine FGAUSS must define the type
TYPE
    glnparam = ARRAY [1..na] OF real;
in the main routine.    *)
VAR
    i,ii: integer; fac,ex,arg: real;
BEGIN
    y := 0.0;
    FOR ii := 1 TO (na DIV 3) DO BEGIN
        i := 3*ii-2; arg := (x-a[i+1])/a[i+2]; ex := exp(-sqr(arg));
        fac := a[i]*ex*2.0*arg; y := y+a[i]*ex; dyda[i] := ex;
        dyda[i+1] := fac/a[i+2]; dyda[i+2] := fac*arg/a[i+2] END
END;
```

Robust Estimation

```
PROCEDURE medfit(x,y: glndata; ndata: integer; VAR a,b,abdev: real);
(* Programs using routine MEDFIT must define the type
TYPE
    glndata = ARRAY [1..ndata] OF real;
in the main routine. MEDFIT also assumes that the instructions
at the beginning of ROFUNC have been carried out so that arrays
x,y and arr, and real variables aa and abdevt exist as globally
defined variables. *)
LABEL 3;
VAR
    j: integer; sy,sxy,sxx,sx,sigb,f2,f1,f: real; del,chisq,bb,b2,b1: real;
BEGIN
    sx := 0.0; sy := 0.0; sxy := 0.0; sxx := 0.0;
    FOR j := 1 TO ndata DO BEGIN
```

```
        sx := sx+x[j]; sy := sy+y[j]; sxy := sxy+x[j]*y[j];
        sxx := sxx+sqr(x[j]) END;
    del := ndata*sxx-sx*sx; aa := (sxx*sy-sx*sxy)/del;
    bb := (ndata*sxy-sx*sy)/del; chisq := 0.0;
    FOR j := 1 TO ndata DO BEGIN
        chisq := chisq+sqr(y[j]-(aa+bb*x[j])) END;
    sigb := sqrt(chisq/del); b1 := bb; f1 := rofunc(b1);
    IF (f1 >= 0.0) THEN BEGIN
        b2 := bb+abs(3.0*sigb) END
    ELSE BEGIN
        b2 := bb-abs(3.0*sigb) END;
    f2 := rofunc(b2);
    WHILE ((f1*f2) > 0.0) DO BEGIN
        bb := 2.0*b2-b1; b1 := b2; f1 := f2; b2 := bb; f2 := rofunc(b2)
    END;
    sigb := 0.01*sigb;
    WHILE (abs(b2-b1) > sigb) DO BEGIN
        bb := 0.5*(b1+b2);
        IF ((bb = b1) OR (bb = b2)) THEN GOTO 3;
        f := rofunc(bb);
        IF ((f*f1) >= 0.0) THEN BEGIN
            f1 := f; b1 := bb END
        ELSE BEGIN
            f2 := f; b2 := bb END END;
3:  a := aa;
    b := bb; abdev := abdevt/ndata
END;

FUNCTION rofunc(b: real): real;
(* Programs using ROFUNC must declare
CONST
    ndata = [the number of data points]
VAR
    aa,abdevt: real;
    x,y,arr: ARRAY [1..ndata] OF real;
in the main routine. *)
VAR
    d,sum: real; j,n1,nmh,nml: integer;
BEGIN
    n1 := ndata+1;
    nml := n1 DIV 2;
    nmh := n1-nml;
    FOR j := 1 TO ndata DO arr[j] := y[j]-b*x[j];
    sort(ndata,arr); aa := 0.5*(arr[nml]+arr[nmh]); sum := 0.0;
    abdevt := 0.0;
    FOR j := 1 TO ndata DO BEGIN
        d := y[j]-(b*x[j]+aa); abdevt := abdevt+abs(d);
        IF (d > 0.0) THEN sum := sum+x[j] ELSE sum := sum-x[j]
    END;
    rofunc := sum
END;
```

Chapter 15. Integration of Ordinary Differential Equations

Runge-Kutta Method

```
PROCEDURE rk4(y,dydx: glnarray; n: integer; x,h: real; VAR yout: glnarray);
(* Programs using routine RK4 must provide a
PROCEDURE derivs(x:real; y:glnarray; VAR dydx:glnarray);
which returns the derivatives dydx at location x, given both x and the
```

```
function values y. The calling program must also define the types
TYPE
    glnarray = ARRAY [1..nvar] OF real;
where nvar is the number of variables y. *)
VAR
    i: integer; xh,hh,h6: real; dym,dyt,yt: glnarray;
BEGIN
    hh := h*0.5; h6 := h/6.0; xh := x+hh;
    FOR i := 1 TO n DO BEGIN
        yt[i] := y[i]+hh*dydx[i] END;
    derivs(xh,yt,dyt);
    FOR i := 1 TO n DO BEGIN
        yt[i] := y[i]+hh*dyt[i] END;
    derivs(xh,yt,dym);
    FOR i := 1 TO n DO BEGIN
        yt[i] := y[i]+h*dym[i]; dym[i] := dyt[i]+dym[i] END;
    derivs(x+h,yt,dyt);
    FOR i := 1 TO n DO BEGIN
        yout[i] := y[i]+h6*(dydx[i]+dyt[i]+2.0*dym[i]) END
END;
```

```
PROCEDURE rkdumb(vstart: glnarray; nvar: integer; x1,x2: real; nstep: integer);
(* Programs using routine RKDUMB must provide a
PROCEDURE derivs(x:real; v:glnarray; VAR dvdx:glnarray);
which returns the derivatives dvdx at location x, given both x and the
function values v. They must also define the type
TYPE
    glnarray = ARRAY [1..nvar] OF real;
where nvar is the number of functions. They must also
declare the variables
VAR
    xx: ARRAY [1..200] OF real;
    y: ARRAY [1..nvar,1..200] OF real;
in the main routine. *)
VAR
    k,i: integer; x,h: real; v,vout,dv: glnarray;
BEGIN
    FOR i := 1 TO nvar DO BEGIN
        v[i] := vstart[i]; y[i,1] := v[i] END;
    xx[1] := x1; x := x1; h := (x2-x1)/nstep;
    FOR k := 1 TO nstep DO BEGIN
        derivs(x,v,dv); rk4(v,dv,nvar,x,h,vout);
        IF (x+h = x) THEN BEGIN
            writeln('pause in routine RKDUMB');
            writeln('stepsize to small'); readln END;
        x := x+h; xx[k+1] := x;
        FOR i := 1 TO nvar DO BEGIN
            v[i] := vout[i]; y[i,k+1] := v[i] END END
END;
```

Adaptive Stepsize Control for Runge-Kutta

```
PROCEDURE rkqc(VAR y,dydx: glarray; n: integer; VAR x: real;
        htry,eps: real; yscal: glarray; VAR hdid,hnext: real);
(* Programs using routine RKQC must provide a
PROCEDURE derivs(x:real; y:glnarray; VAR dydx:glnarray);
which returns the derivatives dydx at location x, given both x and the
function values y. They must also define the type
TYPE
    glarray = ARRAY [1..n] OF real;
in the main routine.    *)
LABEL 1;
```

```
CONST
    pgrow=-0.20; pshrnk=-0.25;
    fcor=0.06666666;    (* 1.0/15.0 *)
    one=1.0; safety=0.9; errcon=6.0e-4;
VAR
    i: integer; xsav,hh,h,temp,errmax: real; dysav,ysav,ytemp: glarray;
BEGIN
    xsav := x;
    FOR i := 1 TO n DO BEGIN
        ysav[i] := y[i]; dysav[i] := dydx[i] END;
    h := htry;
1:  hh := 0.5*h;
    rk4(ysav,dysav,n,xsav,hh,ytemp); x := xsav+hh; derivs(x,ytemp,dydx);
    rk4(ytemp,dydx,n,x,hh,y); x := xsav+h;
    IF (x = xsav) THEN BEGIN
        writeln('pause in routine RKQC');
        writeln('stepsize too small'); readln END;
    rk4(ysav,dysav,n,xsav,h,ytemp); errmax := 0.0;
    FOR i := 1 TO n DO BEGIN
        ytemp[i] := y[i]-ytemp[i]; temp := abs(ytemp[i]/yscal[i]);
        IF (errmax < temp) THEN errmax := temp
    END;
    errmax := errmax/eps;
    IF (errmax > one) THEN BEGIN
        h := safety*h*exp(pshrnk*ln(errmax));
        GOTO 1 END
    ELSE BEGIN
        hdid := h;
        IF (errmax > errcon) THEN BEGIN
            hnext := safety*h*exp(pgrow*ln(errmax)) END
        ELSE BEGIN
            hnext := 4.0*h END END;
    FOR i := 1 TO n DO BEGIN
        y[i] := y[i]+ytemp[i]*fcor END
END;


PROCEDURE odeint(VAR ystart: glnarray; nvar: integer;
          x1,x2,eps,h1,hmin: real; VAR nok,nbad: integer);
(* Programs using routine ODEINT must provide a
PROCEDURE derivs(x:real; y:glnarray; VAR dydx:glnarray);
which returns the derivatives dydx at location x, given both x
and the function values y. They must also define the type
TYPE
    glnarray = ARRAY [1..nvar] OF real;
and must declare the following parameters
VAR
    kmax,kount: integer; dxsav: real;
    xp: ARRAY [1..200] OF real;
    yp: ARRAY [1..10,1..200] OF real;
and must initialize kmax and dxsav in the main routine. *)
LABEL 99;
CONST
    maxstp=10000; two=2.0; zero=0.0; tiny=1.0e-30;
VAR
    nstp,i: integer; xsav,x,hnext,hdid,h: real; yscal,y,dydx: glnarray;
BEGIN
    x := x1;
    IF (x2 >= x1) THEN h := abs(h1) ELSE h := -abs(h1);
    nok := 0; nbad := 0; kount := 0;
    FOR i := 1 TO nvar DO BEGIN
        y[i] := ystart[i] END;
    IF kmax>0 THEN xsav := x-dxsav*two;
    FOR nstp := 1 TO maxstp DO BEGIN
        derivs(x,y,dydx);
```

```
        FOR i := 1 TO nvar DO BEGIN
            yscal[i] := abs(y[i])+abs(dydx[i]*h)+tiny END;
        IF (kmax > 0) THEN BEGIN
            IF (abs(x-xsav) > abs(dxsav)) THEN BEGIN
                IF (kount < kmax-1) THEN BEGIN
                    kount := kount+1; xp[kount] := x;
                    FOR i := 1 TO nvar DO BEGIN
                        yp[i,kount] := y[i] END;
                    xsav := x END END END;
        IF (((x+h-x2)*(x+h-x1)) > zero) THEN h := x2-x;
        rkqc(y,dydx,nvar,x,h,eps,yscal,hdid,hnext);
        IF (hdid = h) THEN BEGIN
            nok := nok+1 END
        ELSE BEGIN
            nbad := nbad+1 END;
        IF (((x-x2)*(x2-x1)) >= zero) THEN BEGIN
            FOR i := 1 TO nvar DO BEGIN
                ystart[i] := y[i] END;
            IF (kmax <> 0) THEN BEGIN
                kount := kount+1; xp[kount] := x;
                FOR i := 1 TO nvar DO BEGIN
                    yp[i,kount] := y[i] END END;
            GOTO 99
        END;
        IF (abs(hnext) < hmin) THEN BEGIN
            writeln('pause in routine ODEINT');
            writeln('stepsize too small'); readln END;
        h := hnext; END;
    writeln('pause in routine ODEINT - too many steps'); readln;
99: END;
```

Modified Midpoint Method

```
PROCEDURE mmid(y,dydx: glnarray; nvar: integer; xs,htot: real;
        nstep: integer; VAR yout: glnarray);
(* Programs using routine MMID must provide a
PROCEDURE derivs(x:real; y:glnarray; VAR dydx:glnarray);
which returns the derivatives dydx at location x, given both x and the
function values y. They must also define the type
TYPE
    glnarray = ARRAY [1..nvar] OF real;
in the main routine. Note that this routine is in
single precision, unlike the FORTRAN version. *)
VAR
    n,i: integer; x,swap,h2,h: real; ym,yn: glnarray;
BEGIN
    h := htot/nstep;
    FOR i := 1 TO nvar DO BEGIN
        ym[i] := y[i]; yn[i] := y[i]+h*dydx[i] END;
    x := xs+h; derivs(x,yn,yout); h2 := 2.0*h;
    FOR n := 2 TO nstep DO BEGIN
        FOR i := 1 TO nvar DO BEGIN
            swap := ym[i]+h2*yout[i]; ym[i] := yn[i]; yn[i] := swap END;
        x := x+h; derivs(x,yn,yout) END;
    FOR i := 1 TO nvar DO BEGIN
        yout[i] := 0.5*(ym[i]+yn[i]+h*yout[i]) END
END;
```

Richardson Extrapolation and the Bulirsch-Stoer Method

```
PROCEDURE bsstep(VAR y: glyarray; dydx: glyarray; nv: integer; VAR x: real;
       htry,eps: real; yscal: glyarray; VAR hdid,hnext: real);
(* Programs using routine BSSTEP must define the type
TYPE
    glyarray = ARRAY [1..nv] OF real;
in the main routine. *)
LABEL 99;
CONST
    imax=11; nuse=7; one=1.0e0; shrink=0.95e0; grow=1.2e0;
VAR
    j,i: integer; xsav,xest,h,errmax: real; ysav,dysav,yseq,yerr: glyarray;
    nseq: ARRAY [1..imax] OF integer;
BEGIN
    nseq[1] := 2; nseq[2] := 4; nseq[3] := 6;
    nseq[4] := 8; nseq[5] := 12; nseq[6] := 16;
    nseq[7] := 24; nseq[8] := 32; nseq[9] := 48;
    nseq[10] := 64; nseq[11] := 96; h := htry; xsav := x;
    FOR i := 1 TO nv DO BEGIN
        ysav[i] := y[i]; dysav[i] := dydx[i] END;
    WHILE true DO BEGIN
        FOR i := 1 TO imax DO BEGIN
            mmid(ysav,dysav,nv,xsav,h,nseq[i],yseq); xest := sqr(h/nseq[i]);
            rzextr(i,xest,yseq,y,yerr,nv,nuse); errmax := 0.0;
            FOR j := 1 TO nv DO BEGIN
                IF (errmax < abs(yerr[j]/yscal[j])) THEN
                    errmax := abs(yerr[j]/yscal[j]) END;
            errmax := errmax/eps;
            IF (errmax < one) THEN BEGIN
                x := x+h; hdid := h;
                IF (i = nuse) THEN hnext := h*shrink
                ELSE IF (i = nuse-1) THEN hnext := h*grow
                ELSE hnext := (h*nseq[nuse-1])/nseq[i];
                GOTO 99
            END END;
        h := 0.25*h;
        IF (((imax-nuse) DIV 2) > 0) THEN BEGIN
            FOR i := 1 TO ((imax-nuse) DIV 2) DO h := h/2
        END;
        IF ((x+h) = x) THEN BEGIN
            writeln('pause in routine BSSTEP');
            writeln('step size underflow'); readln END END;
99: END;
```

```
PROCEDURE rzextr(iest: integer; xest: real; yest: glyarray;
        VAR yz,dy: glyarray; nv,nuse: integer);
(* Programs using routine RZEXTR must declare
TYPE
    glyarray = ARRAY [1..nv] OF real;
CONST
    glimax=11; glnmax=10; glncol=7;
VAR
    glx: ARRAY [1..glimax] OF real;
    gld: ARRAY [1..glnmax,1..glncol] OF real;
in the main routine. *)
CONST
    ncol=7;
VAR
    m1,k,j: integer; yy,v,ddy,c,b1,b: real;
    fx: ARRAY [1..ncol] OF real;
BEGIN
    glx[iest] := xest;
```

```
    IF (iest = 1) THEN BEGIN
        FOR j := 1 TO nv DO BEGIN
            yz[j] := yest[j]; gld[j,1] := yest[j]; dy[j] := yest[j] END END
    ELSE BEGIN
        IF (iest < nuse) THEN m1 := iest ELSE m1 := nuse;
        FOR k := 1 TO m1-1 DO BEGIN
            fx[k+1] := glx[iest-k]/xest END;
        FOR j := 1 TO nv DO BEGIN
            yy := yest[j]; v := gld[j,1]; c := yy; gld[j,1] := yy;
            FOR k := 2 TO m1 DO BEGIN
                b1 := fx[k]*v; b := b1-c;
                IF (b <> 0.0) THEN BEGIN
                    b := (c-v)/b; ddy := c*b; c := b1*b END
                ELSE ddy := v;
                IF k<>m1 THEN v := gld[j,k]; gld[j,k] := ddy;
                yy := yy+ddy END;
            dy[j] := ddy; yz[j] := yy END END
END;
```

```
PROCEDURE pzextr(iest: integer; xest: real; yest: glyarray;
         VAR yz,dy: glyarray; nv,nuse: integer);
(* Programs using routine PZEXTR must declare
TYPE
    glyarray = ARRAY [1..nv] OF real;
CONST
    glimax=11; glnmax=10; glncol=7;
VAR
    glx: ARRAY [1..glimax] OF real;
    glqcol: ARRAY [1..glnmax,1..glncol] OF real;
in the main routine. *)
CONST
    nmax=10;
VAR
    m1,k1,j: integer; q,f2,f1,delta: real;
    d: ARRAY [1..nmax] OF real;
BEGIN
    glx[iest] := xest;
    FOR j := 1 TO nv DO BEGIN
        dy[j] := yest[j]; yz[j] := yest[j] END;
    IF (iest = 1) THEN BEGIN
        FOR j := 1 TO nv DO BEGIN
            glqcol[j,1] := yest[j] END END
    ELSE BEGIN
        IF (iest < nuse) THEN m1 := iest ELSE m1 := nuse;
        FOR j := 1 TO nv DO BEGIN
            d[j] := yest[j] END;
        FOR k1 := 1 TO m1-1 DO BEGIN
            delta := 1.0/(glx[iest-k1]-xest); f1 := xest*delta;
            f2 := glx[iest-k1]*delta;
            FOR j := 1 TO nv DO BEGIN
                q := glqcol[j,k1]; glqcol[j,k1] := dy[j]; delta := d[j]-q;
                dy[j] := f1*delta; d[j] := f2*delta; yz[j] := yz[j]+dy[j]
            END END;
        FOR j := 1 TO nv DO BEGIN
            glqcol[j,m1] := dy[j] END END
END;
```

Chapter 16. Two-Point Boundary Value Problems

The Shooting Method

```
PROCEDURE shoot(nvar: integer; VAR v: gln2array; delv: gln2array;
       n2: integer; x1,x2,eps,h1,hmin: real; VAR f,dv: glnvar);
(* Programs using routine SHOOT must define the types
TYPE
   gln2array = ARRAY [1..n2] OF real;
   glnvar = ARRAY [1..nvar] OF real;
   gln2byn2 = ARRAY [1..n2,1..n2] OF real;
   glinvar = ARRAY [1..nvar] OF integer;
   glnpbynp = gln2byn2;
in the main routine, and set the variable kmax of ODEINT to zero. *)
VAR
   nok,nbad,iv,i: integer; sav,det: real; y: glnvar; dfdv: gln2byn2;
   indx: glinvar;
BEGIN
   load(x1,v,y); odeint(y,nvar,x1,x2,eps,h1,hmin,nok,nbad); score(x2,y,f);
   FOR iv := 1 TO n2 DO BEGIN
      sav := v[iv]; v[iv] := v[iv]+delv[iv]; load(x1,v,y);
      odeint(y,nvar,x1,x2,eps,h1,hmin,nok,nbad); score(x2,y,dv);
      FOR i := 1 TO n2 DO BEGIN
         dfdv[i,iv] := (dv[i]-f[i])/delv[iv] END;
      v[iv] := sav END;
   FOR iv := 1 TO n2 DO BEGIN
      dv[iv] := -f[iv] END;
   ludcmp(dfdv,n2,nvar,indx,det); lubksb(dfdv,n2,nvar,indx,dv);
   FOR iv := 1 TO n2 DO BEGIN
      v[iv] := v[iv] + dv[iv] END
END;
```

Shooting to a Fitting Point

```
PROCEDURE shootf(nvar: integer; VAR v1: gln2array; VAR v2: gln1array;
         delv1: gln2array; delv2: gln1array; n1,n2: integer;
         x1,x2,xf,eps,h1,hmin: real; VAR f: glnvar;
         VAR dv1: gln2array; VAR dv2: gln1array);
(* Programs using routine SHOOTF must define the types
TYPE
   gln1array = ARRAY [1..n1] OF real;
   gln2array = ARRAY [1..n2] OF real;
   glnvar = ARRAY [1..nvar] OF real;
   glnvarbynvar = ARRAY [1..nvar,1..nvar];
   glindx = ARRAY [1..nvar] OF integer;
   glnpbynp = glnvarbynvar;
in the main routine, and set the variable kmax of ODEINT to zero. *)
VAR
   nok,nbad,j,iv,i: integer; sav,det: real; y,f1,f2: glnvar;
   dfdv: glnvarbynvar; indx: glindx;
BEGIN
   load1(x1,v1,y); odeint(y,nvar,x1,xf,eps,h1,hmin,nok,nbad);
   score(xf,y,f1); load2(x2,v2,y);
   odeint(y,nvar,x2,xf,eps,h1,hmin,nok,nbad); score(xf,y,f2); j := 0;
   FOR iv := 1 TO n2 DO BEGIN
      j := j+1; sav := v1[iv]; v1[iv] := v1[iv]+delv1[iv]; load1(x1,v1,y);
      odeint(y,nvar,x1,xf,eps,h1,hmin,nok,nbad); score(xf,y,f);
      FOR i := 1 TO nvar DO BEGIN
         dfdv[i,j] := (f[i]-f1[i])/delv1[iv] END;
```

```
        v1[iv] := sav END;
    FOR iv := 1 TO n1 DO BEGIN
        j := j+1; sav := v2[iv]; v2[iv] := v2[iv]+delv2[iv]; load2(x2,v2,y);
        odeint(y,nvar,x2,xf,eps,h1,hmin,nok,nbad); score(xf,y,f);
        FOR i := 1 TO nvar DO BEGIN
            dfdv[i,j] := (f2[i]-f[i])/delv2[iv] END;
        v2[iv] := sav END;
    FOR i := 1 TO nvar DO BEGIN
        f[i] := f1[i]-f2[i]; f1[i] := -f[i] END;
    ludcmp(dfdv,nvar,nvar,indx,det); lubksb(dfdv,nvar,nvar,indx,f1); j := 0;
    FOR iv := 1 TO n2 DO BEGIN
        j := j+1; v1[iv] := v1[iv]+f1[j]; dv1[iv] := f1[j] END;
    FOR iv := 1 TO n1 DO BEGIN
        j := j+1; v2[iv] := v2[iv]+f1[j]; dv2[iv] := f1[j] END
END;
```

Relaxation Methods

```
PROCEDURE solvde(itmax: integer; conv,slowc: real; scalv: glscalv;
         indexv: glindex; ne,nb,m: integer; VAR y: glyarray;
         nyj,nyk: integer; VAR c: glcarray; nci,ncj,nck: integer;
         VAR s: glsarray; nsi,nsj: integer);
(* Programs using routine SOLVDE must define the types
TYPE
    glindex = ARRAY [1..nyj] OF integer;
    glscalv = ARRAY [1..nyj] OF real;
    glyarray = ARRAY [1..nyj,1..nyk] OF real;
    glcarray = ARRAY [1..nci,1..ncj,1..nck] OF real;
    glsarray = ARRAY [1..nsi,1..nsj] OF real;
in the main routine. *)
LABEL 99;
CONST
    nmax=10;
VAR
    err,errj,fac,vmax,vz: real; ic1,ic2,ic3,ic4,it: integer;
    j,j1,j2,j3,j4,j5,j6,j7,j8,j9: integer;
    jc1,jcf,jv,k,k1,k2,km,kp,nvars: integer;
    ermax: ARRAY [1..nmax] OF real;
    kmax: ARRAY [1..nmax] OF integer;
BEGIN
    k1 := 1; k2 := m; nvars := ne*m; j1 := 1; j2 := nb; j3 := nb+1;
    j4 := ne; j5 := j4+j1; j6 := j4+j2; j7 := j4+j3; j8 := j4+j4;
    j9 := j8+j1; ic1 := 1; ic2 := ne-nb; ic3 := ic2+1; ic4 := ne;
    jc1 := 1; jcf := ic3;
    FOR it := 1 TO itmax DO BEGIN
        k := k1; difeq(k,k1,k2,j9,ic3,ic4,indexv,ne,s,nsi,nsj,y,nyj,nyk);
        pinvs(ic3,ic4,j5,j9,jc1,k1,c,nci,ncj,nck,s,nsi,nsj);
        FOR k := k1+1 TO k2 DO BEGIN
            kp := k-1;
            difeq(k,k1,k2,j9,ic1,ic4,indexv,ne,s,nsi,nsj,y,nyj,nyk);
            red(ic1,ic4,j1,j2,j3,j4,j9,ic3,jc1,jcf,kp,c,nci,ncj,nck,s,nsi,nsj);
            pinvs(ic1,ic4,j3,j9,jc1,k,c,nci,ncj,nck,s,nsi,nsj) END;
        k := k2+1; difeq(k,k1,k2,j9,ic1,ic2,indexv,ne, s,nsi,nsj,y,nyj,nyk);
        red(ic1,ic2,j5,j6,j7,j8,j9,ic3,jc1,jcf,k2, c,nci,ncj,nck,s,nsi,nsj);
        pinvs(ic1,ic2,j7,j9,jcf,k2+1, c,nci,ncj,nck,s,nsi,nsj);
        bksub(ne,nb,jcf,k1,k2,c,nci,ncj,nck); err := 0.0;
        FOR j := 1 TO ne DO BEGIN
            jv := indexv[j]; errj := 0.0; km := 0;
            vmax := 0.0;
            FOR k := k1 TO k2 DO BEGIN
                vz := abs(c[j,1,k]);
```

```
               IF (vz > vmax) THEN BEGIN
                    vmax := vz; km := k END;
               errj := errj+vz END;
           err := err+errj/scalv[jv]; ermax[j] := c[j,1,km]/scalv[jv];
           kmax[j] := km END;
       err := err/nvars; fac := 1.0;
       IF (err > slowc) THEN fac := slowc/err;
       FOR jv := 1 TO ne DO BEGIN
           j := indexv[jv];
           FOR k := k1 TO k2 DO BEGIN
               y[j,k] := y[j,k]-fac*c[jv,1,k] END END;
       writeln;
       writeln('Iter.':8,'Error':9,'FAC':9);
       writeln(it:6,err:12:6,fac:11:6);
       writeln('Var.':8,'Kmax':8,'Max. Error':14);
       FOR j := 1 TO ne DO writeln(indexv[j]:6,
           kmax[j]:9,ermax[j]:14:6);
       IF (err < conv) THEN GOTO 99
   END;
   writeln('pause in routine SOLVDE');
   writeln('too many iterations'); readln;
99: END;


PROCEDURE bksub(ne,nb,jf,k1,k2: integer; VAR c: glcarray;
       nci,ncj,nck: integer);
(* Programs using routine BKSUB must define the type
TYPE
   glcarray = ARRAY [1..nci,1..ncj,1..nck] OF real;
in the main routine. *)
VAR
   nbf,kp,k,j,i,im: integer; xx: real;
BEGIN
   nbf := ne-nb; im := 1;
   FOR k := k2 DOWNTO k1 DO BEGIN
       kp := k+1; IF k=k1 THEN im := nbf+1;
       FOR j := 1 TO nbf DO BEGIN
           xx := c[j,jf,kp];
           FOR i := im TO ne DO BEGIN
               c[i,jf,k] := c[i,jf,k]-c[i,j,k]*xx END END END;
   FOR k := k1 TO k2 DO BEGIN
       kp := k+1;
       FOR i := 1 TO nb DO BEGIN
           c[i,1,k] := c[i+nbf,jf,k] END;
       FOR i := 1 TO nbf DO BEGIN
           c[i+nb,1,k] := c[i,jf,kp] END END
END;


PROCEDURE pinvs(ie1,ie2,je1,jsf,jc1,k: integer; VAR c: glcarray;
       nci,ncj,nck: integer; VAR s: glsarray; nsi,nsj: integer);
(* Programs using routine PINVS must define the types
TYPE
   glcarray = ARRAY [1..nci,1..ncj,1..nck] OF real;
   glsarray = ARRAY [1..nsi,1..nsj] OF real;
in the main routine. *)
CONST
   zero=0.0; one=1.0; nmax=10;
VAR
   js1,jpiv,jp,je2,jcoff,j,irow,ipiv,id,icoff,i: integer;
   pivinv,piv,dum,big: real;
   pscl: ARRAY [1..nmax] OF real;
   indxr: ARRAY [1..nmax] OF integer;
BEGIN
   je2 := je1+ie2-ie1; js1 := je2+1;
```

```
    FOR i := ie1 TO ie2 DO BEGIN
        big := zero;
        FOR j := je1 TO je2 DO IF (abs(s[i,j]) > big) THEN big := abs(s[i,j]);
        IF (big = zero) THEN BEGIN
            writeln('pause in routine PINVS');
            writeln('singular matrix - row all 0'); readln END;
        pscl[i] := one/big; indxr[i] := 0 END;
    FOR id := ie1 TO ie2 DO BEGIN
        piv := zero;
        FOR i := ie1 TO ie2 DO BEGIN
            IF (indxr[i] = 0) THEN BEGIN
                big := zero;
                FOR j := je1 TO je2 DO BEGIN
                    IF (abs(s[i,j]) > big) THEN BEGIN
                        jp := j; big := abs(s[i,j]) END END;
                IF (big*pscl[i] > piv) THEN BEGIN
                    ipiv := i; jpiv := jp; piv := big*pscl[i] END END END;
        IF (s[ipiv,jpiv] = zero) THEN BEGIN
            writeln('pause in routine PINVS');
            writeln('singular matrix'); readln END;
        indxr[ipiv] := jpiv; pivinv := one/s[ipiv,jpiv];
        FOR j := je1 TO jsf DO s[ipiv,j] := s[ipiv,j]*pivinv;
        s[ipiv,jpiv] := one;
        FOR i := ie1 TO ie2 DO BEGIN
            IF (indxr[i] <> jpiv) THEN BEGIN
                IF (s[i,jpiv] <> zero) THEN BEGIN
                    dum := s[i,jpiv];
                    FOR j := je1 TO jsf DO s[i,j] := s[i,j]-dum*s[ipiv,j];
                    s[i,jpiv] := zero END END END END;
    jcoff := jc1-js1; icoff := ie1-je1;
    FOR i := ie1 TO ie2 DO BEGIN
        irow := indxr[i]+icoff;
        FOR j := js1 TO jsf DO c[irow,j+jcoff,k] := s[i,j]
    END
END;
```

```
PROCEDURE red(iz1,iz2,jz1,jz2,jm1,jm2,jmf,ic1,jc1,jcf,kc: integer;
          c: glcarray; nci,ncj,nck: integer;
          VAR s: glsarray; nsi,nsj: integer);
(* Programs using routine RED must define the types
TYPE
    glcarray = ARRAY [1..nci,1..ncj,1..nck] OF real;
    glsarray = ARRAY [1..nsi,1..nsj] OF real;
in the main routine. *)
VAR
    loff,l,j,ic,i: integer; vx: real;
BEGIN
    loff := jc1-jm1; ic := ic1;
    FOR j := jz1 TO jz2 DO BEGIN
        FOR l := jm1 TO jm2 DO BEGIN
            vx := c[ic,l+loff,kc];
            FOR i := iz1 TO iz2 DO s[i,l] := s[i,l]-s[i,j]*vx
        END;
        vx := c[ic,jcf,kc];
        FOR i := iz1 TO iz2 DO s[i,jmf] := s[i,jmf]-s[i,j]*vx;
        ic := ic+1 END
END;
```

A Worked Example: Spheroidal Harmonics

```
PROGRAM sfroid(input,output);
LABEL 99;
CONST
   ne=3; m=41; nb=1; nsi=ne; nyj=ne; nyk=m; nci=ne;
   ncj=3;        (* ncj=ne-nb+1 *)
   nck=42;        (* nck=m+1 *)
   nsj=7;        (* nsj=2*ne+1 *)
TYPE
   glyarray = ARRAY [1..ne,1..m] OF real;
   glcarray = ARRAY [1..nci,1..ncj,1..nck] OF real;
   glsarray = ARRAY [1..nsi,1..nsj] OF real;
   glscalv = ARRAY [1..ne] OF real;
   glindex = ARRAY [1..ne] OF integer;
VAR
   mm,n: integer; i,itmax,k: integer; h,c2,anorm: real;
   conv,deriv,fac1,fac2: real; q1,slowc: real; scalv: glscalv;
   indexv: glindex; y: glyarray; c: glcarray; s: glsarray;
   x: ARRAY [1..m] OF real;
(*$I MODFILE.PAS *)
(*$I PLGNDR.PAS *)
(*$I DIFEQ.PAS *)
(*$I RED.PAS *)
(*$I PINVS.PAS *)
(*$I BKSUB.PAS *)
(*$I SOLVDE.PAS *)
BEGIN
   itmax := 100; c2 := 0.0; conv := 5.0e-6; slowc := 1.0; h := 1.0/(m-1);
   writeln('Enter m n'); readln(mm,n);
   IF (((n+mm) MOD 2) = 1) THEN BEGIN
      indexv[1] := 1; indexv[2] := 2; indexv[3] := 3 END
   ELSE BEGIN
      indexv[1] := 2; indexv[2] := 1; indexv[3] := 3 END;
   anorm := 1.0;
   IF (mm <> 0) THEN BEGIN
      q1 := n;
      FOR i := 1 TO mm DO BEGIN
         anorm := -0.5*anorm*(n+i)*(q1/i); q1 := q1-1.0 END END;
   FOR k := 1 TO (m-1) DO BEGIN
      x[k] := (k-1)*h; fac1 := 1.0-sqr(x[k]);
      fac2 := exp((-mm/2.0)*ln(fac1)); y[1,k] := plgndr(n,mm,x[k])*fac2;
      deriv := -((n-mm+1)*plgndr(n+1,mm,x[k])-
         (n+1)*x[k]*plgndr(n,mm,x[k]))/fac1;
      y[2,k] := mm*x[k]*y[1,k]/fac1+deriv*fac2; y[3,k] := n*(n+1)-mm*(mm+1)
   END;
   x[m] := 1.0; y[1,m] := anorm; y[3,m] := n*(n+1)-mm*(mm+1);
   y[2,m] := (y[3,m]-c2)*y[1,m]/(2.0*(mm+1.0)); scalv[1] := abs(anorm);
   IF (y[2,m] > abs(anorm)) THEN scalv[2] := y[2,m] ELSE scalv[2] := abs(anorm);
   IF (y[3,m] > 1.0) THEN scalv[3] := y[3,m] ELSE scalv[3] := 1.0;
   WHILE true DO BEGIN
      writeln('Enter c**2 or 999 to end.'); readln(c2);
      IF (c2 = 999) THEN GOTO 99;
      solvde(itmax,conv,slowc,scalv,indexv,ne,nb,m,y,nyj,nyk,
         c,nci,ncj,nck,s,nsi,nsj); writeln;
      writeln('m = ',mm:2,' n = ',n:2,' c**2 = ',c2:7:3,
         ' lam = ',y[3,1]+mm*(mm+1):10:6); END;
99: END.
```

```
PROCEDURE difeq(k,k1,k2,jsf,is1,isf: integer; indexv: glindex;
      ne: integer; VAR s: glsarray; nsi,nsj: integer;
      y: glyarray; nyj,nyk: integer);
(* Programs using routine DIFEQ must define the types
```

```
TYPE
    glindex = ARRAY [1..nyj] OF integer;
    glsarray = ARRAY [1..nsi,1..nsj] OF real;
    glyarray = ARRAY [1..nyj,1..nyk] OF real;
and also declare the global quantities
CONST
    m=41;
VAR
    h,c2,anorm: real; mm,n: integer;
    x: ARRAY [1..m] OF real;
in the main routine.   *)
VAR
    temp2,temp: real;
BEGIN
    IF (k = k1) THEN BEGIN
        IF (((n+mm) MOD 2) = 1) THEN BEGIN
            s[3,3+indexv[1]] := 1.0; s[3,3+indexv[2]] := 0.0;
            s[3,3+indexv[3]] := 0.0; s[3,jsf] := y[1,1] END
        ELSE BEGIN
            s[3,3+indexv[1]] := 0.0; s[3,3+indexv[2]] := 1.0;
            s[3,3+indexv[3]] := 0.0; s[3,jsf] := y[2,1] END END
    ELSE IF (k > k2) THEN BEGIN
        s[1,3+indexv[1]] := -(y[3,m]-c2)/(2.0*(mm+1.0));
        s[1,3+indexv[2]] := 1.0; s[1,3+indexv[3]] := -y[1,m]/(2.0*(mm+1.0));
        s[1,jsf] := y[2,m]-(y[3,m]-c2)*y[1,m]/(2.0*(mm+1.0));
        s[2,3+indexv[1]] := 1.0; s[2,3+indexv[2]] := 0.0;
        s[2,3+indexv[3]] := 0.0; s[2,jsf] := y[1,m]-anorm END
    ELSE BEGIN
        s[1,indexv[1]] := -1.0; s[1,indexv[2]] := -0.5*h;
        s[1,indexv[3]] := 0.0; s[1,3+indexv[1]] := 1.0;
        s[1,3+indexv[2]] := -0.5*h; s[1,3+indexv[3]] := 0.0;
        temp := h/(1.0-sqr(x[k]+x[k-1])*0.25);
        temp2 := 0.5*(y[3,k]+y[3,k-1])-c2*0.25*sqr(x[k]+x[k-1]);
        s[2,indexv[1]] := temp*temp2*0.5;
        s[2,indexv[2]] := -1.0-0.5*temp*(mm+1.0)*(x[k]+x[k-1]);
        s[2,indexv[3]] := 0.25*temp*(y[1,k]+y[1,k-1]);
        s[2,3+indexv[1]] := s[2,indexv[1]];
        s[2,3+indexv[2]] := 2.0+s[2,indexv[2]];
        s[2,3+indexv[3]] := s[2,indexv[3]]; s[3,indexv[1]] := 0.0;
        s[3,indexv[2]] := 0.0; s[3,indexv[3]] := -1.0;
        s[3,3+indexv[1]] := 0.0; s[3,3+indexv[2]] := 0.0;
        s[3,3+indexv[3]] := 1.0;
        s[1,jsf] := y[1,k]-y[1,k-1]-0.5*h*(y[2,k]+y[2,k-1]);
        s[2,jsf] := y[2,k]-y[2,k-1]-temp*((x[k]+x[k-1])
            *0.5*(mm+1.0)*(y[2,k]+y[2,k-1])-temp2 *0.5*(y[1,k]+y[1,k-1]));
        s[3,jsf] := y[3,k]-y[3,k-1] END
END;
```

Chapter 17. Introduction to Partial Differential Equations

Relaxation Methods for Boundary Value Problems

```
PROCEDURE sor(a,b,c,d,e,f: gljmax; VAR u: gljmax;
          jmax: integer; rjac: double);
(* Programs using routine SOR must define the type
TYPE
   gljmax = ARRAY [1..jmax,1..jmax] OF double;
in the main routine. *)
LABEL 99;
CONST
```

```
   maxits=1000; eps=1.0e-5; zero=0.0; half=0.5; qtr=0.25; one=1.0;
VAR
   n,l,j: integer; resid,omega,anormf,anorm: double;
BEGIN
   anormf := zero;
   FOR j := 2 TO jmax-1 DO BEGIN
      FOR l := 2 TO jmax-1 DO BEGIN
         anormf := anormf+abs(f[j,l]) END END;
   omega := one;
   FOR n := 1 TO maxits DO BEGIN
      anorm := zero;
      FOR j := 2 TO (jmax-1) DO BEGIN
         FOR l := 2 TO (jmax-1) DO BEGIN
            IF (((j+l) MOD 2) = (n MOD 2)) THEN BEGIN
               resid := a[j,l]*u[j+1,l]+b[j,l]*u[j-1,l]
                  +c[j,l]*u[j,l+1]+d[j,l]*u[j,l-1]
                  +e[j,l]*u[j,l]-f[j,l]; anorm := anorm+abs(resid);
               u[j,l] := u[j,l]-omega*resid/e[j,l] END END END;
      IF (n = 1) THEN BEGIN
         omega := one/(one-half*sqr(rjac)) END
      ELSE BEGIN
         omega := one/(one-qtr*sqr(rjac)*omega) END;
      IF ((n > 1) AND (anorm < (eps*anormf))) THEN GOTO 99
   END;
   writeln('pause in routine SOR');
   writeln('too many iterations'); readln;
99: END;
```

Operator Splitting Methods and ADI

```
PROCEDURE adi(a,b,c,d,e,f,g: gljmax; VAR u: gljmax;
      jmax,k: integer; alpha,beta,eps: double);
(* Programs using routine ADI must define the type
TYPE
   gljmax = ARRAY [1..jmax,1..jmax] OF double;
in the main routine. *)
LABEL 99;
CONST
   jj=50; kk=6;
   nrr=32;        (* nrr=2 to the power (kk-1) *)
   maxits=100; zero=0.0; two=2.0; half=0.5;
TYPE
   gljjarray = ARRAY [1..jj] OF double;
VAR
   i,nr,nits,next,n,l,kits,k1,j,twopwr: integer;
   rfact,resid,disc,anormg,anorm,ab: double; aa,bb,cc,rr,uu: gljjarray;
   psi: ARRAY [1..jj,1..jj] OF double;
   alph,bet: ARRAY [1..kk] OF double;
   r: ARRAY [1..nrr] OF double;
   s: ARRAY [1..nrr,1..kk] OF double;
PROCEDURE tridag(a,b,c,r: gljjarray; VAR u: gljjarray; n: integer);
   (* This is a double precision version of TRIDAG for use with ADI,
      which defines the constant jj and the type gljjarray.    *)
   VAR
      j: integer; bet: double; gam: gljjarray;
   BEGIN
      IF (b[1] = 0.0) THEN BEGIN
         writeln('pause 1 in TRIDAG'); readln END;
      bet := b[1]; u[1] := r[1]/bet;
      FOR j := 2 TO n DO BEGIN
         gam[j] := c[j-1]/bet; bet := b[j]-a[j]*gam[j];
         IF (bet = 0.0) THEN BEGIN
```

```
                writeln('pause 2 in TRIDAG'); readln END;
            u[j] := (r[j]-a[j]*u[j-1])/bet END;
        FOR j := n-1 DOWNTO 1 DO u[j] := u[j]-gam[j+1]*u[j+1]
    END;
BEGIN
    IF (jmax > jj) THEN BEGIN
        writeln('Pause in routine ADI - increase jj'); readln
    END;
    IF (k > (kk-1)) THEN BEGIN
        writeln('Pause in routine ADI - increase kk'); readln
    END;
    k1 := k+1; nr := 1;
    FOR i := 1 TO k DO nr := 2*nr;
    alph[1] := alpha; bet[1] := beta;
    FOR j := 1 TO k DO BEGIN
        alph[j+1] := sqrt(alph[j]*bet[j]); bet[j+1] := half*(alph[j]+bet[j])
    END;
    s[1,1] := sqrt(alph[k1]*bet[k1]);
    FOR j := 1 TO k DO BEGIN
        ab := alph[k1-j]*bet[k1-j]; twopwr := 1;
        FOR i := 1 TO (j-1) DO twopwr := 2*twopwr;
        FOR n := 1 TO twopwr DO BEGIN
            disc := sqrt(sqr(s[n,j])-ab); s[2*n,j+1] := s[n,j]+disc;
            s[2*n-1,j+1] := ab/s[2*n,j+1] END END;
    FOR n := 1 TO nr DO r[n] := s[n,k1];
    anormg := zero;
    FOR j := 2 TO (jmax-1) DO BEGIN
        FOR l := 2 TO (jmax-1) DO BEGIN
            anormg := anormg+abs(g[j,l]); psi[j,l] := -d[j,l]*u[j,l-1]
                +(r[1]-e[j,l])*u[j,l]-f[j,l]*u[j,l+1] END END;
    nits := maxits DIV nr;
    FOR kits := 1 TO nits DO BEGIN
        FOR n := 1 TO nr DO BEGIN
            IF (n = nr) THEN next := 1 ELSE next := n+1;
            rfact := r[n]+r[next];
            FOR l := 2 TO (jmax-1) DO BEGIN
                FOR j := 2 TO jmax-1 DO BEGIN
                    aa[j-1] := a[j,l]; bb[j-1] := b[j,l]+r[n];
                    cc[j-1] := c[j,l]; rr[j-1] := psi[j,l]-g[j,l] END;
                tridag(aa,bb,cc,rr,uu,jmax-2);
                FOR j := 2 TO (jmax-1) DO BEGIN
                    psi[j,l] := -psi[j,l] +two*r[n]*uu[j-1] END END;
            FOR j := 2 TO (jmax-1) DO BEGIN
                FOR l := 2 TO (jmax-1) DO BEGIN
                    aa[l-1] := d[j,l]; bb[l-1] := e[j,l]+r[n];
                    cc[l-1] := f[j,l]; rr[l-1] := psi[j,l] END;
                tridag(aa,bb,cc,rr,uu,jmax-2);
                FOR l := 2 TO (jmax-1) DO BEGIN
                    u[j,l] := uu[l-1]; psi[j,l] := -psi[j,l]+rfact*uu[l-1]
                END END END;
        anorm := zero;
        FOR j := 2 TO (jmax-1) DO BEGIN
            FOR l := 2 TO (jmax-1) DO BEGIN
                resid := a[j,l]*u[j-1,l]+(b[j,l]+e[j,l])*u[j,l]
                    +c[j,l]*u[j+1,l]+d[j,l]*u[j,l-1] +f[j,l]*u[j,l+1]+g[j,l];
                anorm := anorm+abs(resid) END END;
        IF (anorm < (eps*anormg)) THEN GOTO 99
    END;
    writeln('Pause in routine ADI - too many iterations');
99: END;
```

Table of Program Dependencies

The following table lists, in alphabetical order, all the routines in *Numerical Recipes.* The section number in which the routine is listed occurs next, followed by the list of routines (if any) that are necessary for the execution of the routine first named. Note that the dependency list is complete: you do not need to look at other table entries for the additional dependencies of routines listed on the right-hand side of a given entry.

ADI	(§17.6)	TRIDAG			
AMOEBA	(§10.4)				
ANNEAL	(§10.9)	LINK			
AVEVAR	(§13.4)				
BADLUK	(§1.1)	FLMOON	JULDAY		
BALANC	(§11.5)				
BCUCOF	(§3.6)				
BCUINT	(§3.6)	BCUCOF			
BESSI	(§6.5)	BESSI0			
BESSI0	(§6.5)				
BESSI1	(§6.5)				
BESSJ	(§6.4)	BESSJ0	BESSJ1		
BESSJ0	(§6.4)				
BESSJ1	(§6.4)				
BESSK	(§6.5)	BESSK0	BESSK1	BESSI0	BESSI1
BESSK0	(§6.5)	BESSI0			
BESSK1	(§6.5)	BESSI1			
BESSY	(§6.4)	BESSY0	BESSY1	BESSJ0	BESSJ1
BESSY0	(§6.4)	BESSJ0			
BESSY1	(§6.4)	BESSJ1			
BETA	(§6.1)	GAMMLN			
BETACF	(§6.3)				
BETAI	(§6.3)	BETACF	GAMMLN		
BICO	(§6.1)	FACTLN	GAMMLN		
BKSUB	(§16.3)				
BNLDEV	(§7.3)	RAN1	GAMMLN		
BRENT	(§10.2)				
BSSTEP	(§15.4)	MMID	RZEXTR		
CALDAT	(§1.1)				
CEL	(§6.7)				
CHDER	(§5.7)				
CHEBEV	(§5.6)				

CHEBFT	(§5.6)				
CHEBPC	(§5.8)				
CHINT	(§5.7)				
CHSONE	(§13.5)	GAMMQ	GCF	GSER	GAMMLN
CHSTWO	(§13.5)	GAMMQ	GCF	GSER	GAMMLN
CNTAB1	(§13.6)	GAMMQ	GCF	GSER	GAMMLN
CNTAB2	(§13.6)				
CONVLV	(§12.4)	TWOFFT	REALFT	FOUR1	
CORREL	(§12.5)	TWOFFT	REALFT	FOUR1	
COSFT	(§12.3)	REALFT	FOUR1		
COVSRT	(§14.3)				
CRANK	(§13.8)				
CYFUN	(§7.5)				
DBRENT	(§10.3)				
DDPOLY	(§5.3)				
DES	(§7.5)	CYFUN	KS		
DF1DIM	(§10.6)				
DFPMIN	(§10.7)	LINMIN	MNBRAK	BRENT	F1DIM
DIFEQ	(§16.4)				
ECLASS	(§8.5)				
ECLAZZ	(§8.5)				
EIGSRT	(§11.1)				
EL2	(§6.7)				
ELMHES	(§11.5)				
ERF	(§6.2)	GAMMP	GSER	GCF	GAMMLN
ERFC	(§6.2)	GAMMP	GAMMQ	GSER	GCF
		GAMMLN			
ERFCC	(§6.2)				
EULSUM	(§5.1)				
EVLMEM	(§12.8)				
EXPDEV	(§7.2)	RAN1			
F1DIM	(§10.5)				
FACTLN	(§6.1)	GAMMLN			
FACTRL	(§6.1)	GAMMLN			
FGAUSS	(§14.4)				
FIT	(§14.2)	GAMMQ	GCF	GSER	GAMMLN
FIXRTS	(§12.10)	ZROOTS	LAGUER		
FLEG	(§14.3)				
FLMOON	(§1.0)				
FOUR1	(§12.2)				
FOURN	(§12.11)				
FPOLY	(§14.3)				
FRPRMN	(§10.6)	LINMIN	MNBRAK	BRENT	F1DIM
FTEST	(§13.4)	AVEVAR	BETAI	BETACF	GAMMLN
GAMDEV	(§7.3)	RAN1			
GAMMLN	(§6.1)				
GAMMP	(§6.2)	GSER	GCF	GAMMLN	
GAMMQ	(§6.2)	GSER	GCF	GAMMLN	

GASDEV	(§7.2)	RAN1			
GAULEG	(§4.5)				
GAUSSJ	(§2.1)				
GCF	(§6.2)	GAMMLN			
GOLDEN	(§10.1)				
GSER	(§6.2)	GAMMLN			
HQR	(§11.6)				
HUNT	(§3.4)				
INDEXX	(§8.3)				
IRBIT1	(§7.4)				
IRBIT2	(§7.4)				
JACOBI	(§11.1)				
JULDAY	(§1.1)				
KENDL1	(§13.8)	ERFCC			
KENDL2	(§13.8)	ERFCC			
KS	(§7.5)				
KSONE	(§13.5)	SORT	PROBKS		
KSTWO	(§13.5)	SORT	PROBKS		
LAGUER	(§9.5)				
LFIT	(§14.3)	COVSRT	GAUSSJ		
LINK	(§10.9)				
LINMIN	(§10.5)	MNBRAK	BRENT	F1DIM	
LOCATE	(§3.4)				
LUBKSB	(§2.3)				
LUDCMP	(§2.3)				
MDIAN1	(§13.2)	SORT			
MDIAN2	(§13.2)				
MEDFIT	(§14.6)	ROFUNC	SORT		
MEMCOF	(§12.8)				
METROP	(§10.9)	RAN3			
MIDINF	(§4.4)				
MIDPNT	(§4.4)				
MIDSQL	(§4.4)				
MIDSQU	(§4.4)				
MMID	(§15.3)				
MNBRAK	(§10.1)				
MNEWT	(§9.6)	LUDCMP	LUBKSB		
MOMENT	(§13.1)				
MPROVE	(§2.7)	LUBKSB			
MRQCOF	(§14.4)				
MRQMIN	(§14.4)	MRQCOF	COVSRT	GAUSSJ	
ODEINT	(§15.2)	(RKQC and RK4) or (BSSTEP and MMID and RZEXTR)			
PCSHFT	(§5.8)				
PEARSN	(§13.7)	BETAI	BETACF	GAMMLN	ERFCC
PIKSR2	(§8.1)				
PIKSRT	(§8.1)				
PINVS	(§16.3)				

PLGNDR	(§6.6)				
POIDEV	(§7.3)	RAN1	GAMMLN		
POLCOE	(§3.5)				
POLCOF	(§3.5)	POLINT			
POLDIV	(§5.3)				
POLIN2	(§3.6)	POLINT			
POLINT	(§3.1)				
POWELL	(§10.5)	LINMIN	MNBRAK	BRENT	F1DIM
PREDIC	(§12.10)				
PROBKS	(§13.5)				
PZEXTR	(§15.4)				
QCKSRT	(§8.4)				
QGAUS	(§4.5)				
QROMB	(§4.3)	POLINT	TRAPZD		
QROMO	(§4.4)	POLINT			
		MIDPNT or MIDINF or MIDSQL or MIDSQU			
QROOT	(§9.5)	POLDIV			
QSIMP	(§4.2)	TRAPZD			
QTRAP	(§4.2)	TRAPZD			
QUAD3D	(§4.6)	QUAD3F	QUAD3G	QUAD3H	QGAUS
RAN0	(§7.1)	uses system-supplied RAN			
RAN1	(§7.1)				
RAN2	(§7.1)				
RAN3	(§7.1)				
RAN4	(§7.5)	DES	CYFUN	KS	
RANK	(§8.3)				
RATINT	(§3.2)				
REALFT	(§12.3)	FOUR1			
RED	(§16.3)				
RK4	(§15.1)				
RKDUMB	(§15.1)	RK4			
RKQC	(§15.2)	RK4			
ROFUNC	(§14.6)	SORT			
RTBIS	(§9.1)				
RTFLSP	(§9.2)				
RTNEWT	(§9.4)				
RTSAFE	(§9.4)				
RTSEC	(§9.2)				
RZEXTR	(§15.4)				
SCRSHO	(§9.0)				
SFROID	(§16.4)	DIFEQ	SOLVDE	RED	PINVS
		BKSUB	PLGNDR		
SHELL	(§8.1)				
SHOOT	(§16.1)	LUBKSB	LUDCMP		
		ODEINT	RKQC	RK4	
SHOOTF	(§16.2)	LUBKSB	LUDCMP		
		ODEINT	RKQC	RK4	
SIMP1	(§10.8)				

SIMP2	(§10.8)				
SIMP3	(§10.8)				
SIMPLX	(§10.8)	SIMP1	SIMP2	SIMP3	
SINFT	(§12.3)	REALFT	FOUR1		
SMOOFT	(§13.9)	REALFT	FOUR1		
SNCNDN	(§6.7)				
SOLVDE	(§16.3)	RED	PINVS	BKSUB	DIFEQ
SOR	(§17.5)				
SORT	(§8.2)				
SORT2	(§8.2)				
SORT3	(§8.3)	INDEXX			
SPARSE	(§2.10)				
SPCTRM	(§12.7)	FOUR1			
SPEAR	(§13.8)	SORT2	CRANK	BETAI	BETACF
		GAMMLN	ERFCC		
SPLIE2	(§3.6)	SPLINE			
SPLIN2	(§3.6)	SPLINE	SPLINT		
SPLINE	(§3.3)				
SPLINT	(§3.3)				
SVBKSB	(§2.9)				
SVDCMP	(§2.9)				
SVDFIT	(§14.3)	SVDCMP	SVBKSB		
SVDVAR	(§14.3)				
TOEPLZ	(§2.8)				
TPTEST	(§13.4)	AVEVAR	BETAI	BETACF	GAMMLN
TQLI	(§11.3)				
TRAPZD	(§4.2)				
TRED2	(§11.2)				
TRIDAG	(§2.6)				
TTEST	(§13.4)	AVEVAR	BETAI	BETACF	GAMMLN
TUTEST	(§13.4)	AVEVAR	BETAI	BETACF	GAMMLN
TWOFFT	(§12.3)	FOUR1			
VANDER	(§2.8)				
ZBRAC	(§9.1)				
ZBRAK	(§9.1)				
ZBRENT	(§9.3)				
ZROOTS	(§9.5)	LAGUER			

Index

BOOK AND DISKETTE ORDERING INSTRUCTIONS (OUTSIDE U.S.A. & CANADA)

Also published by Cambridge University Press are example books and software that accompany *Numerical Recipes: The Art of Scientific Computing*.

The example books contain FORTRAN and Pascal source programs respectively. These programs exercise and demonstrate all of *Numerical Recipes* subroutines, procedures, and functions. Each program contains comments and is prefaced by a short description of what it does and of which *Numerical Recipes* routines it exercises. In cases where the demonstration programs require input data, that data is supplied. In some cases, sample output is also shown. The example books should be valuable aids to readers wishing to incorporate procedures and subroutines and to conduct simple validation tests.

NUMERICAL RECIPES EXAMPLE BOOK (FORTRAN) ISBN 0-521-31330-9
192 pages
Contains sample program listing written in the FORTRAN language that demonstrate the use of each *Numerical Recipes* subroutine.

NUMERICAL RECIPES EXAMPLE BOOK (PASCAL) ISBN 0-521-30956-5
256 pages
Contains sample program listings written in the Pascal language that demonstrate the use of each *Numerical Recipes* procedure.

The programs listed in *Numerical Recipes: The Art of Scientific Computing* and *Numerical Recipes Example Book* are available in several machine-readable formats and programming languages. The diskettes listed below are available from Cambridge University Press. All versions of the diskettes are 5¼ inch double-sided/double density floppy diskettes. They operate on DOS 2.0/3.0 on IBM PC, XT, AT, and IBM compatible machines. The diskettes can save hours of tedious keyboarding, leaving users free to adapt, modify, or experiment with the programs.

NUMERICAL RECIPES FORTRAN DISKETTE V1.0 ISBN 0-521-30958-1
FORTRAN subroutines as listed in *Numerical Recipes: The Art of Scientific Computing* in machine-readable form.

NUMERICAL RECIPES PASCAL DISKETTE V1.0 ISBN 0-521-30955-7 Pascal procedures as listed in *Numerical Recipes: The Art of Scientific Computing* in machine-readable form.

NUMERICAL RECIPES EXAMPLE DISKETTE (FORTRAN) ISBN 0-521-30957-3
Demonstration programs in the FORTRAN language as listed in *Numerical Recipes Example Book (FORTRAN)* in machine-readable form.

NUMERICAL RECIPES EXAMPLE DISKETTE (PASCAL) ISBN 0-521-30954-9
Demonstration programs in the Pascal language as listed in *Numerical Recipes Example Book (Pascal)* in machine-readable form.

To order the example books or latest version of the diskettes, complete the information below and mail this page or a copy to Cambridge University Press. RESIDENTS OF THE U.S.A. AND CANADA PLEASE USE THE FORM ON THE FOLLOWING PAGES.

Technical questions, corrections, and requests for information on other available formats and software products should be directed to Numerical Recipes Software, P.O. Box 243, Cambridge, MA 02238, U.S.A. Only diskettes with manufacturing defects may be returned to the publisher for replacement (no cash refunds).

TO ORDER SEND THIS FORM TO:
Customer Services Department, Cambridge University Press, Edinburgh Building, Shaftesbury Road, Cambridge CB2 2RU, U.K.

Please send me:

________ 31330-9 *Numerical Recipes Example Book (FORTRAN)* £15.00 each
________ 30956-5 *Numberical Recipes Example Book (Pascal)* £15.00 each
________ 30958-1 NUMERICAL RECIPES FORTRAN DISKETTE V1.1 £15.00 each
________ 30955-7 NUMERICAL RECIPES Pascal DISKETTE V1.1 £15.00 each
________ 30957-3 NUMERICAL RECIPES EXAMPLE DISKETTE (FORTRAN) £15.00 each
________ 30954-9 NUMERICAL RECIPES EXAMPLE DISKETTE (Pascal) £15.00 each

Name ________________________________ (Block capitals please)

Address ________________________________

____

____

**Please accept my payment by cheque or money order in pounds sterling:* I enclose (circle one) a Cheque (made payable to Cambridge University Press)/UK Postal Order/ International Money Order/Bank Draft/Post office Giro.

**Please accept my payment by credit card:* Charge my (circle one) Barclaycard/Visa/ Eurocard/Access/Mastercard/Bank Americard/any other credit card bearing the Interbank symbol (please specify).

Card No. ________________________________

Signature________________________ Expiry date________

Date________

Address as registered by card company:________________________

____

____

All prices exclude VAT and are subject to alteration without prior notice.

Cut along dotted line

To order the example books or latest version of the diskettes, complete the information below and mail this page or a copy to Cambridge University Press. (Alternatively, customers may call the Press at 914/235-0300 [in N.Y. and Canada] or 800/431-1580 [in rest of U.S.] to place an order.) Orders must be accompanied by payment in U.S. funds or the equivalent in Canadian funds. New York and California residents please add appropriate sales tax. Prices are not guaranteed. Ordinary postage for shipping orders paid by the publisher. RESIDENTS OF COUNTRIES OTHER THAN THE U.S.A. AND CANADA PLEASE USE THE FORM ON THE PRECEDING PAGES.

Technical questions, corrections, and requests for information on other available formats and software products should be directed to Numerical Recipes Software, P.O. Box 243, Cambridge, MA 02238, U.S.A. Only diskettes with manufacturing defects may be re-returned to the publisher for replacement (no cash refunds).

TO ORDER SEND THIS FORM TO:
Cambridge University Press, Order Department, 510 North Avenue, New Rochelle, New York 10801.

Rlease indicate method of payment: check _____, Mastercard _____, or Visa _____.

Name ____

Address ____

____

____

Card No. ____

Expiration date __________

Signature _________________________________

Please send me:

________ 31330-9 *Numerical Recipes Example Book (FORTRAN)* $18.95 each
________ 30956-5 *Numerical Recipes Example Book (Pascal)* $18.95 each
________ 30958-1 NUMERICAL RECIPES FORTRAN DISKETTE V1.1 $19.95 each
________ 30955-7 NUMERICAL RECIPES Pascal DISKETTE V1.1 $19.95 each
________ 30957-3 NUMERICAL RECIPES EXAMPLE DISKETTE (FORTRAN) $19.95 each
________ 30954-9 NUMERICAL RECIPES EXAMPLE DISKETTE (Pascal) $19.95 each

________ Please indicate the total number of items ordered,
________ total price,
________ tax, if applicable (NY and CA residents)
________ total enclosed.

Cut along dotted line

BOOK AND DISKETTE ORDERING INSTRUCTIONS (U.S.A. & CANADA)

Also published by Cambridge University Press are example books and software that accompany *Numerical Recipes: The Art of Scientific Computing*.

The example books contain FORTRAN and Pascal source programs respectively. These programs exercise and demonstrate all of *Numerical Recipes* subroutines, procedures, and functions. Each program contains comments and is prefaced by a short description of what it does and of which *Numerical Recipes* routines it exercises. In cases where the demonstration programs require input data, that data is supplied. In some cases, sample output is also shown. The example books should be valuable aids to readers wishing to incorporate procedures and subroutines and to conduct simple validation tests.

NUMERICAL RECIPES EXAMPLE BOOK (FORTRAN) ISBN 0-521-31330-9
192 pages
Contains sample program listings written in the FORTRAN language that demonstrate the use of each *Numerical Recipes* subroutine.

NUMERICAL RECIPES EXAMPLE BOOK (PASCAL) ISBN 0-521-30956-5
256 pages
Contains sample program listings written in the Pascal language that demonstrate the use of each *Numerical Recipes* procedure.

The programs listed in *Numerical Recipes: The Art of Scientific Computing* and *Numerical Recipes Example Book* are available in several machine-readable formats and programming languages. The diskettes listed below are available from Cambridge University Press. All versions of the diskettes are 5¼ inch double-sided/double density floppy diskettes. They operate on DOS 2.0/3.0 on IBM PC, XT, AT, and IBM compatible machines. The diskettes can save hours of tedious keyboarding, leaving users free to adapt, modify, or experiment with the programs.

NUMERICAL RECIPES FORTRAN DISKETTE V1.0 ISBN 0-521-30958-1
FORTRAN subroutines as listed in *Numerical Recipes: The Art of Scientific Computing* in machine-readable form.

NUMERICAL RECIPES PASCAL DISKETTE V1.0 ISBN 0-521-30955-7
Pascal procedures as listed in *Numerical Recipes: The Art of Scientific Computing* in machine-readable form.

NUMERICAL RECIPES EXAMPLE DISKETTE (FORTRAN) ISBN 0-521-30957-3
Demonstration programs in the FORTRAN language as listed in *Numerical Recipes Example Book (FORTRAN)* in machine-readable form.

NUMERICAL RECIPES EXAMPLE DISKETTE (PASCAL) ISBN 0-521-30954-9
Demonstration programs in the Pascal language as listed in *Numerical Recipes Example Book (Pascal)* in machine-readable form.

W9-BRH-904